Nuclear Shell Theory

Nuclear Shell Theory

Amos de-Shalit and Igal Talmi
Weizmann Institute of Science, Rehovot, Israel

Dover Publications, Inc.
Mineola, New York

Bibliographical Note

This Dover edition, first published in 2004, is an unabridged republication of the work first published by Academic Press, New York, 1963.

Library of Congress Cataloging-in-Publication Data

Shalit, Amos de-, 1926–1969.
Nuclear shell theory / Amos de-Shalit and Igal Talmi.—Dover ed.
p. cm.
Originally published: New York : Academic Press, 1963.
ISBN 0-486-43933-X (pbk.)
1. Nuclear shell theory. I. Talmi, Igal, 1925– II. Title.

QC173.S338 2004
539.7'43—dc22

2004056149

Manufactured in the United States of America
Dover Publications, Inc., 31 East 2nd Street, Mineola, N.Y. 11501

לזכרו של

משה תלמי ז"ל

אב, מורה ומחנך

To the Memory of
MOSHE TALMI
Father and Teacher

PREFACE

This volume is intended to fulfill the need for a comprehensive textbook dealing with modern methods of the nuclear shell model. Although these have been extensively used in the last few years, anyone who wants to learn these methods must resort to various articles scattered through the literature.

In the present book the methods of modern spectroscopy are developed in a form suitable for use in the nuclear shell model. Since deeper understanding of these methods involves group theory the correspondence with this theory is pointed out. However, the mathematical machinery of group representations has not been used. The knowledge of group theory is not a prerequisite to studying shell model methods. On the other hand, those familiar with group theory will easily understand the group theoretical meaning of the ideas and formalism. By keeping the mathematics as simple as possible and by emphasizing the physical content of the formalism we hope that the book will be of value for experimentalists and theorists interested in this field.

The book is intended for graduate students who have had a course in elementary quantum mechanics. The necessary additional formalism is developed in the book. In particular, tensor algebra, which is the most important mathematical tool needed, is developed in detail in Part II.

It would be difficult to list here and thank all friends and colleagues from whom the authors received help and encouragement. We would only mention R. Sherr and A. Reiner who read and criticized certain parts of the book. We would like to thank I. Unna and Miss B. Marks for their great help in the preparation of the manuscript.

November, 1962

A. DE-SHALIT
I. TALMI

CONTENTS

INTRODUCTION

In this book we treat the mathematical theory of a system of Fermions in a central field. Such a model is supposed to give a description of systems of *interacting* particles like the electrons in the atom or protons and neutrons in the nucleus. The real problem of n interacting particles has, in general, no exact solutions. There are, however, cases in which the interaction of one particle with all the others can be replaced, to a good approximation, by an average central field. It is then possible to treat the problem by solving, for each particle, the one particle problem in a central field. Once such a solution is obtained, the parameters of the central potential well can be adjusted in order to satisfy some criterion of *self-consistency*. The action of all particles, moving in their assigned orbits, on a given particle, should be equal, as much as possible, to the action of the central field on this particle. If such a self-consistent central field is found, we have still to consider that part of the mutual interaction which is not accounted for by it. It may be possible to treat that part, usually called the *residual* interaction, as a perturbation. We shall develop the methods of constructing the appropriate many particle wave functions for treating such problems. It turns out that it is possible to give a rather detailed description of these wave functions without making specific assumptions on either the central field or the residual interaction. Thus, the methods developed will have general validity and can be applied to many systems which have very different properties.

The description of a system of interacting particles in terms of independent particles moving in a central field is called a *shell model.* The name is derived from the fact that independent Fermions moving in a central field exhibit some shell structure. The single particle energy levels in a central field are discrete and generally degenerate. Thus, the particles move in orbits (defined by the single particle energies) and there is a certain number of identical Fermions that can occupy a given orbit. All orbits that lie close in energy form a major shell. The definition of a shell is rather arbitrary but the idea is that the energy differences between orbits in different shells are much bigger than the difference between orbits in the same shell. When all orbits are filled we speak of a closed shell. There is only one quantum state for a closed shell. Therefore, the system can be excited only by removing one particle to a higher orbit, which lies in the next shell. Therefore, closed shells exhibit extra stability and play an important role in the shell model.

The best example of a shell model is exhibited by the electron system of an atom. A bare nucleus with a charge $+Ze$ produces a Coulomb field and a hydrogen-like series of orbits. As more and more electrons are added to such a nucleus, they occupy these orbits. The spacing and even the order of the orbits in the presence of many electrons are no longer hydrogen-like. The central field is determined by the combined action of the attractive nuclear field and the mutual repulsion of the electrons. Still, the shell structure is well preserved as shown, for instance, by the periodic system of the elements. The actual central field that acts on each electron can be approximately determined by the Thomas-Fermi method or by the more elaborate Hartree approximation.

The atoms with closed shells are those of noble gases. They are the most difficult to excite (and therefore to participate in chemical reactions). The reason is that the "screening" of the nuclear charge by electrons in the same shell is less effective than the screening by electrons in inner shells. Thus, as a shell is being filled the charge of the nucleus increases (for a neutral atom!) while the increased screening does not compensate it. When an electron shell is closed, the only possible excitation is of a single electron to a higher shell. The orbit of the electron in a closed shell has, as described above, a rather low energy whereas in the next orbit the screening is efficient and its energy is much higher. The ionizing energy increases as we proceed in the periodic table and attains maxima for closed shells. Atoms with single electrons outside closed shells have accordingly the lowest ionization energy. The most typical of these are the atoms of the alkali metals which are also very active chemically.

Considering atoms in which the same electronic shell is being filled, other regularities can be observed. There is a certain regularity concerning the ground states of atoms. The total electron spin S in the ground state has its maximum possible value. As will be explained in detail in Part III, this is due to the repulsive interaction of the electrons. The state with maximum S has minimum symmetry in the spatial coordinates and, therefore, minimum repulsion. Also the total orbital angular momentum L attains in the ground state its maximum possible value. As we proceed from the ground state of one atom to the next, we observe an increase in S and L until the orbit is exactly half filled. From this point on there is a gradual decrease in S and L. Only for completely filled orbits do we have in the ground state $S = 0$ and $L = 0$. This behavior is quite different from the one in the nuclear shell model where the residual interaction is essentially attractive and the resulting situation is considerably simpler.

There is another important difference between nuclei and atoms. When we consider the motion of electrons in an atom we have a natural reference point—the nucleus in whose field all the electrons move and whose mass is very large compared to that of the electrons. No such reference point exists in the nucleus, and at first look it seems hard to understand why a central potential may form a good starting point for a nuclear model.

The difficulty can be resolved in the following way. As long as we are interested in the *intrinsic* properties of nuclei we can add to the nuclear Hamiltonian a potential acting on the nucleus' center of mass. This will obviously have the effect of confining the nucleus to a limited portion of space *without* changing its intrinsic structure. It can be expected that such a system, i.e., a nucleus tied to a point in space, would lend itself to an approximation by a central field. Furthermore, if this potential acting on the center of mass is taken to be a harmonic oscillator potential, we can simplify matters very much by using the identity

$$\left(\frac{1}{A}\sum_i \mathbf{r}_i\right)^2 = \frac{1}{A}\sum_i \mathbf{r}_i^2 - \frac{1}{A^2}\sum_{i<j}(\mathbf{r}_i - \mathbf{r}_j)^2.$$

This identity can be interpreted as saying that a harmonic force acting on the center of mass is equivalent to harmonic forces acting on each of the particles (Σr_i^2) plus a correction to the two-body interaction ($-\ \Sigma(\mathbf{r}_i - \mathbf{r}_j)^2$). We therefore see that by tying the center of mass of the nucleus to a fixed point in space one obtains very naturally as a starting point a harmonic central field.

When working with the harmonic oscillator potential well it should be remembered that the center of mass is also in a motion given by a harmonic oscillator potential. However, if all nucleons occupy the lowest possible oscillator levels according to the Pauli principle, the center of mass motion corresponds to the lowest eigenvalue of the oscillator. This can be seen by observing that the center of mass coordinate is symmetric in all nucleon coordinates. Its motion can thus be given by the oscillator wave function with lowest energy without violating the Pauli principle. A center of mass motion in a higher state corresponds to an excited state of the nucleus. Thus, as long as we are dealing with nucleons in the lowest possible oscillator levels the center of mass motion can be ignored. Only if some of the nucleons are excited to higher oscillator levels must the center of mass motion be carefully taken into account. In particular, certain excited states, called "spurious states," have the same intrinsic wave function as the ground state but the center of mass is not in the lowest oscillator level. We shall not encounter such cases in this book and rather refer the reader to the original paper of J. P. Elliott and T. H. R. Skyrme in *Proc. Roy. Soc.* (London) **A232**, 561 (1955).

The application of the shell model to nuclei seemed very doubtful at the beginning. It was stressed that the nuclear interaction is short ranged. Therefore, the whole idea of an average field acting on a given nucleon seemed highly questionable. It was argued that higher approximations to the self-consistent field may be very important to the determination of nuclear properties. There was little hope that a simple shell structure would be observed in nuclei. Nevertheless, there was enough experimental information that could be interpreted in terms of a shell structure. It was discovered in the 1930's that nuclei with special proton or neutron numbers exhibited properties of closed shells. In particular, the first proton (or neutron) beyond such a number was shown to be less bound than preceding protons (or neutrons). Such proton and neutron numbers were called *magic numbers*. Furthermore, quantitative calculations of energies were carried out for light nuclei which gave a fair agreement with experiment (Wigner's supermultiplet theory).

The order of orbits that could give rise to the observed magic numbers was the subject of many discussions. The central potential well is often approximated by a square well. Taking the order of orbits in such a well, the magic numbers 8 and 20 could very well be explained. The higher magic numbers were more puzzling. There was some empirical evidence that 50 and 82 were magic numbers which do not have a natural explanation in terms of a square well. There were some attempts to obtain these numbers by shifting certain single particle orbits much higher. However, no reasonable explanation could be given to this rather *ad hoc* procedure.

Another approach considered the occurrence of the magic numbers 50 and 82 as an indication of the breakdown of the shell model for heavier nuclei. It was argued that in view of the strong correlations between nucleons, due to the short-range nuclear interaction, the first approximation of a central field is not justified. The higher order terms may then determine the nuclear properties in a way which cannot be predicted from the order of orbits. This approach was greatly supported by the theory of nuclear reactions which was based on the assumption of very strong interactions between nucleons in the nucleus.

It was only after 1945 that the existence of higher magic numbers became more evident. More experimental information was gathered which indicated definite, yet not understood, shell structure in nuclei. The effect of long lifetimes of gamma-emitting nuclear states (nuclear isomerism) was studied and it was found that this could furnish valuable information on the parity of nuclear states. Thus, it was finally possible to check the theories which tried to explain the magic numbers by the shifting of single nucleon orbits. The correct description of the magic numbers was given in 1947 by Goeppert-Mayer and independently by Haxel, Jensen, and Suess. These authors tried to see whether there is any evidence for spin-orbit interaction in nuclei. Such an interaction, if strong enough, splits each l-orbit into two orbits, with $j = l + \frac{1}{2}$ and $j = l - \frac{1}{2}$. An appropriate grouping of these j-orbits not only reproduced the magic numbers but also gave a complete description of a host of facts concerning nuclear ground states, magnetic moments, and nuclear isomerism.

The measured spins (total angular momenta) of even nuclei are all zero. This could be simply understood since the nuclear interaction is essentially attractive and states with $J = 0$ allow maximum interaction (as will be discussed in detail in Parts II and III). For odd mass nuclei, with nucleons filling the j-orbit, the measured ground state spin is usually $J = j$. Also this coupling rule can be explained as due to a short-range attractive nuclear interaction. The spins of nuclear ground states thus give directly the j of the orbit which is being filled. The measured magnetic moments of odd mass nuclei are all rather close to the magnetic moment of a single nucleon in the j-orbit occupied by the last odd nucleon. It seems as if the spin and the magnetic moment are due only to the odd proton or neutron and the other nucleons furnish only the central potential.

There is another property which can be attributed to the last odd nucleon. There are low lying excited states in which one nucleon is excited into a higher j'-orbit. To a good approximation, the other nucleons are in a $J = 0$ state so that the total spin is $J = j'$. The presence of the lowest orbits can thus be demonstrated by exciting the odd nucleon into them. The spins and parities of these "single nucleon states" agree very well with those of the orbits assigned by the shell model.

It is clear that an exact treatment must take into account the presence of several nucleons in the j-orbit. For example, in every case the wave functions must be antisymmetric with respect to identical nucleons.

Nevertheless, due to the properties of the nuclear ground states, one could use a single nucleon picture in many cases. The theory of a single particle moving in a central field is discussed in detail in Part I of this book. Some of the properties of the single particle orbits may be applied to the experimental facts with some success (e.g., in the calculation of magnetic moments). The detailed theory of many nucleons in a central field is given in Part III.

The many qualitative results of the shell model are in excellent agreement with the experiment. Thus, the shell model with strong spin-orbit interaction has been an extremely useful guide for the experimental work. In addition to the many qualitative predictions, several quantitative calculations were carried out using the framework of the shell model. Some of these are in excellent agreement with the experimental data. In these more detailed calculations, shell model wave functions are assumed for the nucleons. The energies of nuclear states are calculated from some residual interaction between the nucleons.

Shell model wave functions of nucleons moving independently, cannot give a complete description of the nucleus. There must be short-range correlations between nucleons which are due to the short-range nuclear interaction. The evidence from nucleon scattering concerning that interaction indicates that it has very strong repulsion at close distances (hard core) followed by a very strong attraction. Therefore, no two nucleons can approach each other more than the hard core radius (about 0.4×10^{-13} cm.). The occurrence of a hard core leads to the divergence of matrix elements of the interaction taken with shell model wave functions. It is therefore impossible to treat the nuclear interaction in first order perturbation theory.

The two nucleon correlations discussed above have a range small compared to the average distance between nucleons. These correlations therefore modify the independent nucleons wave functions only at very close distances between nucleons. It is possible to transform the effect of these correlations from the wave functions onto the real Hamiltonian. In some approximation, the "interaction" in the modified Hamiltonian thus obtained will still be a two-body interaction. This modified, or *effective*, interaction, will be regular and it will be possible to use it as a perturbation with shell model wave functions. This is the philosophy that we adopt in the present book. We consider the shell model as a good approximation to nuclei provided we understand the two body interactions between nucleons to be "effective interactions." It is worthwhile to mention that even in the atomic case, energies can be calculated only if the original Coulomb interaction is replaced by a modified interaction (as manifested, for instance, by the appearance of the correction term $\sum_{i<k} (\boldsymbol{l}_i \cdot \boldsymbol{l}_k)$).

The exact form of the effective nuclear interaction is not yet known.

It may well have a strong dependence on the relative orbital angular momenta of the interacting nucleons. However, let us consider the effect of the interaction on the wave function of two nucleons. At distance smaller than the hard core radius, there is a strong repulsion; this is compensated by the strong attraction at larger distances. At the distance r_0, where these two effects are fully compensated, the attractive potential starts to show a "net" effect. We can visualize the effective interaction as due to a potential well which is zero for distances shorter than r_0 and becomes attractive beyond r_0. Thus, the effective interaction is expected to be attractive, rather weak, and long ranged. The cases where this interaction is sufficiently known phenomenologically, support this expectation.

Most of the results described in this book are independent of the exact nature of the interaction. They apply equally well to the atomic case and to the more complicated nuclear case. The description of the interaction in most general form is carried out best by using its expansion in terms of products of irreducible tensors. The treatment of the interaction from this point of view is carried out in Part II. The whole formalism of the algebra of irreducible tensors is first described in detail and then applied to the two particle system. The methods developed there are general and may be found in a number of textbooks. In order to make this book self-contained, we include this material and introduce it in a form which is suitable for use in Part II as well as in Part III.

It is well known that the shell model with a spherically symmetric central field does not give a valid description of all nuclei. In several regions, far away from closed shells, nuclear deformations are indicated by large values of quadrupole moments. Where these deformations are large, in particular in the rare earth and transuranium regions, the observed spectra can be very well described by rotational bands. The energy levels are given in these cases by the expression $CJ(J+1)$ to a high degree of accuracy. There is much additional evidence, mainly from branching ratios of various transitions, that the model of a rotating deformed nucleus as developed by A. Bohr and B. Mottelson, gives a very good description of many nuclear phenomena. In this model, the wave function of the deformed nucleus is taken to be that of independent nucleons moving in a deformed potential. Furthermore, the deformation is determined by the occupation numbers of the nucleons (Nilsson scheme). In this way, an attempt is made at a unified description of nuclear phenomena. It is clear, however, that this model works well where the nuclear deformations are large. Only then may it be a good approximation to have a separation of the motion of nucleons within the deformed potential and the rotation of the deformed nucleus. In the neighborhood of closed shells the assumption of a spherical central field is probably better. In the regions where there are many nucleons outside closed shells, but no rotational spectra are observed, the situation is really complex. There, probably neither of these two models can be directly applied.

In this book we shall not treat at all the collective or unified models for nuclei. A book dealing with *nuclear physics* must of course deal with the collective model in detail. However, we deal here with the mathematical theory of the shell model and make reference to physical facts about nuclei only as illustrations of the formalism. Our whole approach is directed toward applications in nuclear physics but we do not deal here with the vast amount of experimental nuclear data. The shell theory is a well defined subject which can be presented as a simple and exact mathematical formalism. It is remarkable indeed that it can be successfully applied to several problems in nuclear physics. In the case of other problems, the shell model gives qualitative results. In other cases higher order effects must be also considered. The actual discussion of such cases would take much space which would make this volume even bulkier.

Some of the topics dealt with are described in more detail than some others. These are usually the topics on which there is not enough material in the literature. On the other hand, we have not discussed in detail matters on which much has been written. Probably no book can aim at a complete description. We only hope that our book will help those interested in achieving an understanding of the methods of the nuclear shell model.

It would have been impossible to point out in full the numerous publications which are relevant to the material covered by the present book. Rather than do partial justice we decided to omit any references whatsoever (even to papers of the authors!). In doing so we naturally and consciously denounce any claim to originality of ideas or their presentation. We would only like to refer in particular to the work of G. Racah who can be justly called the founder of modern spectroscopy. The authors are greatly indebted to him for having introduced them to his methods. Much of the material presented in this book is due to him directly as well as indirectly.

In the following we give a list of books and some review articles which describe certain topics dealt with in this book and expand them in different directions. This list is not intended to be comprehensive.

Tensor Algebra is considered in the following books:

E. P. Wigner, "Group Theory and Atomic Spectra," Academic Press, New York, 1959.

U. Fano and G. Racah, "Irreducible Tensorial Sets," Academic Press, New York, 1959.

A. R. Edmonds, "Angular Momentum in Quantum Mechanics," Princeton University Press, Princeton, New Jersey, 1957.

M. E. Rose, "Angular Momentum," John Wiley & Sons, New York, 1957.

The Shell Model for nuclei is described in the following:

M. Goeppert-Mayer and J. H. D. Jensen, "Nuclear Shell Structure," John Wiley & Sons, New York, 1955.

E. Feenberg, "Shell Theory of the Nucleus," Princeton University Press, Princeton, 1955.

J. P. Elliott and A. M. Lane, "The Nuclear Shell Model," *in* Vol. XXXIX of the *Encyclopedia of Physics*, Springer-Verlag, Berlin, 1957.

Part I SINGLE PARTICLE SHELL MODEL

It was pointed out in the Introduction that atomic and nuclear states can probably be approximated by wave functions describing the motion of independent particles in a central field. This description is by no means complete and some residual mutual interaction between the particles has to be taken into account. This residual interaction can probably be treated as a perturbation, so that the independent motion of particles in a central field may constitute a good zeroth-order approximation to the actual situation.

In this part of the book we shall be primarily concerned with the motion of a single particle in a central field. The single particle wave functions derived here will become the building stones for the wave functions of many-particle systems which will be considered subsequently. It is therefore desirable to get well acquainted with their properties. It will also turn out, as we shall see in Part III, that for *nuclei*, the single-nucleon wave functions reproduce very closely many of the properties calculated with wave functions of many equivalent nucleons. The reason for it is that in a system containing an odd number of nucleons, pairs of nucleons tend to "saturate" each other, becoming "inert" with respect to some nuclear properties, so that the main contribution to such properties comes from the last odd nucleon. We shall therefore find that the single particle wave functions derived in this part will already be good enough to explain in a general way some nuclear properties such as nuclear magnetic moments.

1. Wave Functions in a Central Field

The problem of the motion of a particle in a central field of force described by the Hamiltonian

$$H = \frac{1}{2m} p^2 + U(r) \tag{1.1}$$

is one of the few problems whose exact general solution is known both in classical and in quantum mechanics. It is described in many textbooks and we shall give here only a brief sketch of it.

The most important feature of the motion in a central field is the existence of three operators $\mathbf{M} = (M_x, M_y, M_z)$, the *orbital angular momentum operators*, which commute with the Hamiltonian, and therefore constitute constants of motion. The definition of these operators in terms of the Cartesian coordinates is given by

$$\mathbf{M} = \mathbf{r} \times \mathbf{p} = \hbar \boldsymbol{l} \tag{1.2}$$

where

$$l_x = -i\left(y \frac{\partial}{\partial z} - z \frac{\partial}{\partial y}\right) \qquad l_y = -i\left(z \frac{\partial}{\partial x} - x \frac{\partial}{\partial z}\right)$$

$$l_z = -i\left(x \frac{\partial}{\partial y} - y \frac{\partial}{\partial x}\right). \tag{1.3}$$

It is easy to see that

$$[\mathbf{p}, r^2] = -2i\hbar\mathbf{r}\,.$$

It therefore follows from (1.2) that

$$[\boldsymbol{l}, r^2] = 0 \tag{1.4}$$

and similarly

$$[\boldsymbol{l}, p^2] = 0\,. \tag{1.5}$$

Therefore, the vector $\boldsymbol{l}$ commutes with the central field Hamiltonian (1.1), and the three components of $\boldsymbol{l}$ are consequently constants of the motion.

It is sometimes convenient to express the three operators $\boldsymbol{l}$ in terms of polar coordinates $(r\ \theta\ \varphi)$ defined by

$$\begin{aligned} x &= r \sin\theta \cos\varphi \\ y &= r \sin\theta \sin\varphi \\ z &= r \cos\theta\,. \end{aligned} \tag{1.6}$$

Straightforward calculation then shows that

$$
\begin{aligned}
l_x &= i\left[\sin\varphi\frac{\partial}{\partial\theta} + \cos\varphi\cot\theta\frac{\partial}{\partial\varphi}\right] \\
l_y &= -i\left[\cos\varphi\frac{\partial}{\partial\theta} - \sin\varphi\cot\theta\frac{\partial}{\partial\varphi}\right] \\
l_z &= -i\frac{\partial}{\partial\varphi} \qquad (1.7) \\
l^2 = l_x^2 + l_y^2 + l_z^2 &= -\left[\frac{1}{\sin\theta}\frac{\partial}{\partial\theta}\left(\sin\theta\frac{\partial}{\partial\theta}\right) + \frac{1}{\sin^2\theta}\frac{\partial^2}{\partial\varphi^2}\right].
\end{aligned}
$$

Thus, the angular momentum operators involve only the angular coordinates and do not depend at all on the radial coordinate r. This, of course, is in agreement with (1.4).

The three angular momentum operators l_x, l_y, and l_z do not commute with each other. They satisfy the following well-known commutation relations, easily derived from (1.3):

$$[l_x, l_y] = il_z \quad [l_y, l_z] = il_x \quad [l_z, l_x] = il_y. \qquad (1.8)$$

The operator describing the square of the total orbital angular momentum

$$l^2 = l_x^2 + l_y^2 + l_z^2 \qquad (1.9)$$

commutes, according to (1.8), with each of the components of $\boldsymbol{l}$:

$$[l^2, \boldsymbol{l}] = 0. \qquad (1.10)$$

Although each one of the components of $\boldsymbol{l}$ commutes with the Hamiltonian (1.1), because of (1.8) they cannot all simultaneously have sharp values in any eigenstate of (1.1). Therefore the most we can do for the definition of quantum numbers for the eigenstates of H is to require that these eigenstates will also be simultaneous eigenstates of l^2 and, say, l_z. The three operators H, l^2, and l_z commute with each other and will therefore furnish us with three quantum numbers necessary for the complete description of the motion of a single particle: the energy, the square of the angular momentum, and the projection of the angular momentum on a given axis.

If $\psi(m)$ is a state of a single particle with a sharp value of l_z, i.e.,

$$l_z\,\psi(m) = m\,\psi(m) \qquad (1.11)$$

then we see from (1.8) that the expectation value of l_x in such a state is zero. In fact

$$\langle m|l_x|m\rangle = -i\,\langle m|[l_y, l_z]|m\rangle = -i(m - m^*)\,\langle m|l_y|m\rangle.$$

Since l_z is a Hermitian operator $m = m^*$, and hence

$$\langle m|l_x|m\rangle = 0 . \tag{1.12}$$

Similarly, also

$$\langle m|l_y|m\rangle = 0 . \tag{1.13}$$

The angular momentum vector $\mathbf{M}$ of a classical system in a central field is fixed in space. Although this implies that its three components are constants of motion, only one of them can actually be used for the description of the motion since their Poisson brackets do not vanish. In quantum mechanics the corresponding commutators do not vanish and therefore the angular momentum vector $\boldsymbol{l}$ precesses around some direction in space. More precisely, in an eigenstate of l_z the angular momentum vector precesses around the z-direction so that its projection on the z-axis remains constant and its projections on the x and y-axes average out to zero. It follows from (1.8) that, during this precession, a certain phase relation exists between the projections on the x and y-axes, because the expectation value of the product $l_x l_y$ does not, generally, vanish. In fact, we see from (1.8) that

$$\langle m|l_x l_y - l_y l_x|m\rangle = im . \tag{1.14}$$

Many methods can be used to derive the eigenvalues and eigenfunctions of $\boldsymbol{l}^2$ and l_z. For instance, from the commutation relations (1.8) alone we can deduce, in the standard way, that the eigenvalues of the operator $\boldsymbol{l}^2$ are the positive numbers $l(l+1)$ where l is an integer or a half-integer, and that the eigenvalues of l_z are real integers or half-integers. Furthermore it follows from the commutation relations that if $\psi(l\, m)$ is a simultaneous eigenfunction of $\boldsymbol{l}^2$ and l_z

$$\begin{aligned} \boldsymbol{l}^2\psi(l\, m) &= l(l+1)\psi(l\, m) \\ l_z\psi(l\, m) &= m\, \psi(l\, m) \end{aligned} \tag{1.15}$$

then the only possible values of m are

$$m = l, l-1, l-2, \cdots, -(l-1), -l . \tag{1.16}$$

For each value of l there are therefore $2l+1$ eigenfunctions of $\boldsymbol{l}^2$, all with the same eigenvalue $l(l+1)$ of $\boldsymbol{l}^2$, but belonging to different eigenvalues of l_z.

We now come back to the explicit definition (1.2) of $\boldsymbol{l}$. It follows from (1.15) and (1.7) that the φ-dependence of $\psi(l\, m)$ is given by $e^{im\varphi}$. Since the angular momentum operators involve only the angular coordinates, we can therefore write

$$\psi(l\,m) = \Theta_{lm}(\theta)\frac{1}{\sqrt{2\pi}}\,e^{im\varphi} \tag{1.17}$$

where $1/\sqrt{2\pi}$ is introduced for the normalization of $e^{im\varphi}$.

It follows from (1.17) that if $\psi(l\,m)$ is to be single valued, m must be an *integer*. This conclusion is reached only on the basis of the explicit dependence (1.7) of l_z on the space coordinates, and is thus valid for eigenfunctions of the *orbital* angular momentum (1.2). There are other types of angular momenta, which are not given in terms of the differential operators (1.3) or (1.7) whose commutation relations are still given by (1.8). Their eigenfunctions $\psi(j\ m)$ will not be given by (1.17) and j as well as m may generally assume half-integral values. Since m is related to l by (1.16), we conclude that *orbital angular momenta have integral values of both l and m.*

To obtain the eigenfunctions $\psi(lm)$ of the orbital angular momentum in a convenient form we can use the Laplace operator which, in polar coordinates, assumes the form:

$$\Delta = \frac{1}{r^2}\frac{\partial}{\partial r}\left(r^2\frac{\partial}{\partial r}\right) - \frac{1}{r^2}\,\mathbf{l}^2\,. \tag{1.18}$$

Here $\mathbf{l}^2$ is just the operator (1.7) of the square of the orbital angular momentum. We now obtain the eigenfunctions of $\mathbf{l}^2$ by considering the harmonic homogeneous polynomials of degree l

$$P_l(x, y, z) = \sum_{\alpha+\beta+\gamma=l} c_{\alpha\beta\gamma}^{(l)}\, x^\alpha\, y^\beta\, z^\gamma \tag{1.19}$$

which satisfy

$$\Delta P_l(x, y, z) = 0\,. \tag{1.20}$$

Since x/r, y/r, and z/r are functions of the angular variables θ and φ only, we see from (1.19) that P_l can be written as a product:

$$P_l(x, y, z) = r^l Q_l(\theta\ \varphi) \tag{1.21}$$

where

$$Q_l(\theta\ \varphi) = \sum c_{\alpha\beta\gamma}^{(l)}\left(\frac{x}{r}\right)^\alpha\left(\frac{y}{r}\right)^\beta\left(\frac{z}{r}\right)^\gamma.$$

Operating with (1.18) on (1.21) we see that, due to (1.20),

$$\mathbf{l}^2 Q_l(\theta\ \varphi) = l(l+1)\,Q_l(\theta\ \varphi)\,. \tag{1.22}$$

Therefore the $Q_l(\theta\ \varphi)$ derived from the homogeneous harmonic polynomials according to (1.21) are eigenfunctions of $\mathbf{l}^2$.

To obtain $Q_l(\theta\ \varphi)$, which is also an eigenfunction of l_z and must therefore have the form (1.17), we note that

$$(x \pm iy) = r \sin\theta\, e^{\pm i\varphi}.$$

Let us perform the linear transformation

$$\xi = x + iy$$
$$\eta = x - iy.$$

The P_l will then become homogeneous polynomials of degree l in ξ, η, and z:

$$P_l(x, y, z) = \sum b^{(l)}_{\alpha\beta\gamma}\ \xi^\alpha\ \eta^\beta\ z^\gamma \tag{1.23}$$

where $b^{(l)}_{\alpha\beta\gamma}$ are complex numbers obtained from the $c^{(l)}_{\alpha\beta\gamma}$ in (1.19). If we impose now the restriction

$$b^{(l)}_{\alpha\beta\gamma} = 0 \quad \textit{unless}\ \alpha - \beta = m \tag{1.24}$$

we see that $P_l(x, y, z)$ can be written as:

$$P_{lm}(x, y, z) = Ar^l\, e^{im\varphi}\, \Theta_{lm}(\theta).$$

We can thus obtain an eigenfunction of l^2 and l_z by constructing a homogeneous harmonic polynomial of degree l in which the difference between the powers of $(x + iy)$ and $(x - iy)$ is a constant m.

There are altogether $\frac{1}{2}(l + 1)\ (l + 2)$ independent coefficients $b^{(l)}_{\alpha\beta\gamma}$ in (1.23). The expression ΔP_l is a homogeneous polynomial of degree $l - 2$, and hence the equation $\Delta P_l = 0$ imposes $\frac{1}{2}l(l - 1)$ conditions on the coefficients $b^{(l)}_{\alpha\beta\gamma}$. This leaves $\frac{1}{2}(l + 1)\ (l + 2) - \frac{1}{2}\, l\, (l - 1) = 2l + 1$ independent solutions of Eq. (1.20). These solutions are then uniquely determined (up to normalization) by choosing a specific value of m according to (1.24). Since $\alpha + \beta + \gamma = l$, m in (1.24) satisfies $-\ l \leq m \leq l$ and can assume exactly $2l + 1$ different values.

There is still a freedom in the choice of the phases and the normalization of the eigenfunctions of l^2 and l_z as derived above. We see from (1.24) that if $m = l$ the homogeneous polynomial reduces to a single term

$$P_{l,m=l} = b^{(l)}_{l00}\xi^l = b^{(l)}_{l00}\, r^l \sin^l\theta\, e^{il\varphi}.$$

Denoting by $Y_{lm}(\theta\ \varphi)$ the normalized eigenfunctions of l^2 and l_z we define their phases so that

$$Y_{ll}(\theta\ \varphi) = (-1)^l \sqrt{\frac{(2l + 1)!}{2}}\, \frac{1}{2^l l!} \sin^l\theta\, \frac{1}{\sqrt{2\pi}}\, e^{il\varphi} \tag{1.25}$$

and

$$(l_x \pm il_y)\, Y_{lm}(\theta\, \varphi) = + \sqrt{l(l+1) - m(m \pm 1)}\; Y_{lm\pm 1}(\theta\, \varphi)\,. \tag{1.26}$$

In (1.26) the equality of the *magnitudes* of both sides follows from the commutation relations of l_x, l_y, and l_z. The requirement on the relative phases of the various Y_{lm} functions is a matter of convention. This is done by the definitions (1.25) and (1.26).

The functions $Y_{lm}(\theta\, \varphi)$ are called *spherical harmonics* of degree l. From their derivation follows that $r^l\, Y_{lm}(\theta\, \varphi)$ is a harmonic homogeneous polynomial in x, y, and z which contains $x + iy$ and $x - iy$ to powers with a constant difference m. This property is often very useful for the simple derivation of various properties of $Y_{lm}(\theta\, \varphi)$. For example, under reflection in the origin, $x \to x' = -x$, $y \to y' = -y$, and $z \to z' = -z$, the polar coordinates undergo the transformation $r \to r' = r$, $\theta \to \theta' = \pi - \theta$, $\varphi' = \pi + \varphi$. A homogeneous polynomial of degree l is multiplied by $(-1)^l$ when we transform from $\mathbf{r}$ to $\mathbf{r}' = -\mathbf{r}$. Hence we conclude that

$$Y_{lm}(\theta'\, \varphi') = Y_{lm}(\pi - \theta,\, \pi + \varphi) = (-1)^l\, Y_{lm}(\theta\, \varphi)\,. \tag{1.27}$$

Similarly, we obtain for $\Theta_{lm}(\theta)$ defined by

$$Y_{lm}(\theta\, \varphi) = \frac{1}{\sqrt{2\pi}}\, \Theta_{lm}(\theta)\, e^{im\varphi} \tag{1.28}$$

that

$$\Theta_{lm}(\theta') = \Theta_{lm}(\pi - \theta) = (-1)^{l+m}\, \Theta_{lm}(\theta)\,. \tag{1.29}$$

Since $\sin(\pi - \theta) = \sin\theta$ and $\cos(\pi - \theta) = -\cos\theta$ we can construct the Table 1.1 for the powers of $\cos\theta$ and $\sin\theta$ in $\Theta_{lm}(\theta)$.

TABLE 1.1

POWERS OF $\cos\theta$ AND $\sin\theta$ IN $\Theta_{lm}(\theta)$

	Powers of $\cos\theta$		Powers of $\sin\theta$	
$l + m$	l even	l odd	l even	l odd
Even	Even	Even	Even	Odd
Odd	Odd	Odd	Odd	Even

From (1.23) and (1.24) it can further be seen that

m is the maximum power of $\sin\theta$ which can be factorized from $Y_{lm}(\theta\, \varphi)$. (1.30)

The first few spherical harmonics are given in the Appendix.

2. Total Angular Momentum of Spin-$\frac{1}{2}$ Particles

The spherical harmonics introduced in the preceding section are the eigenfunctions of the orbital angular momentum. Certain particles, electrons and nucleons in particular, possess also an intrinsic angular momentum which cannot be described in terms of any internal spatial motion. The electron, for instance, is known experimentally to have an intrinsic angular momentum whose z-projection can assume one of the two values $+\frac{1}{2}\hbar$ or $-\frac{1}{2}\hbar$. No internal motion, in the present framework of quantum mechanics, can give rise to half-integral values of an angular momentum. The spin of the electron should therefore be considered as an internal property like its charge or mass. Still, it has the same dimensions and physical meaning as the external (orbital) angular momentum. Associated with the internal spin there is usually also an internal magnetic moment. The spin degree of freedom may therefore be dynamically involved in the equations of motion. It follows that, generally, only the *total* angular momentum, spin plus orbital, is conserved in closed systems.

Particles with discrete internal degrees of freedom can be described by a set of wave functions instead of a single wave function. Thus, the negative electron with spin $\frac{1}{2}$ can be described by two wave functions.

$$\psi = \begin{pmatrix} \psi_{\uparrow}(\mathbf{r}, t) \\ \psi_{\downarrow}(\mathbf{r}, t) \end{pmatrix} \tag{2.1}$$

where $\psi_{\uparrow}$ and $\psi_{\downarrow}$ give the probability amplitudes of finding the electron at $(\mathbf{r}, t)$ with its spin up or down respectively.

In Part II we shall see the connection between the number of wave functions required to describe a given particle and its internal spin. It will there become clear why ascribing to a particle a number of spatial wave functions implies the existence of an internal *angular momentum.* Here it will suffice to note that a particle with an intrinsic spin s, i.e., whose intrinsic angular momentum z-projection can assume the values $-s\hbar$, $-(s-1)\hbar$, $\cdots$, $(s-1)\hbar$, $s\hbar$, should be described by a set of $2s+1$ functions. These functions include, in addition to the space and time coordinates, an index which specifies to which internal state they belong. This index is often called the *spin coordinate*, and the operators which represent the internal spin operate on this spin coordinate. Thus if we denote by $\hbar s_z$ the operator of the z-projection of the electron's spin $\hbar\mathbf{s}$, we have, from (2.1)

$$
\begin{aligned}
s_z \psi_\uparrow(\mathbf{r}, t) &= \tfrac{1}{2} \psi_\uparrow(\mathbf{r}, t) \\
s_z \psi_\downarrow(\mathbf{r}, t) &= -\tfrac{1}{2} \psi_\downarrow(\mathbf{r}, t) .
\end{aligned}
\tag{2.2}
$$

If we adopt the column wave function ψ of (2.1) as the wave function of the electron, we can write down the matrix representing the operator s_z in the scheme defined by $\uparrow$ and $\downarrow$, namely:

$$
\begin{aligned}
&\langle\uparrow|s_z|\uparrow\rangle = \tfrac{1}{2} \qquad \langle\downarrow|s_z|\downarrow\rangle = -\tfrac{1}{2} \\
&\langle\downarrow|s_z|\uparrow\rangle = \langle\uparrow|s_z|\downarrow\rangle = 0
\end{aligned}
\tag{2.3}
$$

or

$$
s_z = \begin{pmatrix} \frac{1}{2} & 0 \\ 0 & -\frac{1}{2} \end{pmatrix} . \tag{2.4}
$$

It should be remembered that s_z operates on a set of coordinates different from $(\mathbf{r}, t)$ and independent of them. To stress this distinction, the operator s_z is said to operate in the *spin space*. In this space we have, for the electron,

$$
\mathbf{s}^2 \psi = s(s+1)\psi = \tfrac{3}{4} \psi \tag{2.5}
$$

or

$$
\mathbf{s}^2 = \begin{pmatrix} \frac{3}{4} & 0 \\ 0 & \frac{3}{4} \end{pmatrix} . \tag{2.6}
$$

The scalar matrix $\mathbf{s}^2$ is diagonal in any scheme, and has always the same eigenvalues. This should be confronted with the matrix of l^2, which in diagonal form is given by

$$
\begin{array}{c} \\ l^2 = \end{array}
\begin{array}{c}
\begin{array}{ccc} l=0 & l=1 & l=2 \end{array} \\
\left[
\begin{array}{c|c|c|c}
0 & & & \\
\hline
0 & \begin{matrix} 2 & 0 & 0 \\ 0 & 2 & 0 \\ 0 & 0 & 2 \end{matrix} & 0 & \\
\hline
0 & & \begin{matrix} 6 & & \\ & \ddots & \\ & & 6 \end{matrix} & \\
\hline
 & & & \ddots
\end{array}
\right]
\end{array}
\tag{2.7}
$$

The matrix of l^2 in a different scheme is not necessarily diagonal, corresponding to the fact that under some conditions the total angular momentum may not be a good quantum number. The operator $\mathbf{s}^2$ is, however, always diagonal, emphasizing the internal nature of the spin, which remains the same in all schemes. Thus an electron plane wave does not represent a state with a sharp value of the orbital angular momentum nor does it have a sharp total angular momentum or a sharp z-projection of $\mathbf{s}$. But the intrinsic angular momentum (squared) is sharp and has the value $\hbar s(s+1) = \frac{3}{4}\hbar$.

It is customary to quantize the orbital and spin angular momenta along the same z-axis. This is not essential in any way since the two quantities refer to different degrees of freedom. However, since under some conditions, only the sum of the two angular momenta is a constant of motion it is convenient to have a common z-axis for both. We shall denote by

$$\psi(s\, l\, m_s m_l) \tag{2.8}$$

a wave function of a particle with internal spin s, orbital angular momentum l, and corresponding z-projections m_s and m_l. We shall be dealing with $s = \frac{1}{2}$ particles and therefore often omit s from ψ in (2.8) and write it as $\psi(l\, m_s m_l)$. The projection m_s can then take on the two values $m_s = +\frac{1}{2}$ and $m_s = -\frac{1}{2}$. If $\psi(s\, l\, m_s m_l)$ is considered as a function of the space and spin coordinates, it is a single wave function and *not* a column. Since the spin and orbital angular momenta refer to different, independent degrees of freedom, it is possible to write ψ of (2.8) also as a product

$$\psi(l\, m_s m_l) = \varphi(l\, m_l)\, \chi(m_s)\,. \tag{2.9}$$

The function $\chi(m_s)$ is then the spin wave function, independent of space and time coordinates, and satisfies, for spin-$\frac{1}{2}$ particles,

$$\begin{aligned} \mathbf{s}^2\, \chi(m_s) &= \tfrac{3}{4}\, \chi(m_s) \\ s_z\, \chi(m_s) &= m_s\, \chi(m_s)\,. \end{aligned} \tag{2.10}$$

The spin functions satisfy also an orthogonality relation. Formally it is written as

$$\sum_{\substack{\text{spin} \\ \text{coordinates}}} \chi^*(m_s)\, \chi(m_s') = \delta_{m_s m_s'}\,. \tag{2.11}$$

Actually this relation is trivial. Introducing the spin coordinate through the arrows $\uparrow$ and $\downarrow$, we have

$$\begin{aligned} \sum_{\substack{\text{spin} \\ \text{coordinates}}} \chi^*(m_s)\, \chi(m_s') &= \chi_\uparrow^*(m_s)\, \chi_\uparrow(m_s') + \chi_\downarrow^*(m_s)\, \chi_\downarrow(m_s') \\ &= \delta_{m_s,1/2}\, \delta_{m_s',1/2} + \delta_{m_s,-1/2}\, \delta_{m_s',-1/2} = \delta_{m_s m_s'}\,. \end{aligned}$$

The wave functions $\psi(l\, m_s m_l)$ satisfy, of course,

$$\begin{aligned} l^2\, \psi(l\, m_s m_l) &= l(l+1)\, \psi(l\, m_s m_l) \\ l_z\, \psi(l\, m_s m_l) &= \quad m_l \quad \psi(l\, m_s m_l) \\ \mathbf{s}^2\, \psi(l\, m_s m_l) &= \quad \tfrac{3}{4} \quad \psi(l\, m_s m_l) \\ s_z\, \psi(l\, m_s m_l) &= \quad m_s \quad \psi(l\, m_s m_l)\,. \end{aligned} \tag{2.12}$$

They also satisfies the orthogonality relations

$$\sum_{\substack{\text{spin}\\ \text{coordinates}}} \int \psi^*(l\, m_s m_l)\, \psi(l\, m_s' m_l')\, d\tau = \delta(m_s\,, m_s')\delta(m_l\,, m_l')\,. \tag{2.13}$$

The three spin operators s_x, s_y and s_z satisfy the commutation relations of the three components of an angular momentum, namely

$$[s_x\,, s_y] = is_z, \quad [s_y\,, s_z] = is_x \quad [s_z\,, s_x] = \; is_y\,. \tag{2.14}$$

Such commutation relations are usually taken as the definition of an angular momentum operator. It was already mentioned that it follows from (2.14) that there are only integral or half-integral angular momenta. The fact that the orbital angular momentum $\boldsymbol{l}$ has only integral values follows, as we have seen, from its explicit definition (1.3) in terms of the space coordinates.

Since the spin operators and the orbital angular momentum operate on different coordinates they commute with each other

$$[s_i\,, l_k] = 0 \quad i, k = 1, 2, 3\,. \tag{2.15}$$

It is a direct result of (1.8) and (2.14) that the operator

$$\mathbf{j} = \boldsymbol{l} + \mathbf{s} \tag{2.16}$$

also satisfies the commutation relations of an angular momentum, namely

$$[j_x\,, j_y] = ij_z \quad \text{and cyclic permutations.} \tag{2.17}$$

It can therefore be called the *operator of total angular momentum.*

It is worthwhile to note that if the components of a certain operator $\mathbf{j}$ satisfy the commutation relations (2.17) of an angular momentum, this will not be true for the operator $\mathbf{j}' = \alpha\mathbf{j}$ with $\alpha \neq 1$. Thus, for instance, $\boldsymbol{l} - \mathbf{s}$ or $\boldsymbol{l} + \frac{1}{4}\mathbf{s}$ are *not* angular momenta, and *only* $\boldsymbol{l} + \mathbf{s}$ is an angular momentum.

The operator of total angular momentum **j**, satisfies, as can be readily seen, the relations

$$[\mathbf{j}^2, \boldsymbol{l}^2] = [\mathbf{j}^2, \mathbf{s}^2] = [j_z, \boldsymbol{l}^2] = [j_z, \mathbf{s}^2] = 0\,. \tag{2.18}$$

On the other hand

$$[\mathbf{j}^2, l_z] = [\boldsymbol{l}^2 + \mathbf{s}^2 + 2(\boldsymbol{l}\cdot\mathbf{s}), l_z] = [2(\boldsymbol{l}\cdot\mathbf{s}), l_z] \neq 0$$

and

$$[\mathbf{j}^2, s_z] = [\boldsymbol{l}^2 + \mathbf{s}^2 + 2(\boldsymbol{l}\cdot\mathbf{s}), s_z] = [2(\boldsymbol{l}\cdot\mathbf{s}), s_z] \neq 0\,.$$

Thus we conclude that in a state of a well-defined total angular momentum $\mathbf{j}^2$, l_z and s_z cannot have sharp values simultaneously. The set of commuting operators can at most consist of four operators, such as $\mathbf{j}^2$, $\mathbf{s}^2$, $\boldsymbol{l}^2$, and j_z.

A simultaneous eigenfunction $\psi(s\,l\,j\,m)$ of these operators is therefore a linear combination of the functions $\psi(l\,m_s m_l)$ with different values of m_s and m_l

$$\psi(s\,l\,j\,m) = \sum_{m_s m_l} (s\,m_s l\,m_l|s\,l\,j\,m)\psi(s\,l\,m_s m_l) \tag{2.19}$$

where $(s\,m_s\,l\,m_l|s\,l\,j\,m)$ are numerical coefficients, known as *Clebsch-Gordan coefficients.* Their properties and determination in the general case will be taken up in Part II. However, for the case $s = \frac{1}{2}$ they can be easily determined. Thus, $\psi(s\,l\,j\,m)$ has to satisfy

$$\begin{aligned} \mathbf{j}^2\,\psi(s\,l\,j\,m) &= j(j+1)\psi(s\,l\,j\,m) \\ j_z\,\psi(s\,l\,j\,m) &= m\,\psi(s\,l\,j\,m)\,. \end{aligned} \tag{2.20}$$

Operating on $\psi(s\,l\,j\,m)$ with j_z we obtain

$$\begin{aligned} j_z\,\psi(s\,l\,j\,m) &= (l_z + s_z) \sum_{m_s m_l} (\tfrac{1}{2}\,m_s l\,m_l|\tfrac{1}{2}\,l\,j\,m)\psi(s\,l\,m_s m_l) \\ &= \sum_{m_s m_l} (m_s + m_l)\,(\tfrac{1}{2}\,m_s l\,m_l|\tfrac{1}{2}\,l\,j\,m)\psi(s\,l\,m_s m_l) \\ &= m\psi(s\,l\,j\,m) = m \sum (\tfrac{1}{2}\,m_s l\,m_l|\tfrac{1}{2}\,l\,j\,m)\psi(s\,l\,m_s m_l)\,. \end{aligned}$$

Using the orthogonality relations (2.13) we obtain

$$(m_s + m_l)\,(\tfrac{1}{2}\,m_s l\,m_l|\tfrac{1}{2}\,l\,j\,m) = m(\tfrac{1}{2}\,m_s l\,m_l|\tfrac{1}{2}\,l\,j\,m)$$

Hence $(\frac{1}{2}\,m_s\,l\,m_l|\frac{1}{2}\,l\,j\,m)$ can be different from zero only if

$$m_s + m_l = m\,.$$

There are therefore only two terms in the summation (2.19)

$$\psi(s\, l\, j\, m) = a\psi_+ + b\psi_-$$

where $a = (\frac{1}{2}, \frac{1}{2}, l, m - \frac{1}{2}|\frac{1}{2}l\, j\, m)$ $\quad b = (\frac{1}{2}, -\frac{1}{2}, l, m + \frac{1}{2}|\frac{1}{2}l\, j\, m)$ and $\psi_\pm = \psi(\frac{1}{2}, l, \pm\frac{1}{2}, m\mp\frac{1}{2})$. The coefficients a and b can now be determined using (2.20).

The operator $\mathbf{j}^2$ can be written as

$$\mathbf{j}^2 = \boldsymbol{l}^2 + \mathbf{s}^2 + (l_x + il_y)(s_x - is_y) + (l_x - il_y)(s_x + is_y) + 2l_z s_z$$

Using (1.26) we obtain

$$\begin{aligned} \mathbf{j}^2 \psi(s\, l\, j\, m) &= j(j+1)\{a\psi_+ + b\psi_-\} \\ &= [l(l+1) + \tfrac{3}{4}]\{a\psi_+ + b\psi_-\} + \sqrt{l(l+1) - (m - \tfrac{1}{2})(m + \tfrac{1}{2})}\, a\psi_- \\ &+ \sqrt{l(l+1) - (m + \tfrac{1}{2})(m - \tfrac{1}{2})}\, b\psi_+ + (m - \tfrac{1}{2})\, a\psi_+ - (m + \tfrac{1}{2})\, b\psi_- . \end{aligned}$$

Hence, using the orthogonality of the $\psi_\pm$ we obtain

$$\begin{aligned} a[j(j+1) - (l + \tfrac{1}{2})^2 - m] - b\sqrt{(l + \tfrac{1}{2})^2 - m^2} &= 0 \\ -a\sqrt{(l + \tfrac{1}{2})^2 - m^2} \qquad + b[j(j+1) - (l + \tfrac{1}{2})^2 + m] &= 0 . \end{aligned} \tag{2.21}$$

The compatibility condition for these two homogeneous equations in the two unknowns a and b requires that

$$j = l \pm \tfrac{1}{2} \tag{2.22}$$

corresponding to the well-known fact that the total angular momentum of the electron is obtained by either adding or subtracting $\frac{1}{2}$ from the orbital angular momentum. For $j = l + \frac{1}{2}$ we have

$$a\sqrt{l + \tfrac{1}{2} - m} \quad - b\sqrt{l + \tfrac{1}{2} + m} = 0$$

and hence

$$a = N\sqrt{l + \tfrac{1}{2} + m} \qquad b = N\sqrt{l + \tfrac{1}{2} - m}$$

where N can be determined by the normalization requirement on $\psi(s\, l\, j\, m)$

$$a^2 + b^2 = 1 .$$

A similar result is obtained for $j = l - \frac{1}{2}$.

Table 2.1 summarizes the normalized Clebsch-Gordan coefficients.

TABLE 2.1

THE CLEBSCH-GORDAN COEFFICIENTS: $(\frac{1}{2}\, m_s l\, m_l | \frac{1}{2}\, l\, j\, m)$

j \ m_s	$+\frac{1}{2}$	$-\frac{1}{2}$
$l+\frac{1}{2}$	$\sqrt{\frac{l+\frac{1}{2}+m}{2l+1}}$	$\sqrt{\frac{l+\frac{1}{2}-m}{2l+1}}$
$l-\frac{1}{2}$	$\sqrt{\frac{l+\frac{1}{2}-m}{2l+1}}$	$-\sqrt{\frac{l+\frac{1}{2}+m}{2l+1}}$

Equations (2.21) determine only the relative phases of the coefficients a and b; the absolute phase is a matter of convention. The phases in Table 2.1 were fixed according to the accepted phase convention to be discussed in Part II.

To summarize this section we have found that the functions

$$\psi(l\,j\,m) = \sum (\tfrac{1}{2}\, m_s\, l\, m_l | \tfrac{1}{2}\, l\, j\, m)\psi(s\, l\, m_s m_l)$$

are eigenfunctions of $\mathbf{j}^2$ and j_z. As such they satisfy the relations corresponding to (1.15) and (1.26)

$$\begin{aligned} \mathbf{l}^2\psi(l\,j\,m) &= l(l+1)\psi(l\,j\,m) \\ \mathbf{s}^2\psi(l\,j\,m) &= s(s+1)\psi(l\,j\,m) \\ \mathbf{j}^2\psi(l\,j\,m) &= j(j+1)\psi(l\,j\,m) \\ j_z\psi(l\,j\,m) &= m\psi(l\,j\,m) \end{aligned} \qquad (2.23)$$

$$(j_x \pm ij_y)\psi(l\,j\,m) = \sqrt{j(j+1) - m(m \pm 1)}\, \psi(l\,j\,m)\,.$$

Explicitly, these functions are given by

$$\psi(l\,j = l \pm \tfrac{1}{2},\, m) = \sqrt{\frac{l+\frac{1}{2}\pm m}{2l+1}}\, \varphi(l, m-\tfrac{1}{2})\chi(\tfrac{1}{2}) \pm \sqrt{\frac{l+\frac{1}{2}\mp m}{2l+1}}\, \varphi(l, m+\tfrac{1}{2})\chi(-\tfrac{1}{2})\,. \qquad (2.24)$$

3. Transformation Properties of Spherical Harmonics

The eigenfunctions of the orbital angular momentum operators $\mathbf{l}^2$ and l_z play a very important role in the atomic and nuclear shell model, as well as in many other problems which involve central fields. Being a complete orthonormal set of functions, which in addition are eigenfunctions of the operator representing a constant of motion in a central field, the spherical harmonics naturally become the most convenient functions to work with in such problems.

There is another point of view from which the properties of the spherical harmonics can be considered. The functional form of the operator $\mathbf{l}^2$, whose eigenfunctions are the spherical harmonics, remains unchanged when we perform a rotation of the frame of reference. Let us perform a rotation R of the frame of reference, given by a matrix a, so that a space point P with coordinates (x_1, x_2, x_3) is given, after the rotation, by its new coordinates (x_1', x_2', x_3'), where

$$x_i' = \sum_k a_{ik} x_k \tag{3.1}$$

and

$$\sum_i a_{ik} a_{il} = \sum_i a_{ki} a_{li} = \delta(k, l) .$$

The operator $\mathbf{l}^2$ is then transformed into

$$(\mathbf{l}')^2 = \frac{1}{\hbar^2} [(\mathbf{r}' \times \mathbf{p}')]^2 = \frac{1}{\hbar^2} [(\mathbf{r} \times \mathbf{p})]^2 = (\mathbf{l})^2 . \tag{3.2}$$

In other words, the operator [cf. (1.7)]

$$\mathbf{l}^2 = -\left[\frac{1}{\sin\theta} \frac{\partial}{\partial\theta} \left(\sin\theta \frac{\partial}{\partial\theta}\right) + \frac{1}{\sin^2\theta} \frac{\partial^2}{\partial\varphi^2}\right] \tag{3.3}$$

goes over into the operator

$$(\mathbf{l}')^2 = -\left[\frac{1}{\sin\theta'} \frac{\partial}{\partial\theta'} \left(\sin\theta' \frac{\partial}{\partial\theta'}\right) + \frac{1}{\sin^2\theta'} \frac{\partial^2}{\partial\varphi'^2}\right] . \tag{3.4}$$

This last operator is constructed from the new coordinates $\theta' \varphi'$ according to exactly the same prescription employed to construct $\mathbf{l}^2$ from the old coordinates. This is a very special property, of a small class of operators, which is essential if we want to give them an invariant meaning as will be explained in Part II. The reader may easily verify that an operator $\Omega = x\, \partial/\partial y$, for instance, does not possess this property:

$$\Omega' = x' \frac{\partial}{\partial y'}$$

is not generally what we obtain by applying the transformation (3.1) to $x\, \partial/\partial y$.

The fact that $\boldsymbol{l}^2$ does have the invariance property exhibited by (3.3) and (3.4) gives rise to important relations among its eigenfunctions. Let $Y_{lm}(\theta'\, \varphi')$ be eigenfunctions of $(\boldsymbol{l}')^2$ and l'_z where the prime indicates that the operators are taken with respect to the rotated frame

$$\begin{aligned} (\boldsymbol{l}')^2\, Y_{lm'}(\theta'\, \psi') &= l(l+1)\, Y_{lm'}(\theta'\, \varphi') \\ l'_z Y_{lm'}(\theta'\, \varphi') &= \quad m' \quad Y_{lm'}(\theta'\, \psi'). \end{aligned} \tag{3.5}$$

Introducing the transformation (3.1) in $Y_{lm}(\theta'\, \varphi')$ it becomes a function of θ and φ

$$Y_{lm'}(\theta'\, \varphi') = Y_{lm'}[\theta'(\theta\, \varphi), \varphi'(\theta\, \varphi)] = f(\theta\, \varphi). \tag{3.6}$$

However, since $(\boldsymbol{l}')^2 = \boldsymbol{l}^2$, we conclude from (3.5) that

$$\boldsymbol{l}^2 f(\theta\, \varphi) = l(l+1) f(\theta\, \varphi).$$

The function $f(\theta\, \varphi)$ can therefore be expanded in terms of the independent functions $Y_{lm}\,(\theta\, \varphi)$. Thus we can write

$$Y_{lm'}(\theta'\, \varphi') = \sum_m Y_{lm}(\theta\, \varphi)\, D^{(l)}_{m\, m'}(R) \tag{3.7}$$

where the expansion coefficients which depend on the rotation R are denoted by $D^{(l)}_{m\, m'}(R)$. The important feature of (3.7) is that $Y_{lm'}(\theta'\, \varphi')$ in the rotated frame is given completely by the spherical harmonics $Y_{lm}(\theta\, \varphi)$, of the *same order* l, in the original frame of reference. This property is often referred to by saying that under rotations the spherical harmonics of order l *transform among themselves*. The physical interpretation of this property is obvious. States with a given angular momentum (squared) relative to one frame have the *same* angular momentum relative to any other rotated frame. It will be noted that this statement is not generally true for a *component* of an angular momentum like l_z. A state with a definite value of the z-projection of $\boldsymbol{l}$ in one frame does not generally represent a definite z-projection of the angular momentum in another frame. Equation (3.7) shows that $Y_{lm'}(\theta'\, \varphi')$ is a *linear combination* of $Y_{lm}(\theta\, \varphi)$ with several *different* values of m.

The coefficients $D^{(l)}_{m\,m'}(R)$ for a given value of l and a given rotation R are the components of a $(2l+1)$ by $(2l+1)$ matrix $D^{(l)}(R)$ whose rows and columns are characterized by the indices m and m'. The matrix $D^{(l)}(R)$ is known as *Wigner's D matrix*, and (3.7) can be written in a matrix notation as:

$$\mathbf{Y}_l(\theta'\ \varphi') = \mathbf{Y}_l(\theta\ \varphi)\ D^{(l)}(R)\,. \tag{3.8}$$

Here $\mathbf{Y}_l$ stands for the $2l+1$-dimensional row $Y_{l,l}, Y_{l,l-1}, \cdots, Y_{l,-l}$.

The D matrix can be readily constructed for some simple cases. For example, consider a rotation R_z around the z-axis by an angle φ_0.

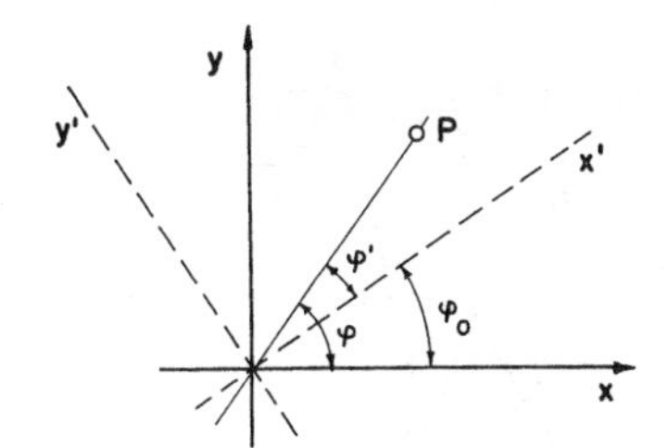

FIG. 3.1 A rotation R_z.

A point P whose coordinates in the original frame were $(r\ \theta\ \varphi)$ will have the new coordinates $(r'\ \theta'\ \varphi')$ where $r' = r$, $\theta' = \theta$, $\varphi' = \varphi - \varphi_0$. Using (1.28) we obtain

$$Y_{lm'}(\theta'\ \varphi') = \frac{1}{\sqrt{2\pi}}\,\Theta_{lm'}(\theta')\,e^{im'\varphi'} = e^{-im'\varphi_0}\,\frac{1}{\sqrt{2\pi}}\,\Theta_{lm'}(\theta)\,e^{im'\varphi}\,.$$

Comparing with (3.7) we see that in this special case

$$D^{(l)}_{m\,m'}(R_z) = e^{-im\varphi_0}\,\delta_{m\,m'} \tag{3.9}$$

Thus, the D-matrix is diagonal for rotations around the z-axis, corresponding to the fact that such rotations *do not* change the z-projection of an angular momentum.

Many properties of the D-matrices can be obtained directly from their definition (3.7) using the known properties of the spherical harmonics. Since the D-matrices transform one orthonormal system into another they must be unitary. This leads to the relation

$$\sum_m D^{(l)*}_{m\,m'}(R)\,D^{(l)}_{m\,m''}(R) = [D^{(l)}(R)^{\dagger}\,D^{(l)}(R)]_{m'\,m''} = \delta_{m'\,m''}\,. \tag{3.10}$$

When it is necessary to write explicitly the rotation R, the Euler angles $(\alpha\ \beta\ \gamma)$ can be used (Fig. 3.2).

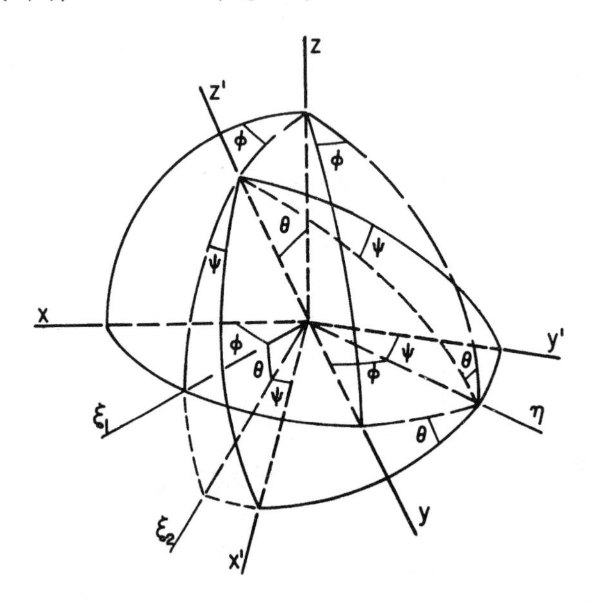

FIG. 3.2. Euler angles between two sets of coordinate axes (x, y, z) and (x', y', z'). The axes ξ_1 and ξ_2 are intermediate between x and x' and η between y and y'.

We can also define integrations over all rotations by

$$\int \dots dR \equiv \int_0^{2\pi}\int_0^{\pi}\int_0^{2\pi} \dots d\alpha \sin\beta d\beta\, d\gamma\,. \tag{3.11}$$

An important relation, the proof of which will not be given here, completes the orthogonality relation (3.10). This relation is

$$\int D^{(l)*}_{m_1 m_2}(R)\ D^{(l')}_{m_1' m_2'}(R)\, dR = \frac{8\pi^2}{2l+1}\,\delta(l, l')\delta(m_1, m_1')\delta(m_2, m_2')\,. \tag{3.12}$$

Another useful relation is obtained if we consider in (3.7) a rotation R given by the Euler angles $\alpha\,\beta\,\gamma$, and a point P whose coordinates in the original frame were $(r\ \theta\ \varphi)$ with $\theta = \beta$ and $\varphi = \alpha$. Clearly the coordinates of this point in the new frame will be $(r\ \theta'\ \varphi')$ with $\theta' = 0$ and φ' undetermined. Using (3.7) we then have

$$Y_{lm'}(0\ \varphi') = \sum_m Y_{lm}(\beta\ \alpha)\, D^{(l)}_{m\,m'}(\alpha\,\beta\,\gamma)$$

and using (3.10)

$$\sum_{m'} D^{(l)*}_{m\,m'}(\alpha\,\beta\,\gamma)\,Y_{lm'}(0\,\varphi') = Y_{lm}(\beta\,\alpha). \tag{3.13}$$

However, from (1.30) we see that, for $\theta = 0$, $Y_{lm}(\theta\,\varphi)$ may be different from zero only if $m = 0$, and, in fact, it can be shown that

$$Y_{lm}(0\,\varphi) = \sqrt{\frac{2l+1}{4\pi}}\,\delta(m,0)\,. \tag{3.14}$$

Substituting (3.14) in (3.13) we find that

$$D^{(l)*}_{m0}(\alpha\,\beta\,\gamma) = \sqrt{\frac{4\pi}{2l+1}}\,Y_{lm}(\beta\,\alpha)\,. \tag{3.15}$$

Thus, the matrix elements $D^{(l)}_{m0}(\alpha\,\beta\,\gamma)$ are independent of γ, and for $\beta = \gamma = 0$, i.e., for a rotation with an angle α around the z-axis, we obtain from (3.15)

$$D^{(l)*}_{m0}(R_z) = \delta(m,0)$$

in agreement with (3.9).

In a similar way it can be shown that

$$D^{(l)*}_{0m}(\alpha\,\beta\,\gamma) = (-1)^m\sqrt{\frac{4\pi}{2l+1}}\,Y_{lm}(\beta\,\gamma)\,. \tag{3.16}$$

If we consider two rotations R_1 and R_2 performed successively, then we can describe their combined effect by a single rotation R_3 which can be written as their product

$$R_3 = R_2\,R_1\,.$$

Denoting the coordinates of a point P after the rotation R_1 by $(r'\,\theta'\,\varphi')$, and the coordinates of P after the combined rotation $R_2\,R_1$ by $(r''\,\theta''\,\varphi'')$ we obtain from (3.7)

$$\begin{aligned} Y_{lm''}(\theta''\,\varphi'') &= \sum_m Y_{lm}(\theta\,\varphi)\,D^{(l)}_{m\,m''}(R_3) \\ &= \sum_{m'} Y_{lm'}(\theta'\,\varphi')D^{(l)}_{m'\,m''}(R_2) = \sum_{m\,m'} Y_{lm}(\theta\,\varphi)D^{(l)}_{m\,m'}(R_1)\,D^{(l)}_{m'\,m''}(R_2). \end{aligned}$$

Hence

$$D^{(l)}_{m\,m''}(R_3) = \sum_{m'} D^{(l)}_{m\,m'}(R_1)\,D^{(l)}_{m'\,m''}(R_2) \tag{3.17}$$

or, in matrix notation,

$$D^{(l)}(R_3) = D^{(l)}(R_1)\,D^{(l)}(R_2)\,. \tag{3.18}$$

It is important to note the order in which R_1 and R_2 appear on the right-hand side of (3.17) or (3.18). This order is the reverse of the conventional one. This is due to our definition (3.7), where the D matrices operate on *rows* to their left rather than on columns to their right.

Due to the relation (3.18) the Wigner $D^{(l)}$ matrices (more precisely $D^{(l)\dagger}$) form a representation of the rotation group, i.e., they satisfy the requirement that the representative matrix of a product of two rotations is the product of the matrices representing each of the rotations. Many properties of the D-matrices and the spherical harmonics can consequently be easily and elegantly obtained by using the powerful group theoretical methods. However, for most of our considerations, we shall try not to make use of these techniques.

4. Radial Functions and Their Properties

The angular part of the wave function of a particle moving in a central field is independent of the special form of the potential. It is determined as shown in the previous sections, by the orbital angular momentum of the particle in that field. The orbital angular momentum can take on a definite set of values, independent of the shape of the central potential. It is enough to know that the field is a central field in order to assure that the orbital angular momentum operator commutes with the energy. The angular dependence of the wave function describing a particle moving in that field is then uniquely fixed. The specific dynamical aspects of various central fields show up only in the *radial* part of the wave function. We do not have therefore any "universal radial functions" like the spherical harmonics for the angular part, and the radial functions have to be calculated separately for every potential.

The situation in most practical cases is even more complex. In treating systems of many particles we can often assume, at least as a first approximation, that each particle moves in a stationary average field created by all other particles. From general symmetry considerations we can then deduce, in some cases, that the average field must have a spherical symmetry, i.e., it must be a central field. In most cases this is about all that can be said about the average field, and we do not even know the detailed radial dependence of this average potential. Thus, the derivation of explicit expressions for the radial wave functions is sometimes not only difficult but actually impossible due to lack of detailed knowledge of the form of the average potential.

Fortunately there are many properties of atoms and nuclei which are not so sensitive to the detailed radial dependence of the wave function. Such properties will generally involve averages of r or r^2 over the radial distribution. On the other hand, energies of systems of particles depend strongly on the radial functions. For such quantities, methods have been developed to reduce all unknown functions to a small number of parameters. This may allow us to draw far-reaching conclusions, as will be shown in more detail in the sequel.

The theory of radial functions is treated in many textbooks and we shall not consider it here in detail. We shall only mention a few properties of the radial functions which will be most useful in the subsequent treatment of the shell theory.

A particle of mass M moving in a central field obeys the Schrödinger equation

$$\left[-\frac{\hbar^2}{2M}\Delta + U(r)\right]\psi(r\ \theta\ \varphi) = E\psi(r\ \theta\ \varphi)\,. \tag{4.1}$$

We shall usually define the zero of the energy by requiring that $U(r) \to 0$ as $r \to \infty$. In any case, the solutions of (4.1) will remain the same if a constant potential U_0 is added to $U(r)$. Only the energy E will change by exactly the same amount U_0. For bound states we have, according to our requirement, $E < 0$ and $\psi(r\ \theta\ \varphi) \to 0$ as $r \to \infty$. Equation (4.1) is separable in the radial and angular coordinates leading, as well known, to

$$\psi_{nlm}(r\ \theta\ \varphi) = \frac{1}{r} R_{nl}(r)\, Y_{lm}(\theta\ \varphi)\,. \tag{4.2}$$

The eigenvalues E_{nl} are independent of m and therefore $(2l + 1)$-fold degenerate.

In (4.2) $R_{nl}(r)$ satisfies the radial Schrödinger equation

$$\frac{\hbar^2}{2M}\frac{d^2R_{nl}(r)}{dr^2} + \left[E_{nl} - U(r) - \frac{\hbar^2}{2M}\frac{l(l+1)}{r^2}\right] R_{nl}(r) = 0 \tag{4.3}$$

with $R_{nl}(r) \to 0$ as $r \to \infty$ (for bound states).

It will be noted that $R_{nl}(r)$ has to vanish at $r = 0$. It may or may not have other zeros (apart from the one at $r = \infty$). The integer n is a quantum number which determines, essentially, the number of these nodes in the radial function. There are various conventions regarding the choice of this quantum number. In atomic spectroscopy it is customary to set the number of nodes (including the one at $r = \infty$ but excluding those at $r = 0$) equal to n-l. States with orbital angular momenta of $l = 0, 1, 2, 3, 4, 5, 6, 7, \ldots$ are denoted respectively by the letters

$$s,\ p,\ d,\ f,\ g,\ h,\ i,\ k,\ \ldots \tag{4.4}$$

(the letter j is not used to avoid confusion).

In the notation of atomic spectroscopy the following single particle (nl) orbits are possible

$$1s,\ 2s,\ 3s,\ \ldots \qquad 2p,\ 3p,\ 4p,\ \ldots \qquad 3d,\ 4d,\ 5d,\ \ldots \text{ etc.}$$

In nuclear spectroscopy it is customary to let n (and not n-l) stand for the number of nodes in the radial functions (including the one at $r = \infty$). The same states will therefore be denoted in nuclear spectroscopy by

$$1s,\ 2s,\ 3s,\ \ldots \qquad 1p,\ 2p,\ 3p,\ \ldots \qquad 1d,\ 2d,\ 3d,\ \ldots \text{ etc.}$$

In this book we shall use the latter notation.

Let us assume for all future considerations that $U(r)$ is an attractive potential, monotonically nonincreasing in its absolute value

$$U(r) \leq 0, \qquad |U(r_1)| \geq |U(r_2)| \qquad \text{for} \qquad r_1 < r_2. \tag{4.5}$$

We shall also assume that $U(r) \to 0$ for $r \to \infty$ and that

$$r^2 \, U(r) \to 0 \qquad \text{for} \qquad r \to 0 \tag{4.6}$$

so that there is a lowest bound state. More singular potentials give rise to infinite binding energies.

For every bound state $E_{nl} < 0$ there is then a point r_0, the *turning point*, for which

$$E_{nl} - U(r_0) - \frac{\hbar^2}{2M} \frac{l(l+1)}{r_0^2} = 0. \tag{4.7}$$

If the value of l is bigger than zero, there are *two* such turning points. We shall denote by r_0 the turning point which is farther away from the origin and by r_0' the other one ($r_0' < r_0$).
It is clear that

$$E_{nl} - U(r) - \frac{\hbar^2}{2M} \frac{l(l+1)}{r^2} > 0 \qquad \text{for} \qquad r_0' < r < r_0 \tag{4.8}$$

and

$$E_{nl} - U(r) - \frac{\hbar^2}{2M} \frac{l(l+1)}{r^2} < 0 \qquad \text{for} \qquad r > r_0. \tag{4.9}$$

In the case (4.8) the radial function $R_{nl}(r)$ is curved towards the r-axis, whereas in the case (4.9) it is curved away from the r-axis (Fig. 4.1).

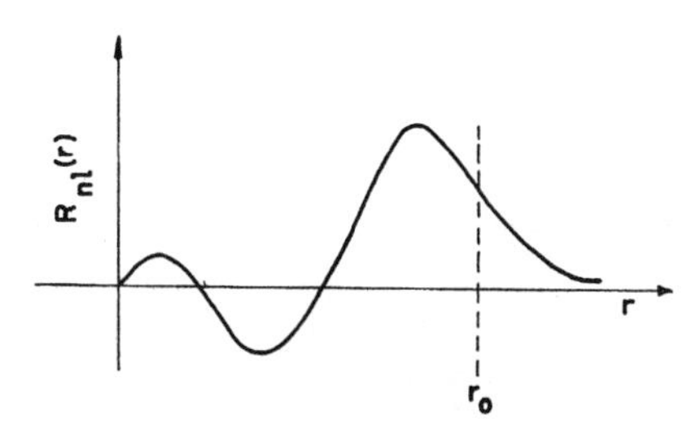

FIG. 4.1 Curvature of the radial function

It follows therefore from the radial equation (4.3) that all the nodes of $R_{nl}(r)$, except the one at infinity, are between $r = 0$ and $r = r_0$. As we see from (4.7) r_0 is within the range of $U(r)$. We therefore conclude that all the nodes of $R_{nl}(r)$ are found within the range of the attractive potential. Outside the range where the potential can be neglected, the wave function decreases exponentially with a characteristic length $\sqrt{\hbar^2/2M|E_{nl}|}$.

The turning point is determined, as we saw in (4.7), by the point at which $R''(r)/R(r) = 0$. It is worthwhile to note that a complete knowledge of a *single* radial function $R_{nl}(r)$ of a given value of l is sufficient to determine $U(r)$ uniquely up to an additive constant. In fact we see from the radial equation (4.3) that

$$U(r) = \frac{\hbar^2}{2M} \frac{R''_{nl}(r)}{R_{nl}(r)} - \frac{l(l+1)}{r^2} + \text{const.}$$

The quantum numbers n and l determine the energy E_{nl} of the corresponding state. The values of the energies E_{nl} depend on the specific central potential considered. There are however a few general features of the relative energies of various states which are true for any central potential.

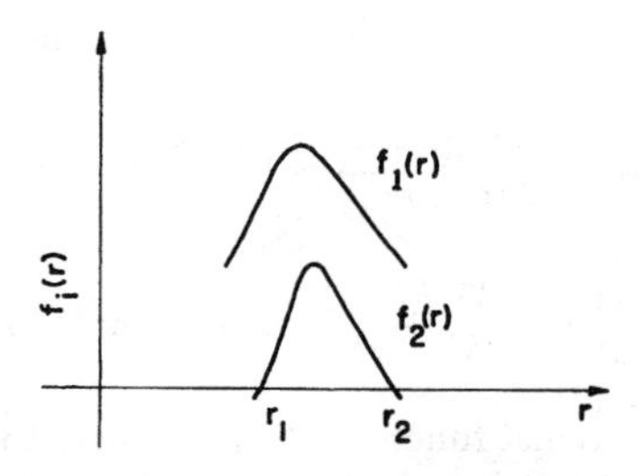

FIG. 4.2 The functions considered in the lemma.

To derive these features we first prove a lemma. Let $f_1(r)$ and $f_2(r)$ be two differentiable real functions, such that $f_1(r) > 0$ for $r_1 \leqslant r \leqslant r_2$, $f_2(r) > 0$ for $r_1 < r < r_2$, $f_2(r_1) = f_2(r_2) = 0$ and $f_2'(r_1) > 0$, $f_2'(r_2) < 0$ (see Fig. 4.2). Then there is at least one point r_0, $r_1 < r_0 < r_2$, such that

$$\frac{f_2''(r_0)}{f_2(r_0)} < \frac{f_1''(r_0)}{f_1(r_0)}. \tag{4.10}$$

To prove the lemma we note that

$$\int_{r_1}^{r_2} [f_1(r) f_2''(r) - f_1''(r) f_2(r)] \, dr = f_1(r) f_2'(r) - f_1'(r) f_2(r) \Big|_{r_1}^{r_2}$$

$$= f_1(r_2) f_2'(r_2) - f_1(r_1) f_2'(r_1) < 0.$$

The last inequality holds even if *one* of the values of $f_1(r_1)$ and $f_1(r_2)$ vanishes. There is therefore at least one point r_0 inside the interval (r_1, r_2) in which the integrand is negative. At this point neither $f_1(r_0)$ nor $f_2(r_0)$ vanish and therefore (4.10) is satisfied.

We can now prove the following theorem

If $U(r)$ is any attractive potential $[U(\infty) = 0]$, *and* $\psi_1 = \psi_{n_1 l}$ *and* $\psi_2 = \psi_{n_2 l}$ *are two bound states with the same orbital angular momentum l, but with* $n_2 > n_1$, *then* $E_{n_2 l} > E_{n_1 l}$. (4.11)

In other words, of two states with the same orbital angular momentum the one with more nodes is less bound. This theorem holds for any potential which need not necessarily be attractive for all values of r. Physically the statement of this theorem is clear. All the nodes of both ψ_1 and ψ_2 must fall within the range of $U(r)$. The effect of the attractive potential is more or less the same in the two states; the kinetic energy, however, is bigger for the function with the bigger number of nodes; hence the result $E_2 > E_1$ $(E_2 = E_{n_2 l}, E_1 = E_{n_1 l})$.

To prove the theorem formally, we note that if $R_2(r)$ (i.e., the radial part of ψ_2 multiplied by r) has more nodes than $R_1(r)$ (i.e., $n_2 > n_1$), then there must be at least two points r_1 and r_2 for which the conditions of the previous lemma are satisfied. These two points are two adjacent nodes of $R_2(r)$ such that $R_1(r)$ does not vanish in the interval (r_1, r_2) and may vanish at most at r_1 or r_2 but not at both points. The phases of R_1 and R_2 are independent of each other, and can therefore be chosen so that both functions are non negative in the above interval. Since the conditions of the lemma are thereby satisfied we conclude that there is a point r_0 at which

$$\frac{R_2''(r_0)}{R_2(r_0)} - \frac{R_1''(r_0)}{R_1(r_0)} < 0. \tag{4.12}$$

Both $R_1(r)$ and $R_2(r)$ satisfy the radial equation (4.3) with the same value of l. We therefore obtain

$$\frac{R_2''(r_0)}{R_2(r_0)} - \frac{R_1''(r_0)}{R_1(r_0)} = \frac{2M}{\hbar}(E_1 - E_2). \tag{4.13}$$

Equation (4.13) together with (4.12) then show that

$$E_1 - E_2 < 0$$

and thereby the theorem is proved.

Another theorem states that

of two states with the same number of nodes the one with the lower value of the orbital angular momentum has a lower energy. (4.14)

The physical interpretation of the theorem is quite obvious. The two states ψ_{nl_1} and ψ_{nl_2} have roughly the same kinetic energy associated with the radial momentum and roughly the same energy in the potential $U(r)$. If, however, $l_2 > l_1$, then the kinetic energy of the transverse motion, given by the centrifugal potential, is higher for the particle in the state ψ_{nl_2}, and it becomes correspondingly less bound.

The theorem can be formally proved as follows. The functions $R_1(r) = R_{nl_1}(r)$ and $R_2(r) = R_{nl_2}(r)$ have, by hypothesis, the same number of nodes. They both vanish for $r = 0$ and $r = \infty$. It follows, therefore, that either all the nodes of R_1 and R_2 coincide or else there are two adjacent nodes of R_2, r_1 and r_2, such that $R_1(r)$ does not vanish in the interval (r_1, r_2) and at most vanishes at r_1 or r_2. The conditions of the previous lemma are satisfied for the latter case, and therefore

$$\frac{R_2''(r_0)}{R_2(r_0)} - \frac{R_1''(r_0)}{R_1(r_0)} < 0. \tag{4.15}$$

Both $R_2(r)$ and $R_1(r)$ satisfy the radial equation (4.3), this time however, with different values of l. We obtain, therefore,

$$\frac{R_2''(r_0)}{R_2(r_0)} - \frac{R_1''(r_0)}{R_1(r_0)} = \frac{1}{r_0^2}\,[l_2(l_2+1) - l_1(l_1+1)] + \frac{2M}{\hbar^2}(E_1 - E_2). \tag{4.16}$$

If, as we assumed, $l_2 > l_1$, we see that (4.15) can be satisfied only if $E_1 - E_2 < 0$. If all nodes of R_1 and R_2 coincide, it is clear that the integral following (4.10) (with R_1 and R_2 replacing f_1 and f_2) must vanish. Its integrand must therefore vanish at least at one point r_0. In this case the right-hand side of (4.16) is equal to zero, which, due to $l_2 > l_1$, implies again $E_2 > E_1$.

It follows from (4.16) that

$$E_1 - E_2 \leq -\frac{\hbar^2}{2M}\frac{1}{r_0^2}\,[l_2(l_2+1) - l_1(l_1+1)]. \tag{4.17}$$

Here r_0 could vary, in principle, between zero and infinity. In some cases, however, more restrictive limits can be put on r_0, leading to a meaningful limitation on the differences $E_{nl_1} - E_{nl_2}$.

From the two theorems proved above we can conclude that in any central potential the lowest bound state is always an s-state (i.e., of zero orbital angular momentum) with just one node at infinity. The next state is either an s-state with two nodes or a p-state with one node, etc. The fact that the ground state in any potential must be an s-state can be seen even in a simpler way. If the wave function for any bound state with energy E is $(1/r)R(r)Y_{lm}(\theta\,\varphi)$ we can consider $(1/r)R(r)$, which is an s state function, as a variation function. Since it has $l = 0$, the expectation value for the central field Hamiltonian calculated with it will be lower than E. Thus, only an s-state can give the lowest energy.

5. Central Potentials in Frequent Use

There are some central potentials which are often used and it is worthwhile to mention some of the properties of the radial functions associated with them.

In atomic spectroscopy the *Coulomb potential* plays the dominant role. For an electron moving in the field of the nucleus it is given by

$$U(r) = -Ze^2 \frac{1}{r} \tag{5.1}$$

where e is the charge of the electron and Z is the atomic number. The eigenvalues of the Hamiltonian for the Coulomb potential are given, as is well known, by

$$E_{nl} = -\frac{MZ^2e^4}{2\hbar^2(n+l)^2} \,. \tag{5.2}$$

Here n stands for the number of nodes in the radial wave function (excluding $r = 0$ and including $r = \infty$). The corresponding normalized radial wave functions are

$$R_{nl}(r) = \sqrt{\frac{Z(n-1)!}{(n+l)^2\,[(n+2l)!]^3 a}}\, \rho^{l+1}\, e^{-\rho/2}\, L^{2l+1}_{n+2l}(\rho) \tag{5.3}$$

where a is given by

$$a = \frac{\hbar^2}{Me^2} \tag{5.4}$$

and ρ is the radial distance measured in units of order a

$$\rho = \frac{2Z}{(n+l)} \frac{r}{a} \,. \tag{5.5}$$

The associated Laguerre polynomial $L^{2l+1}_{n+2l}(x)$ satisfies

$$\int_0^\infty x^m\, e^{-x}\, [L^m_k(x)]^2\, dx = \frac{[\Gamma(k+1)]^3}{(k-m)!}$$

and

$$\int_0^\infty x^{m+1}\, e^{-x}\, [L^m_k(x)]^2\, dx = (2k-m+1)\frac{[\Gamma(k+1)]^3}{(k-m)!} \,. \tag{5.6}$$

The polynomials $L_k^m(x)$ are given explicitly by*

$$L_k^m(x) = (-1)^k \frac{k!}{(k-m)!}$$

$$\left\{ x^{k-m} - \frac{k(k-m)}{1!} x^{k-m-1} + \frac{k(k-1)(k-m)(k-m-1)}{2!} x^{k-m-2} - \dots \right\}.$$

The probability density for a particle in the state ψ_{nlm} to be found in the volume element $r^2 dr d\Omega$ is given by the expression

$$|R_{nl}(r)|^2 \, |Y_{lm}(\theta \, \varphi)|^2 \, dr d\Omega.$$

It follows that the probability for finding the particle anywhere on a *spherical shell* at a distance r from the center of the potential is proportional to

$$P(r)dr = |R_{nl}(r)|^2 dr. \tag{5.7}$$

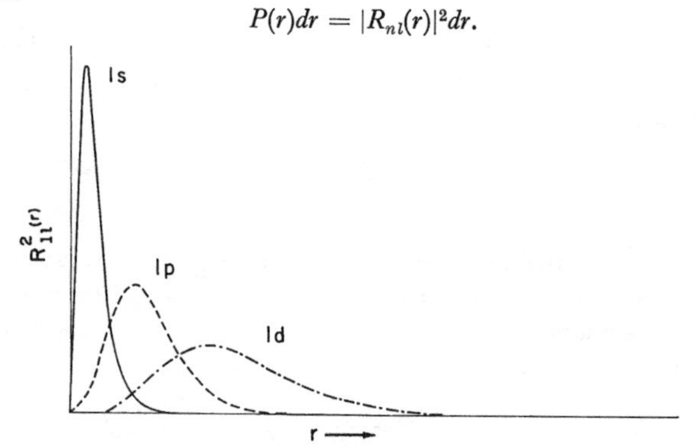

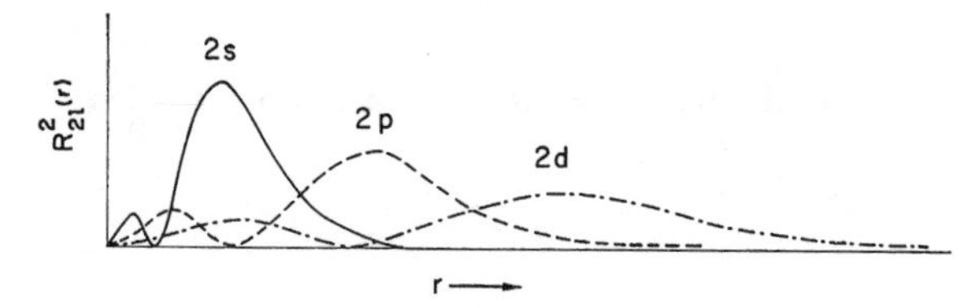

FIG. 5.1 Radial functions in the Coulomb field

Figure 5.1 shows some of these radial distribution functions for the Coulomb porential. Some of the features of these graphs are common to

(*) Some authors use the notation $L_p^m(x)$ with $p = k - m$ to describe the function which we denote as $L_k^m(x)$. In comparing various formulas the reader is urged to make sure which notation is being used.

many potentials. Thus, considering the sequence of states ψ_{nl} with a given value of l, we see that the particle is always most likely to be located just next to the last node. We shall see later that this feature is manifested also in the case of the harmonic oscillator.

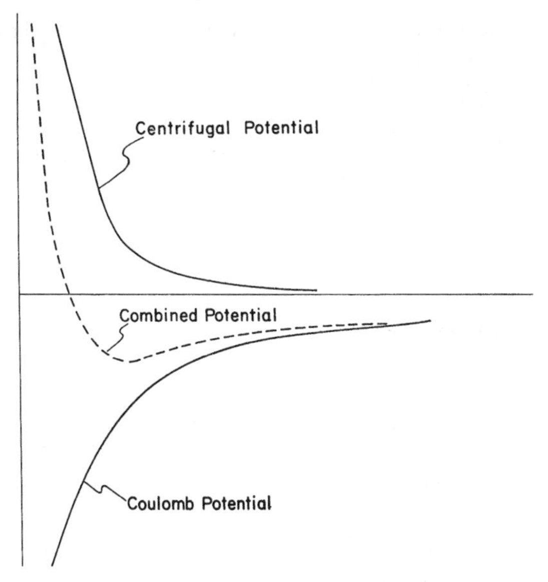

FIG. 5.2 Coulomb and centrifugal potentials

We also see that for a given number of nodes n the most likely distance of the particles from the origin increases with l. This is clearly due to the fact that a particle with higher l tends to stay as far away as possible from the origin so it can still have low transverse momentum. This can be best seen by looking at the effect of the centrifugal forces (see fig. 5.2). The combination of the attractive Coulomb potential $-Ze^2/r$ and the repulsive centrifugal potential $(\hbar^2/2M)[l(l+1)/r^2]$ gives rise to a potential which is repulsive for distances

$$\frac{r}{\hbar/Mc} < \frac{1}{2}\,l(l+1)\cdot\frac{\hbar c}{Ze^2}\,.$$

For higher values of l the radius of the repulsive part of the potential increases, pushing the particle further away from the center.

In nuclear spectroscopy there are several potentials which are frequently used. The most common of them is the *harmonic oscillator potential.* We shall now discuss briefly some of its most important characteristics.

A particle of mass M moving in an attractive harmonic oscillator potential satisfies the Schrödinger equation

$$\left(-\frac{\hbar^2}{2M}\Delta + \frac{1}{2}M\omega^2 r^2\right)\psi(r\,\theta\,\varphi) = E\psi(r\,\theta\,\varphi). \tag{5.8}$$

For obvious reasons the zero of the energy is chosen here so that $U(r) = 0$ for $r = 0$ rather than for $r = \infty$.

The radial functions $R_{nl}(r)$ are given by*

$$R_{nl}(r) = \sqrt{\frac{2(2\nu)^{l+3/2}(n-1)!}{[\Gamma(n+l+\frac{1}{2})]^3}}\, r^{l+1}\, e^{-\nu r^2}\, L^{l+1/2}_{n+l-1/2}(2\nu r^2) \tag{5.9}$$

where

$$\nu = \frac{M\omega}{2\hbar}. \tag{5.10}$$

The energy associated with the eigenfunction $\psi_{nlm}(r\,\theta\,\varphi)$ is, of course, independent of m and is given by

$$E_{nl} = [2(n-1) + l + \tfrac{3}{2}]\,\hbar\omega = [2n + l - \tfrac{1}{2}]\,\hbar\omega. \tag{5.11}$$

The number of nodes (excluding $r = 0$ and including $r = \infty$) in $\psi_{nlm}(r\,\theta\,\varphi)$ is given by n. The quantum number n can take on all positive integer numbers $n = 1, 2, \ldots$†.

(*) Here, and throughout the volume, superscript or subscripts written $l + 3/2$ or $1/2 + l + j_i + 1$ ... etc. should be read as if written thus: $l + \frac{3}{2}$, $\frac{1}{2} + l + j_i + 1$.

(†) Some authors prefer to use the quantum number $n' = n - 1$ instead of our n. The lowest state is then denoted by $0s$ instead of $1s$ in our notation. Equations (5.9) and (5.11) are then modified to

$$R_{n'l}(r) = \sqrt{\frac{2(2\nu)^{l+3/2}n'!}{[\Gamma(n'+l+\frac{3}{2})]^3}}\, r^{l+1}\, e^{-\nu r^2}\, L^{l+1/2}_{n'+l+1/2}(2\nu r^2)$$

and

$$E_{n'l} = (2n' + l + \tfrac{3}{2})\hbar\omega.$$

It is also quite common to find the radial functions expressed in terms of polynomials *proportional* to the associated Laguerre polynomials, and defined by

$$v_{nl}(x) = \frac{2^{2n+l+1}(2l+1)!!\,n!}{(-1)^{l+1/2}\sqrt{\pi}\,[(2n+2l+1)!!]^2}\, L^{l+1/2}_{n+l+1/2}(x)$$

$$= \sum_{k=0}^{n} (-1)^k 2^k \binom{n}{k} \frac{(2l+1)!!}{(2l+2k+1)!!}\, x^k.$$

These polynomials have the advantage that the term independent of x is always 1. The radial functions in terms of $v_{nl}(x)$ are given by

$$R_{n'l}(r) = \sqrt{\frac{2^{l-n'+2}(2\nu)^{l+3/2}(2l+2n'+1)!!}{\sqrt{\pi}\,[(2l+1)!!]^2\,n'!}}\, r^{l+1}\, e^{-\nu r^2}\, v_{n'l}(2\nu r^2).$$

It follows from (5.11) that all states ψ_{nlm} with the same value of $2n + l = N$ are degenerate. Since n can assume only integral values, a set of degenerate states must have all values of l even or all values of l odd. In other words, degenerate states in the harmonic oscillator potential always have the same parity.

Although the harmonic oscillator potential extends to infinity, and even has infinite value at $r \to \infty$, its main effect on particles in low states comes from its part close to the origin. This fact can be nicely brought out by calculating the overlap of eigenfunctions of two different harmonic oscillator potentials. Let $\psi_{nlm}(r\ \theta\ \varphi)$ and $\psi'_{nlm}(r\ \theta\ \varphi)$ be two such eigenfunctions corresponding to the values ν and ν' of the oscillator constant. Then for $n = 1$ we obtain

$$\int \psi^*_{1lm}\, \psi'_{1lm}\, d\tau = \int R_{1l}(r)\, R'_{1l}(r)\, dr$$

$$= [\sqrt{\nu\ \nu'}/\tfrac{1}{2}(\nu + \nu')]^{l+3/2}.$$

We see then that, even if $\nu' = 2\nu$, the overlap between the above two functions is better than 91 % for $l = 0$, 87 % for $l = 1$, 82 % for $l = 2$, etc. Thus the potential at large distances must have little influence on the actual wave functions which are concentrated at the origin.

The shape of the average nuclear potential probably follows the nuclear mass distribution and is therefore not very different from the lower part of a harmonic oscillator potential. As we have just seen, the radial functions are not very sensitive to small variations in the parameters of this potential. We can therefore understand why in many problems the harmonic oscillator potential may constitute a good enough approximation to the actual nuclear potential. In Part II we shall see another feature of the harmonic oscillator wave functions which makes them particularly useful for calculations of nuclear spectra.

Other potentials are also used sometimes, notably the infinite square well $[U(r) = 0$ for $r < r_0$ and $U(r) = \infty$ for $r \geqslant r_0]$ and the finite square well $[U(r) = -U_0$ for $r < r_0$ and $U(r) = 0$ for $r \geqslant r_0]$. We shall not consider here the radial functions for these potentials since they are discussed in detail in many textbooks.

6. Degeneracies in a Central Field

In Section 4 it was shown that the eigenstates of the general central field Hamiltonian can be classified according to the number n of nodes in their radial part and the eigenvalues of $\boldsymbol{l}^2$ and l_z. We further saw that the eigenvalues of a central field Hamiltonian depend only on n and l but not on m. Consequently each eigenvalue E_{nl} is $(2l + 1)$-fold degenerate.

It is instructive to study this degeneracy from a more general point of view. Physically, it results, as well known, from the fact that in a central field there is no preferred direction. Hence any choice of a z-axis along which the angular momentum vector is quantized should lead to the same set of energies. The energies in a central field must consequently be m-independent, leading to the above degeneracy.

From a formal point of view we note that the central-field Hamiltonian commutes with the vector $\boldsymbol{l}$. Hence, if ψ is an eigenfunction which belongs to the eigenvalue E, the function $(a_x l_x + a_y l_y + a_z l_z)\psi$ is also an eigenfunction of H belonging to the same eigenvalue E. The three numbers (a_x, a_y, a_z) can be chosen arbitrarily. There are therefore two possibilities: either it is true that

$$(a_x l_x + a_y l_y + a_z l_z)\psi = \lambda\psi$$

for any choice of (a_x, a_y, a_z) where λ is a space-independent constant (which may, of course, depend on a_x, a_y and a_z), or else the eigenvalue E of H must be degenerate.

The matrices l_x, l_y and l_z do not commute among themselves. Therefore they cannot be diagonalized simultaneously. When l_x and l_y operate on an eigenfunction of $\boldsymbol{l}^2$ and l_z, they give functions which belong to the same l but different values of m (if $l \neq 0$). Therefore H must have at least a $(2l + 1)$-fold degeneracy.

As we shall see in Part II the fact that H commutes with $\boldsymbol{l}$ is closely related to its invariance under rotations. If we perform a rotation R of the frame of reference and call the coordinates in the new frame x', y', and z', then it is easy to see that a central-field Hamiltonian remains invariant under this transformation. In other words, if we take $H(\mathbf{x}, \mathbf{p})$ and substitute in it $\mathbf{x}$ in terms of $\mathbf{x}'$ and $\mathbf{p}$ in terms of $\mathbf{p}'$ we shall generally obtain a *new* function $H'(\mathbf{x}', \mathbf{p}')$ of the new coordinates. However in the case of a central-field Hamiltonian we have

$$H'(\mathbf{x}', \mathbf{p}') = H(\mathbf{x}', \mathbf{p}'). \tag{6.1}$$

Introducing the rotation operator R through the definition

$$Rf(\mathbf{x}) = f(\mathbf{x}(\mathbf{x}')) \tag{6.2}$$

we obtain, using (6.1), for the central-field Hamiltonian

$$RH(\mathbf{x}, \mathbf{p}) = H(\mathbf{x}(\mathbf{x}'), \mathbf{p}(\mathbf{p}')) = H'(\mathbf{x}', \mathbf{p}') = H(\mathbf{x}', \mathbf{p}'). \tag{6.3}$$

We can now apply the operation R to both sides of the Schrödinger equation

$$H(\mathbf{x}, \mathbf{p})\psi(\mathbf{x}) = E\psi(\mathbf{x})$$

and obtain

$$H(\mathbf{x}', \mathbf{p}')\psi'(\mathbf{x}') = E\psi'(\mathbf{x}') \tag{6.4}$$

where

$$\psi'(\mathbf{x}') = \psi(\mathbf{x}(\mathbf{x}')) = R\psi(\mathbf{x}).$$

The functions ψ and ψ' are generally different functions of their arguments. In (6.4) we can now drop the primes from $\mathbf{x}'$ and $\mathbf{p}'$ (this is *not* a transformation!) and obtain

$$H(\mathbf{x}, \mathbf{p})\psi'(\mathbf{x}) = E\psi'(\mathbf{x}).$$

It thus follows from the invariance of H under rotations that, if $\psi(\mathbf{x})$ is an eigenfunction, $\psi'(\mathbf{x})$ is also an eigenfunction belonging to the same eigenvalue. We have already seen that under rotations each $Y_{lm}(\theta\ \varphi)$ is transformed into a linear combination of spherical harmonics of the *same* order l. Therefore, by applying different rotations to the Schrödinger equation we can obtain only $2l + 1$ linearly independent solutions and E_{nl} is consequently $(2l + 1)$-fold degenerate. The invariance under rotations is thus the direct cause for the degeneracy of the eigenvalues.

The advantage of this formulation of the degeneracy problem in the central field lies in its generality. The fact that R is a rotation operation does not come into the considerations in an essential way. The essential point is the existence of an operation Ω on the coordinates and momenta which leaves the Hamiltonian invariant, i.e., if

$$\Omega f(\mathbf{x}, \mathbf{p}) = f(\mathbf{x}(\mathbf{x}', \mathbf{p}'), \mathbf{p}(\mathbf{x}', \mathbf{p}'))$$

then

$$\Omega H(\mathbf{x}, \mathbf{p}) = H(\mathbf{x}(\mathbf{x}', \mathbf{p}'), \mathbf{p}(\mathbf{x}', \mathbf{p}')) = H(\mathbf{x}', \mathbf{p}').$$

As an example of a more general invariance we consider the Hamiltonian of the three-dimensional harmonic oscillator.

We write it in the form*

$$H = \frac{1}{2M} p^2 + \frac{1}{2} M\omega^2 x^2 = \left(\frac{1}{\sqrt{2M}} \mathbf{p} + i\omega \sqrt{\frac{M}{2}} \mathbf{x}\right) \cdot \left(\frac{1}{\sqrt{2M}} \mathbf{p} - i\omega \sqrt{\frac{M}{2}} \mathbf{x}\right)$$

$$= \left(\frac{1}{\sqrt{2M}} \mathbf{p} - i\omega \sqrt{\frac{M}{2}} \mathbf{x}\right)^* \cdot \left(\frac{1}{\sqrt{2M}} \mathbf{p} - i\omega \sqrt{\frac{M}{2}} \mathbf{x}\right). \tag{6.5}$$

We see that this specific Hamiltonian can be written as the squared magnitude of the complex vector $(1/\sqrt{2M})\,\mathbf{p} - i\omega\sqrt{M/2}\,\mathbf{x}$. It is therefore invariant under the three-dimensional unitary transformations on this vector. In fact, let us define $\mathbf{x}'$ and $\mathbf{p}'$ by

$$\left(\frac{1}{\sqrt{2M}} \mathbf{p}' - i\omega \sqrt{\frac{M}{2}} \mathbf{x}'\right) = U \left(\frac{1}{\sqrt{2M}} \mathbf{p} - i\omega \sqrt{\frac{M}{2}} \mathbf{x}\right)$$

where U is a 3×3 unitary matrix $U^\dagger U = 1$. [Such a matrix is often referred to as a matrix of the group $U(3)$.] It is then obvious that H will assume the same form (6.5) in terms of the new variables $\mathbf{x}'$ and $\mathbf{p}'$. The invariance of H with respect to transformations of the group $U(3)$ will therefore result in a degeneracy of its eigenvalues. The three-dimensional rotations $R(3)$ are special cases of $U(3)$, and the degeneracy implied by $R(3)$ is consequently only a part of the degeneracy implied by $U(3)$. This agrees with what we saw in Section 5 about the energy levels in the harmonic oscillator potential: in addition to the $(2l+1)$-fold degeneracy of the eigenstates E_{nl}, we have the degeneracy of all eigenstates E_{nl} for which $2n + l = N$, where N is a fixed integer.

The group-theoretical classification of states uses frequently such considerations. One then starts by studying the most general group of transformations under which a given Hamiltonian is invariant, and then studies this group instead of the Hamiltonian. This procedure very often results in a considerably greater transparency and generality.

(*) The commutator of $\mathbf{p}$ and $\mathbf{x}$, being equal to a constant, is ignored here.

7. Spin-Orbit Interaction and Order of Single Nucleon Levels

In nuclei the order of single particle levels in the central potential is strongly affected by the spin-orbit interaction. We shall therefore start this section with a brief discussion of such interactions.

A particle moving in an electric field feels, in its rest system, also a magnetic field. If it has a magnetic moment $g_s\mathbf{s}$ then the interaction of this moment with the magnetic field is proportional to $(\boldsymbol{l} \cdot \mathbf{s})$. This *spin-orbit interaction* (not to be confused with LS coupling) has long been observed in atoms where it agrees with the order of magnitude expected.

Since nucleons also possess magnetic moments they should also exhibit a spin-orbit interaction. We can get an idea of the order of magnitude of the strength of this spin orbit interaction by evaluating $(\boldsymbol{\mu} \cdot \mathbf{H})$ where $\boldsymbol{\mu}$ is the nucleon's magnetic moment and $\mathbf{H}$ is the magnetic field experienced, in its rest frame, by a nucleon moving in a nucleus. Straightforward calculation then shows that $\boldsymbol{\mu} \cdot \mathbf{H}$ is at most of order 10^3 ev. This is much too small to explain the experimental findings, as will be explained below.

Consider a nucleon moving in a nucleus and is subject to a spin orbit interaction

$$V_s = \zeta(r)\,(\boldsymbol{l} \cdot \mathbf{s}). \tag{7.1}$$

Let us further assume that this interaction is small compared with the difference between single particle levels in the potential. We can then estimate its effects on the energy of a single particle in a definite state by calculating its expectation value in this state. Thus considering a nucleon in the state $\psi(nlj)$ we obtain for its energy shift due to (7.1)

$$\delta E_s = \int \psi^*(n\,l\,j\,m)\,\zeta(r)\,(\boldsymbol{l} \cdot \mathbf{s})\,\psi(n\,l\,j\,m). \tag{7.2}$$

Since

$$(\boldsymbol{l} \cdot \mathbf{s}) = \tfrac{1}{2}\,[(\boldsymbol{l} + \mathbf{s})^2 - \boldsymbol{l}^2 - \mathbf{s}^2] = \tfrac{1}{2}\,[\mathbf{j}^2 - \boldsymbol{l}^2 - \mathbf{s}^2]$$

we have

$$(\boldsymbol{l} \cdot \mathbf{s})\,\psi(n\,l\,j\,m) = \tfrac{1}{2}\,[j(j+1) - l(l+1) - \tfrac{3}{4}]\,\psi(n\,l\,j\,m). \tag{7.3}$$

Substituting (7.3) in (7.2) we obtain

$$\delta E_s = \tfrac{1}{2}\,[j(j+1) - l(l+1) - \tfrac{3}{4}]\,\zeta_{nl}. \tag{7.4}$$

Here

$$\zeta_{nl} = \int \psi^*(n\,l\,j\,m)\;\zeta(r)\,\psi(n\,l\,j\,m) \tag{7.5}$$

is independent of either j or m since $\zeta(r)$ is a function of r only.

In a central field, which is spin-independent, the levels $n, l, j = l + \frac{1}{2}$ and $n, l, j = l - \frac{1}{2}$ are degenerate. It makes no difference in the energy of a level with definite values of n and l of the spin is parallel or anti-parallel to the orbital angular momentum. The spin-orbit interaction, however, does introduce such a difference; if it is attractive for one relative position of l and s, it is repulsive for the other. It thus removes the degeneracy of the above two states in the central field and creates a *doublet*. The origin of this term is from atomic spectroscopy where the spin-orbit interaction is generally weak and its effects are observed as a doubling of each level. In nuclear spectra the doublet splitting may be bigger than the distance between two states with different values of l. For clarity we shall still retain the same terminology. From (7.4) we find that the doublet splitting is given by

$$\begin{aligned} \delta E_s\,(j = l + \tfrac{1}{2}) &= \tfrac{1}{2}\,l\,\zeta_{nl} \\ \delta E_s\,(j = l - \tfrac{1}{2}) &= -\,\tfrac{1}{2}\,(l + 1)\,\zeta_{nl}. \end{aligned} \tag{7.6}$$

The m-degeneracy of the states remains, of course, and is not removed by the spin-orbit interaction. This is obvious physically, since even the inclusion of the spin-orbit interaction in the Hamiltonian does not introduce any preferred direction in space, and hence the energy cannot depend on the choice of the z-axis. Formally, this results from the fact that the eigenvalues of $(\boldsymbol{l} \cdot \mathbf{s})$ are independent of m, as shown by (7.3). The state with $j = l + \frac{1}{2}$ is therefore $2j + 1 = 2l + 2$-fold degenerate and that with $j' = l - \frac{1}{2}$ is $2j' + 1 = 2l$-fold degenerate. It is seen from (7.6) that the "center of mass" of the two components of a doublet remains unshifted by the spin-orbit interaction

$$(2l + 2)\,[\delta E_s\,(j = l + \tfrac{1}{2})] + 2l\,[\delta E_s\,(j = l - \tfrac{1}{2}] = 0.$$

This is a special case of a more general theorem which will be proved in Part II.

We see from (7.6) that the separation between the two components of a doublet is given by

$$\delta E = \delta E_s\,(j = l + \tfrac{1}{2}) - \delta E_s\,(j = l - \tfrac{1}{2}) = \frac{2l + 1}{2}\,\zeta_{nl}.$$

In nuclear spectra δE is observed to be of order 1 to 6 Mev. The spin orbit interaction in nuclei cannot therefore be of electromagnetic origin, since, as we saw above, these energies are expected to be at most of the order of 10^3 ev. The origin of the large nuclear spin-orbit interaction is not yet clearly understood, but the recognition of its existence led to very significant advances in the theory of nuclear structure. It is this large spin-orbit interaction which leads to what is known as the *jj-coupling shell model.* With the help of the resulting *jj*-coupling scheme (to be subsequently described) a great variety of nuclear phenomena could be understood.

The sign of the nuclear spin-orbit interaction is experimentally found to be such that the state with $j = l + \frac{1}{2}$ is *lower* than that with $j = l - \frac{1}{2}$, and furthermore, states with higher values of l seem to have a bigger absolute doublet splitting. This state of affairs has important consequences on the order of single particle levels in the average nuclear potential, as we shall now see.

8. Centrifugal and Coulomb Barriers

The effects of the centrifugal force on the radial function of a particle moving in a central field were studied in Section 4. In nuclear physics the magnitude of the centrifugal potential at the nuclear surface is about $[l(l+1)]/2A^{2/3}$ Mev. It represents therefore a small correction to the nuclear potential at the same point.

Representing the nuclear potential by a square well of radius r_0 we see that the result of adding the centrifugal potential amounts to the addition of a "barrier" at the surface of the nucleus. The height of this barrier increases quadratically with the orbital angular momentum involved and decreases quadratically with the nuclear radius. Of two particles with equal energy the one carrying a higher orbital angular momentum with respect to the nucleus will find it more difficult either to penetrate the nucleus from outside or to leave the nucleus from inside. This is often referred to as the effect of the *centrifugal barrier*.

The Coulomb barrier gives rise to a similar effect. There is good evidence that the nuclear part of the proton-proton interaction is the same as the neutron-neutron and neutron-proton interactions. We can therefore conclude that as far as nuclear interactions are concerned a proton and a neutron are both affected by the *same* average potential when they move in the nucleus. However, a proton is affected in addition by the Coulomb field due to the other protons in the nucleus.

In Fig. 8.1 we have drawn a qualitative picture of the potential acting on a neutron, and the one acting on a proton. It is seen that whereas for a

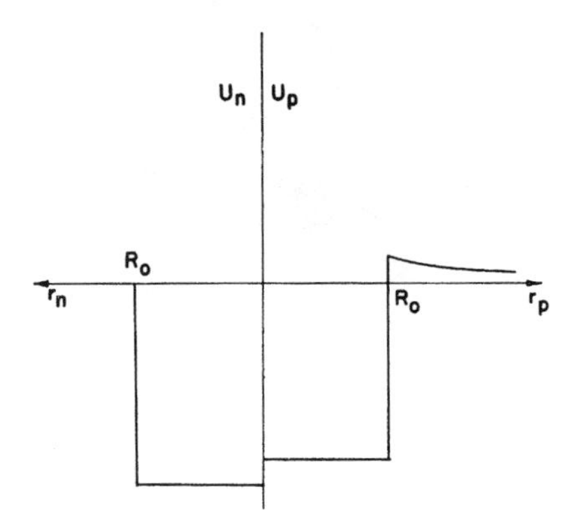

FIG. 8.1 Nuclear and Coulomb potentials

neutron there is an attraction starting at $r = R_0$, a proton which approaches the nucleus from infinity feels first the Coulomb repulsion, and only when it comes within the reach of the nuclear forces at R_0, it is attracted by the nucleus. The depth of the potential acting on a proton inside the nucleus is slightly smaller than that acting on a neutron due to the Coulomb repulsion inside. The actual shape of the combined nuclear and Coulomb potentials inside the nucleus depends on the charge distribution in the nucleus. However, since this effect is small any how, and since the square well potential is in itself an idealization of the actual situation, we can approximate the Coulomb effect on the potential inside the nucleus by a slight decrease in the depth of the nuclear average potential.

Let us put in the nuclear potential Z protons and $N = Z$ neutrons, both protons and neutrons occupying the lowest possible states consistent with the Pauli principle. We see from Fig. 8.1 that the protons will be, on the average, closer to ionization than neutrons. It will thus require less energy to get a proton out of a nucleus with $Z = N$ than to get a neutron out of the same nucleus. This is very nicely born out by experimental data. It takes 18.7 Mev to get a neutron out of C^{12} ($Z = N = 6$), whereas a proton requires only 15.9 Mev, and for Ca^{40} ($Z = N = 20$) they require 16.0 Mev and 8.3 Mev respectively. The difference between these *separation energies* is seen, from Fig. 8.1, to be of order Ze^2/R_0. With $R_0 = 1.2\ A^{1/3}\ 10^{-13}$ cm the fit between this simple minded estimate and experimental data is very good.

The effect described above on self-conjugate ($Z = N$) nuclei is sometimes confusing. A barrier is associated with preventing particles from getting out. In reality, however, the effect of the Coulomb force is to make it "easier" for a proton rather than a neutron to get out of a nucleus. It should be realized that the effect of the Coulomb force is not limited to the "barrier" part ($r > R_0$), but is even stronger inside the nucleus and is manifested by the elevation of the total potential inside the nucleus.

The effects of the outside part of the Coulomb field are best seen when one compares neutron capture to proton capture. Whereas neutrons of the smallest energies are readily captured by nuclei, protons must have an energy of at least a couple of hundreds of kev in order to react with even the lightest elements. This energy is required, of course, to enable the proton to come close enough to the nucleus and to have an appreciable probability of penetrating the Coulomb barrier.

The Coulomb field has interesting effects on bound nucleons in light nuclei. This can be illustrated by examining the level scheme of two conjugate nuclei like C^{13} ($Z = 6$, $N = 7$) and N^{13} ($Z = 7$, $N = 6$). The first few levels of these nuclei are given in Fig. 8.2. The two schemes are plotted so that their ground states coincide. Actually the separation energies of the last neutron in C^{13} is 3 Mev bigger than that of N^{13}. It is seen that, whereas the levels $\frac{1}{2}-$, $\frac{3}{2}-$, $\frac{5}{2}+$ have retained, more or less, their relative separation in passing from C^{13} to N^{13}, the level $\frac{1}{2}+$ is relatively more bound in N^{13} than it is in C^{13}.

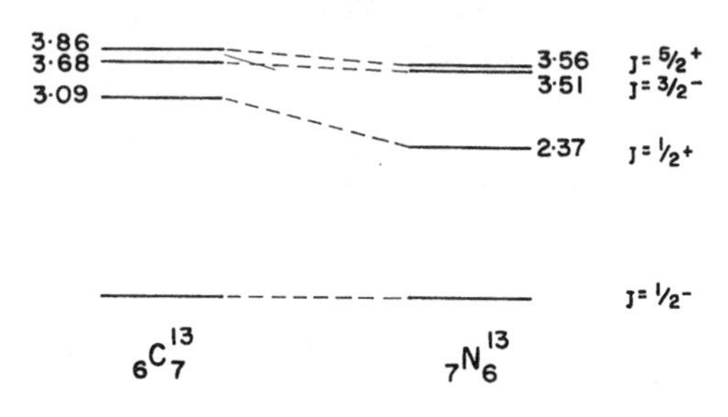

FIG. 8.2 Level schemes of C^{13} and N^{13}.

Neglecting the Coulomb effects we should have expected the same spectrum in C^{13} and N^{13}. Nuclear forces are known to be charge-independent and their effects should be the same for a nucleus consisting of 6 protons and 7 neutrons and for a nucleus consisting of 6 neutrons and 7 protons. The difference in excitation energies of the corresponding states in C^{13} and N^{13} should therefore be searched in the differences in Coulomb energies.

C^{12} represents a doubly-magic nucleus ($1s_{1/2}$ and $1p_{3/2}$ orbits filled for both protons and neutrons). It may therefore be a good approximation to visualize C^{13} as a doubly-magic "core" plus one neutron outside, and N^{13} as composed of the same core plus one proton outside. The first few levels in these nuclei will then represent the single particle excited states. We can therefore assume that the ground state is a $1p_{1/2}$ state, the first excited state—a $2s_{1/2}$, etc. Of these states the $2s_{1/2}$ probably extends most outside the nucleus, since it has no centrifugal barrier. Taking the Coulomb effect as a perturbation, the level shift is given by the expectation value of the Coulomb potential taken with the unperturbed wave function. Such a calculation gives essentially equal Coulomb repulsion in the $1p_{1/2}$ and $2s_{1/2}$ states. The situation becomes different if we consider the change of the wave functions due to the Coulomb potential. The wave functions of states with large separation energies or large values of l will not change appreciably so that the perturbation calculation is essentially exact. However, the $2s_{1/2}$ wave function, with no centrifugal barrier, for a proton will be different from that of a neutron. This modification of the wave function attenuates the effect of the Coulomb repulsion. In comparing N^{13} with

C^{13} we expect a *decrease* in the binding energies of all the single particle states. For the $2s_{1/2}$ state the Coulomb effect is, according to the argument given above, less pronounced. Therefore, relative to the other states, the $2s_{1/2}$ state is more tightly bound in N^{13} than in C^{13}.

9. Magnetic Moments and Beta Decay

We have already mentioned the fact that the single particle wave functions can give a good estimate of some nuclear properties. This is particularly true for the expectation values of sums of single particle operators each of which is proportional to an odd power of the single particle angular momentum. For such operators, the contribution of a particle in a state $\psi(j, m)$ is equal in magnitude and opposite in sign to that of a particle in the state $\psi(j, -m)$. Therefore, saturated pairs, i.e., pairs of particles with opposite directions of their angular momenta, do not contribute to the expectation values of such operators, and the only contribution comes from the unsaturated particles. In many low-lying states of odd A nuclei there is, probably, only one such "unsaturated" particle. Hence the importance of expectation values calculated with single particle wave functions. This approximation of ignoring saturated pairs is often referred to as the *single particle shell model.* More detailed calculations will be given in Parts II and III. Here we shall confine ourselves to the evaluation of matrix elements of some of the single particle operators with single particle wave functions.

We start with the magnetic moment operator. A charged particle moving in a central field interacts with an external magnetic field both via its internal magnetic dipole moment as well as through the current it produces. This interaction energy is given by

$$\delta E = \langle g_l(\boldsymbol{l} \cdot \mathbf{H}) + g_s(\mathbf{s} \cdot \mathbf{H}) \rangle. \tag{9.1}$$

The expectation value is taken with the wave function of the state considered in the limit of a vanishing external magnetic field. The magnetic field $\mathbf{H}$ is assumed to be uniform and small so that δE will, in fact, be linear in $\mathbf{H}$, as (9.1) implies. The g-factors g_l and g_s belong to the orbital motion and the intrinsic spin respectively. For electrons

$$g^{(l)}_{\text{elect.}} = -\frac{e}{2m_e c}\hbar \tag{9.2}$$

where e is the absolute value of the electron charge and m_e is its mass. The electron spin g-factor is (apart from radiative corrections)

$$g^{(s)}_{\text{elect.}} = -\frac{e}{m_e c}\hbar. \tag{9.3}$$

It is customary to measure g-factors in units of *Bohr magnetons*. The electron's orbital g-factor is then 1 Bohr magneton and its spin g-factor is 2 such units.

Nuclear g-factors are measured in units of *nuclear magnetons* which are given by

$$1 \text{ nuclear magneton} = \frac{e\hbar}{2m_p c}. \tag{9.4}$$

Here e is the positive unit of charge and m_p is the *proton* mass. In terms of this unit the spin g factors of the proton and the neutron are

$$g_p^{(s)} = 5.5855 \qquad g_n^{(s)} = -3.8256. \tag{9.5}$$

The orbital g factors are, in the same units,

$$g_p^{(l)} = 1 \qquad g_n^{(l)} = 0. \tag{9.6}$$

We shall now show that it is possible to define a g-factor for a particle in a state with a definite value of j which satisfies, for any orientation of the field $\mathbf{H}$,

$$\delta E = \langle g_l(\boldsymbol{l} \cdot \mathbf{H}) + g_s(\mathbf{s} \cdot \mathbf{H})\rangle = g\langle \mathbf{j} \cdot \mathbf{H}\rangle. \tag{9.7}$$

The magnetic moment of the state under consideration is then defined as

$$\mu = g j. \tag{9.8}$$

Since $\mathbf{H}$ is arbitrary, (9.7) can be satisfied only if we can find a value of g such that

$$g\langle \mathbf{j}\rangle = \langle g_l \boldsymbol{l} + g_s \mathbf{s}\rangle. \tag{9.9}$$

Equation (9.9) is actually a set of three equations, one for each of the components j_x, j_y, and j_z. However, only one equation, namely that for the z-component, is really significant. We have already seen (1.12) that the expectation values of j_x and j_y taken with eigenstates of j_z, vanish identically and therefore the left-hand side of (9.9) vanishes for the x- and y-components. To prove (9.9) we therefore need to show only that the x- and y-components of the right-hand side vanish as well, and then use the z-component for the actual computation of g as a function of g_l and g_s. Actually we shall prove a more general theorem which has many spectroscopic applications.

We note first that $\boldsymbol{l}$ satisfies the following commutation relations with $\mathbf{j}$:

$$\begin{aligned} [j_x, l_y] &= i l_z, \text{ etc.} \\ [j_x, l_x] &= 0, \quad \text{etc} \end{aligned} \tag{9.10}$$

It follows from (9.10) that

$$[\mathbf{j}^2, \boldsymbol{l}] = -2i(\mathbf{j} \times \boldsymbol{l}) - 2\boldsymbol{l}. \tag{9.11}$$

We use here the notation of vector algebra, but it is understood that all components are operators which do not necessarily commute. Therefore their order is important and we define $(j \times l)_z = j_x l_y - j_y l_x$, etc. We now use (9.11) to compute the commutator $[\mathbf{j}^2, (\mathbf{j}^2, \boldsymbol{l})]$ thus obtaining

$$\mathbf{j}^4\boldsymbol{l} - 2\mathbf{j}^2\boldsymbol{l}\mathbf{j}^2 + \boldsymbol{l}\mathbf{j}^4 \equiv [\mathbf{j}^2, (\mathbf{j}^2, \boldsymbol{l})] = 2(\mathbf{j}^2\boldsymbol{l} + \boldsymbol{l}\mathbf{j}^2) - 4\mathbf{j}(\mathbf{j} \cdot \boldsymbol{l}). \quad (9.12)$$

Equation (9.12) is an operator identity. The matrix elements of its left-hand side between the states $\psi(n\,j\,m)$ and $\psi(n'\,j\,m')$ (same j for both functions!) vanish. The right-hand side then gives a very important relation:

$$j(j+1)\,\langle n\,j\,m|\boldsymbol{l}|n'j\,m'\rangle = \langle j\,m|\mathbf{j}|\,j\,m'\rangle\,\langle n\,j\,m'|(\mathbf{j} \cdot \boldsymbol{l})|n'j\,m'\rangle. \quad (9.13)$$

The only property of $\boldsymbol{l}$ used in deriving this relation is its commutation relations with $\mathbf{j}$. We shall have similar relations for *any* set of three operators satisfying the same commutation relations (9.10). In particular the three components of the spin vector $\mathbf{s}$ are easily seen to satisfy

$$[j_x, s_y] = is_z, \qquad [j_x, s_x] = 0, \text{ etc.}$$

and hence

$$j(j+1)\,\langle n\,j\,m|\mathbf{s}|n'j\,m'\rangle = \langle j\,m|\mathbf{j}|\,j\,m'\rangle\,\langle n\,j\,m'|(\mathbf{j} \cdot \mathbf{s})|n'j\,m'\rangle. \quad (9.14)$$

In (9.13) and (9.14) n and n' stand for the radial quantum numbers and any other quantum numbers (such as l for instance) of the states considered. Both $\boldsymbol{l}$ and $\mathbf{s}$ do not connect states with different quantum numbers n. We preferred nevertheless to leave these relations in their general form since it is possible in general to have a vector operator $\mathbf{t}$ which does connect states with different values of n and satisfies

$$[j_x, t_x] = 0 \qquad [j_x, t_y] = it_z, \text{ etc.} \quad (9.15)$$

It is easy to see that the vectors $\mathbf{x}$ and $\mathbf{p}$ satisfy (9.15). In Part II we shall see that any vector operator has to satisfy this relation. It then follows that $\mathbf{t}$ satisfies also

$$j(j+1)\,\langle n\,j\,m|\mathbf{t}|n'j\,m'\rangle = \langle j\,m|\mathbf{j}|\,j\,m'\rangle\,\langle n\,j\,m'|(\mathbf{j} \cdot \mathbf{t})|n'j\,m'\rangle. \quad (9.16)$$

The result (9.16), known as the *generalized Landé formula*, actually states that *within a shell nj* every vector operator is proportional to the angular momentum operator with the proportionality factor

$$\frac{1}{j(j+1)}\,\langle n\,j\,m|(\mathbf{j} \cdot \mathbf{t})|n\,j\,m\rangle$$

which is independent of m. This is, of course, not true for matrix elements between shells with different values of j; an arbitrary $\mathbf{t}$ *need not* be diagonal with respect to $\mathbf{j}^2$, whereas $\mathbf{j}$, of course, is.

Using the results (9.13) and (9.14) it is seen that the right-hand side of (9.9) does in fact vanish whenever the left-hand side does, and furthermore, we obtain, for $j \neq 0$,

$$g = \frac{\langle n\,l\,j\,m|g_l(\mathbf{j}\cdot\mathbf{l}) + g_s(\mathbf{j}\cdot\mathbf{s})|n\,l\,j\,m\rangle}{j(j+1)}. \tag{9.17}$$

To simplify (9.17) we make use of the operator identities

$$(\mathbf{j}\cdot\mathbf{l}) = \tfrac{1}{2}[\mathbf{j}^2 + \mathbf{l}^2 - \mathbf{s}^2]$$
$$(\mathbf{j}\cdot\mathbf{s}) = \tfrac{1}{2}[\mathbf{j}^2 + \mathbf{s}^2 - \mathbf{l}^2].$$

Since the state $\psi(nljm)$ is an eigenstate of $\mathbf{j}^2$, $\mathbf{l}^2$, and $\mathbf{s}^2$, we obtain, using the above identities,

$$g = \frac{1}{2j(j+1)}\{g_l[j(j+1) + l(l+1) - s(s+1)] + g_s[j(j+1) + s(s+1) - l(l+1)]\}. \tag{9.18}$$

The expression (9.18) can be even further simplified if we note that j can assume only one of two values $j = l + \frac{1}{2}$ or $j = l - \frac{1}{2}$. We obtain then by direct substitution

$$\begin{aligned} g &= \frac{(2j-1)g_l + g_s}{2j} \qquad \text{for} \quad j = l + \tfrac{1}{2} \\ g &= \frac{(2j+3)g_l - g_s}{2(j+1)} \qquad \text{for} \quad j = l - \tfrac{1}{2}. \end{aligned} \tag{9.19}$$

For the magnetic moments $\mu = gj$ we obtain therefore

$$\mu = \begin{cases} lg_l + \frac{1}{2}g_s & j = l + \frac{1}{2} \\ \dfrac{j}{j+1}[(l+1)g_l - \frac{1}{2}g_s] & j = l - \frac{1}{2}. \end{cases} \tag{9.20}$$

The values (9.20) for the magnetic moments of a single nucleon are known as the *Schmidt values*, and the lines giving μ as a function of j are known as the *Schmidt lines*. We see that for each type of nucleon (proton or neutron) there are two Schmidt lines, one corresponding to states in which the spin and orbital angular momenta are parallel ($j = l + \frac{1}{2}$) and the other when they are antiparallel ($j = l - \frac{1}{2}$). The line for $j = l + \frac{1}{2}$ is a straight line with a slope g_l, and the line for $j = l - \frac{1}{2}$ is slightly curved tending asymptotically to a line with a slope g_l.

Actual magnetic moment of odd-A nuclei do not generally coincide with the Schmidt values (9.20). They are, however, close enough to them, so that plotted against j, do show a clear breakup into two bands, roughly parallel to the Schmidt lines with $j = l \pm \frac{1}{2}$. The deviations of magnetic moments from the Schmidt values give us an idea of the departure of the actual nuclear wave function from the one we get using some simple models. They have thus been the subject of numerous studies into which we shall not go here.

In some odd-odd nuclei it is presumably possible to characterize the state of the system by the total angular momentum of the protons J_p, that of the neutrons J_n, and the total angular momentum $\mathbf{J} = \mathbf{J}_p + \mathbf{J}_n$. If we know the g-factor of the protons in their state J_p, and that of the neutrons in J_n, we can use the generalized Landé formula to derive the total g-factor. Since $\mathbf{J}_p$ and $\mathbf{J}_n$ both satisfy the commutation relations (9.15) with $\mathbf{J}$, we have

$$\langle g\,\mathbf{J}\rangle = \langle g_p\,\mathbf{J}_p + g_n\,\mathbf{J}_n\rangle$$

and therefore

$$\begin{aligned} g = \frac{1}{2J(J+1)} \{ & g_p[J(J+1) + J_p(J_p+1) - J_n(J_n+1)] \\ & + g_n[J(J+1) + J_n(J_n+1) - J_p(J_p+1)]\} \\ = \tfrac{1}{2}&(g_p + g_n) + (g_p - g_n)\frac{J_p(J_p+1) - J_n(J_n+1)}{2J(J+1)}. \end{aligned} \tag{9.21}$$

As can be expected we find that if $J_n \ll J_p$ then $g \approx g_p$ and if $J_p \ll J_n$ then $g \approx g_n$. We also find the interesting result that whenever $J_p = J_n$ g is given simply by

$$g = \tfrac{1}{2}(g_p + g_n) \qquad \text{for} \qquad J_p = J_n. \tag{9.22}$$

Equation (9.21) holds also for odd-even nuclei provided the state can be characterized by the J values of the protons and the neutrons separately. It is seen that if we have for instance an even number of neutrons and $J_n = 0$ then $g = g_p$. Thus an even group of nucleons whose angular momentum vanishes, does not contribute to the magnetic moment of a nucleus. In particular a closed shell always has an even number of nucleons in it, and their total angular momentum vanishes. It follows, therefore, that closed shells do not contribute to the nuclear magnetic moment.

There are other operators whose matrix elements can be approximately calculated with single nucleon wave functions. We shall consider now the operators which give the probability amplitude for β-decay.

We shall not go into the details of the theory of β-decay. For our purpose it will be enough to note that the interaction leading to β-decay is extremely weak, and that we are therefore well justified in treating it as a small perturbation. The transition probability amplitude is then given by an off-diagonal matrix element of the β-decay interaction Hamiltonian between the initial and final states.

Both the initial and the final states are composed of the corresponding nuclear states ψ_i and ψ_f and the states of the electron-neutrino field (for instance no neutrinos or electrons in the initial state and an antineutrino and an electron in the final state). Here we are concerned only with the nuclear part of the transition amplitude, and we shall confine ourselves to nonrelativistic allowed β-transitions. There are, then, two operators which determine the transition probability, the unit operator 1 and the spin operator $\boldsymbol{\sigma}$. More precisely, if a nucleon in the initial state ψ_i emits an electron plus an antineutrino (or a positron plus a neutrino) and is transformed thereby into a nucleon in the state ψ_f, then the transition probability is proportional to

$$G_F|\langle 1 \rangle|^2 + G_{GT}|\langle \boldsymbol{\sigma} \rangle|^2. \tag{9.23}$$

Here G_F and G_{GT} give the weights of the two possible operators in the actual β-decay interaction Hamiltonian. The symbol $\langle 1 \rangle$ is an accepted notation for the Fermi part of the β-decay interaction. It is defined, in the case of a single nucleon by

$$\langle 1 \rangle = \int \psi_f^* \, \psi_i. \tag{9.24}$$

Similarly $\langle \boldsymbol{\sigma} \rangle$ gives the contribution from the Gamow-Teller part of the β-decay interaction

$$\langle \boldsymbol{\sigma} \rangle = \int \psi_f^* \, \boldsymbol{\sigma} \, \psi_i. \tag{9.25}$$

The expression $\langle \boldsymbol{\sigma} \rangle^2$ in (9.23) stands for the square of (9.25) summed over all $2j_f + 1$ magnetic substates of the final state and averaged over initial states. In (9.24) and (9.25) ψ_f and ψ_i are functions of the space and spin coordinates. They refer always to the states of the particle which undergoes the transformation from a proton to a neutron or vice versa. The description of the change of the nucleon is not included in ψ_f and ψ_i. A more formal way of writing the same thing can be achieved by the use of the isospin formalism as will be discussed in Part II.

Let us now proceed with the evaluation of the matrix elements (9.24) and (9.25). The evaluation of $\langle 1 \rangle$ is straightforward. In fact we see that unless ψ_f and ψ_i are characterized by exactly the same quantum numbers, $\langle 1 \rangle$ will vanish. Thus we obtain the *Fermi selection rule* for allowed β-transitions

The Fermi interaction contributes to allowed β-decay only if a neutron in the nlj-orbit is transformed into a proton in the same orbit, or vice versa. For these cases $\langle 1 \rangle = 1$; otherwise $\langle 1 \rangle = 0$. (9.26)

To evaluate $\langle \boldsymbol{\sigma} \rangle$ we decompose the initial and final wave functions into their spatial and spin-dependent parts

$$\psi_f = \sum_{m_l\, m_s} (\tfrac{1}{2}\, m_s l_f m_l | \tfrac{1}{2}\, l j_f m_f)\, \varphi(n_f l_f m_l)\, \chi(m_s)$$
$$\psi_i = \sum_{m_l'\, m_s'} (\tfrac{1}{2}\, m_s' l_i m_l' | \tfrac{1}{2}\, l j_i m_i)\, \varphi(n_i l_i m_l')\, \chi(m_s') \tag{9.27}$$

where $(\frac{1}{2}\, m_s l m_l | \frac{1}{2}\, l j m)$ is the proper Clebsch-Gordan coefficient introduced in (2.19).

Since the operator $\boldsymbol{\sigma}$ operates on the spin coordinates only, we see that

$$\langle \boldsymbol{\sigma} \rangle = 0 \qquad \text{for} \qquad n_i \neq n_f \qquad \text{and/or} \qquad l_i \neq l_f. \tag{9.28}$$

Using the notation

$$\sigma_1 = -\frac{1}{\sqrt{2}}(\sigma_x + i\sigma_y) \qquad \sigma_0 = \sigma_z \qquad \sigma_{-1} = \frac{1}{\sqrt{2}}(\sigma_x - i\sigma_y)$$

we obtain for $n_i = n_f = n$ and $l_i = l_f = l$,

$$\langle \boldsymbol{\sigma} \rangle^2 = \frac{1}{2j_i + 1} \sum_{m_i,\, m_f,\, \kappa} (-1)^\kappa \langle \sigma_\kappa \rangle \langle \sigma_{-\kappa} \rangle$$
$$= \frac{1}{2j_i + 1} \sum_{\substack{m_s m_s' \varkappa, m_l m_i, m_f \\ \overline{m_s}\, \overline{m_s'}\, \overline{m_l}}} (-1)^\varkappa (\tfrac{1}{2}\, m_s l\, m_l | \tfrac{1}{2}\, l j_f m_f)(\tfrac{1}{2}\, m_s' l\, m_l | \tfrac{1}{2}\, l j_i m_i) \langle m_s | \sigma_\kappa | m_s' \rangle$$
$$\times (\tfrac{1}{2}\, \overline{m_s} l\, \overline{m_l} | \tfrac{1}{2}\, l j_f m_f)(\tfrac{1}{2}\, \overline{m_s'} l\, \overline{m_l} | \tfrac{1}{2}\, l j_i m_i) \langle \overline{m_s} | \sigma_{-\kappa} | \overline{m_s'} \rangle. \tag{9.29}$$

Since the Clebsch-Gordan coefficients in (9.29) vanish unless $j_i = l \pm \frac{1}{2}$ and $j_f = l \pm \frac{1}{2}$, we see that $\langle \boldsymbol{\sigma} \rangle^2$ may be different from zero only if $|j_i - j_f| = 0$ or 1. We thus obtain the Gamow-Teller selection rules for allowed β-transitions

The Gamow-Teller interaction contributes to allowed β-decay only if a neutron in the nlj_i-orbit is transformed into a proton in the $n\,l\,j_f$-orbit (or vice versa) where $|j_i - j_f| = 0$ or 1. (9.30)

The matrix elements of $\boldsymbol{\sigma}$ can be obtained from (2.23), namely

$$\begin{aligned}\sigma_0\chi(m_s) &= 2s_z\chi(m_s) = 2m_s\chi(m_s)\\ \sigma_\pm\chi(m_s) &= \mp\sqrt{2}\,(s_x \pm is_y)\,\chi(m_s) = \mp\sqrt{\tfrac{3}{2} - 2m_s(m_s \pm 1)}\,\chi(m_s \pm 1).\end{aligned} \quad (9.31)$$

This leads to the following matrix representation of the *Pauli matrices* $\boldsymbol{\sigma}$:

$$\sigma_0 = \begin{Vmatrix} 1 & 0 \\ 0 & -1 \end{Vmatrix} \qquad \sigma_{+1} = \begin{Vmatrix} 0 & -\sqrt{2} \\ 0 & 0 \end{Vmatrix} \qquad \sigma_{-1} = \begin{Vmatrix} 0 & 0 \\ \sqrt{2} & 0 \end{Vmatrix}. \quad (9.32)$$

In (9.32) the first row (or column) corresponds to $m_s = +\frac{1}{2}$, and the second one to $m_s = -\frac{1}{2}$. We now use the explicit values of the Clebsch-Gordan coefficients given in Table 2.1 and carry out the summations in (9.29). The results are tabulated in Table 9.1 for the four possible combinations of the initial and final angular momenta.

TABLE 9.1

THE VALUES OF $\langle\boldsymbol{\sigma}\rangle^2$ FOR SINGLE NUCLEON TRANSITIONS

j_f \ j_i	$l + \frac{1}{2}$	$l - \frac{1}{2}$
$l + \frac{1}{2}$	$\frac{2l+3}{2l+1} = \frac{j_f+1}{j_f}$	$4\frac{l+1}{2l+1}$
$l - \frac{1}{2}$	$\frac{4l}{2l+1}$	$\frac{2l-1}{2l+1} = \frac{j_f}{j_f+1}$

These results will be rederived in a more concise form in Part II where simple expressions will be given for sums over magnetic quantum numbers like the one appearing in (9.29).

Part II TENSOR ALGEBRA. TWO-PARTICLE SYSTEMS

The laws of physics are formulated in the same way in all frames of reference with respect to which they are described. The requirement, that the mathematical description of physical phenomena should be independent of the frame of reference, has far-reaching consequences. For instance, if we know that the vector $\mathbf{J}(\mathbf{r})$ describing the flow of heat in an isotropic body, is associated only with the distribution $T(\mathbf{r})$ of the temperature in that body, it can be immediately concluded that $\mathbf{J}(\mathbf{r})$ is in the direction of grad $T(\mathbf{r})$. Any attempt to build a vector $\mathbf{J}(\mathbf{r})$ in a different way, for instance,

$$(J_x, J_y, J_z) = \left(\frac{\partial T}{\partial x}, 2\frac{\partial T}{\partial y}, \frac{\partial T}{\partial z}\right)$$

describes a different flow pattern in different frames of reference—a conclusion which is physically unacceptable.

The limitations imposed on physical laws by requiring that they should be the same for any observer are known as *invariance requirements*. In particular, invariance under rotations of frames of reference in space implies that physical quantities have to be described by scalars, vectors, or higher tensors. It also implies that these quantities can be combined only in covariant forms (scalar product of two vectors, vector product of two vectors, etc.). There are other consequences of invariance requirements. A trivial example is the requirement that the functional dependence of any equation in physics on its arguments will be independent of the units we use. It follows from this requirement that the two sides of an equality must have the same dimensions.

In this book we shall be primarily concerned with the consequences of the invariance of physical laws under rotations. The basic assumption of the shell theories of both atoms and nuclei is the existence of an effective average spherical field in which the electrons or the nucleons move. The spherical symmetry of this field, which is responsible for many features of the shell model, implies covariance of the physical quantities under rotations around the center of the field. We shall therefore discuss in detail the mathematical concepts and apparatus used in the shell model.

10. Irreducible Tensors

We shall use the following notations for a vector leading from the origin to a point P_i in space

$$\mathbf{r}_i \equiv (x_i, y_i, z_i) \equiv [x_1^{(i)}, x_2^{(i)}, x_3^{(i)}].$$

No distinction will be made between covariant and contravariant indices since we shall be using exclusively a metric in which $g_{ik} = \delta_{ik}$. Unless otherwise stated we shall always mean by a *rotation*—rotation of the frame of reference leaving space and physical phenomena in it unaltered. Although this is of course completely equivalent to rotating the whole physical space in the opposite direction, it is more convenient to visualize the whole sequence of physical events in an "absolute" space with frames of reference anchored in it in different orientations.

If a point P in space has the coordinates x_i in the frame of reference σ, the coordinates x_i' of the *same* point P in space referred to the rotated frame σ' are given by

$$x_i' = \sum_k a_{ik}\, x_k. \tag{10.1}$$

In this equation the a_{ik} are the elements of a real orthogonal matrix whose determinant is equal to $+1$.

$$\sum_k a_{ik}\, a_{jk} = \delta_{ij} \qquad \sum_i a_{ik}\, a_{il} = \delta_{kl}$$

$$\det |a_{ik}| = +1. \tag{10.2}$$

Using these orthogonality relations we see that the old coordinates are given in terms of the new ones by

$$x_i = \sum_k a_{ki}\, x_k'. \tag{10.3}$$

It is important to notice that in (10.1) it is the second index of a_{ik} which is being summed, whereas in (10.3) the first index of a_{ik} plays this role.

A vector is defined as a set of three quantities which, under rotations a_{ik} undergo the transformation (10.1). More generally, a set of 3^s numbers is called a *tensor of rank s*, if these numbers $A_{ijk\cdots}$ undergo the following linear transformation under the rotation (10.1)

$$A'_{ijk\cdots} = \sum_{l,m,n\ldots} a_{il}\, a_{jm}\, a_{kn} \ldots A_{lmn\ldots}. \tag{10.4}$$

Each subscript may assume, independently, the values 1, 2, or 3. $A_{lmn}\dots$ has altogether s subscripts, and each term in (10.4) contains a product of s matrix elements a_{il}.

It follows from the linearity of the transformation (10.4) that any linear combination of tensors of a given rank (in which corresponding components are added) is again a tensor of the same rank. Tensors can be conveniently built from products of components of vectors. Thus a simple tensor of the second rank is given by $A_{ij} = x_i x_j$. Another simple tensor of the second rank is given by the linear combination

$$B_{ij} = \xi_i \eta_j - \xi_j \eta_i \tag{10.5}$$

where $\boldsymbol{\xi}$ and $\boldsymbol{\eta}$ are two nonparallel vectors. The linearity of the transformation (10.4) assures, as mentioned above, the tensor character of B_{ij}.

The tensor B_{ij} defined by (10.5) is peculiar in that $B_{ij} = - B_{ji}$, and in particular, three of its components vanish identically: $B_{11} = B_{22} = B_{33} = 0$. It may be noted that there is no vector whose first component, say, vanishes in every frame. The vanishing of the diagonal elements B_{ii} in every frame is thus not an accidental property and represents an intrinsic feature of such a tensor. Indeed, it may be easily verified that if T_{ik} is any tensor of the second rank whose diagonal elements vanish in every frame, then $T_{ik} = - T_{ki}$.

Every tensor T_{ik} of the second rank can be decomposed into symmetric and antisymmetric tensors

$$T_{ik} = S_{ik} + A_{ik}$$

defined by the following linear combinations

$$S_{ik} = S_{ki} = \tfrac{1}{2}(T_{ik} + T_{ki}) \qquad A_{ik} = - A_{ki} = \tfrac{1}{2}(T_{ik} - T_{ki}). \tag{10.6}$$

Under rotations the components of the symmetric part of any tensor transform among themselves and the same goes for the components of the antisymmetric part. The separation of a tensor into its symmetric and antisymmetric parts has therefore a meaning which is independent of the frame of reference.

It can be asked whether the symmetric and the antisymmetric tensors can be further reduced. In other words, can additional conditions be imposed on the symmetric and antisymmetric tensors thereby decomposing them into parts which transform among themselves separately. It is fairly easy to show that this is not possible for the antisymmetric part. An antisymmetric tensor A_{ik} has essentially only three independent components: A_{12}, A_{23}, and A_{31}. We can define three quantities v_1, v_2, and v_3 by

$$v_1 = A_{23} = - A_{32}, \qquad v_2 = A_{31} = - A_{13}, \qquad v_3 = A_{12} = - A_{12}. \tag{10.7}$$

The transformation properties of the v_i are those of an (axial) vector. To see this it is enough to consider the transformation properties of any antisymmetric tensor. If we look at the antisymmetric tensor B_{ij} defined by (10.5), we see that the v_i associated with it are simply the components of the *vector product* of the two vectors ξ and η. Hence, any further condition imposed on the components of the antisymmetric tensor A_{ik} can be formulated as a condition on the components of the vector $\mathbf{v}$. However there is always a rotation which transforms any given vector into any other vector of the same length. It is therefore not possible to find linear combinations of the components of a vector which transform among themselves under rotations. The vector and therefore also the antisymmetric tensor of the second rank are examples of *irreducible tensors*.

As for the symmetric tensors $S_{ik} = S_{ki}$ we notice the invariance of the *trace*

$$\sum_i S'_{ii} = \sum_{i,l,m} a_{il}\, a_{im}\, S_{lm} = \sum_{l,m} \delta_{lm}\, S_{lm} = \sum_m S_{mm}. \tag{10.8}$$

Thus the sum of the diagonal elements of this tensor has the same value irrespective of the frame to which the tensor is referred. Therefore we can reduce this tensor and impose further restrictions on the symmetric part of T_{ik}. We shall require that the trace of the symmetric part vanishes. It will be shown later that no further reduction is possible. Thus the general tensor T_{ik} of the second rank can be decomposed into a sum of three irreducible tensors in the following way:

$$\begin{aligned} T_{ik} &= S_{ik} + A_{ik} + \tau_{ik} \\ \tau_{ik} &= \tfrac{1}{3}[\mathrm{Tr}(T_{ik})]\delta_{ik} = \tfrac{1}{3}[T_{11} + T_{22} + T_{33}]\delta_{ik} = \tfrac{1}{3}\, T\delta_{ik} \\ A_{ik} &= \tfrac{1}{2}(T_{ik} - T_{ki}) \\ S_{ik} &= \tfrac{1}{2}(T_{ik} + T_{ki}) - \tfrac{1}{3}\, T\delta_{ik}. \end{aligned} \tag{10.9}$$

In the symmetric combination, one third of the trace was subtracted from each of the diagonal elements to assure the vanishing of the trace of S.

The single component of the trace, the three independent components of the antisymmetric tensor and five independent components of the symmetric tensor are completely equivalent to the nine components of the original tensor T_{ik}. The advantage of this reduction of the tensor T_{ik} to those three tensors, lies in the fact that under rotations each one of them transforms separately. The trace in the new frame is equal to the trace in the original frame. Each component of the antisymmetric tensor A'_{ik} in the new frame is a linear combination of only the components in the original frame of the antisymmetric tensor A_{ik}. Similarly, each component of the traceless symmetric tensor S'_{ik} in the new frame is a linear combination of the componets in the original frame of the traceless symmetric tensor S_{ik}. It will thus be possible to ascribe an independent physical meaning to the trace, antisymmetric tensor, and traceless symmetric tensors obtained from the general tensor T_{ik}.

To give an illustration of the procedure described above let us consider a simple example which is actually borrowed from the next section. Consider the set of all wave functions which depend on the space coordinates of two particles in a p-orbit. As pointed out in Part I there are three degenerate states for a particle moving in a p-orbit in a central field. These three functions could be written as $xf(r)$, $yf(r)$, and $zf(r)$. Since these three states are degenerate any linear combination of them is also a state with the same energy. We can therefore look upon the three states as forming a vector $f(r)\mathbf{r}$, the three components of which, in any frame, give a possible set of the mutually orthogonal p-states. If there are two particles in the state p there are two such vectors, one for each of them $f(r)\mathbf{r}$ and $f(r')\mathbf{r}'$. In the absence of interaction between the particles the wave function of the composed system of the two particles is obtained by the product of the wave function of the two particles. Since there are three independent functions for each particle, there will be altogether $3 \times 3 = 9$ independent functions for the combined system. As in the case of a single particle we can consider the set of wave functions for the two particle system as a tensor of the second rank $\psi_{ik} = f(r)\, f(r')\, x_i x'_k$. We can now decompose this tensor into its trace, antisymmetric part, and traceless symmetric part. Operating on the components of each one of them with the operator of total angular momentum squared $\mathbf{L}^2 = (\mathbf{l} + \mathbf{l}')^2$, we find by direct calculation

$$\mathbf{L}^2[f(r)\, f(r')\, (xx' + yy' + zz')] = 0$$

$$\mathbf{L}^2[\tfrac{1}{2} f(r)\, f(r')\, (x_i x'_k - x_k x'_i)] = 2\, [\tfrac{1}{2} f(r)\, f(r')\, (x_i x'_k - x_k x'_i)]$$

$$\mathbf{L}^2 f(r)\, f(r') \left[\tfrac{1}{2} \left(x_i x'_k + x_k x'_i \right) - \tfrac{1}{3} \left(\sum_j x_j x'_j \right) \delta_{ik} \right]$$

$$= 6 f(r)\, f(r') \left[\tfrac{1}{2} \left(x_i x'_k + x_k x'_i \right) - \tfrac{1}{3} \left(\sum_j x_j x'_j \right) \delta_{ik} \right].$$

Recalling that the eigenvalues of the operator $\mathbf{L}^2$ are $l(l + 1)$ we see that the trace of the tensor ψ_{ik} describes a state of zero total angular momentum (an S-state). Similarly, each of the three components of the antisymmetric tensor describes a state of total angular momentum 1 (three P-states), and each of the five independent components of the traceless symmetric tensor describes a state of total angular momentum 2 (five independent D-states). Thus, by reducing the tensor ψ_{ik}, we can build the states of a well-defined orbital angular momentum of the system of two particles in p-orbits. The reduction of the second-order tensor in this case has therefore a simple physical meaning.

We can generalize the procedure described above also to tensors of higher rank. Given such a tensor $T_{ijk...}$, which has a certain number of independent components, it might be possible to find a smaller number of linear combinations of these components that will transform among themselves under any rotation. It is then always possible to find another group of linear combinations, independent of the first, that will also transform among themselves. The total number of such independent linear combinations must be equal to the original number of independent components. If this is the case the tensor $T_{ijk...}$ is said to be *reducible*. It might, however, be impossible to find linear combinations of the components such that they form two (or more) groups which transform among themselves. In this case the tensor $T_{ijk...}$ is called an *irreducible tensor*.

The concept of an irreducible tensor is thus defined with respect to rotations of the frames of reference. Each one of the components of an irreducible tensor is involved in an essential way in the transformation of the tensor under general rotations. By definition this property remains valid if an arbitrary linear transformation is performed on the components of an irreducible tensor.

We have actually met such irreducible tensors in Part I when we were considering the spherical harmonics. There we saw that $r^k Y_{k\kappa}(\theta\,\varphi)$, when expressed in terms of x, y, and z, is a homogenous polynomial of degree k. This means that each one of the $(2k+1)$ functions $r^k Y_{k\kappa}(\theta\,\varphi)$ is a linear combination of the components of the special tensor of rank k whose components are

$$x^\alpha y^\beta z^\gamma \qquad \alpha+\beta+\gamma=k \qquad \alpha,\beta,\gamma \text{ non negative integers.} \tag{10.10}$$

The spherical harmonics $r^k Y_{k\kappa}(\theta\,\varphi)$ constitute linear combinations of these components which transform among themselves under rotations. This is evident from (3.7) which can be written as

$$r^k Y_{k\kappa'}(\theta'\,\varphi') = \sum_\kappa r^k Y_{k\kappa}(\theta\,\varphi)\, D^{(k)}_{\kappa\kappa'}(R). \tag{10.11}$$

Here we consider $\theta'\varphi'$, which were obtained from $\theta\varphi$ by a rotation, to be the coordinates of the same point P in a rotated frame. The proof that the $(2k+1)$ spherical harmonics of order k transform *irreducibly* will not be given here. Thus by taking the $2k+1$ special linear combinations of the tensor (10.10) given by $r^k Y_{k\kappa}(\theta\,\varphi)$ we managed to obtain a reduction of this tensor which has $\frac{1}{2}(k+1)(k+2)$ independent components. We may note in passing that other linear combinations which transform among themselves are given by $r^k Y_{k-2,\kappa}(\theta\,\varphi)$, $r^k Y_{k-4,\kappa}(\theta\,\varphi)$, etc. The total number of their components is $[2k+1]+[2(k-2)+1]+\ldots=\frac{1}{2}(k+1)(k+2)$, which is equal to the original number of independent components. These linear combinations are all independent on account of the orthogonality of the spherical harmonics. The set of all the tensors $r^k Y_{k\kappa}(\theta\,\varphi)$, $r^k Y_{k-2,\kappa}(\theta\,\varphi)$, etc. can therefore replace the original tensor (10.10).

The $2k+1$ numbers $Y_{k\kappa}(\theta\,\varphi)$ (for a given value of θ and φ) are the components of a tensor although their law of transformation (10.11) is not formally identical with (10.4). Equation (10.4) determines the transformation of the Cartesian components of tensors, and (10.11) gives the transformation law of $r^k Y_{k\kappa}$, which are linear combinations of the Cartesian components of a tensor. The matrix $D^{(k)}(R)$ can therefore be obtained, in principle, from the matrix a_{ik} which determines the transformation of an ordinary vector. Thus, we can consider (10.11) as an alternative definition of tensor components. Moreover, (10.11) defines a *special* kind of tensor, namely an irreducible one. It is obvious that any set of $2k+1$ quantities $T_\kappa^{(k)}$, $\kappa=k, k-1, \ldots -k$, which transform with the same matrix $D^{(k)}(R)$ according to the law

$$T_{\kappa'}^{(k)\prime}=\sum_{\kappa} T_{\kappa}^{(k)}\, D_{\kappa\kappa'}^{(k)}(R) \tag{10.12}$$

form the components of an irreducible tensor. In (10.12), R is the rotation from the frame σ to the frame σ', $D^{(k)}(R)$ is the matrix which was introduced in (10.11); $T_\kappa^{(k)}$ and $T_\kappa^{(k)\prime}$ are the components of the tensor in the frames σ and σ' respectively. Thus if there were linear combinations of the components $T_\kappa^{(k)}$ which transformed among themselves, then linear combinations, with the same coefficients, of the $Y_{k\kappa}$ should have the same property.

We shall therefore adopt the following general definition:

An irreducible tensor of degree k is a set of $2k+1$ *components* $T_\kappa^{(k)}$ *which transform under rotations according to* (10.12).

It is, of course, implied in this definition that we are given the components of the tensor in every frame of reference so that we can check whether (10.12) holds. It follows from this definition that if $\mathbf{T}^{(k)}$ is an irreducible tensor, then $\lambda T_\kappa^{(k)}$, where λ is any number, independent of κ, also form the components of an irreducible tensor. Furthermore if $\mathbf{U}^{(k)}$ is an irreducible tensor of the same degree, then $V_\kappa^{(k)}=\mu U_\kappa^{(k)}+\lambda T_\kappa^{(k)}$ are again the components of an irreducible tensor of degree k. Clearly, sums of components of irreducible tensors of *different* degrees do not form the components of another irreducible tensor.

A word should be added on notation. We are accustomed to writing tensors by using several indices $T_{ijk\cdots}$, which define its Cartesian components. For the irreducible tensors we borrow the definition of components from that of spherical harmonics since their transformation properties are then written in a simpler way.

This definition of an irreducible tensor by means of (10.12) implies a definite choice of the independent components. It is clear that if the $2k + 1$ components $T_\kappa^{(k)}$ transform irreducibly among themselves then the $2k + 1$ components $\tilde{T}_\lambda^{(k)}$ defined by

$$\tilde{T}_\lambda^{(k)} = \sum T_\kappa^{(k)} U_{\kappa\lambda}$$

will do the same. Here U is any nonsingular matrix. The transformation law of $\tilde{T}_\lambda^{(k)}$ is however different from (10.12). In fact

$$\tilde{T}_\mu^{(k)\prime} = \sum T_{\kappa'}^{(k)\prime} U_{\kappa'\mu} = \sum T_\kappa^{(k)} D_{\kappa\kappa'}^{(k)} U_{\kappa'\mu}$$

$$= \sum T_\kappa^{(k)} U_{\kappa\lambda} (U^{-1})_{\lambda\kappa''} D_{\kappa''\kappa'}^{(k)} U_{\kappa'\mu}$$

$$= \sum \tilde{T}_\lambda^{(k)} \tilde{D}_{\lambda\mu}^{(k)}$$

where

$$\tilde{D}_{\lambda\mu}^{(k)} = \sum (U^{-1})_{\lambda\kappa} D_{\kappa\kappa'}^{(k)} U_{\kappa'\mu}.$$

Thus, the components $\tilde{T}_\lambda^{(k)}$, while transforming irreducibly under rotations, do not transform according to (10.12). In the following we shall assume that the components of all irreducible tensors are chosen in the standard fashion which implies the transformation law (10.12). The fact that every irreducible tensor can be brought to this standard form will not be proved here.

Let us now consider as an example the spherical harmonics with $k = 1$. These are given by

$$rY_{11}(\theta\,\varphi) = -\frac{1}{2}\sqrt{\frac{3}{2\pi}}\,r\sin\theta\,e^{i\varphi} = \frac{1}{2}\sqrt{\frac{3}{\pi}}\left[-\frac{1}{\sqrt{2}}(x+iy)\right]$$

$$rY_{10}(\theta\,\varphi) = \frac{1}{2}\sqrt{\frac{3}{\pi}}\,r\cos\theta = \frac{1}{2}\sqrt{\frac{3}{\pi}}\,z$$

$$rY_{1-1}(\theta\,\varphi) = +\frac{1}{2}\sqrt{\frac{3}{2\pi}}\,r\sin\theta\,e^{-i\varphi} = \frac{1}{2}\sqrt{\frac{3}{\pi}}\left[\frac{1}{\sqrt{2}}(x-iy)\right].$$

From this we see that the standard components of an irreducible tensor of degree 1 (a vector) are given in terms of its Cartesian coordinates by

$$T_1^{(1)} = -\frac{1}{\sqrt{2}}(T_x + iT_y)$$

$$T_0^{(1)} = T_z$$

$$T_{-1}^{(1)} = \frac{1}{\sqrt{2}}(T_x - iT_y).$$

Similarly, the five independent components of the standard irreducible tensor of degree 2 are given in terms of the Cartesian components of the symmetric tensor S_{ij} by using the definition of the corresponding spherical harmonics (given in the Appendix). These are (up to a common numerical factor)

$$T_0^{(2)} = 2S_{33} - S_{11} - S_{22}$$

$$T_{\pm 1}^{(2)} = \mp \sqrt{6}\,(S_{13} \pm iS_{23})$$

$$T_{\pm 2}^{(2)} = \sqrt{\frac{3}{2}}\,(S_{11} - S_{22} \pm 2iS_{12}).$$

The irreducible tensors which were discussed until now were obtained from Cartesian tensors of rank k. By considering special tensors of rank k (10.10) we found linear combinations which form the components of irreducible tensors. The number of components of these irreducible tensors is $2k + 1$ which is always an odd number. We also mentioned that the special tensors (10.10) can be completely reduced to tensors $\mathbf{Y}_l$ where $l = k, k - 2, k - 4, \cdots$. This means that the only irreducible tensors which appear in the reduction of this special tensor of order k have odd numbers of components. One may ask whether other irreducible tensors may appear in the reduction of the most general Cartesian tensor of order k (which does not possess the symmetry properties of our special tensor (10.10)), namely irreducible tensors with an even number of components. In the following we shall see that the answer is negative. The only irreducible tensors which appear in the reduction of the most general Cartesian tensors are those defined by (10.12) with an odd number of components. However, in the next section we shall consider also other irreducible tensors, with an even number of components. If we keep writing the number of components of an irreducible tensor as $2k + 1$, these other tensors will have degrees given by half-integer values of k.

11. Tensor Fields

Until now we have been considering tensors defined at a given point P in space, and investigated the transformation laws of their components under the change of frames of reference. We shall now extend our notions to tensors defined over all points of space, i.e., to *tensor fields* $\mathbf{T}^{(k)}(\mathbf{r})$. A well-known example of such a field is, for instance, the electric or the magnetic field which are fields of vectors ($k = 1$) defined at every point of space. The temperature distribution in a given body is an example of a scalar field ($k = 0$), where a single number—the temperature—is associated with every point in space. Another example of a scalar field is any solution of the ordinary nonrelativistic Schrödinger equation which associates with every point one (complex) number—the probability amplitude for finding the particle there.

We shall define an *irreducible physical tensor field of degree k* (in short: physical tensor field) as a set of $2k + 1$ *functions* $T_\kappa^{(k)}(\mathbf{r})$ which transform under rotations according to the straightforward generalization of (10.12)

$$T_{\kappa'}^{(k)\prime}(\mathbf{r}') = \sum_\kappa T_\kappa^{(k)}(\mathbf{r})\, D_{\kappa\kappa'}^{(k)}(R). \tag{11.1}$$

Here $\mathbf{r}$ and $\mathbf{r}'$ are the coordinates of the *same* point P in space referred to the frames σ and σ' respectively; $T_\kappa^{(k)}(\mathbf{r})$ and $T_{\kappa'}^{(k)\prime}(\mathbf{r}')$ are the components of the tensor $\mathbf{T}^{(k)}$ at P referred to the two frames σ and σ' respectively. It should be stressed that the functional dependence of $T_\kappa^{(k)\prime}(\mathbf{r}')$ on $\mathbf{r}'$ is generally different from the dependence of $T_\kappa^{(k)}(\mathbf{r})$ on $\mathbf{r}$. Thus for $k = 0$ we can have a field defined by $T_0^{(0)}(\mathbf{r}) = \cos\psi$ where ψ is the azimuthal angle of $\mathbf{r}$. In a frame rotated around the z axis by 90^0, the *same physical field* will be described by $\sin\psi'$, and hence $T_0^{(0)\prime}(\mathbf{r}') = \sin\psi'$. The functional dependence of the field functions in the new frame on the new coordinates is given by (11.1) as a linear combination of the components of $\mathbf{T}^{(k)}$ in the original frame at the *same* point P in space. We have described these fields as "physical tensor fields" to stress the fact that they refer to a well-defined field of tensors in space and not necessarily to a special dependence of the field on its arguments.

We already pointed out that the covariant description of physical quantities requires that they be represented by tensors of various ranks. Let us consider this requirement in more detail. The fields encountered in physics obey certain linear differential equations—the *field equations*. The isotropy of physical space implies that the laws of physics are the same for all Cartesian frames of reference. When we go over from one frame of reference to another, the differential operators undergo a certain transformation. Under rotations the functions describing the field, at a certain point P in space—the *field components*—will undergo a certain transformation. Invariance of the field equations implies that, in terms of the new field components, they will be the same as the original equations in the terms of the original components. It is clear that a physical meaning which does not depend on the special choice of a frame of reference can only be given to the totality of field components which transform among themselves under rotations. Thus, the three components of the electric field E_x, E_y, E_z or the three components of the magnetic field H_x, H_y, H_z, can be given a clear physical meaning, but no physical meaning can be attached to the three fields (E_x, H_y, H_z).

A well-known example of a differential equation which determines a physical quantity is the nonrelativistic Schrödinger equation.

$$H(x_i, p_i)\,\psi(x_i) = E\psi(x_i). \tag{11.2}$$

Its solutions are the probability amplitudes for spinless particles. In this equation the Hamiltonian is constructed from the components of the position and momentum vectors according to a certain given prescription. If we now go over to a new frame of reference we require that the Hamiltonian constructed from the components of $\mathbf{x}$ and $\mathbf{p}$ in the new (rotated) frame according to the same prescription—$H(x'_i, p'_i)$—be equal to $H(x_i, p_i)$ at every point P in space (x_i and x'_i are the coordinates of P in σ and σ' respectively). Thus,

$$H(x_i(x'), p_i(p')) \equiv H'(x'_i, p'_i) = H(x'_i, p'_i). \tag{11.3}$$

This equation implies not only that the Hamiltonian is a physical scalar but also that it is the *same function* of the components of the vectors of position and momentum in every frame of reference. This latter property is the essence of the *invariance of the Schrödinger equation under rotations* : the prescription of its construction is independent of the frame of reference.

Let us consider a *definite* physical solution $\psi_0(x_i)$ of (11.2). When we go over to a new frame σ' we obtain from (11.2)

$$H(x_i(x'), p_i(p'))\,\psi_0(x_i(x')) = E_0\psi_0(x_i(x')). \tag{11.4}$$

With the definition $\psi_0(x_i(x')) = \psi'_0(x'_i)$ we obtain, using the invariance (11.3) of H

$$H(x'_i, p'_i)\,\psi'_0(x'_i) = E_0\psi'_0(x'_i). \tag{11.5}$$

Thus $\psi'_0(x'_i)$ is the same physical field as $\psi_0(x_i)$ but expressed as a function of the coordinates in the new frame σ'. This is precisely what we mean by saying that $\psi_0(x)$ is a *physical scalar field.*

There are other fields which do not satisfy scalar equations. The requirement of their invariance under rotations implies a certain specific transformation law of the corresponding physical fields. As an example we treat a simple equation, namely, the *two-component neutrino equation.* Consider the equation

$$H\psi(x) = c(\sigma_1 p_x + \sigma_2 p_y + \sigma_3 p_z)\,\psi(x) = E\psi(x). \tag{11.6}$$

Here $\mathbf{p}$ is the ordinary momentum operator, c the velocity of light, and σ_i are the three Pauli matrices taken in the following representation:

$$\sigma_1 = \begin{pmatrix} 0 & 1 \\ 1 & 0 \end{pmatrix} \quad \sigma_2 = \begin{pmatrix} 0 & -i \\ i & 0 \end{pmatrix} \quad \sigma_3 = \begin{pmatrix} 1 & 0 \\ 0 & -1 \end{pmatrix} \tag{11.7}$$

The solution ψ of (11.6) is, obviously, a column of two functions

$$\psi(x) = \begin{bmatrix} \psi_1(x) \\ \psi_2(x) \end{bmatrix}.$$

Equation (11.6) is actually equivalent to two coupled differential equations in $\psi_1(x)$ and $\psi_2(x)$. In (11.6) the σ_i are numerical matrices which must be the *same* in every frame, in order to keep the invariance of the field equation. In other words the rule (11.7) for the construction of the matrices σ_i is a part of the definition of the specific Hamiltonian (11.6) and as such applies equally well to all observers, irrespective of the frame they use. It is for this reason that we label the matrices as σ_1, σ_2, and σ_3 rather than σ_x, σ_y, and σ_z. Since the ψ obeying (11.6) is evidently not a single physical scalar field we shall investigate in detail the implications of the invariance of this equation. The wave equation (11.6) in terms of coordinates and momenta in the rotated frame σ' has the form

$$c\left[\sigma_1\left(\sum a_{i1}p_i'\right) + \sigma_2\left(\sum a_{i2}p_i'\right) + \sigma_3\left(\sum a_{i3}p_i'\right)\right]\begin{bmatrix} \psi_1(x(x')) \\ \psi_2(x(x')) \end{bmatrix} = E\begin{bmatrix} \psi_1(x(x')) \\ \psi_2(x(x')) \end{bmatrix} \tag{11.8}$$

where a_{ik} is the matrix introduced in (10.1). We rewrite it now in order to see the dependence on the components in the rotated frame, and obtain

$$c\left[\left(\sum \sigma_i a_{1i}\right)p_x' + \left(\sum \sigma_i a_{2i}\right)p_y' + \left(\sum \sigma_i a_{3i}\right)p_z'\right]\begin{bmatrix} \psi_1(x(x')) \\ \psi_2(x(x')) \end{bmatrix} = E\begin{bmatrix} \psi_1(x(x')) \\ \psi_2(x(x')) \end{bmatrix} \tag{11.9}$$

The column function

$$\psi'(x') = \begin{bmatrix} \psi_1'(x') \\ \psi_2'(x') \end{bmatrix},$$

which describes the same physical solution in the new frame should satisfy the equation

$$c(\sigma_1 p'_x + \sigma_2 p'_y + \sigma_3 p'_z)\,\psi'(x') = E\psi'(x') \tag{11.10}$$

There are no primes over the σ_i since they are the same matrices (11.7). Clearly $\begin{bmatrix}\psi_1(x(x'))\\ \psi_2(x(x'))\end{bmatrix}$ which satisfies (11.9) does not in general satisfy (11.10). This means that ψ is not simply a column of two physical scalars. As will now be shown, $\psi_1'(x')$ and $\psi_2'(x')$ in the rotated frame must each be equal to a linear combination of $\psi_1(x)$ and $\psi_2(x)$ in the original frame, all four functions taken at the same space point P. Let us perform a linear transformation with a two by two matrix U,

$$\psi'(x') \equiv \begin{bmatrix}\psi_1'(x')\\ \psi_2'(x')\end{bmatrix} = U\begin{bmatrix}\psi_1(x(x'))\\ \psi_2(x(x'))\end{bmatrix} \equiv U\psi(x(x')). \tag{11.11}$$

U has to be unitary ($U^\dagger U = 1$) if we want to preserve normalization. With this definition (11.9) becomes

$$c\left[\left(\sum \sigma_i a_{1i}\right) p'_x + \left(\sum \sigma_i a_{2i}\right) p'_y + \left(\sum \sigma_i a_{3i}\right) p'_z\right] U^{-1}\psi' = EU^{-1}\psi'. \tag{11.12}$$

We determine U in such a way that ψ' satisfies (11.10). Multiplying (11.12) by U from the left we obtain, observing that $\mathbf{p}$ commutes with U,

$$c\left[U\left(\sum \sigma_i a_{1i}\right) U^{-1}p'_x + U\left(\sum \sigma_i a_{2i}\right) U^{-1}p'_y + U\left(\sum \sigma_i a_{3i}\right) U^{-1}p'_z\right]\psi' = E\psi'. \tag{11.13}$$

In order that (11.13) will be identical with (11.10), U has to satisfy

$$U\left(\sum_j a_{ij}\sigma_j\right) U^{-1} = \sigma_i\,. \tag{11.14}$$

The three matrices $\rho_i = \Sigma_j a_{ij}\sigma_j$ satisfy the same commutation relations as the σ_j. This can be verified by using the orthogonality of the matrix a_{ij}. We know, from the theory of the Pauli matrices, that a *unitary* matrix U which satisfies (11.14) does in fact exist. It is the same matrix, of course, for all values of i and depends only on the rotation considered, i.e., on the matrix a_{ij}. It is worthwhile to note that (11.11) can be written in the form

$$\psi'(P) = U\psi(P).$$

This way of writing stresses the role played by the matrix U. It takes care of the necessary changes of the components of the spinor, at the fixed point P, as the frame to which it is referred is rotated.

A solution of (11.6) is thus an example of an irreducible physical tensor field of degree higher than zero. We have not proved explicitly the irreducibility of this tensor field. It may be easily verified that the assumption of reducibility leads to the simultaneous diagonalization of the three noncommuting Pauli matrices, which, as well known, cannot be performed.

An irreducible field with $2k + 1$ components was called a tensor field of degree k. According to this the solutions of (11.6) are irreducible physical tensor fields of *degree* $k = \frac{1}{2}$. Such irreducible physical tensor fields, with a half-integer k, are called *spinors* since they always appear in the treatment of particles of half-integral spins. Equation (11.1) which determines the transformation of the components of irreducible physical fields will be extended also to half-integral values of k, and we shall come later to the exact definition of the D matrices for these cases. For $k = \frac{1}{2}$ we see from (11.11) that

$$D^{(1/2)}_{\kappa\kappa'}(R)^* = U_{\kappa'\kappa}(R) \tag{11.15}$$

where R is the rotation defined by the matrix a_{ik}.

Equation (11.1) written in matrix form, treats the components of a tensor as a row. It is customary, however, to write the neutrino equation (11.6) with ψ as a complex-conjugated column. This is the reason for the inversion of the order of indices and the appearence of complex conjugation on $D^{(1/2)}(R)$ in (11.15).

Equation (11.11) can be written explicitly in the form

$$\begin{aligned} \psi_1'(x') &= u_{11}\,\psi_1(x) + u_{12}\,\psi_2(x) \\ \psi_2'(x') &= u_{21}\,\psi_1(x) + u_{22}\,\psi_2(x). \end{aligned} \tag{11.16}$$

This is to be understood in the following way. Given a physical situation which is described by

$$\psi(x) = \begin{bmatrix} \psi_1(x) \\ \psi_2(x) \end{bmatrix}$$

in the frame σ, we can obtain its description in the frame σ' by substituting for x its expression in terms of x' and *in addition* taking the linear combinations (11.16). The situation is similar to that of a vector field, where in order to obtain the functional dependence of its components in a rotated frame we have to change the argument from x to x' as well as to take proper linear combinations of the components of the vector in the original frame. This is precisely what is implied by (11.1).

The three spin matrices σ_1, σ_2, and σ_3 are sometimes referred to as the three components of the vector $\boldsymbol{\sigma}$. When this is done, however, the special character of this vector should be kept in mind, namely the independence of its components of the frame of reference. The elements of the matrices σ_i have three indices, one *vector index i* (i.e., stating to which of the three matrices the element belongs) and two *spinor indices* labelling the rows and columns. We can therefore consider $\boldsymbol{\sigma}$ as a triple tensor. Equation (11.14) then guarantees that its elements in the new frame

$$(\sigma_i)'_{\alpha\beta} = \sum_{\mu\lambda} u_{\alpha\mu} \left(\sum_k a_{ik}\sigma_k\right)_{\mu\lambda} (u^{-1})_{\lambda\beta}$$

are exactly equal to $(\sigma_i)_{\alpha\beta}$. Such numerical tensors whose components have the same *value* in every frame are called *isotropic tensors*. The Kronecker δ_{ij} is a trivial example of such a tensor.

We should remark, however, that the three function

$$f_i(x) = \psi^\dagger(x)\sigma_i\varphi(x)$$

where ψ and φ are any two spinors, are the components of a vector. In fact, using (11.11) and the unitarity of U we obtain

$$f_i(x) = \psi^\dagger(x(x'))\sigma_i\varphi(x(x')) = \psi'^\dagger(x')U\sigma_iU^{-1}\varphi'(x')$$
$$= \sum_j \psi'^\dagger(x')\sigma_j\varphi'(x')a_{ji} = \sum a_{ji}f'_j(x')$$

by virtue of (11.14). We see that the f_i are the components of a vector only because we took into account the transformation U induced by the rotation on the spinor functions ψ and φ. It is in this sense that $\boldsymbol{\sigma}$ is often referred to as the spin vector.

The invariance of the field equations under rotations implies certain transformation properties of the field functions at a point P in space. As in (11.1) the functional dependence of the new field components on the new coordinates may be different from that of the original components as functions of the original coordinates. Equations (11.1) determines only the tensorial character of the field considered. Given a definite physical field in one frame we can construct its expression in any other frame. However if the physical field satisfies in addition an invariant field equation we can draw further important relations which go beyond (11.1). These relations result from the fact that the differential operators of the field equations in the rotated frame must have the same dependence on the new coordinates as that of the differential operators in the original frame on the original coordinates.

Let us consider a definite eigenvalue E_0 of (11.2) and write this equation for the point $P' = (x'y'z')$ in the *frame* σ

$$H(x'_i, p'_i)\,\psi_\lambda(x'_i) = E_0\psi_\lambda(x'_i).$$

Here ψ_λ is one of a complete set of eigenfunctions which belong to E_0. The eigenfunction ψ_λ is thus a definite physical scalar field. Every solution of (11.2) which belongs to the same eigenvalue E_0 is a linear combination of the ψ_λ. If $P = (xyz)$ is *another* point in the same frame σ, such that $r^2 = r'^2$, then a rotation matrix a_{ik} exists so that $x'_i = \Sigma a_{ik} x_k$. This matrix is in fact the one which corresponds to the rotation that brings P' to P. We can therefore use the invariance of H under rotations to obtain

$$H(x'_i, p'_i)\psi_\lambda(x'_i(x)) = H(x_i, p_i)\psi_\lambda(x'_i(x)) = E_0\psi_\lambda(x'_i(x)). \tag{11.17}$$

We see therefore that $\psi_\lambda(x'_i(x))$ considered as a function of the x_i, satisfies the original equation (11.2). This solution is obtained by "rotating" $\psi_\lambda(x_i)$ so that the field originally at P' is carried over to the different point P. Since this new solution $\psi_\lambda(x'_i(x))$ belongs to the same eigenvalue E_0, it can be expanded in terms of the functions $\psi_\lambda(x_i)$

$$\psi_\lambda(x'_i(x)) = \sum_\mu c_{\lambda\mu}\psi_\mu(x). \tag{11.18}$$

The relation (11.18) describes the change in the mathematical function $\psi_\lambda(x_i)$ when we go from one point P to another point P' in the same frame. Since the coordinates of P and P' can be related by a matrix of a rotation we can say that the functions ψ_λ transform under rotation according to (11.18). The matrix elements $c_{\lambda\mu}$ which define the transformation, depend on the rotation. Since we agreed to call sets of quantities which transform among themselves under rotation the components of a tensor, all the functions ψ_λ which belong to a given eigenvalue E_0 deserve this name as well. In order to distinguish between the *physical tensor fields* discussed before and such sets of degenerate solutions, we shall call the latter simply *tensors*. Although we previously used the names tensor and irreducible tensor for tensors defined at one point of space only, no confusion will arise since the latter will not appear in the following section.

It is possible that linear combinations of the ψ_λ can be formed in such a way that there will be several subgroups which transform among themselves in the sense of (11.18). If we do this we can form the components of one or more irreducible tensors from the functions ψ_λ. If the number of components of such an irreducible tensor is $2k + 1$ we can choose them so that they will transform with the matrix $D^{(k)}_{\kappa\kappa'}$ introduced before, where $-k \leq \kappa,\ \kappa' \leq k$. The transformation law of these *irreducible tensors* will then be

$$T^{(k)}_\kappa(\mathbf{r}'(\mathbf{r})) = \sum_{\kappa'} T^{(k)}_{\kappa'}(\mathbf{r})\, D^{(k)}_{\kappa'\kappa}(R). \tag{11.19}$$

That this is always possible follows from the group theoretical theorem about the uniqueness (up to equivalence) of the representations of the rotation group of a given order.

Although (11.19) looks very much like (11.1) its meaning is quite different. The $T_\kappa^{(k)}$ which appear in (11.1) are all components of the same physical tensor field. In other words, a particular field (a solution of the field equations), like the scalar solution of the Schrödinger equation or the spinor solution of (11.6), is given only when the values of all the components are given at every point in space. The functional dependence of the field components on the coordinates may be different in different frames of reference. On the other hand, nothing is implied by (11.19) about the functions $T_\kappa^{(k)}$ as physical tensor fields. This is clearly manifested by the example of the Schrödinger equation (11.2). Each one of its solutions $\psi_\kappa^{(k)}(x)$ is a *physical scalar field.* Once a single solution $\psi_\kappa^{(k)}(x)$ is given in one frame of reference, the physical field is uniquely defined in any frame. In fact, the functional dependence of the same physical solution on its arguments in a rotated frame of reference, can be easily found from the original functional form and is generally different from it. For a physical scalar field we saw that in the new frame it is given by $\psi_\kappa^{(k)\prime}(x') = \psi_\kappa^{(k)}(x(x'))$. Thus, using (11.19) for the reverse rotation R^{-1} we obtain

$$\psi_\kappa^{(k)\prime}(x') = \psi_\kappa^{(k)}(x(x')) = \sum_{\kappa'} \psi_{\kappa'}^{(k)}(x')\, D_{\kappa'\kappa}^{(k)}(R^{-1}).$$

From this we see that the function $\psi_\kappa^{(k)\prime}(x')$ is indeed very different from the function $\psi_\kappa^{(k)}(x')$.

On the other hand, (11.19) gives an important relation between the different degenerate physical fields which transform among themselves under rotations. It is seen that it is enough to know these $2k + 1$ independent fields $T_\kappa^{(k)}(\mathbf{r})$ along a single ray in order to determine them over the whole space. This particular feature is a result of the invariance of the field equations. It is clear that the irreducible tensors have all the properties of physical tensors (and tensor fields) that follow from their transformation law. Similar relations to those discussed above exist also between the degenerate solutions of the two-component neutrino equation (11.6). We shall deal with these relations in detail in a subsequent section.

It should be pointed out that the properties defined by Equations (11.1) and (11.19) are *not contradictory*. It may happen that the components of a physical tensor field have the same functional dependence on the coordinates in every frame of reference. The simplest example is the coordinate vector $\mathbf{r}$ itself. Other tensors built from it also have this property. In particular we can consider the spherical harmonics of order k. Equation (10.11), giving their transformation, is obviously identical with (11.19). However, only if the spherical harmonics $Y_{k\kappa}$ are the *components* of a certain physical tensor field of degree k (as is the case for the quadrupole moment operator with $k = 2$), do they obey (11.1) as well. Other examples of physical tensor fields, which at the same time are also tensors in the sense of (11.19) are the vector operator ∇ and other vectors and tensors built from it or from combinations of ∇ and $\mathbf{r}$ such as $\mathbf{L} = -i\mathbf{r} \times \nabla$. In this book we shall usually deal with physical tensor fields that have this property, namely which are also irreducible tensors. Equation (11.1) is in this case mathematically identical with (11.19) which is the reason why we stressed so much the difference in their physical meaning.

In these new irreducible tensors which we introduced the argument of $T^{(k)}(\mathbf{r})$ does not necessarily represent the coordinates of a single point in ordinary space. It could just as well be the coordinates of several points in ordinary space, all simultaneously referred to the frame σ or σ'. The antisymmetric tensor $B_{ij} = \xi_i\eta_j - \xi_j\eta_i$ is an example of a tensor which depends on the coordinates of two points $\boldsymbol{\xi}$ and $\boldsymbol{\eta}$. If $\mathbf{T}^{(k)}(\mathbf{r})$, with integral k, is a function of the coordinates of one point only then we shall now show that (11.19) implies that

$$T_\kappa^{(k)}(\mathbf{r}) = f_k(r)\, Y_{k\kappa}(\theta\, \varphi). \tag{11.20}$$

Since the spherical harmonics form a complete set of functions we can expand any $T_\kappa^{(k)}(\mathbf{r})$

$$T_\kappa^{(k)}(\mathbf{r}) = \sum_{l,\lambda} C_{k\kappa,l\lambda}(r)\, Y_{l\lambda}(\theta\, \varphi)$$

and similarly

$$T_{\kappa'}^{(\kappa)}(\mathbf{r}') = \sum_{l',\lambda'} C_{k\kappa',l'\lambda'}(r)\, Y_{l'\lambda'}(\theta'\, \varphi').$$

Using the transformation properties of $\mathbf{T}^{(k)}$ and $\mathbf{Y}_l$ we find that

$$\sum_\kappa C_{k\kappa,l\lambda}(r)\, D_{\kappa\kappa'}^{(k)}(R) = \sum_{\lambda'} C_{k\kappa',l\lambda'}(r)\, D_{\lambda\lambda'}^{(l)}(R).$$

If we now make use of the orthogonality of the D-functions (3.12) we find that $C_{k\kappa,l\lambda}(r)$ vanishes unless $k = l$ and $\kappa = \lambda$ and that $C_{k\kappa,k\kappa}$ is independent of κ. This proves (11.20). Thus, the concept of irreducible tensors which depend only on the space coordinates of a single particle leads us back to the theory of spherical harmonics. However, the generalization of the concept of irreducible tensors will be of use wherever there are functions of more than one particle, or when they are differential operators operating on the coordinates of one or more particles, as well as in all the cases of half-integral k.

12. Infinitesimal Rotations and Irreducible Tensors

In the preceding section we considered the covariance properties of various quantities under arbitrary rotations. Any rotation in space can be obtained by successive applications of appropriate rotations around the three axes. Every rotation around a given axis can be obtained by successive infinitesimal rotations around that same axis. Thus, every rotation can be obtained by successive applications of the three independent infinitesimal rotations, δR_x, δR_y and δR_z. It will turn out that for many discussions it is enough to consider the behavior of a physical or mathematical quantity under these infinitesimal rotations and there is no necessity of actually treating the finite rotations. In this section we shall therefore study in detail the infinitesimal rotations. This will make it easier to generalize the concept of irreducible tensors in the sense of (11.19) also to nonscalar physical fields.

We saw in the preceding section how to obtain information on the spatial dependence of certain functions from their behavior under rotations. We would now like to derive similar properties using infinitesimal rotations. Let $f(P)$ be a function defined at every point in space (a physical scalar field). In the preceding section we considered the change in certain functions when going from one point P_0 in space to another point P. This change was considered as being induced by a certain rotation—the rotation that brings P to P_0. We shall repeat now these considerations dealing this time with infinitesimal rotations. If P and P_0 are two points close to each other, at the same distance from the origin, we shall therefore interpret the change in $f(P)$

$$\delta f = f(P) - f(P_0) \tag{12.1}$$

as induced by infinitesimal rotations applied to f.

Let the coordinates of P_0 and P in a given frame σ be (x_0, y_0, z_0) and (x, y, z) respectively. A matrix a_{ik} can be found so that

$$x_i = \sum a_{ik} x_{0k} \qquad x_{0i} = \sum (a^{-1})_{ik} x_k. \tag{12.2}$$

It is, in fact, the matrix which rotates the frame in which the point P_0 has the coordinates (x_0, y_0, z_0) to the frame in which this same space point has the coordinates (x, y, z). Thus if f' denotes, as before, the functional dependence of the function $f(P)$ on the coordinates in the rotated frame, (12.1) can be written also as

$$\begin{aligned} \delta f &= f(P) - f(P_0) = f(x, y, z) - f'(x, y, z) \\ &= f(x, y, z) - f(x_0(x, y, z), y_0(x, y, z), z_0(x, y, z)) \end{aligned} \tag{12.3}$$

so that δf can be interpreted as the change in the function f when we pass from a point (x, y, z) in one frame to a point (x, y, z) in a rotated frame. In this sense it describes the behavior of f under rotations. We would like to stress that (12.3) represents a difference between two values of f taken at two different space points P and P_0. In general, $f(x)$ is different from $f(x_0)$, whereas, of course, for the physical scalar field $f(P)$ we have $f'(x) = f(x_0)$. Substituting (12.2) into (12.3) we obtain

$$\delta f = f(x) - f(x_0(x)) = f(x) - f(a^{-1}x). \tag{12.4}$$

Since a will be an infinitesimal rotation, δf in (12.4) will be equal to $f(ax) - f(x)$.

If the matrix a_{ik} corresponds to an infinitesimal rotation $\delta\theta_z$ around the z axis, then $x = x_0 + y_0\delta\theta_z$, $y = y_0 - x_0\delta\theta_z$, $z = z_0$. In other words

$$a = \begin{pmatrix} 1 & +\delta\theta_z & 0 \\ -\delta\theta_z & 1 & 0 \\ 0 & 0 & 1 \end{pmatrix}. \tag{12.5}$$

Therefore, to first order in $\delta\theta_z$,

$$\begin{aligned} \delta_z f &= f(x, y, z) - f(x - y\delta\theta_z, y + x\delta\theta_z, z) \\ &= f(x + y\delta\theta_z, y - x\delta\theta_z, z) - f(x, y, z) \\ &= \delta\theta_z \left(+y\frac{\partial}{\partial x} - x\frac{\partial}{\partial y}\right) f(x, y, z). \end{aligned} \tag{12.6}$$

This way we obtain the change in a given (scalar) function induced by an infinitesimal rotation equal to the infinitesimal change $\delta\theta_z$ multiplied by the result of the application of a differential operator on the function. We notice that the differential operator in (12.6) is proportional to the operator of angular momentum introduced in Part I $l_z = -i(x(\partial/\partial y) - y(\partial/\partial x))$. Hence

$$\delta_z f(x) = -i\,\delta\theta_z l_z f(x).$$

Similar relations hold for rotations around the x and y axes:

$$\begin{aligned} \delta_x f(x) &= -i\,\delta\theta_x l_x f(x) \\ \delta_y f(x) &= -i\,\delta\theta_y l_y f(x). \end{aligned}$$

The change in f due to an arbitrary infinitesimal rotation obtained by successive infinitesimal rotations around the three axes is obtained by

$$\delta f(x) = [(1 - i\delta\theta_z l_z)(1 - i\delta\theta_y l_y)(1 - i\delta\theta_x l_x) f(x)] - f(x).$$

Keeping only linear terms in $\delta\theta_z$, we obtain

$$\delta f(x) = -i(\delta\theta_x l_x + \delta\theta_y l_y + \delta\theta_z l_z)f(x) = -i\,\delta\theta(\mathbf{n}\cdot\boldsymbol{l})f(x)$$

$$\text{where} \qquad \delta\theta = +\sqrt{(\delta\theta_x)^2 + (\delta\theta_y)^2 + (\delta\theta_z)^2} \tag{12.7}$$

and $\mathbf{n}$ is the unit vector $\dfrac{1}{\delta\theta}(\delta\theta_x, \delta\theta_y, \delta\theta_z)$.

Equation (12.7) corresponds to the well-known fact that infinitesimal rotations, unlike finite rotations, *do* combine like vectors. Equation (12.7) describes the change in $f(x)$ induced by a rotation with an angle $\delta\theta$ around the direction $\mathbf{n}$. More precisely, we are considering a point S rigidly connected to the frame of reference and rotating with it. This point coincides with the space point P when the frame is in the original position, and with the space point P_0 when the frame is in the rotated position. The difference between the values of f at these two points is $\delta f = f(P) - f(P_0)$. Alternatively, δf is the change in the function f recorded at a fixed space point P_0, using a fixed frame of reference, when the whole physical field f is rotated rigidly by an angle $+\delta\theta$ around the direction $\mathbf{n}$. It is obvious that the two statements are completely equivalent, and in both cases the answer is given by the application of the differential operator $-i\delta\theta(\mathbf{n}\cdot\boldsymbol{l})$ to the function f at the point (x, y, z).

For physical fields of degrees higher than zero a similar procedure can be adopted. Special care should however be taken in the proper handling of the components of the field in the various frames.

Let us now consider the case of the two component spinor field ψ introduced in the preceding section. Using a given frame of reference we can ask again what is the change in the field ψ recorded at a given space point P_0 when the field is rotated *rigidly* by an angle $\delta\theta$ around the direction $\mathbf{n}$. Equivalently we can consider a point S rigidly attached to the frame at the coordinates (x, y, z). When the frame is in the original position, S coincides with the point P whose coordinates with respect to this original frame are therefore (x, y, z). The spinor field at this point is $\psi(x, y, z)$. At the rotated position, S coincides with the point P_0 and the spinor field there, referred to the *rotated* frame, is $\psi'(x, y, z)$. Here (x, y, z) are the coordinates referred to the rotated frame, i.e., $x = ax_0$. As in (12.3), $\delta\psi$ is given by

$$\delta\psi = \psi(x, y, z) - \psi'(x, y, z). \tag{12.8}$$

It should be emphasized that in (12.8) ψ and ψ', when written in terms of their components, refer to *different* frames and we are thus subtracting components referred to different frames. This, however, is precisely what happens when the field is rotated rigidly and the variations in it are recorded at a given point in space.

Substituting in (11.11) x for x' and x_0 for x we obtain

$$\psi'(x) = U\psi(x_0(x)).$$

Since $x = ax_0$, we can write

$$\psi'(x) = U\psi(a^{-1}x) \tag{12.9}$$

where U is a matrix which satisfies (11.14), i.e.,

$$U\left(\sum a_{ij}\sigma_j\right) U^{-1} = \sigma_i$$

and a_{ik} is the rotation matrix defined in (12.2). Considering again an infinitesimal rotation around the z-axis, for which a_{ik} is given by (12.5) we find that the U corresponding to it should satisfy

$$\begin{aligned} U_z(\sigma_1 + \delta\theta_z\sigma_2)U_z^{-1} &= \sigma_1 \\ U_z(-\delta\theta_z\sigma_1 + \sigma_2)U_z^{-1} &= \sigma_2 \\ U_z\sigma_3U_z^{-1} &= \sigma_3. \end{aligned} \tag{12.10}$$

Equations (12.10) can be solved easily. It is clear that U coincides with the identity 1 for $\delta\theta_z = 0$, and is linear in $\delta\theta_z$ for infinitesimal rotations. It can therefore be written in the form

$$U_z = 1 + \delta\theta_z\Omega_z \qquad U_z^{-1} = 1 - \delta\theta_z\Omega_z$$

where 1 is the unit matrix and Ω_z is a noninfinitesimal matrix. Retaining only linear terms in $\delta\theta_z$ we then obtain from (12.10)

$$\begin{aligned} +\sigma_2 + \Omega_z\sigma_1 - \sigma_1\Omega_z &= 0 \\ -\sigma_1 + \Omega_z\sigma_2 - \sigma_2\Omega_z &= 0 \\ \Omega_z\sigma_3 - \sigma_3\Omega_z &= 0. \end{aligned} \tag{12.11}$$

The solution of (12.11) is readily found from the properties of the spin matrices. We find $\Omega_z = (i/2)\,\sigma_3$ so that

$$U_z = 1 + \frac{i}{2}\,\delta\theta_z\sigma_3.$$

Thus, we obtain $\psi'(x) = (1 + \frac{1}{2}\,i\delta\theta_z\sigma_3)\,\psi(a^{-1}x)$. We express $\psi(a^{-1}x)$ by using (12.6) and find for a rotation $\delta\theta_z$ around the z-axis

$$\begin{bmatrix}\psi_1'(x)\\ \psi_2'(x)\end{bmatrix} = (1 + \tfrac{1}{2}\,i\delta\theta_z\sigma_3)\begin{bmatrix}(1 + i\delta\theta_z l_z)\psi_1(x)\\ (1 + i\delta\theta_z l_z)\psi_2(x)\end{bmatrix}$$

$$= (1 + \tfrac{1}{2}\,i\delta\theta_z\sigma_3)\,(1 + i\delta\theta_z l_z)\begin{bmatrix}\psi_1(x)\\ \psi_2(x)\end{bmatrix}.$$

Retaining only linear terms in $\delta\theta_z$ we finally obtain

$$\delta_z\psi = -\, i\delta\theta_z(l_z + \tfrac{1}{2}\sigma_3)\psi.$$

Similar expressions can be obtained for $\delta_x\psi$ and $\delta_y\psi$ leading to the general relation

$$\delta\psi = -\, i\delta\theta\, \mathbf{n}.(\boldsymbol{l} + \tfrac{1}{2}\boldsymbol{\sigma})\psi \tag{12.12}$$

where the meaning of $\delta\theta$ and $\mathbf{n}$ is as in (12.7) and

$$\mathbf{n}\,.\,\boldsymbol{\sigma} = n_x\sigma_1 + n_y\sigma_2 + n_z\sigma_3.$$

We see from (12.12) that the change affected in a spinor field by going from a point P_0 to a point P has two sources. First, each component is changed by an amount $-\, i\delta\theta(\mathbf{n}\,.\,\boldsymbol{l})\psi_i$. This is a change which each component, being a function of x, undergoes as a result of the rotation, irrespective of whether it is a component of a spinor field or not. In addition to it there is the transformation $-\, i\delta\theta(\mathbf{n}\,.\,\frac{1}{2}\,\boldsymbol{\sigma})\psi$ which expresses each new component of the spinor field in terms of a linear combination of the original components. This additional change stems from the physical nature of the field, i.e., its being a spinor field. It represents an intrinsic property of a particle which requires such a complicated mathematical entity for its description. Thus, for example, whereas there are column functions $\psi(x)$ whose functional dependence on x is such that $\boldsymbol{l}\psi(x) = 0$ (those for which $\psi(x) = \psi(r)$), no spinor field can satisfy $\boldsymbol{\sigma}\psi = 0$.

The difference between (12.12) and (12.7) can be clearly understood. It was shown in the previous section that if $\psi_\lambda(x)$ is an eigenfunction of the rotationally invariant Hamiltonian (11.2) which belongs to the eigenvalue E, then $\psi_\lambda(ax)$, where a_{ik} is any rotation, is also an eigenfunction of H which belongs to the same eigenvalue E. In particular a_{ik} can be an infinitesimal rotation, in which case $\psi_\lambda(ax) = \psi_\lambda(x) + \delta\psi_\lambda(x)$ where $\delta\psi_\lambda(x)$ is given by (12.7). In the case of a spinor field, if $\psi_\lambda(x)$ is a solution which belongs to a given eigenvalue E, only $U^{-1}\psi_\lambda(ax) = \psi_\lambda(x) + \delta\psi_\lambda(x)$ is a solution of the wave equation (11.6) which belongs to the same eigenvalue. Therefore $\delta\psi_\lambda$ in this case is given by (12.12) rather than by (12.7).

If, for convenience of notation, we consider δ as an operator we can write

$$H(1 + \delta)\psi_\lambda = E(1 + \delta)\psi_\lambda = (1 + \delta)E\psi_\lambda = (1 + \delta)H\psi_\lambda. \tag{12.13}$$

Equation (12.13) being valid for any ψ_λ, implies that the Hamiltonian commutes with the operators $(1 + \delta)$, as well as with δ itself. We can therefore use these operators, or functions built from them, for further specification of states which are degenerate with respect to H. The eigenvalues of these operators are constants of the motion. For scalar fields we can use the two commuting operators $l^2 = l_x^2 + l_y^2 + l_z^2$ and l_z which then classify the degenerate states of the Hamiltonian according to the square of the angular momentum l^2 and its z-projection l_z. For spinor fields we can use the two corresponding operators $\mathbf{j}^2 = (\boldsymbol{l} + \frac{1}{2}\boldsymbol{\sigma})^2$ and $j_z = l_z + \frac{1}{2}\sigma_z$, for the classification of the states. These operators are, as is well known, the operators of the square of the total angular momentum $\mathbf{j}^2$ (orbital plus intrinsic angular momentum), and its z-projection j_z. The fact that a certain particle has to be described by two functions indicates that in addition to the spatial degrees of freedom there is another, *space-independent* degree of freedom, which, as we have just seen gives rise to the intrinsic angular momentum. The neutrino, whose spin is $s = \frac{1}{2}$, can assume two $(= 2s + 1)$ independent intrinsic states, and can therefore be described by spinor fields. Other particles whose intrinsic spin may be 1, for instance, have three independent intrinsic states and therefore require a physical *vector* field for their description.

The occurence of the operators of angular momentum in connection with the invariance of the Hamiltonian under rotations is an example of how invariance requirements may result in constants of motion. In general, let O be any operation on the coordinates and momenta of the particles of the system, and let the Hamiltonian be invariant under this operation $H(Ox) = H(x)$. Such an operation can be a rotation of the frame of reference as above, or a permutation of the particles (e.g., $H[O(\mathbf{r}_1, \mathbf{r}_2)] \equiv H(\mathbf{r}_2, \mathbf{r}_1) = H(\mathbf{r}_1, \mathbf{r}_2)$) or any other operation which leaves H invariant. Let the result of such an operation on a function ψ be described by the operator Ω, i.e., $\psi(Ox) = \Omega\psi$. Then, if $H\psi = E\psi$, we obtain $H(Ox)\psi(Ox) = H(\Omega\psi) = E\psi(Ox) = E\Omega\psi$. Thus ψ and $\Omega\psi$ belong to the same eigenvalue E, and Ω commutes with H. The eigenvalues of Ω are therefore constants of the motion and the different states of the system can be chosen to be simultaneous eigenstates of H and of Ω.

Thus we see that, whereas for Hamiltonians of scalar fields the orbital angular momentum $\boldsymbol{l}$ is a constant of the motion, for Hamiltonians of spinor fields the constant of the motion is $\boldsymbol{l} + \frac{1}{2}\boldsymbol{\sigma}$. It is the sum of the orbital $(\boldsymbol{l})$ and the internal $(\frac{1}{2}\boldsymbol{\sigma})$ angular momenta. The orbital angular momentum $\boldsymbol{l}$ alone is *not* a constant of the motion in this case since the effects of a rotation on the two component wave functions is represented by the sum of the two operators $\boldsymbol{l}$ and $\frac{1}{2}\boldsymbol{\sigma}$.

Our considerations up to now were limited to functions of one particle $f(\mathbf{x})$. However they can easily be extended to functions of several particles $f(\mathbf{x}_1, \mathbf{x}_2, ..., \mathbf{x}_n)$. If we consider a change

$$\delta f = f(P_1', P_2', ..., P_n') - f(P_1, P_2, ..., P_n),$$

we obtain in a way similar to that used to derive (12.7), that

$$\delta f = -i[\delta\theta_1(\mathbf{n}_1 \cdot \boldsymbol{l}_1) + \delta\theta_2(\mathbf{n}_2 \cdot \boldsymbol{l}_2) + ... + \delta\theta_n(\mathbf{n}_n \cdot \boldsymbol{l}_n)]\, f. \quad (12.14)$$

Here $\delta\theta_i$ is the infinitesimal rotation around the direction $\mathbf{n}_i$ which brings P'_i to P_i, and $\boldsymbol{l}_i$ is the operator of angular momentum operating on the coordinates x_i of the ith particle.

Although (12.14) is valid for independent infinitesimal rotations applied to each of the particles, it will be of little use in its general form since the Hamiltonian of a system of particles is generally not invariant under such independent rotations. The Hamiltonian will generally remain invariant only if the *same* rotation is applied to all the particles. They are all referred to one single frame before the rotation and to one single rotated frame after the rotation. Denoting now by $\delta\theta$ the angle of rotation of the frame around the direction $\mathbf{n}$ we obtain

$$\delta f = -i\delta\theta(\mathbf{n}\cdot\mathbf{L})f. \tag{12.15}$$

Here

$$\mathbf{L} = \boldsymbol{l}_1 + \boldsymbol{l}_2 + \ldots + \boldsymbol{l}_n \tag{12.16}$$

can be considered as the operator of the total angular momentum of the system. We see from (12.15) and (12.13) that the invariance of H under rotations can only guarantee that $\mathbf{L}$, the total angular momentum, will be a constant of the motion. The individual angular momenta $\boldsymbol{l}_i$ will, in general, not be constants of the motion.

For spinor fields with many particles the situation is similar. The function ψ can be written as

$$\psi = \begin{bmatrix}\psi'_1(x_1)\\ \psi'_2(x_1)\end{bmatrix}\begin{bmatrix}\psi''_1(x_2)\\ \psi''_2(x_2)\end{bmatrix}\begin{bmatrix}\psi'''_1(x_3)\\ \psi'''_2(x_3)\end{bmatrix}\cdots \tag{12.17}$$

and $\delta\psi$ will then be given by

$$\delta\psi = -i\delta\theta\,\mathbf{n}\,.\,(\mathbf{L}+\mathbf{S})\psi = -i\delta\theta(\mathbf{n}\,.\,\mathbf{J})\ . \tag{12.18}$$

Here

$$\mathbf{S} = \tfrac{1}{2}(\boldsymbol{\sigma}_1 + \boldsymbol{\sigma}_2 + \boldsymbol{\sigma}_3 + \ldots) \qquad \text{and} \qquad \mathbf{J} = \mathbf{L} + \mathbf{S} \tag{12.19}$$

and $\boldsymbol{\sigma}_i$ operates on the column $\begin{bmatrix}\psi_1^{(i)}(x_i)\\ \psi_2^{(i)}(x_i)\end{bmatrix}$ only.

The internal degrees of freedom of a particle described by a spinor field are space-independent. Therefore it is convenient to write such states in a form showing explicitly the spatial part of the wave function and the part which refers to the internal degrees of freedom. As a matter of fact this was already done in Part I when we wrote the wave function of a single particle in a definite lj orbit in the form (2.24)

$$\psi_{nljm}(\mathbf{r}) = \sqrt{\frac{l \pm m + \frac{1}{2}}{2l+1}}\,\varphi_{nlm-1/2}(\mathbf{r})\,\chi_{1/2} \pm \sqrt{\frac{l \mp m + \frac{1}{2}}{2l+1}}\,\varphi_{nlm+1/2}(\mathbf{r})\,\chi_{-1/2} \tag{12.20}$$

for $j = l \pm \frac{1}{2}$. The functions $\chi_{1/2}$ and $\chi_{-1/2}$ are simultaneous eigenfunctions of the operators $\mathbf{s}^2$ and s_z where $\mathbf{s} = \frac{1}{2}\boldsymbol{\sigma}$. Written as a column in the scheme in which σ_z is diagonal they are $\chi_{1/2} = \binom{1}{0}$ $\chi_{-1/2} = \binom{0}{1}$. The $\varphi_{nlm\pm 1/2}$ are eigenfunctions of $\boldsymbol{l}^2$ and l_z corresponding to the eigenvalues $l(l+1)$ and $m \pm \frac{1}{2}$ respectively. Equation (12.20) actually tells us how to construct a physical spinor field which will be an eigenstate of $\boldsymbol{l}^2$, $\mathbf{j}^2 = (\boldsymbol{l} + \mathbf{s})^2$ and $j_z = l_z + s_z$. This state is constructed from the physical scalar fields $\varphi_{nlm\pm 1/2}$ which are eigenfunctions of $\boldsymbol{l}^2$ and l_z and the spinors $\chi_{\pm 1/2}$ which are functions of the *internal* degrees of freedom and are eigenfunctions of $\mathbf{s}^2$ and s_z.

To stress the spinor character of the functions (12.20) we can use the representation of $\chi_{\pm 1/2}$ given above and write

$$\psi_{nljm}(\mathbf{r}) = \begin{bmatrix} \sqrt{\dfrac{l \pm m + \frac{1}{2}}{2l+1}}\, \varphi_{nlm-1/2}(\mathbf{r}) \\ \pm \sqrt{\dfrac{l \mp m + \frac{1}{2}}{2l+1}}\, \varphi_{nlm+1/2}(\mathbf{r}) \end{bmatrix} \quad \text{for} \quad j = l \pm \tfrac{1}{2} \tag{12.21}$$

(note that m is a half-integer and therefore m_l of φ will always be equal to $m \pm \frac{1}{2}$). Thus $\psi_{nljm}(\mathbf{r})$ is the two-component function which describes a given physical state for a particle with spin $\frac{1}{2}$.

We saw in the previous section that physical scalar fields satisfy certain mathematical relations (11.19) if they are eigenstates of a Hamiltonian which is invariant under rotations. We are now in the position to discuss similar relations that exist for physical spinor fields which are eigenstates of a rotationally invariant Hamiltonian. Let us consider the eigenspinors of a Hamiltonian which commutes with $\mathbf{j} = \boldsymbol{l} + \mathbf{s} = \boldsymbol{l} + \frac{1}{2}\boldsymbol{\sigma}$. We shall show that a set of degenerate but different physical solutions ψ_{jm} will satisfy a relation similar to (11.19).

Let us consider a spinor Hamiltonian which is invariant under rotations. Such invariant Hamiltonians can be obtained if we add to the ordinary terms (1.1) the combination $\Sigma\sigma_i v_i$ where v_i are the three components of any vector. In fact it follows from (11.14) that for *any* vector $\mathbf{v}$

$$\sum \sigma_i v_i = \sum \sigma_i a_{ki} v_k' = \sum U^{-1}\sigma_k U v_k' = U^{-1}\left(\sum \sigma_k v_k'\right) U. \tag{12.22}$$

Therefore, a rotationally invariant Hamiltonian, to which we add terms of the type $\Sigma\sigma_i v_i$, will continue to be invariant under rotations provided we stipulate, as for the two-component neutrino equation, that the relation between the spinors describing the same physical solution in the two frames is given by (11.11). The Hamiltonian of a nucleon moving in a central field with spin-orbit interaction

$$H_s = \frac{1}{2m}\, p^2 + U(r) + \zeta(r)\,(\boldsymbol{l} \cdot \mathbf{s}) \tag{12.23}$$

is a special case for which $\mathbf{v} = \mathbf{l}$. It is therefore invariant under rotations, and its energy is degenerate with respect to j_z.

Consider the spinor $\psi_{jm}^{(0)}(x')$ referred to the frame σ_0 which is a solution of the Schrödinger equation with H_s of (12.23)

$$H_s(x'p')\,\psi_{jm}^{(0)}(x') = E_0\psi_{jm}^{(0)}(x').$$

Let x' be the coordinates in σ_0 of the point P'. We now consider another point P whose coordinates in σ_0 are given by x, such that $r^2 = r'^2$. As in the derivation of (11.17), we introduce the rotation a_{ik} so that $x' = ax$. This rotation brings σ_0 to the position σ_1. Substituting in (12.23) x' and p' by ax and ap, respectively, we obtain, using (12.22),

$$H_s(x', p') = UH_s(x, p)U^{-1}.$$

Hence we obtain

$$H_s(x, p)U^{-1}\,\psi_{jm}^{(0)}(x'(x)) = E_0U^{-1}\,\psi_{jm}^{(0)}(x'(x)). \tag{12.24}$$

Thus $U^{-1}\psi^{(0)}(x'(x))$ considered as a function of x belongs also to the eigenvalue E_0. It can therefore be expressed as a linear combination of the spinors which belong to the same eigenvalue E_0 of H_s. Apart from accidental degeneracies, which can be removed without destroying the rotational invariance of H_s, these are the $2j + 1$ spinors $\psi_{jm}^{(0)}(x)$. We thus obtain

$$U^{-1}\,\psi_{jm}^{(0)}(x'(x)) = \sum \psi_{jm'}^{(0)}(x)\,D_{m'm}^{(j)}(R) \tag{12.25}$$

where R is the rotation $x' = ax$. We have denoted the matrix which determines the linear combination (12.25) by $D_{m'm}^{(j)}(R)$ just in order to have the same notation as that used for scalar fields. Here, however, j is a half-integer and therefore (12.25) is the *definition* of these $D^{(j)}$. It should be recalled that we have already introduced the matrix $D^{(1/2)}(R)$ in (11.15) when we studied the behavior of a general spinor field under rotations. However, the present definition is consistent with it. The spinor $U^{-1}\psi_{jm}^{(0)}(x'(x))$ represents, in the original frame σ_0, the field obtained by rotating rigidly the field $\psi_{jm}^{(0)}(x)$ so that the field at P' is brought over to P. The factor U^{-1} originates from the fact that in rotating the spinor field rigidly its components along the axes of σ_0 are also changed.

Using (11.11) we see that $U^{-1}\psi_{jm}^{(0)}(x')$ can be interpreted as the spinor $\psi_{jm}^{(0)}(x')$ referred to the frame σ_1 obtained from σ_0 by the rotation R. Let us denote this spinor by $\psi_{jm}^{(1)}(x')$. We then obtain from (12.25)

$$\psi_{jm}^{(1)}(x'(x)) = \sum \psi_{jm'}^{(0)}(x)\,D_{m'm}^{(j)}(R). \tag{12.26}$$

As is well known $\psi_{jj}^{(0)}$ represents a state in which the angular momentum $\mathbf{j}$ is directed along the z-axis of the frame σ_0. Similarly $\psi_{jj}^{(1)}$ obtained from (12.26) is the state in which $\mathbf{j}$ is oriented along the z-axis of the frame σ_1. This can be understood if one recalls that the matrix of the transformation (11.1) of the components of physical tensor fields, as well as (11.11) for spinor fields, is also the matrix which determines the transformations of the unit tensors or spinors. Thus, (12.26) can also be considered as a relation between unit spinors in the two frames which form independent solutions of the corresponding spinor Schrödinger equations.

To derive an expression for the matrices $D^{(j)}(R)$ we make use of (11.15) which, due to the unitarity of U, can be written as

$$D^{(1/2)}(R) = U^{\dagger}(R) = U^{-1}(R). \tag{12.27}$$

To proceed to higher values of j, we make use of the explicit construction (12.21) of ψ_{jm}. Using (12.25) we obtain

$$\sum_{m'} \psi_{jm'}^{(0)}(x)\, D_{m'm}^{(j)}(R) = U^{-1}\, \psi_{jm}^{(0)}(x'(x))$$

$$= D^{1/2}(R) \begin{bmatrix} \sqrt{\dfrac{l \pm m + \frac{1}{2}}{2l+1}}\, \varphi_{nlm-1/2}(ax) \\ \pm \sqrt{\dfrac{l \mp m + \frac{1}{2}}{2l+1}}\, \varphi_{nlm+1/2}(ax) \end{bmatrix}. \tag{12.28}$$

Since φ_{nlm_l} is the component of an irreducible tensor of degree l it satisfies

$$\varphi_{nlm\pm1/2}(ax) = \sum_{m'} \varphi_{nlm'\pm1/2}(x)\, D_{m'\pm1/2,\, m\pm1/2}^{(l)}(R)$$

and therefore, using (12.28), we obtain

$$\sum_{m'} \begin{bmatrix} \sqrt{\dfrac{l \pm m' + \frac{1}{2}}{2l+1}}\, \varphi_{nlm'-1/2}(x) \\ \pm \sqrt{\dfrac{l \mp m' + \frac{1}{2}}{2l+1}}\, \varphi_{nlm'+1/2}(x) \end{bmatrix} D_{m'm}^{(j)}(R)$$

$$= \sum_{m',m''} D^{(1/2)}(R) \begin{bmatrix} \sqrt{\dfrac{l \pm m + \frac{1}{2}}{2l+1}}\, \varphi_{nlm'-1/2}(x)\; D_{m'-1/2,m-1/2}^{(l)}(R) \\ \pm \sqrt{\dfrac{l \mp m + \frac{1}{2}}{2l+1}}\, \varphi_{nlm''+1/2}(x)\, D_{m''+1/2,m+1/2}^{(l)}(R) \end{bmatrix}$$

$$= \sum_{m'm''} \begin{bmatrix} \left[\sqrt{\frac{l \pm m + \frac{1}{2}}{2l+1}} D^{(l)}_{m'-1/2,m-1/2} D^{(1/2)}_{1/2,1/2} \pm \sqrt{\frac{l \mp m + \frac{1}{2}}{2l+1}} D^{(l)}_{m'-1/2,m+1/2} D^{(1/2)}_{1/2,-1/2} \right] \varphi_{nlm'-1/2}(x) \\ \left[\sqrt{\frac{l \pm m + \frac{1}{2}}{2l+1}} D^{(l)}_{m''+1/2,m-1/2} D^{(1/2)}_{-1/2,1/2} \pm \sqrt{\frac{l \mp m + \frac{1}{2}}{2l+1}} D^{(l)}_{m''+1/2,m+1/2} D^{(1/2)}_{-1/2,-1/2} \right] \varphi_{nlm''+1/2}(x) \end{bmatrix}. \tag{12.29}$$

In (12.29) we have changed for convenience summation indices from $m'' + \frac{1}{2}$ to $m' - \frac{1}{2}$ and from $m' - \frac{1}{2}$ to $m'' + \frac{1}{2}$. Comparing coefficients in (12.29) we obtain

$$\sqrt{\frac{l \pm m' + \frac{1}{2}}{2l+1}} D^{(j)}_{m'm}(R) = \sqrt{\frac{l \pm m + \frac{1}{2}}{2l+1}} D^{(l)}_{m'-1/2,m-1/2}(R)\, D^{(1/2)}_{1/2,1/2}(R) \pm \sqrt{\frac{l \mp m + \frac{1}{2}}{2l+1}} D^{(l)}_{m'-1/2,m+1/2}(R)\, D^{(1/2)}_{1/2,-1/2}(R)$$

$$\pm \sqrt{\frac{l \mp m + \frac{1}{2}}{2l+1}} D^{(j)}_{m'm}(R) = \sqrt{\frac{l \pm m + \frac{1}{2}}{2l+1}} D^{(l)}_{m'+1/2,m-1/2}(R)\, D^{(1/2)}_{-1/2,1/2}(R) \pm \sqrt{\frac{l \mp m + \frac{1}{2}}{2l+1}} D^{(l)}_{m'+1/2,m+1/2}(R)\, D^{(1/2)}_{-1/2,-1/2}(R).$$

The matrix elements of $D^{(j)}(R)$ are given by either of these equations. However, we can use both of them to obtain the symmetric form

$$D^{(j)}_{m'm}(R) = \sum_{m_s m_s'} a^{jm'}_{m_l' m_s'} D^{(l)}_{m_l' m_l}(R)\, D^{(1/2)}_{m_s' m_s}(R)\, a^{jm}_{m_l m_s} \tag{12.30}$$

where the coefficients $a^{jm}_{m_l m_s}$ are given by the Table 12.1 and where $m_l = m - m_s$ $m_l' = m' - m'_s$. These coefficients are, just the Clebsch-Gordan coefficients given in Table 2.1 for the addition $\mathbf{l} + \mathbf{s} = \mathbf{j}$.

It does not follow from (12.30) that the values of $D^{(j)}_{m'm}(R)$ depend only on j and are the same for $j = l + \frac{1}{2}$ and $j = l' - \frac{1}{2} = (l + 1) - \frac{1}{2}$. This fact follows from the uniqueness of the representations of the rotation group and will not be shown here.

Equation (12.30) gives the transformation matrices for tensors of half-integral degree, i.e., tensors which have an even number of components. Here we shall accept without proof the fact that this transformation is irreducible and shall extend the definition (11.19) of irreducible tensors $T_\kappa^{(k)}(\mathbf{r})$, for integral values of k, by requiring that, for

TABLE 12.1

$a_{m_l m_s}^{jm}$	$m_s = \frac{1}{2}$	$m_s = -\frac{1}{2}$
$j = l + \frac{1}{2}$	$\sqrt{\frac{l+m+\frac{1}{2}}{2l+1}}$	$\sqrt{\frac{l-m+\frac{1}{2}}{2l+1}}$
$j = l - \frac{1}{2}$	$\sqrt{\frac{l-m+\frac{1}{2}}{2l+1}}$	$-\sqrt{\frac{l+m+\frac{1}{2}}{2l+1}}$

half-integral values of k, they transform under rotations according to (12.26). The D matrix in this case is given by (12.30). As was pointed out after the derivation (12.26), $T_\kappa^{(k)}$ may be a physical spinor, or tensor field. In this case (11.19) is the transformation between unit spinors or tensors in the two frames. Since, however, the transformation matrix is uniquely determined by k, we shall often avoid explicit reference to the physical tensorial character of the field $T_\kappa^{(k)}$. Each eigenfunction ψ_{nljm} describing the motion of a spin $\frac{1}{2}$ particle in a central field with spin-orbit interaction is thus an irreducible physical tensor field of degree $\frac{1}{2}$ and at the same time a component of an irreducible tensor of degree j.

A word should be added about the uniqueness of the definition (12.30) of the D-matrices. The matrix U was defined by (11.14) as that matrix which satisfies

$$U\left(\sum_j a_{ij}\sigma_j\right) U^{-1} = \sigma_i \,.$$

It is immediately seen that, if U is a solution of this equation, $-U$ is also a solution. It is also possible to see that otherwise U is uniquely determined by the rotation matrix a_{ik}. Thus, $D^{(1/2)}(R) = U^\dagger(R)$ is determined only up to a sign, and therefore also $D^{(j)}(R)$ is determined only up to a sign. This indeterminacy cannot be removed by any convention, since it is possible to show that by changing continuously the matrix $a_{ij}^{(0)}$ to $a_{ij}^{(1)}$ and back to $a_{ij}^{(0)}$ it is possible to choose such a continuous sequence of matrices a_{ij}, that the corresponding U will change from $U^{(0)}$ to $-U^{(0)}$. If one tries to adopt any convention and discard half the matrices $D^{(1/2)}(R)$ (e.g., by taking only matrices whose element $D^{(1/2)}_{1/2\;1/2}$ has a given sign), one loses the group property of these matrices, i.e., it will no longer be true that $D^{(1/2)}(R_3) = D^{(1/2)}(R_1)\, D^{(1/2)}(R_2)$ if $R_3 = R_2 R_1$. The doublevaluedness of the $D^{(j)}(R)$ matrices, which constitute representations of the rotation group, is thus an inherent property of the group and cannot be removed by one convention or another. In the following we shall therefore have to take care in dealing with D-matrices of half-integral values of j.

Having introduced irreducible tensors with half-integral values of k, we can proceed to discuss the infinitesimal rotations. In particular we shall study the effects of infinitesimal rotations on irreducible tensors.

The change in an irreducible tensor under a rotation δR is given, according to (11.19), by

$$\begin{aligned} -\delta T_{\kappa}^{(k)}(\mathbf{r}) &= T_{\kappa}^{(k)(1)}(\mathbf{r}') - T_{\kappa}^{(k)(0)}(\mathbf{r}) \\ &= \Big[\sum_{\kappa'} T_{\kappa'}^{(k)}(\mathbf{r})\, D_{\kappa'\kappa}^{(k)}(\delta R)\Big] - T_{\kappa}^{(k)}(\mathbf{r}). \end{aligned} \tag{12.31}$$

On the other hand, by (12.18), $\delta T_{\kappa}^{(k)}$ is given by

$$\delta T_{\kappa}^{(k)}(\mathbf{r}) = -i\delta\theta(\mathbf{n}\cdot\mathbf{J})T_{\kappa}^{(k)}(\mathbf{r}). \tag{12.32}$$

The components of the irreducible tensors we are considering here can be either physical scalars or physical spinors. We shall understand $\mathbf{J}$ to be identical with $\mathbf{L}$ in the first case and with $\mathbf{L}+\mathbf{S}$ in the second. As a matter of fact we shall see later on that $\mathbf{J}=\mathbf{L}+\mathbf{S}$ will represent the infinitesimal variations of an irreducible tensor whose components are physical tensors of higher degree, provided $\mathbf{S}$ is defined properly for each degree of these physical tensors.

Combining (12.31) and (12.32) we can derive an expression for the infinitesimal changes in an irreducible tensor in terms of its components and the derivatives of the components at the point under consideration, namely

$$\sum_{\kappa'} T_{\kappa'}^{(k)}(\mathbf{r})\,[D_{\kappa'\kappa}^{(k)}(\delta R) - \delta_{\kappa'\kappa}] = i\delta\theta(\mathbf{n}\cdot\mathbf{J})T_{\kappa}^{(k)}(\mathbf{r}). \tag{12.33}$$

Here the rotation δR is a rotation by an angle $\delta\theta$ around the direction $\mathbf{n}$.

In Part I we studied the operation of $\boldsymbol{l}$ on special irreducible tensors of an integral degree—namely the spherical harmonics. Similarly we treated the operation of $\mathbf{j}$ on the functions ψ_{jm}. We see from (12.33) that the operation with $\boldsymbol{l}$ or $\mathbf{j}$ is essentially equivalent to an infinitesimal rotation. Since under rotations all irreducible tensors transform in the same manner we can generalize immediately (2.23) and obtain for any irreducible tensor the following relations

$$\begin{aligned} J_z\, T_{\kappa}^{(k)}(\mathbf{r}) &= \kappa\, T_{\kappa}^{(k)}(\mathbf{r}) \\ (J_x \pm iJ_y)T_{\kappa}^{(k)}(\mathbf{r}) &= \sqrt{k(k+1)-\kappa(\kappa\pm 1)}\; T_{\kappa\pm 1}^{(k)}(\mathbf{r}). \end{aligned} \tag{12.34}$$

These are essentially the relations (11.19) written explicitly for special infinitesimal rotations ($\delta\theta$ does not appear explicitly because both sides of the equations are proportional to it). Equation (12.34) is completely equivalent to (11.19) since, as was mentioned, any finite rotation can be obtained by multiplying successively infinitesimal rotations. There are only three independent infinitesimal rotations, and they are given by (12.34). We shall therefore use these relations for the definition of irreducible tensors.

The derivation of (12.34) was made by the use of (12.32). In this relation the irreducible tensor $T_\kappa^{(k)}(\mathbf{r})$ is written in the form of a wave function. If, however, $T_\kappa^{(k)}$ is to be considered as an operator, it should be handled differently.

The rotated operator T' is defined as the operator which, acting on the rotated function ψ', gives the rotated function φ', where $\varphi = T\psi$. The change in $T_\kappa^{(k)}$ is given by

$$\delta T_\kappa^{(k)} = UT_\kappa^{(k)}U^{-1} - T_\kappa^{(k)} = -i\delta\theta[(\mathbf{n}\cdot\mathbf{J}), T_\kappa^{(k)}].$$

Equations (12.34) for irreducible tensor operators will then be written as follows:

$$\begin{aligned}[][J_z, T_\kappa^{(k)}] &= \kappa\, T_\kappa^{(k)} \\ [J_x \pm iJ_y, T_\kappa^{(k)}] &= \sqrt{k(k+1) - \kappa(\kappa \pm 1)}\; T_{\kappa\pm1}^{(k)}.\end{aligned} \qquad (12.35)$$

Equation (12.31), or more generally Equation (11.19), are unaffected by $T_\kappa^{(k)}$ being an operator. Since the transformation properties of irreducible tensors and irreducible tensor operators are the same, we shall, wherever possible, not use a different notation for them.

Equation (12.34) shows that a component $T_\kappa^{(k)}$ of an irreducible tensor of degree k which can be considered as a wave function is always an eigenfunction of the z-component of the total angular momentum belonging to the eigenvalue κ. Furthermore, it can be easily seen from (12.34) that each component is also simultaneously an eigenfunction of $\mathbf{J}^2$ belonging to the eigenvalue $k(k+1)$

$$\mathbf{J}^2 T_\kappa^{(k)} = k(k+1)\, T_\kappa^{(k)}. \qquad (12.36)$$

We saw in (11.20) that, if $T_\kappa^{(k)}$ is a function of the coordinates of just one particle only, then it is proportional to $Y_{k\kappa}(\theta\varphi)$, which is a simultaneous eigenfunction of $\mathbf{l}^2$ and l_z belonging to the eigenvalues $k(k+1)$ and κ respectively. Equations (12.34) and (12.36) are more powerful in the sense that they also describe many-particle wave functions which are simultaneous eigenfunctions of the square of the *total* angular momentum and its z-component. Thus, they can be used for the construction of such wave functions. We have to make sure that the $2k + 1$ functions $T_\kappa^{(k)}$ we construct transform like the components of an irreducible tensor of degree k. This will guarantee that they will be the desired eigenfunctions of $\mathbf{J}^2$ and J_z.

13. Vector Addition Coefficients

In theoretical spectroscopy we have often to consider wave functions of the composite system of two particles. If one particle is in a state of angular momentum j_1 and the other in j_2, we have to construct an eigenstate of $\mathbf{J}^2 = (\mathbf{j}_1 + \mathbf{j}_2)^2$ and $J_z = j_{1z} + j_{2z}$ from the wave functions of the individual particles. From the previous discussion it is seen that this problem is equivalent to that of constructing an irreducible tensor of degree J out of two irreducible tensors of degrees j_1 and j_2. In the following, tensors will be considered as wave functions, but all the results will be derived only from their transformation properties under rotations. Therefore the results obtained will hold for general tensors.

As we saw in Part I there are $2j + 1$ degenerate states for a single particle moving in a central field with a total angular momentum j. These states can be characterized by the projection m of the angular momentum on an arbitrary z axis and are given for spinless particles by

$$\psi_{nlm}(r\,\theta\,\varphi) = \frac{1}{r} R_{nl}(r)\, Y_{lm}(\theta\,\varphi) \qquad m = -l, -l+1, ..., l-1, l$$

where n is the radial quantum number. For particles with spin $\frac{1}{2}$ the spin $\mathbf{s}$ is coupled to $\boldsymbol{l}$ to form a total $\mathbf{j}$ in the manner discussed above and the single particle wave functions can be characterized by the values of $n\,l\,j$ and the projection $m = j_z$.

The wave functions of a system of two noninteracting particles moving in a central field can be obtained by taking products of such wave functions of the single particles

$$\psi_{n_1 j_1 m_1 n_2 j_2 m_2}(\mathbf{r}_1 \mathbf{r}_2) = \psi_{n_1 j_1 m_1}(\mathbf{r}_1)\, \psi_{n_2 j_2 m_2}(\mathbf{r}_2). \tag{13.1}$$

This wave function is an eigenfunction of $J_z = j_{1z} + j_{2z}$, which belongs to the eigenvalue $M = m_1 + m_2$ since

$$\begin{aligned}(j_{1z} + j_{2z})\,\psi_{n_1 j_1 m_1 n_2 j_2 m_2} &= j_{1z}\psi_{n_1 j_1 m_1}\psi_{n_2 j_2 m_2} + j_{2z}\psi_{n_1 j_1 m_1}\psi_{n_2 j_2 m_2}\\ &= m_1\psi_{n_1 j_1 m_1}\psi_{n_2 j_2 m_2} + m_2\psi_{n_1 j_1 m_1}\psi_{n_2 j_2 m_2}\\ &= (m_1 + m_2)\psi_{n_1 j_1 m_1 n_2 j_2 m_2}.\end{aligned}$$

The product (13.1) is generally not an eigenfunction of $\mathbf{J}^2 = (\mathbf{j}_1 + \mathbf{j}_2)^2$. This can be proved readily by noting that (13.1) is also an eigenfunction of j_{1z} (with an eigenvalue m_1). Since j_{1z} does not commute with $\mathbf{J}^2$, (13.1) cannot be always a simultaneous eigenfunction of $\mathbf{J}^2$ and j_{1z}. This fact can be shown by direct calculation.

Thus,

$$\begin{aligned}\mathbf{J}^2 &= \mathbf{j}_1^2 + \mathbf{j}_2^2 + 2(\mathbf{j}_1 \cdot \mathbf{j}_2)\\ &= \mathbf{j}_1^2 + \mathbf{j}_2^2 + [(j_{1x} + ij_{1y})(j_{2x} - ij_{2y}) + (j_{1x} - ij_{1y})(j_{2x} + ij_{2y}) + 2j_{1z}j_{2z}].\end{aligned}$$

Recalling that

$$\mathbf{j}^2\psi_{jm} = j(j+1)\psi_{jm}, \qquad j_z\psi_{jm} = m\psi_{jm}$$

$$(j_x \pm ij_y)\psi_{jm} = \sqrt{j(j+1) - m(m \pm 1)}\, \psi_{jm\pm 1}$$

we see that

$$\begin{aligned}\mathbf{J}^2\psi_{j_1m_1j_2m_2} &= [j_1(j_1+1) + j_2(j_2+1) + 2m_1m_2]\, \psi_{j_1m_1j_2m_2}\\ &+ \sqrt{j_1(j_1+1) - m_1(m_1+1)}\sqrt{j_2(j_2+1) - m_2(m_2-1)}\, \psi_{j_1m_1+1\ j_2m_2-1} \qquad (13.2)\\ &+ \sqrt{j_1(j_1+1) - m_1(m_1-1)}\sqrt{j_2(j_2+1) - m_2(m_2+1)}\, \psi_{j_1m_1-1\ j_2m_2+1}.\end{aligned}$$

Hence, except for the special case in which $m_1 = j_1$ and $m_2 = j_2$ (or $m_1 = -j_1$, $m_2 = -j_2$), $\mathbf{J}^2$ operating on $\psi_{j_1m_1j_2m_2}$ will not be proprotional to the same function, and these functions are therefore not eigenfunctions of the total angular momentum. We can, however, obtain from the functions (13.1) eigenfunctions of $\mathbf{J}^2$ and J_z in the following way.

We first note from (13.2) that the function (13.1) with $m_1 = j_1$ $m_2 = j_2$ is an eigenfunction of $\mathbf{J}^2$ which belongs to the eigenvalue $(j_1 + j_2)(j_1 + j_2 + 1)$. It is obviously also an eigenfunction of J_z belonging to the eigenvalue $M = j_1 + j_2$. Thus

$$\psi(J = j_1 + j_2, M = j_1 + j_2) = \psi_1(j_1, m_1 = j_1)\, \psi_2(j_2, m_2 = j_2) \qquad (13.3)$$

where we introduce the notation $\psi_{j_1m_1}(\mathbf{r}_1) = \psi_1(j_1m_1)$, etc. We now operate on $\psi(JM)$ of (13.3) with $J_x - iJ_y = (j_{1x} - ij_{1y}) + (j_{2x} - ij_{2y})$. According to (12.34) this will give us $\psi(J, M-1)$. Thus

$$\begin{aligned}&\psi(J = j_1 + j_2, M = j_1 + j_2 - 1)\\ &= \frac{1}{\sqrt{J(J+1) - J(J-1)}}(J_x - iJ_y)\, \psi(J = j_1 + j_2, M = j_1 + j_2)\\ &= \frac{1}{\sqrt{2J}}[(j_{1x} - ij_{1y})\, \psi_1(j_1, m_1 = j_1)\, \psi_2(j_2m_2 = j_2) \qquad (13.4)\\ &\qquad + \psi_1(j_1, m_1 = j_1)\, (j_{2x} - ij_{2y})\, \psi_2(j_2, m = j_2)]\\ &= \frac{1}{\sqrt{2J}}[\sqrt{2j_1}\, \psi_1(j_1, m_1 = j_1 - 1)\, \psi_2(j_2, m_2 = j_2)\\ &\qquad + \sqrt{2j_2}\, \psi_1(j_1, m_1 = j_1)\, \psi_2(j_2, m_2 = j_2 - 1)]\\ &= \sum_{m_1m_2} a_{m_1m_2}\, \psi_1(j_1m_1)\psi_2(j_2m_2)\end{aligned}$$

where

$$a_{m_1=j_1-1, m_2=j_2} = \sqrt{\frac{j_1}{j_1+j_2}} \qquad a_{m_1=j_1, m_2=j_2-1} = \sqrt{\frac{j_2}{j_1+j_2}} \tag{13.5}$$

and all other coefficients are equal to zero. By repeated application of the operator $J_x - iJ_y$ we can obtain all the eigenfunctions of $\mathbf{J}^2$ and J_z which belong to the eigenvalues $J = j_1 + j_2$ and $M = J, J-1, ..., -J$.

To obtain an eigenfunction which belongs to the eigenvalues $J = j_1 + j_2 - 1$ and $M = J = j_1 + j_2 - 1$, we note that such an eigenfunction is necessarily a combination of products $\psi_1(j_1m_1)\,\psi_2(j_2m_2)$ with $m_1 + m_2 = j_1 + j_2 - 1$. Thus we can write

$$\psi(J = j_1 + j_2 - 1, M = j_1 + j_2 - 1) = \sum a'_{m_1m_2}\, \psi_1(j_1m_1)\,\psi_2(j_2m_2) \tag{13.6}$$

where $a'_{m_1m_2}$ vanishes unless $m_1 = j_1 - 1$, $m_2 = j_2$ or $m_1 = j_1$, $m_2 = j_2 - 1$. Since there are two linearly independent states with $M = j_1 + j_2 - 1$ (for $j_1 + j_2 \geqslant 1$) and since one such state can be taken as $\psi(J = j_1 + j_2, M = j_1 + j_2 - 1)$ we can obtain just one state $\psi(J = j_1 + j_2 - 1, M = j_1 + j_2 - 1)$. This state will be determined by the condition that it be orthogonal to $\psi(J = j_1 + j_2, M = j_1 + j_2 - 1)$. This requirement yields, for the normalized wave function,

$$a'_{m_1=j_1-1, m_2=j_2} = \sqrt{\frac{j_2}{j_1+j_2}} \qquad a'_{m_1=j_1, m_2=j_2-1} = -\sqrt{\frac{j_1}{j_1+j_2}}. \tag{13.7}$$

Operating now on (13.6) with $J_x - iJ_y$ repeatedly, we can obtain from it all the eigenstates of $\mathbf{J}^2$ and J_z which belong to $J = j_1 + j_2 - 1$ and $M = J, ..., -J$. We can then construct $\psi(J = j_1 + j_2 - 2, M = j_1 + j_2 - 2)$ by noting that it must be given by

$$\psi(J = j_1 + j_2 - 2, M = j_1 + j_2 - 2) = \sum a''_{m_1m_2}\, \psi_1(j_1m_1)\,\psi_2(j_2m_2)$$

with $m_1 + m_2 = j_1 + j_2 - 2$. Using again the orthogonality and normalization requirements we derive the coefficients $a''_{m_1m_2}$ and can proceed in this way until all wave functions have been constructed.

The number of functions of the type (13.1) which have a given value of $M = m_1 + m_2$ is, as can be easily seen, $j_1 + j_2 + 1 - M$ as long as $|M| \geqslant j_1 - j_2$ (assuming $j_1 \geqslant j_2$). For $-(j_1 - j_2) < M < j_1 - j_2$ the number of these functions remains unchanged, namely $2j_2 + 1$. Thus we see that for $|M| < j_1 - j_2$ we shall not be able to find a linear combination of $\psi_1(j_1m_1)\,\psi_2(j_2m_2)$ which will be orthogonal to all the functions $\psi(JM)$ with $J = j_1 - j_2$, $J = j_1 - j_2 + 1$, ... $J = j_1 + j_2$, since there are $2j_2 + 1$ such functions and each one of them is an independent linear combination of the $2j_2 + 1$ functions $\psi_1(j_1m_1)\,\psi_1(j_2m_2)$

with $m_1 + m_2 = M$. This shows that the minimum total angular momentum in our case is $J = j_1 - j_2$, as well known. Thus, in the expression

$$\psi_{JM}(\mathbf{r}_1, \mathbf{r}_2) = \sum a^{JM}_{m_1 m_2} \psi_{j_1 m_1}(\mathbf{r}_1) \psi_{j_2 m_2}(\mathbf{r}_2) \tag{13.8}$$

if the conditions $|j_1 - j_2| \leq J \leq j_1 + j_2$ *and* $M = m_1 + m_2$ *are not satisfied then* $a^{JM}_{m_1 m_2} = 0$.

This method of constructing the eigenfunctions of $\mathbf{J}^2$ and J_z leads to soluble recursion relations between the coefficients $a^{JM}_{m_1 m_2}$. However, we shall find it more convenient to construct the desired eigenfunctions by writing explicitly their transformation properties under finite rotations. This procedure will enable us to obtain directly many of the properties of the coefficients $a^{JM}_{m_1 m_2}$. If we want to construct a wave function which belongs to the eigenvalues $J(J+1)$ and M of $\mathbf{J}^2$ and J_z respectively, the coefficients $a^{JM}_{m_1 m_2}$ in (13.8) should be determined, according to (12.26), by

$$\psi'_{JM}(\mathbf{r}'_1, \mathbf{r}'_2) = \sum_{M'} \psi_{JM'}(\mathbf{r}_1 \mathbf{r}_2) D^{(J)}_{M'M}(R) \tag{13.9}$$

where $\mathbf{r}'$ is the coordinate in the rotated frame of the point P whose coordinate in the original frame was $\mathbf{r}$, and R is an arbitrary rotation. We know the transformation properties of $\psi_{n_1 j_1 m_1}(\mathbf{r}_1)$ and $\psi_{n_2 j_2 m_2}(\mathbf{r}_2)$, since they are eigenstates of the respective single particle angular momentum operators

$$\begin{aligned} \psi'_{n_1 j_1 m_1}(\mathbf{r}'_1) &= \sum_{m'_1} \psi_{n_1 j_1 m'_1}(\mathbf{r}_1) D^{(j_1)}_{m'_1 m_1}(R) \\ \psi'_{n_2 j_2 m_2}(\mathbf{r}'_2) &= \sum_{m'_2} \psi_{n_2 j_2 m'_2}(\mathbf{r}_2) D^{(j_2)}_{m'_2 m_2}(R). \end{aligned} \tag{13.10}$$

Substituting (13.10) into (13.9) we find that

$$\begin{aligned} \psi'_{JM}(\mathbf{r}'_1 \mathbf{r}'_2) &= \sum_{m_1 m_2} a^{JM}_{m_1 m_2} \psi'_{n_1 j_1 m_1}(\mathbf{r}'_1) \psi'_{n_2 j_2 m_2}(\mathbf{r}'_2) \\ &= \sum a^{JM}_{m_1 m_2} \psi_{n_1 j_1 m'_1}(\mathbf{r}_1) \psi_{n_2 j_2 m'_2}(\mathbf{r}_2) D^{(j_1)}_{m'_1 m_1}(R) D^{(j_2)}_{m'_2 m_2}(R) \\ &= \sum_{m'} \psi_{JM'}(\mathbf{r}_1 \mathbf{r}_2) D^{(J)}_{M'M}(R) = \sum a^{JM'}_{m''_1 m''_2} \psi_{n_1 j_1 m''_1}(\mathbf{r}_1) \psi_{n_2 j_2 m''_2}(\mathbf{r}_2) D^{(J)}_{M'M}(R). \end{aligned}$$

Using the orthogonality of $\psi_{n_1j_1m_1}(\mathbf{r}_1)$ and $\psi_{n_2j_2m_2}(\mathbf{r}_2)$ for different values of m we find that the coefficients $a^{JM}_{m_1m_2}$ satisfy the equations

$$\sum_{m_1'm_2'} a^{JM'}_{m_1'm_2'} \overset{(j_1)}{D}_{m_1m_1'}(R) \overset{(j_2)}{D}_{m_2m_2'}(R) = \sum_{M} a^{JM}_{m_1m_2} D^{(J)}_{MM'}(R). \qquad (13.11)$$

Although (13.11) should hold for any rotation R it is clear from the previous discussion that in order to assure its validity for any rotation it is enough that it holds for three independent infinitesimal rotations. We shall however keep the general rotation R since it will turn out to be more convenient for the subsequent treatment.

In order to guarantee that the functions ψ_{JM} in (13.8) will be normalized to unity and orthogonal to each other the coefficients $a^{JM}_{m_1m_2}$ have to satisfy

$$\sum_{m_1m_2} (a^{JM}_{m_1m_2})^* \, a^{J'M'}_{m_1m_2} = \delta_{JJ'} \, \delta_{MM'} \qquad (13.12)$$

$$\sum_{J,M} (a^{JM}_{m_1m_2})^* \, a^{JM}_{m_1'm_2'} = \delta_{m_1m_2} \, \delta_{m_1'm_2'}. \qquad (13.13)$$

Using these orthogonality relations we can transform (13.11) into

$$\overset{(j_1)}{D}_{m_1m_1'}(R) \overset{(j_2)}{D}_{m_2m_2'}(R) = \sum_{J,M',M} a^{JM}_{m_1m_2} \, D^{(J)}_{MM'}(R) \, (a^{JM'}_{m_1'm_2'})^* .$$

This equation can be further simplified by using the orthogonality properties of the D-matrices (3.12) which can be shown to be true for half-integer values of j as well.

We obtain finally

$$a^{JM}_{m_1m_2} (a^{JM'}_{m_1'm_2'})^* = \frac{2J+1}{8\pi^2} \int \overset{(j_1)}{D}_{m_1m_1'}(R) \overset{(j_2)}{D}_{m_2m_2'}(R) \overset{(J)*}{D}_{MM'}(R) \, dR. \qquad (13.14)$$

Inasmuch as the D-matrices are well-defined functions of R, (13.14) determines the a's nearly uniquely as will be presently shown. Taking $M = M'$, $m_1 = m_1'$ and $m_2 = m_2'$ (13.14) yields the magintude of $a^{JM}_{m_1m_2}$. By choosing arbitrarily the phase of one coefficient $a^{J\overline{M}}_{\overline{m}_1\overline{m}_2}$ we can use (13.14) for $m_1' = \overline{m}_1$, $m_2' = \overline{m}_2$, $M' = \overline{M}$ to determine uniquely the magnitude of $a^{JM}_{m_1m_2}$ as well as its phase.

Equation (13.14) can be used to derive many important properties of the coefficients $a^{JM}_{m_1m_2}$. Since (13.14) involves an integration over all rotations R, we can obviously change the arguments of all the D-matrices from R to RR_0 where R_0 is an arbitrary constant rotation. Using the relation

$$D^{(k)}_{\kappa\kappa'}(RR_0) = \sum_{\kappa''} D^{(k)}_{\kappa\kappa''}(R_0) \, D^{(k)}_{\kappa''\kappa'}(R)$$

we then obtain

$$a^{JM}_{m_1m_2}\,(a^{JM'}_{m_1'm_2'})^* = \frac{2J+1}{8\pi^2} \sum_{m_1''m_2''M''} D^{(j_1)}_{m_1m_1''}(R_0)\, D^{(j_2)}_{m_2m_2''}(R_0) D^{(J)*}_{MM''}(R_0)$$

$$\times \int D^{(j_1)}_{m_1''m_1'}(R)\, D^{(j_2)}_{m_2''m_2'}(R)\, D^{(J)*}_{M''M'}(R)\, dR$$

$$= \sum_{m_1''m_2''M''} D^{(j_1)}_{m_1m_1''}(R_0)\, D^{(j_2)}_{m_2m_2''}(R_0) D^{(J)*}_{MM''}(R_0) \cdot a^{JM''}_{m_1''m_2''}\,(a^{JM'}_{m_1'm_2'})^*.$$

If we choose $m_1'm_2'$ and M' so that $a^{JM'}_{m_1'm_2'} \neq 0$, we obtain the relation

$$a^{JM}_{m_1m_2} = \sum_{m_1''m_2''M''} D^{(j_1)}_{m_1m_1''}(R_0)\, D^{(j_2)}_{m_2m_2''}(R_0)\, D^{(J)*}_{MM''}(R_0)\, a^{JM''}_{m_1m_2} \tag{13.15}$$

which is valid for any arbitrary rotation R_0. It should be stressed that the coefficients $a^{JM}_{m_1m_2}$ are independent of R_0. Therefore the right-hand side of (13.15) is a combination of the D-matrices which is independent of the rotation R_0. If we specify R_0 to be a rotation around the z-axis by an angle φ then, according to (3.9),

$$D^{(j)}_{mm'}(z_\varphi) = \delta_{mm'} e^{-im\varphi}$$

and hence

$$a^{JM}_{m_1m_2} = e^{-i(m_1+m_2-M)\varphi}\, a^{JM}_{m_1m_2}.$$

We therefore conclude that

$$a^{JM}_{m_1m_2} = 0 \qquad \text{if} \qquad m_1 + m_2 \neq M. \tag{13.16}$$

Equation (13.16) was already derived in a more elementary way in (13.8). It expresses the obvious physical requirement that the sum of the projections of the two angular momenta $\mathbf{j}_1$ and $\mathbf{j}_2$ on the z-axis be equal to M—the projection on the same axis of the total angular momentum $\mathbf{J}$. From (13.16) it also follows that either j_1, j_2, and J are all integers or *two* of them are half-integers and the third an integer. Otherwise $m_1 + m_2 - M$ will be half-integer and therefore different from zero, and consequently the corresponding coefficient must vanish.

Another simple property of these coefficients is obtained if, in (13.15), we substitute for R_0 a rotation by an angle π around the x-axis. For such rotations (see Appendix)

$$D^{(j)}_{mm'}(x_\pi) = (-1)^j\, \delta(m + m').$$

Hence

$$a^{JM}_{m_1m_2} = (-1)^{j_1+j_2-J}\, a^{J,-M}_{-m_1\,-m_2}. \tag{13.17}$$

From (13.17) we see that, if $j_1 = l_1$, $j_2 = l_2$, and $J = L$ are all integers, and if $m_1 = m_2 = M = 0$, then

$$a^{L0}_{00} = 0 \qquad \text{if} \qquad l_1 + l_2 + L \text{ is odd.} \tag{13.18}$$

If we use (3.16) in conjunction with (13.14) we can derive a relation which will be used frequently in the subsequent sections. Introducing into (13.14) the relation [Cf. (3.16)]

$$D^{(l)}_{0m}(\pi\,\beta\,\gamma) = (-1)^m \sqrt{\frac{4\pi}{2l+1}}\, Y^*_{lm}\;(\beta\,\gamma)$$

and integrating over γ we obtain

$$a^{L0}_{00}\,(a^{LM}_{m_1m_2})^* = \sqrt{4\pi}\sqrt{\frac{2L+1}{(2l_1+1)\,(2l_2+1)}}$$

$$\times \int Y^*_{l_1m_1}(\theta\,\varphi)\; Y^*_{l_2m_2}(\theta\,\varphi)\; Y_{LM}(\theta\,\varphi) \sin\theta\; d\theta d\varphi. \tag{13.19}$$

We see from (13.19) and (13.18) that the integral over a product of three spherical harmonics vanishes unless the sum of their degrees is even. This also follows directly from their definition in terms of homogeneous polynomials.

We noted that (13.14) determines the coefficients $a^{JM}_{m_1m_2}$ up to an arbitrary phase. Although the absolute phase of a wave function has no physical meaning, the relative phases between wave functions are of the greatest physical importance. In fact all interference phenomena are determined by the relative phases of different waves.

Let us therefore review the phase conventions adopted so far. The spherical harmonics $Y_{lm}(\theta\,\varphi)$ were introduced as the normalized solutions of the eigenvalue problem

$$\begin{aligned} \mathbf{l}^2\, Y_{lm}(\theta\,\varphi) &= l(l+1)\; Y_{lm}(\theta\,\varphi) \\ l_z\, Y_{lm}(\theta\,\varphi) &= m\; Y_{lm}(\theta\,\varphi) \\ \int_0^{2\pi}\!\!\int_0^{\pi} |Y_{lm}(\theta\,\varphi)|^2 &\sin\theta\; d\theta d\varphi = 1. \end{aligned} \tag{13.20}$$

These equations determine the functions $Y_{lm}(\theta\,\varphi)$ up to a phase. To fix a definite phase we adopted the convention that

$$Y_{ll}(\theta\,\varphi) = \frac{(-1)^l}{\sqrt{2\pi}}\sqrt{\frac{(2l+1)!}{2}}\,\frac{1}{2^l l!}\sin^l\theta\, e^{il\varphi} \tag{13.21}$$

and that the Y_{lm} satisfy the relation

$$(l_x \pm il_y)Y_{lm}(\theta\,\varphi) = \sqrt{(l\mp m)\,(l+m-1)}\; Y_{l,m\pm1}(\theta\,\varphi) \tag{13.22}$$

with the *positive sign* of the square root.

The relations (13.20)-(13.22) determine the spherical harmonics uniquely, along with their phases. An important corollary of them, which we shall now prove, is the relation

$$Y^*_{lm}(\theta\,\varphi) = (-1)^m\, Y_{l,-m}(\theta\,\varphi). \tag{13.23}$$

To obtain this relation we note that $\boldsymbol{l}^* = -\boldsymbol{l}$ (the star denotes complex conjugation, *not* Hermitian conjugation). Hence, from $l_z Y_{lm} = mY_{lm}$ we find that $l_z Y^*_{lm} = -mY^*_{lm}$ so that $Y^*_{lm} = e^{i\delta}Y_{l,-m}$. Then, by taking the complex conjugate of (13.22), we find that

$$\frac{Y^*_{lm}}{Y_{l,-m}} = -\frac{Y^*_{l\,m+1}}{Y_{l,-(m+1)}}.$$

Hence $Y^*_{lm} = e^{i\delta'}(-1)^m Y_{l,-m}$ where δ' is another phase, independent of m. Finally, since

$$l_x - il_y = -e^{-i\varphi}\left[\frac{\partial}{\partial\theta} + \cot\theta\, i\,\frac{\partial}{\partial\varphi}\right]$$

it is clear from (13.21) and (13.22) that $Y_{l0}(\theta\,\varphi)$ is real, and hence the result (13.23).

As for the Wigner matrices $D^{(k)}$ with an integer k, their definition (3.7) leaves no ambiguity in their phases once the Y_{lm} are unambiguously defined. From (13.23) and (3.7) we conclude furthermore that

$$D^{(k)}_{\kappa,\kappa'}(R) = (-1)^{\kappa-\kappa'}\, D^{(k)}_{-\kappa,-\kappa'}(R). \tag{13.24}$$

Equation (13.24) gives us the effect of complex conjugation on the D-matrices with their phases chosen according to the above convention. We have proved (13.24) only for integral values of k. It can be shown, by using the construction (12.30) of $D^{(j)}$ that the same relation holds also for half-integral values of j. We can now use this equation to define the phases of the coefficients $a^{JM}_{m_1m_2}$. Going back to (13.14) and using (13.24) we see that

$$\left[\int D^{(j_1)}_{m_1m_1'}(R)\, D^{(j_2)}_{m_2m_2'}(R)\, D^{(J)*}_{MM'}(R)\, dR\right]^*$$

$$= (-1)^{(m_1+m_2-M)-(m_1'+m_2'-M')} \int D^{(j_1)}_{-m_1,-m_1'}(R)\, D^{(j_2)}_{-m_2,-m_2'}(R)\, D^{(J)*}_{-M,-M'}(R)\, dR.$$

We use the fact that the variable of integration can be changed from R to R_0R or even to R_0RR_1 where R_0 and R_1 are any fixed rotations. We use the relation $D^{(j)}_{mm'}(x_\pi) = (-1)^j\, \delta(m+m')$ where x_π is a rotation through 180° around the x-axis, and choose $R_0 = R_1 = x_\pi$ to obtain

$$\left[\int D^{(j_1)}_{m_1m_1'}(R)\, D^{(j_2)}_{m_2m_2'}(R)\, D^{(J)*}_{MM'}(R)\, dR\right]^* = \int D^{(j_1)}_{m_1m_1'}(R)\, D^{(j_2)}_{m_2m_2'}(R)\, D^{(J)*}_{MM'}(R)\, dR. \tag{13.25}$$

In deriving this relation we used the fact that these integrals vanish if $m_1 + m_2 - M \neq 0$ or if $m_1' + m_2' - M' \neq 0$, and that $2(j_1 + j_2 + J)$ is always an even number. We conclude that our phase conventions on the spherical harmonics made all the integrals (13.14) *real.*

Thus, if we choose the phase of one coefficient $a^{JM}_{m_1m_2}$ so that it becomes real, and this we are still free to do, all other coefficients with the same J, j_1, and j_2 will also be real. This becomes self-evident if we consider the straightforward construction of the coefficients $a^{JM}_{m_1m_2}$ as in (13.5)-(13.7). We shall therefore adopt as our first convention on the phases of these coefficients the requirement

$$a^{JM}_{m_1m_2} = (a^{JM}_{m_1m_2})^*. \tag{13.26}$$

We are thus left only with the question of determining the actual sign of the coefficients. Since, as we saw before, for $J = j_1 + j_2$, $M = j_1 + j_2$ there is only one nonvanishing coefficient namely that with $m_1 = j_1$ and $m_2 = j_2$ we define

$$a^{J=j_1+j_2, M=j_1+j_2}_{m_1=j_1, m_2=j_2} = +1. \tag{13.27}$$

Equation (13.14) then fixes the sign of all other coefficients with the same J, j_1, and j_2. For other values of J we fix the sign of the coefficients by requiring that

$$(j_1j_2J\,M \mid j_{1z} \mid j_1j_2J-1, M) = \int \psi^*_{JM}(\mathbf{r}_1\mathbf{r}_2) j_{1z} \psi_{J-1,M}(\mathbf{r}_1\mathbf{r}_2) > 0. \tag{13.28}$$

It can be shown that, due to (12.34), (13.28) will hold for all values of M if it holds for one specific value of M. It can further be shown that for permissible values of J and $(J-1)$ (13.28) is always different from zero. Substituting for ψ the expression (13.8) in the last equation we obtain

$$\sum_{m_1 m_2} m_1 a_{m_1 m_2}^{JM} a_{m_1 m_2}^{J-1,M} > 0. \tag{13.29}$$

It should be pointed out that the last convention introduces an asymmetry between $\mathbf{j}_1$ and $\mathbf{j}_2$ which is not inherent in the problem. Thus it is clear that

$$(j_1 j_2 J\, M | j_{1z} + j_{2z} | j_1 j_2 J - 1\, M) = (j_1 j_2 J\, M | J_z | j_1 j_2 J - 1\, M) = 0$$

and hence if

$$(j_1 j_2 J\, M \mid j_{1z} \mid j_1 j_2 J - 1\, M) \geqslant 0$$

it follows that

$$(j_1 j_2 J\, M | j_{2z} | j_1 j_2 J - 1\, M) \leq 0.$$

Of course, this mathematical asymmetry does not represent any physical asymmetry between the two angular momenta. Whenever the phase of a function is of any importance it will be the *relative* phase of two states which matters. If this relative phase can be measured it is unaffected by any phase convention. The purpose of having a definite phase convention is to facilitate actual computations of different phenomena and has, of course, nothing to do with physical properties of the systems under consideration.

The phase convention we have introduced is known as the *Condon and Shortley phase convention.* The coefficients $a_{m_1 m_2}^{JM}$ so defined are called *Clebsch-Gordan coefficients.* We shall use the following notation for Clebsch-Gordan coefficients

$$a_{m_1 m_2}^{JM} = (j_1 m_1 j_2 m_2 | j_1 j_2 J\, M).$$

The explicit dependence of the coefficients on j_1 and j_2 is noted, and $j_1 j_2$ is repeated on the right half of the coefficients to stress that these are coefficients for the transformation from the scheme in which $\mathbf{j}_1^2$, j_{1z}, $\mathbf{j}_2^2$, and j_{2z} are diagonal to the scheme in which $\mathbf{j}_1^2$, $\mathbf{j}_2^2$, $\mathbf{J}^2$, and J_z are diagonal. With the new notation we write

$$\psi_{JM}(\mathbf{r}_1 \mathbf{r}_2) = \sum_{m_1 m_2} \psi_{j_1 m_1}(\mathbf{r}_1)\, \psi_{j_2 m_2}(\mathbf{r}_2)\, (j_1 m_1 j_2 m_2 | j_1 j_2 J\, M) \tag{13.30}$$

or alternatively

$$(j_1 m_1 j_2 m_2 | j_1 j_2 J\, M) = \int \psi_{j_1 m_1}^*(\mathbf{r}_1)\, \psi_{j_2 m_2}^*(\mathbf{r}_2)\, \psi_{JM}(\mathbf{r}_1 \mathbf{r}_2)\, d\mathbf{r}_1\, d\mathbf{r}_2. \tag{13.31}$$

The function ψ_{JM} will be referred to as *the state in which* $\mathbf{j}_1$ *and* $\mathbf{j}_2$ *are coupled to* $\mathbf{J}$. The Clebsch-Gordan coefficients are also called *vector addition* (*or coupling*) *coefficients*, since they are the coefficients used in coupling two states with definite angular momenta to a state with a definite total angular momentum.

There are several useful relations satisfied by the phases of various Clebsch-Gordan coefficients which can be deduced from the conventions adopted. From (12.34) and (13.27) we notive that for any value of m_1m_2 and $M = m_1 + m_2$ we have

$$(j_1m_1j_2m_2|j_1j_2\ j_1 + j_2M) \geqslant 0. \tag{13.32}$$

Another relation can be obtained by using the equation

$$(J_x + iJ_y)\psi_{JJ} = 0. \tag{13.33}$$

Recalling that

$$\psi_{JJ} = \sum_{m_1'm_2'} (j_1m_1'j_2m_2'|j_1j_2J\ J)\ \psi_{j_1m'}(1)\ \psi_{j_2m_2'}(2)$$

and

$$(J_x + iJ_y) = (j_{1x} + ij_{1y}) + (j_{2x} + ij_{2y})$$

we obtain from (13.33) and (12.34)

$$\sqrt{j_1(j_1+1) - m_1(m_1+1)}\,(j_1m_1j_2\ m_2 + 1|j_1j_2J\ J)$$
$$+ \sqrt{j_2(j_2+1) - m_2(m_2+1)}\ (j_1\ m_1 + 1\ j_2m_2|j_1j_2J\ J) = 0$$

and hence

$$(j_1m_1j_2\ m_2 + 1|j_1j_2J\ J)(j_1\ m_1 + 1\ j_2m_2|j_1j_2J\ J) \leq 0. \tag{13.34}$$

Equation (13.34), which is valid irrespective of the phase conventions, can be generalized by multiplying a sequence of the relations (13.34) for decreasing values of m_1.

$$(-1)^{j_1-m_1}(j_1m_1j_2m_2|j_1j_2J\ J)(j_1j_1j_2m_2'|j_1j_1J\ J) \geqslant 0 \tag{13.35}$$

where, of course,

$$m_1 = j_1 + m_2' - m_2 = J - m_2.$$

A third relation can be obtained from the consideration of the matrix of the product

$$\begin{aligned}&(j_1j_2J\ J|(J_x + iJ_y)j_{1z}|j_1j_2J - 1\ J - 1)\\ &= (j_1j_2J\ J|(J_x + iJ_y)|j_1j_2J\ J-1)\ (j_1j_2J\ J-1|j_{1z}|j_1j_2J - 1\ J - 1).\end{aligned} \tag{13.36}$$

By our phase convention (13.28) and, using (12.34), the right-hand side of (13.36) is positive. Hence

$$(j_1j_2J\,J|(J_x+iJ_y)j_{1z}|j_1j_2J-1\;J-1)\geqslant 0. \tag{13.37}$$

Using the commutation relations

$$(J_x+iJ_y)j_{1z}=-(j_{1x}+ij_{1y})+j_{1z}(J_x+iJ_y)$$

and the relation

$$(j_1j_2J\,J|j_{1z}(J_x+iJ_y)|j_1j_2J-1\;J-1)=0$$

we obtain from (13.37)

$$(j_1j_2J\,J|j_{1x}+ij_{1y}|j_1j_2J-1\;J-1)\leq 0. \tag{13.38}$$

Expanding both $\psi(j_1j_2J\,J)$ and $\psi(j_1j_2J-1\;J-1)$ in (13.38) with Clebsch-Gordan coefficients, and using (12.34) we find that

$$(j_1\,m_1+1\,j_2m_2|j_1j_2J\,J)\,(j_1m_1j_2m_2|j_1j_2J-1\;J-1)\leq 0. \tag{13.39}$$

Combining (13.39) with (13.35) we finally obtain

$$(j_1j_1j_2m_2|j_1j_2J\,J)\,(j_1j_1j_2m_2'|j_1j_2J-1\;J-1)\geqslant 0.$$

Since we adopted the convention (13.27) that

$$(j_1j_1j_2j_2|j_1j_2,j_1+j_2,j_1+j_2)=+1$$

we conclude that for any J

$$(j_1j_1j_2m_2|j_1j_2J\,J)\geqslant 0. \tag{13.40}$$

Equations (13.40) and (13.38) both coincide with (13.27) for $M=J=j_1+j_2$. By combining (13.40) and (13.35) we obtain

$$(-1)^{j_1-m_1}(j_1m_1j_2m_2|j_1j_2J\,J)\geqslant 0. \tag{13.41}$$

The Clebsch-Gordan coefficients were introduced in order to obtain a state with $\mathbf{J}=\mathbf{j}_1+\mathbf{j}_2$. Instead, we can present a considerably more symmetric treatment of this problem by looking for the coefficients which appear in the addition of three angular momenta $\mathbf{j}_1$, $\mathbf{j}_2$ and $\mathbf{j}_3$ to zero. The two problems are related, since if $\mathbf{J}^2=(\mathbf{j}_1+\mathbf{j}_2)^2=j_3(j_3+1)$ we can combine this $\mathbf{J}$ and $\mathbf{j}_3$ to zero. Thus we are interested in the coefficients $a^{j_1j_2j_3}_{m_1m_2m_3}$ defined by requiring that

$$\psi_{00}(\mathbf{r}_1,\mathbf{r}_2,\mathbf{r}_3)=\sum_{m_i}a^{j_1j_2j_3}_{m_1m_2m_3}\,\psi_{j_1m_1}(\mathbf{r}_1)\,\psi_{j_2m_2}(\mathbf{r}_2)\,\psi_{j_3m_3}(\mathbf{r}_3) \tag{13.42}$$

is a state of zero total angular momentum.

We can obtain a possible state $\psi'_{00}(\mathbf{r}_1, \mathbf{r}_2, \mathbf{r}_3)$ of this type by using the Clebsch-Gordan coefficients. In fact all we have to do is to couple $\psi_{j_1m_1}(\mathbf{r}_1)$ and $\psi_{j_2m_2}(\mathbf{r}_2)$ to $\psi_{j_3m_3'}(\mathbf{r}_1, \mathbf{r}_2)$ and then couple the resulting state with $\psi_{j_3m_3}(\mathbf{r}_3)$ to $\psi'_{00}(\mathbf{r}_1, \mathbf{r}_2, \mathbf{r}_3)$. Thus

$$\psi'_{00} = \sum_{m_3m_3'} (j_3m_3'j_3m_3|j_3j_30\,0)\Big[\sum_{m_1m_2} (j_1m_1j_2m_2|j_1j_2j_3m_3')\,\psi_1(j_1m_1)\,\psi_2(j_2m_2)\Big]\,\psi_3(j_3m_3). \tag{13.43}$$

Comparing (13.43) and (13.42) we see that

$$a^{j_1j_2j_3}_{m_1m_2m_3} = \sum_{m_3'} (j_1m_1j_2m_2|j_1j_2j_3m_3')\,(j_3m_3'j_3m_3|j_3j_30\,0). \tag{13.44}$$

The Clebsch-Gordan coefficient $(j_3m_3'j_3m_3|j_3j_30\,0)$ is particularly simple and can be calculated by introducing $J = M = M' = 0$ into (13.14) and using (13.24). We then have:

$$\begin{aligned}(j_1m_1j_2m_2|j_1j_20\,0)\,(j_1m_1'j_2m_2'|j_1j_20\,0) &= \frac{1}{8\pi^2}\int D^{(j_1)}_{m_1m_1'}(R)\,D^{(j_2)}_{m_2m_2'}(R)\,dR \\ &= \frac{(-1)^{m_2-m_2'}}{8\pi^2}\int D^{(j_1)}_{m_1m_1'}(R)\,D^{(j_2)*}_{-m_2,-m_2'}(R)\,dR \\ &= (-1)^{m_2-m_2'}\frac{1}{2j_1+1}\,\delta_{j_1j_2}\,\delta_{m_1,-m_2}\,\delta_{m_1',-m_2'}.\end{aligned} \tag{13.45}$$

Hence

$$(j_1m_1j_2m_2|j_1j_20\,0)^2 = \frac{1}{2j_1+1}\,\delta_{j_1j_2}\,\delta_{m_1,-m_2}$$

and because of the phase in (13.45) we conclude that

$$(j\,m\,j-m\,|\,j\,j\,0\,0) = E(j)\cdot(-1)^m\frac{1}{\sqrt{2j+1}}$$

where $E(j)$ includes the possible dependence of the phase on j. This relation could, in fact, be as easily obtained by expanding $(J_x + iJ_y)\,\psi(j\,j\,0\,0) = 0$ in terms of Clebsch-Gordan coefficients. Since for $m = j$ (13.40) indicated that the coefficient has to be positive we conclude that

$$(j\,m\,j-m'\,|\,j\,j\,0\,0) = (-1)^{j-m}\frac{1}{\sqrt{2j+1}}\,\delta_{mm^1}. \tag{13.46}$$

Substituting (13.46) into (13.44) we obtain

$$a^{j_1j_2j_3}_{m_1m_2m_3} = \frac{(-1)^{j_3+m_3}}{\sqrt{2j_3+1}}(j_1m_1j_2m_2|j_1j_2j_3-m_3). \tag{13.47}$$

With these values of the coefficients, (13.42) assumes the form

$$\psi'_{00}(\mathbf{r}_1, \mathbf{r}_2, \mathbf{r}_3) = \sum_{m_1 m_2 m_3} \frac{(-1)^{j_3+m_3}}{\sqrt{2j_3+1}} (j_1 m_1 j_2 m_2 | j_1 j_2 j_3 - m_3)$$

$$\times \psi_{j_1 m_1}(\mathbf{r}_1)\, \psi_{j_2 m_2}(\mathbf{r}_2)\, \psi_{j_3 m_3}(\mathbf{r}_3). \qquad (13.48)$$

Equation (13.48) describes a possible state of the three particles which has a zero total angular momentum. It is clear that in order to get a zero total angular momentum $\mathbf{J}^2 = (\mathbf{j}_1 + \mathbf{j}_2)^2$ must equal $\mathbf{j}_3^2$. Since there is only one way of coupling $\mathbf{j}_1$ to $\mathbf{j}_2$ which will give $\mathbf{J}^2 = \mathbf{j}_3^2$ there is only one state of zero total angular momentum of three particles. Therefore any such state which is obtained by another order of coupling may differ from ψ'_{00} of (13.48) at most by a phase. In particular, we may construct a state $\psi''_{00}(\mathbf{r}_1, \mathbf{r}_2, \mathbf{r}_3)$ by coupling $\mathbf{j}_2$ and $\mathbf{j}_3$ to $\mathbf{J}'$ where $\mathbf{J}'^2 = \mathbf{j}_1^2$, and then $\mathbf{J}'$ and $\mathbf{j}_1$ to $\mathbf{J} = 0$. Obviously, we shall then obtain

$$\psi'_{00}(\mathbf{r}_2, \mathbf{r}_3, \mathbf{r}_1) = \psi''_{00}(\mathbf{r}_1, \mathbf{r}_2, \mathbf{r}_3) = \sum_{m_1 m_2 m_3} \frac{(-1)^{j_1+m_1}}{\sqrt{2j_1+1}} (j_2 m_2 j_3 m_3 | j_2 j_3 j_1 - m_1)$$

$$\times \psi_{j_1 m_1}(\mathbf{r}_1)\, \psi_{j_2 m_2}(\mathbf{r}_2)\, \psi_{j_3 m_3}(\mathbf{r}_3). \qquad (13.49)$$

Comparing (13.48) and (13.49) we conclude that

$$\frac{(-1)^{j_1+m_1}}{\sqrt{2j_1+1}} (j_2 m_2 j_3 m_3 | j_2 j_3 j_1 - m_1) = E(j_1 j_2 j_3) \frac{(-1)^{j_3+m_3}}{\sqrt{2j_3+1}} (j_1 m_1 j_2 m_2 | j_1 j_2 j_3 - m_3).$$

where E is a phase which may depend on j_1, j_2, and j_3 but not on m_1, m_2, or m_3. By repeated cyclic permutations we obtain

$$E(j_1 j_2 j_3) E(j_2 j_3 j_1) E(j_3 j_1 j_2) = +1.$$

Using the phase conventions introduced previously we can show that $E(j_1 j_2 j_3) = (-1)^{2j_1}$. Therefore, since $m_1 + m_2 + m_3 = 0$,

$$(j_2 m_2 j_3 m_3 | j_2 j_3 j_1 - m_1) = (-1)^{j_3 - j_1 - m_2} \sqrt{\frac{2j_1+1}{2j_3+1}}\, (j_1 m_1 j_2 m_2 | j_1 j_2 j_3 - m_3). \quad (13.50)$$

We can now use (13.50) to define the phase of the state ψ_{00} in such a way that cyclic permutations of the particles will leave it invariant. In other words we are looking for a phase $\eta(j_1 j_2 j_3)$ such that

$$\eta(j_1 j_2 j_3) \psi'_0(\mathbf{r}_1, \mathbf{r}_2, \mathbf{r}_3) = \eta(j_2 j_3 j_1) \psi'_0(\mathbf{r}_2, \mathbf{r}_3, \mathbf{r}_1) = \eta(j_3 j_1 j_2) \psi'_0(\mathbf{r}_3, \mathbf{r}_1, \mathbf{r}_2). \qquad (13.51)$$

Substituting from (13.48), (13.49), and (13.50) into (13.51) we see that this requirement is equivalent to

$$\eta(j_1j_2j_3) = (-1)^{2j_1}\,\eta(j_2j_3j_1) = (-1)^{2j_3}\,\eta(j_3j_1j_2).$$

The simplest solution of this equation is readily given by

$$\eta(j_1j_2j_3) = (-1)^{j_1-j_2+j_3}.$$

We thus introduce new coefficients defined by

$$\begin{pmatrix} j_1 & j_2 & j_3 \\ m_1 & m_2 & m_3 \end{pmatrix} = (-1)^{j_1-j_2+j_3}\, a^{j_1j_2j_3}_{m_1m_2m_3}$$

$$= \frac{(-1)^{j_1-j_2-m_3}}{\sqrt{2j_3+1}}\,(j_1m_1j_2m_2|j_1j_2j_3-m_3) \tag{13.52}$$

and call them *Wigner coefficients* or *3-j symbols*. For typographic convenience we may also use the notation

$$\begin{pmatrix} j_1 & j_2 & j_3 \\ m_1 & m_2 & m_3 \end{pmatrix} = V(j_1j_2j_3|m_1m_2m_3) \tag{13.53}$$

and shall often refer to a Wigner coefficient simply as "a V". In terms of the Wigner coefficients ψ_{00} is given by

$$\psi_{00}(\mathbf{r}_1, \mathbf{r}_2, \mathbf{r}_3) = \sum_{m_1m_2m_3} \begin{pmatrix} j_1 & j_2 & j_3 \\ m_1 & m_2 & m_3 \end{pmatrix} \psi_{j_1m_1}(\mathbf{r}_1)\,\psi_{j_2m_2}(\mathbf{r}_2)\,\psi_{j_3m_3}(\mathbf{r}_3). \tag{13.54}$$

It has the property

$$\psi_{00}(\mathbf{r}_1, \mathbf{r}_2, \mathbf{r}_3) = \psi_{00}(\mathbf{r}_2, \mathbf{r}_3, \mathbf{r}_1) = \psi_{00}(\mathbf{r}_3, \mathbf{r}_1, \mathbf{r}_2).$$

Consequently the Wigner coefficients have the symmetry property

$$\begin{pmatrix} j_1 & j_2 & j_3 \\ m_1 & m_2 & m_3 \end{pmatrix} = \begin{pmatrix} j_2 & j_3 & j_1 \\ m_2 & m_3 & m_1 \end{pmatrix} = \begin{pmatrix} j_3 & j_1 & j_2 \\ m_3 & m_1 & m_2 \end{pmatrix}. \tag{13.55}$$

Equations (13.55) can be compared with their equivalent (13.50) expressed in terms of the Clebsch-Gordan coefficients. The simpler symmetry of the Wigner coefficients is due to the fact that they are the coefficients for the addition $\mathbf{j}_1 + \mathbf{j}_2 + \mathbf{j}_3 = 0$ which is completely invariant even under cyclic permutations. The two types of coefficients are obviously closely related as, indeed, is shown by (13.52).

Another symmetry property of the Clebsch-Gordan and the Wigner coefficients can be obtained by changing the order of coupling of $\mathbf{j}_1$ and $\mathbf{j}_2$. It follows from (13.41) that

$$(j_1m_1j_2m_2|j_1j_2j_3j_3) = (-1)^{j_1+j_2-j_3}(j_2m_2j_1m_1|j_2j_1j_3j_3)$$

since obviously the magnitude of the two coefficients must be equal. It follows then from (13.14) that this phase relation holds for every value of m_3. Therefore

$$(j_2m_2j_1m_1|j_2j_1j_3m_3) = (-1)^{j_1+j_2-j_3}(j_1m_1j_2m_2|j_1j_2j_3m_3). \tag{13.56}$$

Using (13.52) we can express this last symmetry property of the Clebsch-Gordan coefficients in terms of a corresponding symmetry property of the Wigner coefficients by

$$\begin{pmatrix} j_1 & j_2 & j_3 \\ m_1 & m_2 & m_3 \end{pmatrix} = (-1)^{j_1+j_2+j_3}\begin{pmatrix} j_2 & j_1 & j_3 \\ m_2 & m_1 & m_3 \end{pmatrix}. \tag{13.57}$$

Equation (13.17) can also be rewritten in terms of Wigner coefficients as follows:

$$\begin{pmatrix} j_1 & j_2 & j_3 \\ -m_1 & -m_2 & -m_3 \end{pmatrix} = (-1)^{j_1+j_2+j_3}\begin{pmatrix} j_1 & j_2 & j_3 \\ m_1 & m_2 & m_3 \end{pmatrix}. \tag{13.58}$$

We thus conclude that a cyclic permutation of the columns in a Wigner coefficient leaves it unchanged while an odd permutation of the columns or a change in the sign of all the m values multiplies the coefficient by $(-1)^{j_1+j_2+j_3}$.

It will be usefull to have some of the properties of the Clebsch-Gordan coefficients written in terms of the Wigner coefficients. The orthogonality relations (13.12) and (13.13) can be rewritten, using (13.25), in the form

$$\sum_{m_1m_2}\begin{pmatrix} j_1 & j_2 & j_3 \\ m_1 & m_2 & m_3 \end{pmatrix}\begin{pmatrix} j_1 & j_2 & j_3' \\ m_1 & m_2 & m_3' \end{pmatrix} = \delta_{j_3j_3'}\,\delta_{m_3m_3'}\frac{1}{2j_3+1} \tag{13.59}$$

and

$$\sum_{j_3,m_3}(2j_3+1)\begin{pmatrix} j_1 & j_2 & j_3 \\ m_1 & m_2 & m_3 \end{pmatrix}\begin{pmatrix} j_1 & j_2 & j_3 \\ m_1' & m_2' & m_3 \end{pmatrix} = \delta_{m_1m_1'}\,\delta_{m_2m_2'}. \tag{13.60}$$

Since in (13.59) the sum on the left-hand side is independent of m_3 we can perform the summation over m_3 simply by multiplying both sides by $2j_3+1$. Hence

$$\sum_{m_1m_2m_3}\begin{pmatrix} j_1 & j_2 & j_3 \\ m_1 & m_2 & m_3 \end{pmatrix}\begin{pmatrix} j_1 & j_2 & j_3 \\ m_1 & m_2 & m_3 \end{pmatrix} = 1. \tag{13.61}$$

The phase conventions (13.27) and (13.29) give for the Wigner coefficients the following relations

$$\begin{pmatrix} j_1 j_2 & j_1 + j_2 \\ j_1 j_2 & -(j_1 + j_2) \end{pmatrix} = \frac{(-1)^{2j_1}}{\sqrt{2j+1}} \tag{13.62}$$

$$\sum_{m_i} (2j_3 + 1)\, m_1 \begin{pmatrix} j_1 & j_2 & j_3 \\ m_1 & m_2 & m_3 \end{pmatrix} \begin{pmatrix} j_1 & j_2 & j_3' \\ m_1 & m_2 & m_3 \end{pmatrix} \geqslant 0. \tag{13.63}$$

An important relation involving the Wigner coefficients and the Wigner D matrices can be obtained from (13.15)

$$\sum_{m_i'} \begin{pmatrix} j_1 & j_2 & j_3 \\ m_1' & m_2' & m_3' \end{pmatrix} D^{(j_1)}_{m_1' m_1}(R)\, D^{(j_2)}_{m_2' m_2}(R)\, D^{(j_3)}_{m_3' m_3}(R) = \begin{pmatrix} j_1 & j_2 & j_3 \\ m_1 & m_2 & m_3 \end{pmatrix}. \tag{13.64}$$

Since the D-matrices are the matrices of the transformation in the space of irreducible tensors induced by the rotation R, it follows from (13.64) that the Wigner coefficients can be considered as isotropic tensors with three indices $T_{m_1 m_2 m_3}$. They are given by the same set of numbers in every frame, and their transformation is governed by the product of three transformation matrices, very much like (10.4).

In subsequent applications we shall very often come across problems of calculating quantities, like energies, lifetimes, etc., whose values do not depend on the frame of reference. In as much as such calculations will involve the use of Wigner coefficients, it is clear from (13.64) that they will have to appear in the form of contracted products. The properties of these invariant products of Wigner coefficients will be treated in subsequent sections.

14. Tensor Products. The Wigner-Eckart Theorem

The construction of a state with a definite total angular momentum from two states, each of which has a well-defined angular momentum, is a special case of the reduction of a product of any two irreducible tensors. Given two such irreducible tensors $T^{(k_1)}_{\kappa_1}$ and $U^{(k_2)}_{\kappa_2}$ we can form $(2k_1 + 1)(2k_2 + 1)$ products and obtain a new tensor, their *external product*, $P^{(k_1k_2)}_{\kappa_1\kappa_2} = T^{(k_1)}_{\kappa_1}U^{(k_2)}_{\kappa_2}$. The tensor P, however, will generally not be irreducible. In the applications of tensor algebra to spectroscopy we are often faced with the problem of reducing this tensor P to a linear combination of irreducible tensors. One example of this procedure was given in Section 10 when we discussed the reduction of the second rank tensor. In the preceding section we saw another example. The construction of a state of a given total angular momentum J out of $\psi_{j_1m_1}$ and $\psi_{j_2m_2}$ is equivalent to the reduction of the tensor $P^{(j_1j_2)}_{m_1m_2} = \psi_{j_1m_1}\psi_{j_2m_2}$ to irreducible parts picking the one which transforms like a tensor of degree J.

Since any irreducible tensor of degree k always transforms under rotations with Wigner's matrix $D^{(k)}(R)$, we see that the reduction of the tensor $P^{(k_1k_2)}_{\kappa_1\kappa_2}$ amounts to finding coefficients $a^{k\kappa}_{\kappa_1\kappa_2}$ such that

$$S^{(k)}_{\kappa} = \sum_{\kappa_1\kappa_2} a^{k\kappa}_{\kappa_1\kappa_2} P^{(k_1k_2)}_{\kappa_1\kappa_2} = \sum a^{k\kappa}_{\kappa_1\kappa_2} T^{(k_1)}_{\kappa_1} U^{(k_2)}_{\kappa_2}$$

will transform under rotations with the matrix $D^{(k)}(R)$. This problem is thus identical to that of the coupling of two angular momenta as in (13.8). We therefore conclude that the coefficients $a^{k\kappa}_{\kappa_1\kappa_2}$ are just the Clebsch-Gordan coefficients

$$a^{k\kappa}_{\kappa_1\kappa_2} = (k_1\kappa_1k_2\kappa_2|k_1k_2k\,\kappa).$$

This result which follows only from the transformation properties, remains valid irrespective of whether the tensors T and U refer to the same particle or to different particles. We conclude that given any two irreducible tensors $T^{(k_1)}_{\kappa_1}$ and $T^{(k_2)}_{\kappa_2}$ of degrees k_1 and k_2 we can obtain another irreducible tensor of degree k_3 by forming the bilinear combination

$$T^{(k_3)}_{\kappa_3} = \sum_{\kappa_1\kappa_2} (k_1\kappa_1k_2\kappa_2|k_1k_2k_3\kappa_3)\; T^{(k_1)}_{\kappa_1}\; T^{(k_2)}_{\kappa_2}. \tag{14.1}$$

The tensor $\mathbf{T}^{(k_3)}$ is called the tensor product of degree k_3 of $\mathbf{T}^{(k_1)}$ and $\mathbf{T}^{(k_2)}$ and will be denoted by

$$\mathbf{T}^{(k_3)} = [\mathbf{T}^{(k_1)} \times \mathbf{T}^{(k_2)}]^{(k_3)}. \tag{14.2}$$

Equation (14.1) can be used to give $T^{(k_1)}_{\kappa_1} T^{(k_1)}_{\kappa_2}$ in terms of $T^{(k_3)}_{\kappa_3}$. Using the orthogonality properties of the Clebsch-Gordan coefficients we obtain from (14.1)

$$T^{(k_1)}_{\kappa_1} T^{(k_2)}_{\kappa_2} = \sum_{k_3\kappa_3} (k_1\kappa_1k_2\kappa_2|k_1k_2k_3\kappa_3)\, T^{(k_3)}_{\kappa_3}. \tag{14.3}$$

There are numerous examples of tensor products in addition to that of the coupling of two angular momenta mentioned above. Thus, for instance, the ordinary vector product is a special case of (14.1) with $k_1 = k_2 = k_3 = 1$. In particular the orbital angular momentum operator $\boldsymbol{l}$ is proportional to the tensor product of degree 1 of the two tensors of the first degree $\mathbf{r}$ and $\mathbf{p}$, $\boldsymbol{l} = -i\sqrt{2}\,[\mathbf{r} \times \mathbf{p}]^{(1)}$. The Legendre polynomial of order l is proportional to the tensor product of degree zero of two spherical harmonics of degree l:

$$P_l(\cos\omega_{12}) = \frac{4\pi(-1)^l}{\sqrt{2l+1}}\,[\mathbf{Y}^{(l)}(\theta_1\,\varphi_1) \times \mathbf{Y}^{(l)}(\theta_2\,\varphi_2)]^{(0)}_0 \tag{14.4}$$

where ω_{12} is the angle between the directions $(\theta_1\varphi_1)$ and $(\theta_2\varphi_2)$. Noting that by (13.46)

$$(l\,m_1\,l\,m_2|l\,l\,0\,0) = \frac{(-1)^{l-m_1}}{\sqrt{2l+1}}\,\delta_{m_1,-m_2}$$

we can also write (14.4) in the form

$$\begin{aligned} P_l(\cos\omega_{12}) &= \frac{4\pi}{2l+1}\sum_m (-1)^m\, Y_{l,-m}(\theta_1\,\varphi_1)\, Y_{l,m}(\theta_2\,\varphi_2) \\ &= \frac{4\pi}{2l+1}\sum_m Y^*_{lm}(\theta_1\,\varphi_1)\, Y_{lm}(\theta_2\,\varphi_2). \end{aligned} \tag{14.5}$$

Equation (14.5) is known as the *addition theorem for spherical harmonics*. Another, slightly more complicated, example of products of irreducible tensors is offered by the expression of the interaction of two magnetic dipoles $g_1\mathbf{s}_1$ and $g_2\mathbf{s}_2$. As is well known the interaction energy when the two dipoles are a distance $\mathbf{r}$ apart is given by a function of r multiplied by

$$\left[\frac{(\mathbf{s}_1\cdot\mathbf{r})\,(\mathbf{s}_2\cdot\mathbf{r})}{r^2} - \frac{1}{3}\,(\mathbf{s}_1\cdot\mathbf{s}_2)\right]. \tag{14.6}$$

Like every interaction energy, (14.6) is, of course, a scalar built from the scalars $\mathbf{r}^2$, $(\mathbf{s}_1 \cdot \mathbf{r})$, $(\mathbf{s}_2 \cdot \mathbf{r})$, and $(\mathbf{s}_1 \cdot \mathbf{s}_2)$. We can build scalars out of $\mathbf{s}_1$, $\mathbf{s}_2$, and $\mathbf{r}$ in other ways as well. Thus for instance,

$$V_1 = [[\mathbf{s}_1 \times \mathbf{s}_2]^{(1)} \times \mathbf{r}]_0^{(0)} \tag{14.7}$$

or

$$V_2 = [[\mathbf{s}_1 \times \mathbf{s}_2]^{(2)} \times [\mathbf{r} \times \mathbf{r}]^{(2)}]_0^{(0)} \tag{14.8}$$

are examples of such scalars obtained by first forming a tensor of the first or second degree out of $\mathbf{s}_1$ and $\mathbf{s}_2$, and then forming the scalar product of this tensor with a tensor of the corresponding degree built from the coordinate vector.

Noting that $r_{\pm 1}^{(1)} = \mp (1/\sqrt{2})\,(x \pm iy)$, $r_0^{(1)} = z$, and using the Clebsch-Gordan coefficients given in the Appendix for the addition of two vectors to form a tensor of the second degree, we can verify that (14.8) is equal to

$$V_2 = \frac{r^2}{\sqrt{5}} \left[\frac{(\mathbf{s}_1 \cdot \mathbf{r})\,(\mathbf{s}_2 \cdot \mathbf{r})}{r^2} - \frac{1}{3}\,(\mathbf{s}_1 \cdot \mathbf{s}_2) \right]. \tag{14.9}$$

Comparing (14.9) with (14.6) we see that the dipole-dipole interaction is proportional to the scalar product of two second degree tensors. It is therefore called a *tensor interaction*. Similarly, (14.7) is called a *vector interaction*. In the following we shall treat such interactions in more detail. It is worthwhile to notice that using the spin operator $\mathbf{s}$ of a single particle, no tensors of degree higher than 1 can be constructed. In fact, the vector product of $\mathbf{s}$ with itself yields $i\mathbf{s}$ due to the commutation relations (2.14). The only other possible tensor product of $\mathbf{s}$ with itself, of degree 2, vanishes due to the well-known anticommuting of the spin matrices.

The concept of tensor products can be applied to obtain an important theorem on the matrix elements of irreducible tensor operators $T_\kappa^{(k)}$ between two states of given angular momenta ψ_{JM} and $\psi_{J'M'}$. In the definition of the tensor product (14.1), $\mathbf{T}^{(k_1)}$ and $\mathbf{T}^{(k_2)}$ may be any two arbitrary irreducible tensors which may have very different physical properties. In particular, $\mathbf{T}^{(k_2)}$ can be a set of $2k_2 + 1$ functions with angular momentum k_2 and $\mathbf{T}^{(k_1)}$ can be a tensor operator which is operating on these functions. With this in mind, let us consider the matrix element

$$\langle J M | T_\kappa^{(k)} | J'M' \rangle = \int \psi_{JM}^* T_\kappa^{(k)} \psi_{J'M'} \, d\tau. \tag{14.10}$$

Here the ψ are wave functions of any number of particles in a state of well-defined total angular momentum but otherwise unrestricted. Thus, for instance, ψ_{JM} may or may not refer to a definite configuration. In (14.10), $T_\kappa^{(k)}$ is any irreducible tensor operating on all or part of the coordinates appearing in ψ. The operation of $T_\kappa^{(k)}$ gives some wave function $\psi' = T_\kappa^{(k)} \psi_{J'M'}$ of the same physical nature as that of $\psi_{J'M'}$. Thus, if ψ_{JM} and $\psi_{J'M'}$ are physical scalar fields, so is also $\psi' = T_\kappa^{(k)} \psi_{J'M'}$. If ψ_{JM} and $\psi_{J'M'}$ are both physical spinor fields, so is also $\psi' = T_\kappa^{(k)} \psi_{J'M'}$, etc. Therefore, if $\psi_{J'M'}$ is generally a physical tensor field of degree s, $T_\kappa^{(k)}$ is a $(2s + 1) \times (2s + 1)$ matrix, and the product $T_\kappa^{(k)} \psi_{J'M'}$ is to be understood as the multiplication of the matrix $[T_\kappa^{(k)}]_{\sigma\sigma'}$ by the column $(\psi_{J'M'})_{\sigma'}$ where $\sigma, \sigma' = -s, -s+1, ..., +s$. Similarly, $\psi_{JM}^*(T_\kappa^{(k)} \psi_{J'M'})$ is to be understood as the matrix multiplication of the row $(\psi_{JM}^*)_\sigma$ by the column

$$(T_\kappa^{(k)}\psi_{J'M'})_\sigma = \sum_{\sigma'} [T_\kappa^{(k)}]_{\sigma\sigma'} (\psi_{J'M'})_{\sigma'} .$$

Thus, irrespective of the physical nature of the ψ, the integrand in (14.10) is always a physical scalar field. It is a function of the space coordinates of the particles only and of no other internal degrees of freedom which the system may possess.

Before evaluating the matrix element (14.10) we shall prove a useful lemma:

If $T_\kappa^{(k)}(\mathbf{r}_1, \mathbf{r}_2, ...)$ is an irreducible tensor of degree $k \neq 0$ which is a physical scalar field of the coordinates only, then

$$F_\kappa^{(k)} = \int T_\kappa^{(k)} (\mathbf{r}_1, \mathbf{r}_2, ...)\, d\tau_1\, d\tau_2 \,... = 0 \qquad \text{if} \qquad k \neq 0. \tag{14.11}$$

To prove this lemma we note that the integration is taken over all space. Therefore the value of the integral is not affected if we change the integrand into $T_\kappa^{(k)}(\mathbf{r}_1', \mathbf{r}_2' ...)$ where $\mathbf{r}_i' = a\mathbf{r}_i$ and a is any fixed rotation independent of i. Using the transformation properties of irreducible tensors (11.19) we then obtain

$$\begin{aligned} F_\kappa^{(k)} &= \int T_\kappa^{(k)} (\mathbf{r}_1, \mathbf{r}_2 ...)\, d\tau_1\, d\tau_2 \,... = \int T_\kappa^{(k)}(\mathbf{r}_1'(\mathbf{r}_1), \mathbf{r}_2'(\mathbf{r}_2), ...)\, d\tau_1\, d\tau_2 \,... \\ &= \sum D_{\kappa'\kappa}^{(k)}(R) \int T_{\kappa'}^{(k)} (\mathbf{r}_1, \mathbf{r}_2, ...)\, d\tau_1\, d\tau_2 \,... = \sum F_{\kappa'}^{(k)} D_{\kappa'\kappa}^{(k)}(R). \end{aligned} \tag{14.12}$$

Thus $F_\kappa^{(k)}$ is an irreducible tensor of the type discussed in Section 10 (*not* a tensor field). The $F_\kappa^{(k)}$ are just the $2k + 1$ components of a tensor which according to (14.12) have the same numerical values in all frames. Such a tensor must vanish identically if $k > 0$ due to the irreducibility of the transformation carried out by the D matrices. To illustrate this fact we can choose for R a rotation by an angle φ around the z axis, for which $D_{\kappa'\kappa}(R) = e^{-i\kappa\varphi}\delta_{\kappa',\kappa}$ and obtain $F_\kappa^{(k)} = e^{-i\kappa\varphi}F_\kappa^{(k)}$. Thus, for $\kappa \neq 0$ $F_\kappa^{(k)} = 0$. Since this relation holds in every frame, it is clear that also $F_0^{(k)} = 0$, and the lemma is thus proved.

Consider now the matrix elements (14.10). The ψ_{JM}, being a state of a definite total angular momentum, is a component of an irreducible tensor of degree J, i.e., it satisfies

$$\psi'_{JM}(\mathbf{r}') = \sum_{M'} \psi'_{JM'}(\mathbf{r})\, D^{(J)}_{M'M}(R).$$

Since we know that

$$D^{(J)}_{M'M}(R)^* = (-1)^{M-M'}\, D^{(J)}_{-M',-M}(R) = (-1)^{(J-M)-(J-M')}\, D^{(J)}_{-M',-M}(R)$$

we see that

$$(-1)^{J+M}\, \psi^*_{J,-M}(\mathbf{r}') = \sum_{M'} [(-1)^{J+M'}\, \psi^*_{J,-M'}(\mathbf{r})]\, D^{(J)}_{M'M}(R).$$

Therefore, also $(-1)^{J+M}\psi^*_{J,-M}$ is the M component of an irreducible tensor of degree J. The integrand in (14.10) is thus proportional to the product of components of three irreducible tensors. By using (14.3) twice we can express this integrand as a sum of irreducible tensors multiplied by the proper Clebsch-Gordan coefficients

$$\int \psi^*_{JM} T^{(k)}_{\kappa} \psi_{J'M'}$$

$$= \int (-1)^{J-M}\, [(-1)^{J+(-M)}\, \psi^*_{J,-(-M)}] \sum_{k_1\kappa_1} (k\kappa J'M'|k\, J'k_1\kappa_1)[\mathbf{T}^{(k)} \times \psi_{J'}]^{(k_1)}_{\kappa_1}$$

$$= \int (-1)^{J-M} \sum_{k_1k_2\kappa_1\kappa_2} (J\, -M\, k_1\kappa_1|J\, k_1k_2\kappa_2)\, (k\, \kappa\, J'M'|k\, J'k_1\kappa_1)$$

$$\times\, [\varphi_J \times [\mathbf{T}^{(k)} \times \psi_{J'}]^{(k_1)}]^{(k_2)}_{\kappa_2} \qquad (14.13)$$

where

$$\varphi_{JM} = (-1)^{J+M}\, \psi^*_{J,-M}. \qquad (14.14)$$

Since the integrand in (14.10) is a function of the space coordinates only, we can now use the lemma (14.11) and conclude that on carrying out the integration in (14.13) the only contribution will come from the term with $k_2 = \kappa_2 = 0$. Hence

$$\int \psi^*_{JM} T^{(k)}_{\kappa} \psi_{J'M'} = (-1)^{J-M} \sum_{k_1\kappa_1} (J\, -M\, k_1\kappa_1|J\, k_1 0\, 0)\, (k\, \kappa\, J'M'|k\, J'k_1\kappa_1)$$

$$\times \int [\varphi_J \times [\mathbf{T}^{(k)} \times \psi_{J'}]^{(k_1)}]^{(0)}_0$$

$$= (-1)^{2J}\, (2J+1)^{-1/2}\, (k\, \kappa\, J'M'|k\, J'J\, M)\, \times$$

$$\times \int [\varphi_J \times [\mathbf{T}^{(k)} \times \psi_{J'}]^{(J)}]^{(0)}_0.$$

Introducing now a Wigner coefficient instead of the Clebsch-Gordan coefficient we find that

$$\langle JM|T_\kappa^{(k)}|J'M'\rangle = \int \psi_{JM}^* T_\kappa^{(k)} \psi_{J'M'} = (-1)^{J-M} \begin{pmatrix} J & k & J' \\ -M & \kappa & M' \end{pmatrix} (J||\mathbf{T}^{(k)}||J') \tag{14.15}$$

where

$$(J||\mathbf{T}^{(k)}||J') = (-1)^{J-k+J'} \int [\varphi_J \times [\mathbf{T}^{(k)} \times \psi_{J'}]^{(J)}]_0^{(0)}$$

$$= \sum_{M\kappa M'} (-1)^{J-M} \begin{pmatrix} J & k & J' \\ -M & \kappa & M' \end{pmatrix} \int \psi_{JM}^* T_\kappa^{(k)} \psi_{J'M'}. \tag{14.16}$$

Equation (14.15), which is known as the *Wigner-Eckart theorem*, is a basic theorem in the algebra of irreducible tensor operators. It expresses the matrix elements of any component of a tensor operator as the product of two factors. The first factor, $(-1)^{J-M} V(J\,k\,J'|-M\,\kappa\,M')$, takes care of the geometrical properties of the tensor and the states considered. It is generally different for different components of the tensor and for different magnetic substates. This factor is a well-defined number, *independent* of the physical contents of the tensor $\mathbf{T}^{(k)}$ provided we know that $T_\kappa^{(k)}$ is an *irreducible* tensor. The second factor, $(J||\mathbf{T}^{(k)}||J')$, known as the *reduced matrix element of* $\mathbf{T}^{(k)}$, includes all the specific physical information which is contained in the tensor $\mathbf{T}^{(k)}$. The reduced matrix elements are obviously independent of either M, κ, or M' as can readily be seen from (14.16). It is in these matrix elements that the difference in the physical meaning of different tensors shows itself up. Thus, for instance, the total angular momentum operator $\mathbf{J}$ and the magnetic moment operator $\boldsymbol{\mu}$ are both vectors, i.e., tensors of degree 1. When we consider their matrix elements, the M and κ dependence will be the same. However, the reduced matrix elements will be very different, reflecting the physical difference in these two operators. In fact, $(J||\mathbf{J}||J') = 0$ if $J \neq J'$ whereas $(J||\boldsymbol{\mu}||J')$ can be different from zero for $J \neq J'$.

The generality of (14.15) should be pointed out. Nothing was assumed about the wave functions ψ_{JM}, $\psi_{J'M'}$, and the irreducible tensor operator $T_\kappa^{(k)}$ except their transformation properties. They can be functions and operators of a single particle or of many particles, belonging to a well-defined configuration or not. The ψ can be physical scalar fields, or spinor fields or higher tensor physical fields. As long as the ψ refer to definite *total* angular momenta and $\mathbf{T}^{(k)}$ is an irreducible tensor operator (14.15) holds.

It may be worthwhile to see in detail the form which the change-of-variables-transformation (14.12) assumes in the case of a spinor field.

Consider, for instance, the function $\psi^*_{jm}(\mathbf{r})\, s_i\, \psi_{j_1m_1}(\mathbf{r})$ where $s_i = \frac{1}{2}\sigma_i$ are the spin matrices. The integral of this function over all space is equal to that of $\psi^*_{jm}(\mathbf{r}')s_i\,\psi_{j_1m_1}(\mathbf{r}')$ with $\mathbf{r}' = a\mathbf{r}$ where a is a fixed rotation. According to (12.25) we obtain

$$\psi^*_{jm}(\mathbf{r}')s_i\,\psi_{j_1m_1}(\mathbf{r}') = \Big[\sum_{m'} \psi_{jm'}(\mathbf{r})\, D^{(j)}_{m'm}(a)\Big]^* \; U^{-1}s_iU\Big[\sum_{m_1'} \psi_{j_1m_1'}(\mathbf{r})\, D^{(j_1)}_{m_1'm_1}(a)\Big].$$

Recalling (11.14), this expression becomes equal to

$$\Big[\sum_{m'} \psi_{jm'}(\mathbf{r})\, D^{(j)}_{m'm}(a)\Big]^* \; \Big[\sum_{k} a_{ik}\, s_k\Big]\Big[\sum_{m_1'} \psi_{j_1m_1'}(\mathbf{r})\, D^{(j_1)}_{m_1'm_1}(a)\Big]$$

Thus, due to the transformation properties of the spinors ψ_{jm}, the spin matrices are transformed as the components of a vector. Therefore, in order for the integral of such a function not to vanish, a zero degree tensor should be built of the irreducible tensors φ_{jm} [defined in (14.14)], $\psi_{j_1m_1}$ and the *vector* $\mathbf{s}$.

Although the reduced matrix element of a given tensor contains very important information on that tensor, it will turn out, as we shall see in the following, that many conclusions can be drawn just from the tensor character of physical quantities and from the M-dependence of their matrix elements. Thus, for instance, we see from (14.15) that the matrix elements of any irreducible tensor of degree k between states of total angular momenta J and J' will vanish unless $|J - J'| \leq k \leq J + J'$ and unless $M = \kappa + M'$. It follows then that the matrix elements of any conceivable tensor operator of degree higher than 1 in the spin space of one $s = \frac{1}{2}$ particle must vanish. There are therefore no other such tensors apart from 1 and $\mathbf{s}$. Another simple result of the Wigner-Eckart theorem is the vanishing of the quadrupole moment in a system with $J < 1$. The quadrupole moment is an irreducible tensor of degree 2, and if $J = 0$ or $\frac{1}{2}$ the triangular condition cannot be satisfied.

It also follows that in the ordinary applications of spectroscopy, where the angular momenta of the initial and final state are both integers or both half-integers, only tensor operators with integral values of k can appear. This will usually be the case in the following.

A less trivial example is the theorem which states that the matrix elements of a scalar, i.e., an irreducible tensor of degree zero, are independent of M. In fact, for $k = \kappa = 0$

$$\langle J\,M|T^{(0)}_0|J'M'\rangle = (-1)^{J-M}\begin{pmatrix} J & 0 & J' \\ -M & 0 & M' \end{pmatrix}(J||\mathbf{T}^{(0)}||J')$$

$$= \frac{1}{\sqrt{2J+1}}(J||\mathbf{T}^{(0)}||J')\,\delta_{JJ'}\,\delta_{MM'}. \tag{14.17}$$

Here we have used (13.52) and (13.46) to compute the value of $V(J\,O\,J'|-M\,0\,M') = (-1)^{j+j'}V(J J' 0|-M\,M'\,0)$.

Since the M-dependence of the matrix elements of any two tensor operators $\mathbf{T}^{(k)}$ and $\mathbf{U}^{(k)}$ of the same degree is the same, we conclude that in as much as $\langle J\,M|U_\kappa^{(k)}|J'M'\rangle \neq 0$

$$\langle J\,M|T_\kappa^{(k)}|J'M'\rangle = A_{JJ'}\langle J\,M|U_\kappa^{(k)}|J'M'\rangle \tag{14.18}$$

where the proportionality constant $A_{JJ'}$ which is the ratio of the two reduced matrix elements will generally depend on J and J' and on the physical nature of $\mathbf{T}^{(k)}$ and $\mathbf{U}^{(k)}$, but not on M and M'. In particular, if we consider matrix elements diagonal in J, i.e., $J = J'$, and take for $\mathbf{U}^{(k)}$ the vector $\mathbf{J}$ itself, then for any vector $\mathbf{T}^{(1)} \equiv \mathbf{T}$ we have

$$\langle J\,M|\mathbf{T}|J\,M'\rangle = A\langle J\,M|\mathbf{J}|J\,M'\rangle. \tag{14.19}$$

To obtain an expression for A we calculate the diagonal matrix elements of the scalar product $(\mathbf{T}\cdot\mathbf{J})$. Noting that the operator $\mathbf{J}$ is diagonal in the quantum numbers J, we obtain

$$\langle J\,M|(\mathbf{T}\cdot\mathbf{J})|J\,M\rangle = \sum_{M'}(\langle J\,M|\mathbf{T}|J\,M'\rangle\cdot\langle J\,M'|\mathbf{J}|J\,M\rangle)$$

$$= A\sum_{M'}(\langle J\,M|\mathbf{J}|J\,M'\rangle\cdot\langle J\,M'|\mathbf{J}|J\,M\rangle) = A\langle J\,M|\mathbf{J}^2|J\,M\rangle = AJ(J+1).$$

Hence the value of A is determined and we obtain from (14.19)

$$\langle J\,M|\mathbf{T}|J\,M'\rangle = \frac{\langle J\,M|(\mathbf{T}\cdot\mathbf{J})|J\,M\rangle}{J(J+1)}\langle J\,M|\mathbf{J}|J\,M'\rangle. \tag{14.20}$$

Equation (14.20) is the *Landé formula* introduced in (9.16) where we saw some of its applications.

It should be pointed out that in the generalized Landé formula (14.20) the matrix elements of the vector $\mathbf{T}$ are taken between two states of the *same* total angular momentum J. Since $\mathbf{J}$ is diagonal in the quantum numbers J, and an arbitrary vector need not necessarily be diagonal in J, a relation of the type (14.20) cannot hold for the elements $\langle J\,M|\mathbf{T}|J'M'\rangle$ with $J' \neq J$.

It will be useful to have the expression for the reduced matrix element of the angular momentum vector $\mathbf{j}$. Using the Wigner-Eckart theorem we have, since $\mathbf{j}$ is a tensor of degree 1,

$$m = \langle j\,m|j_z|j\,m\rangle = (-1)^{j-m}\begin{pmatrix} j & 1 & j \\ -m & 0 & m \end{pmatrix}(j||\mathbf{j}||j).$$

Using now the relation (c.f. Appendix)

$$\begin{pmatrix} j & 1 & j \\ -m & 0 & m \end{pmatrix} = (-1)^{j-m} \frac{m}{\sqrt{j(2j+1)(j+1)}}.$$

We obtain

$$(j||\mathbf{j}||j) = \sqrt{j(j+1)(2j+1)}. \tag{14.21}$$

Although we have derived (14.21) using j_z, it is clear that we would have obtained the same result had we used j_+ or j_-. Using j_z is most convenient for obvious reasons.

It is often required to calculate matrix elements of tensor products. We shall now derive relations between the reduced matrix elements of tensor products $T_\kappa^{(k)} = [\mathbf{T}^{(k_1)} \times \mathbf{T}^{(k_2)}]_\kappa^{(k)}$ and those of $T_{\kappa_1}^{(k_1)}$ and $T_{\kappa_2}^{(k_2)}$. We shall be interested, in particular, in the case where $T_{\kappa_1}^{(k_1)}$ are tensors operating on one set of coordinates of the system and $T_{\kappa_2}^{(k_2)}$ are tensors operating on the other set. These sets may each consist of the space coordinates of some of the particles or the spin coordinates of some particles etc. The case in which $\mathbf{T}^{(k_1)}$ and $\mathbf{T}^{(k_2)}$ operate on the same system will be considered in the next section. The wave functions of the whole system are built from wave functions of the two parts. Such wave functions will therefore be eigenfunctions of $\mathbf{j}_1^2$, $\mathbf{j}_2^2$, $\mathbf{J}^2 = (\mathbf{j}_1 + \mathbf{j}_2)^2$, and J_z where $\mathbf{j}_1$ and $\mathbf{j}_2$ are the angular momenta of the first and second parts of the system respectively. In addition to j_1 and j_2 we shall use, when necessary, additional quantum numbers α_1 and α_2 to specify uniquely the states of the two parts of the system. We shall thus derive relations between $(\alpha_1 j_1 \alpha_2 j_2 J||\mathbf{T}^{(k)}||\alpha_1' j_1' \alpha_2' j_2' J')$ and $(\alpha_1 j_1||\mathbf{T}^{(k_1)}||\alpha_1' j_1')$ and $(\alpha_2 j_2||\mathbf{T}^{(k_2)}||\alpha_2' j_2')$. Let us consider the irreducible tensor operator

$$T_\kappa^{(k)} = [\mathbf{T}^{(k_1)} \times \mathbf{T}^{(k_2)}]_\kappa^{(k)} = \sum (k_1\kappa_1 k_2\kappa_2|k_1k_2k\,\kappa)\, T_{\kappa_1}^{(k_1)}\, T_{\kappa_2}^{(k_2)} \tag{14.22}$$

and the two states

$$\begin{aligned} \psi_{\alpha_1 j_1 \alpha_2 j_2 JM}(1, 2) &= \sum (j_1m_1j_2m_2|j_1j_2J\,M)\,\psi_{\alpha_1 j_1 m_1 \alpha_2 j_2 m_2}(1, 2) \\ \psi_{\alpha_1' j_1' \alpha_2' j_2' J'M'}(1, 2) &= \sum (j_1'm_1'j_2'm_2'|j_1'j_2'J'M')\,\psi_{\alpha_1' j_1' m_1' \alpha_2' j_2' m_2'}(1, 2). \end{aligned} \tag{14.23}$$

Using the Wigner-Eckart theorem we obtain

$$\langle j_1j_2J\,M|T_\kappa^{(k)}|j_1'j_2'J'M'\rangle = (-1)^{J-M} \begin{pmatrix} J & k & J' \\ -M & \kappa & M' \end{pmatrix} (j_1j_2J||\mathbf{T}^{(k)}||j_1'j_2'J'). \tag{14.24}$$

On the other hand, if we use the explicit expansions (14.22) and (14.23) we obtain

$$\langle \alpha_1 j_1 \alpha_2 j_2 J\, M | T_\kappa^{(k)} |\, \alpha_1' j_1' \alpha_2' j_2' J' M' \rangle$$

$$= \sum_{\substack{m_1, m_2, m_1' m_2' \\ \kappa_1 \kappa_2}} (j_1 m_1 j_2 m_2 | j_1 j_2 J\, M)\, (k_1 \kappa_1 k_2 \kappa_2 | k_1 k_2 k\, \kappa)\, (j_1' m_1' j_2' m_2' | j_1' j_2' J' M')$$

$$(-1)^{j_1 - m_1} \begin{pmatrix} j_1 & k_1 & j_1' \\ -m_1 & \kappa_1 & m_1' \end{pmatrix} (-1)^{j_2 - m_2} \begin{pmatrix} j_2 & k_2 & j_2' \\ -m_2 & \kappa_2 & m_2' \end{pmatrix} (\alpha_1 j_1 || \mathbf{T}^{(k_1)} || \alpha_1' j_1')$$

$$\times\, (\alpha_2 j_2 || \mathbf{T}^{(k_2)} || \alpha_2' j_2'.) \qquad (14.25)$$

Comparing (14.24) and (14.25) and using the orthogonality relations of the Wigner coefficients we obtain an expression which contains products of six V-coefficients. We write it in the form

$$(\alpha_1 j_1 \alpha_2 j_2 J || \mathbf{T}^{(k)} || \alpha_1' j_1' \alpha_2' j_2' J') = \sqrt{(2J+1)(2k+1)(2J'+1)} \begin{Bmatrix} j_1 & j_2 & J \\ j_1' & j_2' & J' \\ k_1 & k_2 & k \end{Bmatrix}$$

$$\times\, (\alpha_1 j_1 || \mathbf{T}^{(k_1)} || \alpha_1' j_1')\, (\alpha_2 j_2 || \mathbf{T}^{(k_2)} || \alpha_2' j_2'). \qquad (14.26)$$

Here we introduced the notation

$$\sqrt{(2J+1)(2k+1)(2J'+1)} \begin{Bmatrix} j_1 & j_2 & J \\ j_1' & j_2' & J' \\ k_1 & k_2 & k \end{Bmatrix}$$

$$= \sqrt{(2J+1)(2k+1)(2J'+1)}\; X(j_1 j_2 J | j_1' j_2' J' | k_1 k_2 k)$$

$$= \sum_{\substack{m_1 m_2 M \\ m_1' m_2' M' \\ \kappa_1 \kappa_2 \kappa}} (-1)^{J-M} \begin{pmatrix} J & k & J' \\ -M & \kappa & M' \end{pmatrix} (j_1 m_1 j_2 m_2 | j_1 j_2 J\, M)\, (k_1 \kappa_1 k_2 \kappa_2 | k_1 k_2 k\, \kappa)$$

$$(14.27)$$

$$\times\, (j_1' m_1' j_2' m_2' | j_1' j_2' J' M')\, (-1)^{j_1 - m_1} \begin{pmatrix} j_1 & k_1 & j_1' \\ -m_1 & \kappa_1 & m_1' \end{pmatrix} (-1)^{j_2 - m_2} \begin{pmatrix} j_2 & k_2 & j_2' \\ -m_2 & \kappa_2 & m_2' \end{pmatrix}.$$

The reduced matrix element of the tensor product of $\mathbf{T}^{(k_1)}$ and $\mathbf{T}^{(k_2)}$ is thus proportional to the product of their reduced matrix elements. The proportionality factor is generally dependent on all angular momenta and degrees of the tensors involved.

We note that this result will hold also if $\mathbf{T}^{(k_1)}$ is an irreducible tensor which operates on the space coordinates of a given particle, and $\mathbf{T}^{(k_2)}$ operates on the spin coordinates of the *same* particle. In this case $\mathbf{j}_1$ will be $\boldsymbol{l}$—the orbital angular momentum, $\mathbf{j}_2$ will be $\mathbf{s}$—the spin angular momentum, and $\mathbf{J}$ will be $\mathbf{j}$ the total angular momentum of the particle considered.

The coefficient

$$X(j_1j_2J|j_1'j_2'J'|k_1k_2k) \equiv \begin{Bmatrix} j_1 & j_2 & J \\ j_1' & j_2' & J' \\ k_1 & k_2 & k \end{Bmatrix},$$

whose explicit expression in terms of Clebsch-Gordan and Wigner coefficients is given by (14.27), is called a *9-j symbol* or an *X-coefficient*. It includes all the *geometric* relations involved in taking matrix elements of a product of two irreducible tensors. It is thus a universal coefficient which is independent of the physical nature of the tensors $\mathbf{T}^{(k_1)}$ and $\mathbf{T}^{(k_2)}$. As in the derivation of the Wigner-Eckart theorem, we find that using well-defined transformation properties for the construction of tensor products, the relation between the corresponding reduced matrix elements becomes a universal one, related to the geometric configurations involved.

The expression (14.27) for the 9-j symbol can be brought into a considerably more symmetrical form by changing all Clebsch-Gordan coefficients into Wigner coefficients, rearranging them slightly and changing some summation indices (e.g., from M to $-M$). Using the symmetry properties of the Wigner coefficients we then obtain

$$\begin{Bmatrix} j_1 & j_2 & J \\ j_1' & j_2' & J' \\ k_1 & k_2 & k \end{Bmatrix} = \sum_{\substack{m_1m_2M \\ m_1'm_2'M' \\ \kappa_1\kappa_2\kappa}} \begin{pmatrix} j_1 & j_2 & J \\ m_1 & m_2 & M \end{pmatrix} \begin{pmatrix} j_1' & j_2' & J' \\ m_1' & m_2' & M' \end{pmatrix} \begin{pmatrix} k_1 & k_2 & k \\ \kappa_1 & \kappa_2 & \kappa \end{pmatrix}$$

$$\times \begin{pmatrix} j_1 & j_1' & k_1 \\ m_1 & m_1' & \kappa_1 \end{pmatrix} \begin{pmatrix} j_2 & j_2' & k_2 \\ m_2 & m_2' & \kappa_2 \end{pmatrix} \begin{pmatrix} J & J' & k \\ M & M' & \kappa \end{pmatrix}. \qquad (14.28)$$

Remembering that a Wigner coefficient $V(j_1j_2j_3|m_1m_2m_3)$ behaves, in a sense, like a third rank (isotropic) tensor with indices m_1 m_2 and m_3, we see that a 9-j symbol is an invariant function built by contracting completely the product of six Wigner coefficients. The upper rows of the six Wigner coefficients are identical with the three rows and the three columns of the 9-j symbol. Thus, each summation index appears once in a Wigner coefficient which is obtained from a row and a second time in a Wigner coefficient obtained from a column of the X-coefficient.

There are some important symmetry and orthogonality properties of the 9-j symbols which we shall now proceed to derive. In the first place it is immediately seen that

a 9-j symbol vanishes identically unless each row and each column satisfies the triangular condition. (14.29)

Since rows and columns play an equivalent role in (14.28) we have

$$\begin{Bmatrix} j_1 & j_2 & J \\ j_1' & j_2' & J' \\ k_1 & k_2 & k \end{Bmatrix} = \begin{Bmatrix} j_1 & j_1' & k_1 \\ j_2 & j_2' & k_2 \\ J & J' & k \end{Bmatrix}. \tag{14.30}$$

Interchanging two rows in a 9-j symbol does not affect three of the coefficients in (14.28), but causes a noncyclic permutation in each of the other three Wigner coefficients. Using (13.57) we therefore obtain

$$\begin{Bmatrix} j_1' & j_2' & J' \\ j_1 & j_2 & J \\ k_1 & k_2 & k \end{Bmatrix} = (-1)^s \begin{Bmatrix} j_1 & j_2 & J \\ j_1' & j_2' & J' \\ k_1 & k_2 & k \end{Bmatrix}, \tag{14.31}$$

$$s = j_1 + j_2 + J + j_1' + j_2' + J' + k_1 + k_2 + k.$$

Thus, an odd permutation of the rows or columns of a 9-j symbol multiplies it by $(-1)^s$, and an even permutation leaves it unchanged. We can also conclude from this that

A 9-j symbol with two equal rows or columns vanishes if s is odd. (14.32)

The 9-j symbol can also be visualized as a transformation matrix. Thus, in (14.24) we actually started from a scheme in which $\mathbf{j}_1$ and $\mathbf{j}_2$ were coupled to $\mathbf{J}$, $\mathbf{j}_1'$ and $\mathbf{j}_2'$ were coupled to $\mathbf{J}'$, and then $\mathbf{J}$ and $\mathbf{J}'$ were coupled to a tensor of degree k in order to obtain the nonvanishing contribution to the integral $\int \psi_{JM}^* T_\kappa^{(k)} \psi_{J'M'}$ (compare the derivation of the Wigner-Eckart theorem). In (14.25) we used another scheme for the coupling of the various tensors: $\mathbf{j}_1$ and $\mathbf{j}_1'$ were first coupled to an irreducible tensor of degree k_1, $\mathbf{j}_2$ and $\mathbf{j}_2'$ were coupled to an irreducible tensor of degree k_2, and $\mathbf{T}^{(k_1)}$ and $\mathbf{T}^{(k_2)}$ were coupled to a tensor of degree k. The 9-j symbol serves to carry out the transformation between these two schemes as indicated by (14.26).

This is an example of a more general problem, i.e., that of the *change of coupling scheme*. Let us take four independent systems with angular momenta $\mathbf{j}_1$, $\mathbf{j}_2$, $\mathbf{j}_3$, and $\mathbf{j}_4$ and consider the construction of a state with a given total angular momentum $\mathbf{J}^2$ and $\mathbf{J}_z$. This problem has generally more than one solution. Using Clebsch-Gordan coefficients we can first couple $\mathbf{j}_1$ and $\mathbf{j}_2$ to $\mathbf{J}_{12}$, then couple $\mathbf{j}_3$ and $\mathbf{j}_4$ to $\mathbf{J}_{34}$, and finally couple $\mathbf{J}_{12}$ and $\mathbf{J}_{34}$ to $\mathbf{J}$. The resulting state will be

$$\psi[j_1 j_2(J_{12}) j_3 j_4(J_{34}) J\, M]$$

$$= \sum_{m_1 m_2 m_3 m_4 M_{12} M_{34}} (j_1 m_1 j_2 m_2 | j_1 j_2 J_{12} M_{12})\, (j_3 m_3 j_4 m_4 | j_3 j_4 J_{34} M_{34}) \tag{14.33}$$

$$\times\, (J_{12} M_{12} J_{34} M_{34} | J_{12} J_{34} J\, M)\, \psi_{j_1 m_1}(1)\, \psi_{j_2 m_2}(2)\, \psi_{j_3 m_3}(3)\, \psi_{j_4 m_4}(4).$$

The angular momenta J_{12} and J_{34} are arbitrary except for the triangular conditions which they have to satisfy in order that (14.33) will not vanish identically. We already see that different choices of J_{12} and J_{34}, though they lead to the same J and M, will generally result in different functions $\psi[j_1 j_2(J_{12}) j_3 j_4(J_{34}) J\, M]$. In fact it is easily seen that

$$\int \psi^*[j_1 j_2(J_{12}) j_3 j_4(J_{34}) J\, M]\, \psi[j_1 j_2(J'_{12}) j_3 j_4(J'_{34}) J\, M] = \delta_{J_{12} J'_{12}}\, \delta_{J_{34} J'_{34}}.$$

Moreover, we know that by taking all the possible values of J_{12} (i.e., $J_{12} = |j_1 - j_2|,\ |j_1 - j_2| + 1,\ \ldots,\ j_1 + j_2$) we exhaust all the possible $(2j_1 + 1)\,(2j_2 + 1)$ states for the pair 1, 2, and similarly for the pair 3,4. Thus, by taking all the values of J_{12} and J_{34} which are compatible with the triangular conditions in (14.33) we obtain a complete basis for the description of the four systems with j_1, j_2, j_3, and j_4 coupled to a total angular momentum J.

We can construct states of total angular momentum J also by other prescriptions. We can first couple $\mathbf{j}_1$ and $\mathbf{j}_3$ to $\mathbf{J}_{13}$, $\mathbf{j}_2$ and $\mathbf{j}_4$ to $\mathbf{J}_{24}$, and then couple $\mathbf{J}_{13}$ and $\mathbf{J}_{24}$ to $\mathbf{J}$. The resulting wave function will be

$$\psi[j_1 j_3(J_{13}) j_2 j_4(J_{24}) J\, M]$$

$$= \sum (j_1 m_1 j_3 m_3 | j_1 j_3 J_{13} M_{13})\, (j_2 m_2 j_4 m_4 | j_2 j_4 J_{24} M_{24})\, (J_{13} M_{13} J_{24} M_{24} | J_{13} J_{24} J\, M)$$

$$\times\, \psi_{j_1 m_1}(1)\, \psi_{j_2 m_2}(2)\, \psi_{j_3 m_3}(3)\, \psi_{j_4 m_4}(4). \tag{14.34}$$

However, since the set of functions (14.33) forms a complete basis for the states with a total angular momentum J, built from the states of the four systems, we conclude that a unitary transformation exists so that

$$\psi[j_1 j_3(J_{13}) j_2 j_4(J_{24}) J\, M]$$

$$= \sum_{J_{12}, J_{34}} \langle j_1 j_3(J_{13}) j_2 j_4(J_{24}) J | j_1 j_2(J_{12}) j_3 j_4(J_{34}) J \rangle\, \psi[j_1 j_2(J_{12}) j_3 j_4(J_{34}) J\, M]. \tag{14.35}$$

Equation (14.35) is an expansion of an irreducible tensor of degree J (i.e., $\psi[j_1 j_3(J_{13}) j_2 j_4(J_{24}) J\, M]$) in a sum of other irreducible tensors of the same degree (i.e., $\psi[j_1 j_2(J_{12}) j_3 j_4(J_{34}) J\, M]$). It follows that the transformation matrix, which will be concisely written as $\langle J_{13}J_{24} | J_{12}J_{34}\rangle$, does not depend on M.

We can obtain an expression for the transformation matrix $\langle J_{13}J_{24} | J_{12}J_{34}\rangle$ by direct substitution of (14.33) and (14.34) into (14.35). Using the orthogonality of the functions $\psi_{j_k m_k}(k)$ and introducing Wigner coefficients instead of the Clebsch-Gordan coefficients, we obtain

$$(-1)^{(j_3-j_1-M_{13})+(j_4-j_2-M_{24})+(J_{24}-J_{13}-M)} \begin{pmatrix} j_1 & j_3 & J_{13} \\ m_1 & m_3 & -M_{13} \end{pmatrix} \begin{pmatrix} j_2 & j_4 & J_{24} \\ m_2 & m_4 & -M_{24} \end{pmatrix}$$

$$\times \begin{pmatrix} J_{13} & J_{24} & J \\ M_{13} & M_{24} & -M \end{pmatrix} \sqrt{(2J_{13}+1)(2J_{24}+1)(2J+1)}$$

$$= \sum_{J_{12}, J_{34}} \langle J_{13}J_{24} | J_{12}J_{34}\rangle (-1)^{(j_2-j_1-M_{12})+(j_4-j_3-M_{34})+(J_{34}-J_{12}-M)} \begin{pmatrix} j_1 & j_2 & J_{12} \\ m_1 & m_2 & -M_{12} \end{pmatrix}$$

$$\times \begin{pmatrix} j_3 & j_4 & J_{34} \\ m_3 & m_4 & -M_{34} \end{pmatrix} \begin{pmatrix} J_{12} & J_{34} & J \\ M_{12} & M_{34} & -M \end{pmatrix} \sqrt{(2J_{12}+1)(2J_{34}+1)(2J+1)} \qquad (14.36)$$

Using the orthogonality relations of the V-coefficients we obtain, after slight rearrangements,

$$\langle J_{13}J_{24} | J_{12}J_{34}\rangle$$
$$= \sqrt{(2J_{13}+1)(2J_{24}+1)(2J_{12}+1)(2J_{34}+1)} \begin{Bmatrix} j_1 & j_2 & J_{12} \\ j_3 & j_4 & J_{34} \\ J_{13} & J_{24} & J \end{Bmatrix}. \qquad (14.37)$$

Therefore (14.35) can be rewritten in the form

$$\psi[j_1 j_3(J_{13}) j_2 j_4(J_{24}) J\, M] = \sum_{J_{12} J_{34}} \sqrt{(2J_{13}+1)(2J_{24}+1)(2J_{12}+1)(2J_{34}+1)}$$

$$\times \begin{Bmatrix} j_1 & j_2 & J_{12} \\ j_3 & j_4 & J_{34} \\ J_{13} & J_{24} & J \end{Bmatrix} \psi[j_1 j_2(J_{12}) j_3 j_4(J_{34}) J\, M]. \qquad (14.38)$$

Thus, we see that the 9-j symbols are the elements of the transformation matrix between two schemes for the coupling of four angular momenta to a given total angular momentum. Although, as we shall see later, questions of different coupling schemes and the transformations between them occur already for the addition of three angular momenta, we have preferred to treat the case of four angular momenta first. This we did since this problem has a higher degree of symmetry and compactness. The important case of three angular momenta will be treated in detail below, as a special case.

The fact that the 9-j symbols define a unitary transformation gives rise to important relations between them. The unitarity of the transformation matrix can be written as follows:

$$\sum_{J_{13}J_{24}} \langle J_{12}J_{34} | J_{13}J_{24} \rangle \langle J_{13}J_{24} | J'_{12}J'_{34} \rangle = \delta_{J_{12}J'_{12}} \delta_{J_{34}J'_{34}}.$$

Using (14.37) we can express this relation in terms of the 9-j symbols and obtain the orthogonality theorem

$$\sum_{J_{13}J_{24}} (2J_{13}+1)(2J_{24}+1) \begin{Bmatrix} j_1 & j_2 & J_{12} \\ j_3 & j_4 & J_{34} \\ J_{13} & J_{24} & J \end{Bmatrix} \begin{Bmatrix} j_1 & j_2 & J'_{12} \\ j_3 & j_4 & J'_{34} \\ J_{13} & J_{24} & J \end{Bmatrix} = \frac{\delta_{J_{12}J'_{12}} \delta_{J_{34}J'_{34}}}{(2J_{12}+1)(2J_{34}+1)}. \tag{14.39}$$

The product of two transformations of coupling schemes is again such a transformation, namely

$$\sum_{J_{13}J_{24}} \langle J_{12}J_{34} | J_{13}J_{24} \rangle \langle J_{13}J_{24} | J_{14}J_{23} \rangle = \langle J_{12}J_{34} | J_{14}J_{23} \rangle.$$

From this we obtain

$$\sum_{J_{13}J_{24}} (2J_{13}+1)(2J_{24}+1) \begin{Bmatrix} j_1 & j_3 & J_{13} \\ j_2 & j_4 & J_{24} \\ J_{12} & J_{34} & J \end{Bmatrix} \begin{Bmatrix} j_1 & j_4 & J_{14} \\ j_3 & j_2 & J_{23} \\ J_{13} & J_{24} & J \end{Bmatrix} (-1)^{(j_2+j_4+J_{24})+(j_2+j_3+J_{23})}$$

$$= \begin{Bmatrix} j_1 & j_4 & J_{14} \\ j_2 & j_3 & J_{23} \\ J_{12} & J_{34} & J \end{Bmatrix} (-1)^{(j_3+j_4+J_{34})}.$$

Here we have to introduce the extra phases since in $\langle J_{13}J_{24} | J_{14}J_{23} \rangle$ for instance, systems 2 and 3 *in this order* are coupled to J_{23}, whereas in $X(j_1j_4J_{14} | j_3j_2J_{23} | J_{13}J_{24}J)$ they are coupled in the reversed order: j_3 and j_2 to J_{23}. The change in the corresponding V coefficient thus introduces the phases $(-1)^{j_2+j_3+j_{23}}$, and similarly for the other phases. Rearranging slightly we therefore obtain

$$\sum_{J_{13}J_{24}} (-1)^{2j_2+J_{23}+J_{24}-J_{34}} (2J_{13}+1)(2J_{24}+1) \begin{Bmatrix} j_1 & j_3 & J_{13} \\ j_2 & j_4 & J_{24} \\ J_{12} & J_{34} & J \end{Bmatrix} \begin{Bmatrix} j_1 & j_4 & J_{14} \\ j_3 & j_2 & J_{23} \\ J_{13} & J_{24} & J \end{Bmatrix}$$

$$= \begin{Bmatrix} j_1 & j_4 & J_{14} \\ j_2 & j_3 & J_{23} \\ J_{12} & J_{34} & J \end{Bmatrix}. \tag{14.40}$$

Using (14.36) we can derive further useful relations involving both

9-j symbols and V-coefficients. Thus, if we multiply both sides of (14.36) by

$$V(j_1 j_2 J'_{12} | m_1 m_2 - M'_{12}) \, V(j_3 j_4 J'_{34} | m_3 m_4 - M'_{34})$$

and sum over m_1, m_2, m_3 and m_4 only, we obtain, after slight rearrangements and taking into account also (14.37)

$$\begin{Bmatrix} j_1 & j_2 & J_{12} \\ j_3 & j_4 & J_{34} \\ J_{13} & J_{24} & J \end{Bmatrix} \begin{pmatrix} J_{12} & J_{34} & J \\ M_{12} & M_{34} & M \end{pmatrix}$$

$$= \sum_{m_1 m_2 m_3 m_4} \begin{pmatrix} j_1 & j_2 & J_{12} \\ m_1 & m_2 & M_{12} \end{pmatrix} \begin{pmatrix} j_3 & j_4 & J_{34} \\ m_3 & m_4 & M_{34} \end{pmatrix} \begin{pmatrix} j_1 & j_3 & J_{13} \\ m_1 & m_3 & M_{13} \end{pmatrix} \begin{pmatrix} j_2 & j_4 & J_{24} \\ m_2 & m_4 & M_{24} \end{pmatrix} \begin{pmatrix} J_{13} & J_{24} & J \\ M_{13} & M_{24} & M \end{pmatrix}. \tag{14.41}$$

Multiplying both sides by $(2J + 1) \, V(J_{13} J_{24} J | M_{13} M_{24} M)$ and summing over J we obtain

$$\sum_{J,M} (2J+1) \begin{pmatrix} J_{12} & J_{34} & J \\ M_{12} & M_{34} & M \end{pmatrix} \begin{pmatrix} J_{13} & J_{24} & J \\ M_{13} & M_{24} & M \end{pmatrix} \begin{Bmatrix} j_1 & j_2 & J_{12} \\ j_3 & j_4 & J_{34} \\ J_{13} & J_{24} & J \end{Bmatrix}$$

$$= \sum_{m_1 m_2 m_3 m_4} \begin{pmatrix} j_1 & j_2 & J_{12} \\ m_1 & m_2 & M_{12} \end{pmatrix} \begin{pmatrix} j_3 & j_4 & J_{34} \\ m_3 & m_4 & M_{34} \end{pmatrix} \begin{pmatrix} j_1 & j_3 & J_{13} \\ m_1 & m_3 & M_{13} \end{pmatrix} \begin{pmatrix} j_2 & j_4 & J_{24} \\ m_2 & m_4 & M_{24} \end{pmatrix}. \tag{14.42}$$

By a similar procedure more V-coefficients can be transferred from the right-hand side to the left-hand side of (14.42), thus yielding further identities.

The introduction of the various symbols and coefficients enables us to separate the geometrical aspects of various problems from their physical contents. The calculations then become considerably more transparent and clearly exhibit the full symmetry of the problem. In addition, the various relations satisfied by these coefficients enable a great saving in actual numerical calculations.

15. Racah Coefficients

We saw in Section 14 how to form new irreducible tensors from the products of any two irreducible tensors, and derived the relations between the corresponding reduced matrix elements. By far the most important case of these products is that of the *scalar product* of two tensors $\mathbf{T}^{(k)}$ and $\mathbf{U}^{(k)}$ with integral value k. It is convenient to define a scalar product not by putting $k_3 = \kappa_3 = 0$ in (14.1), but rather by an expression proportional to it, namely

$$(\mathbf{T}^{(k)} \cdot \mathbf{U}^{(k)}) = (-1)^k \sqrt{2k+1}\, [\mathbf{T}^{(k)} \times \mathbf{U}^{(k)}]^{(0)}_0. \tag{15.1}$$

We shall stress the different definition by using the ordinary notation of a scalar product of two vectors [which is a special case of (15.1) with $k = 1$]. Using (14.1) we obtain as an expression for the scalar product of two irreducible tensors of degree k

$$(\mathbf{T}^{(k)} \cdot \mathbf{U}^{(k)}) = (-1)^k \sum (k\,\kappa_1\,k\,\kappa_2|k\,k\,0\,0)\, \mathbf{T}^{(k)}_{\kappa_1} \mathbf{U}^{(k)}_{\kappa_2} \sqrt{2k+1}$$

$$= \sum (-1)^\kappa\, T^{(k)}_\kappa\, U^{(k)}_{-\kappa}\,. \tag{15.2}$$

The various relations which we derived in the previous section on the matrix elements of tensor products hold, of course, also in the special case of the scalar product of two tensors. The fact, however, that the combined tensor has the degree zero enables us to simplify some of the relations and to abbreviate some of the notations. Also historically the tensor algebra was developed by Racah starting from the scalar product of tensors rather than from the more general tensor product. Therefore the symbols and coefficients introduced were chosen to fit best this special case rather than the more general one. We shall proceed now to study this case in more detail.

Using (14.26) we can write the reduced matrix element of the scalar product of two tensors $\mathbf{T}^{(k)}(1)$ and $\mathbf{U}^{(k)}(2)$ operating on two independent sets of coordinates 1 and 2 as

$$(j_1 j_2 J||\mathbf{T}^{(k)}(1) \cdot \mathbf{U}^{(k)}(2)||j_1' j_2' J') = (-1)^k \sqrt{(2J+1)(2J'+1)(2k+1)} \begin{Bmatrix} j_1 & j_2 & J \\ j_1' & j_2' & J' \\ k & k & 0 \end{Bmatrix}$$

$$\times\, (j_1||\mathbf{T}^{(k)}||j_1')\,(j_2||\mathbf{U}^{(k)}||j_2').$$

The 9-j symbol $X(j_1j_2J|j_1'j_2'J'|kk0)$ vanishes unless $J=J'$. This reflects the fact that $(\mathbf{T}^{(k)}(1)\cdot\mathbf{U}^{(k)}(2))$, being a scalar, cannot have nonvanishing matrix elements between two states with different values of J. This 9-j symbol therefore depends on six independent numbers only, namely, j_1j_2, $j_1'j_2'$ k, and J. We shall therefore define the *6-j symbol* by

$$W(j_1j_2J|j_2'j_1'k) \equiv \begin{Bmatrix} j_1 & j_2 & J \\ j_2' & j_1' & k \end{Bmatrix} = (-1)^{j_2+J+j_1'+k}\sqrt{(2J+1)(2k+1)}\begin{Bmatrix} j_1 & j_2 & J \\ j_1' & j_2' & J \\ k & k & 0 \end{Bmatrix} \quad (15.3)$$

where the special phase and normalization factor were chosen for reasons which will become clear later. This 6-j symbol, originally introduced by Racah and independently by Wigner, is more commonly called a *Racah coefficient* or a W-coefficient. Using the definition (15.3), we can write the reduced matrix elements of the scalar product $(\mathbf{T}^{(k)}\cdot\mathbf{U}^{(k)})$ as

$$(j_1j_2J||\mathbf{T}^{(k)}(1)\cdot\mathbf{U}^{(k)}(2)||j_1'j_2'J')$$

$$= (-1)^{j_2+J+j_1'}\sqrt{2J+1}\begin{Bmatrix} j_1 & j_2 & J \\ j_2' & j_1' & k \end{Bmatrix}(j_1||\mathbf{T}^{(k)}||j_1')(j_2||\mathbf{U}^{(k)}||j_2')\,\delta_{JJ'}. \quad (15.4)$$

Since the matrix elements of a scalar operator do not depend on the values of J_z we can conveniently use the Wigner-Eckart theorem and obtain for the ordinary (not reduced) matrix elements of $(\mathbf{T}^{(k)}\cdot\mathbf{U}^{(k)})$

$$\langle j_1j_2J\,M|(\mathbf{T}^{(k)}(1)\cdot\mathbf{U}^{(k)}(2))|j_1'j_2'J'M'\rangle$$

$$= (-1)^{J-M}\begin{pmatrix} J & 0 & J' \\ -M & 0 & M' \end{pmatrix}(j_1j_2J||(\mathbf{T}^{(k)}(1)\cdot\mathbf{U}^{(k)}(2))||j_1'j_2'J')$$

$$= (-1)^{j_1'+j_2+J}\begin{Bmatrix} j_1 & j_2 & J \\ j_2' & j_1' & k \end{Bmatrix}(j_1||\mathbf{T}^{(k)}||j_1')(j_2||\mathbf{U}^{(k)}||j_2')\,\delta_{JJ'}\,\delta_{MM'}. \quad (15.5)$$

Equation (15.5) often serves as the starting point for the definition of the Racah coefficients. It is one of the most fundamental relations in tensor algebra and is used very often in various applications. For instance, we shall see that any two-body interaction can be decomposed into a sum of scalar products of irreducible tensors. Equation (15.5) will then determine the matrix elements of the two-body interaction and will therefore be of frequent use in the determination of nuclear spectra.

Although all the symmetry properties and orthogonality relations of the Racah coefficients can be derived from those of the 9-j symbols, it will be convenient to write them down explicitly since we shall have to make very frequent reference to them.

First we note that a Racah coefficient $W(j_1j_2j_3|l_1l_2l_3)$ wanishes unless each of the following four triads satisfy the triangular condition

$$(j_1j_2j_3),\ (j_1l_2l_3),\ (l_1j_2l_3),\ (l_1l_2j_3). \tag{15.6}$$

Furthermore, from (14.31) and (15.3) it follows that

$$\begin{Bmatrix} j_1 & j_2 & j_3 \\ l_1 & l_2 & l_3 \end{Bmatrix} = (-1)^{j_2+j_3+l_2+l_3}\sqrt{(2l_3+1)(2j_3+1)}\begin{Bmatrix} j_1 & j_2 & j_3 \\ l_2 & l_1 & j_3 \\ l_3 & l_3 & 0 \end{Bmatrix}$$

$$= (-1)^{j_1+j_3+l_1+l_3}\sqrt{(2l_3+1)(2j_3+1)}\begin{Bmatrix} j_2 & j_1 & j_3 \\ l_1 & l_2 & j_3 \\ l_3 & l_3 & 0 \end{Bmatrix} = \begin{Bmatrix} j_2 & j_1 & j_3 \\ l_2 & l_1 & l_3 \end{Bmatrix} \tag{15.7}$$

where we have used the fact that $(-1)^{2j_2+2l_2+2j_3+2l_3} = +1$. The relation (15.7) explains why the special phase in the definition (15.3) of the 6-j symbol was chosen. It is only this phase which allows the 6-j symbol to have the simple symmetry property (15.7) and similar other properties. The coefficients originally defined by Racah are related to the 6-j symbol by

$$W(j_1j_2l_2l_1;j_3l_3) = (-1)^{j_1+j_2+l_1+l_2}\begin{Bmatrix} j_1 & j_2 & j_3 \\ l_1 & l_2 & l_3 \end{Bmatrix} \tag{15.8}$$

[note the difference between $W(j_1j_2l_2l_1;\,j_3l_3)$ and $W(j_1j_2j_3|l_1l_2l_3)$]. Its symmetry properties involve different phases for different permutations. It is therefore more convenient to work with the modified Racah coefficient as defined by (15.3). By permuting proper rows and columns in the 9-j symbol in (15.3) it can be easily shown, like in (15.7), that

$$\begin{Bmatrix} j_1 & j_2 & j_3 \\ l_1 & l_2 & l_3 \end{Bmatrix} = \begin{Bmatrix} j_2 & j_1 & j_3 \\ l_2 & l_1 & l_3 \end{Bmatrix} = \begin{Bmatrix} j_1 & j_3 & j_2 \\ l_1 & l_3 & l_2 \end{Bmatrix} = \begin{Bmatrix} l_1 & l_2 & j_3 \\ j_1 & j_2 & l_3 \end{Bmatrix} = \begin{Bmatrix} j_1 & l_2 & l_3 \\ l_1 & j_2 & j_3 \end{Bmatrix}. \tag{15.9}$$

Thus any permutation of the columns of the 6-j symbol does not affect its value. Also any two numbers from the upper row can be interchanged with the corresponding numbers in the lower row without affecting the value of the 6-j symbol.

The orthogonality relations of the 9-j symbols can be used to obtain orthogonality relations for the Racah coefficients. Thus, from (14.39) it follows that

$$\sum_j (2j+1)\begin{Bmatrix} j_1 & j_2 & j \\ j_3 & j_4 & j' \end{Bmatrix}\begin{Bmatrix} j_1 & j_2 & j \\ j_3 & j_4 & j'' \end{Bmatrix}$$

$$= \sum_j (-1)^{2j_2+2j_4+2j+j'+j''}(2j+1)(2j+1)\sqrt{(2j'+1)(2j''+1)}$$

$$\times \begin{Bmatrix} j_1 & j_2 & j \\ j_4 & j_3 & j \\ j' & j' & 0 \end{Bmatrix}\begin{Bmatrix} j_1 & j_2 & j \\ j_4 & j_3 & j \\ j'' & j'' & 0 \end{Bmatrix} = \frac{\delta_{j'j''}}{2j'+1}. \tag{15.10}$$

This relation also explains why the special normalization (15.3) was chosen. Using the relation (14.40) we obtain in a similar way

$$\sum_{j} (-1)^{j+j'+j''} (2j+1) \begin{Bmatrix} j_1 j_2 j' \\ j_3 j_4 j \end{Bmatrix} \begin{Bmatrix} j_1 j_3 j'' \\ j_2 j_4 j \end{Bmatrix} = \begin{Bmatrix} j_1 j_2 j' \\ j_4 j_3 j'' \end{Bmatrix}. \qquad (15.11)$$

We can also use (14.28) to obtain an expression for the Racah coefficient in terms of Wigner coefficients. Using

$$V(j\,j\,0|m\,m'0) = \frac{(-1)^{j-m}}{\sqrt{2j+1}}\,\delta_{m,-m'}$$

we obtain

$$\begin{Bmatrix} j_1 j_2 j_3 \\ l_1\, l_2\, l_3 \end{Bmatrix} = \sum_{m_1 m_2 m_3, m_1' m_2' m_3'} (-1)^{j_1+j_2+j_3+l_1+l_2+l_3+m_1+m_2+m_3+m_1'+m_2'+m_3'}$$

$$\times \begin{pmatrix} j_1 & j_2 & j_3 \\ m_1 & m_2 & m_3 \end{pmatrix} \begin{pmatrix} j_1 & l_2 & l_3 \\ -m_1 & m_2' & -m_3' \end{pmatrix} \begin{pmatrix} l_1 & j_2 & l_3 \\ -m_1' & -m_2 & m_3' \end{pmatrix} \begin{pmatrix} l_1 & l_2 & j_3 \\ m_1' & -m_2' & -m_3 \end{pmatrix}. \qquad (15.12)$$

As in the case of the 9-j symbols, (15.12) can be written in several equivalent forms in which one or more Wigner coefficients are shifted from the right-hand side to the left-hand side. Thus, from (14.41) and (14.42) we can readily obtain

$$\begin{pmatrix} j_1 & j_2 & j_3 \\ m_1 & m_2 & m_3 \end{pmatrix} \begin{Bmatrix} j_1 j_2 j_3 \\ l_1\, l_2\, l_3 \end{Bmatrix} \qquad (15.13)$$

$$= \sum_{m_1' m_2' m_3'} (-1)^{l_1+l_2+l_3+m_1'+m_2'+m_3'} \begin{pmatrix} j_1 & l_2 & l_3 \\ m_1 & m_2' & -m_3' \end{pmatrix} \begin{pmatrix} l_1 & j_2 & l_3 \\ -m_1' & m_2 & m_3' \end{pmatrix} \begin{pmatrix} l_1 & l_2 & j_3 \\ m_1' & -m_2' & m_3 \end{pmatrix}$$

and

$$\sum_{m_3} \begin{pmatrix} j_1 & j_2 & j_3 \\ m_1 & m_2 & m_3 \end{pmatrix} \begin{pmatrix} l_1 & l_2 & j_3 \\ m_1' & m_2' & -m_3 \end{pmatrix}$$

$$= \sum_{l_3 m_3'} (-1)^{l_3+j_3+m_1+m_1'} \begin{pmatrix} j_1 & l_2 & l_3 \\ m_1 & m_2' & m_3' \end{pmatrix} \begin{pmatrix} l_1 & j_2 & l_3 \\ m_1' & m_2 & -m_3' \end{pmatrix} \begin{Bmatrix} j_1 j_2 j_3 \\ l_1\, l_2\, l_3 \end{Bmatrix}. \qquad (15.14)$$

There are several other useful identities satisfied by the Racah coefficients, which can be deduced directly from the previous identities. Thus, if in (15.12) we take $j_3 = 0$ we obtain

$$\begin{Bmatrix} j_1 j_2\, 0 \\ l_1\, l_2\, l_3 \end{Bmatrix} = \sum_{m,m'} (-1)^{j_1+j_2+l_1+l_2+l_3+m_1'+m_2'} \begin{pmatrix} j_1 & j_2 & 0 \\ m_1 & m_2 & 0 \end{pmatrix} \begin{pmatrix} j_1 & l_2 & l_3 \\ -m_1 & m_2' & -m_3' \end{pmatrix}$$

$$\times \begin{pmatrix} l_1 & j_2 & l_3 \\ -m_1' & -m_2 & m_3' \end{pmatrix} \begin{pmatrix} l_1 & l_2 & 0 \\ m_1' & -m_2' & 0 \end{pmatrix}$$

$$= (-1)^{l_3+l_1+j_1} \sum \begin{pmatrix} j_1 & l_1 & l_3 \\ -m_1 & m_1' & -m_3' \end{pmatrix} \begin{pmatrix} l_1 & j_1 & l_3 \\ -m_1' & m_1 & m_3' \end{pmatrix} \frac{\delta_{j_1 j_2}\delta_{l_1 l_2}}{\sqrt{(2j_1+1)(2l_1+1)}}.$$

Hence, taking into account the symmetry properties and the orthogonality of the V-coefficients, we obtain

$$\begin{Bmatrix} j_1 j_1' \, 0 \\ j_2 j_2' j_3 \end{Bmatrix} = \frac{(-1)^{j_1+j_2+j_3}}{\sqrt{(2j_1+1)(2j_2+1)}} \delta_{j_1 j_1'} \delta_{j_2 j_2'}. \tag{15.15}$$

This result can be obtained directly from the definition of the Racah coefficients in terms of the elements of a transformation matrix which will be considered later. Putting in (15.30) below $j_3 = 0$, the matrix element becomes unity and thus (15.15) follows. Using (15.15) and putting $j'' = 0$ in (15.10) we obtain a very useful relation

$$\sum_j (-1)^{j_1+j_2+j} (2j+1) \begin{Bmatrix} j_1 j_1 j' \\ j_2 j_2 j \end{Bmatrix} = \sqrt{(2j_1+1)(2j_2+1)}\, \delta_{j'0}. \tag{15.16}$$

Inserting $j'' = 0$ in (15.11) and using (15.15) results in

$$\sum_j (2j+1) \begin{Bmatrix} j_1 j_2 j \\ j_1 j_2 j' \end{Bmatrix} = (-1)^{2(j_1+j_2)}. \tag{15.17}$$

As could be expected, there are several identities satisfied by combinations of Racah coefficients and 9-j symbols. To derive them we can start from (14.42) and use (15.14). Thus,

$$\sum_{m_1 m_2 m_3 m_4} \begin{pmatrix} j_1 & j_2 & J_{12} \\ m_1 & m_2 & M_{12} \end{pmatrix} \begin{pmatrix} j_3 & j_4 & J_{34} \\ m_3 & m_4 & M_{34} \end{pmatrix} \begin{pmatrix} j_1 & j_3 & J_{13} \\ m_1 & m_3 & -M_{13} \end{pmatrix} \begin{pmatrix} j_2 & j_4 & J_{24} \\ m_2 & m_4 & -M_{24} \end{pmatrix}$$

$$= \sum_{J,M} (2J+1) \begin{pmatrix} J_{12} & J_{34} & J \\ M_{12} & M_{34} & M \end{pmatrix} \begin{pmatrix} J_{13} & J_{24} & J \\ -M_{13} & -M_{24} & M \end{pmatrix} \begin{Bmatrix} j_1 & j_2 & J_{12} \\ j_3 & j_4 & J_{34} \\ J_{13} & J_{24} & J \end{Bmatrix}$$

$$= \sum_{J,J''M''} (2J+1)(-1)^{(J+J''+M_{12}+M_{13})+(J_{13}+J_{24}+J)} \begin{pmatrix} J_{12} & J_{24} & J'' \\ M_{12} & M_{24} & M'' \end{pmatrix}$$

$$\times \begin{pmatrix} J_{13} & J_{34} & J'' \\ M_{13} & M_{34} & -M'' \end{pmatrix} \begin{Bmatrix} J_{12} & J_{34} & J \\ J_{13} & J_{24} & J'' \end{Bmatrix} \begin{Bmatrix} j_1 & j_2 & J_{12} \\ j_3 & j_4 & J_{34} \\ J_{13} & J_{24} & J \end{Bmatrix}.$$

If we multiply now both sides by

$$\begin{pmatrix} J_{12} & J_{24} & J' \\ M_{12} & M_{24} & M' \end{pmatrix} \begin{pmatrix} J_{13} & J_{34} & J' \\ M_{13} & M_{34} & -M' \end{pmatrix}$$

and sum over M_{12} M_{24} M_{13}, and M_{34}, we obtain, by using (15.13) and rearranging slightly

$$\sum_J (2J+1) \begin{Bmatrix} J_{13} & J_{24} & J \\ J_{12} & J_{34} & J' \end{Bmatrix} \begin{Bmatrix} J_{13} & J_{24} & J \\ j_1 & j_2 & J_{12} \\ j_3 & j_4 & J_{34} \end{Bmatrix} = (-1)^{2J'} \begin{Bmatrix} j_1 & j_2 & J_{12} \\ J_{24} & J' & j_4 \end{Bmatrix} \begin{Bmatrix} j_3 & j_4 & J_{34} \\ J' & J_{13} & j_1 \end{Bmatrix}. \tag{15.18}$$

We can now use the orthogonality of the Racah coefficients (15.10) to express the 9-j symbol as a sum over products of three Racah coefficients. Multiplying both sides of (15.18) by $(2J'+1)W(J_{13}J_{24}J_0|J_{12}J_{34}J')$ and summing over J' we obtain

$$\begin{Bmatrix} J_{13} & J_{24} & J_0 \\ j_1 & j_2 & J_{12} \\ j_3 & j_4 & J_{34} \end{Bmatrix} = \sum_{J'} (-1)^{2J'} (2J'+1) \begin{Bmatrix} J_{13} & J_{24} & J_0 \\ J_{12} & J_{34} & J' \end{Bmatrix} \begin{Bmatrix} j_1 & j_2 & J_{12} \\ J_{24} & J' & j_4 \end{Bmatrix} \begin{Bmatrix} j_3 j_4 & J_{34} \\ J' & J_{13} j_1 \end{Bmatrix}. \tag{15.19}$$

The Racah coefficients can be used to derive an expression for the reduced matrix elements of the tensor product of two tensors operating on the coordinates of the *same* system. Consider

$$T^{(k)}_\kappa = [\mathbf{T}^{(k_1)} \times \mathbf{T}^{(k_2)}]^{(k)}_\kappa. \tag{15.20}$$

Using the Wigner-Eckart theorem we have, on the one hand,

$$\langle \alpha\, j\, m | T^{(k)}_\kappa | \alpha' j' m' \rangle = (-1)^{j-m} \begin{pmatrix} j & k & j' \\ -m & \kappa & m' \end{pmatrix} (\alpha j || \mathbf{T}^{(k)} || \alpha' j'). \tag{15.21}$$

On the other hand we have

$$\langle \alpha\, j\, m | T^{(k)}_\kappa | \alpha' j' m' \rangle$$

$$= \langle \alpha\, j\, m | \sum_{\kappa_1 \kappa_2} (k_1 \kappa_1 k_2 \kappa_2 | k_1 k_2 k\, \kappa) T^{(k_1)}_{\kappa_1} T^{(k_2)}_{\kappa_2} | \alpha' j' m' \rangle$$

$$= \sum_{\kappa_1 \kappa_2} (k_1 \kappa_1 k_2 \kappa_2 | k_1 k_2 k\, \kappa) \sum_{\alpha'' j'' m''} \langle \alpha\, j\, m | T^{(k_1)}_{\kappa_1} | \alpha'' j'' m'' \rangle \langle \alpha'' j'' m'' | T^{(k_2)}_{\kappa_2} | \alpha' j' m' \rangle. \tag{15.22}$$

Using again the Wigner-Eckart theorem in (15.22) and comparing with (15.21) we find, after slight rearrangements,

$$(\alpha j || \mathbf{T}^{(k)} || \alpha' j') = \sum_{\alpha'' j''} (\alpha j || \mathbf{T}^{(k_1)} | \alpha'' j'') (\alpha'' j'' || \mathbf{T}^{(k_2)} || \alpha' j') (-1)^{k+j+j'} \sqrt{2k+1}$$

$$\times \Bigg[\sum_{\substack{\kappa_1 \kappa_2 \kappa \\ m m' m''}} (-1)^{k_1+k_2+k+j'+j''+j+m'+m''+m} \begin{pmatrix} k_1 & k_2 & k \\ \kappa_1 & \kappa_2 & \kappa \end{pmatrix} \begin{pmatrix} k_1 & j & j'' \\ -\kappa_1 & m & -m'' \end{pmatrix}$$

$$\times \begin{pmatrix} j' & k_2 & j'' \\ -m' & -\kappa_2 & m'' \end{pmatrix} \begin{pmatrix} j' & j & k \\ m' & -m & -\kappa \end{pmatrix} \Bigg].$$

Comparing the expression in the square brackets with (15.12) we finally obtain for the tensor product of two operators operating on the same coordinates

$$(\alpha j||\mathbf{T}^{(k)}||\alpha' j')$$
$$= (-1)^{j+k+j'} \sqrt{2k+1} \sum_{\alpha'' j''} (\alpha j||\mathbf{T}^{(k_1)}||\alpha'' j'')(\alpha'' j''||\mathbf{T}^{(k_2)}||\alpha' j') \begin{Bmatrix} k_1 & k_2 & k \\ j' & j & j'' \end{Bmatrix}. \tag{15.23}$$

Another important relation involving the Racah coefficient is obtained if we consider the matrix element of an operator $\mathbf{T}^{(k)}(1)$, operating on the variables of system 1, between the states $\psi(j_1 j_2 J M)$ and $\psi(j_1' j_2' J' M')$. Here, as before, j_1 is the angular momentum of system 1, j_2 is that of system 2, and J is their total angular momentum. To derive an expression for $(j_1 j_2 J||\mathbf{T}^{(k)}(1)||j_1' j_2' J')$ we can consider $\mathbf{T}^{(k)}(1)$ as the tensor product of $\mathbf{T}^{(k)}(1)$ and the unit operator 1 operating on the system 2. Since the unit operator is a tensor operator of degree zero we obtain, using (14.26)

$$(j_1 j_2 J||\mathbf{T}^{(k)}(1)||j_1' j_2' J')$$
$$= \sqrt{(2J+1)(2k+1)(2J'+1)} \begin{Bmatrix} j_1 & j_2 & J \\ j_1' & j_2' & J' \\ k & 0 & k \end{Bmatrix} (j_1||\mathbf{T}^{(k)}||j_1')(j_2||1||j_2'). \tag{15.24}$$

Obviously we have $\langle j_2 m_2|1|j_2' m_2\rangle = \delta_{j_2 j_2'}$. Since, however,

$$V(j_2 0\, j_2|m_2 0\, m_2) = (-1)^{j_2 - m_2} \frac{1}{\sqrt{2j_2+1}},$$

we find that

$$(j||1||j') = \sqrt{2j+1}\, \delta_{jj'}. \tag{15.25}$$

Using now (15.7) to reduce the 9-j symbol in (15.24) to the corresponding Racah coefficient we obtain finally

$$(j_1 j_2 J||\mathbf{T}^{(k)}(1)||j_1' j_2' J')$$
$$= (-1)^{j_1+j_2+J'+k} \sqrt{(2J+1)(2J'+1)} \begin{Bmatrix} j_1 & j_1' & k \\ J' & J & j_2 \end{Bmatrix} (j_1||\mathbf{T}^{(k)}(1)||j_1')\delta_{j_2 j_2'}. \tag{15.26}$$

Similarly we obtain for the reduced matrix elements of a tensor $\mathbf{T}^{(k)}(2)$, operating on the coordinates of system 2

$$(j_1 j_2 J||\mathbf{T}^{(k)}(2)||j_1' j_2' J')$$
$$= (-1)^{j_1+j_2'+J+k} \sqrt{(2J+1)(2J'+1)} \begin{Bmatrix} j_2 & j_2' & k \\ J' & J & j_1 \end{Bmatrix} (j_2||\mathbf{T}^{(k)}(2)||j_2')\delta_{j_1 j_1'}. \tag{15.27}$$

We wish to draw attention to the phases appearing in (15.26) and (15.27). They result from our phase conventions regarding the addition of angular momenta.

The Racah coefficients, like the 9-j symbols, are also the elements of a transformation matrix from one coupling scheme to another. Thus, if in (14.38) we put $J = 0$ we obtain

$$\psi[j_1 j_3(J') j_2 j_4(J') 0\, 0]$$

$$= \sum_{J''} (-1)^{j_2+j_3+J'+J''} \sqrt{(2J'+1)(2J''+1)} \begin{Bmatrix} j_1 & j_2 & J'' \\ j_4 & j_3 & J' \end{Bmatrix} \psi[j_1 j_2(J'') j_3 j_4(J'') 0\, 0] \tag{15.28}$$

where, since $J = 0$, we put $J_{13} = J_{24} = J'$, $J_{12} = J_{34} = J''$. Thus, a Racah coefficient, multiplied by a proper factor, is a matrix element of a transformation between two coupling schemes of *four* irreducible tensors which are coupled to a tensor of zero degree. Alternatively it can, of course, be considered as the transformation matrix between two coupling schemes of any *three* irreducible tensors. Inserting $j_4 = 0$ (and therefore $J_{34} = j_3$ and $J_{24} = j_2$) in (14.38) we obtain, using (15.7),

$$\psi[j_1 j_3(J_{13}) j_2 J\, M]$$

$$= \sum_{J_{12}} (-1)^{j_2+j_3+J_{12}+J_{13}} \sqrt{(2J_{12}+1)(2J_{13}+1)} \begin{Bmatrix} j_1 & j_2 & J_{12} \\ J & j_3 & J_{13} \end{Bmatrix} \psi[j_1 j_2(J_{12}) j_3 J\, M]. \tag{15.29}$$

Or, introducing the matrix $\langle j_1 j_3(J_{13}) j_2 J | j_1 j_2(J_{12}) j_3 J \rangle$ for the transformation from the scheme in which j_1 and j_2 are coupled to J_{12}, J_{12} and j_3 being coupled to J, to the scheme in which j_1 and j_3 are coupled to J_{13}, J_{13} and j_2 being coupled to J, we obtain

$$\langle j_1 j_3(J_{13}) j_2 J | j_1 j_2(J_{12}) j_3 J \rangle$$

$$= (-1)^{j_2+j_3+J_{12}+J_{13}} \sqrt{(2J_{12}+1)(2J_{13}+1)} \begin{Bmatrix} j_1 & j_2 & J_{12} \\ J & j_3 & J_{13} \end{Bmatrix} \tag{15.30}$$

Various identities satisfied by the Racah coefficients can be easily derived from their interpretation as transformation matrices. Thus, the orthogonality relations of the Racah coefficients is just an expression of the unitarity of the transformation matrix (15.30). The relation (15.11) demonstrates the fact that the product of two consecutive transformations of the type (15.30) is another such transformation.

A useful relation, known as the *Biedenharn-Elliot identity*, can be derived by considering two different sets of successive transformations from one specific scheme to another specific one. These two schemes are the one in which j_1 and j_2 are coupled to J_{12}, j_3 and j_4 are coupled to J_{34}, J_{12} and J_{34} are coupled to J, and the one in which j_3 and j_1 are coupled to J_{31}, J_{31} and j_2 are coupled to J_{312}, and J_{312} and j_4 are coupled to J. Thus, this transformation can be decomposed according to either one of the two ways

$$
\begin{aligned}
&\langle j_3 j_1(J_{31}) j_2(J_{312}) j_4 J | j_1 j_2(J_{12}) j_3 j_4(J_{34}) J\rangle \\
&= \langle j_3 j_1(J_{31}) j_2(J_{312}) j_4 J | j_1 j_2(J_{12}) j_3(J_{312}) j_4 J\rangle \\
&\times \langle j_1 j_2(J_{12}) j_3(J_{312}) j_4 J | j_1 j_2(J_{12}) j_3 j_4(J_{34}) J\rangle \\
&= \sum_{J_{314}} \langle j_3 j_1(J_{31}) j_2(J_{312}) j_4 J | j_3 j_1(J_{31}) j_4(J_{314}) j_2 J\rangle \\
&\times \langle j_3 j_1(J_{31}) j_4(J_{314}) j_2 J | j_3 j_4(J_{34}) j_1(J_{314}) j_2 J\rangle \\
&\times \langle j_3 j_4(J_{34}) j_1(J_{314}) j_2 J | j_1 j_2(J_{12}) j_3 j_4(J_{34}) J\rangle. \qquad (15.31)
\end{aligned}
$$

Each of the transformation matrices in the decomposition (15.31) is a transformation involving three angular momenta only. It can therefore be expressed with the help of a corresponding Racah coefficient. The identity of the two decompositions then leads to the following identity satisfied by the Racah coefficients

$$
\begin{aligned}
&(-1)^{j_1+j_2+j_1'+j_2'+j_1''+j_2''+j} \begin{Bmatrix} j_1 & j_2 & j \\ j_1'' & j_2'' & k_1 \end{Bmatrix} \begin{Bmatrix} j_1' & j_2' & j \\ j_1'' & j_2'' & k_2 \end{Bmatrix} \\
&= \sum_k (-1)^{k_1+k_2+k}(2k+1) \begin{Bmatrix} j_1 & j_1' & k \\ j_2' & j_2 & j \end{Bmatrix} \begin{Bmatrix} k_1 & k_2 & k \\ j_1' & j_1 & j_2'' \end{Bmatrix} \begin{Bmatrix} k_1 & k_2 & k \\ j_2' & j_2 & j_1'' \end{Bmatrix}. \qquad (15.32)
\end{aligned}
$$

In (15.32) we have given the various angular momenta new names in order to exhibit the symmetry of the relation.

We shall now illustrate the use of the Racah coefficients by an example taken from Part I. There, in (8.29) we calculated the matrix element

$$\langle\sigma\rangle^2 = \frac{1}{2j_i+1} \sum_{\kappa m_i m_f} (-1)^\kappa \langle\sigma_\kappa\rangle \langle\sigma_{-\kappa}\rangle$$

where

$$\langle\sigma_\kappa\rangle = \langle \tfrac{1}{2} l\, j_f m_f | \sigma_\kappa | \tfrac{1}{2} l\, j_i m_i\rangle.$$

Using (15.26) and the Wigner-Eckart theorem (14.15) we obtain

$$
\begin{aligned}
\langle\sigma_\kappa\rangle &= (-1)^{1/2+l+j_i+1} \sqrt{(2j_i+1)(2j_f+1)} \begin{Bmatrix} \frac{1}{2} & \frac{1}{2} & 1 \\ j_i & j_f & l \end{Bmatrix} (\tfrac{1}{2}||\sigma||\tfrac{1}{2}) \\
&\times (-1)^{j_f-m_f} \begin{pmatrix} j_f & 1 & j_i \\ -m_f & \kappa & m_i \end{pmatrix} \\
\langle\sigma_{-\kappa}\rangle &= (-1)^{1/2+l+j_f+1} \sqrt{(2j_i+1)(2j_f+1)} \begin{Bmatrix} \frac{1}{2} & \frac{1}{2} & 1 \\ j_f & j_i & l \end{Bmatrix} (\tfrac{1}{2}||\sigma||\tfrac{1}{2}) \\
&\times (-1)^{j_i-m_i} \begin{pmatrix} j_i & 1 & j_f \\ -m_i & -\kappa & m_f \end{pmatrix}.
\end{aligned}
$$

Hence

$$\langle\sigma\rangle^2 = \sum_{\kappa m_i m_f} (2j_f+1)\begin{Bmatrix}\frac{1}{2} & \frac{1}{2} & 1\\ j_i & j_f & l\end{Bmatrix}^2 (\tfrac{1}{2}||\sigma||\tfrac{1}{2})^2(-1)^{-m_f+\kappa+m_i}$$

$$\times\begin{pmatrix} j_f & 1 & j_i\\ -m_f & \kappa & m_i\end{pmatrix}\begin{pmatrix} j_f & 1 & j_i\\ -m_f & \kappa & m_i\end{pmatrix} = (2j_f+1)\begin{Bmatrix}\frac{1}{2} & \frac{1}{2} & 1\\ j_i & j_f & l\end{Bmatrix}^2 (\tfrac{1}{2}||\sigma||\tfrac{1}{2})^2.$$

Using (14.21) we see that $(\frac{1}{2}||\sigma||\frac{1}{2}) = 2(\frac{1}{2}||\mathbf{s}||\frac{1}{2}) = \sqrt{6}$, and hence

$$\langle\sigma\rangle^2 = 6(2j_f+1)\begin{Bmatrix}\frac{1}{2} & \frac{1}{2} & 1\\ j_i & j_f & l\end{Bmatrix}^2. \tag{15.33}$$

Substitution of the Racah coefficients from the Appendix reproduces Table 9.1 in Part I.

We may note that in many calculations we are required to sum transition probabilities over all final states and average over initial states in order to obtain the total transition probability. If the operator operating between these two states is a component of a tensor, say $T_\kappa^{(k)}$, we have then to calculate

$$P_\kappa = \frac{1}{2j_i+1}\sum_{m_i m_f} |\langle j_i m_i|T_\kappa^{(k)}|j_f m_f\rangle|^2.$$

Using the Wigner-Eckart theorem we obtain then

$$P_\kappa = \frac{1}{2j_i+1}\sum_{m_i m_f}\begin{pmatrix} j_i & k & j_f\\ -m_i & \kappa & m_f\end{pmatrix}^2 |(j_i||\mathbf{T}^{(k)}||j_f)|^2$$

$$= \frac{1}{(2j_i+1)(2k+1)}|(j_i||\mathbf{T}^{(k)}||j_f)|^2. \tag{15.34}$$

In case we have to sum also over all components κ of $T_\kappa^{(k)}$ [i.e., if the transition in question can proceed through any one of the components of $\mathbf{T}^{(k)}$] we obtain

$$P = \sum P_\kappa = \frac{1}{2j_i+1}|(j_i||\mathbf{T}^{(k)}||j_f)|^2. \tag{15.35}$$

This result is very useful in computing transition probabilities and is very often used in the literature.

Calculations with central field wave functions always involve certain combinations of Wigner coefficients. The introduction of special abbreviations—the 6-j and the 9-j symbols—for these special combinations, has greatly facilitated the handling of many problems. Instead of carrying whole sets of Wigner coefficients all through the calculation, it became possible to reduce considerably the complexity of the problem by using the high symmetry and the various orthogonality relations of these coefficients. Relatively complicated expressions can be summed to yield few 6-j or 9-j coefficients. These coefficients can be evaluated numerically once and for all.

The numerical evaluation of the various coefficients can be done directly from their definition in terms of Clebsch-Gordan coefficients. Since, as we shall see below, closed formulas are available for the Clebsch-Gordan coefficients, it is possible to get also closed formulas for the other coefficients. These formulas are very practical for the calculations of individual coefficients. In many cases, when a whole series of coefficients has to be calculated, it is found to be considerably simpler to calculate the coefficients using various recursion relations. Such relations can be derived from (14.40) or (15.11) or other similar relations. We shall not go into the details of these calculations here, since extensive tables of such coefficients already exist. A practical point which can often serve for checking numerical calculations should, however, be mentioned. The various coefficients are, as we shall see, square roots of rational numbers. In a given problem which has to do with angular momenta j_i and tensors of degrees k_i these rational numbers will be formed from products and quotients of j_i, k_i, $2j_i + 1$, $2k_i + 1$, etc. Thus the various coefficients which are involved in a given calculation will generally have many common prime factors. Therefore, if the calculation is carried out with rational fractions, it is very often easy to spot a numerical error. For this reason it is also advantageous to have the coefficients tabulated as rational fractions (or rather as square roots thereof).

The first closed expression for the V-coefficients was given by Wigner himself who used, for this purpose, their relation to the D matrices. Racah obtained a similar formula by solving the recursion relations which these coefficients satisfy. Racah's result can be written in the form

$$\begin{aligned}\begin{pmatrix} j_1 & j_2 & j_3 \\ m_1 & m_2 & m_3 \end{pmatrix} &= \delta(m_1 + m_2 + m_3)\sqrt{\frac{(j_1+j_2-j_3)!\,(j_2+j_3-j_1)!\,(j_3+j_1-j_2)!}{(j_1+j_2+j_3+1)!}} \\ &\times \sqrt{(j_1+m_1)!\,(j_1-m_1)!\,(j_2+m_2)!\,(j_2-m_2)!\,(j_3+m_3)!\,(j_3-m_3)!} \\ &\times \sum_t (-1)^{j_1-j_2-m_3+t} \\ &\times \frac{1}{t!\,(j_1+j_2-j_3-t)!\,(j_3-j_2+m_1+t)!\,(j_3-j_1-m_2+t)!} \\ &\times \frac{1}{(j_1-m_1-t)!\,(j_2+m_2-t)!}\,. \end{aligned} \tag{15.36}$$

The summation over t is intended to extend over all integers with the convention that $(-m)! = \infty$ when $m > 0$. In other words, t actually runs over such integers for which all the expressions appearing in the denominators of (15.36), i.e., $(j_1 + j_2 - j_3 - t)$, $(j_3 - j_2 + m_1 + t)$, etc., are non-negative.

Equation (15.36) becomes especially simple for some special coefficients. Thus, if j_1, j_2, and j_3 are all integers and if $2g = j_1 + j_2 + j_3$ is even, then

$$\begin{pmatrix} j_1 & j_2 & j_3 \\ 0 & 0 & 0 \end{pmatrix} = (-1)^g \sqrt{\frac{(2g-2j_1)!\,(2g-2j_2)!\,(2g-2j_3)!}{(2g+1)!}} \times \frac{g!}{(g-j_1)!\,(g-j_2)!\,(g-j_3)!}. \tag{15.37}$$

We have already seen that $V(j_1 j_2 j_3 | 0\,0\,0) = 0$ if $2g$ is odd.

The explicit dependence of the Wigner coefficients on the various j's and m's given by (15.36) was used by Racah to derive a closed formula for the W-coefficients

$$\begin{aligned} \begin{Bmatrix} j_1 & j_2 & j_3 \\ l_1 & l_2 & l_3 \end{Bmatrix} &= \Delta(j_1 j_2 j_3)\,\Delta(j_1 l_2 l_3)\,\Delta(l_1 j_2 l_3)\,\Delta(l_1 l_2 j_3) \\ &\times \sum_t (-1)^t (t+1)!\,\{[t-(j_1+j_2+j_3)]!\,[t-(j_1+l_2+l_3)]! \\ &\times [t-(l_1+j_2+l_3)]!\,[t-(l_1+l_2+j_3)]!\,(j_1+j_2+l_1+l_2-t)! \\ &\times (j_2+j_3+l_2+l_3-t)!\,(j_3+j_1+l_3+l_1-t)!\}^{-1} \end{aligned} \tag{15.38}$$

where

$$\Delta(a\,b\,c) = \sqrt{\frac{(a+b-c)!\,(b+c-a)!\,(c+a-b)!}{(a+b+c+1)!}}. \tag{15.39}$$

A different, but equivalent expression can be derived as follows*. Let us denote the four triangular sums which appear in a Racah coefficient by

$$2p_1 = j_1 + l_2 + l_3 \quad 2p_2 = l_1 + j_2 + l_3 \quad 2p_3 = l_1 + l_2 + j_3 \quad 2p = j_1 + j_2 + j_3$$

and the corresponding "areas" of the triangles by S_1, S_2, S_3, and S where

$$S(a, b, c) = \sqrt{(a+b+c+1)!\,(a+b-c)!\,(b+c-a)!\,(c+a-b)!}$$

Let us also define

$$q_1 = j_2 + l_2 + j_3 + l_3, \qquad q_2 = j_3 + l_3 + j_1 + l_1, \qquad q_3 = j_1 + l_1 + j_2 + l_2,$$

(*) Y. Lehrer, private communication.

then

$$\begin{Bmatrix} j_1 & j_2 & j_3 \\ l_1 & l_2 & l_3 \end{Bmatrix} = \frac{S_1 S_2 S_3 S}{(2q_1+1)(2q_2+1)(2q_3+1)}$$

$$\times \sum_t (-1)^{t+1} \binom{t}{2p_1+1} \binom{t}{2p_2+1} \binom{t}{2p_3+1} \binom{t}{2p+1}$$

$$\times \binom{2q_1+1}{t} \binom{2q_2+1}{t} \binom{2q_3+1}{t}. \tag{15.40}$$

Both Eqs. (15.38) and (15.40) exhibit the great symmetry of the Racah coefficients, and both can be used to derive further properties of these coefficients.

We conclude this section by considering some classical limits of the quantities we have been dealing with. This will enable us to have a more intuitive picture of some of the relations in which they are involved.

We first consider a wave function ψ_{nlm} of a single particle in a central field. For $m = l$ the angular dependence of this function

$$\psi_{nlm=l}(r\, \theta\, \varphi) \textit{ is proportional to } \sin^l\theta e^{il\varphi}. \tag{15.41}$$

Thus, when $l \to \infty$ the state $\psi_{nlm=l}(r\, \theta\, \varphi)$ describes a particle confined to the plane perpendicular to $\boldsymbol{l}$ [i.e., the plane given by $\theta = (\pi/2)$]. This indeed is always the case for a classical particle moving in a central field, for which $(\mathbf{r} \cdot \boldsymbol{l}) = 0$.

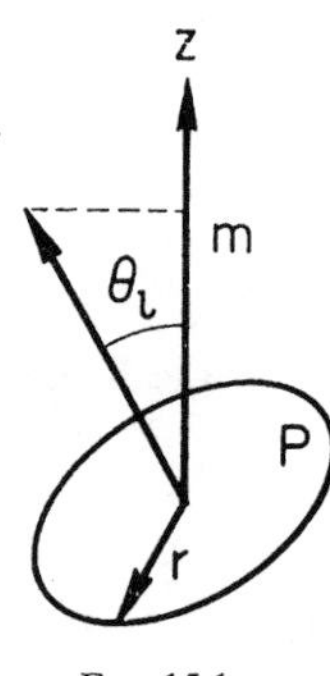

FIG. 15.1

For $m \neq l$ we cannot expect such a simple result to hold since ψ_{nlm} describes a state of a given z-projection of $\boldsymbol{l}$ but undetermined x and y projections of the same vector. Rather, for $m \neq l$, ψ_{nlm} describes a distribution which is averaged over all the values of the x and y projections of $\boldsymbol{l}$. If θ_l and φ_l are the angles defining the direction of the vector $\boldsymbol{l}$, then $\psi_{nlm}(\mathbf{r})$ corresponds to a distribution obtained by keeping θ_l fixed and averaging over φ_l. Since in the classical limit, $l \to \infty$, the particles are confined to a plane perpendicular to $\boldsymbol{l}$, we see that we obtain the classical limit of the distribution described by ψ_{nlm} by taking the distribution of ψ_{nll} in a plane P perpendicular to $\boldsymbol{l}$ and rotating this plane around the z axis keeping a fixed angle between them.

We shall now obtain the classical limit of the coefficients by considering the average value of a known angular function when $l \to \infty$. Take the function $Y_{k0}(\theta\, \varphi)$ and let us consider its expectation value in the state ψ_{nlm}. Since for $l \to \infty$ we know that the density distribution of the particle is predominantly in the plane perpendicular to the direction of $\mathbf{l}$, it will be convenient to refer also the function $Y_{k0}(\theta\,\varphi)$ to this direction. Denoting the angular coordinates of the point $\mathbf{r}$ with respect to $\mathbf{l}$ by θ_0 and φ_0, we obtain, using the addition theorem of spherical harmonics (14.5),

$$Y_{k0}(\theta\,\varphi) = \sqrt{\frac{2k+1}{4\pi}}\,P_k(\cos\theta) = \sqrt{\frac{4\pi}{2k+1}}\sum Y^*_{k\kappa}(\theta_0\,\varphi_0)\,Y_{k\kappa}(\theta_l\,\varphi_l).$$

For $l \to \infty$ we evaluate this expectation value with $\psi_{nll}(\theta_0\,\varphi_0)$, keeping θ_l constant and averaging over φ_l. Thus, the only contribution comes from $\theta_0 = (\pi/2)$ and therefore

$$\begin{aligned}\langle lm|Y_k(\theta\,\varphi)|lm\rangle &\approx \sqrt{\frac{4\pi}{2k+1}}\,\langle Y^*_{k0}(\pi/2,\varphi_0)\rangle_{\varphi_0}\,\langle Y_{k0}(\theta_l\,\varphi_l)\rangle_{\varphi_l}\\ &= \sqrt{\frac{2k+1}{4\pi}}\,P_k(0)\,P_k(\cos\theta_l).\end{aligned} \tag{15.42}$$

Substituting for θ_l its approximate value, $\cos\theta_l = (m/l)$, and using the Wigner-Eckart theorem, we obtain

$$(-1)^{l-m}(l||\mathbf{Y}_k||l)\begin{pmatrix} l & k & l\\ -m & 0 & m\end{pmatrix} \approx \sqrt{\frac{2k+1}{4\pi}}\,P_k(0)\,P_k\left(\frac{m}{l}\right), \qquad l\to\infty. \tag{15.43}$$

For the particular value $m = l$ we see from (15.36) that

$$\begin{pmatrix} l & k & l\\ -l & 0 & l\end{pmatrix} \approx \sqrt{\frac{1}{2l+1}} \qquad \text{for} \qquad l\to\infty\ (k\ \text{finite}). \tag{15.44}$$

We now divide (15.43) by its value for $m = l$ and obtain for the classical limit of the Wigner coefficient, if $P_k(0) \neq 0$, i.e., if k is even,

$$\begin{pmatrix} l & k & l\\ -m & 0 & m\end{pmatrix} \approx \frac{(-1)^{l-m}}{\sqrt{2l+1}}\,P_k\left(\frac{m}{l}\right), \qquad l\to\infty. \tag{15.45}$$

Thus, the classical limit of $V(l\,k\,l|-m\,0\,m)$ is proportional to the Legendre polynomial P_k whose argument is $\cos\theta = (m/l)$.

We can use these results to obtain also the classical limit of some special Racah coefficients. Consider the expression

$$\begin{aligned}&\langle l_1l_2L\,M|P_k(\cos\omega_{12})|l_1l_2L\,M\rangle\\ &= \frac{4\pi}{2k+1}\,\langle l_1l_2L\,M|(\mathbf{Y}_k(1)\cdot\mathbf{Y}_k(2))|l_1l_2L\,M\rangle\\ &= \frac{4\pi}{2k+1}(-1)^{l_1+l_2-L}(l_1||\mathbf{Y}_k||l_1)\,(l_2||\mathbf{Y}_k||l_2)\begin{Bmatrix} l_1 & l_2 & L\\ l_2 & l_1 & k\end{Bmatrix}\end{aligned} \tag{15.46}$$

where ω_{12} is the angle between the directions of $\mathbf{r}_1$ and $\mathbf{r}_2$. It is clear that when both l_1 and l_2 are large, particles 1 and 2 will be concentrated in the planes perpendicular to $\boldsymbol{l}_1$ and $\boldsymbol{l}_2$ respectively. The average value of $P_k(\cos\omega_{12})$ can then be conveniently calculated by taking the z-axis along $\boldsymbol{l}_1$. We obtain then, using (15.42),

$$\langle l_1 l_2 L\, M | P_k(\cos\omega_{12}) | l_1 l_2 L\, M\rangle$$

$$= \frac{4\pi}{2k+1} \langle l_1 l_2 L\, M | (\mathbf{Y}_k(1)\cdot\mathbf{Y}_k(2)) | l_1 l_2 L\, M\rangle$$

$$= \frac{4\pi}{2k+1} Y_{k0}\left(\frac{\pi}{2}, 0\right) \overline{Y_{k0}(\theta_2\,\varphi_2)}$$

$$= \frac{4\pi}{2k+1} Y_{k0}\left(\frac{\pi}{2}, 0\right) \sqrt{\frac{2k+1}{4\pi}}\, P_k(0)\, P_k(\cos\omega_{12})$$

$$= [P_k(0)]^2\, P_k(\cos\omega_{12}).$$

Recalling that $\cos\omega_{12} = [(\boldsymbol{l}_1\cdot\boldsymbol{l}_2)/l_1 l_2] = (1/2l_1 l_2)(\mathbf{L}^2 - l_1^2 - l_2^2)$ we obtain, by substituting into (15.46), using also (15.43) and (15.44),

$$[P_k(0)]^2\, P_k\left(\frac{\mathbf{L}^2 - l_1^2 - l_2^2}{2l_1 l_2}\right)$$

$$\approx \frac{4\pi}{2k+1}(-1)^{l_1+l_2-L}\left[\sqrt{\frac{2k+1}{4\pi}}\, P_k(0)\sqrt{2l_1+1}\right]$$

$$\times\left[\sqrt{\frac{2k+1}{4\pi}}\, P_k(0)\sqrt{2l_2+1}\right]\begin{Bmatrix} l_1 & l_2 & L \\ l_2 & l_1 & k\end{Bmatrix}.$$

Hence, for even k for which $P_k(0) \neq 0$ and in the limit $l_1 \to \infty$ $l_2 \to \infty$

$$\begin{Bmatrix} l_1 & l_2 & L \\ l_2 & l_1 & k\end{Bmatrix} = (-1)^{l_1+l_2-L}\frac{P_k[(\mathbf{L}^2 - l_1^2 - l_2^2)/2l_1 l_2]}{\sqrt{(2l_1+1)(2l_2+1)}}. \tag{15.47}$$

For $k = 0$ (15.47) holds also for finite l_1 and l_2 since $P_0(x) = 1$.

Thus the classical limit of the Racah coefficient $W(l_1 l_2 L | l_2 l_1 k)$ is proportional to the Legendre polynomial P_k of the cosine of the angle between the two vectors $\boldsymbol{l}_1$ and $\boldsymbol{l}_2$.

16. Multipole Expansion of the Electromagnetic Field

Atoms and nuclei in excited states are very often de-excited by the interaction with the electromagnetic field. As a matter of fact, in many cases this is the only possibility for their de-excitation. Thus, isolated atoms in excited states can decay to their ground states only by the emission of electromagnetic radiation, or, what is essentially the same thing, the ejection of Auger electrons. Practically all our information on atoms comes from their electromagnetic interactions. On the other hand, nuclei in unstable states can emit α-particles, neutrons, protons, etc. They can also undergo β-decay or fission, in addition to the possible emission of electromagnetic radiation. Whereas the interactions responsible for the other modes of nuclear decay are not so well known, the electromagnetic interaction is very well understood. Therefore, a large amount of our information on nuclei is deduced from their electromagnetic interactions.

Due to the weakness of the coupling of nucleons to the electromagnetic field the lifetimes for the emission of electromagnetic radiation are relatively long. This accounts, to some extent, for the thoroughness with which such transitions have been studied experimentally. The presence of the atomic electrons very close to the nucleus makes the variety of the phenomena associated with the interactions of nucleons with the electromagnetic field even bigger. It allows, for instance, for an appreciable "internal conversion" of the γ-rays emitted by the nucleus. This effect has been widely used to determine various properties of nuclei.

The electromagnetic transitions in atoms and nuclei are also easy to analyze theoretically since the electromagnetic field is so well known. They furnish us with very reliable and straightforward information on the physical system which does not involve any assumption about unknown features of other forces like nuclear forces. We shall therefore devote this section to the study of electromagnetic fields.

In atomic physics the electromagnetic field most frequently encountered, is that of an electric dipole. Contributions from higher multipoles are very rare. In nuclear physics, on the other hand, higher multipoles are very common. This situation is due to the different energies and dimensions involved in the two cases. As we shall see, the probability of emitting a radiation of wave number k from a 2^l pole of linear dimensions R is proportional to $(kR)^{2l+1}$. Whereas kR is of order 10^{-4} in atoms, it is of order 10^{-2} for nuclei. Atoms in states which cannot be de-excited by a dipole radiation can usually be de-excited by other, faster mechanisms (e.g., inelastic collisions with other atoms). We are therefore rarely confronted in practice with long-lived atomic metastable states and most of the observed atomic radiations are associated with electric dipoles. No analogous mechanism exists for nuclei at normal temperatures, and lifetimes for observed electromagnetic transitions in nuclei range all the way from 10^{-14} sec and less to several months and more. The electromagnetic radiations from nuclei are thus associated with various multipoles of the nulcear charge and current distributions and supply a considerably richer information on the nucleus than that obtained on the atomic structure from atomic radiation.

Since electromagnetic radiation is always accompanied by a transition of the emitting system from one state of a definite total angular momentum to another such state, we must expect that the electromagnetic radiation carries with it the balance as required by conservation of angular momentum. We are therefore interested in constructing electromagnetic fields which carry a well-defined angular momentum, as well as a well-defined energy and eventually also a well-defined parity.

We have already seen that by decomposing a scalar field, like the wave function of a spinless particle, into irreducible tensors of degree k we obtained fields which were eigenstates of the orbital angular momentum operator $\boldsymbol{l}^2$. We have also seen that the irreducible tensors obtained from physical spinor fields were eigenstates of $\mathbf{j}^2 = (\boldsymbol{l} + \mathbf{s})^2$. The electromagnetic field is a physical vector field. As can be expected the corresponding irreducible tensors will turn out to be eigenstates of $\mathbf{j}^2 = (\boldsymbol{l} + \mathbf{s})^2$ where $\mathbf{s}$ will be eventually interpreted as the internal angular momentum (spin) of the photon (divided by $\hbar$) with the eigenvalue $\mathbf{s}^2 = s(s+1) = 2$.

Let $\mathbf{A}(\mathbf{r})$ be a physical vector field with components $A_\kappa(\mathbf{r})$. If we perform a rotation R of the frame of reference then, from (11.1), we obtain

$$A'_{\kappa'}(\mathbf{r}) = \sum_\kappa A_\kappa(a^{-1}\mathbf{r})\, D^{(1)}_{\kappa\kappa'}(R) \tag{16.1}$$

where $\mathbf{A}'$ is the field referred to the *rotated* frame and a_{ik} is the matrix which determines the rotation R. We see from (16.1), as in the case of a spinor field, that the rotation of a vector field has a double effect on it. Firstly, the numerical values of the coordinates of each point are changed and therefore, to get the same physical field, its functional dependence on the coordinates must generally change from $\mathbf{A}(\mathbf{r})$ to $\mathbf{A}'(\mathbf{r}')$. Secondly, the vector at each point in the vector field is referred, after the rotation, to a new frame of reference. Hence, another transformation, $D^{(1)}(R)$ on the components of the vector field as referred to the original frame is required in order to obtain the components referred to the new frame. As in the case of the spinor field, it is especially instructive to consider infinitesimal transformations. For such a rotation δR by an angle $\delta\theta$ around the direction $\mathbf{n}$ we know, from (12.4) and (12.7), that

$$A_\kappa(a^{-1}\mathbf{r}) = [1 + i\delta\theta(\mathbf{n}\cdot\boldsymbol{l})]\, A_\kappa(\mathbf{r}).$$

We now define the three 3×3 spin matrices for a vector field by the relation

$$[D^{(1)}(\delta R)]_{\kappa\kappa'} = \delta_{\kappa'\kappa} + i\delta\theta[n_x(s_x)_{\kappa'\kappa} + n_y(s_y)_{\kappa'\kappa} + n_z(s_z)_{\kappa'\kappa}] \qquad (16.2)$$

[note that the element $(\kappa\kappa')$ of $D^{(1)}$ defines the element $(\kappa'\kappa)$ of $\mathbf{s}$ as in (11.15)]. We then obtain, to first order in $\delta\theta$,

$$A'_{\kappa'}(\mathbf{r}) = \sum_\kappa \{(\delta_{\kappa\kappa'} + [i\delta\theta\,\mathbf{n}\cdot(\boldsymbol{l}+\mathbf{s})]_{\kappa'\kappa}\}\, A_\kappa(\mathbf{r}).$$

Therefore as in (12.3) and in (12.8)

$$\delta A_{\kappa'}(\mathbf{r}) = A_{\kappa'}(\mathbf{r}) - A'_{\kappa'}(\mathbf{r}) = -\,i\delta\theta \sum_\kappa [\mathbf{n}\cdot(\boldsymbol{l}+\mathbf{s})]_{\kappa'\kappa}\, A_\kappa(\mathbf{r}). \qquad (16.3)$$

We can easily derive a representation of the spin matrices s_i by considering a special vector field which is constant in space. The effect of rotation on such a field is then due only to the fact that after rotation the vector is referred to the new frame of reference. This effect is represented by the operation of the spin matrices on the vector field, since $\boldsymbol{l}$ applied to a constant field gives zero. In this case it can be verified that

$$\delta A_\kappa = A_\kappa - A'_\kappa = (\mathbf{n}\times\mathbf{A})_\kappa\,\delta\theta. \qquad (16.4)$$

Comparing (16.3) and (16.4) we obtain

$$(\mathbf{n}\times\mathbf{A})_\kappa = -\,i\sum_{\kappa'}(\mathbf{n}\cdot\mathbf{s})_{\kappa\kappa'}\, A_{\kappa'}. \qquad (16.5)$$

If in (16.5) we take for $\mathbf{n}$ unit vectors in the x, y, and z directions, we readily obtain

$$s_x = \begin{pmatrix} 0 & 0 & 0 \\ 0 & 0 & -i \\ 0 & +i & 0 \end{pmatrix} \quad s_y = \begin{pmatrix} 0 & 0 & +i \\ 0 & 0 & 0 \\ -i & 0 & 0 \end{pmatrix} \quad s_z = \begin{pmatrix} 0 & -i & 0 \\ +i & 0 & 0 \\ 0 & 0 & 0 \end{pmatrix}. \qquad (16.6)$$

In Section 12, following (12.13), we discussed the constants of motion of rotationally invariant Hamiltonians. We see now that, since the equations of the electromagnetic field are invariant under rotations, the vector $\mathbf{j} = \boldsymbol{l} + \mathbf{s}$ will be a constant of the motion. The angular momenta $\boldsymbol{l}$ alone or $\mathbf{s}$ alone will not, generally, commute with the Hamiltonian and will therefore not be constants of the motion.

The matrices s_x, s_y, and s_z can be seen to satisfy the commutation relations of angular momentum

$$[s_x, s_y] = is_z, \qquad [s_y, s_z] = is_x, \qquad [s_z, s_x] = is_y. \tag{16.7}$$

The conserved vector $\mathbf{j} = \boldsymbol{l} + \mathbf{s}$ can therefore be interpreted as the total angular momentum of the system. The vector $\mathbf{s}$ will then represent an internal angular momentum of the vector field in the same sense as $\frac{1}{2}\boldsymbol{\sigma}$ was the internal angular momentum of the spinor field. No matter what vector field we choose, it will always belong to the eigenvalue $\mathbf{s}^2 = s(s+1) = 2$ of the intrinsic spin (squared), i.e., it represents an internal angular momentum of magnitude 1. Different vector fields may belong to different eigenvalues of $\boldsymbol{l}^2$, but $\mathbf{s}^2$, being an internal property of a vector field, remains always the same. It is generally true that an irreducible physical tensor field of degree k (integer or half-integer) represents particles with an intrinsic angular momentum of magnitude k. The scalar, spinor, and vector fields are the special cases discussed above.

We shall now show, using the Maxwell equations, how the operator $\mathbf{j} = \boldsymbol{l} + \mathbf{s}$ is related to the physical angular momentum of the electromagnetic field as defined classically. Let us make some transformations on the expression of this quantity expressed as follows in terms of the fields

$$\mathbf{G} = \frac{1}{4\pi c} \int [\mathbf{r} \times (\mathbf{E} \times \mathbf{H})] dV.$$

We first assume that the fields $\mathbf{E}$ and $\mathbf{H}$ oscillate with a fixed frequency ω (corresponding to a collection of photons with the same energy $\hbar\omega$).

$$\begin{aligned} \mathbf{E}(\mathbf{x}, t) &= \mathbf{E}(\mathbf{x})\, e^{-i\omega t} + \mathbf{E}^*(\mathbf{x})\, e^{+i\omega t} \\ \mathbf{H}(\mathbf{x}, t) &= \mathbf{H}(\mathbf{x})\, e^{-i\omega t} + \mathbf{H}^*(\mathbf{x})\, e^{+i\omega t}. \end{aligned} \tag{16.8}$$

Using Maxwell's equation $\operatorname{curl} \mathbf{E} + (1/c)\,(\partial \mathbf{H}/\partial t) = 0$ we obtain the relation

$$\operatorname{curl} \mathbf{E}(\mathbf{x}) - \frac{i\omega}{c} \mathbf{H}(\mathbf{x}) = 0 \tag{16.9}$$

and similarly

$$\operatorname{curl} \mathbf{H}(\mathbf{x}) + \frac{i\omega}{c} \mathbf{E}(\mathbf{x}) = 0. \tag{16.10}$$

Averaging over a time $T \gg (1/\omega)$ we obtain for the angular momentum per photon *

$$\frac{\mathbf{G}}{N} = \frac{\hbar\omega}{c} \frac{\int \mathbf{r} \times (\mathbf{E}^* \times \mathbf{H} + \mathbf{E} \times \mathbf{H}^*) dV}{\int (\mathbf{E}^*\mathbf{E} + \mathbf{H}^*\mathbf{H}) dV}. \tag{16.11}$$

(*) From now on we shall denote by **E** and **H** the spatial parts of the fields. This will cause no confusion since we deal with fields with a definite frequency.

Here we have taken the number of photons N equal to the total energy of the field divided by the energy $\hbar\omega$ of each photon.

Introducing (16.9) and (16.10) into (16.11) we obtain

$$\frac{\mathbf{G}}{N} = \frac{\hbar}{i} \frac{\int \mathbf{r} \times (\mathbf{E}^* \times \operatorname{curl} \mathbf{E} - \operatorname{curl} \mathbf{H} \times \mathbf{H}^*) dV}{\int (\mathbf{E}^*\mathbf{E} + \mathbf{H}^*\mathbf{H}) dV}. \tag{16.12}$$

We now use well-known identities from vector analysis

$$[\mathbf{r} \times (\mathbf{F}^* \times \operatorname{curl} \mathbf{F})]_k = i(\mathbf{F}^* \cdot l_k \mathbf{F}) + (\mathbf{F}^* \times \mathbf{F})_k - (\mathbf{F}^* \cdot \operatorname{grad}) [\mathbf{r} \times \mathbf{F}]_k$$

where l_k is the kth component of the operator of orbital angular momentum $\boldsymbol{l} = -i(\mathbf{r} \times \nabla)$ and $\mathbf{F}$ is an arbitrary complex vector field. From (16.6) we see that $(\mathbf{F}^* \times \mathbf{F})_k = i(\mathbf{F}^* \cdot s_k \mathbf{F})$, and hence

$$[\mathbf{r} \times (\mathbf{F}^* \times \operatorname{curl} \mathbf{F})]_k = i[\mathbf{F}^* \cdot (l_k + s_k)\mathbf{F}] - (\mathbf{F}^* \cdot \operatorname{grad}) [\mathbf{r} \times \mathbf{F}]_k. \tag{16.13}$$

Equation (16.12) is symmetric in $\mathbf{E}$ and $\mathbf{H}$, and both terms can be transformed by using (16.13) yielding

$$\frac{G_k}{N} = \hbar \frac{\int (\mathbf{E}^* \cdot j_k \mathbf{E} + \mathbf{H}^* \cdot j_k \mathbf{H}) dV}{\int (\mathbf{E}^* \cdot \mathbf{E} + \mathbf{H}^* \cdot \mathbf{H}) dV}, \qquad \mathbf{j} = \boldsymbol{l} + \mathbf{s}. \tag{16.14}$$

Here we have used the relation

$$\int (\mathbf{E}^* \cdot \operatorname{grad}) (\mathbf{r} \times \mathbf{E}) dV = \oint (\mathbf{r} \times \mathbf{E}) (\mathbf{E}^* \cdot \mathbf{n}) dS - \int (\mathbf{r} \times \mathbf{E}) \operatorname{div} \mathbf{E}^* dV = 0$$

which follows from Maxwell's equations, from $\operatorname{div} \mathbf{E}^* = 0$ and from the vanishing of the surface integral for a big enough sphere. Similar considerations hold for the integral which involves the magnetic field $\mathbf{H}$.

Equation (16.14) shows that $\mathbf{j}$ is in fact the angular momentum operator for the electromagnetic field and that if $\mathbf{E}$ and $\mathbf{H}$ are both eigenfields of, say, j_z belonging to the same eigenvalue m, then each photon in this field carries a z-component of angular momentum which is equal to m. As a matter of fact it is enough that only $\mathbf{E}$ will be an eigenfield of j_z. As we shall see below, it then follows from Maxwell's equations that $\mathbf{H}$ is automatically an eigenfield of j_z belonging to the same eigenvalue.

Thus, we see here again, as in the case of the spinless electron or the two-component neutrino, that the operator which determines the transformation of the vector field under infinitesimal rotations is the operator of the total angular momentum built from an orbital angular momentum $\boldsymbol{l}$ and a spin angular momentum $\mathbf{s}$.

We shall now actually construct an irreducible tensor, whose components will be eigenfunctions of $\mathbf{j}^2$ and j_z. Each component of these irreducible tensors will be a physical vector field. The situation is completely analogous to that of the spinor field where we constructed an irreducible tensor, each component of which was a two-rowed physical spinor field. Since $\mathbf{j} = \boldsymbol{l} + \mathbf{s}$ we know we may do so by taking a tensor product of eigenfunctions of $\boldsymbol{l}^2$, l_z and eigenfunctions of $\mathbf{s}^2$, s_z. The former are simply the ordinary spherical harmonics $Y_{lm}(\theta\,\varphi)$. On the other hand the eigenfunctions of $\mathbf{s}^2$, s_z are vectors which need not have any dependence on the space coordinates. Since the operator $\mathbf{s}^2$ in our case has only one eigenvalue $s(s+1) = 2$, and since s_z has therefore only three eigenvalues $m_s = \pm 1, 0$ there are altogether three simultaneous eigenstates of the operators $\mathbf{s}^2$ and s_z: χ_1, χ_0, and χ_{-1}. Each one of these eigenfunctions is a vector, i.e., has three components, and is independent of the coordinates $\mathbf{x}$. The three vectors χ_κ are defined to satisfy

$$(s_x \pm is_y)\,\chi_\kappa = \sqrt{2 - \kappa(\kappa \pm 1)}\,\chi_{\kappa\pm1} \tag{16.15}$$

$$s_z\,\chi_\kappa = \kappa\chi_\kappa$$

and therefore

$$\mathbf{s}^2\,\chi_\kappa = 2\chi_\kappa. \tag{16.16}$$

We can conclude, taking the representation (16.6) of s_z, that the components of χ_κ are given by

$$\chi_1 = \frac{1}{\sqrt{2}}\begin{pmatrix}-1\\-i\\0\end{pmatrix} \qquad \chi_0 = \begin{pmatrix}0\\0\\1\end{pmatrix} \qquad \chi_{-1} = \frac{1}{\sqrt{2}}\begin{pmatrix}+1\\-i\\0\end{pmatrix}. \tag{16.17}$$

It follows from (16.15) that the three vectors χ_κ are the three components of an irreducible tensor. Under rotations these unit vectors transform according to (11.19) with $k = 1$, yielding the unit vectors in the rotated frame. The index κ signifies the component of the tensor $\chi_\kappa^{(1)}$ in a space which is called the *spin space.* We have added the superscript 1 to emphasize this tensorial character. Each of the components $\chi_\kappa^{(1)}$ is, of course, a vector in ordinary space. We shall not refer to the components in ordinary space explicitly.

Using the vectors $\chi_\kappa^{(1)}$ we construct now the irreducible tensors in the vector field, also known as the *vector spherical harmonics* $\mathbf{Y}_m^{j(l1)}$, defined by

$$\mathbf{Y}_m^{j(l1)}(\theta\,\varphi) = [\mathbf{Y}_l(\theta\,\varphi) \times \chi^{(1)}]_m^{(j)}. \tag{16.18}$$

Obviously, $\mathbf{Y}_m^{j(l1)}$ is a physical vector field. It is clear from their construction that the vector spherical harmonics are eigenfunctions of $\boldsymbol{l}^2$, $\mathbf{s}^2$, $\mathbf{j}^2$, and j_z satisfying

$$\begin{aligned} \boldsymbol{l}^2\mathbf{Y}_m^{j(l1)} &= l(l+1)\,\mathbf{Y}_m^{j(l1)} & \mathbf{s}^2\mathbf{Y}_m^{j(l1)} &= 2\,\mathbf{Y}_m^{j(l1)} \\ \mathbf{j}^2\mathbf{Y}_m^{j(l1)} &= j(j+1)\,\mathbf{Y}_m^{j(l1)} & j_z\mathbf{Y}_m^{j(l1)} &= m\,\mathbf{Y}_m^{j(l1)}. \end{aligned} \tag{16.19}$$

Since we have taken the $\chi_\kappa^{(1)}$ to be normalized vectors, i.e.,

$$(\chi_\kappa^{(1)*} \cdot \chi_{\kappa'}^{(1)}) = \delta_{\kappa\kappa'},$$

it follows from their definition (16.18) that the vector spherical harmonics are also normalized,

$$\int [\mathbf{Y}_m^{j(l1)}(\theta\,\varphi)]^\dagger \cdot \mathbf{Y}_{m'}^{j'(l'1)}(\theta\,\varphi) \sin\theta\, d\theta d\varphi = \delta_{jj'}\, \delta_{ll'}\, \delta_{mm'}. \qquad (16.20)$$

The special vector spherical harmonics with $l = j$, i.e., $\mathbf{Y}_m^{j(j1)}(\theta\,\varphi)$, are particularly useful. It is easy to verify that

$$\mathbf{Y}_m^{j(j1)}(\theta\,\varphi) = \frac{1}{\sqrt{j(j+1)}}\, \boldsymbol{l}\, Y_{jm}(\theta\,\varphi) \qquad (16.21)$$

where $\boldsymbol{l}$ is the differential operator $\boldsymbol{l} = -i(\mathbf{r} \times \nabla)$. Although $\boldsymbol{l}$ is, of course, the operator of orbital angular momentum, it will be more appropriate to consider it here just as a vector differential operator in the sense that $\boldsymbol{l}\, Y_{jm}$ is the vector $(l_x Y_{jm}, l_y Y_{jm}, l_z Y_{jm})$. From (16.6) we see that

$$s_z(\boldsymbol{l}\, Y_{jm}) = \begin{pmatrix} 0 & -i & 0 \\ i & 0 & 0 \\ 0 & 0 & 0 \end{pmatrix} \begin{pmatrix} l_x Y_{jm} \\ l_y Y_{jm} \\ l_z Y_{jm} \end{pmatrix} = \begin{pmatrix} -i\, l_y Y_{jm} \\ i\, l_x Y_{jm} \\ 0 \end{pmatrix}.$$

Hence

$$j_z(\boldsymbol{l}\, Y_{jm}) = (l_z + s_z)(\boldsymbol{l}\, Y_{jm}) = \begin{pmatrix} l_z(l_x Y_{jm}) \\ l_z(l_y Y_{jm}) \\ l_z(l_z Y_{jm}) \end{pmatrix} + \begin{pmatrix} -i\, l_y\, Y_{jm} \\ i\, l_x\, Y_{jm} \\ 0 \end{pmatrix}$$

$$= \begin{pmatrix} l_x\, l_z\, Y_{jm} \\ l_y\, l_z\, Y_{jm} \\ l_z\, l_z\, Y_{jm} \end{pmatrix} = m(\boldsymbol{l}\, Y_{jm}).$$

Thus, the vector field $\boldsymbol{l}[Y_{jm}(\theta\,\varphi)]$ is an eigenfield of the operator j_z belonging to the eigenvalue m. In a similar way it can be shown that

$$\mathbf{j}^2[\boldsymbol{l}\, Y_{jm}(\theta\,\varphi)] = j(j+1)\,[\boldsymbol{l}\, Y_{jm}(\theta\,\varphi)].$$

Since the operators l_x, l_y, and l_z do not connect spherical harmonics of different degrees, we conclude that

$$\boldsymbol{l}\, Y_{jm}(\theta\,\varphi) = \alpha\, \mathbf{Y}_m^{j(j1)}(\theta\,\varphi)$$

where α is a proportionality factor.

To determine α we calculate the normalization integral and find

$$\int [\boldsymbol{l}\, Y_{jm}(\theta\,\varphi)]^\dagger \cdot [\boldsymbol{l}\, Y_{jm}(\theta\,\varphi)] = \int Y^*_{jm}(\theta\,\varphi)\boldsymbol{l} \cdot \boldsymbol{l}\, Y_{jm}(\theta\,\varphi)$$

$$= \int Y^*_{jm}(\theta\,\varphi)\boldsymbol{l}^2\, Y_{jm}(\theta\,\varphi) = j(j+1)$$

and hence the result (16.21).

Since the operator $\boldsymbol{l}$ is an angular operator, i.e., it commutes with r, it follows from (16.21) that $\mathbf{Y}_m^{j(j1)}(\theta\,\varphi)$ is a transverse field. This means that

$$\mathbf{r} \cdot \mathbf{Y}_m^{j(j1)}(\theta\,\varphi) = 0. \tag{16.22}$$

The transversality of the fields $\mathbf{Y}_m^{j(j1)}(\theta\,\varphi)$ accounts for their great use in analyzing the electromagnetic fields.

Consider in fact the field $\mathbf{H}(\mathbf{r})$ given by

$$\mathbf{H}_{jm}(\mathbf{r}) = f(r)\, \mathbf{Y}_m^{j(j1)}(\theta\,\varphi). \tag{16.23}$$

We shall prove that with a proper choice of $f(r)$ it satisfies Maxwell's equations for a magnetic field in vacuum. First we notice that

$$\operatorname{div} \mathbf{H}_{jm}(\mathbf{r}) = \operatorname{grad} f(r) \cdot \mathbf{Y}_m^{j(j1)}(\theta\,\varphi) + f(r) \operatorname{div} \mathbf{Y}_m^{j(j1)}(\theta\,\varphi).$$

Since grad $f(r)$ is in the direction of $\mathbf{r}$, the first term vanishes on account of (16.22). For the second term we obtain

$$\operatorname{div} \mathbf{Y}_m^{j(j1)}(\theta\,\varphi) = \operatorname{div} \boldsymbol{l}\, Y_{jm}(\theta\,\varphi) = -i\mathbf{r} \cdot (\nabla \times \nabla)\, Y_{jm}(\theta\,\varphi)$$

$$= -i(\nabla \times \mathbf{r}) \cdot [\nabla Y_{jm}(\theta\,\varphi)]_c - i\,\mathbf{r} \cdot (\nabla \times \nabla)\, Y_{jm}(\theta\,\varphi).$$

Here $[\;]_c$ denotes that the expression in the brackets behaves like a constant with respect to the differential operator which precedes it. Since $(\nabla \times \mathbf{r}) = -(\mathbf{r} \times \nabla)$, and since $\nabla \times \nabla$ is identically zero we see that div $\mathbf{Y}_m^{j(j1)}(\theta\,\varphi) = 0$. Thus for any choice of $f(r)$, $\mathbf{H}_{jm}(\mathbf{r})$ satisfies div $\mathbf{H}_{jm}(\mathbf{r}) = 0$. In order to describe an electromagnetic field in free space, $\mathbf{H}_{jm}(\mathbf{r})$ has to satisfy in addition the wave equation

$$\left[\Delta + \left(\frac{\omega}{c}\right)^2\right] \mathbf{H}(\mathbf{r}) = 0.$$

This leads to the following equation for $f(r)$

$$\left[\frac{d^2}{dr^2} - \frac{j(j+1)}{r^2} + \left(\frac{\omega}{c}\right)^2\right] [r\, f(r)] = 0.$$

The solutions to the last equation are the well-known Bessel functions of half-integral order. Therefore, with the choice

$$f(r) = \frac{1}{r}\sqrt{\frac{\omega r}{c}}\, Z_{j+1/2}\left(\frac{\omega}{c} r\right) \tag{16.24}$$

the field $\mathbf{H}_{jm}(\mathbf{r})$ defined by (16.23) satisfies Maxwell's equations in free space. To obtain the corresponding electric field we use (16.10)

$$\mathbf{E}_{jm}(\mathbf{r}) = \frac{ic}{\omega} \operatorname{curl} \mathbf{H}_{jm}(\mathbf{r}) = \frac{ic}{\omega} \operatorname{curl} [f(r)\, \mathbf{Y}_m^{j(j1)}(\theta\, \varphi)]. \tag{16.25}$$

Since the operator $\mathbf{j}$ induces the infinitesimal change in a vector field, and the operator curl commutes with rotations, we conclude that

$$j_k[\operatorname{curl} \mathbf{Y}_m^{j(l1)}] = \operatorname{curl} [j_k\, \mathbf{Y}_m^{j(l1)}]. \tag{16.26}$$

By construction, $\mathbf{H}_{jm}(\mathbf{r})$ is an eigenfield of $\mathbf{j}^2$ and j_z belonging to the eigenvalues $j(j+1)$ and m respectively. We therefore conclude from (16.25) and (16.26) that $\mathbf{E}_{jm}(\mathbf{r})$ is also an eigenfield of $\mathbf{j}^2$ and j_z belonging to the same eigenvalues.

The electromagnetic field in free space defined by (16.23) and (16.25) is known as the field of an *electric multipole of order j*. It is given by

$$\begin{aligned} \mathbf{E}_{jm}^{(\mathrm{el})}(\mathbf{r}, t) &= \mathbf{E}_{jm}^{(\mathrm{el})}(\mathbf{r})\, e^{-i\omega t} + \mathrm{c.c.} \\ \mathbf{H}_{jm}^{(\mathrm{el})}(\mathbf{r}, t) &= \mathbf{H}_{jm}^{(\mathrm{el})}(\mathbf{r})\, e^{-i\omega t} + \mathrm{c.c.} \end{aligned} \tag{16.27}$$

where

$$\begin{aligned} \mathbf{E}_{jm}^{(\mathrm{el})}(\mathbf{r}) &= \frac{ic}{\omega} \operatorname{curl} [f(r)\, \mathbf{Y}_m^{j(j1)}(\theta\, \varphi)] \\ \mathbf{H}_{jm}^{(\mathrm{el})}(\mathbf{r}) &= f(r)\, \mathbf{Y}_m^{j(j1)}(\theta\, \varphi). \end{aligned} \tag{16.28}$$

This field corresponds to photons of energy $\hbar\omega$ and angular momentum j.

There is also another solution of Maxwell's equations which has the same angular momentum, the *magnetic multipole field of order j*, given by

$$\begin{aligned} \mathbf{E}_{jm}^{(\mathrm{mag})}(\mathbf{r}, t) &= \mathbf{E}_{jm}^{(\mathrm{mag})}(\mathbf{r})\, e^{-i\omega t} + \mathrm{c.c.} \\ \mathbf{H}_{jm}^{(\mathrm{mag})}(\mathbf{r}, t) &= \mathbf{H}_{jm}^{(\mathrm{mag})}(\mathbf{r})\, e^{-i\omega t} + \mathrm{c.c.} \end{aligned}$$

where now

$$\mathbf{E}_{jm}^{(\mathrm{mag})}(\mathbf{r}) = f(r)\, \mathbf{Y}_m^{j(j1)}(\theta\, \varphi) \tag{16.29a}$$

and because of Maxwell's equation (16.9)

$$\mathbf{H}^{(\text{mag})}_{jm}(\mathbf{r}) = -\frac{ic}{\omega} \operatorname{curl} f(r)\, \mathbf{Y}^{j(j1)}_{m}(\theta\, \varphi). \tag{16.29b}$$

The difference between the solutions (16.28) and (16.29) is that the role played by the magnetic field in one is taken over by the electric field in the other and vice versa. Whether a system emits one radiation or the other can be determined in the following way.

We know that the probability amplitude for the emission of an electromagnetic radiation accompanied by a transition of the emitting system from ψ_a to ψ_b is given by

$$\int \psi_b^* (\mathbf{x})\, (\mathbf{j}(\mathbf{x}) \cdot \mathbf{A}(\mathbf{x}))\, \psi_a(\mathbf{x})\, d^3\mathbf{x}. \tag{16.30}$$

Here $\mathbf{j}(\mathbf{x})$ is the current operator of the emitting system operating on the coordinate $\mathbf{x}$ and $\mathbf{A}(\mathbf{x})$ is the electromagnetic potential. The integral in (16.30) extends over all space. We can therefore make the substitution $\mathbf{y} = -\mathbf{x}$ without changing the limits of the integration. We obtain then

$$\int \psi_b^*(\mathbf{x})\, (\mathbf{j}(\mathbf{x}) \cdot \mathbf{A}(\mathbf{x}))\, \psi_a(\mathbf{x}) d^3\mathbf{x} = \int \psi_b^*(-\mathbf{y})\, (\mathbf{j}(-\mathbf{y}) \cdot \mathbf{A}(-\mathbf{y}))\, \psi_a(-\mathbf{y}) d^3\mathbf{y}. \tag{16.31}$$

If the states ψ_a and ψ_b have definite parities $(-1)^{P_a}$ and $(-1)^{P_b}$ respectively, then

$$\psi_a(-\mathbf{y}) = (-1)^{P_a}\, \psi(\mathbf{y}) \qquad \psi_b(-\mathbf{y}) = (-1)^{P_b}\, \psi(\mathbf{y}). \tag{16.32}$$

Furthermore, $\mathbf{j}$ is a polar vector operator, and therefore

$$\mathbf{j}(-\mathbf{y}) = -\mathbf{j}(\mathbf{y}) \tag{16.33}$$

If, in addition, the field $\mathbf{A}$ satisfies

$$\mathbf{A}(-\mathbf{y}) = (-1)^P\, \mathbf{A}(\mathbf{y}) \tag{16.34}$$

then we obtain, by substituting into (16.31) that

$$\int \psi_b^*(\mathbf{x})\, (\mathbf{j} \cdot \mathbf{A})\, \psi_a(\mathbf{x}) d^3\mathbf{x} = (-1)^{P_a+P_b+1+P} \int \psi_b^*(\mathbf{y})\, (\mathbf{j} \cdot \mathbf{A})\, \psi_a(\mathbf{y}) d^3\mathbf{y}.$$

Hence

$$\int \psi_b(\mathbf{x})\, (\mathbf{j} \cdot \mathbf{A})\, \psi_a(\mathbf{x}) d^3\mathbf{x} = 0 \qquad \text{if} \qquad (-1)^{P_a+P_b+P+1} = -1. \tag{16.35}$$

Equation (16.35) represents the result of parity conservation in electromagnetic transitions. It states that the electromagnetic field emitted in the transition from ψ_a to ψ_b must be such that its vector potential **A** has a parity $(-1)^P$ satisfying $(-1)^{P+1} = (-1)^{P_a+P_b}$. It is convenient to define the *parity of an electromagnetic field* P_{ph} by $P_{ph} = P + 1$ where P is the parity of the vector potential field associated with it. The result (16.35) can then be rewritten in the form:

Electromagnetic transitions between the states a and b are allowed only if $(-1)^{P_a+P_b+P_{ph}} = 1.$ (16.36)

For any electromagnetic field **E** has the same parity as **A** whereas **H** is proportional to curl **A**. It follows therefore that the parity of the electric field **E** is equal to that of **A**, namely $(-1)^P$, whereas the parity of **H** is equal to $(-1)^{P+1}$. Thus, our convention about the parity of an electromagnetic field can be formulated in the following way:

The parity P_{ph} *of an electromagnetic field described by the electric field* **E** *and the magnetic field* **H** *is defined as the parity of* **H**, *i.e.,* P_{ph} *satisfies*

$$\mathbf{H}(-\mathbf{x}, t) = (-1)^{P_{ph}}\mathbf{H}(\mathbf{x}, t). \tag{16.37}$$

It should be noted that the parity of the electromagnetic field is not associated with the question of whether the vector $\mathbf{H}(\mathbf{x}_0, t_0)$, at given point $\mathbf{x}_0$ in space, is a polar or an axial vector. This latter question has to do with the "intrinsic" parity of the photon, whereas it is clear from the derivation of (16.35) that what comes into the parity selection rule is the total parity of the vector potential **A**. Since the intrinsic parity of the polar vector **A** is fixed (-1), the total parity is determined by the "orbital" parity of the electromagnetic field, i.e., the relation between the field at the point **x** and the field at the point $-\mathbf{x}$. The distinction between intrinsic and orbital parities is similar to the corresponding distinction between intrinsic and orbital angular momenta. The intrinsic parity of a field is determined by its physical nature $(+1$ for axial vector and -1 for polar vector fields). The orbital parity, on the other hand, depends on the particular solution considered.

These considerations can be visualized by comparing a magnetic dipole field and an electric dipole field. Consider first a static magnetic dipole D at the origin. The magnetic field at A has the same direction and intensity as that at the point B, which is the mirror image of A relative to the origin. If now the magnetic dipole oscillates, the fields at A and at B will oscillate in the same phase. Thus the magnetic field of the radiation of an oscillating magnetic dipole has a posivite parity. The electric field of the radiation of an oscillating magnetic dipole must have a negative parity since the vector, $\mathbf{E} \times \mathbf{H}$, which describes the direction of energy flow must point in opposite directions at A and at B (energy is radiated *away* from the dipole). If, now, D is taken to be an *electric* dipole, then the electric field of the electric dipole radiation has a positive parity and consequently the magnetic field must have a negative parity.

Coming back to the two solutions of Maxwell's equations in free space, (16.28) and (16.29) we note that since $\mathbf{Y}_m^{j(j1)}$ equals $[1/\sqrt{j(j+1)}]\, lY_{jm}(\theta\,\varphi)$ the parity of the vector spherical harmonics $\mathbf{Y}_m^{j(j1)}(\theta\,\varphi)$ is the same as that of the ordinary spherical harmonics $Y_{jm}(\theta\,\varphi)$. Since the parity of $Y_{jm}(\theta\,\varphi)$ is $(-1)^j$, we conclude that

$$\begin{array}{l}\textit{The parity of an electric multipole field of order } j \textit{ is } (-1)^j;\\ \textit{the parity of a magnetic multipole field of order } j \textit{ is } (-1)^{j+1}.\end{array} \tag{16.38}$$

We can draw the following conclusion from (16.38) and (16.36). If in the transition from ψ_a to ψ_b the system emits a multipole radiation of order j (i.e., if because of conservation of angular momentum the electromagnetic field carries with it j units of angular momentum), then the radiation is that of an electric or magnetic multipole according to whether $(-1)^{P_a+P_b}$ is equal to $(-1)^j$ or $(-1)^{j+1}$ respectively. Electric dipole radiation ($E1$) for which $j=1$ requires different parity for the initial and final states of the system. Electric quadrupole ($E\,2$) radiation has $j=2$ and involves no change in parity, etc. Magnetic dipole radiation ($M\,1$) has $j=1$ and involves no change in parity, whereas magnetic quadrupole radiation ($M\,2$) has $j=2$ and does involve a change in parity, etc.

It should be pointed out that the amount of angular momentum carried by the radiation is very often not uniquely determined by the initial and final states of the system. Only if the angular momentum of the initial or the final states is zero will the angular momentum of the radiation be uniquely determined. For example, if the excited state of the system has $J_i = 2^+$ (i.e., its angular momentum is 2 and its parity positive), and the ground state has $J_f = 0^+$, then the radiation emitted must have $j=2$ in order to conserve angular momentum. Since the parity of the two states is the same, the radiation will be that of an *electric multipole* of order 2 and will be given by (16.28) with $j=2$. On the other hand if the ground state has $J_f = 1^-$, conservation of angular momentum only tells us that $|J_i - J_f| \leq j \leq J_i + J_f$, so that j could take on the values 1, 2, or 3. Since the transition in this example involves a change of the parity of the system, the radiations emitted could be either $E\,1(j=1)$, $M\,2(j=2)$, or $E\,3(j=3)$, or of course, any mixture of these. We shall see in the following section that the emission probability generally decreases with increasing multipolarity. Therefore, in the example above we can generally expect the radiation to be predominantly that of an electric dipole.

Before closing this section we would like also to derive the angular distribution of the radiations of the various multipoles. We have already pointed out that the field $\mathbf{Y}_m^{j(j1)}(\theta\,\varphi)$ is transversal. It follows that for an electric multipole field the magnetic field is always transversal; the electric field, however, is generally not purely transversal for such a multipole. In fact we obtain from (16.28)

$$\mathbf{E}_{jm}^{(\mathrm{el})}(\mathbf{x}) = \frac{ic}{\omega}\,[\operatorname{grad} f(r) \times \mathbf{Y}_m^{j(j1)}(\theta\,\varphi)] + \frac{ic}{\omega} f(r) \operatorname{curl} \mathbf{Y}_m^{j(j1)}(\theta\,\varphi).$$

The first term, being the vector product of a transversal field and a vector in the $\mathbf{r}$-direction, is again transversal. The second term has, however, a nonvanishing component along $\mathbf{r}$. In fact

$$\mathbf{r}\cdot\operatorname{curl}\mathbf{Y}_m^{j(j1)}(\theta\,\varphi) = \mathbf{r}\cdot\boldsymbol{\nabla}\times\left(\frac{\mathbf{l}}{\sqrt{j(j+1)}}\,Y_{jm}(\theta\,\varphi)\right)$$

$$= i\,\frac{\mathbf{l}^2}{\sqrt{j(j+1)}}\,Y_{jm}(\theta\,\varphi) = i\sqrt{j(j+1)}\;Y_{jm}(\theta\,\varphi).$$

We see that the ratio of the radial component of the electric field to its transversal component decreases as $1/r$ when $r \to \infty$. Hence for $kr \gg 1$ both the magnetic and the electric fields are transversal, as, of course, should be expected at regions where the radiation field is essentially a plane wave. The angular distribution of the radiation is therefore proportional to the distribution of energy density of that radiation on a given sphere. If the sphere is large enough the radiation at each point can be approximated by a plane wave and therefore $E \approx H$. Thus, the energy density on the surface of a large sphere is proportional to $E^*E \approx H^*H$. It is therefore the same for both electric and magnetic multipole radiations of order j and is given by

$$Q_{jm}(\theta\,\varphi) = [\mathbf{Y}_m^{j(j1)}(\theta\,\varphi)]^\dagger\cdot\mathbf{Y}_m^{j(j1)}(\theta\,\varphi)$$

$$= \sum (j\lambda 1\mu|j1jm)\,(j\lambda' 1\mu'|j1jm)\;Y_{j\lambda}^*(\theta\,\varphi)\,Y_{j\lambda'}(\theta\,\varphi)\,(\chi_\mu^\dagger\cdot\chi_{\mu'}).$$

However, since $(\chi_\mu^\dagger\cdot\chi_{\mu'}) = \delta_{\mu\mu'}$ we obtain the angular distribution

$$Q_{jm}(\theta) = [\mathbf{Y}_m^{j(j1)}(\theta\,\varphi)]^\dagger\cdot\mathbf{Y}_m^{j(j1)}(\theta\,\varphi) = \sum_{\lambda\mu} (j\lambda 1\mu|j1jm)^2|Y_{j\lambda}(\theta\,\varphi)|^2. \tag{16.39}$$

It follows that $Q_{jm}(\theta)$ is independent of φ and thus the radiation emitted has a cylindrical symmetry around the z-axis. Inserting the values of the Clebsch-Gordan coefficients in (16.39) (see Appendix) we obtain explicitly

$$Q_{jm}(\theta) = \frac{(j+m)(j-m+1)}{2j(j+1)} |Y_{jm-1}(\theta\,\varphi)|^2$$

$$+ \frac{m^2}{j(j+1)} |Y_{jm}(\theta\,\varphi)|^2 + \frac{(j-m)(j+m+1)}{2j(j+1)} |Y_{jm+1}(\theta\,\varphi)|^2. \tag{16.40}$$

Since $Y_{jm}(\theta\,\varphi)$ vanishes for $\theta = 0$ unless $m = 0$, we see from (16.40) that a multipole field has a nonvanishing intensity in the z-direction only if $m = \pm 1$. For radiation emitted in this direction the z-projection of the orbital angular momentum vanishes. Thus, the only possible projection of the total angular momentum comes from the intrinsic angular momentum. This can have the values ± 1 only, and not $m = 0$, due to the transversality of the electromagnetic field. By conservation of angular momentum we know that m, the z-component of the photon angular momentum, is equal to the change in the z-component of the angular momentum of the emitting system. We conclude therefore from (16.40) that a system can emit radiation in the z-direction only if in that transition $\Delta M = \pm 1$ where $\Delta M = M_i - M_f$. As mentioned before we cannot distinguish between an electric and magnetic multipole radiation by measuring their angular distributions. The angular distribution determines only the angular momentum j and its z-component m, carried by the radiation.

There is another interesting conclusion which can be drawn from (16.39) concerning the angular correlation of two successive radiations. Let us first consider a system in an initial state $|J_i M_i\rangle$ radiating a given multipole field $\mathbf{F}_{jm}$ and going over to a final state $|J_f M_f\rangle$, where of course, $M_i = M_f + m$ and $|J_i - J_f| \leq j \leq J_i + J_f$. It is well known [cf. (15.34)] that, if the initial system has equal probability of being in any of the substates M_i, then the total probability of emitting the field $\mathbf{F}_{jm}$ from the initial states $|J_i M_i\rangle$ ($M_i = J_i, J_i - 1, \ldots, -J_i$) is independent of m. If we allow the system to emit all possible fields $\mathbf{F}_{jm}$, with $m = j, j-1, \ldots, -j$, we shall obtain for the angular distribution of the total radiation the expression $\Sigma Q_{jm}(\theta)$. Since

$$\sum_{m\mu} |(j\,\lambda\,1\,\mu | j\,1\,j\,m)|^2 = 1,$$

and since by the addition theorem of spherical harmonics

$$\sum_{\lambda} |Y_{j\lambda}(\theta\,\varphi)|^2 = \frac{2j+1}{4\pi} P_j(1) = \frac{2j+1}{4\pi},$$

we see that the angular distribution of the total radiation, emitted from a system uniformly distributed among all the states M_i, is uniform.

This should have been expected since an equal probability of finding the initial system in any of the substates M_i means that there is no preferred direction in the initial system. Consequently there can be no preferred direction in the subsequent emission of the radiation.

Let us consider the situation where a system initially in a state of $J = 0$, emits an electromagnetic radiation R_1 in the transition to a state of $J = 1$. This state then further decays to the ground state with $J = 0$, emitting the radiation R_2. If the radiation R_1 is observed in a certain direction, we choose this to be the z-axis. We know that the value of the z-projection of the angular momentum of the system could then change only by ± 1 (otherwise there is no intensity in the z-direction!). Thus, the substates of the intermediate states of those nuclei, whose first radiation was observed, are not equally populated. The states with $M = 1$ and $M = -1$ are (equally) populated, and the state with $M = 0$ is not populated at all. Thus, if we look at the second radiation which is emitted from those nuclei whose first radiation was detected in a given direction, its distribution in space will not generally be uniform. In the specific case of this example we see from (16.40) that it will be given by

$$Q_{1,1}(\theta) + Q_{1,-1}(\theta) = [|Y_{1,1}(\theta\,\varphi)|^2 + |Y_{1,0}(\theta\,\varphi)|^2] = \tfrac{3}{8\pi}(1 + \cos^2\theta).$$

Although each radiation in itself is emitted uniformly in all directions, an angular correlation between two successive radiations exists, and is usually not uniform.

17. Electromagnetic Transitions in Nuclei

So far we have treated only the electromagnetic field in free space and have disregarded its relation to the sources which produce it. We did use, to be sure, the requirement that the electromagnetic field carries with it whatever angular momentum is lost by the system in the process of emitting the radiation. However, since the angular momentum of the field can be determined from its asymptotic behavior, it was enough, for our purposes, to consider the solutions in free space only.

When we come to the question of evaluating the transition probability for a given system emitting electromagnetic radiation, we have to consider the solutions of Maxwell's equations also in those regions of space which are not free of charges and currents.

Considering classical systems first, Maxwell's equations with sources are

$$\begin{aligned} &\operatorname{curl} \mathbf{E} + \frac{1}{c}\frac{\partial}{\partial t}(\mathbf{H} + 4\pi\, \mathbf{m}_0) = 0 \qquad && \operatorname{div} \mathbf{E} = 4\pi\, \rho_0 \\ &\operatorname{curl} \mathbf{H} - \frac{1}{c}\frac{\partial E}{\partial t} - \frac{4\pi}{c}\mathbf{j}_0 = 0 && \operatorname{div} \mathbf{H} = -4\pi \operatorname{div} \mathbf{m}_0. \end{aligned} \tag{17.1}$$

Here $\mathbf{m}_0(\mathbf{x}, t)$ is the density of magnetization (such as that arising from intrinsic magnetic moments of the particles in the system), $\rho_0(\mathbf{x}, t)$ the charge density, and $\mathbf{j}_0(\mathbf{x}, t)$ the current density; ρ_0 and $\mathbf{j}_0$ satisfy the continuity equation

$$\operatorname{div} \mathbf{j}_0 + \frac{\partial \rho_0}{\partial t} = 0. \tag{17.2}$$

We choose again fields of a definite frequency and expand the sources according to the frequency of oscillation. The part which has the frequency ω is given by

$$\begin{aligned} \mathbf{m}_0(\mathbf{x}, t) &= \mathbf{m}(\mathbf{x})e^{-i\omega t} + \mathbf{m}^\dagger(\mathbf{x})e^{i\omega t} \\ \rho_0(\mathbf{x}, t) &= \rho(\mathbf{x})e^{-i\omega t} + \rho^\dagger(\mathbf{x})e^{i\omega t} \\ \mathbf{j}_0(\mathbf{x}, t) &= \mathbf{j}(\mathbf{x})e^{-i\omega t} + \mathbf{j}^\dagger(\mathbf{x})e^{i\omega t}. \end{aligned}$$

We then obtain from Maxwell's equations (17.1)

$$\begin{aligned} \left[\operatorname{curl}\operatorname{curl} - \left(\frac{\omega}{c}\right)^2\right] \mathbf{E}(\mathbf{x}) &= \omega\,\frac{4\pi i}{c^2}\left[\mathbf{j}(\mathbf{x}) + c\operatorname{curl}\mathbf{m}(\mathbf{x})\right] \\ \left[\operatorname{curl}\operatorname{curl} - \left(\frac{\omega}{c}\right)^2\right] \mathbf{H}(\mathbf{x}) &= \frac{4\pi}{c}\left[\frac{\omega^2}{c}\,\mathbf{m}(\mathbf{x}) + \operatorname{curl}\mathbf{j}(\mathbf{x})\right]. \end{aligned} \tag{17.3}$$

Equation (17.3) can be solved by assuming, for example, $\mathbf{H}(\mathbf{r}) = (1/r) f(r) \mathbf{Y}_m^{j(j1)}(\theta\,\varphi)$. This special choice will guarantee that the electromagnetic field carries a definite angular momentum as before. The presence of the sources, however, will affect the function $f(r)$ and we shall now get a radial dependence different from that of the free field case. Electric multipole radiations will again be given by electromagnetic fields whose *magnetic* field is $\mathbf{H}_{jm}^{(\mathrm{el})}(\mathbf{r}) = (1/r) f^{(\mathrm{el})}(r)\, \mathbf{Y}_m^{j(j1)}(\theta\,\varphi)$, and magnetic multipole radiations will be given by electromagnetic fields whose *electric* field is $\mathbf{E}_{jm}^{(\mathrm{mag})}(\mathbf{r}) = (1/r) f^{(\mathrm{mag})}(r)\, \mathbf{Y}_m^{j(j1)}(\theta\,\varphi)$. Under the above conditions $\operatorname{div} \mathbf{H}_{jm}^{(\mathrm{el})}(\mathbf{r}) = 0$ and $\operatorname{div} \mathbf{E}_{jm}^{(\mathrm{mag})}(\mathbf{r}) = 0$. For any electromagnetic field $\operatorname{div} \mathbf{H} = -4\pi \operatorname{div} \mathbf{m}$ and $\operatorname{div} \mathbf{E} = 4\pi\rho$. We therefore see that a system will emit a pure electric multipole radiation only if $\operatorname{div} \mathbf{m} = 0$, and a pure magnetic multipole radiation only if $\rho = 0$. This explains why we have called an electromagnetic field with angular momentum j an *electric*-multipole radiation ($\operatorname{div} \mathbf{m} = 0$) if its parity is $(-1)^j$, and a *magnetic*-multipole radiation ($\rho = 0$) if its parity is $(-1)^{j+1}$.

To obtain an equation for the radial dependence of an electric multipole field of degree j, we multiply Maxwell's equation (17.3) for $\mathbf{H}_{jm}^{(\mathrm{el})}(\mathbf{r}) = (1/r) f(r)\, \mathbf{Y}_m^{j(j1)}(\theta\,\varphi)$ by $[\mathbf{Y}_m^{j(j1)}(\theta\,\varphi)]^\dagger$ and integrate over all angles. Since, as can be verified easily

$$\operatorname{curl} \operatorname{curl} \left[\frac{1}{r} f(r)\, \mathbf{Y}_m^{j(j1)}(\theta\,\varphi)\right] = -\frac{1}{r}\left[\frac{d^2}{dr^2} - \frac{j(j+1)}{r^2}\right] f(r)\, \mathbf{Y}_m^{j(j1)}(\theta\,\varphi)$$

we obtain this way the equation

$$\left[\frac{d^2}{dr^2} - \frac{j(j+1)}{r^2} + \frac{\omega^2}{c^2}\right] f(r)$$
$$= -\frac{4\pi}{c} \int [\mathbf{Y}_m^{j(j1)}(\theta\,\varphi)]^\dagger \left[\frac{\omega^2}{c} \mathbf{m}(\mathbf{r}) + \operatorname{curl} \mathbf{j}(\mathbf{r})\right] \sin\theta\, d\theta d\varphi. \tag{17.4}$$

The right-hand side is, of course, a function of r only. Similar equations can be obtained in the case of magnetic multipole radiations.

The right-hand side of (17.4) is actually the coefficient of the expansion of $(4\pi/c)\,[(\omega^2/c)\, \mathbf{m}(\mathbf{r}) + \operatorname{curl} \mathbf{j}(\mathbf{r})]$ in its multipoles. Thus, as can be expected, the amplitude of $f(r)$ right outside the emitting system, which determines the strength of the particular multipole radiation, depends on the extent to which this particular multipole order exists in the distribution of magnetization $\mathbf{m}(\mathbf{r})$ and of $\operatorname{curl} \mathbf{j}(\mathbf{r})$ in the source. Furthermore, one sees from (17.4) that for very small values of r, where $[j(j+1)]/r^2$ is the dominant r-dependent coefficient, $f(r)$ behaves like $(\omega r/c)^{(j+1)}$. Thus, higher multipoles have smaller amplitudes in the emitting system (as long as $\omega r/c < 1$). This qualitative result explains why the probability of emitting a radiation of a given frequency *decreases* with increasing multipolarity, and why the probability of emitting a radiation of a given multipolarity *increases* with increasing frequency ω.

In a quantum-mechanical treatment of a radiating system, the multipole moments of the charge, current, and magnetization of the emitting system are replaced by matrix elements of corresponding operators between the initial and final states of the system. We shall not discuss here the derivation of the formulas for the transition probability, which are found in a number of standard textbooks. We shall rather write them down and concentrate on the evaluation of the matrix elements involved.

In the "long wave approximation" for which $c/\omega \gg R$ (R is the linear dimension of the emitting system), the transition probability per unit time is given by

$$T(L) = 8\pi c \frac{e^2}{\hbar c} \frac{(L+1)}{L[(2L+1)!!]^2} k^{2L+1} B(L) \qquad k = \frac{\omega}{c} \ll \frac{1}{R}. \tag{17.5}$$

Here the *reduced transition rate* B is given for electric multipoles by

$$B(\text{el}, L) = \sum_{M,M_f} |\langle J_f M_f | \frac{1}{e} \int \rho(\mathbf{r}) r^L\, Y_{LM}(\theta\, \varphi)\, d^3x | J_i M_i \rangle|^2 \tag{17.6}$$

and for magnetic multipoles by

$$B(\text{mag}, L) = \sum_{M,M_f} |\langle J_f M_f | \frac{1}{L+1} \frac{1}{ec} \int [\text{grad}\,(r^L\, Y_{LM}(\theta\, \varphi)] \cdot \{\mathbf{r} \times [\mathbf{j}(\mathbf{r}) + c\, \text{curl}\, \mathbf{m}(\mathbf{r})]\}\, d^3x | J_i M_i \rangle|^2. \tag{17.7}$$

$J_i M_i$ and $J_f M_f$ characterize the initial and final states respectively. We have denoted the degree of the multipole radiation by L rather than by j, as was done previously, in order to conform with accepted notation. It should however be remembered that this L always stands for the sum of an "orbital" and "intrinsic" angular momenta. We have also introduced the wave number $k = \omega/c$ to simplify notations. The $\rho(\mathbf{r})$ and $\mathbf{j}(\mathbf{r})$ are the operators for the charge density and the current density of the emitting system respectively. The nonrelativistic approximation was assumed for the nucleons. For a system of particles of masses m_i, carrying charges e_i and magnetic moments $\mu_i \boldsymbol{\sigma}_i$ we have

$$\begin{aligned} \rho(\mathbf{r}) &= \sum e_i \delta(\mathbf{r} - \mathbf{r}_i) \\ \mathbf{j}(\mathbf{r}) &= \sum \frac{e_i}{m_i} \mathbf{p}_i \delta(\mathbf{r} - \mathbf{r}_i) \\ \mathbf{m}(\mathbf{r}) &= \sum_i \mu_i \boldsymbol{\sigma}_i \delta(\mathbf{r} - \mathbf{r}_i). \end{aligned} \tag{17.8}$$

Since $\rho(\mathbf{r})r^L Y_{LM}(\theta\,\varphi)$ is an irreducible tensor of degree L we see from (17.6) that $B(\text{el}, L)$, and therefore also the corresponding transition probability, vanishes unless L satisfies the triangular condition $|J_f - J_i| \leq L \leq J_f + J_i$. Similarly, from (17.7) we see that, since $r^L Y_{LM}$ is a homogeneous polynomial of degree L, grad $[r^L Y_{LM}(\theta\,\varphi)]$ contains at most tensors of degree $L - 1$. But as $[\mathbf{r} \times (\mathbf{j} + c \operatorname{curl} \mathbf{m})]$ is a tensor of degree 1, the scalar product of $[\mathbf{r} \times (\mathbf{j} + c \operatorname{curl} \mathbf{m})]$ and grad $[r^L Y_{LM}(\theta\,\varphi)]$ contains at most tensors of degree L (later we shall see that it contains only tensors of degree L). The transition is allowed only if angular momentum is conserved, i.e., if

$$|J_i - J_f| \leq L \leq J_i + J_f.$$

Substituting (17.8) in (17.6) and carrying out the integration d^3x we obtain, according to (15.35)

$$\begin{aligned} B(\text{el}, L) &= \sum_{M,M_f} \left\langle J_f M_f \left| \frac{1}{e} \sum_i e_i r_i^L Y_{LM}(\theta_i\,\varphi_i) \right| J_i M_i \right\rangle^2 \\ &= \sum_{M,M_f} \begin{pmatrix} J_f & L & J_i \\ -M_f & M & M_i \end{pmatrix}^2 \left(J_f \left\| \frac{1}{e} \sum e_i r_i^L \mathbf{Y}_L(\theta_i\,\varphi_i) \right\| J_i \right)^2 \\ &= \frac{1}{2J_i + 1} \left(J_f \left\| \frac{1}{e} \sum e_i r_i^L \mathbf{Y}_L(\theta_i\,\varphi_i) \right\| J_i \right)^2 . \end{aligned} \tag{17.9}$$

The matrix elements in (17.9) are those of an irreducible tensor operator. Their evaluation was therefore straightforward. For magnetic transitions some transformations are required to simplify the expression. Introducing (17.8) into (17.7) we obtain for the matrix elements

$$\left\langle J_f M_f \left| \frac{1}{L+1} \frac{1}{ec} \int \left(\operatorname{grad} r^L Y_{LM} \cdot [\mathbf{r} \times (\mathbf{j} + c \operatorname{curl} \mathbf{m})] \right) d^3x \right| J_i M_i \right\rangle$$

$$= \frac{1}{(L+1)ec} \left\langle J_f M_f \left| \sum_i \left(\operatorname{grad} r_i^L Y_{LM}(\theta_i\,\varphi_i) \cdot \frac{e_i}{m_i} \hbar \mathbf{l}_i \right) \right| J_i M_i \right\rangle$$

$$+ \frac{1}{e(L+1)} \left\langle J_f M_f \left| \int \left(\operatorname{grad} r^L Y_{LM} \cdot \left\{ \mathbf{r} \times \sum_i \mu_i \operatorname{curl} [\boldsymbol{\sigma}_i \delta(\mathbf{r} - \mathbf{r}_i)] \right\} \right) d^3x \right| J_i M_i \right\rangle.$$

By partial integration we obtain for the operator in the second matrix element

$$\int \left(\operatorname{grad} r^L Y_{LM} \cdot \{ \mathbf{r} \times \operatorname{curl} [\boldsymbol{\sigma}_i \delta(\mathbf{r} - \mathbf{r}_i)] \} \right)$$

$$\begin{aligned} = -\int \delta(\mathbf{r} - \mathbf{r}_i) \Big[& (\boldsymbol{\sigma}_i \cdot \mathbf{r})\, \Delta(r^L Y_{LM}) - (\boldsymbol{\sigma}_i \cdot \operatorname{grad} r^L Y_{LM}) \\ & - (\boldsymbol{\sigma}_i \cdot \operatorname{grad} (\mathbf{r} \cdot \operatorname{grad} r^L Y_{LM})) \Big]. \end{aligned}$$

However, $\Delta(r^L Y_{LM}) = 0$ and $\mathbf{r} \cdot \text{grad}\, r^L Y_{LM} = L r^L Y_{LM}$ as can be easily seen by noting that $r^L Y_{LM}$ is a homogeneous polynomial of degree L in x, y, and z. Adding all terms together we therefore obtain

$$B(\text{mag}, L) = \frac{1}{2J_i + 1}$$
$$\times \left| \left(J_f \left\| \sum_i [\text{grad}\,(r_i^L Y_{LM}(\theta_i\, \varphi_i)] \cdot \left[\frac{1}{ec(L+1)} \frac{e_i \hbar}{m_i} \mathbf{l}_i + \frac{1}{e} \mu_i \boldsymbol{\sigma}_i\right] \right\| J_i\right)\right|^2 . \tag{17.10}$$

Equations (17.9) and (17.10) assume a simple form when the system is taken to be that of a single particle moving in a central field. It will be shown in Part III that the calculation of operators like (17.9) or (17.10) can always be reduced to the calculation of single particle matrix elements. Let us denote the quantum numbers of the initial and final single particle states by $n_i l_i j_i$ and $n_f l_f j_f$ respectively. If the single particle is a proton with $e_i = e$, $m_i = m_p$ and $\mu_i = \mu_p\,(e\hbar/2m_p c)$ we obtain

$$B_{\text{s.p.}}(\text{el}, L) = \frac{1}{2j_i + 1} |(n_f l_f j_f || r^L Y_L(\theta\, \varphi) || n_i l_i j_i)|^2 \tag{17.11}$$

$$B_{\text{s.p.}}(\text{mag}, L) = \frac{1}{2j_i + 1}$$
$$\times \left| \left(n_f l_f j_f \left\| \text{grad}\,[r^L Y_{LM}(\theta\, \varphi)] \cdot \left[\frac{\hbar}{m_p c} \frac{\mathbf{l}}{L+1} + \frac{\hbar}{2m_p c} \mu_p \boldsymbol{\sigma}\right] \right\| n_i l_i j_i\right)\right|^2 . \tag{17.12}$$

These equations can be further simplified by the use of tensor algebra. In (17.11) we note that the radial part of the matrix element is separable from its angular part. We also note that the operator $r^L Y_{LM}(\theta\, \varphi)$ does not operate on the spin coordinates of the proton. We can therefore use (15.26) and obtain

$$B_{\text{s.p.}}(\text{el}, L) = (2j_f + 1)\, (l_f || \mathbf{Y}_L || l_i)^2 \begin{Bmatrix} l_f & j_f & \frac{1}{2} \\ j_i & l_i & L \end{Bmatrix}^2 \left| \int_0^\infty R^*_{n_f l_f}(r)\, r^L R_{n_i l_i}(r)\, dr \right|^2 . \tag{17.13}$$

We saw in (13.19) that

$$\int Y^*_{l_1 m_1}(\theta\, \varphi)\, Y_{LM}(\theta\, \varphi)\, Y_{l_2 m_2}(\theta\, \varphi) \sin\theta\, d\theta d\varphi$$

$$= \frac{1}{\sqrt{4\pi}} \sqrt{\frac{(2L+1)(2l_2+1)}{(2l_1+1)}}\, (L\, M l_2 m_2 | L\, l_2 l_1 m_1)\, (L\, 0 l_2 0 | L\, l_2 l_1 0)$$

$$= \frac{(-1)^{m_1}}{\sqrt{4\pi}} \sqrt{(2L+1)(2l_1+1)(2l_2+1)} \begin{pmatrix} l_1 & L & l_2 \\ -m_1 & M & m_2 \end{pmatrix} \begin{pmatrix} l_1 & L & l_2 \\ 0 & 0 & 0 \end{pmatrix} .$$

Using the Wigner-Eckart theorem we find that

$$(l_1||\mathbf{Y}_L||l_2) = \sqrt{\frac{2L+1}{4\pi}} \cdot (-1)^{l_1} \sqrt{(2l_1+1)(2l_2+1)} \begin{pmatrix} l_1 & L & l_2 \\ 0 & 0 & 0 \end{pmatrix}. \tag{17.14}$$

Substituting (17.14) in (17.13) we obtain an explicit form more convenient for direct calculations

$$B_{\text{s.p.}}(\text{el}, L) = \frac{(2j_f+1)(2l_f+1)(2L+1)(2l_i+1)}{4\pi} \times \begin{pmatrix} l_f & L & l_i \\ 0 & 0 & 0 \end{pmatrix}^2 \begin{Bmatrix} l_f & j_f & \frac{1}{2} \\ j_i & l_i & L \end{Bmatrix}^2 \langle r^L \rangle^2 \tag{17.15}$$

where

$$\langle r^L \rangle = \int_0^\infty R^*_{n_f l_f}(r)\, r^L R_{n_i l_i}(r)\, dr. \tag{17.16}$$

The reduction of $B_{\text{s.p.}}(\text{mag}, L)$ is less simple. To carry it through we first derive an expression for grad $[r^L Y_{LM}(\theta\,\varphi)]$. Using (16.6) and the explicit expression for $\boldsymbol{l}$ we derive the commutators

$$[((l_x + s_x) \pm i(l_y + s_y)), \text{grad}\,(r^L Y_{LM}(\theta\,\varphi))]$$

$$= \sqrt{L(L+1) - M(M \pm 1)}\, \text{grad}\,(r^L Y_{LM\pm 1}(\theta\,\varphi))$$

$$[(l_z + s_z), \text{grad}\,(r^L Y_{LM}(\theta\,\varphi))] = M\, \text{grad}\,(r^L Y_{LM}(\theta\,\varphi)).$$

Comparing this result with (12.35) we see that grad $[r^L Y_{LM}(\theta\,\varphi)]$ is an irreducible tensor of degree L. Furthermore, since it is obtained from the grad operation on a homogeneous polynomial of degree L it can only be built from the spherical harmonics of degree $L-1$. The only irreducible tensor of degree L built from spherical harmonics of degree $L-1$, each component of which is a vector field, is the vector spherical harmonic $\mathbf{Y}_m^{L(L-1,1)}(\theta\,\varphi)$. Thus we conclude that

$$\text{grad}\,[r^L Y_{LM}(\theta\,\varphi)] = \alpha\, r^{L-1}\, \mathbf{Y}_M^{L(L-1,1)}(\theta\,\varphi) \tag{17.17}$$

where α is a proportionality factor. To obtain α we evaluate by partial integration the integral over a volume within the sphere S of radius R

$$\int_S \{\text{grad}\,[r^L Y_{LM}(\theta\,\varphi)]\}^\dagger \cdot \{\text{grad}\,[r^L Y_{LM}(\theta\,\varphi)]\}\, d^3x$$

$$= \int_{r=R} [\text{grad}\,(r^L Y_{LM}(\theta\,\varphi))]^\dagger \cdot \frac{\mathbf{r}}{r}\, [r^L Y_{LM}(\theta\,\varphi)]\, dS$$

$$- \int_S [r^L Y_{LM}(\theta\,\varphi)]^\dagger\, \Delta [r^L Y_{LM}(\theta\,\varphi)]\, d^3x.$$

The volume integral on the right-hand side vanishes since

$$\Delta[r^L Y_{LM}(\theta\,\varphi)] = 0.$$

As to the surface integral, we have already seen that

$$\mathbf{r} \cdot \text{grad}\,[r^L Y_{LM}(\theta\,\varphi)] = L r^L Y_{LM}(\theta\,\varphi).$$

Hence, using (17.17) and the orthogonality of the vector spherical harmonics, we obtain

$$|\alpha|^2 \int_0^R r^{2L} dr = L R^{2L+1}$$

and hence

$$\alpha = \pm \sqrt{L(2L+1)}.$$

The actual sign can be easily determined by taking, for instance, the spherical harmonic with maximum M in (17.17). We thus obtain

$$\text{grad}\,[r^L Y_{LM}(\theta\,\varphi)] = \sqrt{L(2L+1)}\, r^{L-1}\, \mathbf{Y}_M^{L(L-1,1)}(\theta\,\varphi). \tag{17.18}$$

A useful relation valid for any vector $\mathbf{v}$ can be obtained from (17.18),

$$\text{grad}\,[r^L Y_{LM}(\theta\,\varphi)] \cdot \mathbf{v} = \sqrt{L(2L+1)}\, r^{L-1} [\mathbf{Y}_{L-1}(\theta\,\varphi) \times \mathbf{v}]_M^{(L)} \tag{17.19}$$

where the components of the irreducible tensor (17.19) are physical *scalars*. To prove this relation we note that from (16.17) it follows that for any vector $\mathbf{v}$

$$(\boldsymbol{\chi}_\kappa \cdot \mathbf{v}) = v_\kappa. \tag{17.20}$$

Hence

$$\text{grad}\,[r^L Y_{LM}(\theta\,\varphi)] \cdot \mathbf{v} = \sqrt{L(2L+1)}\, r^{L-1} (\mathbf{Y}_M^{L(L-1,1)}(\theta\,\varphi) \cdot \mathbf{v})$$

$$= \sqrt{L(2L+1)}\, r^{L-1} \sum_{\mu,\kappa} (L-1, \mu\, 1\, \kappa | L-1, 1\, L\, M)\, Y_{L-1,\mu} (\boldsymbol{\chi}_\kappa \cdot \mathbf{v})$$

$$= \sqrt{L(2L+1)}\, r^{L-1} \sum_{\mu,\kappa} (L-1, \mu\, 1\, \kappa | L-1, 1\, L\, M)\, Y_{L-1,\mu}\, v_\kappa$$

$$= \sqrt{L(2L+1)}\, r^{L-1} [\mathbf{Y}_{L-1}(\theta\,\varphi) \times \mathbf{v}]_M^{(L)}.$$

We can now use (17.19) to transform the matrix element in (17.12) and obtain:

$$\left(n_f l_f j_f \left\| \left(\operatorname{grad}\left[r^L Y_{LM}(\theta\,\varphi)\right] \cdot \left[\frac{\hbar}{m_p c}\frac{1}{L+1}\boldsymbol{l} + \frac{\hbar}{2m_p c}\mu_p \boldsymbol{\sigma}\right]\right) \right\| n_i l_i j_i\right)$$

$$= \frac{\hbar}{m_p c}\frac{1}{L+1}\left(n_f l_f j_f \|\left(\operatorname{grad}\left[r^L Y_{LM}(\theta\,\varphi)\right] \cdot \mathbf{j}\right)\| n_i l_i j_i\right)$$

$$+ \frac{\hbar}{2m_p c}\left(\mu_p - \frac{1}{L+1}\right)\left(n_f l_f j_f \|(\operatorname{grad}\left[r^L Y_{LM}(\theta\,\varphi)\right] \cdot \boldsymbol{\sigma})\| n_i l_i j_i\right)$$

$$= \frac{\hbar}{m_p c}\frac{1}{L+1}\left(n_f l_f j_f \| r^{L-1}\left[\mathbf{Y}_{L-1}(\theta\,\varphi) \times \mathbf{j}\right]^{(L)} \| n_i l_i j_i\right)\sqrt{L(2L+1)}$$

$$+ \frac{\hbar}{2m_p c}\left(\mu_p - \frac{1}{L+1}\right)\left(n_f l_f j_f \| r^{L-1}\left[\mathbf{Y}_{L-1}(\theta\,\varphi) \times \boldsymbol{\sigma}\right]^{(L)} \| n_i l_i j_i\right)\sqrt{L(2L+1)} \tag{17.21}$$

where $\mathbf{j} = \boldsymbol{l} + \frac{1}{2}\boldsymbol{\sigma}$ is the operator of total angular momentum of the proton. The right-hand side of (17.21) can be evaluated by the use of tensor algebra. Observing that $\mathbf{j}$ is diagonal in the quantum numbers l and j, we obtain for the first term, using (15.23),

$$\left(n_f l_f j_f \| r^{L-1}\left[\mathbf{Y}_{L-1}(\theta\,\varphi) \times \mathbf{j}\right]^{(L)} \| n_i l_i j_i\right)$$

$$= \langle r^{L-1}\rangle \left(l_f j_f \| \mathbf{Y}_{L-1} \| l_i j_i\right)\left(l_i j_i \| \mathbf{j} \| l_i j_i\right)\sqrt{2L+1}\,(-1)^{j_f+j_i+L}\begin{Bmatrix} j_f & L & j_i \\ 1 & j_i & L-1 \end{Bmatrix}. \tag{17.22}$$

The expectation value $\langle r^{L-1}\rangle$ is defined in (17.16).

The second term in (17.21) can be evaluated by using (14.26). We obtain

$$\left(n_f l_f j_f \| r^{L-1}\left[\mathbf{Y}_{L-1}(\theta\,\varphi) \times \boldsymbol{\sigma}\right]^{(L)} \| n_i l_i j_i\right)$$

$$= \langle r^{L-1}\rangle \left(l_f \| \mathbf{Y}_{L-1} \| l_i\right)\left(\tfrac{1}{2}\|\boldsymbol{\sigma}\|\tfrac{1}{2}\right)\sqrt{(2j_f+1)(2L+1)(2j_i+1)}\begin{Bmatrix} l_f & \frac{1}{2} & j_f \\ l_i & \frac{1}{2} & j_i \\ L-1 & 1 & L \end{Bmatrix}. \tag{17.23}$$

The reduced matrix elements of $\boldsymbol{\sigma}$ and $\mathbf{j}$ are given by (14.21) and in particular

$$\left(\tfrac{1}{2}\|\boldsymbol{\sigma}\|\tfrac{1}{2}\right) = 2\left(\tfrac{1}{2}\|\mathbf{s}\|\tfrac{1}{2}\right) = \sqrt{6}. \tag{17.24}$$

Using the relation (15.26) we obtain

$$\left(l_f j_f \| \mathbf{Y}_{L-1} \| l_i j_i\right)$$

$$= (-1)^{l_f+1/2+j_i+L-1}\left(l_f \| \mathbf{Y}_{L-1} \| l_i\right)\sqrt{(2j_i+1)(2j_f+1)}\begin{Bmatrix} l_f & j_f & \frac{1}{2} \\ j_i & l_i & L-1 \end{Bmatrix}. \tag{17.25}$$

We can add up all contributions and obtain for (17.12) the expression

$$B_{\text{s.p.}}(\text{mag. } L) = \left(\frac{\hbar}{m_p c}\right)^2 \langle r^{L-1}\rangle^2 (l_f||Y_{L-1}||l_i)^2 (2j_f+1)\frac{(2L+1)^2}{L}$$

$$\times\left[\frac{(-1)^{j_f+l_f+1/2}L}{L+1}\sqrt{(2j_i+1)j_i(j_i+1)}\begin{Bmatrix} l_f & j_f & \frac{1}{2} \\ j_i & l_i & L-1\end{Bmatrix}\begin{Bmatrix} j_f & L & j_i \\ 1 & j_i & L-1\end{Bmatrix}\right.$$

$$\left. + \sqrt{\tfrac{3}{2}}\left(L\mu_p - \frac{L}{L+1}\right)\begin{Bmatrix} l_f & \frac{1}{2} & j_f \\ l_i & \frac{1}{2} & j_i \\ L-1 & 1 & L\end{Bmatrix}\right]^2. \qquad (17.26)$$

Equation (17.26) is considerably simplified if $|j_f - j_i| = L$, i.e., if L is the minimum angular momentum which must be carried away by the radiation when the proton jumps from j_i to j_f. In this case

$$\begin{Bmatrix} l_f\, j_f & \frac{1}{2} \\ j_i\, l_i\, L-1\end{Bmatrix} = 0$$

since j_f, j_i and $L-1$ do not satisfy the triangular condition. Thus we are left only with

$$B_{\text{s.p.}}(\text{mag}, L) = \left(\frac{\hbar}{m_p c}\right)^2 \langle r^{L-1}\rangle^2 (l_f||\mathbf{Y}_{L-1}||l_i)^2 (2j_f+1)\frac{(2L+1)^2}{L}$$

$$\times \tfrac{3}{2}\left(L\mu_p - \frac{L}{L+1}\right)^2 \begin{Bmatrix} l_f & \frac{1}{2} & j_f \\ l_i & \frac{1}{2} & j_i \\ L-1 & 1 & L\end{Bmatrix}^2 \quad \text{for} \quad |j_i - j_f| = L. \qquad (17.27)$$

Equations (17.5), (17.13), and (17.26) give the rate of emission of electromagnetic radiation by a proton moving in a central field. This rate depends on the initial and final angular momenta, the multipolarity of the radiation, the dimensions of the emitting system, and the energy available for the transition. Combining these equations we obtain for electric multipole radiation

$$T_{\text{s.p.}}(\text{el}, L) = c\frac{e^2}{\hbar c}k^{2L+1}\langle r^L\rangle^2 \frac{2(L+1)}{L[(2L+1)!!]^2}S(\text{el}) \qquad (17.28)$$

and for magnetic multipole radiation with $|j_i - j_f| = L$

$$T_{\text{s.p.}}(\text{mag}, L) = c\frac{e^2}{\hbar c}k^{2L+1}\left(\frac{\hbar}{m_p c}\right)^2\langle r^{L-1}\rangle^2$$

$$\times\left(L\mu_p - \frac{L}{L+1}\right)^2 \frac{2(L+1)}{L[(2L+1)!!]^2}S(\text{mag}) \qquad (17.29)$$

where

$$S(\text{el}) = (2j_f + 1) \begin{Bmatrix} l_f & j_f & \frac{1}{2} \\ j_i & l_i & L \end{Bmatrix}^2 4\pi |(l_f||\mathbf{Y}_L||l_i)|^2 \tag{17.30}$$

$$S(\text{mag}) = (2j_f + 1) \frac{(2L+1)^2}{L} \frac{3}{2} \begin{Bmatrix} l_f & \frac{1}{2} & j_f \\ l_i & \frac{1}{2} & j_i \\ L-1 & 1 & L \end{Bmatrix}^2 4\pi |(l_f||\mathbf{Y}_{L-1}||l_i)|^2. \tag{17.31}$$

The factors S, which are often referred to as the *statistical factors*, are of order unity as can be verified readily. The main determining factors for the transition rates are therefore the multipolarity L, the energy of the radiation $\hbar\omega = c\hbar k$, and the dimensions of the system which enter through $\langle (kr)^L \rangle$. For a given multipolarity L, the transition probability increases with energy as $(\hbar\omega)^{2L+1}$, corresponding to the fact that more energetic γ rays have a bigger amplitude inside the emitting system. For a given energy $\hbar\omega$, or wave number k, the transition probability decreases like $(kR)^{2L}$ where R is the linear dimension of the emitting system, and $kR \ll 1$. This result corresponds to the fact that higher multipole fields have a lower amplitude inside the emitting system, i.e., near $r = 0$.

As a direct consequence of these considerations we can conclude that of all the multipole radiations of a definite type (electric or magnetic) which are allowed by conservation of angular momentum, the one with smallest L is the most probable.

A comparison of (17.28) and (17.29) shows that in single proton transitions the minimum L is *always* favored, irrespective of whether this radiation turns out to be electric or magnetic. For transitions of a single neutron moving in a central field the rate for magnetic radiation is given by (17.29) with $(L\mu_N)^2$ replacing $[L\mu_p - (L/L+1)]^2$. Electric multipoles cannot be radiated by neutron transitions in the approximation which was used to derive (17.28). In this derivation we neglected the contribution of the magnetic moment of the proton to the radiation of electric multipoles since it was of order $(v/c)^2$ compared to the contribution of the proton charge to this same radiation, where v is the velocity of the proton in the nucleus. For neutron transitions, however, this is the only possible source for the radiation of electric multipoles. Since, as will be pointed out below, such processes are of little practical interest, we shall not derive an explicit formula for them.

Although (17.28) and (17.29) and their analogs for neutron transitions are strictly valid only for *single* nucleons, we shall see in Part III that in the shell model practically the same transition rates are obtained also when more complicated configurations are involved. The reason for this is that the interaction of a nucleus, composed of many nucleons, with the electromagnetic field is the sum of the interactions of each of the nucleons with that field, as assumed explicitly in (17.8). Therefore, the probability amplitude for the emission of an electromagnetic radiation from a system of many nucleons is the sum of the probability amplitudes for that radiation being emitted by each of the nucleons. In the shell model, which is an independent particle model, interference phenomena hardly occur. The process is thus essentially a single particle transition with the rest of the nucleons modifying only the overlap of the initial and final states and occasionally the statistical factor S.

This modification may sometimes be quite appreciable. Thus, for instance, consider a transition between any state of the j_1^2 configuration and a state of the j_2^2 configuration with $j_1 \neq j_2$. Here no first-order effect is possible at all, since the overlap of the initial and final wave functions of the nonradiating nucleon vanishes. However, in many cases of practical interest the two levels involved differ, at most, by the quantum numbers of one nucleon. Therefore, we can then expect for the radiation rates, at least as far as orders of magnitude are concerned, some values in the vicinity of those given by (17.28) and (17.29). Other models may not have the nucleons moving in well-defined orbits, in which case many contributions may have to be added up together possibly showing coherence effects. These may therefore lead to appreciably different rates.

In comparing (17.28) and (17.29) with the experimental data it is better to concentrate first on general qualitative agreement and then discuss actual quantitative agreement. As was pointed out in Part I, the shell model does not only assert that each nucleon moves in a well-defined orbit but also gives the sequence of levels occupied by successive nucleons added to the nucleus. Since this sequence also determines generally the first few excited states of the nucleus, we see that the shell model determines the expected multipolarities of radiations emitted by various nuclei. Thus, for instance, in the nuclear shell model just below $Z = 50$ or $N = 50$, odd protons or neutrons, respectively, occupy either the level $p_{1/2}$ or $g_{9/2}$. As a result, the lowest single nucleon excitations can be expected to be from $p_{1/2}$ to $g_{9/2}$ or from $g_{9/2}$ to $p_{1/2}$, orbits. Hence, the predominant radiation emitted from these excited states should be a magnetic $L = 4$ ($M4$) radiation. This, indeed, is very convincingly confirmed by experiment. Similar general predictions of the nuclear shell model are experimentally confirmed over the whole range of the periodic table.

Some other predictions of the theory presented above are, however, not found to agree with experimental findings. We have already noted that electric multipole radiations by neutrons should proceed at a considerably slower rate than those emitted by a proton. The former can occur only due to relativistic effects, which are small for the nucleons in a nucleus. No such difference between neutrons and protons is observed even in cases when we have every reason to believe that the radiating nucleon is definitely either a neutron or a proton (e.g., for transitions in odd-A nuclei in which either the protons or the neutrons fill a closed shell). Quite generally no systematic difference could be found between transitions which are believed to involve radiating protons and transitions in which a neutron is believed to be responsible for the radiation.

We also see from (17.31) that for magnetic dipole radiations, $L = 1$, l_f must be equal to l_i, otherwise the 9-j symbol, and therefore $S(\text{mag})$, vanishes. Nevertheless such "l-forbidden" transitions, e.g., magnetic dipole radiations in transitions between a $d_{3/2}$-state and an $s_{1/2}$-state, are observed to proceed at a rate comparable to other transitions.

The actual quantitative agreement between the predicted rates and observed values for γ-ray emission from nuclei is quite poor. It is considerably better in atoms, but even there it is much worse than the agreement between theory and experiment on energy spacings. It can be expected that the very much idealized picture offered by the shell model does not represent exactly the actual situation in nuclei. Even in atoms, where it has a much more sound justification, it needs a number of corrections. The inadequacy of the shell model to describe the actual situation exactly can be formulated by saying that the actual orbits ascribed to each particle have to be corrected. This will naturally result in it being impossible to ascribe unique occupation number to the different orbits. Some operators, like those which give the rate of electromagnetic radiation, may be affected considerably by such modifications. On the other hand, the energy operator has a stationary value for a given state. It will consequently be less affected by the deviations of the actual states of the shell model. Thus, in using the shell model for predicting various properties of atoms or nuclei we have to keep in mind the possibility that it may be a better model for some properties and less good for others.

18. Two Interacting Particles in a Central Field

In Part I we considered the nucleus as a system of nucleons moving in a common potential well. This picture was motivated by the similar picture used to describe the atom, where, to a first approximation, the electrons are visualized as moving independently of each other in the common field created by the charge of the nucleus. Many properties of such idealized systems can be obtained from the study of the motion of one particle, the other particles showing their effect through the Pauli principle only, namely, in forbidding the occupation of some states by the particle considered.

Obviously a common single particle central interaction cannot replace completely the actual mutual interactions in a system of many particles. Under certain circumstances it can do so to a good approximation, and the whole philosophy of the shell model is built on the idea that such is the case in atoms and in nuclei. However, to be more realistic we have to consider corrections to the central field. These will naturally take the form of effective residual interactions among the particles. Thus, the motion is no more that of independent particles.

We are therefore interested in solving the Schrödinger equation

$$(H_0 + H_1)\,\psi(\mathbf{r}_1,\mathbf{r}_2, \ldots, \mathbf{r}_A) = E\psi(\mathbf{r}_1, \mathbf{r}_2, \ldots, \mathbf{r}_A) \tag{18.1}$$

where H_0 is the Hamiltonian of the central field

$$H_0 = \sum [T_i + U(r_i)] \tag{18.2}$$

and H_1 is whatever has to be added to it in order to make it more realistic. It may include corrections to the single particle potential $U(r_i)$ as well as additional interactions between the particles. The wave function $\psi(\mathbf{r}_1, \mathbf{r}_2, \ldots, \mathbf{r}_A)$ is antisymmetric in the coordinates of the A particles. Since the main term in H_1 is most probably an effective *two-body* interaction, we shall devote the rest of Part II to the study of systems of two particles in a central field.

There are very few cases in which the Schrödinger equation (18.1) can be solved exactly, and in most cases we have to use approximation methods to solve it even for two particles. Although several approaches are known today for the approximate solution of (18.1), we shall confine ourselves in this book to its solution by means of perturbation theory, and in most cases shall be satisfied with the lowest approximation. There is not as yet any theoretical justification for doing so in treating nuclei. Therefore we shall develop this approach in its utmost generality so as to make it applicable also for other methods of approximation. This will be achieved by making no specific assumptions on the residual interactions between particles.

The states of two noninteracting particles moving in a central field are generally degenerate. Thus, the energy of a particle in the orbit (l, m_s, m_l) where m_l is the z-projection of the orbital angular momentum and m_s that of the spin, does not depend on either m_l or m_s. Therefore all the *m-scheme states*

$$\varphi(l_1 m_{s1} m_{l1}, l_2 m_{s2} m_{l2}) = \varphi_{n_1 l_1 m_{s1} m_{l1}}(\mathbf{r}_1)\, \varphi_{n_2 l_2 m_{s2} m_{l2}}(\mathbf{r}_2) \tag{18.3}$$

which are characterized by putting each particle in a state with definite values of m_l and m_s, are degenerate. The wave function $\varphi_{nlm_s m_l}(\mathbf{r})$ satisfies

$$H_0 \varphi_{nlm_s m_l}(\mathbf{r}) = (T + U)\varphi_{nm_s m_l}(\mathbf{r}) = E_{nl} \varphi_{nlm_s m_l}(\mathbf{r}) \tag{18.4}$$

and all the states $\varphi(l_1 m_{s1} m_{l1}, l_2 m_{s2} m_{l2})$ in (18.3) belong to the same eigenvalue $E_{n_1 l_1} + E_{n_2 l_2}$ of H_0. According to perturbation theory we obtain the first-order corrections to the energies of the two-particle system by diagonalizing the submatrix of the perturbation H_1 defined by the space of the degenerate states. Thus, strictly speaking, we should construct the matrix

$$\langle n_1 l_1 m_{s1} m_{l1}, n_2 l_2 m_{s2} m_{l2} | H_1 | n_1 l_1 m'_{s1} m'_{l1}, n_2 l_2 m'_{s2} m'_{l2} \rangle \tag{18.5}$$

and diagonalize it. The approximate nature of the answer obtained this way is stressed in (18.5) by the fact that the quantum numbers $n_1 l_1$ and $n_2 l_2$ are the same on both sides. The exact solution would have involved the diagonalization of the complete matrix

$$\langle n_1 l_1 m_{s1} m_{l1}, n_2 l_2 m_{s2} m_{l2} | H_0 + H_1 | n'_1 l'_1 m'_{s1} m'_{l1}, n'_2 l'_2 m'_{s2} m'_{l2} \rangle .$$

By choosing specific linear combinations of the functions $\varphi(l_1 m_{s1} m_{l1}, l_2 m_{s2} m_{l2})$, defined in (18.3), we can reduce considerably the problem of diagonalizing H_1 in (18.5). Thus, we know that H_1 is a scalar function of the coordinates (including spins) of the two particles. It therefore commutes with the total angular momentum $\mathbf{J}$ of the two particles and has no matrix elements between states of different values of J and M. We therefore construct the states

$$\varphi(l_1 l_2 \alpha\, J M) = \sum a^{(\alpha JM)}_{m_{s1} m_{l1} m_{s2} m_{l2}}\, \varphi(l_1 m_{s1} m_{l1}, l_2 m_{s2} m_{l2}) \tag{18.6}$$

so that

$$
\begin{aligned}
(\boldsymbol{l}_1 + \mathbf{s}_1 + \boldsymbol{l}_2 + \mathbf{s}_2)^2 \varphi(l_1 l_2 \alpha\, J\, M) &= J(J+1)\, \varphi(l_1 l_2 \alpha\, J\, M) \\
(l_{1z} + s_{1z} + l_{2z} + s_{2z})\, \varphi(l_1 l_2 \alpha\, J\, M) &= M \varphi(l_1 l_2 \alpha\, J\, M).
\end{aligned}
\tag{18.7}
$$

Here α is an additional quantum number required for specifying different states with the same values of $n_1 l_1 n_2 l_2$ and J. H_1 has then no matrix elements between states of different values of J and M, and the problem of diagonalizing H_1 in (18.5) reduces to the problem of diagonalizing the set of matrices

$$
\langle \alpha | H_1^{(JM)} | \alpha' \rangle \equiv \langle l_1 l_2 \alpha\, J\, M | H_1 | l_1 l_2 \alpha'\, J\, M \rangle \tag{18.8}
$$

where, for each matrix, l_1, l_2, J and M are kept constant. Actually we even know that since H_1 is a scalar function of the coordinates of the two particles, its eigenvalues do not depend on M at all, so that it is enough to diagonalize only those matrices (18.8) which are obtained for different values of J and take one specific values of M, say $M = J$, in each case.

The order of the matrices which have to be diagonalized in (18.8) is in general considerably smaller than that of the matrix in (18.5). For each value of J this order is given by the number of possible independent states with total angular momentum J in the configuration $l_1 l_2$. We can easily find this number by constructing these states according to the following prescription. Couple first $\boldsymbol{l}_1$ and $\boldsymbol{l}_2$ to $\mathbf{L}$, $\mathbf{s}_1$ and $\mathbf{s}_2$ to $\mathbf{S}$, and then $\mathbf{S}$ and $\mathbf{L}$ to $\mathbf{J}$. S can obviously have the values 0 and 1 only. For $S = 0$ we must have $L = J$, thus obtaining the state $\varphi(l_1 l_2\, S = 0\, L = J\, J\, M)$. For $S = 1$, L can take on the values $J + 1$, J, and $J - 1$, so that we get the three states $\varphi(l_1 l_2\, S = 1\, L = J \pm 1\, J\, M)$ and $\varphi(l_1 l_2\, S = 1\, L = J\, J\, M)$. Therefore, there are at most 4 states of a given total angular momentum J in the $l_1 l_2$ configuration and the matrices in (18.8) are at most of order 4.

In many cases there are even less states of a given J and M. For example, if we are dealing with two equivalent particles, for which $n_1 = n_2 = n$ and $l_1 = l_2 = l$, we find that

$$
\begin{aligned}
\varphi_{12}(l\, l\, S\, L\, J\, M) &= \sum (l\, m_{l1} l\, m_{l2} | l\, l\, L\, M_L)\, (s\, m_{s1} s\, m_{s2} | s\, s\, S\, M_S) \\
&\quad \times (S\, M_S L\, M_L | S\, L\, J\, M)\, \varphi_1(l\, m_{l1})\, \varphi_2(l\, m_{l2})\, \chi_1(m_{s1})\, \chi_2(m_{s2}) \\
&= \sum (S\, M_S L\, M_L | S\, L\, J\, M)\, \varphi_{12}(l\, l\, L\, M_L)\, \chi_{12}(s\, s\, S\, M_S)
\end{aligned}
\tag{18.9}
$$

where

$$
\varphi_{12}(l\, l\, L\, M_L) = \sum (l\, m\, l\, m' | l\, l\, L\, M_L)\, \varphi_1(l\, m)\, \varphi_2(l\, m') \tag{18.10}
$$

and similarly

$$
\chi_{12}(s\, s\, S\, M_S) = \sum (s\, m_s s\, m_s' | s\, s\, S\, M_S)\, \chi_1(m_s)\, \chi_2(m_s'). \tag{18.11}
$$

Using the symmetry properties of the V-coefficients (13.57) it is easily seen from (18.10) that

$$\varphi_{21}(l\,l\,L\,M_L) = \sum (l\,m\,l\,m'|l\,l\,L\,M_L)\,\varphi_2(l\,m)\,\varphi_1(l\,m') = (-1)^{2l-L}\,\varphi_{12}(l\,l\,L\,M_L) \tag{18.12}$$

and similarly

$$\chi_{21}(s\,s\,S\,M_S) = (-1)^{2s-S}\,\chi_{12}(s\,s\,S\,M_S). \tag{18.13}$$

Since l is an integer and s a half-integer, we see that $\varphi_{12}(l\,l L\,M_L)$ is a symmetric or antisymmetric function of the space coordinates of the two particles according to whether L is even or odd. On the other hand $\chi_{12}(s\,s\,S\,M_S)$ is a symmetric or antisymmetric function of the spin coordinates of the two particles according to whether S is odd or even. Since the Pauli principle holds for two identical fermions, the wave function $\varphi_{12}(l\,l\,S\,L\,J\,M)$ in (18.9) should be antisymmetric with respect to the simultaneous exchange of space and spin coordinates. Bearing in mind the results (18.12) and (18.13) we see that this is possible only if $L + S$ is even. Thus for antisymmetric wave functions we obtain:

$$\varphi(l^2 S\,L\,J\,M) = 0 \qquad \text{for} \qquad S + L \text{ odd}. \tag{18.14}$$

If J is odd, there is therefore only one possible nonvanishing, antisymmetric state, namely,

$$\varphi(l^2\ S = 1\ L = J\ J\ M) \qquad J \text{ odd}. \tag{18.15}$$

If J is even, there are generally only three possible states

$$\varphi(l^2\ S = 0\ L = J\ J\ M), \quad \varphi(l^2\ S = 1\ L = J \pm 1\ J\ M) \qquad J \text{ even}. \tag{18.16}$$

We therefore find that for two equivalent identical particles in a state of odd angular momentum the energy shift due to the interaction H_1 is given by the only element of the corresponding matrix (18.8)

$$\Delta E(l^2 J) = (l^2\ S = 1\ L = J\ J\ M|H_1|l^2\ S = 1\ L = J\ J\ M) \quad J \text{ odd}. \tag{18.17}$$

For even J, it is given by the eigenvalues of the matrix

$$\begin{Vmatrix} (0\ J\ \ J|H_1|0\ J\ J) & (0\ J\ \ J|H_1|1\ J-1\ J) & (0\ J\ \ J|H_1|1\ J+1\ J) \\ (1\ J-1\ J|H_1|0\ J\ J) & (1\ J-1\ J|H_1|1\ J-1\ J) & (1\ J-1\ J|H_1|1\ J+1\ J) \\ (1\ J+1\ J|H_1|0\ J\ J) & (1\ J+1\ J|H_1|1\ J-1\ J) & (1\ J+1\ J|H_1|1\ J+1\ J) \end{Vmatrix} \tag{18.18}$$

where we used the notation

$$(1\ J-1\ J|H_1|0\ J\ J) \equiv (l^2\ S = 1\ L = J - 1\ J\ M|H_1|l^2\ S = 0\ L = J\ J\ M),$$

etc. In as much as the three eigenvalues of (18.18) are different, we conclude that the three possible states with that particular value of J in the l^2 configuration are affected differently by the interaction H_1.

It may seem that since H_1 is at most of order 4 its diagonalization poses no special problems. However, it is difficult to make general statements about the eigenvalues of H_1 without further specifying it, and furthermore we cannot construct the general eigenfunctions of the system unless we have further information on H_1.

If H_1 does not contain the spin coordinates at all, it commutes with $\mathbf{S}$. Since under this condition it is a scalar built of the space coordinates only, it commutes also with $\mathbf{L}$. If for example we consider again equivalent nucleons in the l^2 configuration we see from (18.18) that, for such an interaction, H_1 is already diagonal in the scheme used. In general, for perturbations H_1 which do not depend on the spin coordinates of the particles we thus know how to construct wave functions with which H_1 already assumes a diagonal form within the $l_1 l_2$. configuration These wave functions are

$$\varphi(l_1 l_2 S\, L\, J\, M) = \sum (l_1 m_{l1} l_2 m_{l2} | l_1 l_2 L\, M_L)\, (s_1 m_{s1} s_2 m_{s2} | s\, s\, S\, M_S)$$

$$\times\, (S\, M_S L\, M_L | S\, L\, J\, M)\, \varphi(l_1 m_{s1} m_{l1},\, l_2 m_{s2} m_{l2}). \qquad (18.19)$$

The wave function (18.19) is obtained by coupling first $\boldsymbol{l}_1$ and $\boldsymbol{l}_2$ to yield $\mathbf{L}$, $\mathbf{s}_1$, and $\mathbf{s}_2$ to $\mathbf{S}$, and then $\mathbf{S}$ and $\mathbf{L}$ to $\mathbf{J}$.

H_1 need not be spin-independent in order to be diagonal in the wave functions (18.19). It is sufficient, for this purpose, that H_1 will commute with both $\mathbf{S}$ and $\mathbf{L}$. This is obviously true for any interaction which is spin-independent, i.e., for $H_1 = V(|\mathbf{r}_1 - \mathbf{r}_2|)$, but also for interactions of the type $H_1 = (\mathbf{s}_1 \cdot \mathbf{s}_2) V(|\mathbf{r}_1 - \mathbf{r}_2|)$. The forces due to such interactions, which commute with both $\mathbf{S} = \mathbf{s}_1 + \mathbf{s}_2$ and $\mathbf{L} = \boldsymbol{l}_1 + \boldsymbol{l}_2$, are called *scalar forces*. This is not to be confused with the fact that any interaction is a scalar, i.e., commutes with the *total* angular momentum $\mathbf{J} = \mathbf{S} + \mathbf{L}$. The term scalar forces refers to the fact that these interactions which commute with $\mathbf{S}$ and $\mathbf{L}$, are scalars both in ordinary and in spin space.

The wave functions (18.19) constitute a definite scheme for the coupling of the four angular momenta $\boldsymbol{l}_1$, $\boldsymbol{l}_2$, $\mathbf{s}_1$, and $\mathbf{s}_2$ to a total $\mathbf{J}$. It is generally known as the *LS-coupling scheme* or *Russel-Saunders coupling.* It gives the "correct" linear combinations of the two particle functions $\varphi(l_1 m_{s1} m_{l1}, l_2 m_{s2} m_{l2})$ for scalar forces H_1. The matrix $\langle l_1 l_2 S\, L\, J\, M | H_1 | l_1 l_2 S' L' J\, M\rangle$ in this case is diagonal in L and S and is therefore diagonal in this scheme. Thus, for perturbations H_1, which commute with both $\mathbf{S}$ and $\mathbf{L}$, the states of the $l_1 l_2$ configuration are conveniently characterized by the values of S and L which, together, play the role of α in (18.6). The energy shift of the state $\psi(l_1 l_2 S\, L\, J\, M) \equiv \psi(^{2S+1}L_J)$ is given by

$$\Delta E(^{2S+1}L_J) = \langle l_1 l_2 S\, L\, J\, M | H_1 | l_1 l_2 S\, L\, J\, M\rangle. \qquad (18.20)$$

An important property of $\Delta E(^{2S+1}L_J)$ for the special interactions we have been considering, is that for a given value of S and L it is independent of J. If H_1 commutes with $\mathbf{S}$ and $\mathbf{L}$ separately it is necessarily a function of operators which are scalars with respect to S and L. Such operators can be any central two-body interaction $V(|\mathbf{r}_1 - \mathbf{r}_2|)$ and the operator $(\mathbf{s}_1 \cdot \mathbf{s}_2)$. It can therefore be written in the form

$$H_1 = \sum_{k=0,1} f_k V_k(|\mathbf{r}_1 - \mathbf{r}_2|)\,(\mathbf{s}_1 \cdot \mathbf{s}_2)^k \tag{18.21}$$

where f_k are real numbers. In (18.21) k does not exceed 1 since higher powers of $(\mathbf{s}_1 \cdot \mathbf{s}_2)$ can be expressed as linear combinations of 1 and $(\mathbf{s}_1 \cdot \mathbf{s}_2)$. The matrix element (18.20) is, according to (18.21), a sum of matrix elements of products of two scalars, one operating on the space coordinates and the other on the spin coordinates. Using (15.4) and (15.15) we see that (18.20) is independent of J. This result can be understood by noting that, since H_1 commutes with $\mathbf{S}$ and $\mathbf{L}$ separately, its eigenvalues must be independent of M_S and M_L. Therefore states with definite S and L must have the same energy irrespective of the value of J.

The procedure used above for the construction of the "correct linear combinations" in the case of two particles with scalar forces can be extended to more than two particles. The crucial point is that H_1 for scalar forces commutes with both $\mathbf{S} = \sum \mathbf{s}_i$ and $\mathbf{L} = \sum \mathbf{l}_i$. Therefore, by constructing states with a definite value of S and L in the $l_1 l_2 \ldots l_n$ configuration, we automatically obtain a partial diagonalization of H_1. However, with more than two particles, S and L alone are not generally sufficient to specify the states uniquely. Thus, a general function of n particles in a central field in LS coupling will be denoted by $\psi(l_1 l_2 \ldots l_n \alpha SLJM)$ where α is a set of additional quantum numbers.

Now coming back to two-particle configurations we can see what happens when the interaction H_1 does not commute with both $\mathbf{L}$ and $\mathbf{S}$. Clearly the matrix (18.8) of $H_1(J\,M)$ for definite values of J and M will not be diagonal any more in the LS-coupling scheme. However, the off-diagonal elements might still be small, in the sense of perturbation theory. In this case, we have for every value of S, L, S', and L'

$$\begin{aligned} |\langle l_1 l_2 S\,L\,J\,M|H_1|l_1 l_2 S'L'J\,M\rangle| \ll |&\langle l_1 l_2 S\,L\,J\,M|H_1|l_1 l_2 S\,L\,J\,M\rangle \\ &- \langle l_1 l_2 S'L'J\,M|H_1|l_1 l_2 S'L'J\,M\rangle|. \end{aligned} \tag{18.22}$$

The eigenvalues of $H_1(J\,M)$ will then differ only slightly from its diagonal elements. In fact, in such cases we obtain the correction up to the second order, to the actual eigenvalue $\Delta E(^{2S+1}L_J)$ as

$$\Delta E(^{2S+1}L_J) = \langle H_1\rangle_{SLJ} + \sum_{S'L' \neq SL} \frac{|\langle l_1 l_2\, S\,L\,J|H_1|l_1 l_2 S'L'J\rangle|^2}{\langle H_1\rangle_{SLJ} - \langle H_1\rangle_{S'L'J}} + \ldots . \tag{18.23}$$

In (18.23) we use the notation

$$\langle H_1 \rangle_{SLJ} \equiv \langle l_1 l_2 S\, L\, J | H_1 | l_1 l_2 S\, L\, J \rangle.$$

Hence we conclude that, as long as the part of H_1 which is non-diagonal in S and L is small compared to the appropriate differences of the diagonal elements (18.22), the LS-coupling scheme offers a good zeroth-order approximation.

An important interaction which is not diagonal in S and L is the well-known spin-orbit interaction (7.1)

$$U_{so} = \sum_i \zeta(r_i)\,(\boldsymbol{l}_i \cdot \mathbf{s}_i). \tag{18.24}$$

The matrix elements of the operator $\zeta(r)\,(\boldsymbol{l} \cdot \mathbf{s})$ in a single particle state were calculated in (9.4) and are given by

$$\langle n\, l\, j\, m | \zeta(r)\,(\boldsymbol{l} \cdot \mathbf{s}) | n\, l\, j\, m \rangle = \zeta_{nl} \frac{j(j+1) - l(l+1) - \frac{3}{4}}{2} \tag{18.25}$$

where

$$\zeta_{nl} = \int_0^\infty R^2_{nl}(r)\, \zeta(r) dr.$$

It is instructive to calculate also this matrix element using the formalism of tensor algebra. We notice that $(\boldsymbol{l} \cdot \mathbf{s})$ is a scalar product of two tensor operators of degree 1 (vectors) $\boldsymbol{l}$ and $\mathbf{s}$ operating on different degrees of freedom of the particle. Using (15.5) we therefore obtain

$$\langle n\, l\, j\, m | \zeta(r)\,(\boldsymbol{l} \cdot \mathbf{s}) | n\, l\, j\, m \rangle = (-1)^{l+1/2+j} (l || \boldsymbol{l} || l)\, (\tfrac{1}{2} || \mathbf{s} || \tfrac{1}{2}) \begin{Bmatrix} l & \frac{1}{2} & j \\ \frac{1}{2} & l & 1 \end{Bmatrix} \zeta_{nl}.$$

Introducing the values (14.21) for the reduced matrix elements we find that

$$\langle n\, l\, j\, m | \zeta(r)\,(\boldsymbol{l} \cdot \mathbf{s}) | n\, l\, j\, m \rangle = \zeta_{nl} (-1)^{l+1/2+j} \sqrt{(2l+1)\, l(l+1)\, \tfrac{3}{2}} \begin{Bmatrix} l & \frac{1}{2} & j \\ \frac{1}{2} & l & 1 \end{Bmatrix}.$$

We can now use the explicit form of the W-coefficient involved here (see Appendix) in order to obtain the simple result (18.25).

In two-particle configurations the matrix elements of

$$U_{so} = \zeta(r_1)\,(\boldsymbol{l}_1 \cdot \mathbf{s}_1) + \zeta(r_2)\,(\boldsymbol{l}_2 \cdot \mathbf{s}_2)$$

can be most conveniently calculated in a scheme in which we couple $\mathbf{s}_1$ and $\boldsymbol{l}_1$ to $\mathbf{j}_1$, $\mathbf{s}_2$ and $\boldsymbol{l}_2$ to $\mathbf{j}_2$, and then $\mathbf{j}_1$ and $\mathbf{j}_2$ to $\mathbf{J}$. This coupling scheme is known as the *jj-coupling scheme*. It is the natural scheme for calculating matrix elements of spin-orbit interactions because the spin-orbit interaction (18.24) commutes with $\mathbf{l}_1^2$, $\mathbf{l}_2^2$, $\mathbf{j}_1^2$, $\mathbf{j}_2^2$, and $\mathbf{J}$. The jj-coupling wave functions are thus defined by

$$\psi(l_1 j_1 l_2 j_2 J\, M) = \sum (s_1 m_{s1} l_1 m_{l1} | s_1 l_1 j_1 m_1)\,(s_2 m_{s2} l_2 m_{l2} | s_2 l_2 j_2 m_2) \times (j_1 m_1 j_2 m_2 | j_1 j_2 J M)\, \varphi(l_1 m_{s1} m_{l1}, l_2 m_{s2} m_{l2}). \tag{18.26}$$

With those functions we obtain

$$\langle l_1 j_1 l_2 j_2 J\, M | \sum_i \zeta(r_i)\,(\mathbf{l}_i \cdot \mathbf{s}_i) | l_1 j_1' l_2 j_2' J\, M \rangle = \left[\sum \zeta_{n_i l_i} \frac{j_i(j_i+1) - l_i(l_i+1) - \frac{3}{4}}{2}\right] \delta_{j_1 j_1'}\, \delta_{j_2 j_2'}. \tag{18.27}$$

In order to calculate the matrix elements of the spin-orbit interaction in the LS-coupling scheme, we express a wave function in the LS-coupling scheme in terms of wave functions in the jj-coupling scheme. This is a change of coupling transformation which is given by (14.38)

$$\varphi(l_1 l_2 S\, L\, J\, M) = \varphi[s_1 s_2(S) l_1 l_2(L) J\, M] = \sum_{j_1 j_2} \sqrt{(2S+1)(2L+1)(2j_1+1)(2j_2+1)} \times \begin{Bmatrix} \frac{1}{2} & l_1 & j_1 \\ \frac{1}{2} & l_2 & j_2 \\ S & L & J \end{Bmatrix} \psi(l_1 j_1 l_2 j_2 J\, M). \tag{18.28}$$

We can now use (18.25) to express the matrix elements of the spin-orbit interaction in the LS-coupling scheme in terms of those in the jj-coupling scheme. It is possible to arrive at the same result in a more straightforward manner. Thus, consider the matrix element

$$\langle l_1 l_2 S\, L\, J\, M | \zeta(r_1)\,(\mathbf{s}_1 \cdot \mathbf{l}_1) | l_1 l_2 S' L' J\, M \rangle. \tag{18.29}$$

This is the matrix element of a scalar product of two vectors, $\mathbf{s}_1$ and $\mathbf{l}_1$, multiplied by $\zeta_{n_1 l_1}$. We can therefore use (15.5) to express (18.29) in terms of $(S||\mathbf{s}_1||S')$ and $(L||\mathbf{l}_1||L')$. These reduced matrix elements can further be simplified using (15.26) and be expressed in terms of $(\frac{1}{2}||\mathbf{s}_1||\frac{1}{2})$ and $(l_1||\mathbf{l}_1||l_1)$. We thus obtain

$$\langle l_1 l_2 S\, L\, J\, M | \zeta(r_1)\,(\mathbf{l}_1 \cdot \mathbf{s}_1) + \zeta(r_2)\,(\mathbf{l}_2 \cdot \mathbf{s}_2) | l_1 l_2 S' L' J\, M \rangle$$
$$= \left[(-1)^{L-L'} \zeta_{n_1 l_1} (l_1||\mathbf{l}_1||l_1) \begin{Bmatrix} l_1 & L & l_2 \\ L' & l_1 & 1 \end{Bmatrix} + (-1)^{S-S'} \zeta_{n_2 l_2} (l_2||\mathbf{l}_2||l_2) \begin{Bmatrix} l_2 & L & l_1 \\ L' & l_2 & 1 \end{Bmatrix}\right]$$
$$\times (-1)^{l_1+l_2+J+1} \sqrt{(2S+1)(2S'+1)(2L+1)(2L'+1)}$$
$$\times (\tfrac{1}{2}||\mathbf{s}||\tfrac{1}{2}) \begin{Bmatrix} \frac{1}{2} & S & \frac{1}{2} \\ S & \frac{1}{2} & 1 \end{Bmatrix} \begin{Bmatrix} S & L & J \\ L' & S' & 1 \end{Bmatrix}. \tag{18.30}$$

Equation (18.30) assumes a simple form in the important case of the diagonal elements ($S = S'$, $L = L'$) in configurations of equivalent particles. Noting that (see Appendix)

$$\begin{Bmatrix} l & L & l \\ L & l & 1 \end{Bmatrix} = (-1)^{L+1} \sqrt{\frac{L(L+1)}{4l(l+1)(2l+1)(2L+1)}}$$

and using (14.21) we obtain in this special case

$$\langle l^2 S\, L\, J\, M | \sum_i \zeta(r_i)\, (\boldsymbol{l}_i \cdot \mathbf{s}_i) | l^2 S\, L\, J\, M \rangle$$

$$= \zeta_{nl}(-1)^{L-J}(2S+1)\sqrt{(2L+1)L(L+1)}\,(\tfrac{1}{2}||\mathbf{s}||\tfrac{1}{2}) \begin{Bmatrix} \frac{1}{2} & S & \frac{1}{2} \\ S & \frac{1}{2} & 1 \end{Bmatrix} \begin{Bmatrix} S & L & J \\ L & S & 1 \end{Bmatrix}$$

$$= \tfrac{1}{2}\zeta_{nl}[J(J+1) - L(L+1) - S(S+1)]. \tag{18.31}$$

The J, L, and S dependence of (18.31) is quite general and is valid for the diagonal elements of $\Sigma_i\, \zeta(r_i)\, (\boldsymbol{l}_i \cdot \mathbf{s}_i)$ for any number of particles and any configuration. This can be verified easily from the above derivation. It is due solely to the fact that the spin-orbit interaction is a scalar built from two vectors, one operating on the space coordinates and the other on the spin coordinates, and that the matrix elements are taken in a scheme in which $\mathbf{L}^2$ and $\mathbf{S}^2$ are diagonal.

We can now see the effect of the addition of a spin-orbit interaction to the H_1 with scalar forces discussed above. With H_0 alone, all the states of the $l_1 l_2$ configuration are degenerate. Introducing the interaction H_1 each state $\varphi(l_1 l_2 S\, L\, J\, M)$ is shifted by an amount $\langle H_1 \rangle_{SLJ}$ which may be different for different values of S and L, but for given S and L is independent of J. The previously degenerate states are therefore split into a number of degenerate *multiplets*. Each multiplet comprises all the states of different J and M but the same S and L. If either $S = 0$ or $L = 0$, the multiplet is composed of just one state $\varphi(l_1 l_2 S\, L\, J)$; if $L \neq 0$ and $S = 1$, it is composed of the three states $\varphi(l_1 l_2\; S = 1\; L\; J = L - 1)$, $\varphi(l_1 l_2\; S = 1\; L\; J = L)$, and $\varphi(l_1 l_2\; S = 1\; L\; J = L + 1)$. The introduction of a spin-orbit interaction will split the different J states within a multiplet. Even if the off-diagonal elements of the spin-orbit interaction are small compared to the differences of the diagonal elements of H_1, we see from (18.30) that the spin-orbit interaction will have different contributions to the diagonal elements of H_1 for different values of J. We can illustrate the situation by the following example.

Let us consider the *pd* configuration of two particles with $l_1 = 1$ $l_2 = 2$. Before the introduction of H_1 there are altogether

$$2(2l_1 + 1) \times 2(2l_2 + 1) = 60$$

degenerate states in this configuration [the factor 2 is the number of possible spin orientations, $(2l + 1)$—that of different values of m_l]. S can assume two values, 0 and 1, and L can assume any integral values consistent with $|l_1 - l_2| \leq L \leq l_1 + l_2$, i.e., $L = 1$, 2, and 3. All possible combinations of S, L, and J are therefore

$$\begin{array}{cccc} {}^1P_1 & {}^3P_0 & {}^3P_1 & {}^3P_2 \\ {}^1D_2 & {}^3D_1 & {}^3D_2 & {}^3D_3 \\ {}^1F_3 & {}^3F_2 & {}^3F_3 & {}^3F_4 \end{array}$$

where in ${}^{2S+1}L_J$ the P, D, and F stand for $L = 1$, 2, and 3 respectively. Each symbol denotes the $(2J + 1)$ states obtained by taking different values of M.

It has already been pointed out that all the states of the multiplet ${}^{2S+1}L_J$ are degenerate in the absence of H_1. Upon the introduction of the perturbation H_1 the various multiplets are separated from each other. The states ${}^3P_0$, ${}^3P_1$, and ${}^3P_2$ still remain degenerate after the introduction of the scalar force, and are split only if the spin-orbit interaction is included.

If $\zeta_{n_1l_1}$ and $\zeta_{n_2l_2}$ are small compared to the separation induced by the scalar force between different multiplets, we expect the spectrum of energy levels of the l_1l_2 configuration to be composed of several clusters of levels. Each such cluster—a multiplet—is characterized by definite values of S and L. These clusters have a *fine structure* which is due to the splitting of their levels according to the different values of J.

We can draw some general conclusions about the structure of such multiplets from (18.30). In fact, if we ignore the contribution of the off-diagonal elements of the spin-orbit interaction, then the shift of the state with total angular momentum J within a given multiplet is given by

$$\delta E_J = c(l_1l_2\, S\, L)\, [J(J + 1) - L(L + 1) - S(S + 1)]. \tag{18.32}$$

Here, c is a certain function of l_1, l_2, S, and L (but not of J). The explicit expression for $W(S\, L\, J|L\, S\, 1)$ given in the Appendix was used in deriving (18.32) [as well as in (18.31) above].

It follows from (18.32) that the shift δE_J increases quadratically with J. Therefore the intervals between two levels are proportional to J

$$\delta E_{J+1} - \delta E_J = 2c(l_1l_2S\, L)\, (J + 1). \tag{18.33}$$

Equation (18.33) is known as the *Landé interval rule*. It is obeyed relatively well in many atomic spectra where the spin-orbit interaction is indeed very small compared to the splitting between two multiplets.

Another general conclusion can be drawn by noting that the only J dependence of (18.30) is given by

$$(-1)^J \begin{Bmatrix} S & L & J \\ L' & S' & 1 \end{Bmatrix}.$$

Noting that the fine-structure splitting ΔE_J is given by the diagonal elements of (18.32) we obtain, using (15.16)

$$\sum (2J+1)\, \Delta E_J = 0. \tag{18.34}$$

Since each level with a definite J is $(2J+1)$-fold degenerate, (18.34) means that the *center of mass* of a multiplet coincides with the level into which the multiplet would have degenerated in the absence of spin-orbit interaction.

The preceding discussion on the multiplet structure of the $l_1 l_2$ configuration is valid only if the spin-orbit interaction is small compared to the interaction which splits the multiplets. Only in this approximation it is justified to neglect off-diagonal elements of the spin-orbit interaction in the LS-coupling scheme and to consider contributions to the diagonal elements only. If the spin-orbit interaction is not so small we have to calculate the complete matrix of H_1 in the LS-coupling scheme and diagonalize it. Alternatively, we can try to use another coupling, scheme, such that the spin-orbit interaction is diagonal in it, and consider the rest of H_1 as a perturbation. We have already seen how to construct such a scheme—the *jj-coupling* scheme given by (18.26). The four possible states $\psi(l_1 j_1 l_2 j_2 J M)$ of the $l_1 l_2$ configuration belonging to the angular momentum J and to $J_z = M$ will now be characterized by the four combinations of the quantum numbers j_1 and j_2, namely: $j_1 = l_1 + \frac{1}{2}$, $j_2 = l_2 + \frac{1}{2}$; $j_1 = l_1 + \frac{1}{2}$, $j_2 = l_2 - \frac{1}{2}$; $j_1 = l_1 - \frac{1}{2}$, $j_2 = l_2 + \frac{1}{2}$; $j_1 = l - \frac{1}{2}$, $j_2 = l - \frac{1}{2}$. The spin-orbit interaction is diagonal in this scheme as shown in (18.27). However, a two-body interaction like $V(|\mathbf{r}_1 - \mathbf{r}_2|)$ will have nonvanishing off-diagonal matrix elements. In order that the jj-coupling scheme be a good first approximation, the off-diagonal elements in $H_1(JM)$ must be small compared to the corresponding differences of the diagonal elements

$$|\langle l_1 j_1 l_2 j_2 J\, M | H_1 | l_1 j_1' l_2 j_2' J\, M\rangle| \ll |\langle l_1 j_1 l_2 j_2 J\, M | H_1 | l_1 j_1 l_2 j_2 J\, M\rangle - \langle l_1 j_1' l_2 j_2' J\, M | H_1 | l_1 j_1' l_2 j_2' J\, M\rangle|. \tag{18.35}$$

Taking for H_1 the expression

$$H_1 = \sum \zeta(r_i)\, (\boldsymbol{l}_i \cdot \mathbf{s}_i) + V_{12} \tag{18.36}$$

and using (18.27), (18.35) becomes

$$|\langle l_1 j_1 l_2 j_2 J\, M | V_{12} | l_1 j_1' l_2 j_2' J\, M\rangle| \ll |\tfrac{1}{2}\zeta_{n_1 l_1}[j_1(j_1+1) - j_1'(j_1'+1)] + \tfrac{1}{2}\zeta_{n_2 l_2}[j_2(j_2+1) - j_2'(j_2'+1)] + \langle l_1 j_1 l_2 j_2 J\, M | V_{12} | l_1 j_1 l_2 j_2 J\, M\rangle - \langle l_1 j_1' l_2 j_2' J M | V_{12} | l_1 j_1' l_2 j_2' J\, M\rangle| \quad \text{for} \quad jj\text{-coupling.} \tag{18.37}$$

With a two-body interaction that gives rise to scalar forces the conditions for the validity of the LS-coupling scheme are given by rewriting (18.22), in the form

$$|\langle l_1 l_2 S\, L\, J| \sum \zeta(r_i)\,(\boldsymbol{l}_i \cdot \mathbf{s}_i)| l_1 l_2 S' L' J\rangle| \ll |V_{SLJ} - V_{S'L'J} + \langle l_1 l_2 S\, L\, J| \sum \zeta(r_i)\,(\boldsymbol{l}_i \cdot \mathbf{s}_i)| l_1 l_2 S\, L\, J\rangle - \langle l_1 l_2 S' L' J| \sum \zeta(r_i)\,(\boldsymbol{l}_i \cdot \mathbf{s}_i)| l_1 l_2 S' L' J\rangle| \quad \text{for} \quad LS\text{-coupling.} \tag{18.38}$$

We see that LS-coupling is a good approximation if ζ_{nl} is considerably smaller than an average difference between two V_{SLJ}. The jj-coupling scheme becomes a good approximation in the other extreme.

It is possible to derive more precise conditions for the validity of jj-coupling or LS-coupling within a given configuration by using the numerical values of the coefficients of the transformation (18.28) from LS- to jj-coupling. We shall not enter here into a full discussion of this point. It was pointed out in Part I that there are several indications that in nuclei the spin-orbit interaction is considerably bigger than the residual two-body force. Therefore, in nuclei, the jj-coupling scheme is a considerably better approximation than LS-coupling. Most of our subsequent considerations and examples will be therefore limited to the jj-coupling scheme.

In jj-coupling the spin-orbit interaction splits the $l_1 l_2$ configuration into four *jj-coupling* $(l_1 j_1 l_2 j_2)$ *configurations* obtained by taking the four possible combinations of $j_1 = l_1 \pm \frac{1}{2}$ and $j_2 = l_2 \pm \frac{1}{2}$. Each jj-coupling configuration is composed of states of different values of J and M with $|j_1 - j_2| \leq J \leq j_1 + j_2$ and $M = J,\ J-1,\ \ldots,\ -J$. As long as we do not introduce any further interactions, these states are degenerate. Two-particle interactions can remove this degeneracy and the shift of each state is given, to a first approximation, by the diagonal elements of the interaction

$$\Delta E_J = \langle l_1 j_1 l_2 j_2 J\, M | V_{12} | l_1 j_1 l_2 j_2 J\, M\rangle. \tag{18.39}$$

The calculation of two-particle spectra in the *jj*-coupling scheme amounts therefore to the evaluation of the matrix elements (18.39). In the following we shall very often speak of the "j_1j_2 configuration," and consider matrix elements $\langle j_1j_2J\,M|V_{12}|j_1j_2J\,M\rangle$, where l_1 and l_2 are omitted. We shall always mean by that the *jj-coupling* $l_1j_1l_2j_2$ configuration and the matrix element (18.39). Also, when we speak of the *orbit* of a particle in *jj*-coupling, we mean the state with the function $\varphi(l\,j\,m)$ with definite *l and j*.

19. Antisymmetric Wave Functions and Isospin

So far we have been considering general two-particle states given by

$$\varphi(l_1 j_1 l_2 j_2 J\,M) = \sum (j_1 m_1 j_2 m_2 | j_1 j_2 J\,M)\, \varphi_1(l_1 j_1 m_1)\, \varphi_2(l_2 j_2 m_2). \tag{19.1}$$

In dealing with nuclei it will be convenient to consider separately states of two identical nucleons (i.e., two protons or two neutrons) and two different nucleons (i.e., a proton and a neutron). For two identical particles the correct wave functions $\varphi_a(l_1 j_1 l_2 j_2 J\,M)$ should be antisymmetric

$$\varphi_a(l_a j_a l_b j_b J\,M)$$
$$= N \sum_{m_a, m_b} (j_a m_a j_b m_b | j_a j_b J\,M)\, [\varphi_1(l_a j_a m_a)\, \varphi_2(l_b j_b m_b) - \varphi_1(l_b j_b m_b)\, \varphi_2(l_a j_a m_a)]. \tag{19.2}$$

Here we denoted the quantum numbers by indices a and b since it is no longer possible to say that particle 1 is in one definite state and particle 2 is in another definite state. N is a normalization factor which can be calculated by requiring that $\int |\psi_a|^2 d\mathbf{r}_1 d\mathbf{r}_2 = 1$. If $(l_a j_a) \neq (l_b j_b)$ we readily obtain $N = (1/\sqrt{2})$. If however, $(l_a j_a) = (l_b j_b) = (l\,j)$ we obtain, using the symmetry properties of the V-coefficients (13.56),

$$\psi_a[(lj)^2 J\,M] = N \sum_{m,m'} (j\,m\,j\,m' | j\,j\,J\,M)\, [\varphi_1(l\,j\,m)\, \varphi_2(l\,j\,m') - \varphi_1(l\,j\,m')\, \varphi_2(l\,j\,m)]$$

$$= N \sum_{m,m'} [(j\,m\,j\,m' | j\,j\,J\,M) - (j\,m'\,j\,m | j\,j\,J\,M)]\, \varphi_1(l\,j\,m)\, \varphi_2(l\,j\,m')$$

$$= N[1 - (-1)^{2j-J}] \sum_{m,m'} (j\,m\,j\,m' | j\,j\,J\,M)\, \varphi_1(l\,j\,m)\, \varphi_2(l\,j\,m'). \tag{19.3}$$

Since j is a half-integer we see that

In configurations of two equivalent particles $[(l_a j_a) = (l_b j_b)]$ *in jj-coupling the only antisymmetric states are those having* even *total angular momentum. These are the only allowed states for identical fermions.* (19.4)

For even values of J and equivalent particles we obtain for the normalization factor in (19.3) $N = \frac{1}{2}$. The antisymmetric normalized function is thus

$$\psi_a[(lj)^2 J\,M] = \sum_{m,m'} (j\,m\,j\,m' | j\,j\,J\,M)\, \varphi_1(l\,j\,m)\, \varphi_2(l\,j\,m') \qquad J \text{ even}. \tag{19.5}$$

This restriction to even values of J in jj-coupling is similar to the restrictions on the antisymmetric states of equivalent particles in LS-coupling, (18.14). It is instructive to see how we can obtain one restriction from the other. Using the transformation from LS-coupling to jj-coupling (18.28) we can write

$$\psi_a[(lj)^2 J\,M] = \sum_{S,L} \sqrt{(2S+1)(2L+1)}\,(2j+1)\begin{Bmatrix} \frac{1}{2} & l & j \\ \frac{1}{2} & l & j \\ S & L & J \end{Bmatrix} \varphi_a(l^2 S\,L\,J\,M). \tag{19.6}$$

The 9-j symbol in (19.6) vanishes if $(-1)^{2(1/2+l+j)+S+L+J} = -1$, i.e., if $S+L+J$ is odd [cf. (14.32)]. The LS-coupling wave function $\varphi_a(l^2 S\,L\,J\,M)$ vanishes if $S+L$ is odd. Thus, both $S+L$ and $S+L+J$ have to be even in order to get a nonvanishing jj-coupling wave functions. It follows that $\psi_a[(l\,j)^2 J\,M]$ does not vanish only for even values of J.

For two nonidentical particles, i.e., a proton and a neutron, there is no restriction on the symmetry of the wave function and we could describe a system of a proton and a neutron moving in a central field simply by (19.1). A complication arises, however, due to the great similarity between the neutron and the proton. Consider the state $\varphi_{pn}(j_1 j_2 J\,M)$ defined by (19.1), in which the proton and the neutron are in j_1 and j_2-orbits respectively, and the state $\varphi_{np}(j_1 j_2 J\,M)$ in which their orbits are interchanged. Since the masses of the neutron and the proton are very close to each other, the total kinetic energy in both states is practically the same. The potential energy depends on the central field in which these two particles move. Strictly speaking a proton in a nucleus sees a different average field than that seen by a neutron, since the proton is also affected by the electric repulsion of all the other protons in the nucleus. However, the energies associated with Coulomb forces, especially in light nuclei, are small compared to those associated with nuclear forces. Since we know that nuclear forces are the same between two protons, two neutrons, or a neutron and a proton (in the same state), we conclude that the two states we considered above have also practically the same potential energy in the central field. Thus $\varphi_{pn}(j_1 j_2 J\,M)$ and $\varphi_{np}(j_1 j_2 J_2 M)$ are degenerate states. This degeneracy remains even if we take into account the diagonal elements of the interaction V_{pn}. Therefore, the residual interaction V_{pn} has to be diagonalized in the subspace of these two states. To obtain the first-order correction to the energies of the two degenerate states $\varphi_{pn}(j_1 j_2 J\,M)$ and $\varphi_{np}(j_1 j_2 J\,M)$, we have then to evaluate the eigenvalues of the matrix

$$\begin{pmatrix} \langle p\,n|V_{pn}|p\,n\rangle & \langle p\,n|V_{pn}|n\,p\rangle \\ \langle n\,p|V_{pn}|p\,n\rangle & \langle n\,p|V_{pn}|n\,p\rangle \end{pmatrix} \tag{19.7}$$

where

$$\langle p\, n|V_{pn}|n\, p\rangle = \int \varphi^*_{pn}(j_1 j_2 J\, M) V_{pn} \varphi_{np}(j_1 j_2 J\, M) d\mathbf{r}_p d\mathbf{r}_n$$

etc.

Since $V_{pn} = V_{np}$, the diagonal elements in (19.7) are equal. Therefore, the eigenstates of (19.7) are the symmetric and antisymmetric combinations of φ_{pn} and φ_{np}. We define accordingly

$$\psi^{(\pm)}_{pn}(j_a j_b\, J\, M) \equiv \psi^{(\pm)}_{pn}(l_a j_a l_b j_b\, J\, M)$$

$$= N \sum (j_a m_a\, j_b m_b | j_a j_b\, J\, M)\, [\varphi_p(l_a j_a m_a)\, \varphi_n(l_b j_b m_b) \pm \varphi_n(l_a j_a m_a)\, \varphi_p(l_b j_b m_b)]. \tag{19.8}$$

These functions satisfy

$$\psi^{(\pm)}_{pn} = \pm\, \psi^{(\pm)}_{np} \tag{19.9}$$

and therefore

$$\int \psi^{(\pm)*}_{pn}\, V_{pn}\, \psi^{(\mp)}_{pn} = 0. \tag{19.10}$$

In general, an interaction V_{pn} will therefore split the two degenerate linear combinations $\psi^{(+)}_{pn}$ and $\psi^{(-)}_{pn}$ of (19.8). The energy shift of each of these states is given by

$$\Delta E^{(\pm)}_{JM} = \int \psi^{(\pm)*}_{pn}(j_a j_b\, J\, M) V_{pn} \psi^{(\pm)}_{pn}(j_a j_b\, J\, M). \tag{19.11}$$

Since we assumed that the dependence of $\varphi(l\, j\, m)$ on the spin and space coordinates is the same for both a neutron and a proton, it is convenient to write the nucleon wave function in the form of a product of two functions

$$\varphi_N(l j\, m) = \varphi(l j\, m)\, \xi(N). \tag{19.12}$$

Here $\varphi(l\, j\, m)$ depends only on the space and spin coordinates, and $\xi(N)$ depends only on the charge coordinates of the *nucleon* (either a proton or a neutron). To make the definition of $\xi(N)$ more precise we introduce an operator Q which measures the charge of the nucleon. This operator is defined by the requirement that it has two eigenvalues 1 and 0, with corresponding two eigenfunctions $\xi(p)$ and $\xi(n)$

$$Q\xi(p) = \xi(p) \qquad Q\xi(n) = 0. \tag{19.13}$$

The matrix representation of Q is of order 2, and $\xi(N)$ can be considered as a two-rowed column:

$$Q = \begin{pmatrix} 1 & 0 \\ 0 & 0 \end{pmatrix}, \qquad \xi(p) = \begin{pmatrix} 1 \\ 0 \end{pmatrix}, \qquad \xi(n) = \begin{pmatrix} 0 \\ 1 \end{pmatrix}, \tag{19.14}$$

Such column functions were encountered before when we discussed the spin eigenfunctions. Therefore the mathematical apparatus introduced there can be used in treating the charge functions $\xi(N)$. We shall introduce the operator τ_3, which operates on the charge functions $\xi(N)$, and whose representation in the scheme (19.14) is given by

$$\tau_3 = \begin{pmatrix} 1 & 0 \\ 0 & -1 \end{pmatrix}. \tag{19.15}$$

The operator Q then becomes $Q = \frac{1}{2}(1 + \tau_3)$, and both $\xi(p)$ and $\xi(n)$ are seen to be eigenfunctions of τ_3 belonging to the eigenvalues $+1$ and -1 respectively. (Sometimes another convention is used ascribing to the proton the eigenvalue -1 of τ_3. The two conventions are obviously equivalent.) Using (19.12) and (19.14) we can write the wave function of a proton in the state $\varphi(l\,j\,m)$ as

$$\varphi_p(l\,j\,m) = \begin{bmatrix} \varphi(l\,j\,m) \\ 0 \end{bmatrix} \tag{19.16}$$

and similarly for a neutron in the same orbit

$$\varphi_n(l\,j\,m) = \begin{bmatrix} 0 \\ \varphi(l\,j\,m) \end{bmatrix}. \tag{19.17}$$

The form (19.16) and (19.17) of the nucleon wave functions is that of a spinor field whose spinor indices refer to components in a mathematical space—the *charge space*. Actually, to assure that $\varphi_N(l\,j\,m)$ is a spinor in such a space we have to know its transformation properties under rotations in this space. Since, however, this is a hypothetical new space which has nothing to do with actual space we can *define* the rotations in this space so as to make $\varphi_N(l\,j\,m)$ a spinor field. The situation here can be confronted with that of the ordinary spin. There we saw that in order to preserve the invariance of certain equations—such as the two-component neutrino equation—we had to impose on the two components of the field describing the neutrino certain transformation laws associated with rotations in real space. Although the spin indices of the neutrino referred to components in a different space—the spin space—their transformation was tightly linked to rotations in ordinary space. Each rotation in the latter space induced a well-defined rotation in the former space. The rotations in the charge-space we are considering here are entirely independent of the rotations in ordinary space and are introduced, at this stage at least, as a formal device only, which considerably simplifies the treatment of systems with many particles.

To distinguish this new spinor field from the ordinary spins, we call it an *isospinor* field, (also called isotopic spinor or isobaric spinor). Its components are called *isospinor components*, and the charge-space is often referred to as the *isospin space* or *isospace*. Comparing (19.15) and (11.7) we see that τ_3 is most conveniently identified with the operator which gives the effect of infinitesimal rotations around the z-axis in the isospin space. More precisely, an infinitesimal rotation by an angle $\delta\theta_3$ around this axis induces a change in φ_N given by

$$\delta\varphi_N = -\tfrac{1}{2} i \delta\theta_3 \, \tau_3 \varphi_N.$$

It is therefore more appropriate to consider ξ_p and ξ_n as eigenstates of $\frac{1}{2}\tau_3$, rather than τ_3, with eigenvalues $+\frac{1}{2}$ and $-\frac{1}{2}$. The nucleon is then visualized as a particle with *isospin* $\mathbf{t}$ with $t = \frac{1}{2}$ (not to be confused with its normal spin which also happens to be $s = \frac{1}{2}$). The proton is the state with a "z-projection" (in isospin space!) $t_3 = +\frac{1}{2}$ and the neutron—the state with $t_3 = -\frac{1}{2}$.

We can also introduce the operators for infinitesimal rotations around the other two axes in isospace, τ_1 and τ_2, defined in the above scheme by

$$\tau_1 = \begin{pmatrix} 0 & 1 \\ 1 & 0 \end{pmatrix} \qquad \tau_2 = \begin{pmatrix} 0 & -i \\ i & 0 \end{pmatrix}. \tag{19.18}$$

It is readily seen, by operating with the matrices (19.18) on the isospinors (19.16) and (19.17) that

$$\begin{aligned} \tfrac{1}{2}(\tau_1 + i\tau_2)\varphi_p &= 0 & \tfrac{1}{2}(\tau_1 + i\tau_2)\varphi_n &= \varphi_p \\ \tfrac{1}{2}(\tau_1 - i\tau_2)\varphi_p &= \varphi_n & \tfrac{1}{2}(\tau_1 - i\tau_2)\varphi_n &= 0. \end{aligned} \tag{19.19}$$

Hence $\frac{1}{2}(\tau_1 + i\tau_2)$ is the operator which increases the nucleon charge by one unit and $\frac{1}{2}(\tau_1 - i\tau_2)$ decreases it by one unit.

Coming back now to (19.8) we see that it can be written in the form

$$\begin{aligned} &\psi_{pn}^{(\pm)}(j_a j_b J M) \\ &= N \sum (j_a m_a j_b m_b | j_a j_b J M) \, [\varphi_1(a)\varphi_2(b) \pm \varphi_2(a)\varphi_1(b)] \, \xi_1(\tfrac{1}{2})\xi_2(-\tfrac{1}{2}). \end{aligned} \tag{19.20}$$

Here the notation $\varphi_1(a) \equiv \varphi_1(l_a j_a m_a)$ implies that particle no. 1 is in the state $(l_a j_a m_a)$, whereas $\xi_1(\frac{1}{2})$ then assures that this particle is a proton. We could have obviously changed the role of 1 and 2 without changing at all the physical contents of the wave function. In other words, the wave functions

$$\begin{aligned} \psi_{pn}^{(\pm)}(j_a j_b J M) = N \sum (j_a m_a j_b m_b | j_a j_b J M) \\ \times [\varphi_2(a)\varphi_1(b) \pm \varphi_1(a)\varphi_2(b)] \, \xi_2(\tfrac{1}{2}) \, \xi_1(-\tfrac{1}{2}) \end{aligned} \tag{19.21}$$

describe exactly the same physical states as (19.20). We are therefore free to describe the physical states considered by any linear combination of (19.20) and (19.21). Using this freedom it is possible to arrive at a unified treatment of two nucleon systems, irrespective of whether they are composed of two neutrons, two protons, or a neutron and a proton. We make the following choice for reasons to become clear later.

$$\psi_{pn}^{(+)}(j_a j_b J M) = N \sum (j_a m_a j_b m_b | j_a j_b J M)$$
$$\times [\varphi_1(a)\varphi_2(b) + \varphi_2(a)\varphi_1(b)] [\xi_1(\tfrac{1}{2})\xi_2(-\tfrac{1}{2}) - \xi_1(-\tfrac{1}{2})\xi_2(\tfrac{1}{2})] \tag{19.22}$$

$$\psi_{pn}^{(-)}(j_a j_b J M) = N \sum (j_a m_a j_b m_b | j_a j_b J M)$$
$$\times [\varphi_1(a)\varphi_2(b) - \varphi_2(a)\varphi_1(b)] [\xi_1(\tfrac{1}{2})\xi_2(-\tfrac{1}{2}) + \xi_1(-\tfrac{1}{2})\xi_2(\tfrac{1}{2})]. \tag{19.23}$$

Both these functions are linear combinations of (19.20) and (19.21), and it can be verified directly that they are both eigenfunctions of $\mathbf{l}_1^2$, $\mathbf{l}_2^2$, $\mathbf{j}_1^2$, $\mathbf{j}_2^2$, $\mathbf{J}^2$, and J_z. They both correspond to a state of a neutron plus a proton in the sense that

$$(Q_1 + Q_2)\psi_{pn}^{(\pm)} = (+1)\psi_{np}^{(\pm)}.$$

The exchange of the proton and the neutron in (19.22) and (19.23) can be carried out by exchanging the charge on particles 1 and 2, i.e., by the transformation

$$[\xi_1(\tfrac{1}{2})\xi_2(-\tfrac{1}{2}) \mp \xi_1(-\tfrac{1}{2})\xi_2(\tfrac{1}{2})] \to [\xi_1(-\tfrac{1}{2})\xi_2(\tfrac{1}{2}) \mp \xi_1(\tfrac{1}{2})\xi_2(-\tfrac{1}{2})].$$

This leads to

$$\psi_{np}^{(\pm)}(j_a j_b J M) = \mp \psi_{pn}^{(\pm)}(j_a j_b J M).$$

The special combinations (19.22) and (19.23) have, however, an additional symmetry. Both of them are antisymmetric with respect to the exchange of all the coordinates (including the isospin) of particles 1 and 2. Nucleons 1 and 2 therefore appear, with this particular choice of wave functions, as two identical fermions whose wave function should be antisymmetric to obey the Pauli exclusion principle. This choice of the two degenerate wave functions for the system of a neutron and a proton therefore enables us to give a uniform treatment to a system of any two nucleons. Writing in fact the antisymmetric states of two protons and two neutrons in a similar way we have

$$\psi_{pp}(j_a j_b J M) = N \sum (j_a m_a j_b m_b | j_a j_b J M)$$
$$\times [\varphi_1(a)\varphi_2(b) - \varphi_2(a)\varphi_1(b)]\, \xi_1(\tfrac{1}{2})\xi_2(\tfrac{1}{2}) \tag{19.24}$$

$$\psi_{nn}(j_a j_b J M) = N \sum (j_a m_a j_b m_b | j_a j_b J M)$$
$$\times [\varphi_1(a)\varphi_2(b) - \varphi_2(a)\varphi_1(b)]\, \xi_1(-\tfrac{1}{2})\xi_2(-\tfrac{1}{2}). \tag{19.25}$$

The charge-dependent parts of (19.22)-(19.25) have a simple interpretation in the isospin formalism. We define a total isospin **T** by

$$\mathbf{T} = \mathbf{t}^{(1)} + \mathbf{t}^{(2)} = \tfrac{1}{2}[\boldsymbol{\tau}^{(1)} + \boldsymbol{\tau}^{(2)}] \tag{19.26}$$

where $\boldsymbol{\tau}^{(1)} = (\tau_1^{(1)}, \tau_2^{(1)}, \tau_3^{(1)})$ are the three isospin matrices operating on nucleon 1. Consequently, for two-particle systems the eigenfunctions of the total isospin (squared) and its third component are given by

$$\xi(\tfrac{1}{2}\tfrac{1}{2}T\,M_T) = \sum_{m_t, m'_t} (\tfrac{1}{2}m_t\,\tfrac{1}{2}m'_t|\tfrac{1}{2}\,\tfrac{1}{2}T\,M_T)\,\xi_1(m_t)\xi_2(m'_t) \tag{19.27}$$

where m_t and m'_t take on the values $\pm\,\frac{1}{2}$. It will be recognized now that the four functions (19.22)-(19.25) can be written as follows

$$\begin{aligned}
\psi_{pn}^{(+)}(j_a j_b J\,M) &= \psi_s(j_a j_b J\,M)\,\xi(\tfrac{1}{2}\,\tfrac{1}{2}\,0\,0)\\
\psi_{pn}^{(-)}(j_a j_b J\,M) &= \psi_a(j_a j_b J\,M)\,\xi(\tfrac{1}{2}\,\tfrac{1}{2}\,1\,0)\\
\psi_{pp}(j_a j_b J\,M) &= \psi_a(j_a j_b J\,M)\,\xi(\tfrac{1}{2}\tfrac{1}{2}\,1\,1)\\
\psi_{nn}(j_a j_b J\,M) &= \psi_a(j_a j_b J\,M)\,\xi(\tfrac{1}{2}\,\tfrac{1}{2}\,1\,-1)
\end{aligned} \tag{19.28}$$

where

$$\begin{aligned}
\psi_s(j_a j_b J\,M) &= \frac{1}{\sqrt{2}}\sum (j_a m_a\,j_b m_b|\,j_a j_b J\,M)\\
&\qquad\times[\varphi_1(j_a m_a)\varphi_2(j_b m_b) + \varphi_1(j_b m_b)\varphi_2(j_a m_a)]\\
\psi_a(j_a j_b J\,M) &= \frac{1}{\sqrt{2}}\sum (j_a m_a\,j_b m_b|\,j_a j_b J\,M)\\
&\qquad\times[\varphi_1(j_a m_a)\varphi_2(j_b m_b) - \varphi_1(j_b m_b)\varphi_2(j_a m_a)].
\end{aligned} \tag{19.29}$$

In other words, the four possible states for the two-nucleon $j_a j_b$ configuration for each state of given $J\ M$ values can be divided into two groups. The first group contains just one state which is characterized by total isospin $T = 0$. It is antisymmetric in its charge coordinates and symmetric with respect to the simultaneous exchange of the space and spin coordinates. It always corresponds to a state of one neutron and one proton, since $T = 0$ implies $M_T = 0$ and hence the total charge of such a state is just one unit. The second group contains three states all characterized by the same total isospin $T = 1$ and the different values of M_T. These values are $M_T = +1$ (hence $Q = 2$, i.e., two-proton system), $M_T = 0$ ($Q = 1$, i.e., a proton and a neutron), and

$M_T = -1$ ($Q = 0$, i.e., two neutrons). All these states are symmetric in the charge coordinates and antisymmetric in the space and spin coordinates.

It is now obvious why we have made the choice (19.22) and (19.23). Starting with the wave function (19.24) of two protons and operating on it with $T_1 - iT_2$ we obtain directly the state (19.23). The state (19.22) is the state orthogonal to it. The physical basis of choosing functions like (19.22) and (19.23) rather than the function (19.20) is *charge independence*. It means that we consider, on the basis of experimental facts, the proton and neutron to be the same particle in two different charge states. As long as we consider ordinary interactions V_{pn}, the charge serves merely as a label and choosing (19.22) and (19.23) is just a matter of convenience. However, the charges can also appear dynamically in the Hamiltonian, as in the case of charge exchange interactions or beta decay. In this case, we *must* use functions of the charge coordinates that are either symmetric or antisymmetric. The requirement of antisymmetry of the wave function under exchange of all coordinates (including the charge coordinates) follows by comparison with the case of two protons or two neutrons ($M_T = +1$ or $M_T = -1$).

The notion of isospin can be extended to any number of nucleons. By requiring that the states of A nucleons be antisymmetric with respect to the simultaneous exchange of any two sets of space, spin, and charge coordinates of two nucleons, we form a complete basis for all the allowed A-nucleon states in a given $j_1 j_2 \dots j_A$ configuration. The use of isospin with the generalized Pauli principle does not give any new wave function, neither does it forbid any of the wave functions which would have been allowed otherwise. It is merely a convenient formalism for classifying wave functions of many nucleons. As we saw the introduction of the isospin coordinates leads to a bigger number of wave functions, which, however, may be different only formally and describe no new physical states. The generalized Pauli principle eliminates this redundancy and makes each wave function describe a distinct physical state. At the same time it expresses these states in terms of functions which have convenient properties in the isospin formalism.

For configurations of two equivalent nucleons $\psi(j^2 J M)$ is antisymmetric or symmetric according to whether J is even or odd. Denoting by $\psi(j_a j_b \dots T M_T J M)$ a wave function of nucleons in the orbits $j_a j_b, \dots$ coupled to total angular momentum $J(J_z = M)$ and to total isospin $T(T_3 = M_T)$, we see that $\psi(j^2 T M_T J M)$ may be different from zero only if $J + T$ is odd. Similar conclusions can be drawn for two-nucleon wave functions in LS-coupling. In fact $\psi(l^2 T M_T S L J M)$ may be different from zero only if $S + L + T$ is odd. The group of states $\psi(l^2 T M_T S L J M)$ with given fixed values of T, S, and L and all possible values of M_T, J, and M may be called a *supermultiplet* and is denoted by the symbol ${}^{2T+1,2S+1}L$. Thus, for instance, ${}^{3\,1}P$ stands for the group of all states (of a given two-nucleon configuration) with $T = 1$, $S = 0$, and $L = 1$. Since different values of M_T imply a different charge, different members of the same supermultiplet may belong to different nuclei. The atomic weight of all members of a supermultiplet is, however, the same. A more precise definition of a supermultiplet will be given in Section 36.

20. The Two-Particle Interaction

The first-order correction to the energy of a state $\psi(l_1l_2S\,L\;J\,M)$ in LS-coupling or $\psi(l_1j_1l_2j_2J\;M)$ in jj-coupling of two particles moving in a central field due to the mutual interaction V_{12} is given by

$$\varDelta E_J = \int \psi^*(J\,M)V_{12}\psi(J\,M). \tag{20.1}$$

This section will be devoted to the discussion of some general properties of $\varDelta E_J$.

Let us first consider two identical particles in LS-coupling. If $l_1 \neq l_2$ the antisymmetric wave function $\psi_a(l_1l_2S\,L\;J\,M)$ is given by

$$\psi_a(l_1l_2S\,L\;J\,M) = \frac{1}{\sqrt{2}}\sum (S\,M_SL\,M_L|S\,L\;J\,M)$$
$$\times\,[\varphi_{12}(l_1l_2L\,M_L)\,\chi_{12}(s_1s_2S\,M_S) - \varphi_{21}(l_1l_2L\,M_S)\,\chi_{21}(s_1s_2\;S\,M_S)] \tag{20.2}$$

in analogy to (19.3). Using the symmetry properties of the V-coefficients we obtain

$$\psi_a(l_1l_2S\,L\;J\,M) = \frac{1}{\sqrt{2}}\sum_{M_SM_L}(S\,M_S\,L\,M_L|S\,L\;J\,M)$$
$$\times\,[\varphi(l_1l_2L\,M_L) + (-\,1)^{l_1+l_2-L+S}\,\varphi(l_2l_1L\,M_L)]\,\chi(s_1s_2S\,M_S) \tag{20.3}$$

where now

$$\varphi(l_il_jL\,M_L) = \sum_{m_im_j}(l_im_il_jm_j|l_il_jL\,M_L)\,\varphi_1(l_im_i)\,\varphi_2(l_jm_j) \tag{20.4}$$

and

$$\chi(s_1s_2S\,M_S) = \sum_{m_sm_s'}(\tfrac{1}{2}\,m_s\,\tfrac{1}{2}\,m_s'|\tfrac{1}{2}\;\tfrac{1}{2}\,S\,M_S)\,\chi_1(m_s)\,\chi_2(m_s'). \tag{20.5}$$

Thus $\varphi(l_2l_1L\,M_L)$ stands for the wave function in which particle 1 is in the l_2-orbit, particle 2 is in the l_1-orbit, and l_2 and l_1 are coupled to L in this order. In this way we replace an antisymmetrized wave function by a linear combination of different functions in which the particles appear in a definite order. This is necessary for the convenient use of tensor algebra. *This convention will be adhered to throughout, so that the whole formalism of tensor algebra can be applied in a straightforward manner.*

Substituting (20.3) in (20.1) we obtain

$$\Delta E_{SLJ} = \tfrac{1}{2}\{\langle l_1 l_2 S\,L\;J|V_{12}|l_1 l_2 S\,L\;J\rangle + \langle l_2 l_1 S\,L\;J|V_{12}|l_2 l_1 S\,L\;J\rangle$$

$$+ (-1)^{l_1+l_2-L+S}[\langle l_1 l_2 S\,L\;J|V_{12}|l_2 l_1 S\,L\;J\rangle + \langle l_2 l_1 S\,L\;J|V_{12}|l_1 l_2 S\,L\;J\rangle]\}. \tag{20.6}$$

It should be stressed that the matrix elements in (20.6) have their usual meaning, i.e.,

$$\langle l_a l_b S\,L\;J\,M|V_{12}|l_c l_d S'L'\,J'M'\rangle$$

$$= \sum (S\,M_S L\,M_L|S\,L\,J\,M)\,(S'M'_S L'\,M'_L|S'L'\,J'M')\,(s_1 m_{s1} s_2 m_{s2}|s_1 s_2 S M_S)$$

$$\times\,(s'_1 m'_{s1} s'_2 m'_{s2}|s'_1 s'_2 S' M'_S)\,(l_a m_a l_b m_b|l_a l_b L M_L)\,(l_c m_c l_d m_d|l_c l_d L' M'_L)$$

$$\times \int [\varphi_1(a)\,\chi_1(m_{s1})\,\varphi_2(b)\,\chi_2(m_{s2})]^* V_{12}[\varphi_1(c)\,\chi_1(m'_{s1})\,\varphi_2(d)\,\chi_2(m'_{s2})]\;d\mathbf{r}_1 d\mathbf{r}_2. \tag{20.7}$$

Thus, in the matrix element $\langle l_a l_b S\,L\;J\,M|V_{12}|l_c l_d S'L'\,J'M'\rangle$, l_a and l_c are the quantum numbers of particle 1, l_b, and l_d are those of particle 2; l_a and l_b (in this order!) are coupled to L, l_c, and l_d (again in this order) are coupled to L'.

Since $V_{12} = V_{21}$ it follows from (20.7) that

$$\begin{aligned}\langle l_1 l_2 S\,L\;J\,M|V_{12}|l_1 l_2 S\,L\;J\,M\rangle &= \langle l_2 l_1 S\,L\;J\,M|V_{12}|l_2 l_1 S\,L\;J\,M\rangle\\ \langle l_1 l_2 S\,L\;J\,M|V_{12}|l_2 l_1 S\,L\;JM\rangle &= \langle l_2 l_1 S\,L J\,M|V_{12}|l_1 l_2 S\,L J\,M\rangle.\end{aligned} \tag{20.8}$$

Hence

$$\Delta E_{SLJ} = \langle l_1 l_2 S\,L\;J\,M|V_{12}|l_1 l_2 S\,L\;J\,M\rangle + (-1)^{l_1+l_2-L+S}\langle l_1 l_2 S\,L\;J\,M|V_{12}|l_2 l_1 S\,L\;J\,M\rangle. \tag{20.9}$$

Similar considerations for jj coupling show that, for $j_1 \neq j_2$,

$$\Delta E_{j_1 j_2 J} = \langle l_1\,j_1 l_2\,j_2 J\,M|V_{12}|l_1\,j_1 l_2\,j_2 J\,M\rangle - (-1)^{j_1+j_2-J}\langle l_1\,j_1 l_2\,j_2 J\,M|V_{12}|l_2\,j_2 l_1\,j_1 J\,M\rangle. \tag{20.10}$$

Thus, in both cases the first-order correction to the energy of a state of given J and M contains two terms: the *direct term* $\langle l_1 l_2 S\,L\;J|V_{12}|l_1 l_2 S\,L\;J\rangle$ or $\langle j_1 j_2 J\,M|V_{12}|j_1 j_2 J\,M\rangle$ and the *exchange term* $\langle l_1 l_2 S\,L\,J|V_{12}|l_2 l_1 S\,L\,J\rangle$ or $\langle j_1 j_2 J\,M|V_{12}|j_2 j_1 J\,M\rangle$. The direct term is identical with the matrix element obtained with nonantisymmetrized wave functions. The exchange term appears as a result of using antisymmetrized wave functions. If the two orbits l_1 and l_2 are very different from each other, i.e., if their overlap in space is very small, the contribution of the exchange term will also become very small. The integral in the exchange matrix element contains the product $\varphi_1^*(l_1)\varphi_1(l_2)$, and a small overlap of the two orbits just means that this product is small compared to both $\varphi_1^*(l_1)\varphi_1(l_1)$ and to $\varphi_1^*(l_2)\varphi_1(l_2)$. Thus, for completely nonoverlapping orbits the contribution of the exchange terms vanishes and we are left only with the direct terms. This should be expected since the antisymmetrization becomes less and less effective as we increase the spatial separation between the two particles.

For orbits which overlap each other considerably, the contribution of the exchange terms may be quite important. We see from (20.9) that for spin-independent interactions the contribution of the exchange terms has *different signs* for states with $S = 0$ and $S = 1$. This result may seem at first somewhat surprizing since we have assumed that the interaction is spin-independent. However, we have to remember that we are dealing with antisymmetric wave functions. The spin-dependent part of the wave function is symmetric or antisymmetric according to whether $S = 0$ or $S = 1$. Therefore, for $S = 0$ the wave function must be symmetric with respect to the exchange of the space coordinates alone, whereas for $S = 1$ it is antisymmetric in the space coordinates. Thus, although V_{12} may be spin-independent, the choice of $S = 0$ or 1 indirectly determines the spatial properties of the state considered, and therefore affects the average value of V_{12}, i.e., of ΔE_{SLJ}.

The effects of the spatial symmetry of the wave function can best be visualized by choosing for V_{12} a very short-range interaction. Putting actually $V_{12} = \delta(\mathbf{r}_1 - \mathbf{r}_2)$, states antisymmetric in their space coordinates will contribute nothing to the average value of this interaction. Since for $S = 1$ the state $\psi_a(l_1l_2S\,L\,J\,M)$ is antisymmetric in its space coordinates, we have

$$\Delta E_{SLJ}(\delta) = 0 \qquad \text{for} \qquad V_{12} = \delta(\mathbf{r}_1 - \mathbf{r}_2) \qquad \text{and} \qquad S = 1. \tag{20.11}$$

This same result can also be obtained in a more formal way. We notice by using the symmetry properties of the V-coefficients that for *δ-forces*

$$\begin{aligned}&\langle l_1l_2S\,L\,J\,M|\delta(\mathbf{r}_1 - \mathbf{r}_2)|l_2l_1S\,L\,J\,M\rangle \\ &= (-1)^{l_1+l_2-L}\,\langle l_1l_2S\,L\,J\,M|\delta(\mathbf{r}_1 - \mathbf{r}_2)|l_1l_2S\,L\,J\,M\rangle.\end{aligned} \tag{20.12}$$

Substituting (20.12) into (20.9) we therefore obtain

$$\Delta E_{SLJ}(\delta) = [1 + (-1)^S]\,\langle l_1l_2S\,L\,J\,M|\delta(\mathbf{r}_1 - \mathbf{r}_2)|l_1l_2S\,L\,J\,M\rangle \tag{20.13}$$

and hence the result (20.11).

For the singlet states, $S = 0$, the wave function is symmetric in its space coordinates. Unlike the case of $S = 1$ this information does not suffice to draw conclusions about the behavior of the wave function for $|\mathbf{r}_1 - \mathbf{r}_2| \approx 0$. Antisymmetric functions of $\mathbf{r}_1$ and $\mathbf{r}_2$ *must* vanish when $\mathbf{r}_1 = \mathbf{r}_2$, whereas symmetric functions of $\mathbf{r}_1$ and $\mathbf{r}_2$ may or may not vanish when $\mathbf{r}_1 = \mathbf{r}_2$.

Let us in fact, consider the *symmetric* function

$$\varphi_s(l_1 l_2 L\, M) = \sum_{m_1 m_2} (l_1 m_1 l_2 m_2 | l_1 l_2 L\, M) \times [\varphi_1(l_1 m_1)\, \varphi_2(l_2 m_2) + \varphi_1(l_2 m_2)\, \varphi_2(l_1 m_1)]. \quad (20.14)$$

Since the m-dependence of the function $\varphi(l\, m)$ is wholly contained in their angular parts, it will be convenient to separate the angular and radial dependence of $\varphi(l\, m)$ by writing

$$\varphi(n\, l\, m) = \frac{1}{r} R_{nl}(r)\, Y_{lm}(\Omega) \qquad \Omega \equiv (\theta\, \varphi). \quad (20.15)$$

Equation (20.14) can then be rewritten as

$$\varphi_s(l_1 l_2 L\, M) = [R^{(+)}_{l_1 l_2}(r_1, r_2)\, \mathscr{Y}^{(+)}_{LM}(\Omega_1, \Omega_2) + R^{(-)}_{l_1 l_2}(r_1, r_2)\, \mathscr{Y}^{(-)}_{LM}(\Omega_1, \Omega_2)] \quad (20.16)$$

where

$$R^{(\pm)}_{l_1 l_2}(r_1, r_2) = \frac{1}{r_1 r_2 \sqrt{2}} [R_{n_1 l_1}(r_1)\, R_{n_2 l_2}(r_2) \pm R_{n_2 l_2}(r_1)\, R_{n_1 l_1}(r_2)]$$

$$\mathscr{Y}^{(\pm)}_{LM}(\Omega_1, \Omega_2) = \frac{1}{\sqrt{2}} \sum (l_1 m_1 l_2 m_2 | l_1 l_2 L\, M) \times [Y_{l_1 m_1}(\Omega_1)\, Y_{l_2 m_2}(\Omega_2) \pm Y_{l_2 m_2}(\Omega_1)\, Y_{l_1 m_1}(\Omega_2)]. \quad (20.17)$$

These functions have the following symmetry properties

$$R^{(\pm)}_{l_1 l_2}(r_1, r_2) = \pm R^{(\pm)}_{l_1 l_2}(r_2, r_1) \qquad \mathscr{Y}^{(\pm)}(\Omega_1, \Omega_2) = \pm \mathscr{Y}^{(\pm)}(\Omega_2, \Omega_1). \quad (20.18)$$

We see that the symmetric function $\varphi_s(l_1 l_2 L\, M)$ can be decomposed into a sum of two products; one product of two symmetric functions and one product of two antisymmetric functions. We can therefore derive the energies in the singlet case with δ-forces by direct calculation.

In order to do so, we shall evaluate in a straightforward manner the expression

$$\Delta E_L(\delta) = \sum (l_1 m_1 l_2 m_2 | l_1 l_2 L\, M)\, (l_1 m_1' l_2 m_2' | l_1 l_2 L\, M) \times \int \varphi_1^*(l_1 m_1)\, \varphi_2^*(l_2 m_2)\, \delta(\mathbf{r}_1 - \mathbf{r}_2)\, \varphi_1(l_1 m_1')\, \varphi_2(l_2 m_2')\, d\mathbf{r}_1 d\mathbf{r}_2.$$

From (20.13) it follows that $\Delta E_{SLJ}(\delta) = 2\Delta E_L(\delta)$ for $S = 0$. Using the explicit expression

$$\delta(\mathbf{r}_1 - \mathbf{r}_2) = \frac{1}{r_1 r_2}\,\delta(r_1 - r_2)\,\delta(\cos\theta_1 - \cos\theta_2)\,\delta(\varphi_1 - \varphi_2) \tag{20.19}$$

we obtain

$$\Delta E_L(\delta) = \int \frac{1}{r^2}\,[R_{n_1l_1}(r)\,R_{n_2l_2}(r)]^2\,dr \sum_{m_1m_2m_1'm_2'} (l_1m_1l_2m_2|l_1l_2L\,M)$$

$$\times\,(l_1m_1'l_2m_2'|l_1l_2L\,M)\int Y^*_{l_1m_1}(\theta\,\varphi)\,Y^*_{l_2m_2}(\theta\,\varphi)\,Y_{l_1m_1'}(\theta\,\varphi)\,Y_{l_2m_2'}(\theta\,\varphi)\,d\Omega. \tag{20.20}$$

Using (13.19) and (17.14) we can expand a product of two spherical harmonics of the same angles as follows

$$Y_{l_1m_1}(\theta\,\varphi)\,Y_{l_2m_2}(\theta\,\varphi) = \sum_{lm}(-1)^{l-m}\begin{pmatrix} l & l_1 & l_2\\ -m & m_1 & m_2\end{pmatrix}(l||\mathbf{Y}_{l_1}||l_2)\,Y_{lm}(\theta\,\varphi). \tag{20.21}$$

We therefore obtain for (20.20)

$$\Delta E_L(\delta) = \sum_{\substack{m_1m_2m_1'm_2'\\ l\;m\;l'\;m'}} (l_1m_1l_2m_2|l_1l_2L\,M)\begin{pmatrix} l_1 & l_2 & l\\ m_1 & m_2 & -m\end{pmatrix}(l_1m_1'l_2m_2'|l_1l_2L\,M)\begin{pmatrix} l_1 & l_2 & l'\\ m_1' & m_2' & -m'\end{pmatrix}$$

$$\times\,(-1)^{(l+l')-(m+m')}\,(l||\mathbf{Y}_{l_1}||l_2)\,(l'||\mathbf{Y}_{l_1}||l_2)\int\frac{1}{r^2}\,[R_{n_1l_1}(r)\,R_{n_2l_2}(r)]^2dr$$

$$\times\int Y^*_{lm}(\theta\,\varphi)\,Y_{l'm'}(\theta\,\varphi)\,d\Omega = \frac{1}{2L+1}\,|(L||\mathbf{Y}_{l_1}||l_2)|^2\int\frac{1}{r^2}\,[R_{n_1l_1}(r)\,R_{n_2l_2}(r)]^2dr$$

$$= \frac{1}{2L+1}\,|(l_1||\mathbf{Y}_L||l_2)|^2\int\frac{1}{r^2}\,[R_{n_1l_1}(r)\,R_{n_2l_2}(r)]^2dr. \tag{20.22}$$

For numerical calculations it is convenient to express the reduced matrix element in (20.22) in terms of a special V-coefficient. We then obtain, using (17.14),

$$\Delta E_L(\delta) = \frac{1}{4\pi}\int\frac{1}{r^2}\,[R_{n_1l_1}(r)\,R_{n_2l_2}(r)]^2dr\,(2l_1+1)\,(2l_2+1)\begin{pmatrix} l_1 & L & l_2\\ 0 & 0 & 0\end{pmatrix}^2 \tag{20.23}$$

Since $V(l_1Ll_2\,0\,0\,0)$ vanishes unless $l_1 + L + l_2$ is even [cf. (13.18)] we conclude that $\Delta E_L(\delta)$ vanishes for $(-1)^{l_1+l_2-L} = -1$. We can now supplement the rule (20.11) for the triplet states by

$$\Delta E_{SLJ}(\delta) = 0 \quad\text{for}\quad V_{12} = \delta(\mathbf{r}_1 - \mathbf{r}_2)\quad\text{if}\quad S = 1\quad\text{and}\quad(-1)^{l_1+l_2-L} = \pm 1$$

$$\text{or} \tag{20.24}$$

$$\text{if}\quad S = 0\quad\text{and}\quad(-1)^{l_1+l_2-L} = -1$$

Thus, of all the degenerate states of the $l_1 l_2$ configuration in a central field, only those for which $S = 0$ and $l_1 + l_2 - L$ is even will be affected by a δ-force. The other states are left unaffected. Therefore, if we are considering an *attractive δ-force* in LS-coupling, the ground state of the $l_1 l_2$ configuration of identical particles will be a singlet state ($S = 0$) with an even or odd value of L according to whether $l_1 + l_2$ is even or odd. Since $(-1)^{l_1+l_2}$ is also the parity of each of the states of the $l_1 l_2$ configuration, we can say that under the above conditions the L of the ground state is even or odd according to whether the parity of the ground state is even or odd.

Equation (20.24) actually amounts to saying that the wave function $\psi_a(l_1 l_2 S L J M)$ vanishes for $\mathbf{r}_1 = \mathbf{r}_2$ if $S = 1$ or even if $S = 0$ provided $(-1)^{l_1+l_2-L} = -1$. It generally does not vanish for $\mathbf{r}_1 = \mathbf{r}_2$ if $S = 0$ and $(-1)^{l_1+l_2-L} = +1$, since in that case (20.24) assumes a finite value. Suppose now that V_{12} is an attractive interaction, not necessarily a δ-function but monotonically decreasing as $V_{12} \to \infty$, and let us consider a state (SLJ) of the $l_1 l_2$ configuration for which $(-1)^{l_1+l_2-L} = +1$. Since $\psi_a(l_1 l_2 S L J M)$ vanishes for $\mathbf{r}_1 = \mathbf{r}_2$ if $S = 1$ and is different from zero if $S = 0$, we can expect that for even values of $l_1 + l_2 - L$, $|\Delta E_{S=0LJ}|$ will be bigger than $|\Delta E_{S=1LJ}|$, and thus the level ${}^1L_J$ will be lower than the levels ${}^3L_J$, (the order being reversed for repulsive interactions).

We have already seen that if $(-1)^{l_1+l_2-L} = -1$ then the corresponding wave functions vanish for $\mathbf{r}_1 = \mathbf{r}_2$ both for the singlets and the triplets. Therefore, for short-range attractive interactions we can expect both triplets and singlets with odd values of $l_1 + l_2 - L$ to lie higher (i.e., be less bound) than the singlets with even values of $l_1 + l_2 - L$. Whether the singlets with odd values of $l_1 + l_2 - L$ are higher than the triplets or vice versa cannot be generally decided without more details on the interaction.

One case in which we can go further with our general considerations is that of the $l\,l$ configuration, i.e., a configuration of two particles in orbits of equal angular momenta $l_1 = l_2 = l$ but different radial quantum numbers $n_1 \neq n_2$ (if also $n_1 = n_2$ we refer to the configuration as being the l^2 configuration).

Using the decomposition (20.16) for the wave functions in this case we see that

$$\mathscr{Y}_{LM}^{(\pm)}(\Omega_1, \Omega_2)$$

$$= \frac{1}{\sqrt{2}} \sum (l\, m_1 l m_2 | l l L\, M) \, [Y_{lm_1}(\Omega_1)\, Y_{lm_2}(\Omega_2) \pm Y_{lm_2}(\Omega_1)\, Y_{lm_1}(\Omega_2)]$$

$$= \frac{1}{\sqrt{2}} [1 \pm (-1)^{2l-L}] \sum (l\, m_1 l\, m_2 | l l L\, M) \, Y_{lm_1}(\Omega_1)\, Y_{lm_2}(\Omega_2) \tag{20.25}$$

Thus, if $(-1)^{2l-L} = +1$, i.e., if L is even, the antisymmetric angular wave function vanishes identically

$$\mathscr{Y}^{(-)}_{LM}(\Omega_1, \Omega_2) = 0 \qquad \text{for even } L \tag{20.26}$$

and similarly

$$\mathscr{Y}^{(+)}_{LM}(\Omega_1, \Omega_2) = 0 \qquad \text{for odd } L. \tag{20.27}$$

Therefore, the wave function $\varphi_s(l\,l\,L\,M)$ for even values of L will be a product of a *symmetric* angular wave function $\mathscr{Y}^{(+)}_{LM}(\Omega_1, \Omega_2)$ and a *symmetric* radial wave function $R^{(+)}_{ll}(r_1 r_2)$. For odd values of L it will be a product of the *two* antisymmetric functions $R^{(-)}_{ll}(r_1 r_2)\,\mathscr{Y}^{(-)}_{LM}(\Omega_1, \Omega_2)$. The function $\varphi_a(l\,l\,L\,M)$, on the other hand, is *always* a product of one antisymmetric function and one symmetric function:

$$\varphi_a(l\,l\,L\,M) = \begin{cases} R^{(-)}_{ll}(r_1 r_2)\,\mathscr{Y}^{(+)}_{LM}(\Omega_1 \Omega_2) & \text{for even } L \\ R^{(+)}_{ll}(r_1 r_2)\,\mathscr{Y}^{(-)}_{LM}(\Omega_1 \Omega_2) & \text{for odd } L \end{cases} \tag{20.28}$$

We can therefore expect that for short-range attractive interactions in the $l\,l$ configuration the states with even values of L will have their singlets lower than their triplets, whereas the states with odd values of L, which are altogether higher than the states with even values of L, will have their triplets somewhat lower than their singlets.

Thus, for instance, in the $d\,d$ configuration ($l = 2$), L can assume all values from 0 to 4. With a δ-interaction between the particles the lowest states will be ^1S, ^1D, and ^1G. All other states remain unaffected by the δ-interaction. Increasing slightly the range of the interaction, keeping it still attractive, the effect will be bigger on ^3S, ^3D, ^3G, ^3P, and ^3F and smaller on ^1P and ^1F.

These configurations, as well as others, will be considered later, after developing some methods for the calculation of energy levels for arbitrary potentials. For the time being it is enough to notice, using (20.23) and the explicit expression (15.37) for $V(l_1 L\, l_2|0\,0\,0)$, that for an attractive δ-interaction the levels in the $l\,l$ configuration for small values of l are usually arranged in increasing order of L: ^1S being the lowest, ^1D next, etc.

It is also possible to give a physical picture of why the S state in the $l\,l$ configuration becomes the lowest one when we introduce an attractive δ-interaction between the particles. As discussed in Section 15, the mass distribution of a single particle with an orbital angular momentum $\mathbf{l}$ is concentrated around the plane perpendicular to $\mathbf{l}$. Therefore, to obtain the biggest possible effect from the attractive interaction the two particles should have the best possible overlap of their mass distributions. The optimum situation is obtained when l_1 and l_2 point at opposite directions, yielding a total $L = 0$, i.e., an S-state.

The general results on the structure of the energy levels in the $l_1 l_2$ configuration obtained above can be used to derive similar relations for the jj-coupling $j_1 j_2$ configurations. Using the transformation (18.28) between the two schemes we obtain

$$\psi(l_1 j_1 l_2 j_2 J M)$$

$$= \sum_{S,L} \sqrt{(2S+1)(2L+1)(2j_1+1)(2j_2+1)} \begin{Bmatrix} \frac{1}{2} & l_1 & j_1 \\ \frac{1}{2} & l_2 & j_2 \\ S & L & J \end{Bmatrix} \psi(l_1 l_2 S L J M). \tag{20.29}$$

Since this relation is based on pure geometrical properties of the wave functions (i.e., on the coupling scheme of the angular momenta involved), it holds for antisymmetric wave functions as well. We therefore obtain from (20.1)

$$\Delta E_J = \Delta E[l_1 j_1 l_2 j_2 J]$$

$$= (2j_1+1)(2j_2+1) \sum_{S,L} (2S+1)(2L+1) \begin{Bmatrix} \frac{1}{2} & l_1 & j_1 \\ \frac{1}{2} & l_2 & j_2 \\ S & L & J \end{Bmatrix}^2 \Delta E_{SLJ}(l_1 l_2) \tag{20.30}$$

where $\Delta E_{j_1 j_2 J}$ is given by (20.10) and $\Delta E_{SLJ}(l_1 l_2)$ is given by (20.9).

Taking now the special case of a δ-interaction, ΔE_{SLJ} vanishes for $S = 1$, hence the summation in (20.30) is reduced to the one term only with $S = 0$ and $L = J$. We thus obtain

$$\Delta E_J(\delta) = (2j_1+1)(2j_2+1)(2J+1) \begin{Bmatrix} \frac{1}{2} & l_1 & j_1 \\ \frac{1}{2} & l_2 & j_2 \\ 0 & J & J \end{Bmatrix}^2 \Delta E_{SLJ}(\delta) \text{ with } S = 0\ L = J \tag{20.31}$$

Since $\Delta E_{SLJ}(\delta)$ vanishes also for $S = 0$ if $(-1)^{l_1+l_2-L} = -1$, we see that in jj-coupling

$$\Delta E_J(\delta) = 0 \quad \text{if} \quad (-1)^{l_1+l_2-J} = -1. \tag{20.32}$$

Thus, for an attractive δ-interaction the lowest states of the $j_1 j_2$ configuration have an even or odd value of J according to whether the state has an even or odd *parity*. This general conclusion can be expected to hold also for other short-range attractive interactions and some of its features may hold also for configurations of more than two identical nucleons. We can therefore expect to find that low-lying levels in even-even nuclei have positive parity if J is even and negative parity if J is odd. It is found experimentally that lowest lying odd parity states in even-even nuclei have generally also an odd value of J. Since odd parity states of an even number of particles necessarily involve at least two different orbits, say configurations like j_1j_2, $j_1^3j_2$, $j_1j_2^3$, etc., both even and odd values of J are allowed. The fact that when the parity is negative, odd J states are the lowest is in a nice agreement with the above expectation.

The relation (20.31) can be simplified by substituting for $\Delta E_{SLJ}(\delta)$ its value from (20.23). We then obtain, recalling that $\Delta E_{SLJ} = 2\Delta E_L$,

$$\Delta E_J(\delta) = \tfrac{1}{2}(2j_1+1)(2j_2+1)\begin{Bmatrix} l_1 & l_2 & J \\ j_2 & j_1 & \tfrac{1}{2} \end{Bmatrix}^2 2(2l_1+1)(2l_2+1)\begin{pmatrix} l_1 & J & l_2 \\ 0 & 0 & 0 \end{pmatrix}^2 F_0 \tag{20.33}$$

where

$$F_0 = \frac{1}{4\pi}\int \frac{1}{r^2}[R_{n_1l_1}(r)\,R_{n_2l_2}(r)]^2\,dr.$$

Using (15.13) we can transform

$$\left[\begin{Bmatrix} l_1 & l_2 & J \\ j_2 & j_1 & \tfrac{1}{2} \end{Bmatrix}\begin{pmatrix} l_1 & l_2 & J \\ 0 & 0 & 0 \end{pmatrix}\right]^2$$

for $\quad (-1)^{l_1+l_2-J} = 1 \quad$ into $\quad \dfrac{1}{(2l_1+1)(2l_2+1)}\begin{pmatrix} j_1 & j_2 & J \\ \tfrac{1}{2} & -\tfrac{1}{2} & 0 \end{pmatrix}^2.$

Hence

$$\Delta E_J(\delta) = (2j_1+1)(2j_2+1)\begin{pmatrix} j_1 & j_2 & J \\ \tfrac{1}{2} & -\tfrac{1}{2} & 0 \end{pmatrix}^2 F_0. \tag{20.34}$$

Equation (20.34) is a closed formula which gives the energy levels of the antisymmetric states in the j_1j_2 configuration with a δ-interaction, for those values of J for which the δ-interaction has a nonvanishing contribution. The appearance of a special V-coefficient in (20.34) is just a convenient way of writing the dependence of $\Delta E_J(\delta)$ on j_1, j_2, and J. We see that the energy $\Delta E_J(\delta)$ does not depend explicitly on l_1 and l_2 (except through F_0). Implicitly it does, however, depend strongly on l_1 and l_2, or rather on $l_1 + l_2$, since $l_1 + l_2$ determines whether (20.34) should be taken for odd values of J or for even values of J. We can also obtain the lowest state among those for which $(-1)^{l_1+l_2-J} = 1$. By taking the numerical values of $V(j_1j_2J|\tfrac{1}{2} - \tfrac{1}{2}\,0)$ from the Appendix we obtain the following result for small values of j. If $j_1 = l_1 \pm \tfrac{1}{2}$ $j_2 = l_2 \pm \tfrac{1}{2}$ [the states with $(-1)^{l_1+l_2-J} = +1$ then have $J = |j_1 - j_2|$, $J = |j_1 - j_2| + 2$, etc.], the state with $J = |j_1 - j_2|$ is the lowest. If $j_1 = l_1 \pm \tfrac{1}{2}$ $j_2 = l_2 \mp \tfrac{1}{2}$ [the states with $(-1)^{l_1+l_2-J} = +1$ then have $J = j_1 + j_2$, $J = j_1 + j_2 - 2$, etc.], the state with $J = j_1 + j_2$ is the lowest. Thus, in jj-coupling, the ground state of the $j_1 j_2$ configuration is obtained by coupling $\mathbf{j}_1$ and $\mathbf{j}_2$ so that the spins $\mathbf{s}_1$ and $\mathbf{s}_2$ form a singlet state. The total angular momentum J is given by $J = |j_1 - j_2|$ if the spins are both "parallel" or both "antiparallel" to l in both j_1 and j_2 $(j_1 = l_1 \pm \frac{1}{2}, j_2 = l_2 \pm \frac{1}{2})$, and $J = j_1 + j_2$ if the spin is parallel for one j and antiparallel for the other.

Some of the formulas derived above have to be modified for equivalent particles. Thus, in (20.2) the normalization factor has to be changed from $1/\sqrt{2}$ to $\frac{1}{2}$ if $n_1 = n_2 = n$ and $l_1 = l_2 = l$. Using the symmetry properties of the V-coefficients we can write in this case

$$\psi_a(l^2 S\, L\, J\, M) = \sum\ (S\, M_S L\, M_L | S\, L\, J\, M)\,(l\, m_{l1} l\, m_{l2} | l\, l\, L\, M_L)$$

$$\times\ (s\, m_{s1} s\, m_{s2} | s\, s\, S\, M_S)\, \varphi_1(l\, m_{l1})\, \varphi_2(l\, m_{l2})\, \chi_1(m_{s1})\, \chi_2(m_{s2}) \qquad \text{for} \qquad S + L \text{ even.} \tag{20.35}$$

Similarly, in jj-coupling

$$\psi_a(j^2 J\, M) = \sum (j\, m_1\, j\, m_2 |\, j\, j\, J\, M)\, \varphi_1(j\, m_1)\, \varphi_2(j\, m_2) \qquad \text{for} \qquad J \text{ even.} \tag{20.36}$$

For equivalent particles the exchange term is equal in magnitude to the direct term and there is no need to distinguish between them. Taking into account the proper normalization we obtain

$$\Delta E(l^2 S\, L\, J) = \langle l^2 S\, L\, J\, M | V_{12} | l^2 S\, L\, J\, M \rangle$$
$$\Delta E(j^2 J) = \langle j^2 J\, M | V_{12} | j^2 J\, M) \rangle. \tag{20.37}$$

In the case of the δ-interaction, $\Delta E(l^2 SLJ)$ is given by (20.23), with $l_1 = l_2 = l$. Similarly, $\Delta E(j^2 J)$ is given by (20.34), with $j_1 = j_2 = j$, divided by 2.

The functions $\psi_s(l_1 l_2 S\, L\, M)$ required for the construction of states with $T = 0$ according to (19.29) can be obtained in analogy with (20.2)

$$\psi_s(l_1 l_2 S\, L\, J\, M) = \frac{1}{\sqrt{2}} \sum (S\, M_S L\, M_L | S\, L\, J\, M)$$

$$\times\ [\varphi_{12}(l_1 l_2 L\, M_L)\, \chi_{12}(s_1 s_2 S\, M_S) + \varphi_{21}(l_1 l_2 L\, M_L)\, \chi_{21}(s_1 s_2 S\, M_S)]. \tag{20.38}$$

In jj-coupling, the symmetric functions are given by

$$\psi_s(j_1 j_2 J\, M) = \frac{1}{\sqrt{2}} \sum (j_1 m_1\, j_2 m_2 |\, j_1 j_2 J\, M)$$

$$\times\ [\varphi_1(j_1 m_1)\, \varphi_2(j_2 m_2) + \varphi_1(j_2 m_2)\, \varphi_2(j_1 m_1)]. \tag{20.39}$$

For equivalent particles, the analogs of (20.35) and (20.36) are

$$\psi_s(l^2 S\,L\,J\,M) = \sum (S\,M_S L\,M_L|S\,L\,J\,M)\,\varphi_{12}(l\,l\,L\,M_L)\,\chi_{12}(s\,s\,S\,M_S) \quad \text{for} \quad S + L \text{ odd} \tag{20.40}$$

in LS-coupling, and

$$\psi_s(j^2 J\,M) = \sum (j\,m\,j\,m'|j\,j\,J\,M)\,\varphi_1(j\,m)\,\varphi_2(j\,m') \quad \text{for} \quad J \text{ odd} \tag{20.41}$$

in jj-coupling.

Using these wave functions, similar considerations about the energies of two-nucleon systems with $T = 0$ can be made. These energies, however, depend, strongly on the exchange character of the two-body interaction, even for short-range forces. We shall therefore proceed now to discuss exchange interactions in general.

The existence of internal degrees of freedom, such as the spin of a particle or its charge, gives rise to a certain classification of quantum-mechanical two-body scalar forces. We have so far considered mainly *ordinary interactions* which do not affect the internal degrees of freedom of the interacting particles, but arise solely from a momentum exchange between the two particles. We shall now consider also *exchange interactions* for which an exchange of spin or isospin may take place. Ordinary forces derived from a potential are often called *Wigner forces* (W), a typical example being the Coulomb force. The exchange interactions are classified into *Bartlett forces* (B) if they involve only spin exchange, *Heisenberg forces* (H) if they involve only charge exchange, and *Majorana forces* (M) if they involve both spin and charge exchange.

More precisely when these forces are derivable from potentials, the four possible potentials are given by

$$\begin{aligned}
V_W &= f_W(|\mathbf{r}_1 - \mathbf{r}_2|)\\
V_B &= \tfrac{1}{2}[1 + (\boldsymbol{\sigma}_1 \cdot \boldsymbol{\sigma}_2)]\,f_B(|\mathbf{r}_1 - \mathbf{r}_2|)\\
V_H &= -\tfrac{1}{2}[1 + (\boldsymbol{\tau}_1 \cdot \boldsymbol{\tau}_2)]\,f_H(|\mathbf{r}_1 - \mathbf{r}_2|)\\
V_M &= -\tfrac{1}{4}[1 + (\boldsymbol{\sigma}_1 \cdot \boldsymbol{\sigma}_2)]\,[1 + (\boldsymbol{\tau}_1 \cdot \boldsymbol{\tau}_2)]\,f_M(|\mathbf{r}_1 - \mathbf{r}_2|).
\end{aligned} \tag{20.42}$$

Here the $f(|\mathbf{r}_1 - \mathbf{r}_2|)$ are any functions of the relative distance between particles 1 and 2, not necessarily the same for the different interactions.

The operators $\boldsymbol{\tau}_1$ and $\boldsymbol{\tau}_2$ are the isospin operators introduced in the preceding section. For two identical particles $T = 1$, where $\mathbf{T} = \frac{1}{2}\boldsymbol{\tau}_1 + \frac{1}{2}\boldsymbol{\tau}_2$. Therefore the expectation value of $(\boldsymbol{\tau}_1 \cdot \boldsymbol{\tau}_2)$ between states of two protons or of two neutrons is

$$\langle T = 1\,M_T|(\boldsymbol{\tau}_1 \cdot \boldsymbol{\tau}_2)|T = 1\,M_T\rangle = 2[T(T+1) - 2\cdot\tfrac{3}{4}] = 1. \tag{20.43}$$

Similarly for $T = 0$

$$\langle T = 0\ M_T = 0\ |(\boldsymbol{\tau}_1 \cdot \boldsymbol{\tau}_2)|\ T = 0\ M_T = 0\rangle = -3. \tag{20.44}$$

The operator

$$P_\tau = \tfrac{1}{2}[1 + (\boldsymbol{\tau}_1 \cdot \boldsymbol{\tau}_2)] \tag{20.45}$$

has therefore the eigenvalues $+1$ and -1 for states symmetric ($T = 1$) and antisymmetric ($T = 0$) in the two isospin coordinates. P_τ is therefore often called the *charge exchange operator*. Similarly, the *spin exchange operator*

$$P_\sigma = \tfrac{1}{2}[1 + (\boldsymbol{\sigma}_1 \cdot \boldsymbol{\sigma}_2)] \tag{20.46}$$

leaves invariant a state symmetric in the two spins ($S = 1$) and multiplies by -1 any antisymmetric state ($S = 0$).

We can also define a *space exchange operator* P_x by

$$P_x \psi(x_1, x_2) = \psi(x_2, x_1) \tag{20.47}$$

so that its eigenvalues are $+1$ and -1 corresponding respectively to space-symmetric and space-antisymmetric functions. However, since the complete wave functions are always antisymmetric with respect to the simultaneous exchange of the two space, spin, and charge coordinates of the particles, the three exchange operators obey the relation

$$P_\sigma P_\tau P_x \psi_a(1\ 2) = -\psi_a(1\ 2).$$

Therefore, in as much as only totally antisymmetric states are involved, we have

$$P_x = -P_\sigma P_\tau. \tag{20.48}$$

Thus, the Majorana force is equivalent to a space-exchange force.

For systems of identical particles there are only two types of scalar forces. In fact, we see from (20.43) that for such systems Heisenberg forces reduce to Wigner forces and Majorana forces reduce to Bartlett forces.

The right-hand side of (20.43) is independent of M_T. This is evident since $(\boldsymbol{\tau}_1 \cdot \boldsymbol{\tau}_2)$ is a scalar product in isospace and as such its expectation values do not depend on the choice of the z-axis in this space. Thus, the expectation value of the Heisenberg interaction is the same for two neutrons ($M_T = -1$), two protons ($M_T = 1$), or a neutron and a proton in a state with $T = 1$ ($M_T = 0$), provided the spatial dependence of their wave-functions is the same. This property of the interaction is often referred to as *charge independence*. Obviously the Wigner, Bartlett, and Majorana forces are charge-independent. More precisely, an interaction is said to be charge-independent if it commutes with the total isospin $\mathbf{T}$. It is readily seen that the only independent two-particle isospin operators which have this property are 1 and $(\boldsymbol{\tau}_1 \cdot \boldsymbol{\tau}_2)$. This is similar to the situation with the ordinary spin where 1 and $(\boldsymbol{\sigma}_1 \cdot \boldsymbol{\sigma}_2)$ are the only independent operators which commute with **S**. The four interactions in (20.42) are therefore the most general potentials which can give rise to charge-independent scalar forces.

It is very well established that nuclear forces are charge independent. There are, however, numerous indications that they are not scalar forces. Nuclear interactions probably do not commute with either **S** or **L**. The most obvious indication in this direction is rendered by the existence of a quadrupole moment of the deuteron. From the magnetic moment of the deuteron we know that the spatial part of the relative motion of the proton and the neutron is predominantly an S-state. This is in agreement with Section 4 where we saw that the lowest state of a two-particle system with a well-behaved (local) attractive potential is a state with a zero relative angular momentum. On the other hand, an S-state of the proton-neutron system cannot give rise to any quadrupole moment. The quadrupole moment operator is a tensor operator of the second degree in the space coordinates. Its expectation value in a state of zero orbital angular momentum therefore vanishes according to the Wigner-Eckart theorem. We are therefore forced to conclude that the ground state of the deuteron contains both an S-state and a state with a higher value of the relative orbital angular momentum.

Thus, at least the proton-neutron interaction does not commute with **L**. Scattering experiments also indicate quite clearly the existence of spin-orbit forces which do not commute with either **S** or **L**. We must therefore consider interactions more general than those given by (20.42). These interactions must, of course, commute with **J** but should not do so with either **S** or **L**. Such interactions must therefore be built from tensors of degrees higher than zero in spin space or in ordinary space.

Since the spin operators are tensors of the first degree we can obtain from $\boldsymbol{\sigma}_1$ and $\boldsymbol{\sigma}_2$ only a scalar $(\boldsymbol{\sigma}_1 \cdot \boldsymbol{\sigma}_2)$, a vector, e.g. $[\boldsymbol{\sigma}_1 \times \boldsymbol{\sigma}_2]^{(1)}$, or a tensor of the second degree $[\boldsymbol{\sigma}_1 \times \boldsymbol{\sigma}_2]^{(2)}$. There are therefore the following three possibilities for the spin and space dependence of two-particle interactions.

Scalar forces, which are obtained by taking either a scalar function of $\mathbf{r} = \mathbf{r}_1 - \mathbf{r}_2$ or a product of the scalar $(\boldsymbol{\sigma}_1 \cdot \boldsymbol{\sigma}_2)$ and a scalar function of $\mathbf{r}$. The most general potential of this type is

$$V_0(1\,2) = f_0(r) + (\boldsymbol{\sigma}_1 \cdot \boldsymbol{\sigma}_2) f(r). \tag{20.49}$$

Vector forces, which can be obtained by taking the scalar product of the vector $[\boldsymbol{\sigma}_1 \times \boldsymbol{\sigma}_2]^{(1)}$ and a vector function of $\mathbf{r} = \mathbf{r}_1 - \mathbf{r}_2$. Excluding for the moment differential operators, the only possibility for such a potential is

$$V_1(1\,2) = ([\boldsymbol{\sigma}_1 \times \boldsymbol{\sigma}_2]^{(1)} \cdot \mathbf{r}) f(r). \tag{20.50}$$

Tensor forces, which are obtained by taking a scalar product of the second degree tensor $[\boldsymbol{\sigma}_1 \times \boldsymbol{\sigma}_2]^{(2)}$ and a tensor of the second degree suitably constructed from the space coordinates. Excluding derivatives we obtain

$$V_2(1\,2) = ([\boldsymbol{\sigma}_1 \times \boldsymbol{\sigma}_2]^{(2)} \cdot [\mathbf{r} \times \mathbf{r}]^{(2)}) f(r). \tag{20.51}$$

Recalling that, for charge-independent interactions, the charge dependence can be introduced only through a factor $a + b(\boldsymbol{\tau}_1 \cdot \boldsymbol{\tau}_2)$, we finally obtain for the most general charge-independent two-body interaction, which does not include derivatives, the following expression:

$$\begin{aligned} V(1\,2) = V_0(r) &+ P_\sigma V_{\sigma 0}(r) + P_\tau V_{\tau 0}(r) + P_\sigma P_\tau V_{\sigma\tau 0}(r) \\ &+ ([\boldsymbol{\sigma}_1 \times \boldsymbol{\sigma}_2]^{(1)} \cdot \mathbf{r}) V_1(r) + P_\tau ([\boldsymbol{\sigma}_1 \times \boldsymbol{\sigma}_2]^{(1)} \cdot \mathbf{r}) V_{\tau 1}(r) \\ &+ ([\boldsymbol{\sigma}_1 \times \boldsymbol{\sigma}_2]^{(2)} \cdot [\mathbf{r} \times \mathbf{r}]^{(2)}) V_2(r) \\ &+ P_\tau ([\boldsymbol{\sigma}_1 \times \boldsymbol{\sigma}_2]^{(2)} \cdot [\mathbf{r} \times \mathbf{r}]^{(2)}) V_{\tau 2}(r). \end{aligned} \tag{20.52}$$

The indices 0, 1, and 2 stand for the scalar, vector, and tensor forces respectively, and the various functions $V_0(r)$, $V_{\sigma 0}(r)$, etc., may be independent of each other. We note that we do not gain anything by adding a term obtained by multiplying a vector or a tensor force with the spin-exchange operator P_σ. For instance, if we take

$$(\boldsymbol{\sigma}_1 \cdot \boldsymbol{\sigma}_2)([\boldsymbol{\sigma}_1 \times \boldsymbol{\sigma}_2]^{(2)} \cdot [\mathbf{r} \times \mathbf{r}]^{(2)})$$

we can combine the two spin-dependent factors. However, since one of them is a scalar and the other is a tensor, their product is again a tensor, and the *only* second degree tensor we can construct from $\boldsymbol{\sigma}_1$ and $\boldsymbol{\sigma}_2$ is $[\boldsymbol{\sigma}_1 \times \boldsymbol{\sigma}_2]^{(2)}$. We are therefore back to the ordinary tensor force. For similar reasons we do not gain anything by constructing a product of vector and tensor forces:

$$([\boldsymbol{\sigma}_1 \times \boldsymbol{\sigma}_2]^{(1)} \cdot \mathbf{r})([\boldsymbol{\sigma}_1 \times \boldsymbol{\sigma}_2]^{(2)} \cdot [\mathbf{r} \times \mathbf{r}]^{(2)}).$$

Equation (20.52) is therefore the most general charge-independent two-particle interaction which does not involve differential operators.

Having thus explicitly indicated the spin and isospin dependence of the interaction we are able to discuss the contribution of the various terms in (20.52) to the expectation value of the interaction $V(12)$ in any given state of the two-particle system. We shall not do it here in great detail, since the calculations are straightforward. We shall, however, indicate some general properties of some of these interactions.

From the properties of the charge-exchange operator we see that whenever an interaction is proprotional to P_τ it is either repulsive for isotriplets and attractive for iso-singlets or vice versa. A charge-independent interaction which is either attractive or repulsive for *both* iso-triplets and iso-singlets *must* therefore contain a term with no charge operators, such as $V_0(r)$ or $P_\sigma V_{\sigma 0}(r)$. In a similar way, $P_\sigma V_{\sigma 0}(r)$ gives rise to a force which is either attractive for singlets and repulsive for triplets or vice versa [depending on the sign of $V_{\sigma 0}(r)$].

The expectation value of the vector interaction $([\boldsymbol{\sigma}_1 \times \boldsymbol{\sigma}_2]^{(1)} \cdot \mathbf{r})V_1(r)$ vanishes for any two-particle configuration, both in LS-coupling and in jj-coupling. This happens because that particular interaction changes sign under space reflections, whereas $\psi^*(\mathbf{r})\,\psi(\mathbf{r})$ always has an even parity. The above vector force does have nonvanishing, nondiagonal elements and may give rise to higher order corrections to the energy. It should be realized, however, that this is a parity nonconserving interaction, and therefore probably does not appear in the actual strong nuclear interaction. We can consider another type of two-particle vector interaction which is parity-conserving, similar to the single particle spin-orbit interaction (7.1)

$$V_1(1\,2) = ((\boldsymbol{\sigma}_1 + \boldsymbol{\sigma}_2) \cdot \boldsymbol{l})\, V(|\mathbf{r}_1 - \mathbf{r}_2|). \tag{20.53}$$

Here $\boldsymbol{l} = \frac{1}{2}(\mathbf{r}_1 - \mathbf{r}_2) \times (\mathbf{p}_1 - \mathbf{p}_2)$ is the operator of the relative angular momentum which, like $\boldsymbol{\sigma}$, is an axial vector. The interaction $V_1(1\,2)$ is thus a scalar (not a pseudoscalar) as required. It is a spin-orbit interaction which depends on the relative angular momentum of the two particles rather than on the angular momentum of each one of them in the central field. Since $V_1(1\,2)$ is a tensor of the first degree in both spin and space coordinates, its expectation value vanishes for states of zero total spin or zero total orbital angular momentum, i.e., for all singlets and for all S states. It is also obvious that the off-diagonal elements of $V_1(1\,2)$ between singlet and triplet states vanish indentically unlike the case with the single particle interaction (7.1). This is so because in $\langle S = 1\, |\boldsymbol{\sigma}_1 + \boldsymbol{\sigma}_2|\, S = 0\rangle$, the $S = 1$ state and $\boldsymbol{\sigma}_1 + \boldsymbol{\sigma}_2$ are both symmetric in the two spins, and therefore the function $(\boldsymbol{\sigma}_1 + \boldsymbol{\sigma}_2)|S = 1\rangle$ is orthogonal to the antisymmetric state $|S = 0\rangle$.

The existence of an interaction of the type (20.53) is suggested by nuclear scattering experiments. We shall also see later that for a single particle outside closed shells (20.53) gives rise to an ordinary spin-orbit interaction. It may therefore be, at least partially, responsible for the spin-orbit interaction observed in nuclei.

The tensor interaction $([\boldsymbol{\sigma}_1 \times \boldsymbol{\sigma}_2)^{(2)} \cdot [\mathbf{r} \times \mathbf{r}]^{(2)})\, V_2(r)$ is parity-conserving and is most probably part of the actual nuclear interaction. As we saw in Section 14 it is proportional to the interaction between two magnetic moments $\mu_1\boldsymbol{\sigma}_1$ and $\mu_2\boldsymbol{\sigma}_2$ at a distance $\mathbf{r}$ apart. Since both the spin part and the space part of the tensor interactions are tensors of the second degree, its expectation values vanish for all states with zero total spin or zero total orbital angular momentum, i.e., for all singlets and for all S-states. The tensor interaction has also vanishing off-diagonal elements between singlets and triplets, because the triangular condition between S, k, and S' cannot be satisfied with $S = 1$, $k = 2$, and $S' = 0$.

Taking expectation values of the above interactions with antisymmetric wave functions, it may happen that two interactions which look very different will actually turn out to be completely equivalent. A simple example is the δ interaction $V_0(r) = g\delta(\mathbf{r})$. Let us compare it with the special Bartlett interaction $V_B(r) = -\frac{1}{2}g(1 + \boldsymbol{\sigma}_1 \cdot \boldsymbol{\sigma}_2)\,\delta(\mathbf{r})$, and assume we are dealing with identical particles. Because of the occurrence of $\delta(\mathbf{r})$ we shall obtain contributions, in both cases, only from the space-symmetric part of the wave function. However, due to the antisymmetry of the wave function of two identical particles, its space-symmetric part is always multiplied by a spin-antisymmetric part, i.e., by a singlet spin function. Since for singlets the expectation value of $-\frac{1}{2}(1 + \boldsymbol{\sigma}_1 \cdot \boldsymbol{\sigma}_2)$ is $+1$, we obtain the general result independent of the coupling scheme

$$\langle l_1 l_2 \alpha\, J\, M|\delta(\mathbf{r})|l_1 l_2 \alpha\, J\, M\rangle_a = \langle l_1 l_2 \alpha\, J\, M| -\tfrac{1}{2}(1 + \boldsymbol{\sigma}_1 \cdot \boldsymbol{\sigma}_2)\delta(\mathbf{r})|l_1 l_2 \alpha\, J\, M\rangle_a \tag{20.54}$$

for antisymmetric states.

In the general case of any two nucleons, the Majorana interaction $P_x\delta(\mathbf{r})$ is equivalent to the ordinary interaction $\delta(\mathbf{r})$.

There is not yet conclusive evidence on the exact form of nuclear forces. We know that they are short-range and charge-independent forces, certainly involving the spin operators and probably containing vector and tensor forces. Despite this rather scant information we shall see that many nuclear properties can be predicted and some regularities can be explained, using only the general results of the shell model of the nucleus.

21. Tensor-Expansion of the Interaction

In the preceding sections the spacing of the energy levels in a given configuration was discussed, and calculated for a specific interaction (the δ-force). Some general conclusions about the structure of such configurations were deduced from the known exchange properties of the interaction. In order to be able to treat in detail arbitrary interactions we have to make full use of the powerful methods of tensor algebra. To this end we shall find it useful to express the interaction in terms of scalar products of irreducible tensor operators. Due to the Wigner-Eckart theorem it will turn out that only a small number of these tensor operators affect a given configuration. Such a method, first introduced by Slater for the Coulomb potential, thus allows the determination of energy levels of a given configuration, with an *arbitrary* interaction, in terms of a small number of parameters. In many cases, especially in the well-studied configurations of atomic spectroscopy, the number of levels whose calculated energy can be compared with experiment is bigger than the number of parameters. It was also found possible to apply such methods to some nuclear spectra. Thus, at least the internal consistency of the theory can be checked.

Let us consider first an interaction $V(|\mathbf{r}_1 - \mathbf{r}_2|)$ and its matrix elements in the $l_1 l_2$ configuration, $\int \psi^*(l_1 l_2 L\, M) V(|\mathbf{r}_1 - \mathbf{r}_2|) \psi(l_1 l_2 L\, M)$. The evaluation of these integrals is complicated because V is a function of $|\mathbf{r}_1 - \mathbf{r}_2|$ whereas $\psi^*\psi$ is essentially a product of a function of $\mathbf{r}_1$ and a function of $\mathbf{r}_2$. The evaluation of these integrals will become considerably simpler if we manage to separate $V(|\mathbf{r}_1 - \mathbf{r}_2|)$ into a product of functions of $\mathbf{r}_1$ and $\mathbf{r}_2$. This can be achieved to a certain extent in the following way. The distance $|\mathbf{r}_1 - \mathbf{r}_2|$ is a function of r_1, r_2, and the angle ω_{12} between $\mathbf{r}_1$ and $\mathbf{r}_2$

$$\cos \omega_{12} = \frac{(\mathbf{r}_1 \ \mathbf{r}_2)}{r_1 r_2}. \tag{21.1}$$

We can therefore expand $V(|\mathbf{r}_1 - \mathbf{r}_2|)$ in a series of Legendre polynomials of $\cos \omega_{12}$

$$V(|\mathbf{r}_1 - \mathbf{r}_2|) = \sum_{k=0}^{\infty} v_k(r_1, r_2)\, P_k(\cos \omega_{12}) \tag{21.2}$$

where

$$v_k(r_1, r_2) = \frac{2k+1}{2} \int V(|\mathbf{r}_1 - \mathbf{r}_2|)\, P_k(\cos \omega_{12})\, d(\cos \omega_{12}). \tag{21.3}$$

In (21.2) we prefer to expand $V(|\mathbf{r}_1 - \mathbf{r}_2|)$ in Legendre polynomials of $\cos\omega_{12}$ rather than in powers of $\cos\omega_{12}$ because $P_k(\cos\omega_{12})$ can be expressed in a simple way in terms of irreducible tensor operators. This is not true of $\cos^n\omega_{12}$. In fact, using the addition theorem of spherical harmonics (14.5) we obtain

$$V(|\mathbf{r}_1 - \mathbf{r}_2|) = \sum_{k,\kappa} \frac{4\pi}{2k+1} v_k(r_1, r_2)\, Y^*_{k\kappa}(\Omega_1)\, Y_{k\kappa}(\Omega_2). \tag{21.4}$$

Although $V(|\mathbf{r}_1 - \mathbf{r}_2|)$ is not fully decomposed into a sum of products of functions of $\mathbf{r}_1$ and functions of $\mathbf{r}_2$, we did manage to do so in (21.4) for the angular coordinates. However, in $\psi(l_1l_2L\,M)$ the dependence on r_1 and r_2 is the *same* for all allowed values of L and M. The only difference between the functions $\psi(l_1l_2L\,M)$ for different values of L and M is their dependence on the angles of the two particles. Thus, the decomposition (21.4) will nevertheless be very useful. Writing explicitly

$$\psi(l_1l_2L\,M) = \frac{1}{r_1r_2} R_{n_1l_1}(r_1)\, R_{n_2l_2}(r_2)\, \varphi_{\rm ang}(l_1l_2L\,M) \tag{21.5}$$

where $\varphi_{\rm ang}$ depends only on θ_1, φ_1, θ_2, and φ_2

$$\varphi_{\rm ang}(l_1l_2L\,M) = \sum (l_1m_1l_2m_2|l_1l_2L\,M)\, Y_{l_1m_1}(\theta_1\varphi_1)\, Y_{l_2m_2}(\theta_2\varphi_2) \tag{21.6}$$

we obtain from (21.4)

$$\langle l_1l_2L\,M|V(|\mathbf{r}_1 - \mathbf{r}_2|)|l_1l_2L\,M\rangle$$

$$= \sum_{k=0}^{\infty} F^k \langle l_1l_2L\,M| \frac{4\pi}{2k+1} (\mathbf{Y}_k(\Omega_1)\cdot\mathbf{Y}_k(\Omega_2))|l_1l_2L\,M\rangle. \tag{21.7}$$

Here the *radial integrals* or *Slater integrals* F^k are given by

$$F^k = F^k(n_1l_1n_2l_2) = \int |R_{n_1l_1}(r_1)\, R_{n_2l_2}(r_2)|^2\, v_k(r_1, r_2)\, dr_1dr_2 \tag{21.8}$$

and the matrix elements of the spherical harmonics are to be taken with the angular part of the wave function (21.6). Since the operator $[4\pi/(2k+1)]\,(\mathbf{Y}_k(\Omega_1)\cdot\mathbf{Y}_k(\Omega_2))$ is a scalar product of two tensor operators

$$C^k_\kappa(\Omega) = \sqrt{\frac{4\pi}{2k+1}}\, Y_{k\kappa}(\Omega) \tag{21.9}$$

we can use (15.5) directly and obtain for the angular matrix element in (21.7)

$$f_k = \langle l_1 l_2 L\, M \; \frac{4\pi}{2k+1} (\mathbf{Y}_k(\Omega_1) \cdot \mathbf{Y}_k(\Omega_2)) \; l_1 l_2 L\, M \rangle$$

$$= (-1)^{l_1+l_2+L} (l_1 || \mathbf{C}^k || l_1)(l_2 || \mathbf{C}^k || l_2) \begin{Bmatrix} l_1 & l_2 & L \\ l_2 & l_1 & k \end{Bmatrix} \tag{21.10}$$

where the reduced matrix elements can be obtained from (17.14).

Considering the W-coefficient in (21.10) we see that the angular matrix elements vanish unless $0 \leq k \leq 2l_1$, $0 \leq k \leq 2l_2$. Since $(l || \mathbf{C}^{(k)} || l)$ vanishes for odd values of k due to (13.18), only even values of k can appear in the sum (21.7). Thus we finally obtain

$$\langle l_1 l_2 L\, M | V(|\mathbf{r}_1 - \mathbf{r}_2|) | l_1 l_2 L\, M \rangle = \sum_{k \text{ even}} f_k F^k \tag{21.11}$$

where f_k is defined in (21.10) and the summation over k goes from $k = 0$ to $k = 2 \min(l_1, l_2)$.

Equation (21.11) is the expression of the energies of all the $2l_2 + 1$ levels of the $l_1 l_2$ configuration ($l_2 \leq l_1$) in terms of $l_2 + 1$ parameters F^k, for $0 \leq k \leq 2l_2$. It is important to note that in (21.11) the geometrical factors are separated from the physical factors in the determination of the structure of the $l_1 l_2$ configuration. The numbers f_k depend only on the quantum numbers l_1, l_2, and L, and are thus completely determined by the spatial configuration of the three vectors $\mathbf{l}_1$, $\mathbf{l}_2$, and $\mathbf{L}$. The f_k are *independent* of the particular interaction $V(|\mathbf{r}_1 - \mathbf{r}_2|)$ under consideration, nor do they depend on the shape of the central field. The particular r-dependence of the central field affects only the *radial* parts of the single particle wave functions. Thus, the detailed form of $V(|\mathbf{r}_1 - \mathbf{r}_2|)$, as well as that of the central potential, affects only the values of the F^k. The expansion (21.7) can be generalized to any two-particle interaction, including spins, isospins, and noncentral forces. We can fit (21.11) to experimental data by properly choosing the values of the parameters F^k. *By doing this we are automatically relieved of the necessity of making any detailed assumption about either the central field in which the two particles move or on their mutual interaction.* The only assumption which we then make is that the major part of the interactions is absorbed in the central field and therefore the residual interaction can be treated as a perturbation.

The treatment given above can be extended to antisymmetric states. In (20.9) we obtained for identical particles

$$\Delta E_{SLJ} = \langle l_1 l_2 S\, L\, J\, M | V_{12} | l_1 l_2 S\, L\, J\, M \rangle$$

$$+ (-1)^{l_1+l_2-L+S} \langle l_1 l_2 S\, L\, J\, M | V_{12} | l_2 l_1 S\, L\, J\, M \rangle.$$

Taking $V_{12} = V(|\mathbf{r}_1 - \mathbf{r}_2|)$ and using the expansion (21.4) we obtain

$$\Delta E_{SLJ} = \sum f_k F^k + (-1)^{l_1+l_2+L+S} \sum g_k G^k \tag{21.12}$$

where f_k and F^k are given by (21.10) and (21.8). The corresponding coefficients and *exchange integrals* in ΔE_{SLJ} are defined by

$$G^k(n_1 l_1 n_2 l_2) = \int R_{n_1 l_1}(r_1) R_{n_2 l_2}(r_1) R_{n_1 l_1}(r_2)\, R_{n_2 l_2}(r_2)\, v_k(r_1, r_2)\, dr_1 dr_2 \tag{21.13}$$

and

$$g_k = (-1)^L\, (l_1||\mathbf{C}^k||l_2)\, (l_2||\mathbf{C}^k||l_1) \begin{Bmatrix} l_1 & l_2 & L \\ l_1 & l_2 & k \end{Bmatrix}. \tag{21.14}$$

We note that, for equivalent particles, $(n_1, l_1) = (n_2, l_2)$, G^k as given by (21.13) and F^k given by (21.8) are identical. Observing also that the normalization factor for equivalent particles has to be changed from $1/\sqrt{2}$ to $\frac{1}{2}$, we obtain for this case that ΔE_{SLJ} is given simply by (21.11). This is simply due to the fact that for equivalent particles the state $\psi(l^2 S\, L\, J\, M)$ is automatically antisymmetric if $L + S$ is even, so that formally no exchange integrals show up.

Before we proceed to the more general case of spin-dependent interactions we shall discuss some simple examples of the formalism developed above. Consider, for instance, the simple $l\,s$ configuration, i.e., $l_1 = l$ (arbitrary) and $l_2 = 0$. There are only two multiplets in this configuration, namely ^1L and ^3L with $L = l$. From (21.10) and (21.14) we see that the only coefficients which are different from zero are f_0 and g_l. Introducing

$$F_0 = f_0 F^0 \qquad G_l = (-1)^l g_l G^l$$

we obtain

$$\begin{aligned} \Delta E(^1L) &= F_0 + G_l \\ \Delta E(^3L) &= F_0 - G_l. \end{aligned} \tag{21.15}$$

Thus, the separation between the two multiplets is $2G_l$ and can be estimated from (21.13) if the central potential and the residual interaction $V(|\mathbf{r}_1 - \mathbf{r}_2|)$ are known.

Another simple example, is offered by the p^2 configuration, i.e., $n_1 = n_2 = n$, $l_1 = l_2 = 1$. It is clear from the previous discussion that only $k = 0$ and $k = 2$ terms will give nonvanishing contributions and that the results will contain only the direct integrals $F^{(k)}$.

An actual computation gives

$$\begin{aligned} \Delta E(p^2, {}^1S) &= F^{(0)} + \tfrac{10}{25} F^{(2)} \\ \Delta E(p^2, {}^3P) &= F^{(0)} - \tfrac{5}{25} F^{(2)} \\ \Delta E(p^2, {}^1D) &= F^{(0)} + \tfrac{1}{25} F^{(2)}. \end{aligned} \tag{21.16}$$

The position of the three multiplets is determined by two parameters $F^{(0)}$ and $F^{(2)}$. We can therefore deduce from (21.16) one relation involving the spacings between the multiplets which is independent of $F^{(0)}$ and $F^{(2)}$. In fact, we conclude that

$$\frac{\Delta E({}^1S) - \Delta E({}^1D)}{\Delta E({}^1D) - \Delta E({}^3P)} = \frac{{}^1S - {}^1D}{{}^1D - {}^3P} = \frac{3}{2} \quad \text{for} \quad p^2. \tag{21.17}$$

Here ${}^1S - {}^1D$ stands for the spacing between the level ${}^1S$ and the level ${}^1D$, etc. The relation (21.17) is valid for *any* central field and *any* interaction $V(|\mathbf{r}_1 - \mathbf{r}_2|)$. This example shows the power of the method we have discussed and the sort of predictions it can make, even in cases where the central potential and the two-body interactions are not explicitly known.

So far, we have been considering only interactions of the type $V(|\mathbf{r}_1 - \mathbf{r}_2|)$. The method can, however, be easily generalized to spin-dependent interactions as well, since we know already the most general expressions through which the spin dependence can come into the picture. Consider the interaction $(\boldsymbol{\sigma}_1 \cdot \boldsymbol{\sigma}_2)V(|\mathbf{r}_1 - \mathbf{r}_2|)$. Using the expansion (21.4) we can write

$$(\boldsymbol{\sigma}_1 \cdot \boldsymbol{\sigma}_2)\, V(|\mathbf{r}_1 - \mathbf{r}_2|) = \sum_k v_k(r_1, r_2)\, (\boldsymbol{\sigma}_1 \cdot \boldsymbol{\sigma}_2)\, \frac{4\pi}{2k+1} (\mathbf{Y}_k(\Omega_1) \cdot \mathbf{Y}_k(\Omega_2)). \tag{21.18}$$

Taking the matrix elements of this interaction in the state $\psi(l_1 l_2 S L\, J\, M)$ the radial integrals can be separated off and we obtain

$$\langle l_1 l_2 S L\, J\, M | (\boldsymbol{\sigma}_1 \cdot \boldsymbol{\sigma}_2) V(|\mathbf{r}_1 - \mathbf{r}_2|) | l_1 l_2 S\, L\, J\, M\rangle = \sum f'_k F^k. \tag{21.19}$$

Now,

$$\begin{aligned} f'_k &= \langle l_1 l_2 S\, L\, J\, M \Big| (\boldsymbol{\sigma}_1 \cdot \boldsymbol{\sigma}_2) \frac{4\pi}{2k+1} (\mathbf{Y}_k(\Omega_1) \cdot \mathbf{Y}_k(\Omega_2)) \Big| l_1 l_2\, S\, L\, J\, M\rangle \\ &= \sum_{M_S, M_L} \langle \tfrac{1}{2}\, \tfrac{1}{2}\, S\, M_S | (\boldsymbol{\sigma}_1 \cdot \boldsymbol{\sigma}_2) |\, \tfrac{1}{2}\, \tfrac{1}{2}\, S\, M_S\rangle \\ &\times \langle l_1 l_2 L\, M_L \Big| \frac{4\pi}{2k+1} (\mathbf{Y}_k(\Omega_1) \cdot \mathbf{Y}_k(\Omega_2)) \Big| l_1 l_2 L\, M_L\rangle (S\, M_S L\, M_L | S\, L\, J\, M)^2 \\ &= 2[S(S+1) - \tfrac{3}{2}]\, (-1)^{l_1 + l_2 + L} (l_1 || \mathbf{C}^k || l_1)\, (l_2 || \mathbf{C}^k || l_2) \begin{Bmatrix} l_1 & l_2 & L \\ l_2 & l_1 & k \end{Bmatrix}. \end{aligned} \tag{21.20}$$

To obtain similar relations in jj-coupling we define a tensor $\mathbf{T}^{(1k)r}$ by the tensor product of $\boldsymbol{\sigma}$ and $\mathbf{C}^{(k)} = \sqrt{4\pi/(2k+1)}\,\mathbf{Y}_k$

$$T^{(1k)r}_\rho = [\boldsymbol{\sigma} \times \mathbf{C}^{(k)}]^{(r)}_\rho = \sum_{\mu,\kappa} (1\mu k\kappa|1kr\rho)\,\sigma_\mu C^{(k)}_\kappa. \tag{21.21}$$

Using the transformation (15.28) for tensor products with $j_1 = j_3 = 1$, $j_2 = j_4 = k$, $J' = 0$, observing the convention (15.1) for scalar products, we then obtain

$$(\boldsymbol{\sigma}_1 \cdot \boldsymbol{\sigma}_2)(\mathbf{C}_k(1) \cdot \mathbf{C}_k(2)) = \sum_r (-1)^{k+r+1} (\mathbf{T}^{(1k)r}(1) \cdot \mathbf{T}^{(1k)r}(2)). \tag{21.22}$$

The tensors $\mathbf{T}^{(1k)r}$ depend only on the spin and angular coordinates of one particle, but not on its radial coordinates. Using (21.22) we can therefore rewrite (21.18) in the form

$$(\boldsymbol{\sigma}_1 \cdot \boldsymbol{\sigma}_2)\, V(|\mathbf{r}_1 - \mathbf{r}_2|) = \sum_{k,r} (-1)^{k+r+1} v_k(r_1, r_2)\,(\mathbf{T}^{(1k)r}(1) \cdot \mathbf{T}^{(1k)r}(2)). \tag{21.23}$$

In general, any interaction (not necessarily a scalar force) which depends on the spin and space coordinates of particles 1 and 2 can be written in the form

$$V_{12} = \sum_{\substack{skr\\ s'k'}} v_{ss'kk',r}(r_1, r_2)\,(\mathbf{T}^{(s\,k)r}(1) \cdot \mathbf{T}^{(s'k')r}(2)). \tag{21.24}$$

Here $\mathbf{T}^{(sk)r}$ is an irreducible tensor operator of degree r which is built out of a tensor of degree s of the spin coordinates and a tensor of degree k of the angular coordinates. Taking matrix elements of (21.24) in jj-coupling we obtain the separation of the radial integrals from the angular and spin integrals. Thus

$$\Delta E_J = \langle j_1 j_2 J\,M|V_{12}|\,j_1 j_2 J\,M\rangle = \sum f_r F^r \tag{21.25}$$

where for convenience we have omitted the indices (sk), and f_r is now given by

$$\begin{aligned} f_r &= \langle j_1 j_2 J\,M|(\mathbf{T}^{(r)}(1) \cdot \mathbf{T}^{(r)}(2))|\,j_1 j_2 J\,M\rangle \\ &= (-1)^{j_1+j_2+J} (j_1||\mathbf{T}^{(r)}||j_1)\,(j_2||\mathbf{T}^{(r)}||\,j_2) \begin{Bmatrix} j_1 & j_2 & J \\ j_2 & j_1 & r \end{Bmatrix}. \end{aligned} \tag{21.26}$$

The summation over r in (21.25) is limited to $0 \leq r \leq 2j_1$, $0 \leq r \leq 2j_2$. Thus, the energy levels of the $j_1 j_2$ configuration are given in terms of a small number of parameters. The reduced matrix elements of $\mathbf{T}^{(r)}$ can be further decomposed into the product of the reduced matrix elements of the spin operator $\mathbf{\Sigma}^{(s)}$ and the space operator $\mathbf{U}^{(k)}$ by using (14.26)

$$(j_1||\mathbf{T}^{(r)}||j_1) = (\tfrac{1}{2}l_1 j_1||[\mathbf{\Sigma}^{(s)} \times \mathbf{U}^{(k)}]^{(r)}||\tfrac{1}{2}l_1 j_1)$$

$$= \sqrt{2r+1}\,(2j_1+1)\begin{Bmatrix} \frac{1}{2} & l_1 & j_1 \\ \frac{1}{2} & l_1 & j_1 \\ s & k & r \end{Bmatrix}(\tfrac{1}{2}||\mathbf{\Sigma}^{(s)}||\tfrac{1}{2})(l_1||\mathbf{U}^{(k)}||l_1). \quad (21.27)$$

From (14.32) we see that $(j_1||\mathbf{T}^{(r)}(1)||j_1)$ vanishes unless $2l_1 + 2j_1 + 1 + s + k + r$ is even. Since $2l_1$ is even and $2j_1$ is odd, the requirement is that $s + k + r$ be even. We can therefore rewrite (21.26) using (21.27) and obtain

$$f_r = (-1)^{j_1+j_2+J}(2r+1)(2j_1+1)(2j_2+1)\begin{Bmatrix} j_1 & j_2 & J \\ j_2 & j_1 & r \end{Bmatrix}\begin{Bmatrix} \frac{1}{2} & l_1 & j_1 \\ \frac{1}{2} & l_1 & j_1 \\ s & k & r \end{Bmatrix}\begin{Bmatrix} \frac{1}{2} & l_2 & j_2 \\ \frac{1}{2} & l_2 & j_2 \\ s' & k' & r \end{Bmatrix}$$

$$\times (\tfrac{1}{2}||\mathbf{\Sigma}^{(s)}||\tfrac{1}{2})(\tfrac{1}{2}||\mathbf{\Sigma}^{(s')}||\tfrac{1}{2})(l_1||\mathbf{U}^{(k)}||l_1)(l_2||\mathbf{U}^{(k')}||l_2). \quad (21.28)$$

We take for $s = 0$, $\mathbf{\Sigma}^{(0)} = 1$ whereas for $s = 1$ we take $\mathbf{\Sigma}^{(1)} = \boldsymbol{\sigma}$. The corresponding reduced matrix elements are then given according to (14.21) by

$$(\tfrac{1}{2}||\mathbf{\Sigma}^{(s)}||\tfrac{1}{2}) = \sqrt{2(2s+1)}.$$

An irreducible tensor $\mathbf{U}^{(k)}$ which depends on the coordinates only is necessarily proportional to $\mathbf{C}^{(k)} = \sqrt{4\pi/2k+1)}\,\mathbf{Y}_k$. We can therefore choose $\mathbf{U}^{(k)} = \mathbf{C}^{(k)}$. Since $s + k + r$ must be even and since $(l||\mathbf{C}^{(k)}||l)$ vanishes for odd values of k we see that

$$\begin{array}{lll} \text{if} & s \neq s', & f_r \text{ vanishes for any value of } r \\ \text{if} & s = s' = 0, & f_r \text{ vanishes for odd values of } r \\ \text{if} & s = s' = 1, & f_r \text{ vanishes for even values of } r. \end{array} \quad (21.29)$$

It will be shown later that interactions whose expansions (21.24) contain only tensors with odd values of r play an important role in nuclear spectroscopy. It is worthwhile to note here that because of (21.29) such interactions are equivalent to spin-dependent forces.

The coefficients f_r in (21.25) turn out to be rational numbers

$$f_r = \frac{C_r}{D_r} \qquad C_r \text{ and } D_r\text{—both integers}$$

It is therefore convenient to define new parameters

$$F_r = \frac{1}{D_r} F^r.$$

In terms of these parameters the energy is given by

$$\Delta E_J = \sum C_r F_r$$

where the C_r are all integers. In this book we shall not use this convention.

The simple dependence of f_r, and consequently of ΔE_J, on J enables us to draw some general conclusions on the two-particle spectra. Consider the expression

$$\overline{\Delta E} = \sum_J (2J+1)\, \Delta E_J \Big/ \sum_J (2J+1) \tag{21.30}$$

which gives the average shift of all the levels of the $j_1 j_2$ configuration. Using (21.26) and (15.16) we obtain

$$\sum_J (2J+1)\, \Delta E_J = \sum_r F^r \sum_J (2J+1) f_r(J)$$

$$= \sum_r F^r (j_1 || \mathbf{T}^{(r)} || j_1)\, (j_2 || \mathbf{T}^{(r)} || j_2) \sqrt{(2j_1+1)(2j_2+1)}\, \delta_{r,0}$$

$$= F^0 (j_1 || \mathbf{T}^{(0)} || j_1)\, (j_2 || \mathbf{T}^{(0)} || j_2) \sqrt{(2j_1+1)(2j_2+1)}. \tag{21.31}$$

Since

$$\sum_J (2J+1) = (2j_1+1)(2j_2+1)$$

we obtain

$$\overline{\Delta E} = \frac{F^0}{\sqrt{(2j_1+1)(2j_2+1)}} (j_1 || \mathbf{T}^{(0)} || j_1)\, (j_2 || \mathbf{T}^{(0)} || j_2). \tag{21.32}$$

However, the reduced matrix element of $T^{(0)}$ can be simply obtained from (21.27). Because of (21.29) and since in addition s, k, and r have to satisfy the triangular condition we must have $s = k = 0$ and hence

$$(j_1 || \mathbf{T}^{(sk)0} || j_1)$$

$$= (2j_1+1) \begin{Bmatrix} \frac{1}{2} & l_1 & j_1 \\ \frac{1}{2} & l_2 & j_2 \\ 0 & 0 & 0 \end{Bmatrix} (\tfrac{1}{2} || \mathbf{\Sigma}^{(0)} || \tfrac{1}{2})\, (l_1 || \mathbf{U}^{(0)} || l_1) \delta_{s,0} \delta_{k,0}$$

$$= \sqrt{\frac{2j_1+1}{2l_1+1}}\, (l_1 || \mathbf{C}^{(0)} || l_1) \delta_{s,0} \delta_{k,0} = \sqrt{2j_1+1}\, \delta_{s,0} \delta_{k,0}. \tag{21.33}$$

Inserting (21.33) into (21.32) we obtain

$$\overline{\Delta E} = F^{(00)0} \tag{21.34}$$

The physical interpretation of (21.34) is quite simple. We note from (21.3) that

$$V_0(r_1, r_2) = \tfrac{1}{2} \int V(|\mathbf{r}_1 - \mathbf{r}_2|)\, d(\cos \omega_{12}) = \overline{V(|\mathbf{r}_1 - \mathbf{r}_2|)} \qquad (21.35)$$

where $\overline{V}$ stands for the average over all angles ω_{12}. Thus, $F^{(00)0}$ is the expectation value of the average interaction $\overline{V}$ in the given configuration, and by (21.34) is equal to the average shift of the levels.

Other simple conclusions can be obtained if we consider a configuration of two equivalent particles in jj-coupling. We obtain, then, from (21.25) and (21.26)

$$\Delta E(j^2 J) = (-1)^{J+1} \sum_r (j||\mathbf{T}^{(r)}||j)^2 \begin{Bmatrix} j & j & J \\ j & j & r \end{Bmatrix} F^r. \qquad (21.36)$$

For identical particles, only even values of J are allowed, and (21.36) gives then the energy, including the exchange energy, as explained above. For nonidentical particles, a proton in a j-orbit and a neutron in a j-orbit, both even and odd values of J are allowed and the energy is again given by (21.36). The states with $T = 1$ have even values of J and those with $T = 0$ have odd values of J. For $J = 0$ we obtain

$$\Delta E(j^2, J = 0) = \frac{1}{2j+1} \sum_r (-1)^r (j||\mathbf{T}^{(r)}||j)^2 F^r. \qquad (21.37)$$

The average shift of all states, with $T = 1$ and $T = 0$, is given by (21.32) as

$$\overline{\Delta E} = \frac{(j||T^{(0)}||j)^2}{2j+1} F^0. \qquad (21.38)$$

To evaluate the average shift of the $T = 1$ states for two equivalent particles, identical or nonidentical, we have to sum (21.36) over even values of J only. Thus, using (15.16) and (15.17) we obtain

$$\sum_{J\ \text{even}} (2J+1)\, \Delta E(j^2 J)$$

$$= - \sum_r (j||\mathbf{T}^{(r)}||j)^2 F^r \Big(\sum_J \tfrac{1}{2} [1 + (-1)^J]\, (2J+1) \begin{Bmatrix} j & j & J \\ j & j & r \end{Bmatrix} \Big)$$

$$= -\tfrac{1}{2} \sum_r (j||\mathbf{T}^{(r)}||j)^2 F^r [1 - (2j+1)\, \delta_{r,0}]. \qquad (21.39)$$

Since

$$\sum_{J\ \text{even}} (2J+1) = 1 + 5 + \ldots + (4j-1) = j(2j+1) \qquad (21.40)$$

we obtain, using (21.33)

$$\varDelta\overline{E}\,(T=1)=\frac{-1}{2j(2j+1)}\sum(j||\mathbf{T}^{(r)}||j)^2F^r+\frac{1}{2j}(j||\mathbf{T}^{(0)}||j)^2F^0$$
$$=\frac{2j+1}{2j}F^0-\frac{1}{2j(2j+1)}\sum(j||\mathbf{T}^{(r)}||j)^2F^r. \tag{21.41}$$

Some of the relations obtained above can still be simplified if the interaction $V(1\,2)$ contains only odd tensors. For such *odd-tensor interactions* we have from (21.37)

$$\varDelta E(j^20)=-\frac{1}{2j+1}\sum_{r\text{ odd}}(j||\mathbf{T}^{(r)}||j)^2F^r \tag{21.42}$$

and from (21.39) we obtain for the states with $T=1$

$$\sum_{J\text{ even}}(2J+1)\,\varDelta E(j^2J)=-\tfrac{1}{2}\sum_{r\text{ odd}}(j||\mathbf{T}^{(r)}||j)^2F^r.$$

Hence for odd tensor interactions

$$\sum_{J\text{ even}}(2J+1)\,\varDelta E(j^2J)=\frac{2j+1}{2}\,\varDelta E(j^20) \tag{21.43}$$

or more explicitly

$$\sum_{J>0\text{ even}}(2J+1)\,\langle j^2J\,M|V_{12}{}^{(\text{odd})}|j^2J\,M\rangle=\frac{2j-1}{2}\langle j^20\,0|V_{12}{}^{(\text{odd})}|j^20\,0\rangle. \tag{21.44}$$

Considering all states we obtain from (21.36) and (15.17)

$$\sum(-1)^J(2J+1)\,\varDelta E(j^2J)=-\sum_r(j||\mathbf{T}^{(r)}||j)^2F^r.$$

Hence, for odd tensor interactions, we obtain, using (21.42)

$$\sum(-1)^J(2J+1)\,\varDelta E(j^2J)=(2j+1)\,\varDelta E(j^20) \tag{21.45}$$

or, more explicitly,

$$\sum_{J>0}(-1)^J(2J+1)\,\langle j^2J\,M|V_{12}^{(\text{odd})}|j^2J\,M\rangle=2j\langle j^20\,0|V_{12}^{(\text{odd})}|j^20\,0\rangle. \tag{21.46}$$

Odd tensor interactions play an important role in nuclear spectroscopy. We shall discuss now some interactions which involve odd tensors. We first recall that, due to (21.29), the interactions (21.23), within the l^2-configuration contain only odd tensors.

Let us now consider the δ-force. As a function of $\mathbf{r}_1 - \mathbf{r}_2$ its expansion in the form (21.11) contains only even tensors. However, as will be now shown, the δ-force for identical nucleons is equivalent to an odd tensor interaction.

In fact we have already seen (20.54) that for antisymmetric wave functions

$$\langle l_1 l_2 \alpha \, J \, M | \delta(\mathbf{r}_1 - \mathbf{r}_2) | l_1 l_2 \alpha \, J \, M \rangle_a$$
$$= - \tfrac{1}{2} \langle l_1 l_2 \alpha \, J \, M | (1 + \boldsymbol{\sigma}_1 \cdot \boldsymbol{\sigma}_2) \, \delta(\mathbf{r}_1 - \mathbf{r}_2) | l_1 l_2 \alpha \, J \, M \rangle_a .$$

Hence we obtain by slight rearrangement

$$\langle l_1 l_2 \alpha \, J \, M | \delta(\mathbf{r}_1 - \mathbf{r}_2) | l_1 l_2 \alpha \, J \, M \rangle_a$$
$$= - \tfrac{1}{3} \langle l_1 l_2 \alpha \, J \, M | (\boldsymbol{\sigma}_1 \cdot \boldsymbol{\sigma}_2) \, \delta(\mathbf{r}_1 - \mathbf{r}_2) | l_1 l_2 \alpha \, J \, M \rangle_a \tag{21.47}$$

However, $(\boldsymbol{\sigma}_1 \cdot \boldsymbol{\sigma}_2)\delta(\mathbf{r}_1 - \mathbf{r}_2)$ is just an interaction of the type (21.23). By (21.29) and for $l_1 = l_2$ the only contribution to its expectation value in the state $\psi_a(l^2 \alpha \, J \, M)$ comes from tensors with odd values of r.

The matrix elements of the δ-interaction, which were evaluated in Section 20 by a direct calculation, can, of course, be calculated also by the method presented here. As is well known,

$$\delta(\mathbf{r}_1 - \mathbf{r}_2) = \frac{1}{r_1 r_2} \, \delta(r_1 - r_2) \frac{1}{2\pi} \, \delta(\cos \omega_{12} - 1). \tag{21.48}$$

Using the completeness of the Legendre polynomials we can expand $\delta(\cos \omega_{12} - 1)$ and obtain

$$\delta(\cos \omega_{12} - 1) = \sum_k \frac{2k+1}{2} P_k(\cos \omega_{12}) \, P_k(1) = \sum_k \frac{2k+1}{2} P_k(\cos \omega_{12}).$$

Hence,

$$\delta(\mathbf{r}_1 - \mathbf{r}_2) = \sum_k \frac{2k+1}{4\pi} \cdot \frac{\delta(r_1 - r_2)}{r_1 r_2} \, P_k(\cos \omega_{12}). \tag{21.49}$$

Comparing this result with the Slater expansion (21.2) we see that for the δ-force $v_k(r_1, r_2)$ is given by

$$v_k(r_1, r_2) = \frac{2k+1}{4\pi} \frac{\delta(r_1 - r_2)}{r_1 r_2} . \tag{21.50}$$

The Slater integrals F^k are therefore

$$F^k = (2k+1) \frac{1}{4\pi} \int \frac{1}{r^2} \, |R_{n_1 l_1}(r) \, R_{n_2 l_2}(r)|^2 \, dr = (2k+1) \, F^0. \tag{21.51}$$

Using (21.10) and (21.11) we obtain for the δ-force

$$\Delta E_L(\delta) = \langle l_1 l_2 L\, M | \delta(\mathbf{r}_1 - \mathbf{r}_2) | l_1 l_2 L\, M \rangle$$

$$= \sum_k (-1)^{l_1+l_2+L} (l_1 || \mathbf{C}^{(k)} || l_1)\, (l_2 || \mathbf{C}^{(k)} || l_2) \begin{Bmatrix} l_1 & l_2 & L \\ l_2 & l_1 & k \end{Bmatrix} (2k+1)\, F^0$$

$$= (2l_1 + 1)(2l_2 + 1)\, F^0 \sum (-1)^L (2k+1)$$

$$\times \begin{pmatrix} l_1 & k & l_1 \\ 0 & 0 & 0 \end{pmatrix} \begin{pmatrix} l_2 & k & l_2 \\ 0 & 0 & 0 \end{pmatrix} \begin{Bmatrix} l_1 & l_2 & L \\ l_2 & l_1 & k \end{Bmatrix}.$$

Since $V(l_1 k l_1 | 0\,0\,0)$ vanishes for odd values of k we can multiply each term in the summation by $(-1)^k$ without affecting the value of the sum. We can then use (15.14) to obtain

$$\Delta E_L(\delta) = (2l_1 + 1)(2l_2 + 1) \begin{pmatrix} l_1 & l_2 & L \\ 0 & 0 & 0 \end{pmatrix} \begin{pmatrix} l_2 & l_1 & L \\ 0 & 0 & 0 \end{pmatrix} F^0$$

$$= (2l_1 + 1)(2l_2 + 1) \begin{pmatrix} l_1 & L & l_2 \\ 0 & 0 & 0 \end{pmatrix}^2 F^0. \qquad (21.52)$$

This result is identical with (20.23) which was obtained by direct evaluation of the integrals involved.

Another interaction which contains in its Slater-expansion odd tensors only is the tensor interaction introduced in (20.51). As we saw in (14.8) and (14.9) this interaction can be written also in the form

$$V_2(1\,2) = \left[\frac{(\boldsymbol{\sigma}_1 \cdot (\mathbf{r}_1 - \mathbf{r}_2))\, (\boldsymbol{\sigma}_2 \cdot (\mathbf{r}_1 - \mathbf{r}_2))}{r^2} - \tfrac{1}{3} (\boldsymbol{\sigma}_1 \cdot \boldsymbol{\sigma}_2)\right] \frac{r^2}{\sqrt{5}} f(r). \qquad (21.53)$$

The expression (21.53), or (20.51), can be expanded in the form (21.24). Only terms with $s = s' = 1$ occur in this expansion; however, terms with both $k = k'$ and $k = k' \pm 2$ will be included. Obviously, within the l^2 configuration both k and k' must be even. The expansion (21.24) will, therefore, contain in this case, only tensors with odd degrees r. We shall not discuss the detailed expansion here. Starting from (20.51) and the Slater expansion of $f(|\mathbf{r}_1 - \mathbf{r}_2|)$, we can use successively the recoupling transformations (14.38) and (15.28) to obtain the desired expansion.

So far, our considerations have been limited to either two non-identical particles or two particles in the same orbit. In both cases there was no need of introducing the exchange term. In general the inclusion of the exchange terms results, for LS-coupling, in the relation (21.12)

$$\Delta E_{SLJ} = \sum f_k F^k + (-1)^{l_1+l_2+L+S} \sum g_k G^k.$$

Here G^k are the radial integrals (21.13) and g_k is given by (21.14) as

$$g_k = (-1)^L (l_1||\mathbf{C}^{(k)}||l_2)\,(l_2||\mathbf{C}^{(k)}||l_1) \begin{Bmatrix} l_1 & l_2 & L \\ l_1 & l_2 & k \end{Bmatrix}.$$

It is possible to reinterpret g_k as the f_k of a modified interaction, i.e., to consider the exchange term as the direct term of a modified interaction. This is required for considering many particles as well as for the transition to jj-coupling. To achieve this end we first introduce *unit tensor operators.*

The *unit tensor operator* of degree r is defined as the tensor $\mathbf{u}^{(r)}$ whose reduced matrix elements are given by

$$(\alpha j||\mathbf{u}^{(r)}||\alpha' j') = \delta_{\alpha\alpha'}\delta_{jj'}. \tag{21.54}$$

Here α and α' are any quantum numbers which are required to define the single particle states apart from its angular momentum j. These tensors operate on the coordinates of the single particle. Equation (21.54) defines them uniquely, since, together with the Wigner-Eckart theorem, it gives all their matrix elements.

Consider, for the sake of simplicity, spin-independent interactions. Using (15.11) we obtain from (21.14)

$$g_k = (-1)^L (l_1||\mathbf{C}^{(k)}||l_2)\,(l_2||\mathbf{C}^{(k)}||l_1) \sum_r (-1)^{r+L+k}(2r+1) \begin{Bmatrix} l_1 & l_2 & L \\ l_2 & l_1 & r \end{Bmatrix} \begin{Bmatrix} l_1 & l_2 & k \\ l_2 & l_1 & r \end{Bmatrix}$$

$$= (-1)^{k+l_1+l_2+L}(l_1||\mathbf{C}^{(k)}||l_2)\,(l_2||\mathbf{C}^{(k)}||l_1) \tag{21.55}$$

$$\times \langle l_1 l_2 L\, M \Big| \sum_r \begin{Bmatrix} l_1 & l_2 & k \\ l_2 & l_1 & r \end{Bmatrix} (-1)^r (2r+1)\,(\mathbf{u}^{(r)}(1)\cdot\mathbf{u}^{(r)}(2)) \Big| l_1 l_2 L\, M\rangle .$$

We can therefore consider the exchange term $(-1)^{l_1+l_2+L+S}\sum g_k G^k$ in (21.12) as being the direct term of an interaction V', whose expansion is determined by (21.55).

We can now use the fact that

$$(-1)^S = -\langle \tfrac{1}{2}\,\tfrac{1}{2}\,S\,M_S|\tfrac{1}{2}[1+(\sigma_1\cdot\sigma_2)]|\tfrac{1}{2}\,\tfrac{1}{2}\,S\,M_S\rangle$$

to obtain the exchange term in (21.12) as the direct term of the interaction

$$V'(1\,2) = \sum_{r,k} (-1)^{r+1}(2r+1)G^k(l_1||\mathbf{C}^{(k)}||l_2)\,(l_2||\mathbf{C}^{(k)}||l_1)$$

$$\times \begin{Bmatrix} l_1 & l_2 & k \\ l_2 & l_1 & r \end{Bmatrix} (\mathbf{u}^{(r)}(1)\cdot\mathbf{u}^{(r)}(2)) \left[\frac{1+(\sigma_1\cdot\sigma_2)}{2}\right]. \tag{21.56}$$

The expansion of (21.56) in the general form (21.24) is now straightforward.

It is important to note that although we started from an ordinary (nonexchange) interaction $V(|\mathbf{r}_1 - \mathbf{r}_2|)$, the equivalent interaction V' given by (21.56) has an exchange character. It is generally nonlocal (i.e., cannot be expressed in terms of $|\mathbf{r}_1 - \mathbf{r}_2|$), but nevertheless it is *rotationally* invariant since it involves only scalar products of irreducible tensors. A similar treatment can be given to the exchange term in *jj*-coupling.

The expansion of the interaction in terms of scalar products of irreducible tensors is one of the strongest tools of atomic and nuclear spectroscopy. It enables us to derive all the consequences resulting from the basic assumptions of the shell model without making any unnecessary, and usually unreliable, assumptions about particular types of potentials and interactions. The assumptions which enter into this method are only that the residual effective interaction, which is considered perturbation-wise, is a two-body interaction, and the adequacy of *jj*-coupling or *LS*-coupling as the case may be. This method is essentially equivalent to expressing the energies in the *n*-particle configurations in terms of those of corresponding two-particle configurations. Therefore, the method can become useful only if we consider spectra of more than 2 particles.

22. Closed Shells. Particles and Holes

Until now we have considered only two-particle configurations. The treatment of n-particle configurations with $n > 2$ will be carried out in Part III. We can still, however, treat some simple cases with the formalism developed so far. The simplest of them is that in which all the particles are in a closed shell.

Let us consider a closed jj-coupling shell of identical particles, (similar considerations hold also for closed shells in LS-coupling). Since we can put $2j + 1$ identical particles in the j-orbit, the configuration we are dealing with is j^{2j+1}. Due to the Pauli principle there is only *one* such state in which the $2j + 1$ particles can be put into the j-orbit. This state has $M = 0$ and therefore its total angular momentum is $J = 0$. This is also evident upon inspecting the only possible antisymmetric wave function $\psi(j^{2j+1})$ of $2j + 1$ particles in the orbit characterized by the single-particle wave functions $\varphi(j, m)$. This antisymmetric function can be written in the form of a *Slater determinant*

$$\psi(j^{2j+1}) = \frac{1}{\sqrt{(2j+1)!}} \begin{vmatrix} \varphi_1(j,j) & \varphi_1(j,j-1) & \cdots & \varphi_1(j,-j) \\ \varphi_2(j,j) & \varphi_2(j,j-1) & \cdots & \varphi_2(j,-j) \\ \vdots & \vdots & & \vdots \\ \varphi_{2j+1}(j,j) & \varphi_{2j+1}(j,j-1) & \cdots & \varphi_{2j+1}(j,-j) \end{vmatrix} . \tag{22.1}$$

On performing a rotation, each $\varphi(j, m)$ goes over into a linear combination of $\varphi(j, m')$ with the same j. Since all values of m appear in (22.1) in a symmetric way, $\psi(j^{2j+1})$ is transformed under rotations into an antisymmetric function of the same configuration and therefore into itself. This shows that the state considered has zero total angular momentum.

To evaluate the interaction of the particles in the j shell, we have to calculate the expectation value

$$\Delta E(j^{2j+1}\, J = 0) = \tfrac{1}{2} \int \psi^*(j^{2j+1}) \Big(\sum_{i \neq k} V_{ik}\Big) \psi(j^{2j+1})\, d(1)d(2) \ldots d(2j+1). \tag{22.2}$$

In (22.2) i and k run over all the $2j + 1$ particles and $d(i)$ indicates integration over space coordinates and summation over spin coordinates. All the particles considered are completely equivalent to each other and the wave function $\psi(j^{2j+1})$ is antisymmetric in all of them. We can therefore obtain $\Delta E(j^{2j+1},\ J = 0)$ by evaluating the integral in (22.2) for just one pair, V_{12}, and multiplying the result by the number of pairs in j^{2j+1}, namely $\frac{1}{2}(2j+1)\,2j = j(2j+1)$. Thus we obtain

$$\Delta E(j^{2j+1}\; J=0) = j(2j+1) \int \psi^*(j^{2j+1})\, V_{12}\, \psi(j^{2j+1})\, d(1) \ldots d(2j+1). \quad (22.3)$$

We can now expand the determinant in (22.1) into a sum of products of the determinants of the first two rows and their respective minors. Substituting this expansion into (22.3) we can integrate over particles 3 to $2j+1$ and obtain

$$\Delta E(j^{2j+1}\; J=0) = \sum_{m>m'} \int \varphi_{12}^*(j\,m, j\,m')\, V_{12}\, \varphi_{12}(j\,m, j\,m')\, d(1)d(2)$$

$$= \tfrac{1}{2} \sum_{m,m'} \int \varphi_{12}^*(j\,m, j\,m')\, V_{12}\, \varphi_{12}(j\,m, j\,m')\, d(1)d(2) \quad (22.4)$$

where

$$\varphi_{12}(j\,m, j\,m') = \frac{1}{\sqrt{2}} \begin{vmatrix} \varphi_1(j\,m) & \varphi_1(j\,m') \\ \varphi_2(j\,m) & \varphi_2(j\,m') \end{vmatrix}. \quad (22.5)$$

The summation over m in (22.4) can be transformed into a more familiar expression. If $\psi_{12}(j^2 J\, M)$ is the antisymmetric two-particle wave function in the j^2 configuration, then

$$\psi_{12}(j^2 J\, M) = \frac{1}{\sqrt{2}} \sum (j\,m\,j\,m' \mid j\,j\,J\,M)\, \varphi_{12}(j\,m, j\,m'). \quad (22.6)$$

Note the factor $1/\sqrt{2}$ that stems from the fact that $\varphi_{12}(j\,m, j\,m')$ is the determinant (22.5) and *not* the product $\varphi_1(j\,m)\varphi_2(j\,m')$.

Using the orthogonality of the Clebsch-Gordan coefficients we obtain from (22.6)

$$\varphi_{12}(j\,m, j\,m') = \sqrt{2} \sum_{JM} (j\,m\,j\,m' \mid j\,j\,J\,M)\, \psi_{12}(j^2 J\, M). \quad (22.7)$$

Introducing (22.7) into (22.4) we recall that V_{12} commutes with $\mathbf{J}_{12}$. V_{12} is therefore diagonal with respect to J and independent of M. Hence,

$$\Delta E(j^{2j+1}\; J=0) = \sum_{JM} \langle j^2 J\, M | V | j^2 J\, M \rangle = \sum_{J\ \text{even}} (2J+1)\, \Delta E_J \quad (22.8)$$

Here the summation is limited to even values of J only, since ψ_{12} in (22.6) vanishes for odd values of J. The sum (22.8) was calculated in the previous section. Substituting from (21.39) and (21.33) we obtain

$$\Delta E(j^{2j+1}\; J=0) = \tfrac{1}{2} \left\{ (2j+1)^2 F^{(00)0} - \sum_r (j \| \mathbf{T}^{(r)} \| j)^2 F^r \right\}. \quad (22.9)$$

For odd tensor interactions we use (21.43) to obtain

$$\Delta E_{\text{odd}}(j^{2j+1}J=0)=\frac{2j+1}{2}\,\Delta E_{\text{odd}}(j^2J=0). \tag{22.10}$$

Equation (22.10) can be interpreted by saying that, for odd tensor interactions, the total energy seems to be that of $(2j+1)/2$ "saturated pairs" with no further interaction between them. This point will be considered in greater detail in Part III.

Another case which can also be treated simply is that of a single particle outside a closed shell. Here we have to do with a configuration $j^{2j+1}j'$, which contains $2j+2$ identical particles and we are interested in calculating

$$\Delta E(J)=\langle j^{2j+1}j'JM\left|\tfrac{1}{2}\sum_{i\neq k}V_{ik}\right|j^{2j+1}j'JM\rangle_a. \tag{22.11}$$

In (22.11) the total angular momentum J must equal j', since, as we have already seen, the $2j+1$ equivalent particles can yield only a zero total angular momentum. The antisymmetric state $\psi_a(j^{2j+1}j'JM)$ is therefore again given by a single Slater determinant

$$\psi_a(j^{2j+1}j'\;J=j',\;M=m')$$
$$=\frac{1}{\sqrt{(2j+2)!}}\begin{vmatrix}\varphi_1(j,j) & \varphi_1(j,j-1) & \dots & \varphi_1(j,-j) & \varphi_1(j'm')\\ \varphi_2(j,j) & \varphi_2(j,j-1) & \dots & \varphi_2(j,-j) & \varphi_2(j'm')\\ \vdots & \vdots & & \vdots & \vdots\\ \varphi_{2j+2}(j,j) & \varphi_{2j+2}(j,j-1) & \dots & \varphi_{2j+2}(j,-j) & \varphi_{2j+2}(j'm')\end{vmatrix} \tag{22.12}$$

Since the absolute value squared of (22.12) is a symmetric function it is clear that we can limit the summation in (22.11) to V_{12} and multiply the result by $\frac{1}{2}(2j+2)(2j+1)$ (the total number of pairs in the $j^{2j+1}j'$ configuration). We then obtain, as in (22.4),

$$\Delta E(j^{2j+1}j')=\frac{1}{2}\sum_{m_1m_2}\int\begin{vmatrix}\varphi_1^*(j\,m_1) & \varphi_1^*(j\,m_2)\\ \varphi_2^*(j\,m_1) & \varphi_2^*(j\,m_2)\end{vmatrix}V_{12}\begin{vmatrix}\varphi_1(j\,m_1) & \varphi_1(j\,m_2)\\ \varphi_2(j\,m_1) & \varphi_2(j\,m_2)\end{vmatrix}d(1)d(2)$$
$$+\frac{1}{2}\sum_{m}\int\begin{vmatrix}\varphi_1^*(j\,m) & \varphi_1^*(j'm')\\ \varphi_2^*(j\,m) & \varphi_2^*(j'm')\end{vmatrix}V_{12}\begin{vmatrix}\varphi_1(j\,m) & \varphi_1(j'm')\\ \varphi_2(j\,m) & \varphi_2(j'm')\end{vmatrix}d(1)d(2) \tag{22.13}$$

The first term on the right-hand side of (22.13) is exactly equal to $\Delta E(j^{2j+1}\;J=0)$ given by (22.4). The antisymmetrized normalized function $\psi_a(jj'JM)$ can be defined by

$$\psi_a(jj'JM)=\sum_{mm'}(j\,m\,j'm'\mid jj'JM)\frac{1}{\sqrt{2}}\begin{vmatrix}\varphi_1(j\,m) & \varphi_1(j'm')\\ \varphi_2(j\,m) & \varphi_2(j'm')\end{vmatrix}. \tag{22.14}$$

We can then rewrite (22.13), and obtain

$$\Delta E(j^{2j+1}j')$$
$$= \Delta E(j^{2j+1}\, J = 0) + \sum_{mJM} |(j\, m\, j'm' \mid j\, j' J\, M)|^2 \langle j\, j' J\, M | V | j\, j' J\, M\rangle_a$$
$$= \Delta E(j^{2j+1}\, J = 0) + \frac{1}{2j'+1} \sum_J (2J+1)\, \Delta E(j\, j' J) \tag{22.15}$$

where

$$\Delta E(j\, j' J) = \langle j\, j' J\, M | V | j\, j' J\, M\rangle_a . \tag{22.16}$$

In deriving (22.15) we have used the fact that (22.16) is independent of M.

The sum over J in (22.15) can be carried out in terms of the Slater integrals. The direct integral is given by (21.31) and (21.33). For the exchange integral we obtain, using (15.4) and (15.17).

$$\sum_J (2J+1)(-1)^{j+j'+J} \langle j\, j' J\, M | V | j' j\, J\, M\rangle$$
$$= \sum_{Jr} (2J+1)(-1)^{j-j'} G^r (j || \mathbf{T}^{(r)} || j')(j' || \mathbf{T}^{(r)} || j) \begin{Bmatrix} j & j' & J \\ j & j' & r \end{Bmatrix}$$
$$= \sum_r G^r (j || \mathbf{T}^{(r)} || j')(j' || \mathbf{T}^{(r)} || j)(-1)^{j-j'}$$
$$= \sum_r G^r |(j || \mathbf{T}^{(r)} || j')|^2 . \tag{22.17}$$

Here we used the fact that Hermitian operators $\mathbf{T}^{(r)}$ satisfy the relation

$$(j || \mathbf{T}^{(r)} || j') = (-1)^{j-j'} \overline{(j' || \mathbf{T}^{(r)} || j)} . \tag{22.18}$$

This relation can be obtained directly from the definition of the reduced matrix elements.

We thus obtain the following expression for the interaction in the $j^{2j+1}j'$ configuration

$$\Delta E(j^{2j+1}j') = \Delta E(j^{2j+1}\, J = 0)$$
$$+ \left\{ (2j+1)\, F^{(00)0}(n\, l, n'l') - \frac{1}{2j'+1} \sum_r |(j || \mathbf{T}^{(r)} || j')|^2\, G^r(n\, l, n'l') \right\} . \tag{22.19}$$

Here $F^{(00)0}(n\, l, n'l')$ is given by (21.8) with $v_0(r_1, r_2) = v_{00,0}(r_1, r_2)$ defined by (21.24) and $G^r(n\, l, n'l')$ is similarly given by (21.13).

Equation (21.19) shows that the interaction in the $j^{2j+1}\, j'$ configuration is composed of two parts: one is the interaction of the particles in the closed shell—$\Delta E(j^{2j+1},\ J=0)$—and the other part is the interaction of the single j' particle with all the $2j+1$ particles in the closed shell. This latter interaction is of course independent of m' since the $2j+1$ particles in j^{2j+1} are coupled to the spherically symmetric state with $J=0$. Putting it in another way, since the z-projection M of the total angular momentum J of the $j^{2j+1}j'$ configuration is necessarily equal to m', and since the total interaction is a scalar, it is independent of M and therefore of m'. Thus, *the interaction of a particle with closed shells is equivalent to an interaction with a central field.*

Although our conclusions were derived under the assumption that the particles in the j-shell and the j'-shell are identical, they hold, *a fortiori*, when the j'-particle is different from those in the j-shell. Assuming that the particles in the j-shell are protons and the particle in j' is a neutron, it can be seen that (22.15) holds with the modification

$$\Delta E(j_p j'_n J) = \langle j_p j'_n J\, M | V_{pn} | j_p j'_n J\, M\rangle. \tag{22.20}$$

Here, the *ordinary* (nonantisymmetric) matrix element has to be taken, and J can assume all values consistent with $|j-j'| \leq J \leq j+j'$. Consequently (22.19) holds as well with the exchange integrals G^r being put equal to zero.

We thus obtain

$$\Delta E(j_p^{2j+1} j'_n) = \Delta E(j_p^{2j+1}\ J=0) + (2j_p+1)\, F^{(00)0}. \tag{22.21}$$

We can easily understand why only F^0 appears in (22.21). The interaction of the particles in the closed shells with the outside particle can be written as a sum of terms of the type

$$\left(\left[\sum_p \mathbf{T}^{(r)}_{(p)}\right] \cdot \mathbf{T}^{(r)}_{(n)}\right).$$

The fact that the protons fill a closed shell and therefore have $J=0$ makes all contributions from $\Sigma_p\, \mathbf{T}^{(r)}_{(p)}$ with $r \neq 0$ vanish. We are thus left with the contribution from $\mathbf{T}^{(0)}$ only. Therefore contributions from odd tensor interactions, like the tensor forces, vanish in this case.

In the case of identical particles, similar considerations hold when we replace the exchange term by its equivalent direct term of V'. The only contribution to the exchange term in (22.19) comes from the unit tensor of zero degree.

Equation (22.19) for identical and nonidentical nucleons has an important consequence for the treatment of particles in given shells. Let us consider for instance, the $j^{2j+1}\, j'^n$, configuration in which n is any number such that $1 \leq n \leq 2j'+1$. This configuration will generally have many states with different values of J, M, and perhaps other quantum numbers α, $\psi(j^{2j+1} j'^n \alpha\, J M)$. These different states are obtained by choosing different relative $\mathbf{j}$ orientations of the n particles in the j'-orbit.

Since the interaction of a particle in the j'-orbit with the $2j + 1$ particles in the j-shell is independent of its m'-value, we see that the following result holds quite generally

$$\begin{aligned} \Delta E(j^{2j+1} j'^n \alpha\, J M) &= \Delta E(j^{2j+1}\ J = 0) + n \Big\{ (2j+1)\, F^{(00)0}(n\, l\, n' l') \\ &\quad - \frac{1}{2j'+1} \sum_r |(j||\mathbf{T}^{(r)}||j')|^2\, G^r(n\, l, n' l') \Big\} + E(j'^n \alpha\, J M) \\ &= \Delta E_0 + \Delta E(j'^n \alpha\, J M). \end{aligned} \tag{22.22}$$

Here ΔE_0 is independent of α, J, and M and is equal to the interaction of the particles in the j'-orbit with the closed shell plus the interaction in the closed shell. Thus, in as much as we are interested only in the spacings between levels with different values of J (or α), we can ignore the existence of closed shells and treat only the particles in unfilled shells. This result is responsible for the great simplification offered by the shell model in calculating energies, and other quantities as well, in systems of many particles. For practically all purposes the particles in the closed shells can be ignored completely, or at most their influence can be very simply estimated. As we saw their main effect on the particles in the unfilled shells can be taken into account as a modification of the central field.

It is interesting to note that the interaction within a closed shell, (22.9), can be obtained from (22.19). In fact, we see from (22.19) that the interaction of a j-particle with a closed j-shell is given by

$$(2j+1)\, F^0(n\, l, n\, l) - \frac{1}{2j+1} \sum |(j||\mathbf{T}^{(r)}||j)|^2\, F^r(n\, l, n\, l) \tag{22.23}$$

(the interaction of the specific particle with itself vanishes because of the antisymmetry). To obtain the total interaction in the j-shell we multiply (22.23) by the number of particles there, i.e., $2j + 1$, and divide by 2 since each pair was counted twice. The resulting expression is identical with (22.9).

A simple example of the use of (22.15) or (22.21) is offered by the vector force introduced in (20.53). For the sake of simplicity let us assume that we are dealing with a configuration of $2j + 1$ protons in the j-shell and a neutron in the j'-shell, with a neutron-proton interaction given by

$$V_{pn} = [(\boldsymbol{\sigma}_p + \boldsymbol{\sigma}_n) \cdot \boldsymbol{l}_{pn}]. \tag{22.24}$$

Here $\boldsymbol{l}_{pn}$ is the relative angular momentum.

$$\boldsymbol{l}_{pn} = \tfrac{1}{2}(\mathbf{r}_p - \mathbf{r}_n) \times (\mathbf{p}_p - \mathbf{p}_n) = \tfrac{1}{2}(\boldsymbol{l}_p + \boldsymbol{l}_n) - \tfrac{1}{2}[(\mathbf{r}_n \times \mathbf{p}_p) + (\mathbf{r}_p \times \mathbf{p}_n)]. \tag{22.25}$$

A distance dependence $f(|\mathbf{r}_p - \mathbf{r}_n|)$ could have been introduced into (22.24), but it would have complicated slightly the calculations without modifying in any substantial way the nature of the conclusions.

We first notice that if the protons form a closed LS-coupling shell ($4l + 2$ protons in the l orbit) the interaction (22.24) leads to simple results. From (18.31) we see that in this case the dependence of the energy on S, L, and J is given by the operator $(\mathbf{S} \cdot \mathbf{L})$. In the case considered here, the protons' spin is $\mathbf{S}_p = 0$ and similarly $\mathbf{L}_p = 0$. Hence, $(\mathbf{S} \cdot \mathbf{L})$ reduces to $(\mathbf{s}_n \cdot \boldsymbol{l}_n)$. The energy has therefore the form of an ordinary spin-orbit interaction of the outside particle.

In jj-coupling the situation is more complicated. However, we shall derive also in this case a similar result.

Introducing (22.25) into (22.24) we obtain

$$V_{pn} = \tfrac{1}{2}(\boldsymbol{\sigma}_p \cdot \boldsymbol{l}_p + \boldsymbol{\sigma}_n \cdot \boldsymbol{l}_n) + V'_{pn} \tag{22.26}$$

where

$$V'_{pn} = \tfrac{1}{2}\{(\boldsymbol{\sigma}_p \boldsymbol{l}_n + \boldsymbol{\sigma}_n \boldsymbol{l}_p) - (\boldsymbol{\sigma}_p + \boldsymbol{\sigma}_n) \cdot [(\mathbf{r}_n \times \mathbf{p}_p) + (\mathbf{r}_p \times \mathbf{p}_n)]\}. \tag{22.27}$$

Thus, the interaction (22.24) is equivalent to an ordinary spin-orbit interaction $\tfrac{1}{2}(\boldsymbol{\sigma}_p \cdot \boldsymbol{l}_p + \boldsymbol{\sigma}_n \cdot \boldsymbol{l}_n)$, plus another interaction V'_{pn}. We shall now prove that the expectation value of the interaction V'_{pn}, summed over all the protons in a closed shell, vanishes. Hence, the interaction (22.24) of a neutron with a closed proton shell is equivalent to a simple spin-orbit interaction of that neutron.

To prove this it is convenient to express V'_{pn} as a sum of scalar products of a vector operator operating on the proton coordinates and a vector operator operating on the neutron coordinates. For instance

$$[\boldsymbol{\sigma}_p \cdot (\mathbf{r}_n \times \mathbf{p}_p)] = -(\boldsymbol{\sigma}_p \times \mathbf{p}_p) \cdot \mathbf{r}_n.$$

We can therefore write

$$V'_{pn} = \sum_{ik} (\mathbf{T}_i^{(1)}(p) \cdot \mathbf{T}_k^{(1)}(n))$$

where $\mathbf{T}_i^{(1)}$ are irreducible vector operators like $\boldsymbol{\sigma}$, $\boldsymbol{\sigma} \times \mathbf{p}$, $\boldsymbol{\sigma} \times \mathbf{r}$, and $\boldsymbol{l}$. We thus see that in this expansion of V'_{pn} there are no tensors of zero degree of the proton space and spin coordinates. Since the protons are in a closed shell with $J = 0$, we see from (22.21) that the expectation value of $\sum_p V'_{pn}$ vanishes.

Therefore when the protons fill a closed shell the only contribution from (22.24) comes from the single particle spin-orbit term $\frac{1}{2}[(\sigma_p \cdot l_p) + (\sigma_n \cdot l_n)]$.

The fact that closed shells can be ignored in most calculations has wider applications. We shall now show that there is a very close relation between the energies of the j^{2j+1-n} configuration and the j^n configuration. It will turn out that an *almost closed shell* is equivalent, in a sense, to an almost empty shell. For the sake of convenience we shall call the j^{2j+1-n} configuration an *n-hole configuration*, and the j^n configuration its *conjugate n*-particle configuration.

In the first place we observe that the hole configuration and its conjugate particle configuration contain the same number of states. The enumeration of the states is most easily done in the m-scheme where each antisymmetric state is characterized by a set of occupied m-states, so that

$$\psi(m_1 m_2 \dots m_k) = \frac{1}{\sqrt{k!}} \begin{vmatrix} \psi_1(j\, m_1) & \psi_1(j\, m_2) & \dots & \psi_1(j\, m_k) \\ \psi_2(j\, m_1) & \psi_2(j\, m_2) & \dots & \psi_2(j\, m_k) \\ \vdots & \vdots & & \vdots \\ \psi_k(j\, m_1) & \psi_k(j\, m_2) & \dots & \psi_k(j\, m_k) \end{vmatrix} . \tag{22.28}$$

In order to establish a definite phase for this wave function we shall adopt the convention that the m values are arranged in a descending order, i.e., $m_1 > m_2 > \dots > m_k$. Instead of specifying a state by a set of $2j + 1 - n$ occupied m-states, we can specify it by the set of n unoccupied m-states, i.e., by the set of n holes $m_1^{-1}\, m_2^{-1} \dots m_n^{-1}$ [m_i^{-1} stands for a hole in the state $\psi(jm_i)$]. Thus, a one-to-one correspondence can be established between each hole state and a conjugate particle state. The state in the j^{2j+1-n} configuration in which $m_1\, m_2 \dots m_n$ are not occupied corresponds to the particle state in the j^n configuration in which $m_1\, m_2 \dots m_n$ *are* occupied. Thus, the number of states in the n-hole and n-particle configurations is necessarily the same.

We shall now show that the energies of the states of the hole configuration are equal, up to an additive constant, to those in the conjugate particle configuration. To do so we shall calculate the matrices of the interaction $\Sigma\, V_{ij}$ in the m-scheme for both configurations. Starting with the particle configuration, each row and column in the matrix of $\Sigma\, V_{ij}$ is characterized by a set of the n numbers $m_1 > m_2 > \dots > m_n$. Since V_{ij} is a two-particle operator, $\langle m_1\, m_2 \dots m_n | \Sigma\, V_{ik} | m_1'\, m_2' \dots m_n' \rangle_a$ is different from zero only if the set m_i' differs from the set m_i by *at most* two m values. Moreover, since V_{ik} is diagonal with respect to $m_i + m_k$, the set m_i' is either identical with the set m_i or there are exactly *two* numbers in the set m_i' which differ from the corresponding numbers in the set m_i.

For the diagonal elements of the n-particle configuration we therefore obtain by expanding (22.28) in the derivation of (22.4),

$$\langle m_1\, m_2 \ldots m_n \left| \tfrac{1}{2} \sum_{i,k=1}^{n} V_{ik} \right| m_1\, m_2 \ldots m_n \rangle_a = \sum_{m>m'} \langle m\, m' | V_{12} | m\, m' \rangle_a$$

$$m,\, m' = m_1,\, m_2,\, \ldots\, m_n \tag{22.29}$$

where the summation extends over the whole set $m_1\, m_2 \ldots m_n$. For the nondiagonal elements we obtain

$$\langle m_1\, m_2\, m_3 \ldots m_n \left| \sum V_{ik} \right| m_1'\, m_2'\, m_3 \ldots m_n \rangle_a = \langle m_1\, m_2 | V | m_1'\, m_2' \rangle_a. \tag{22.30}$$

Equation (22.30) was obtained for the case in which the set m_i' differs from the set m_i in the values of m_1 and m_2. Similar relations hold in other cases. Labeling the states of the n-hole configuration by the *unoccupied* values of m as explained before, we obtain for the hole configuration

$$\langle m_1^{-1}\, m_2^{-1} \ldots m_n^{-1} \left| \tfrac{1}{2} \sum_{i,k=n+1}^{2j+1} V_{ik} \right| m_1^{-1}\, m_2^{-1} \ldots m_n^{-1} \rangle_a = \sum_{m>m'} \langle m\, m' | V_{12} | m\, m' \rangle_a$$

$$m,\, m' \neq m_1,\, m_2,\, \ldots,\, m_n. \tag{22.31}$$

Here the summation extends over all values of m *different* from m_1, m_2, ... m_n, i.e., over all the occupied m-states in the hole state considered. We have, however,

$$\sum_{\substack{m>m' \\ m,m'\neq m_1,\ldots m_n}} \langle m\, m' | V_{12} | m\, m' \rangle_a = \tfrac{1}{2} \sum_{\substack{m,m' \\ mm'\neq m_1,\ldots m_n}} \langle m\, m' | V_{12} | m\, m' \rangle_a$$

$$= \tfrac{1}{2}\Big\{ \sum_{\substack{m,m' \\ m,m'=j,j-1,\ldots,-j}} \langle m\, m' | V_{12} | m\, m' \rangle_a - \sum_{i=1}^{n} \sum_{m'=j,\ldots,-j} \langle m_i m' | V_{12} | m_i\, m' \rangle_a$$

$$- \sum_{i=1}^{n} \sum_{m=j,\ldots,-j} \langle m\, m_i | V_{12} | m\, m_i \rangle_a \Big\} + \tfrac{1}{2} \sum_{\substack{m,m' \\ m,m'=m_1,\ldots m_n}} \langle m\, m' | V_{12} | m\, m' \rangle_a$$

$$= E_0 + \sum_{\substack{m>m' \\ m,m'=m_1,\ldots m_n}} \langle m\, m' | V_{12} | m\, m' \rangle_a. \tag{22.32}$$

The expression $\sum_{m'=j,j-1,\ldots-j} \langle m_i m' | V_{12} | m_i m' \rangle_a$ is independent of m_i; we see that E_0, given by the expression in the curly brackets in (22.32), is independent of the specific values of $m_1, m_2, \ldots m_n$. Comparing (22.32) with (22.29) we obtain for the diagonal elements

$$\langle m_1^{-1}\, m_2^{-1} \ldots m_n^{-1} \left| \tfrac{1}{2} \sum_{i,k=n+1}^{2j+1} V_{ik} \right| m_1^{-1}\, m_2^{-1} \ldots m_n^{-1} \rangle_a$$

$$= E_0 + \langle m_1\, m_2 \ldots m_n \left| \tfrac{1}{2} \sum_{i,k=1}^{n} V_{ik} \right| m_1\, m_2 \ldots m_n \rangle_a. \qquad (22.33)$$

A similar calculation shows that for the nondiagonal elements we obtain, for instance,

$$\langle m_1^{-1}\, m_2^{-1}\, m_3^{-1} \ldots m_n^{-1} \left| \tfrac{1}{2} \sum_{i,k=n+1}^{2j+1} V_{ik} \right| m_1'^{-1}\, m_2'^{-1}\, m_3^{-1} \ldots m_n^{-1} \rangle_a$$

$$= \langle m_1'\, m_2'\, m_3 \ldots m_n \left| \tfrac{1}{2} \sum_{i,k=1}^{n} V_{ik} \right| m_1\, m_2\, m_3 \ldots m_n \rangle_a. \qquad (22.34)$$

Equation (22.33) holds for any phase convention adopted for the hole configuration. On the other hand, the signs on the right-hand side of (22.34) *depend* on the phase convention. However, the eigenvalues of the interaction matrix are obviously independent of such a choice. It can be easily shown that (22.34) is a permissible choice.

Thus, the whole matrix of the interaction $\frac{1}{2} \sum V_{ik}$ in a hole configuration, taken in the m-scheme, is equal to a constant E_0 plus the transposed matrix, in the same scheme, of the conjugate particle configuration. The eigenvalues of the matrix $\frac{1}{2} \sum V_{ik}$ taken in the j^{2j+1-n} configuration are therefore equal to E_0 plus the eigenvalues of the matrix $\frac{1}{2} \sum V_{ik}$ taken in the j^n-configuration. The *spacings* of the levels in the two configurations are therefore equal.

We shall not enter here into a discussion of the quantum numbers which characterize conjugate states in conjugate configurations. This point will be considered in Part III. We would, however, like to point out here that the allowed angular momenta in the n-hole configuration are also allowed in the n-particle configuration, and that states which have the same energy in the two configurations (apart from the constant E_0) also have the same total angular momentum.

Let us now consider matrix elements of tensor operators in hole configurations

$$T_\kappa^{(k)} = \sum_{i=n+1}^{2j+1} T_\kappa^{(k)}(i). \qquad (22.35)$$

Using again the m-scheme and labeling the various states by the m-values of the unoccupied levels we obtain for the diagonal elements

$$\langle m_1^{-1} m_2^{-1} \ldots m_n^{-1} \left| \sum_{i=n+1}^{2j+1} T_\kappa^{(k)}(i) \right| m_1^{-1} m_2^{-1} \ldots m_n^{-1} \rangle_a$$

$$= (2j+1-n) \langle m_1^{-1} m_2^{-1} \ldots m_n^{-1} | T_\kappa^{(k)}(1) | m_1^{-1} m_2^{-1} \ldots m_n^{-1} \rangle_a. \qquad (22.36)$$

Expanding the Slater determinants of the hole configuration (of order $2j + 1 - n$) in the first row, we obtain for (22.36)

$$\sum_{m \neq m_1, m_2, \ldots, m_n} \langle j\, m | T_\kappa^{(k)} | j\, m \rangle$$

$$= \sum_{m=-j,\ldots,j} \langle j\, m | T_\kappa^{(k)} | j\, m \rangle - \sum_{m=m_1,\ldots,m_n} \langle j\, m | T_\kappa^{(k)} | j\, m \rangle. \qquad (22.37)$$

A similar expansion of (22.1) yields

$$\langle j^{2j+1}\, J=0 \left| \sum_{i=1}^{2j+1} T_\kappa^{(k)}(i) \right| j^{2j+1}\, J=0 \rangle = \sum_{m=-j}^{j} \langle j\, m | T_\kappa^{(k)} | j\, m \rangle$$

and therefore

$$\sum_{m=-j}^{j} \langle j\, m | T_\kappa^{(k)} | j\, m \rangle = \sqrt{2j+1}\, \delta_{k,0} (j || \mathbf{T}^{(0)} || j). \qquad (22.38)$$

Hence we find the following connection between the diagonal elements of $\sum T_\kappa^{(k)}(i)$ in a hole configuration and its conjugate particle configuration

$$\langle m_1^{-1} m_2^{-1} \ldots m_n^{-1} \left| \sum_{i=n+1}^{2j+1} T_\kappa^{(k)}(i) \right| m_1^{-1} m_2^{-1} \ldots m_n^{-1} \rangle$$

$$= - \langle m_1\, m_2 \ldots m_n \left| \sum_{i=1}^{n} T_\kappa^{(k)}(i) \right| m_1\, m_2 \ldots m_n \rangle \qquad \text{for} \quad k > 0. \qquad (22.39)$$

In (22.39) the matrix elements on both sides vanish if $\kappa \neq 0$ due to the Wigner-Eckart theorem. For $\kappa = 0$, $T_0^{(k)} = \sum_i T_0^{(k)}(i)$ is diagonal in the m-scheme and it has no nonvanishing elements in the j^n configuration other than those in (22.39). Thus, we conclude that the relation (22.39) holds also in any other scheme, and in particular in the scheme with definite total J and total M. However, we must be careful in identifying the corresponding states. In (22.39) the total M of the hole configuration is

$$M_{\text{hole}} = \sum_{m \neq m_1, m_2, \ldots m_n} m = \Big(\sum_{m=j,\ldots,-j} m \Big) - \Big(\sum_{m=m_1, m_2, \ldots m_n} m \Big)$$

$$= - \sum_{m=m_1, m_2, \ldots m_n} m. \tag{22.40}$$

For the particle configuration, on the other hand

$$M_{\text{part.}} = \sum_{m=m_1, m_2, \ldots m_n} m. \tag{22.41}$$

Hence, the correspondence established in (22.39) is between hole states with a given $M_{\text{hole}} = M$ and a particle state with $M_{\text{part.}} = -M$. Thus we conclude that

$$\langle j^{2j+1-n} J\, M \Big| \sum_{i=n+1}^{2j+1} T_0^{(k)}(i) \Big| j^{2j+1-n} J\, M \rangle$$

$$= -\langle j^n J, -M \Big| \sum_{i=1}^{n} T_0^{(k)}(i) \Big| j^n\, J, -M \rangle \qquad \text{for} \qquad k > 0.$$

By the Wigner-Eckart theorem

$$\langle J, -M | T_\kappa^{(k)} | J, -M' \rangle = (-1)^{J+M} \begin{pmatrix} J & k & J \\ M & \kappa & -M' \end{pmatrix} (J || \mathbf{T}^{(k)} || J)$$

$$= (-1)^{J-M+k} \begin{pmatrix} J & k & J \\ -M & -\kappa & M' \end{pmatrix} (J || \mathbf{T}^{(k)} || J)$$

$$= (-1)^k \langle J\, M | T_{-\kappa}^{(k)} | J\, M' \rangle.$$

Hence the final result

$$\langle j^{2j+1-n} J\, M \Big| \sum_{i=n+1}^{2j+1} T_0^{(k)}(i) \Big| j^{2j+1-n} J\, M \rangle$$

$$= (-1)^{k+1} \langle j^n J\, M \Big| \sum_{i=1}^{n} T_0^{(k)}(i) \Big| j^n J\, M \rangle \qquad \text{for} \qquad k > 0 \tag{22.42}$$

or, equivalently

$$\left(j^{2j+1-n}J\left\|\sum_{i=n+1}^{2j+1}\mathbf{T}^{(k)}(i)\right\|j^{2j+1-n}J\right)=(-1)^{k+1}\left(j^nJ\left\|\sum_{i=1}^{n}\mathbf{T}^{(k)}(i)\right\|j^nJ\right)$$

$$\text{for } k>0. \tag{22.43}$$

For $k = 0$ we have, of course,

$$\left(j^{2j+1-n}J\left\|\sum_{i=n+1}^{2j+1}\mathbf{T}^{(0)}(i)\right\|j^{2j+1-n}J\right)=\frac{2j+1-n}{n}\left(j^nJ\left\|\sum_{i=1}^{n}\mathbf{T}^{(0)}(i)\right\|j^nJ\right).$$

The relation (22.43) has many applications. For example, the multipole moment of order L of a system of particles is defined by an irreducible tensor operator of degree L. We see, therefore, that the odd multipoles, such as the dipole, octopole, etc., have the same value in a given state of a particle configuration and in the corresponding state of the conjugate hole configuration. For even multipoles, on the other hand, the *magnitudes* of the moments are the same, but their *signs are reversed.* Since for half a shell the hole configuration and the particle configuration are identical, we can conclude, from (22.43), that even multipole moments vanish for all the states of a half-filled shell.

These general features are nicely observed in the experiment. Magnetic moments of nuclei do not show any systematic changes from the beginning of a shell to its end. Quadrupole moments, on the other hand, are always negative just after a magic number and always positive just before a magic number. The detailed treatment of the behavior of these moments must, however, wait until we have developed the formalism for handling many particle configurations, which will be treated in Part III.

Another useful application of (22.43) is found in treating particle-hole configurations. Consider the $j^{2j}j'$ configuration consisting of $2j$ protons in the j-orbit (i.e., a proton hole in the j-orbit) and a neutron in the j' orbit. Expanding the interaction V_{pn} between the protons and the neutron as before,

$$V_{pn}=\sum_k v_k(r_p,r_n)\,\mathbf{T}^{(k)}(p)\cdot\mathbf{T}^{(k)}(n)$$

we obtain

$$\langle j^{2j}j'JM\left|\sum_p V_{pn}\right|j^{2j}j'JM\rangle$$

$$=\sum_p\sum_k F^k\langle j^{2j}j'JM|(\mathbf{T}^{(k)}(p)\cdot\mathbf{T}^{(k)}(n))|j^{2j}j'JM\rangle$$

$$=\sum_k F^k\langle j^{2j}j'JM|(\mathbf{T}^{(k)}(P)\cdot\mathbf{T}^{(k)}(n))|j^{2j}j'JM\rangle. \tag{22.44}$$

Here we introduced the notation

$$\mathbf{T}^{(k)}(P) = \sum_{p} \mathbf{T}^{(k)}(p) \tag{22.45}$$

so that $\mathbf{T}^{(k)}(P)$ is a tensor operator of degree k operating on all the $2j$ protons. The j^{2j} configuration of identical particles can have for its total angular momentum only $J = j$. Therefore, in (22.44) the $2j$ protons are in a state of a definite total angular momentum. We can therefore use (15.4) and obtain

$$\begin{aligned} \Delta E(j^{2j}j'J) &= \langle j^{2}{}_{j}j'JM \Big| \sum_{p} V_{pn} \Big| j^{2j}j'JM\rangle \\ &= \sum_{k} (-1)^{j+j'+J} (j^{2j}j||\mathbf{T}^{(k)}(P)||j^{2j}j)(j'||\mathbf{T}^{(k)}(n)||j') \begin{Bmatrix} j & j' & J \\ j' & j & k \end{Bmatrix} F^{k}. \end{aligned} \tag{22.46}$$

We can now use (22.43) to express the reduced matrix elements $(j^{2j}j||\mathbf{T}^{(k)}(P)||j^{2j}j)$ in terms of those of the conjugate configuration, namely $(j||\mathbf{T}^{(k)}(p)||j)$. We thus obtain

$$\begin{aligned} \Delta E(j^{2j}j'J) = \sum_{k} (-1)^{j+j'+J+k+1} (j||\mathbf{T}^{(k)}(p)||j)(j'||\mathbf{T}^{(k)}(n)||j') \begin{Bmatrix} j & j' & J \\ j' & j & k \end{Bmatrix} F^{k} \\ + \sqrt{\frac{2j+1}{2j'+1}} (j||\mathbf{T}^{(0)}(p)||j)(j'||\mathbf{T}^{(0)}(n)||j') F^{0} \end{aligned} \tag{22.47}$$

where for $k = 0$ there is an additional term.

However, we have found in (21.26) that

$$\Delta E(jj'J) = \sum_{k} (-1)^{j+j'+J} (j||\mathbf{T}^{(k)}(p)||j)(j'||\mathbf{T}^{(k)}(n)||j') \begin{Bmatrix} j & j' & J \\ j' & j & k \end{Bmatrix} F^{k}. \tag{22.48}$$

Comparing (22.48) and (22.47) we see that if the interaction contains tensors of only even degrees k, then

$$\Delta E(j^{-1}j'J) = \text{const.} - \Delta E(jj'J) \qquad \text{for even tensor interactions.} \tag{22.49}$$

The constant in (22.49), which is independent of J, arises from $\mathbf{T}^{(0)}$. The energy levels in a hole-particle configuration are then obtained by changing the sign of the interaction in the corresponding particle-particle interaction. On the other hand, for odd tensor interactions, the order and their spacings are the same in the two conjugate configurations

$$\Delta E(j^{-1}j'J) = \Delta E(jj'J) \qquad \text{for odd tensor interactions.} \tag{22.50}$$

These results hold for nonidentical particles as well as for the direct term in the interaction of identical particles.

In the general case, where both odd and even tensors have to be included, we can derive a relation between energies in the two conjugate configuration in the following way. Using the orthogonality of the W-coefficients we obtain from (22.48)

$$(2k+1)\sum_{J'}(-1)^{J'}(2J'+1)\begin{Bmatrix} j & j' & J' \\ j' & j & k \end{Bmatrix} \Delta E(jj'J')$$

$$= (-1)^{j+j'}(j||\mathbf{T}^{(k)}(p)||j)(j'||\mathbf{T}^{(k)}(n)||j')F^k. \qquad (22.51)$$

Equation (22.51) gives the strengths of the various irreducible tensors in the expansion of an interaction in terms of the expectation values of this interaction. Substituting from (22.51) into (22.47) we obtain

$$\Delta E(j^{2j}j'J) = -\sum_{k,J'}(-1)^{k+J+J'}\begin{Bmatrix} j & j' & J \\ j' & j & k \end{Bmatrix}(2J'+1)(2k+1)\begin{Bmatrix} j & j' & J' \\ j' & j & k \end{Bmatrix}\Delta E(jj'J')$$

$$+\frac{1}{2j'+1}\sum(2J+1)\Delta E(jj'J)$$

$$= -\sum_{J'}(2J'+1)\Delta E(jj'J')\sum_{k}(-1)^{k+J+J'}(2k+1)\begin{Bmatrix} j & j' & J \\ j' & j & k \end{Bmatrix}\begin{Bmatrix} j & j' & J' \\ j' & j & k \end{Bmatrix}$$

$$+\frac{1}{2j'+1}\sum(2J+1)\Delta E(jj'J)$$

$$= -\sum(2J'+1)\begin{Bmatrix} j & j' & J' \\ j & j' & J \end{Bmatrix}\Delta E(jj'J') + (2j+1)\overline{\Delta E(jj')} \qquad (22.52)$$

where we have used the definition (21.30) of $\overline{\Delta E}$. For the special case of equal energies $\Delta E(jj'J) = \overline{\Delta E}$, we obtain from (22.52) $\Delta E(j^{2j}j'J) = 2j\,\overline{\Delta E}$ as it should be.

We have thus succeeded in expressing the expectation values of the interaction in the particle-hole configuration in terms of the energies of the corresponding particle-particle configuration. These relations are valid under the most general conditions. We have used the expansion of the proton-neutron interaction in a sum of products of irreducible tensors only in order to be able to use (22.43). We did not have to assume anything about the tensors $\mathbf{T}^{(k)}(p)$ and $\mathbf{T}^{(k)}(n)$, and they have all been completely eliminated from the final result (22.52). In fact, it is possible to obtain (22.52) without using the expansion of the interaction in scalar products of irreducible tensors. That proof makes use of the methods which will be introduced in Part III and therefore will not be given here.

Due to the generality of (22.52) we expect it to hold in actual nuclei if the basic assumptions of the shell model are valid. We may note that the relations between the particle-hole configuration and the particle-particle configuration are different for *LS*-coupling and *jj*-coupling. Thus, the comparison of (22.52) with experiment may also indicate whether *jj*-coupling is a good approximation or not.

One of the most striking examples of the use of (22.52) is offered by the two nuclei Cl^{38} and K^{40}. In Cl^{38} there is a $d_{3/2}$-proton and $f_{7/2}$-neutron outside closed shells. In K^{40} there are 3 $d_{3/2}$-protons, i.e., a $d_{3/2}$-proton hole, and an $f_{7/2}$-neutron outside closed shells. The energy levels in these two nuclei should therefore be related according to (22.52). In fact, we have the results given in Table 22.1.

TABLE 22.1

J	K^{40}	Cl^{38} Calculated from K^{40} using (22.52)	Cl^{38} Experimental
2	0.80	0	0
3	0.03	0.75	0.762
4	0	1.32	1.312
5	0.89	0.70	0.672

The energies (in Mev) are referred to the ground states in both cases since the same change in all $\Delta E(jj'J')$ contributes only to the last, J-independent term of (22.52). The agreement between the calculated levels in Cl^{38} and the observed ones is extremely good. It thus lends great support to the basic assumptions underlying the derivation of (22.52), namely, the validity of the *jj*-coupling shell model in nuclei with two-body effective forces. The good agreement obtained indicates also that the two-particle matrix elements of the effective interaction do not change appreciably when going from one nucleus to the other.

23. Two Particles in the Harmonic Oscillator Potential

The Slater expansion which we have used in the preceding sections is valid for arbitrary central potentials and arbitrary two-body interactions. If the central potential can be approximated by a harmonic oscillator potential, another method can be used which often results in great simplification of the formalism.

The harmonic oscillator potential has the property

$$\tfrac{1}{2}m\omega^2(r_1^2 + r_2^2) = \tfrac{1}{2}\left(\frac{m}{2}\right)\omega^2 r^2 + \tfrac{1}{2}(2m)\omega^2 R^2 \tag{23.1}$$

where

$$\mathbf{r} = \mathbf{r}_2 - \mathbf{r}_1 \quad \text{and} \quad \mathbf{R} = \tfrac{1}{2}(\mathbf{r}_1 + \mathbf{r}_2).$$

Thus, the potential acting on two particles moving in a central harmonic well is equivalent to a harmonic potential acting between them [with the reduced mass $(m/2)$], plus a harmonic potential acting on their center of mass (mass $2m$). It follows that an (unperturbed) eigenfunction $\psi(\mathbf{r}_1, \mathbf{r}_2)$ of two particles in a harmonic well can be written as a product of two normalized functions $\psi(\mathbf{R})$ describing the center-of-mass motion, and $\varphi(\mathbf{r})$ describing the relative motion

$$\psi(\mathbf{r}_1, \mathbf{r}_2) = \psi(\mathbf{R})\varphi(\mathbf{r}). \tag{23.2}$$

It can be shown that no potential other than the harmonic oscillator potential has this property.

Consider now an interaction V_{12}, which depends on the relative coordinates of two particles. Taking its diagonal matrix elements we obtain

$$\int \psi^*(\mathbf{r}_1, \mathbf{r}_2)V_{12}\psi(\mathbf{r}_1, \mathbf{r}_2)d\mathbf{r}_1 d\mathbf{r}_2 = \int \varphi^*(\mathbf{r})V_{12}(\mathbf{r})\varphi(\mathbf{r})d\mathbf{r}. \tag{23.3}$$

Thus the function $\psi(\mathbf{R})$ drops out from the final expression and we are left only with the relatively simple integral (23.3). If we consider a state in which the two particles are in definite orbits, n_1l_1 and n_2l_2, of the oscillator potential, its wave function will, in general, not have the form (23.2). It will be, however, a linear combination of a limited number of functions (23.2) which belong to the same eigenvalue of the oscillator Hamiltonian. Thus, matrix elements of $V_{12}(\mathbf{r})$ between such states will be given by linear combinations of a limited number of integrals (23.3).

In the preceding sections we saw that, in order to diagonalize the interaction matrix (in the subspace of the degenerate states of a given configuration), we should use functions with a definite total angular momentum. We are therefore interested in decomposing the special function $\psi_{12}(n_1 l_1 n_2 l_2 L\, M)$ into a sum of products of two functions as in (23.2). Since $V_{12}(r)$ is a scalar function in the relative coordinate, it commutes with the relative angular momentum and it is therefore convenient to introduce a complete set of functions $\varphi_{nlm}(\mathbf{r})$ of the relative coordinate. These functions have definite values of the relative angular momentum l and m, and a relative radial quantum number n. Similarly, we can take for the center-of-mass functions a complete set $\psi_{N\Lambda M_\Lambda}(\mathbf{R})$ with radial quantum number N, angular momentum Λ, and a z projection M_Λ. We have then

$$\psi_{12}(n_1 l_1 n_2 l_2 L\, M)$$

$$= \sum_{\substack{n,l,N,\Lambda, \\ m,M_\Lambda}} a_{nlN\Lambda}^{n_1 l_1 n_2 l_2, L} (l\, m\, \Lambda\, M_\Lambda | l\, \Lambda\, L\, M)\, \psi_{N\Lambda M_\Lambda}(\mathbf{R})\, \varphi_{nlm}(\mathbf{r}). \tag{23.4}$$

Here we used explicitely the Clebsch-Gordan coefficients since by definition we have

$$\mathbf{l}_1 + \mathbf{l}_2 = \mathbf{l} + \mathbf{\Lambda} = \mathbf{L}. \tag{23.5}$$

Consequently, if ψ_{12} has a definite total angular momentum $\mathbf{L}$, we must have the same total angular momentum also on the right-hand side of (23.4).

Although the summation in (23.4) extends over all values of n, l, N, and Λ we can easily see that it is actually limited to quite a small number of these parameters. According to (5.11), the energy, in the central field, of the state $\psi_{12}(n_1 l_1 n_2 l_2 L\, M)$ is $(2n_1 + l_1 - \frac{1}{2} + 2n_2 + l_2 - \frac{1}{2})\hbar\omega$. On the other hand the energies of the states $\psi_{N\Lambda M_\Lambda}(\mathbf{R})$ and $\varphi_{nlm}(\mathbf{r})$ are $(2N + \Lambda - \frac{1}{2})\hbar\omega$ and $(2n + l - \frac{1}{2})\hbar\omega$ respectively. The n, l, N, and Λ should therefore satisfy the relation

$$(2n + l) + (2N + \Lambda) = (2n_1 + l_1) + (2n_2 + l_2).$$

Thus,

$$a_{nlN\Lambda}^{n_1 l_1 n_2 l_2, L} = 0 \qquad \textit{unless} \qquad 2n + l + 2N + \Lambda = 2n_1 + l_1 + 2n_2 + l_2$$

and

$$|l_1 - l_2| \leq L \leq l_1 + l_2, \qquad |l - \Lambda| \leq L \leq l + \Lambda. \tag{23.6}$$

We are thus left with a limited number of nonvanishing coefficients in (23.4). Some of these coefficients are given in the Appendix. Introducing

(23.4) into (23.3), and noting that $V_{12}(r)$ depends on the *magnitude* of $\mathbf{r}$ only, we can carry out both the $\mathbf{R}$ integration and the integration over the relative angular coordinates and obtain:

$$\Delta E(l_1 l_2 L) = \langle n_1 l_1 n_2 l_2 L\, M | V_{12}(r) | n_1 l_1 n_2 l_2 L\, M \rangle$$

$$= \sum_{nlN\Lambda} |a^{n_1 l_1 n_2 l_2, L}_{nlN\Lambda}|^2 \int |R_{nl}(r)|^2\, V(r) dr. \tag{23.7}$$

Here $R_{nl}(r)$ is the normalized radial part of $\varphi_{nlm}(\mathbf{r})$

$$\varphi_{nlm}(\mathbf{r}) = \frac{1}{r} R_{nl}(r)\, Y_{lm}(\theta\, \varphi).$$

Introducing the notation

$$b^L_{nl} = \sum_{N,\Lambda} |a^{n_1 l_1 n_2 l_2, L}_{nlN\Lambda}|^2 \tag{23.8}$$

we obtain

$$\Delta E(l_1 l_2 L) = \sum_{nl} b^L_{nl} \int |R_{nl}(r)|^2\, V(r) dr. \tag{23.9}$$

The b^L_{nl} are universal coefficients which do not depend on the interaction $V(r)$ and can therefore be calculated once for all interactions. However, we can simplify (23.9) even further. We have seen in Section 5 that for the harmonic oscillator potential, $R_{nl}(r)$ is given by

$$R_{nl}(r) = N_{nl} e^{-\nu r^2} r^{l+1} L^{l+1/2}_{n+l-1/2}(2\nu r^2)$$

where N_{nl} is a normalization factor, $\nu = (m\omega/2\hbar)$ and $L^m_k(x)$ is a polynomial of degree $k - m$ in x. Thus $|R_{nl}(r)|^2$ is given by the product of $e^{-2\nu r^2}$ and a polynomial of degree $2[2n + l - 1]$ in r. It is therefore clear that we can find numerical coefficients $c^{(nl)}_{l'}$ such that

$$|R_{nl}(r)|^2 = \sum_{l'} c^{(nl)}_{l'} |R_{1l'}(r)|^2 \tag{23.10}$$

where l' now goes generally from $l' = l$ up to $l' = l + 2n - 2$.

We therefore define

$$\alpha^L_l = \sum_{n'l'} b^L_{n'l'} c^{(n'l')}_l \tag{23.11}$$

where the sum extends over all possible values of n' and l' determined by $n_1 l_1$ and $n_2 l_2$ according to (23.6). Introducing the integrals

$$I_l = \int |R_{1l}(r)|^2\, V(r) dr \tag{23.12}$$

we obtain for (23.9) the expression

$$\Delta E(l_1 l_2 L) = \sum_l \alpha_l^L I_l. \tag{23.13}$$

Here the α_l^L are universal constants, independent of the particular interaction $V(r)$. The integrals I_l on the other hand are averages of the interaction $V(r)$ taken with the weight function $|R_{1l}(r)|^2$. The sum extends over a finite number of values of l, and it can be verified easily that

$$0 \leq l \leq 2n_1 + l_1 + 2n_2 + l_2. \tag{23.14}$$

The method described above is, in a sense, complementary to Slater's expansion. In the latter the interaction V_{12} is decomposed essentially into a product of a function of $\mathbf{r}_1$ and a function of $\mathbf{r}_2$, thus enabling us to carry out the integrations over $d\Omega_1$ and $d\Omega_2$ separately. Here the particle density $\psi_{12}^* \psi_{12}$ is decomposed into a product of a function of $\mathbf{r}$ and a function of $\mathbf{R}$, enabling us now to carry out the two integrations $d\mathbf{r}$ and $d\mathbf{R}$ separately. Noncentral forces can also be treated with this method. The integrals (23.3) become then slightly more complicated. This method is particularly convenient when $V(r)$ is a short-range interaction. In this case, I_l is a strongly decreasing function of l, and in many practical cases we can put $I_l = 0$ if $l > l_0$ where l_0 is a small integer. In particular for δ-forces only I_0 is different from zero and we find that

$$\Delta E_{l_1 l_2 L}(\delta) = \alpha_0^L I_0. \tag{23.15}$$

Comparing (23.15) with (20.23) we can conclude that

$$\alpha_0^L = \lambda(2l_1 + 1)(2l_2 + 1) \begin{pmatrix} l_1 & l_2 & L \\ 0 & 0 & 0 \end{pmatrix}^2 \tag{23.16}$$

where λ is a constant, independent of L, which can be fixed by considering the normalizations of the various functions involved. Although we have derived (23.16) from the study of the energy levels for the δ-forces, it is clear from the foregoing that this expression is valid for any interaction.

The convenience of working with harmonic oscillator wave functions can be illustrated by calculating the Coulomb energies in a given configuration. Using (23.13) we see that in the $l_1 l_2$ configuration the Coulomb energy is given by

$$E(l_1 l_2 L) = \sum_l \alpha_l^L I_l \tag{23.17}$$

where

$$I_l = e^2 \int \frac{1}{r} |R_{1l}|^2 \, dr. \tag{23.18}$$

Substituting for R_{1l} from (5.9) we find that

$$I_l = e^2 \frac{2\sqrt{2\nu}}{\Gamma(l+\frac{3}{2})} \int_0^\infty x^{2l+1} e^{-x^2} \, dx. \tag{23.19}$$

Thus the Coulomb interaction between two particles described by harmonic oscillator wave functions is given by $e^2 \sqrt{\nu}$ multiplied by a universal, easily calculable function of l_1, l_2, and L. As we shall see in Part III, energies in the n-body configurations can be expressed in terms of the energies in corresponding two-body configurations. Thus Coulomb energies in the harmonic oscillator potential will always be given in terms of $e^2 \sqrt{\nu}$ multiplying a universal function of the angular momenta involved and the number of particles. This result is also obvious from dimensional analysis, since the Coulomb energy is proportional to e^2, and the characteristic length in this problem is $\nu^{-1/2}$.

It is also clear from the above remark that the nuclear radius, defined by $\sqrt{\overline{r^2}}$, is proportional to $\nu^{-1/2}$. Here $\overline{r^2}$ is the average value of r^2 taken over all the occupied states. In the harmonic oscillator potential the expectation value of r_i^2 is proportional to the energy of the ith particle, and hence to $\hbar\omega = (2\hbar^2/m)\nu$. It follows that r^2 is proportional to $\nu^{-1/2}$. Thus the measured Coulomb energies of nuclei can serve to determine the nuclear radii, or more specifically the *Coulomb radius* of the nucleus. Radii defined in terms of other averaged properties can differ, of course, from those determined from the Coulomb energies, although they all come fairly close to each other.

Part III THREE- OR MORE-PARTICLE SYSTEMS

The treatment of the motion of electrons in an atom or of nucleons in a nucleus is an example of a *many-body problem*. In such systems each particle interacts with all the others. If we know the forces acting between the particles we can write down the Schrödinger equation of their motion. However, if there are more than two particles there is generally no analytical method to solve this equation. Even in classical mechanics there is no general solution of the problem of three bodies moving under the influence of mutual gravitational forces.

Various approximation methods have been developed for the solution of the atomic many-body problem. The characteristic feature of all of them is that they are individual particle models. The motion of a particle under the influence of all the others is approximated by its motion in a self-consistent central field of force.

There is enough empirical information on nuclear structure to justify the use of the shell model for nuclei. We have therefore ignored in this book the problem of finding a thorough theoretical foundation for this approximation.

As mentioned in the Introduction, approximation methods, more sophisticated than the ordinary perturbation theory, have been developed for the treatment of nuclei. On the basis of these methods it seems that, at least as far as calculations of the energy are concerned, we can use independent particle wave functions provided we take a properly chosen effective two-body interaction. Such effective interaction may well have little resemblance to the original free nucleon interaction. We therefore primarily discuss methods which are independent of the particular form of the interaction.

In the following we shall describe the various methods of constructing antisymmetric wave functions of several nucleons moving in a central field. Using these wave functions we shall calculate interaction energies and matrix elements of single particle operators without making specific assumptions on the form of the operators involved.

24. The *m*-Scheme

In the central field approximation we use shell model wave functions to calculate expectation values of the Hamiltonian

$$H = \sum_i T_i + \sum_i U_i + \sum_{i<j} V_{ij}. \tag{24.1}$$

Here $T_i = (1/2m)\, p_i^2$ is the kinetic energy of the ith particle, U_i is its potential energy in the central field, and V_{ij} is the interaction between the ith and jth particles. We shall proceed with the calculation of matrix elements of (24.1) using shell model wave functions.

A *configuration* is the group of states which can be formed by particles in orbits specified by n (principal quantum number) and l (orbital angular momentum) in LS-coupling, or n, l, and j (total angular momentum of the individual particle) in the case of jj-coupling. A configuration in LS-coupling is thus specified by the values of n_i and l_i of each particle. In jj-coupling a configuration is specified by the values of n_i, l_i, and j_i. Each configuration has usually many different states. The reason for grouping them under one name lies in the fact that in the absence of the mutual interaction they are all degenerate. Thus, in addition to specifying a state by the configuration to which it belongs, we shall need additional quantum numbers for its complete specification.

We shall deal first with the simpler case of one kind of particles obeying the Pauli principle, such as electrons in an atom. We can characterize the various shell model wave functions of a certain configuration by the values of the z-projections of the individual $\mathbf{l}_i$ and $\mathbf{s}_i$ (in LS-coupling) or $\mathbf{j}_i$ (in jj-coupling). In this case the wave function is a product of single nucleon wave functions $\psi_{nlm_sm_l}$ or ψ_{nljm_j}. Let us write a wave function of this type as

$$\psi_{a_1}(x_1)\, \psi_{a_2}(x_2) \dots \psi_{a_n}(x_n) \tag{24.2}$$

where x_i stands for all space and spin coordinates of the ith particle. The numbers a_i stand for the single particle quantum numbers n_i and l_i and either $m_{l_i} m_{s_i}$ or $j_i m_{j_i}$. This wave function is not yet antisymmetric. In order to antisymmetrize it we construct the expression

$$\psi = \sum_P (-1)^P\, P\psi_{a_1}(x_1) \dots \psi_{a_n}(x_n). \tag{24.3}$$

Here P is an operator which permutes the space and spin coordinates of the particles. The permutation $(1, 2, \dots, n) \rightarrow (i_1 i_2, \dots i_n)$ changes

$\psi_{a_1}(x_1) \ldots \psi_{a_n}(x_n)$ into $\psi_{a_1}(x_{i_1}) \ldots \psi_{a_n}(x_{i_n})$. The phase $(-1)^P$ is $+1$ or -1 according to whether the permutation P is even or odd. The summation is extended over all $n!$ possible permutations P.

The normalized antisymmetrized wave function (24.3) is by definition equal to a *Slater determinant*

$$\psi = \frac{1}{\sqrt{n!}} \begin{vmatrix} \psi_{a_1}(x_1) & \psi_{a_1}(x_2) & \ldots & \psi_{a_1}(x_n) \\ \psi_{a_2}(x_1) & \psi_{a_2}(x_2) & \ldots & \psi_{a_2}(x_n) \\ \vdots & \vdots & & \vdots \\ \psi_{a_n}(x_1) & \psi_{a_n}(x_2) & \ldots & \psi_{a_n}(x_n) \end{vmatrix} . \tag{24.4}$$

Slater determinants were already used in Section 22. For the sake of completeness, we shall include here some calculations and results which were already obtained there. The normalization of ψ follows very simply from the fact that all terms in (24.3) are orthogonal to one another, (since no two particles can have the same quantum numbers).

We now calculate the expectation value of H with these wave functions. It is immediately seen that the kinetic energy, being a sum of single particle operators, gives rise to an expectation value which is also a sum of single particle terms. The expectation value of the kinetic energy $\Sigma\, T_i$ is the sum of the expectation values of the single particle kinetic energies in the various occupied orbits. This is true both for the wave function (24.2) as well as for the normalized antisymmetric wave function (24.4). The kinetic energy operator of a single nucleon, which is the scalar $(1/2\, m)p^2$, operates only on the spatial part of the single particle wave function. It is independent of m_s and its expectation value depends on n and l, but is independent of the orientation of l in space, namely of m_l. Consequently the expectation value of T_i does not depend on the z-projection of $\mathbf{j}_i$. If the potential is spin-dependent (e.g., includes a spin-orbit term) the radial functions may, however, depend on the value of j. We therefore have for the kinetic energy

$$\langle \sum_i T_i \rangle = \sum_i \langle T_i \rangle_{n_i l_i} \qquad \text{or in } jj\text{-coupling} \qquad \langle \sum_i T_i \rangle = \sum_i \langle T_i \rangle_{n_i l_i j_i}.$$

The potential energy in a central field has also the same property. It is a sum of single particle terms, the expectation values of which depend on n_i and l_i. In the case of a spin-independent central field the single particle expectation value does not depend on either the value of m_s or the spatial orientation of l_i, i.e., on the value of m_l. In this case

$$\langle \sum U_i \rangle = \sum \langle U_i \rangle_{n_i l_i}.$$

However, if there is a single particle spin-orbit term in the central field, the value of the single nucleon expectation value depends on m_l

and m_s. The jj-coupling scheme has been introduced to handle this particular case. In this scheme the potential energy in the central field has always the simple expression

$$\langle \sum U_i \rangle = \sum \langle U_i \rangle_{n_i l_i j_i}.$$

As pointed out in Part I, the single nucleon expectation values are obviously independent of the orientation of $\mathbf{j}$, i.e., of m_j. The expectation value of the interaction energy is more complicated and we shall look into it in more detail.

As mentioned before, all states of the same configuration are degenerate in the central field. Therefore, if we consider the effective interaction V_{ij} as a perturbation, we have to diagonalize it within each configuration. In other words, if $|\alpha\rangle$ and $|\alpha'\rangle$ are two states of the *same* configuration we have to diagonalize the matrix $\langle \alpha | \sum_{i<j} V_{ij} | \alpha' \rangle$ in order to obtain the first-order corrections due to the mutual interaction. We are free to choose as basic states of a given configuration any set of independent combinations of wave functions (24.3) [the Hamiltonian (24.1), being symmetric in all particles, transforms an antisymmetric function into an antisymmetric function]. It is better to choose such combinations that diagonalize $\sum_{i<j} V_{ij}$ as much as possible.

The invariance properties of the interactions determine to a large extent the proper linear combinations. Since V_{ij} is assumed to be invariant under rotations in ordinary and spin spaces, the total angular momentum $\mathbf{J}$ commutes with the interaction. Therefore, the matrix of V_{ij} will not have nondiagonal elements between states with different values of J. Hence, we choose linear combinations of wave functions (24.4) which are eigenfunctions of $\mathbf{J}^2$.

Two remarks should be added. First, the requirement that the functions be eigenfunctions of $\mathbf{J}^2$ is usually not sufficient to specify uniquely the linear combinations. There may be two or more linearly independent combinations which are eigenstates of $\mathbf{J}^2$ with the same eigenvalue J. In these cases, some additional quantum numbers will be needed for the specification of the states. This point will be considered in much detail later on. Meanwhile, we shall limit ourselves to simple cases where this problem does not arise.

The second point that should be kept in mind is that the matrix of V_{ij} is not diagonal in our scheme. Only the *submatrix* of V_{ij} defined by all states of the same configuration can be made diagonal by the choice made above for wave functions. The matrix V_{ij} has nondiagonal elements between states with the same J which belong to different configurations. It is clear that the smaller these nondiagonal elements, the better the approximation of the shell model. These nondiagonal elements of V_{ij} are responsible for *configuration interaction*. They will be discussed only at the end of this part. Here, we restrict ourselves to the submatrices which belong to a single configuration. In other words, we have to calculate matrix elements of V_{ij} with wave functions of a certain configuration.

The difficulty of having to deal with several states with the same value of J occurs in LS-coupling even in rather simple configurations. This is not so in simple jj-coupling configurations. The reason is that an LS-coupling configuration has as many states as several jj-coupling configurations. In LS-coupling states are characterized by the total $\mathbf{L} = \Sigma \mathbf{l}_i$ and the total $\mathbf{S} = \Sigma \mathbf{s}_i$. The angular momenta $\mathbf{S}$ and $\mathbf{L}$ commute with the Hamiltonian only if the latter is invariant under rotations carried out independently on the space coordinates and on the spin coordinates of the particles. Therefore, this scheme diagonalizes the interaction matrix of scalar forces only. In the case of more complicated forces, like spin-orbit interaction and tensor forces, neither $\mathbf{S}$ nor $\mathbf{L}$ commute with the Hamiltonian. In the general case there are non-vanishing nondiagonal matrix elements of the interaction even in the two-particle configurations (i.e., matrix elements between states $|S\,L\,J\rangle$ and $|S'L'J\rangle$). However, in simple cases where the quantum numbers S and L specify the various states uniquely, the LS-coupling scheme can be conveniently used even if the interaction is not diagonal with respect to L and S.

In Part II, matrix elements of various interactions were calculated in detail for two-particle configurations. We now consider many-particle configurations with *two-body forces only*. In this case,

expectation values, as well as nondiagonal matrix elements, in the many-particle configuration can always be expressed as linear combinations of matrix elements in corresponding two-particle configurations. (24.5)

Our problem will be thus reduced to expressing matrix elements of $\Sigma\, V_{ik}$ in the many-particle configuration in terms of matrix elements of V_{ik} in two-particle configurations. If the mutual interaction V_{ik} is given, we can make use of the results of Part II for the matrix elements in the two-particle configurations. If the detailed form of V_{ik} is not known, we can use the general theorem (24.5) to correlate various experimental data as shall be discussed in detail later on.

We can demonstrate the trivial, yet important, result (24.5) by calculating matrix elements of $\Sigma_{i<k}\, V_{ik}$ in the original scheme of Slater determinants (24.4). We shall call this scheme the *m-scheme* since the individual values of m are specified in it. There are $n(n-1)/2$ interaction terms V_{ik}. Each one of them gives rise to the same contribution to any matrix element, since from the antisymmetry it follows that

$\psi_A^*(x_1, ..., x_n)\psi_B(x_1, ..., x_n)$ is invariant under permutations of the particle numbers. Let us therefore calculate the contribution of V_{12}. In the matrix element

$$\langle A|V_{12}|B\rangle = \int \psi_A^*(x_1, ... x_n) V_{12} \psi_B(x_1, ..., x_n) d(1) ... d(n) \qquad (24.6)$$

we can first integrate over the particle coordinates other than 1 and 2. We see at once that the single particle wave functions in which particle coordinates with $i > 2$ appear must be the same in ψ_A and ψ_B. Otherwise the integral (24.6) vanishes because of the orthogonality of the single particle wave functions. This means that in order to obtain a nonvanishing result in (24.6) *ψ_A and ψ_B must differ at most by the quantum numbers of two particles.* Since V_{12} commutes with J_z, $\Sigma_A m_i = \Sigma_B m_i$ for the nonvanishing matrix elements (24.6). Therefore, ψ_A and ψ_B must have either the same set of quantum numbers m_i or two sets differing by *two* values of m_i.

We can now develop the determinants (24.4) of ψ_A and ψ_B with respect to the first two columns. Equation (24.6) will then become a sum of several integrals. This expansion contains terms which are products of

$$[\psi_{a_i}(x_1) \psi_{a_k}(x_2) - \psi_{a_k}(x_1) \psi_{a_i}(x_2)]$$

and a minor determinant of order $n - 2$ [obtained by striking off the ith and kth rows and the first two columns in (24.4)]. It is easy to see that to each minor in the expansion of ψ_A there is at most *one* equal minor in the expansion of ψ_B. All other minors are different and therefore orthogonal to it. Carrying out the integration over particle coordinates with $i > 2$, each term is multiplied by $(n - 2)!$ and the final result is

$$\langle A \Big| \sum_{i<k} V_{ik} \Big| B\rangle = \frac{n(n-1)}{2} \frac{(n-2)!}{n!} \sum_{abcd} \int [\psi_a^*(x_1) \psi_b^*(x_2) - \psi_b^*(x_1) \psi_a^*(x_2)]$$

$$\times V_{12}[\psi_c(x_1) \psi_d(x_2) - \psi_d(x_1) \psi_c(x_2)] \, d(1)d(2). \qquad (24.7)$$

The range of a, b, c, and d depends on the quantum numbers of ψ_A and ψ_B. If the sets A and B differ by *two* quantum numbers then ab of A must be the two quantum numbers which are changed in B to c and d. The result is then simply

$$\langle A \Big| \sum_{i<j} V_{ij} \Big| B\rangle = \int \frac{1}{\sqrt{2}} [\psi_a^*(x_1) \psi_b^*(x_2) - \psi_b^*(x_1) \psi_a^*(x_2)] \, V_{12}$$

$$\times \frac{1}{\sqrt{2}} [\psi_c(x_1) \psi_d(x_2) - \psi_d(x_1) \psi_c(x_2)] \, d(1)d(2)$$

$$= \int \psi_a^*(x_1)\,\psi_b^*(x_2)\,V_{12}\,\psi_c(x_1)\,\psi_d(x_2)\,d(1)d(2)$$

$$- \int \psi_a^*(x_1)\,\psi_b^*(x_2)\,V_{12}\,\psi_d(x_1)\,\psi_c(x_2)\,d(1)d(2) \qquad (24.8)$$

where the four integrals could be reduced to two due to the symmetry $V_{12} = V_{21}$.

If the difference between A and B is in one quantum number, a in A is changed to c in B, there are $n - 1$ terms in (24.7). Each has the form (24.8) with $d = b$ and b goes over all single particle quantum numbers other than a. Such cases, however, do not appear within a given configuration where the sets $n_i l_i$ (or also j_i) are the same in ψ_A and ψ_B. As already mentioned the sets m_i in ψ_A and ψ_B cannot differ by one value of m_i. If sets A and B are identical, there is a sum of $n(n - 1)/2$ terms of the form

$$\sum_{ab} \int \frac{1}{\sqrt{2}} [\psi_a^*(x_1)\,\psi_b^*(x_2) - \psi_b^*(x_1)\,\psi_a^*(x_2)]\,V_{12}$$

$$\times \frac{1}{\sqrt{2}} [\psi_a(x_1)\,\psi_b(x_2) - \psi_b(x_1)\,\psi_a(x_2)]\,d(1)d(2) \qquad (24.9)$$

with a and b going over all quantum numbers which appear in $\psi_A = \psi_B$. Each term (24.9) is equal to the difference of the *direct* term

$$\int \psi_a^*(x_1)\,\psi_b^*(x_2)\,V_{12}\,\psi_a(x_1)\,\psi_b(x_2)\,d(1)d(2) \qquad (24.10)$$

and the *exchange* term

$$\int \psi_a^*(x_1)\,\psi_b^*(x_2)\,V_{12}\,\psi_b(x_1)\,\psi_a(x_2)\,d(1)d(2). \qquad (24.11)$$

Within a specified configuration, the quantum numbers ab and cd can differ at most by the projections of the $\mathbf{l}_i$ and $\mathbf{s}_i$ (or the projections of $\mathbf{j}_i$ in *jj*-coupling). In this case, the expressions (24.8) [or (24.9)] are simply the (properly normalized) matrix elements of V_{12} in the two-nucleon configuration $n_a l_a n_b l_b$ (or $n_a l_a j_a n_b l_b j_b$).

In *jj*-coupling we can simplify matters even further and replace all these matrix elements in the two-particle configuration by the expectation values of V_{12} in the states with a definite total J. This is due to the fact that in the two-particle *jj*-coupling configuration there is only one state for each value of J. Thus, in this scheme there are no nondiagonal matrix elements between any two states in the $j_1 j_2$ configuration. All matrix elements $\langle n_1 l_1 j_1 m_1, n_2 l_2 j_2 m_2 | V_{12} | n_1 l_1 j_1 m_1', n_2 l_2 j_2 m_2' \rangle$ can be expressed in terms of expectation values $\langle n_1 l_1 j_1, n_2 l_2 j_2 J M | V_{12} | n_1 l_1 j_1, n_2 l_2 j_2 J M \rangle$. Thus,

In jj-coupling all matrix elements of two-body interactions in the many-particle configuration can always be expressed as linear combinations of expectation values in two-particle configurations. (24.12)

As mentioned above, the submatrix of the mutual interaction which belongs to a definite configuration is not diagonal in the m-scheme. In order to diagonalize it we have to build linear combinations with definite eigenvalues of $\mathbf{J}^2$ and J_z. This procedure reduces the energy submatrix which belongs to the given configuration, into submatrices characterized by definite values of J. The construction of these better wave functions will be considered in Section 26 by using the powerful methods of tensor algebra. In this section we shall see what can be done with the m-scheme.

The m-scheme was very popular in the older days of spectroscopy. Using this scheme the antisymmetrization could be carried out in an elementary and straightforward way by the use of Slater determinants. The calculation of two-body matrix elements could also be done easily by (24.7). The calculation of expectation values and matrix elements of single particle operators can be carried out in an even easier way.

Consider the operator Σf_i where f_i operates on the coordinates of the ith particle. Such operators define for instance, the magnetic moment or quadrupole moment. Let us take its matrix element between states of type (24.4), ψ_A and ψ_B. Due to the orthogonality of the single particle wave functions, the sets A and B can differ at most by the quantum numbers of a single particle for nonvanishing matrix elements. The contribution of each term, f_i to the matrix element is the same. We can therefore calculate the expectation value of f_1 and multiply it by n. Thus,

$$\int \psi_A^*(x_1, \ldots, x_n) \left(\sum f_i\right) \psi_B(x_1, \ldots, x_n)\, d(1) \ldots d(n)$$

$$= n \int \psi_A^*(x_1, \ldots, x_n) f_1 \psi_B(x_1, \ldots, x_n)\, d(1) \ldots d(n). \tag{24.13}$$

We now develop the determinants ψ_A and ψ_B with respect to the first column. We thus obtain linear combinations of $\psi_{a_i'}(x_1)$ multiplied by the corresponding minor determinants of order $n-1$. For each minor of ψ_A there is at most one minor of ψ_B which is equal to it. All others are orthogonal to it. We carry out the integration in (24.13) over all particle coordinates with $i > 1$. The integral of the product of two equal minors is simply $(n-1)!$ and the final result is therefore

$$\int \psi_A^*(x_1, \ldots, x_n) \left(\sum f_i\right) \psi_B(x_1, \ldots, x_n)\, d(1) \ldots d(n)$$

$$= n \frac{(n-1)!}{n!} \sum_{ab} \int \psi_a^*(x_1) f_1 \psi_b(x_1)\, d(1). \tag{24.14}$$

The ranges of a and b are determined by the sets A and B. If they differ by one single quantum number, the subscript a must be the set of quantum numbers replaced by b in B. There is only one term in this case, namely

$$\int \psi_a^*(x) f \psi_b(x). \tag{24.15}$$

If A and B are identical then (24.13) is an expectation value and the result is

$$\sum_a \int \psi_a^*(x) f_1 \psi_a(x) \tag{24.16}$$

with the subscript a going over all quantum numbers of the set A. Thus, also the matrix elements of single particle operators can be easily calculated in the m-scheme.

Before proceeding to calculate eigenvalues of operators using the m-scheme let us simplify the summations which appear in (24.9) and (24.16). In these expressions the various quantum numbers belong to *individual occupied orbits* rather than to individual particles. We can therefore discuss the contributions from various groups of occupied orbits. If there are particles in closed shells and also *extra particles*—outside closed shells—there are three groups of interactions. The first is the interaction of particles in closed shells. This was calculated in Section 22. There is only one state for particles in closed shells, namely the state with $J = 0$ (also with $S = 0$ and $L = 0$ in LS-coupling). The various states of a configuration are therefore determined by the particles outside closed shells only. The interaction of the particles in closed shells contributes the same energy to any of these states. This contribution can therefore be calculated separately and then added to the remaining terms.

The second group consists of the interactions of the extra particles with those in closed shells. The interaction of one extra particle with the closed shells was calculated in Section 22. As we saw there, this interaction has the same properties as the potential energy of a single particle in a central field. In jj-coupling it depends on n, l, and j of the extra particle but not on the projection m. If there are several extra particles their interaction with the closed shells is just the sum of the single particle interactions. In LS-coupling the situation may be more complex since the interaction with the closed shells of a single extra particle may well depend on m_s and m_l. In fact, if V_{ik} contains a mutual spin-orbit interaction, then the interaction of a single particle with the closed shells contains a single particle spin-orbit term. Its contribution depends on m_s and m_l and is therefore generally different in the various states of the configuration. It is obviously not diagonal in the LS-coupling scheme. This part should be considered along with the spin-orbit terms in U_i. Tensor forces, on the other hand, do not contribute to the interaction of an extra particle with closed shells [see (22.21) and the discussion which follows].

The third group is that of interactions between the extra particles outside closed shells. This is the only part which should be calculated for each state of the configuration. When calculating this part, the individual quantum numbers of occupied closed shells can be omitted. We shall now show how the eigenvalues of this part of the interaction can be calculated in the m-scheme.

When we go over from the m-scheme to the scheme in which $\mathbf{J}^2$ is diagonal, the submatrix of the interaction undergoes a canonical transformation. The trace of the matrix remains therefore invariant. Thus, the sum of the expectation values in the two schemes remains the same. This in itself is not enough for the calculation of each of the expectation values. However, the transformation from the m-scheme to the J-scheme operates only between states with the same z-projection M of the total $\mathbf{J}$. Unlike J, M is a good quantum number both in the J-scheme and in the m-scheme where it is given by

$$M = \sum_i m_{j_i} \text{ in } jj\text{-coupling} \qquad \text{or} \qquad M = \sum_i m_{l_i} + \sum_i m_{s_i} \text{ in } LS\text{-coupling.}$$

Therefore, in this transformation the submatrices characterized by the values of M undergo separate transformations. As a result, the traces of these submatrices—i.e., the sums of the expectation values—are the same in the two schemes. This fact is utilized in the *sum method* as follows.

Let us consider a j^n configuration in jj-coupling. In the m-scheme we consider first the states with the maximum value of M allowed by the Pauli principle. The corresponding state can be denoted by the series of m values (which must all be different) $(j, j-1, j-2, \ldots, j-n+1)$. Obviously, there is only one such state with the maximum value of $M_{\max} = [n(2j+1-n)]/2$. Therefore, this state is also an eigenstate of $\mathbf{J}^2$ with $J = M_{\max} = \frac{1}{2}n(2j+1-n)$. The expectation value of the energy in this state can be thus calculated in the m-scheme directly.

It can be easily verified that there is also only one state with $M = M_{\max} - 1$. It is given by the set of m-values $(j, j-1, j-2, ..., j-n+2, j-n)$. Since we have seen that there is a state with $J = M_{\max}$, $M = M_{\max}$, there must also be a state with $J = M_{\max}$ and $M = M_{\max} - 1$. However as there is only one state with $M = M_{\max} - 1$, *it* must be this very same state. The energy of a state with a definite J is independent of the M-value. Thus we see that there are two possible ways to calculate it directly in the m-scheme: either with the aid of the wave function corresponding to $M = M_{\max}$ or with that corresponding to $M = M_{\max} - 1$.

We now consider all the states with $M = M_{\max} - 2$. It is easy to see that if $M > 0$ there are exactly two states with this value of M. They are given by the sets $(j, j-1, ..., j-n+2, j-n-1)$ and $(j, j-1, ..., j-n+3, j-n+1, j-n)$. Since there must be one state with $J = M_{\max}$, $M = M_{\max} - 2$, this state must be a linear combination of these two states in the m-scheme. The other independent linear combination must belong to another state of definite $\mathbf{J}^2$ and can therefore only be the state with $J = M_{\max} - 2$, $M = M_{\max} - 2$. The invariance of the trace implies then

$$\langle 1|H|1\rangle + \langle 2|H|2\rangle = E_{J=M_{\max}} + E_{J=M_{\max}-2}$$

where $|1\rangle$ and $|2\rangle$ are the two m-scheme states considered above. Since we already know the value of $E_{J=M_{\max}}$ we can immediately find the value of $E_{J=M_{\max}-2}$.

When we consider lower values of M, there appear more states with possible new values of J. If there are no two states with the same J, each step of decreasing M can introduce at most one new value of J. In this case all the E_J can be calculated successively in terms of expectation values in the m-scheme. They are given by

$$E_{J=M} = \mathrm{Tr}\, H_M - \mathrm{Tr}\, H_{M+1} \tag{24.17}$$

provided the order of the submatrix H_{M+1} is smaller than that of H_M by one. Here H_M is the submatrix of $\Sigma\, V_{ik}$ in the m-scheme which corresponds to a given value of M. It is obvious that if two (or more) states with the same value of J exist, only the sum of their energies is given by the sum method. This procedure should be continued to the point where the next step would give a negative value of M. From this point on, no new information will be obtained.

Along with the sum method, we have actually described here a simple and straightforward way to enumerate all the states with the various J

values. This method of enumeration is the only elementary way in the case of *equivalent particles*, i.e., particles with the same values of n, l, and j.

As an illustration let us consider the case of $j = \frac{5}{2}$, $n = 3$. The enumeration of states in the m-scheme is given in Table 24.1. We see that

TABLE 24.1

M	Possible states
$\frac{9}{2}$	$(\frac{5}{2}\,\frac{3}{2}\,\frac{1}{2})$
$\frac{7}{2}$	$(\frac{5}{2}\,\frac{3}{2}\,-\frac{1}{2})$
$\frac{5}{2}$	$(\frac{5}{2}\,\frac{1}{2}\,-\frac{1}{2})\,(\frac{5}{2}\,\frac{3}{2}\,-\frac{3}{2})$
$\frac{3}{2}$	$(\frac{5}{2}\,\frac{1}{2}\,-\frac{3}{2})\,(\frac{5}{2}\,\frac{3}{2}\,-\frac{5}{2})\,(\frac{3}{2}\,\frac{1}{2}\,-\frac{1}{2})$
$\frac{1}{2}$	$(\frac{5}{2}\,\frac{1}{2}\,-\frac{5}{2})\,(\frac{3}{2}\,\frac{1}{2}\,-\frac{3}{2})\,(\frac{5}{2}\,-\frac{1}{2}\,-\frac{3}{2})$

the J-values allowed by the Pauli principle are $J = \frac{9}{2}$, $J = \frac{5}{2}$, and $J = \frac{3}{2}$. There is no state with $J = \frac{1}{2}$ since the number of independent states in the m-scheme with $M = \frac{1}{2}$ is equal to that with $M = \frac{3}{2}$.

The following relations exist between expectation values:

$$\begin{aligned}
E_{J=9/2} &= \langle \tfrac{5}{2}\,\tfrac{3}{2}\quad \tfrac{1}{2}\,|H|\,\tfrac{5}{2}\,\tfrac{3}{2}\quad \tfrac{1}{2}\rangle \\
E_{J=9/2} &= \langle \tfrac{5}{2}\,\tfrac{3}{2}-\tfrac{1}{2}\,|H|\,\tfrac{5}{2}\,\tfrac{3}{2}-\tfrac{1}{2}\rangle \\
E_{J=5/2}+E_{J=9/2} &= \langle \tfrac{5}{2}\,\tfrac{1}{2}-\tfrac{1}{2}\,|H|\,\tfrac{5}{2}\,\tfrac{1}{2}-\tfrac{1}{2}\rangle \\
&\quad + \langle \tfrac{5}{2}\,\tfrac{3}{2}-\tfrac{3}{2}\,|H|\,\tfrac{5}{2}\,\tfrac{3}{2}-\tfrac{3}{2}\rangle \\
E_{J=3/2}+E_{J=5/2}+E_{J=9/2} &= \langle \tfrac{5}{2}\,\tfrac{1}{2}-\tfrac{3}{2}\,|H|\,\tfrac{5}{2}\,\tfrac{1}{2}-\tfrac{3}{2}\rangle \\
&\quad + \langle \tfrac{5}{2}\,\tfrac{3}{2}-\tfrac{5}{2}\,|H|\,\tfrac{5}{2}\,\tfrac{3}{2}-\tfrac{5}{2}\rangle \\
&\quad + \langle \tfrac{3}{2}\,\tfrac{1}{2}-\tfrac{1}{2}\,|H|\,\tfrac{3}{2}\,\tfrac{1}{2}-\tfrac{1}{2}\rangle \\
E_{J=3/2}+E_{J=5/2}+E_{J=9/2} &= \langle \tfrac{5}{2}\,\tfrac{1}{2}-\tfrac{5}{2}\,|H|\,\tfrac{5}{2}\,\tfrac{1}{2}-\tfrac{5}{2}\rangle \\
&\quad + \langle \tfrac{3}{2}\,\tfrac{1}{2}-\tfrac{3}{2}\,|H|\,\tfrac{3}{2}\,\tfrac{1}{2}-\tfrac{3}{2}\rangle \\
&\quad + \langle \tfrac{5}{2}-\tfrac{1}{2}-\tfrac{3}{2}\,|H|\,\tfrac{5}{2}-\tfrac{1}{2}-\tfrac{3}{2}\rangle .
\end{aligned}$$

From these equations the values of $E_J = \frac{9}{2}$, $E_J = \frac{5}{2}$, and $E_J = \frac{3}{2}$ can be determined (in more than one way) in terms of expectation values in the m-scheme.

Let us consider the case of LS-coupling next. With the exception of some trivial cases (e.g., a single particle outside closed shells), a configuration always has several states with the same value of J which may have different values of S and L. This fact makes it impossible to use directly the sum method for a general two-body interaction. Still, for that part of the interaction which is diagonal in the LS-coupling scheme, i.e., the scalar two-body forces, the sum method may be used. Such is the case in atomic spectroscopy for the electrostatic interaction between electrons.

In the transformation from the m-scheme to the LS-coupling scheme, states with various sets of m_{li} and m_{si} are combined to form states with definite values of S and L, M_S, and M_L. In order to obtain states with a definite total J, these wave functions are combined with the appropriate vector addition coefficients. However, for the purpose of using the sum method, it is convenient not to make the last step. The functions with definite S, L, M_S, and M_L can be used because the expectation values of central interactions depend rather strongly on S and L but do not depend at all on their orientation in space. This means that with such interactions all J states, obtained by combining a definite S and definite L, are degenerate. In this group of states, which is called a ***multiplet***, we are free to use wave functions with definite M_S and M_L for calculating the central interaction energy. The removal of the degeneracy within a multiplet can be obtained only if noncentral two-body forces or single particle spin-orbit interactions are also taken into account.

The fact that the interaction energy of scalar forces does not depend on the value of J in a multiplet (with given S and L) has been demonstrated in Part II. It is due to the fact that the corresponding Hamiltonian is invariant under independent rotations of the space coordinates and of the spin coordinates. This invariance is obviously maintained also in the case of more than two particles.

In the transformation from the m-scheme to the scheme characterized by definite S, M_S, L, and M_L only those states which have the same values of M_S and M_L are combined. These, unlike S and L, are good quantum numbers also in the m-scheme where they are given by

$$M_S = \sum_{i=1}^{n} m_{si} \qquad M_L = \sum_{i=1}^{n} m_{li}.$$

Because of the invariance of the traces of the submatrices characterized by M_S and M_L, the sum of expectation values in the LS-scheme is equal to the corresponding sum in the m-scheme. This fact may be used for the calculation of expectation values in the LS-scheme in a way similar to that employed in the simpler case of jj-coupling.

We start by listing the wave functions of the m-scheme according to their values of M_S and M_L. Starting with the maximum values of M_S and M_L allowed by the Pauli principle we take them in descending order (as long as M_S and M_L are positive) getting sets of equations between expectation values in the two schemes. Using the fact that the interaction energy is independent of M_S and M_L, we can obtain from these equations the energies of the various states in the LS-scheme. In cases when there are two or more states with the same values of S and L, we obtain by this method only the sum of expectation values in these states. This procedure gives in all cases a simple and direct enumeration of the various states of an LS-configuration allowed by the Pauli principle.

A simple example can best illustrate the way the enumeration of states is carried out, as well as the method for the calculation of expectation values. Let us consider the p^3 configuration of three particles with orbital angular momentum $l = 1$ with the same principal quantum number. The maximum M_L allowed by the Pauli principle can be obtained with $m_{l_1} = 1\ m_{s_1} = \frac{1}{2}$, $m_{l_2} = 1\ m_{s_2} = -\frac{1}{2}$, $m_{l_3} = 0\ m_{s_3} = \frac{1}{2}$. The corresponding state can be denoted by $(1^+1^-0^+)$ where the superscripts $+$ or $-$ refer to the values $+\frac{1}{2}$ and $-\frac{1}{2}$ of m_s respectively. This particular state has $M_L = 2$ and $M_S = \frac{1}{2}$. Since there are no possible states with higher M_L and M_S, this state must have $L = 2\ S = \frac{1}{2}$ and is denoted by ^2D where D stands for $L = 2$ and $2 = 2S + 1$ is the *multiplicity* of the state.

For $M_L = 1$ there are two states, $(1^+0^+0^-)$ and $(1^+1^- -1^+)$, both with $M_S = \frac{1}{2}$. One linear combination of these states is the ^2D state wave function with $M_L = 1\ M_S = \frac{1}{2}$. Another independent linear combination must be a state with $L = 1\ S = \frac{1}{2}$, namely ^2P.

For $M_L = 0$ there is one state, $(1^+0^+ -1^+)$, with $M_S = \frac{3}{2}$. Since there is no state with $M_S = \frac{3}{2}$ and $M_L > 0$ this must be a state with $S = \frac{3}{2}$ and $L = 0$, namely ^4S. There are three states with $M_L = 0$ and $M_S = \frac{1}{2}$. These are $(1^+0^+ -1^-)$, $(1^-0^+ -1^+)$, and $(1^+0^- -1^+)$. Since we have already established the existence of ^2D, ^2P, and ^4S and since they all have states $M_L = 0$ and $M_S = \frac{1}{2}$, we conclude that there is no additional state with $L = 0$, $S = \frac{1}{2}$.

TABLE 24.2

M_S \ M_L	2	1	0
$\frac{3}{2}$	—	—	$(1^+0^+ -1^+)$
$\frac{1}{2}$	$(1^+1^-0^+)$	$(1^+0^+0^-)$ $(1^+1^- -1^+)$	$(1^+0^+ -1^-)$ $(1^-0^+ -1^+)$ $(1^+0^- -1^+)$

This procedure is described in Table 24.2. Using the sum method for this table we can write down the equations

$$E(^2D) = \langle 1^+1^-0^+|H|1^+1^-0^+\rangle$$

$$E(^2P) + E(^2D) = \langle 1^+0^+0^-|H|1^+0^+0^-\rangle + \langle 1^+1^- - 1^+|H|1^+1^- - 1^+\rangle$$

$$E(^4S) = \langle 1^+0^+ - 1^+|H|1^+0^+ - 1^+\rangle$$

$$E(^4S) + E(^2P) + E(^2D) = \langle 1^+0^+ - 1^-|H|1^+0^+ - 1^-\rangle$$
$$+ \langle 1^-0^+ - 1^+|H|1^-0^+ - 1^+\rangle + \langle 1^+0^- - 1^+|H|1^+0^- - 1^+\rangle.$$

We did not specify the $M_S M_L$ values of the LS-states since the energy is independent of them. From the first three equations, $E(^2D)$, $E(^2P)$, and $E(^4S)$ can be determined. The last equation can then serve for checking the results.

In order to calculate the energies in LS-coupling in the general case we must discuss the expectation values as well as nondiagonal matrix elements of the noncentral interactions. Since there is no simple way to do it in the m-scheme we shall postpone this matter to the following sections where the powerful methods of modern spectroscopy will be introduced.

The sum method may be used also for the calculation of expectation values of operators other than the energy. Simple operators of this kind are the magnetic moment and the electric quadrupole moment. When using the sum method it should be remembered that the expectation values of these operators do depend on M. However, the Wigner-Eckart theorem defines a simple relation between the expectation values in states with the same J and different M. The sum method gives equations expressing these in terms of expectation values in the m-scheme. By means of the Wigner-Eckart relations all the unknown expectation values can be expressed in terms of those in states with definite J and $M = J$. Consequently, the equations can be solved and expectation values in the J-scheme can be expressed in terms of those in the m-scheme.

The calculation of the magnetic moment of a group of identical particles is straightforward both in jj-coupling and in LS-coupling. Therefore we shall demonstrate here the calculation of the quadrupole moment in the j^n configuration.

The quadrupole moment is the $\kappa = 0$ component of a tensor operator of degree 2. The quadrupole moment in a state with a total angular momentum J is defined as the expectation value of the operator in the state with $M = J$. Using the Wigner-Eckart theorem and the values of the V-coefficients (given in the Appendix) we see that the dependence of the expectation value on M is given by

$$Q(JM) = \langle JM|Q_0^{(2)}|JM\rangle = q(J)\,[3M^2 - J(J+1)]. \qquad (24.18)$$

As in the case of the energy, we enumerate the states in the *m*-scheme according to the values of M. The calculation of the expectation values in the *m*-scheme is straightforward. According to (24.16) the expectation value in a given state is a sum over the contributions of the occupied single particle states. Each such contribution is of the form (24.18)

$$\langle j\, m | q_0^{(2)} | j\, m \rangle = q[3m^2 - j(j+1)].$$

The invariance of the trace implies that the sum of the expectation values of all states in the *m*-scheme which belong to $M = M_0$ is equal to the sum of the expectation values in the states allowed by the Pauli principle with values of $J \geqslant M_0$ and $M = M_0$. This latter sum is equal to $\Sigma'_{J \geqslant M_0}\, q(J)\, [3M_0^2 - J(J+1)]$ where the dash signifies summation only over the states allowed by the Pauli principle. From these equations, $q(J)$, and therefore $Q(J) = Q(J, M = J)$, can be determined.

As an illustration we shall now discuss the case of the $(7/2)^3$ configuration. The states in the *m*-scheme listed according to the value of M are given in Table 24.3. From this table we see that the allowed states

TABLE 24.3

M	States in the *m*-scheme
$\frac{15}{2}$	$(\frac{7}{2}\,\frac{5}{2}\,\frac{3}{2})$
$\frac{13}{2}$	$(\frac{7}{2}\,\frac{5}{2}\,\frac{1}{2})$
$\frac{11}{2}$	$(\frac{7}{2}\,\frac{5}{2} - \frac{1}{2})\,(\frac{7}{2}\,\frac{3}{2}\,\frac{1}{2})$
$\frac{9}{2}$	$(\frac{7}{2}\,\frac{5}{2} - \frac{3}{2})\,(\frac{7}{2}\,\frac{3}{2} - \frac{1}{2})\,(\frac{5}{2}\,\frac{3}{2}\,\frac{1}{2})$
$\frac{7}{2}$	$(\frac{7}{2}\,\frac{5}{2} - \frac{5}{2})\,(\frac{7}{2}\,\frac{3}{2} - \frac{3}{2})\,(\frac{7}{2}\,\frac{1}{2} - \frac{1}{2})\,(\frac{5}{2}\,\frac{3}{2} - \frac{1}{2})$
$\frac{5}{2}$	$(\frac{7}{2}\,\frac{5}{2} - \frac{7}{2})\,(\frac{5}{2}\,\frac{3}{2} - \frac{3}{2})\,(\frac{5}{2}\,\frac{1}{2} - \frac{1}{2})\,(\frac{7}{2}\,\frac{1}{2} - \frac{3}{2})\,(\frac{7}{2}\,\frac{3}{2} - \frac{5}{2})$
$\frac{3}{2}$	$(\frac{7}{2}\,\frac{3}{2} - \frac{7}{2})\,(\frac{5}{2}\,\frac{3}{2} - \frac{5}{2})\,(\frac{3}{2}\,\frac{1}{2} - \frac{1}{2})$ $(\frac{5}{2}\,\frac{1}{2} - \frac{3}{2})\,(\frac{7}{2}\,\frac{1}{2} - \frac{5}{2})\,(\frac{7}{2} - \frac{3}{2} - \frac{1}{2})$
$\frac{1}{2}$	$(\frac{7}{2}\,\frac{1}{2} - \frac{7}{2})\,(\frac{5}{2}\,\frac{1}{2} - \frac{5}{2})\,(\frac{3}{2}\,\frac{1}{2} - \frac{3}{2})$ $(\frac{5}{2}\,\frac{3}{2} - \frac{7}{2})\,(\frac{7}{2} - \frac{1}{2} - \frac{5}{2})\,(\frac{5}{2} - \frac{3}{2} - \frac{1}{2})$

have the following J values: $\frac{15}{2}, \frac{11}{2}, \frac{9}{2}, \frac{7}{2}, \frac{5}{2}, \frac{3}{2}$. The first equation, for $M = \frac{15}{2}$, is

$$q\,(\tfrac{15}{2})\,[3\,(\tfrac{15}{2})^2 - \tfrac{15}{2} \times \tfrac{17}{2}] = 105q\,(\tfrac{15}{2}) = Q\,(\tfrac{15}{2})$$
$$= q\,[3\,(\tfrac{7}{2})^2 + 3\,(\tfrac{5}{2})^2 + 3\,(\tfrac{3}{2})^2 - 3\,\tfrac{7}{2} \times \tfrac{9}{2}] = 15q$$

which gives $Q(\frac{15}{2}) = 15q$. The second equation, for $M = \frac{13}{2}$, can serve only for checking the value of $Q(\frac{15}{2})$. The third equation, for $M = \frac{11}{2}$, is

$$q(\tfrac{15}{2})[3(\tfrac{11}{2})^2 - \tfrac{255}{4}] + q(\tfrac{11}{2})[3(\tfrac{11}{2})^2 - \tfrac{11}{2} \times \tfrac{13}{2}]$$
$$= Q(\tfrac{15}{2})\tfrac{27}{105} + Q(\tfrac{11}{2})$$
$$= q[3\{(\tfrac{7}{2})^2 + (\tfrac{5}{2})^2 + (-\tfrac{1}{2})^2 + (\tfrac{7}{2})^2 + (\tfrac{1}{2})^2 + (\tfrac{3}{2})^2\} - 6\tfrac{7}{2} \times \tfrac{9}{2}] = 6q.$$

Using the value of $Q(\frac{15}{2})$ given above, we obtain $Q(\frac{11}{2}) = \frac{15}{7}q$.

In the same way all the other expectation values can be obtained. It may be verified that these values are $Q(\frac{9}{2}) = -\frac{105}{11}q$, $Q(\frac{7}{2}) = 7q$, $Q(\frac{5}{2}) = \frac{39}{2}q$, and $Q(\frac{3}{2}) = -\frac{63}{5}q$. The last equation, for $M = \frac{1}{2}$, does not give any new information and can be used for checking the results.

25. The Transformation from the m-Scheme to the J-Scheme

The sum method enables us to obtain expectation values of the energy and other operators in states with definite J and M in terms of expectation values in the m-scheme. If there are several states with the same value of J, the sum method is not adequate. In such cases we have to construct the explicit wave functions with definite J and M as linear combinations of wave functions in the m-scheme. These wave functions are required also for the calculation of nondiagonal matrix elements, as in the computation of transition probabilities. Before considering straightforward methods to build these wave functions which make use of tensor algebra, let us give a short description of what could be done within the frame-work of the m-scheme.

One method could be that of direct diagonalization. The submatrix of the mutual interaction within a configuration is diagonal with respect to J and M. Therefore, the mutual interaction matrix can be calculated in the m-scheme and then diagonalized. The diagonalization would give *all* the eigenvalues of the interaction matrix (also in cases where the sum method fails) and the corresponding eigenfunctions. Since M is a good quantum number in the m-scheme we have to diagonalize only the submatrices defined by the various values of M.

Such a diagonalization can be carried out for any given mutual interaction. If there is only one state of a given J, it will necessarily be the same, irrespective of the interaction we use. It will then be determined uniquely provided the eigenvalues of the interaction are not degenerate.

Since we are primarily interested in building wave functions with definite values of J, we can diagonalize any interaction with nondegenerate eigenvalues. If there are several states with the same J, the diagonalization of the interaction will give us a possible set of mutually orthogonal states with that J. The eigenstates of any other interaction belonging to the same J would then be linear combinations of these states. The actual form of these linear combinations depends, of course, on the particular form of the interaction.

In jj-coupling a simple interaction which can be used for this purpose is

$$\sum_{k<l} V_{kl} = \sum_{k<l} (\mathbf{j}_k \cdot \mathbf{j}_l) = \tfrac{1}{2}\Big(\mathbf{J}^2 - \sum_k \mathbf{j}_k^2\Big) .$$

Since we are interested in the matrix elements of this interaction within a given configuration, where each particle is in an orbit of a well-defined j, we can replace $\mathbf{j}_k^2$ by its eigenvalue $j_k(j_k + 1)$.

We thus obtain

$$\sum_{k<l} V_{kl} = \frac{1}{2}\left[\mathbf{J}^2 - \sum_k j_k(j_k+1)\right].$$

The eigenvalues of this interaction are obviously different for different values of J. Since these eigenvalues are given explicitly as a function of J, there is no need to solve the secular equation. The diagonalization is therefore straightforward.

In LS-coupling it is simpler to diagonalize the m-scheme submatrices with definite M_L and M_S. Simple interactions can be found, the diagonalization of which gives wave functions with definite values of S and L. These interactions are

$$\sum_{k<l} (\mathbf{s}_k \cdot \mathbf{s}_l) = \frac{1}{2}\left(\mathbf{S}^2 - \sum_k \mathbf{s}_k^2\right) = \frac{1}{2}\left(\mathbf{S}^2 - \frac{3}{4}n\right)$$

and

$$\sum_{i<j} (l_i \cdot l_j) = \frac{1}{2}\left(\mathbf{L}^2 - \sum_i l_i^2\right) = \frac{1}{2}\left[\mathbf{L} - \sum_i l_i(l_i+1)\right].$$

A somewhat simpler method for building the wave functions with definite J and M is suggested by the method used for construction of such wave functions of two particles (Section 13). We start with the wave function in the m-scheme which has the maximum values of $M = M_{\max}$ allowed by the Pauli principle. This wave function is an eigenfunction of $\mathbf{J}^2$ with $J = M_{\max}$ and $M = M_{\max}$. We can obtain from this the wavefunction with the same $J = M_{\max}$ and $M = M_{\max} - 1$ by operating on it with the operator $J_- = \Sigma_k j_-^{(k)}$. This results in $J_-\psi_{JM} = \sqrt{(J+M)(J-M+1)}\,\psi_{JM-1}$. Applied to a wave function in the m-scheme, J_- will give the following expression:

$$J_-\,\psi(j_1m_1, j_2m_2, ..., j_nm_n)$$

$$= \sqrt{(j_1+m_1)(j_1-m_1+1)}\,\psi(j_1m_1-1,\ j_2m_2, ..., j_nm_n)$$

$$+ \sqrt{(j_2+m_2)(j_2-m_2+1)}\,\psi(j_1m_1, j_2m_2-1, ..., j_nm_n) + ...$$

$$+ \sqrt{(j_n+m_n)(j_n-m_n+1)}\,\psi(j_1m_1, j_2m_2, ..., j_nm_n-1).$$

The resulting wave function has the value $M = \Sigma_i m_i = M_{\max} - 1$.

We already mentioned that if in the m scheme there are two independent states with $M = M_{\text{max}} - 2$, then one linear combination is the state with $J = M_{\text{max}}$, $M = M_{\text{max}} - 2$ and the other is the state with $J = M_{\text{max}} - 2$, $M = M_{\text{max}} - 2$. The first state can be obtained by applying twice the operator J_-. The second state is then the combination of the m-scheme states orthogonal to the first one. We can proceed in this manner for lower values of M. As before, if there are several states with the same J, we can only obtain a possible set of independent linear combinations which must be orthogonal to those with higher values of J.

In the methods discussed so far, we antisymmetrize wave functions in the m-scheme and then proceed to construct from them linear combinations with definite values of J and M. On the other hand, equipped as we are with the knowledge of tensor algebra, we can easily construct wave functions with definite values of J and M by straightforward addition of the angular momenta of the different particles. Thus, we can add $\mathbf{j}_1$ and $\mathbf{j}_2$ to $\mathbf{J}_2$, then add $\mathbf{J}_2$ and $\mathbf{j}_3$ to $\mathbf{J}_3$, and so on, until we obtain a function of all the n particles with total angular momentum equal to $\mathbf{J}$. However, such wave functions are usually not antisymmetric and we have to antisymmetrize them. They then may lose a great deal of their simplicity. Still, with the methods of tensor algebra developed previously, they can be more efficiently handled than the functions in the m-scheme.

It should be recalled that in the m-scheme the effect of antisymmetrization is very simple. We start with wave functions (24.2) in which the single particle quantum numbers are all different from each other. The antisymmetrization of such a function always gives a nonvanishing antisymmetric function. The expectation values of single particle operators (24.16) calculated with the antisymmetrized wave functions (24.4) are the same as those calculated with the original wave function (24.2). Also nondiagonal matrix elements of such operators (24.15) are not affected by the antisymmetrization. The effect of antisymmetrization on the expectation values of two-body interactions (24.9) is that of replacing the direct term (24.10) by the difference of the direct term and the exchange term (24.11). This is also true for nondiagonal elements of such operators (24.8). Putting $f_i = 1$ in (24.13) for all i we obtain that any two orthogonal wave functions (24.2) which do not have the same set of single particle quantum numbers, give rise to two orthogonal antisymmetric wave functions (24.4).

Thus we can summarize by saying that in the m-scheme it is possible to take the antisymmetrization into account only in the last step of the actual evaluation of the matrix elements of two-particle operators. Otherwise, properly chosen nonantisymmetrized wave functions in the m-scheme may serve just as well.

On the other hand, if we build wave functions with a definite value of J by a successive coupling of the $\mathbf{j}_i$, the antisymmetrized wave functions bear no such simple relation to the original functions. The wave function of n particles occupying orbits characterized by the j values j^a, j^b, j^c, ... built by successive couplings is

$$\psi[j_1^a j_2^b(J_2) j_3^c(J_3) \ldots; J\, M]$$

$$= \sum_{m^a m^b M_2 m^c M_3 \ldots} (j^a m^a j^b m^b \mid j^a j^b J_2 M_2)\,(J_2 M_2 j^c m^c \mid J_2 j^c J_3 M_3) \ldots$$

$$\times\, \psi_{j^a m^a}(x_1)\, \psi_{j^b m^b}(x_2)\, \psi_{j^c m^c}(x_3) \ldots . \qquad (25.1)$$

It is obtained in a unique way by coupling $\mathbf{j}^a$ and $\mathbf{j}^b$ to $\mathbf{J}_2$, coupling $\mathbf{J}_2$ and $\mathbf{j}^c$ to $\mathbf{J}_3$, etc., arriving finally at the total $\mathbf{J}$. The notation we use in (25.1) implies that particle 1 is in the j^a orbit, particle 2 in the j^b orbit, etc. This clearly indicates that this wave function is not yet antisymmetric. The value of J, as well as the intermediate sums of angular momenta J_2, J_3, ..., J_{n-1}, characterize uniquely the wave functions (25.1). Furthermore, two wave functions (25.1) with different values of at least one of the J_k, $k = 2, \ldots, n-1$ (or J or M), are orthogonal to each other.

As an indication of the complications involved in antisymmetrizing the wave functions (25.1), we note that two such orthogonal non-antisymmetrized wave functions may become nonorthogonal after antisymmetrization. The reason for this becomes clear when we consider the relation between the wave functions (25.1) and the wave functions in the m-scheme. In (25.1) each of the wave functions is expressed as a linear combination of (not antisymmetrized) wave functions in the m-scheme $\psi(j_1^a m_1^a, j_2^b m_2^b, \ldots)$. The coefficients of these linear combinations are products of Clebsch-Gordan coefficients determined by the specific way of coupling. However, some of the m-scheme wave functions appearing in the expansion of (25.1) may have two or more identical single particle quantum numbers and vanish after antisymmetrization. This difficulty will not occur only if the expansion (25.1) will contain no such m-scheme wave functions, forbidden by the Pauli principle. Only in this case will all the simple features of the antisymmetrization in the m-scheme hold also for the antisymmetrization of the wave functions (25.1).

This is certainly the case if the various j values which appear in (25.1) (j^a, j^b, ...) are all different from each other. In this case, the single particle quantum numbers are all different irrespective of the values of m^a, m^b, ... We can then use our previous results about the orthogonality of the antisymmetrized wave functions in the m-scheme. The simple properties of the matrix elements of single particle and two-particle operators will then hold also for the wave function (25.1).

The antisymmetric wave function obtained by applying the operator

$(1/\sqrt{n!})\ \Sigma_P\,(-1)^P P$ to the function (25.1) with different values of j^a, j^b ... will be denoted by

$$\psi[j^a j^b(J_2)\,j^c(J_3)\ldots;\,JM] = \frac{1}{\sqrt{n!}}\sum_P(-1)^P P\,\psi[j_1^a j_2^b(J_2)\,j_3^c(J_3)\ldots;\,JM]. \tag{25.2}$$

All antisymmetric wave functions (25.2) characterized by all possible values of $J_2\ \ldots\ J_{n-1}JM$ are allowed by the Pauli principle. Different values of these intermediate sums of angular momenta, as well as of J or M, characterize *orthogonal* wave functions. Therefore, these quantum numbers, in addition to J and M, can be used to label the various antisymmetric wave functions (25.2). Matrix elements of single particle operators calculated with antisymmetrized wave functions (25.2) are equal to those calculated with the wave functions (25.1)

$$\langle j^a j^b(J_2)\,j^c(J_3)\ldots;\,JM\,\Big|\sum_{i=1}^{n} f_i\Big|\,j^a j^b(J_2')\,j^c(J_3')\ldots;\,J'M'\rangle$$

$$= \langle j_1^a j_2^b(J_2)\,j_3^c(J_3)\ldots;\,JM\,\Big|\sum_{i=1}^{n} f_i\Big|\,j_1^a j_2^b(J_2')\,j_3^c(J_3')\ldots;\,J'M'\rangle\,. \tag{25.3}$$

Similarly, we can simply calculate matrix elements of two-particle operators taken with the antisymmetrized wave functions (25.2)

$$\langle j^a j^b(J_2)\,j^c(J_3)\ldots;\,JM\,\Big|\sum_{i<k} g_{ik}\Big|\,j^a j^b(J_2')\,j^c(J_3')\ldots;\,J'M'\rangle\,. \tag{25.4}$$

We first express (25.4) in terms of matrix elements of two-particle configurations using the wave functions (25.1). Then we replace each two-particle matrix element by the difference of the direct and exchange terms. Thus, the antisymmetrization with respect to pairs of particles can be carried out only in the last step of evaluating the matrix elements with two-particle wave functions.

As indicated above, the very simple procedure for carrying out the antisymmetrization is a direct consequence of the fact that $j^a j^b$, etc., are all different. When several particles are in the same orbit (i.e., with the same values of n, l, and j) the situation becomes complicated.

In the general case, there are n_a particles in the j_a-orbit, n_b particles in the j_b-orbit, etc. This configuration is denoted by $j_a^{n_a} j_b^{n_b}$ .... We consider states of the whole system in which each configuration of equivalent particles is in a definite antisymmetric state with definite total angular momenta (J_a, J_b, ...). If there are several states with the same eigenvalue of $\mathbf{J}^2$ within a configuration $j_a^{n_a}$ of equivalent particles, there will be some additional quantum numbers $\alpha_a, \alpha_b, \ldots$. The problem of constructing the wave functions with the allowed values of $J_a, J_b, \ldots$ within each configuration, as well as the discussion of the additional necessary quantum numbers $\alpha_a, \alpha_b, \ldots$, will be treated only in the following sections. However, once these are given, a state which belongs to an eigenvalue of the total $\mathbf{J}^2$ of the whole system can be obtained by coupling successively the vectors $\mathbf{J}_a, \mathbf{J}_b, \ldots$ to form a total $\mathbf{J}$. If there are several ways to obtain the same value of the total J, each one of them defines an independent state of the whole system. They can be distinguished by the values of the *intermediate* sums of angular momenta $J_a\, J_b\, (J_{ab})\, J_c\, (J_{abc}), \ldots$. Obviously, states with different sets of intermediate angular momenta or total J and M, are orthogonal. The wave functions built in this way will be denoted by

$$\psi[j_a^{n_a}(1, \ldots, n_a; \alpha_a J_a) j_b^{n_b}(n_a + 1, \ldots, n_a + n_b; \alpha_b J_b) J_{ab}$$
$$j_c^{n_c}(n_a + n_b + 1, \ldots, n_a + n_b + n_c; \alpha_c J_c)\, J_{abc} \ldots; JM]. \qquad (25.5)$$

This notation implies that particles 1 to n_a, all occupying the orbit j_a, are coupled to form the *antisymmetric state* characterized by $\alpha_a J_a$; particles $n_a + 1$ to $n_a + n_b$, all occupying the orbit j_b, are coupled to form the antisymmetric state $\alpha_b J_b$, etc. The angular momenta $\mathbf{J}_a$ and $\mathbf{J}_b$ are then coupled to $\mathbf{J}_{ab}$, $\mathbf{J}_{ab}$ and $\mathbf{J}_c$ are coupled to $\mathbf{J}_{abc}$, etc., adding up to $\mathbf{J}$. The wave function (25.5) is not completely antisymmetric. It is antisymmetric within each group of equivalent particles but is not antisymmetric with respect to exchange of two particles from different groups.

When we antisymmetrize these wave functions the situation is very similar to the previous case where all the j_i were different. The wave-functions (25.5) can be expanded as linear combinations of m-scheme wave functions, in which all the single particle quantum numbers are different. This is certainly true for the quantum numbers of particles from different groups, i.e., with different j-values. It is also true for the quantum numbers of particles within each group of equivalent particles since they already form an antisymmetric state. Therefore, the intermediate sums of angular momenta $J_{ab}, J_{abc}, \ldots$ can serve as quantum numbers along with $J_a J_b \ldots$ and $\alpha_a \alpha_b \ldots$ also for the antisymmetric wave functions in this case. Antisymmetric wave functions with different sets of these quantum numbers are orthogonal to each other. Furthermore, we can use the wave functions (25.5) rather than the fully antisymmetric wave functions, in order to obtain matrix elements of single particle operators Σf_i. The matrix elements of two-particle operators g_{ik} can also be calculated with the help of the wave function (25.5). To obtain the correct results we first express these matrix elements as linear combinations of matrix elements of g_{ik} in two-particle configurations. We then antisymmetrize the two-particle wave functions appearing in this calculation if i and k refer to particles in different groups.

We may add a remark about the physical applications of this formalism. The wave functions (25.5) characterized by definite states of the groups of equivalent particles, form a complete scheme of orthogonal wave functions. However, these schemes do not diagonalize the interaction matrix (even within a given configuration). There are generally nondiagonal matrix elements of the interaction between such wave functions with the same value of J. If the states within the first group of equivalent particles are appropriately chosen, the scheme of wave functions $\psi(j_a^{n_a}; \alpha_a J_a)$ diagonalizes the submatrix of the interaction within the group $j_a^{n_a}$. However, the interaction between particles from different groups has nondiagonal elements between states with different sets of values of J_a, J_b..., α_a, α_b..., J_{ab}, J_{abc}, ... In such cases the energy matrix can be calculated in the above scheme of wave functions and then diagonalized. Since the nondiagonal matrix elements depend on the interaction and on the complexity of the given configuration, there is no general prescription for this diagonalization. In actual applications many particles are in closed shells and their contribution can be simply calculated as shown in Section 22. In this book we are usually dealing with only one group of equivalent particles outside closed shells. In a few cases we deal with two such groups at most.

Although our previous considerations were carried out in the jj-coupling scheme, the situation in LS-coupling is the same. Let there be n_a particles in l_a orbits, n_b particles in l_b orbits, etc. We build wave functions of the whole configuration $l_a^{n_a} l_b^{n_b}$ ... starting from definite antisymmetric states of the groups of equivalent particles. These states are characterized by the spins S_a, S_b, ... the orbital angular momenta L_a, L_b, ..., and possibly by additional necessary quantum numbers α_a, α_b, ... We now couple successively the spins: $\mathbf{S}_a$ is coupled to $\mathbf{S}_b$ to form $\mathbf{S}_{ab}$; the intermediate sum $\mathbf{S}_{ab}$ is coupled to $\mathbf{S}_c$ to form $\mathbf{S}_{abc}$; etc., until the total $\mathbf{S}$ is obtained. Similarly, the orbital angular momenta are successively coupled to form the total $\mathbf{L}$. The total angular momentum is obtained by coupling $\mathbf{S}$ and $\mathbf{L}$ to $\mathbf{J} = \mathbf{L} + \mathbf{S}$. These wave functions are completely and uniquely characterized by the set of values S_a, S_b, ..., L_a, L_b, ..., and α_a, α_b, ... as well as by the intermediate sums of angular momenta S_{ab}..., L_{ab}..., the total S, L, and J. These wave functions, not antisymmetric with respect to exchange of particles from different groups, may be used for actual calculations instead of the fully antisymmetric wave functions. Matrix elements of single particle operators can be directly calculated with these wave functions. In the calculation of matrix elements of two-particle operators g_{ik}, the antisymmetrization may be introduced only at the last step of the evaluation of the matrix elements in the two-particle configurations.

26. Coefficients of Fractional Parentage

In the preceding sections we constructed wave functions by successive couplings of the single particle angular momenta. It was pointed out that when such wave functions are antisymmetrized various complications might occur. Thus, wave functions with different intermediate sums of angular momenta, which are orthogonal to each other, may lose this orthogonality after antisymmetrization. Such complications arise if there are several equivalent particles. We shall now turn to the antisymmetrization of such wave functions and first treat the case of three particles in *jj*-coupling.

Consider three particles in orbits characterized by j^a, j^b, and j^c. As before, we couple $\mathbf{j}^a$ and $\mathbf{j}^b$ to form $\mathbf{J}_2$ which is then coupled to $\mathbf{j}^c$ to form the total $\mathbf{J}$. If particle 1 is in the j^a orbit, and particle 2 in the j^b orbit, we obtain a wave function which is obviously not antisymmetric

$$\psi[j_1^a j_2^b(J_2) j_3^c JM]. \tag{26.1}$$

Antisymmetrizing this wave function by applying to it the operator $\Sigma_P (-1)^P P$, we obtain

$$\sum_P (-1)^P P \, \psi[j_1^a j_2^b(J_2) j_3^c JM]$$

$$= \psi[j_1^a j_2^b(J_2) j_3^c JM] - \psi[j_2^a j_1^b(J_2) j_3^c JM] - \psi[j_1^a j_3^b(J_2) j_2^c JM]$$

$$- \psi[j_3^a j_2^b(J_2) j_1^c JM] + \psi[j_2^a j_3^b(J_2) j_1^c JM] + \psi[j_3^a j_1^b(J_2) j_2^c JM]. \tag{26.2}$$

It is clear that, if j^a, j^b, and j^c are all different, any two of the six wave functions which appear in (26.2) are orthogonal to each other. Therefore, the normalization of the wave function (26.2) is $(1/\sqrt{3!}) = \sqrt{\tfrac{1}{6}}$. All other simple features of the antisymmetrized wave function follow simply in a similar way.

However, if two of the three values j^a, j^b, and j^c are equal (and belong to orbits with the same n and l), the various terms in (26.2) are no longer orthogonal to each other. For example, if $j^a = j^b = j$ it follows that

$$\psi[j_2 j_1(J_2) j_3^c J\, M] = -(-1)^{J_2} \psi[j_1 j_2(J_2) j_3^c J\, M].$$

This implies that (26.2) vanishes for odd values of J_2 which are forbidden by the Pauli principle. For the allowed even values of J_2, (26.2) becomes twice the following sum

$$\psi[j_{12}^2(J_2)\, j_3^c J\, M] - \psi[j_{13}^2(J_2)\, j_2^c J\, M] + \psi[j_{23}^2(J_2)\, j_1^c J\, M]. \tag{26.3}$$

In this notation, j_{12}^2 means that particles 1 and 2 are in the j-orbit and $j_1 = j$ is coupled to $j_2 = j$ to form J_2. If j^c is different from j, it is clear that the three terms of (26.3) are orthogonal to each other. However, if j^c is also equal to j, i.e., if there are three equivalent particles, the three terms of (26.3) are not orthogonal as shown below.

Let us consider the antisymmetric wave functions (26.3) for three equivalent particles, i.e., $j^c = j$. It has then the form

$$\psi[j_{12}^2(J_2)\, j_3 J\, M] - \psi[j_{13}^2(J_2)\, j_2 J\, M] + \psi[j_{23}^2(J_2)\, j_1 J\, M]. \tag{26.4}$$

It is easy to see that the second term is not orthogonal to the first term. In fact, the integral

$$\int \psi^*[j_{12}^2(J_2)\, j_3 J\, M]\, \psi[j_{13}^2(J_2)\, j_2 J\, M]\, d(1)d(2)d(3) \tag{26.5}$$

is precisely the matrix element $\langle j_1 j_3(J_2) j_2 J | j_1 j_2(J_2) j_3 J\rangle$ of the transformation (15.29) between two coupling schemes.

The transformation matrix is independent of M and is given in terms of Racah coefficients by (15.30)

$$\langle j_1 j_3(J_2)\, j_2 J |\, j_1 j_2(J_1)\, j_3 J\rangle = (-1)^{j_2+j_3+J_1+J_2} \sqrt{(2J_1+1)(2J_2+1)} \begin{Bmatrix} j_1 & j_2 & J_1 \\ J & j_3 & J_2 \end{Bmatrix}.$$

The functions $\psi[j_{12}^2(J_1) j_3 J\, M]$ and $\psi[j_{12}^2(J_2) j_3 J\, M]$ with $J_1 \neq J_2$ are of course orthogonal. Equation (15.29) enables us to write the functions $\psi[j_{13}^2(J_2) j_2 J\, M]$ and $\psi[j_{23}^2(J_2) j_1 J\, M]$ in terms of the functions $\psi[j_{12}^2(J_1) j_3 J\, M]$ (with various values of J_1). We can therefore obtain an expansion of the antisymmetric wave function (26.4) in terms of orthogonal wave functions by applying the transformation (15.30) to the last two terms in (26.4). We also observe that since $j_1 = j_2 = j_3 = j$ we can always drop the subscripts 1, 2, 3 in the matrix element (15.30). Carrying out this transformation, (26.4) assumes the form

$$\begin{aligned}\psi(j^3 J\, M) = \psi[j_{12}^2(J_2)\, j_3\, JM] &- \sum_{J_1} \langle j\, j(J_2)\, j\, J | \, j\, j(J_1)\, j\, J\rangle\, \psi[j_{12}^2(J_1)\, j_3 J\, M] \\ &+ \sum_{J_1} \langle j\, j(J_2)\, j\, J | \, j\, j(J_1)\, j\, J\rangle\, \psi[j_{21}^2(J_1)\, j_3 J\, M].\end{aligned}$$

Using the relation

$$\psi[j_{21}^2(J_1)\,j_3 J\,M] = -(-1)^{J_1}\,\psi[j_{12}^2(J_1)\,j_3 J\,M],$$

we can add all terms and obtain

$$\psi(j^3 J\,M) = \psi[j_{12}^2(J_2)\,j_3 J\,M] - \sum_{J_1} \langle j\,j(J_2)\,j\,J | j\,j(J_1)\,j\,J\rangle$$

$$\times [1 + (-1)^{J_1}]\,\psi[j_{12}^2(J_1)\,j_3 J\,M]. \tag{26.6}$$

In the summation (26.6) all terms with odd values of J_1, forbidden by the Pauli principle, disappear as should be expected. We thus obtain, using (15.30),

$$\psi(j^3 J\,M) = \sum_{J_1 \text{ even}} [\delta_{J_1 J_2} - 2\langle j\,j(J_2)\,j\,J | j\,j(J_1)\,j\,J\rangle]\,\psi[j_{12}^2(J_1)\,j_3 J\,M]$$

$$= \sum_{J_1 \text{ even}} \left[\delta_{J_1 J_2} + 2\sqrt{(2J_1+1)(2J_2+1)} \begin{Bmatrix} j & j & J_1 \\ J & j & J_2 \end{Bmatrix}\right] \psi[j_{12}^2(J_1)\,j_3 J\,M]. \tag{26.7}$$

For values of J in the j^3 configurations which are forbidden by the Pauli principle, the wave function (26.7) vanishes identically. This requirement is equivalent to certain relations satisfied by the W-functions (e.g., the vanishing of some of them even when the triangular conditions are not violated). These relations are obviously connected with the symmetry properties of the transformation (15.29).

The wave function (26.7) is fully antisymmetric, yet it is given as a linear combination of wave functions $\psi[j_{12}(J_1)j_3 J\,M]$ which are not antisymmetric. It is not surprizing that it is possible to express a completely antisymmetric wave function of 3 particles in terms of wave functions in which particle 3 is so clearly distinguished from the other two. The functions $\psi[j_{12}^2(J_1)j_3 J\,M]$, with different values of J_1, constitute a complete set for three equivalent particles in the j-orbit. Therefore any function of these three particles, including the antisymmetric wave function, can be expanded in terms of the functions $\psi[j_{12}^2(J_1)j_3 J\,M]$.

All terms of (26.7) with different J_1 are orthogonal to each other. Therefore the use of this expansion for antisymmetric wave functions is very simple and straightforward. In the case of particles in different j orbits, matrix elements could be calculated simply with a single non-antisymmetric wave function (26.1). On the other hand, here we have to use a combination of several such functions. However, in doing so we obtain a fully antisymmetric wave function and no special precautions should be taken about the exchange integrals. They are automatically included in our results, since for equivalent particles there is no way to distinguish them from the direct integrals.

Since all terms of (26.7) are orthogonal to each other, the calculation of the normalization factor is straightforward. The integral of the square of the wave function (26.7) is equal to

$$\sum_{J_1 \text{ even}} \left[\delta_{J_1 J_2} + 2\sqrt{(2J_1+1)(2J_2+1)} \begin{Bmatrix} j & j & J_1 \\ J & j & J_2 \end{Bmatrix}\right]^2 = 3 + 6(2J_2+1) \begin{Bmatrix} j & j & J_2 \\ J & j & J_2 \end{Bmatrix}. \tag{26.8}$$

Here we have used Eqs (15.16) and (15.17) to carry out the summation over even values of J_1.

Thus, in order to be normalized to unity, the wave function (26.7) should be multiplied by

$$\left[3 + 6(2J+1) \begin{Bmatrix} j & j & J_2 \\ J & j & J_2 \end{Bmatrix}\right]^{-1/2}. \tag{26.9}$$

The expansion of the normalized antisymmetric wave function built this way is generally written in the form

$$\psi(j^3 \alpha\, J M) = \sum_{J_1 \text{ even}} [j^2(J_1)\, j\, J| \} j^3 \alpha\, J]\, \psi[j_{12}^2(J_1)\, j_3 J\, M]. \tag{26.10}$$

Here α is an additional quantum number which is necessary if there are several antisymmetric states of the j^3 configuration with the same value of J. The coefficients of the linear combinations are written as elements of a transformation matrix which leads from the scheme of wave functions $\psi[j_{12}^2(J_1) j_3 J\, M]$ to the antisymmetric wave functions. The transformation coefficients are obviously independent of M which is therefore omitted. This transformation is only a part of a unitary transformation and therefore cannot be inverted. This fact is indicated by the peculiar sign $|\}$ inserted in the bracket. The coefficients in (26.10) are called *coefficients of fractional parentage* (in short c.f.p.) since each state $\psi(j_{12}^2 J_1)$ can be regarded as a (fractional) "parent" of the antisymmetric state $\psi(j^3 \alpha\, J)$.

If there is only one antisymmetric state of the j^3 configuration with a given value of J, there is no need for the quantum number α. In this case it is clear that starting from *any* allowed value of J_2, say $J_2 = J_0$, we should always obtain the same antisymmetric wave function $\psi(j^3 J)$, except for a possible phase factor. In this case the coefficients of fractional parentage are

$$[j^2(J_1) j J| \} j^3 J]$$

$$= \left[\delta_{J_1 J_0} + 2\sqrt{(2J_0+1)(2J_1+1)} \begin{Bmatrix} j & j & J_1 \\ J & j & J_0 \end{Bmatrix}\right] \left[3 + 6(2J_0+1) \begin{Bmatrix} j & j & J_0 \\ J & j & J_0 \end{Bmatrix}\right]^{-1/2} \tag{26.11}$$

If $J = j$ we can start from $J_0 = 0$. For $j \leq \frac{7}{2}$, there is only one antisymmetric state with $J = j$ in the j^3 configuration. For higher values of j this is no longer the case. However, even for $j > \frac{7}{2}$, the choice $J_0 = 0$ leads to a well-defined state with $J = j$. From (26.11) we obtain, for $J_0 = 0$,

$$[j^2(J_1)j\, J = j \}j^3\, J = j] = \left[\delta_{J_1 0} + 2\sqrt{2J_1+1} \begin{Bmatrix} j & j & J_1 \\ j & j & 0 \end{Bmatrix}\right] \left[3 + 6 \begin{Bmatrix} j & j & 0 \\ j & j & 0 \end{Bmatrix}\right]^{-1/2}$$

Inserting the values of the W-coefficients we obtain

$$\begin{aligned} [j^2(0)j\, J = j \}j^3\, J = j] &= \sqrt{\frac{2j-1}{3(2j+1)}} \\ [j^2(J_1)j\, J = j \}j^3\, J = j] &= -\frac{2\sqrt{2J_1+1}}{\sqrt{3(2j-1)(2j+1)}} \qquad J_1 > 0 \qquad J_1 \text{ even.} \end{aligned} \tag{26.12}$$

Therefore

$$\psi(j^3[J_0 = 0]\, J = j) = \sqrt{\frac{2j-1}{3(2j+1)}}\, \psi[j_{12}^2(J_1 = 0)\, j_3\, J = j] - \frac{2}{\sqrt{3(2j-1)(2j+1)}} \times \sum_{\substack{J_1 > 0 \\ \text{even}}} \sqrt{2J_1+1}\, \psi[j_{12}^2(J_1)\, j_3\, J = j]. \tag{26.13}$$

The value $J_0 = 0$ has been put in square brackets in $\psi(j^3 J)$ to remind us how this particular antisymmetric state was obtained. In the case of $j = \frac{5}{2}$, (26.13) readily gives

$$\psi\{(\tfrac{5}{2})^3\, [J_0 = 0]\, J = \tfrac{5}{2}\} = \tfrac{1}{3}\sqrt{2}\, \psi\, [(\tfrac{5}{2})_{12}^2\, (0)\, (\tfrac{5}{2})_3\, J = \tfrac{5}{2}] - \tfrac{1}{3}\sqrt{\tfrac{5}{2}}\, \psi\, [(\tfrac{5}{2})_{12}^2\, (2)\, (\tfrac{5}{2})_3\, J = \tfrac{5}{2}] - \tfrac{1}{\sqrt{2}}\, \psi\, [(\tfrac{5}{2})_{12}^2\, (4)\, (\tfrac{5}{2})_3\, J = \tfrac{5}{2}].$$

It may be verified that starting from $J_0 = 2$ or $J_0 = 4$ and inserting in (26.11) the appropriate values of the Racah coefficients, the same wave function multiplied by -1 will be obtained. This will not be the case for $j = \frac{9}{2}$ in the $(\frac{9}{2})^3$ configuration. There, starting with $J_0 = 0$ or with $J_0 = 2$ we obtain *different* functions. Such cases will be treated subsequently. The coefficients (26.11) always vanish if J is not allowed by the Pauli principle. Thus, for the $(\frac{5}{2})^3$ configuration, starting with $J_0 = 2$ or $J_0 = 4$, we find that (26.11) vanishes for $J = \frac{1}{2}, \frac{7}{2}, \frac{11}{2}$, or $\frac{13}{2}$. These values of J are indeed forbidden by the Pauli principle as can be easily verified.

This method of constructing antisymmetric states of three particles can be extended to configurations with a larger number of particles. We can start from the antisymmetric states of three particles and build antisymmetric wave functions of four particles. Generally, we can start from the states of the j^{n-1} configuration and obtain antisymmetric states of the j^n configuration. We define the coefficients of fractional parentage in the general case by the following expansion of a normalized antisymmetric wave function

$$\psi(j^n \alpha\, J\, M) = \sum_{\alpha_1 J_1} [j^{n-1}(\alpha_1 J_1) j\, J \| j^n \alpha\, J]\, \psi[j_{1,\ldots,n-1}^{n-1}(\alpha_1 J_1)\, j_n J\, M]. \qquad (26.14)$$

In (26.14) we have a linear combination of tensors of degree J, i.e., the functions $\psi[j_{1,\ldots,n-1}^{n-1}(\alpha_1 J_1) j_n J\, M]$ with the $2J + 1$ values of M, which form another irreducible tensor of the same degree $\psi(j^n \alpha\, J\, M)$. The coefficients of such a combination are obviously independent of M. The quantum numbers α and α_1 are additional quantum numbers which are necessary if there are several states with the same value of J. Since all the terms of (26.14) are orthogonal to each other, the normalization of the wave function imposes a single condition on the c.f.p. This normalization and the orthogonality of wave functions with different values of α can be expressed by the following relation:

$$\sum_{\alpha_1 J_1} [j^n \alpha\, J \{| j^{n-1}(\alpha_1 J_1) j\, J]\, [j^{n-1}(\alpha_1 J_1) j\, J |\} j^n \alpha'\, J] = \delta_{\alpha\alpha'}. \qquad (26.15)$$

The orthogonality of two functions with different values of J is automatically assured by the coupling of angular momenta $\mathbf{J}_1$ and $\mathbf{j}_n$. It therefore imposes no further conditions on the corresponding c.f.p. The coefficient $[j^n \alpha\, J \{| j^{n-1}(\alpha_1 J_1) j\, J]$ is the complex conjugate of $[j^{n-1}(\alpha_1 J_1) j\, J |\} j^n \alpha\, J]$. The c.f.p. are obtained by using the transformation (15.30) whose matrix elements are real. Therefore, the c.f.p. are also real numbers and there is no distinction between the two coefficients $[\ldots|\}\ldots]$ and $[\ldots\{|\ldots]$.

Every antisymmetric wave function of the j^n configuration can be expressed in the form (26.14). In fact, the wave functions

$$\psi[j_{1,\ldots,n-1}^{n-1}(\alpha_1 J_1)\, j_n J\, M] \qquad (26.16)$$

with all allowed values of J_1 and α_1 clearly constitute a complete basis of all functions of n particles which are antisymmetric in the coordinates of the first $n - 1$ particles. Every wave function which has this property can be expressed as a linear combination of these basis wave functions. In particular, the wave functions which are antisymmetric in *all* particle coordinates can be expressed in the form (26.14).

The transformation from the wave functions (26.16) to the antisymmetric functions (26.14) can be considered as a *projection* operation. This interpretation is very useful for dealing with coefficients of fractional parentage. In fact, the wave functions which are antisymmetric in *all* the n particles constitute a subspace of the space spanned by the functions (26.16). The antisymmetrization of $\psi[j^{n-1}(\alpha_0 J_0) j_n J\, M]$ is therefore a projection of this function on the subspace of fully antisymmetric wave functions. This projection is then expressed as in (26.14), in terms of the basis wave functions

$$\psi(j^n[\alpha_0 J_0] J M) = \sum_{\alpha_1 J_1} [j^{n-1}(\alpha_1 J_1) j J|\} j^n[\alpha_0 J_0] J] \, \psi[j^{n-1}(\alpha_1 J_1) j_n J M]. \tag{26.17}$$

The quantum numbers α_0 and J_0 are put in square brackets in order to indicate that the function (26.17) was obtained by antisymmetrization of $\psi[j^{n-1}(\alpha_0 J_0) j_n J M]$. The state with $\alpha_0 J_0$ of the j^{n-1} configuration is accordingly called the ***principal parent*** of the state (26.17). From this point of view, it is clear that a nonvanishing projected wave function must have a component proportional to the original wave function. Therefore, the c.f.p. belonging to the principal parent is different from zero in the case of a nonvanishing antisymmetric function.

If $\mathscr{P}$ is the operator which carries out this projection, applying it again will not change the wave function. Thus,

$$\mathscr{P}^2 \psi[j^{n-1}(\alpha_1 J_1) j_n J M] = \mathscr{P} \psi[j^{n-1}(\alpha_1 J_1) j_n J M]. \tag{26.18}$$

Clearly, antisymmetrization of an antisymmetric wave function does not change it, except for a possible multiplication by a numerical factor. In order to utilize (26.18) we have to define $\mathscr{P}$ so that this factor will be unity. In order to antisymmetrize the function (26.16) we can introduce a new operator $\mathscr{A}$. This operator exchanges only particle n with the first $n - 1$ particles. We thus define

$$\mathscr{A}\psi[j^{n-1}_{1,\dots,n-1}(\alpha_0 J_0) j_n J M]$$

$$= \psi[j^{n-1}_{1,\dots,n-1}(\alpha_0 J_0) j_n J M] - \sum_k \psi[j^{n-1}_{1,\dots,k-1,n,k+1,\dots,n-1}(\alpha_0 J_0) j_k J M]. \tag{26.19}$$

Since $\psi(j^{n-1} \alpha_0 J_0)$ is antisymmetric in the $n - 1$ particles, the function (26.19) is fully antisymmetric. When the antisymmetrizer $\mathscr{A}$ is applied to an antisymmetric wave function, each of its n permutations reproduces the antisymmetric function. Thus, the operator $\mathscr{P}$, in (26.18), is equal to $\mathscr{A}/n$. The wave function (26.19) is obviously not normalized. We introduce the normalization factor $N_{\alpha_0 J_0 J}$ and, comparing it to (26.17), we obtain

$$\mathscr{P}\psi[j^{n-1}(\alpha_0 J_0) j_n J M] = \frac{\mathscr{A}}{n} \psi[j^{n-1}(\alpha_0 J_0) j_n J M]$$

$$= \sum_{\alpha_1 J_1} \frac{[j^{n-1}(\alpha_1 J_1) j J|\} j^n[\alpha_0 J_0] J]}{n N_{\alpha_0 J_0 J}} \psi[j^{n-1}(\alpha_1 J_1) j_n J M]$$

$$= \sum_{\alpha_1 J_1} \langle \alpha_0 J_0 | \mathscr{P} | \alpha_1 J_1 \rangle \, \psi[j^{n-1}(\alpha_1 J_1) j_n J M]. \tag{26.20}$$

Equation (26.20) defines the matrix elements of $\mathscr{P}$. The matrix $\mathscr{P}$ projects every wave function $\psi[j^{n-1}(\alpha_1 J_1) j_n J\, M]$ into an antisymmetric wave function. Therefore, the number of independent antisymmetric states with given J and M in the j^n configuration is equal to the rank of this matrix. Since $\mathscr{P}^2 = \mathscr{P}$, the eigenvalues of $\mathscr{P}$ are either zero or unity. The number of independent antisymmetric states is equal to the multiplicity of the eigenvalue 1 of the matrix $\mathscr{P}$ and therefore to its trace. The interpretation of $\mathscr{P}$ as a projection operator (in a unitary space) implies that the matrix in (26.20) is Hermitian. Since it is real, it is a symmetric matrix. Thus, the c.f.p. satisfy the relation

$$\frac{[j^{n-1}(\alpha_1 J_1) j\, J|\}j^n[\alpha_2 J_2]J]}{N_{\alpha_2 J_2 J}} = \frac{[j^{n-1}(\alpha_2 J_2) j\, J|\}j^n[\alpha_1 J_1]J]}{N_{\alpha_1 J_1 J}}. \tag{26.21}$$

We now obtain a further simple result due to the fact that $(1/n)\mathscr{A}$ is an idempotent projection operator. Let us consider the c.f.p. of the principal parent, given by

$$[j^{n-1}(\alpha_0 J_0) j\, J|\}j^n[\alpha_0 J_0]J] = \int \psi^*(j^n[\alpha_0 J_0]J\, M)\, \psi(j^{n-1}(\alpha_0 J_0) j_n J\, M). \tag{26.22}$$

Since $\psi^*(j^n[\alpha_0 J_0]J\, M)$ is antisymmetric, we can apply to either function in the integrand the operator $(1/n)\mathscr{A}$. We thus obtain

$$[j^{n-1}(\alpha_0 J_0) j\, J|\}j^n[\alpha_0 J_0]J]$$

$$= \frac{1}{n} \int \psi^*(j^n[\alpha_0 J_0]J\, M)\, \mathscr{A}\psi[j^{n-1}(\alpha_0 J_0) j_n J\, M]$$

$$= \frac{1}{n}\frac{1}{N_{\alpha_0 J_0 J}} \int \psi^*(j^n[\alpha_0 J_0]J\, M)\, \psi(j^n[\alpha_0 J_0]J\, M) = \frac{1}{nN_{\alpha_0 J_0 J}}. \tag{26.23}$$

Thus, the normalization factor is given directly in terms of the c.f.p. of the principal parent by

$$nN_{\alpha_0 J_0 J}\, [j^{n-1}(\alpha_0 J_0) j\, J|\}j^n[\alpha_0 J_0]J] = 1. \tag{26.24}$$

If we adopt the convention that the normalization factor is positive, we see that also the c.f.p. of the principal parent is always positive. A special case of (26.24) is the relation between the normalization factor (26.9) and the c.f.p. (26.11) in the case $n = 3$.

The actual construction of wave function (26.14), i.e., the calculation of the c.f.p., can be carried out as in the case of the three particles. The procedure is, however, generally more complicated. We start with an antisymmetric state of the first $n - 1$ particles $\psi(j^{n-1}_{1,\ldots,n-1}\alpha_0 J_0)$. To this state we add the last particle by coupling $\mathbf{J}_0$ to $\mathbf{j}_n = \mathbf{j}$ to form a total $\mathbf{J}$. This wave function can now be antisymmetrized by applying the operator $\mathscr{A}$ thus obtaining (26.19).

To the various terms of (26.19), which are antisymmetric in $n-1$ particles, we now apply the permutations P_k which will bring the nth particle to the last position. We thus obtain

$$\psi[j^{n-1}_{1,\ldots,n-1}(\alpha_0 J_0)\, j_n J\, M] - \sum_k (-1)^{P_k}\, \psi[j^{n-1}_{1,\ldots,k-1,k+1,\ldots,n-1,n}(\alpha_0 J_0)\, j_k J\, M]. \tag{26.25}$$

The antisymmetric wave function $\psi(j^{n-1}\alpha_0 J_0)$ can be expressed in terms of $\psi[j^{n-2}_{1,\ldots,n-2}(\alpha_2 J_2) j_{n-1}\alpha_0 J_0]$ by using the appropriate c.f.p. The expression (26.25) will then assume the form

$$\psi[j^{n-1}_{1,\ldots,n-1}(\alpha_0 J_0)\, j_n J\, M] - \sum_{\alpha_2 J_2} [j^{n-2}(\alpha_2 J_2)\, j\, J_0 \| j^{n-1}\alpha_0 J_0]$$

$$\times \sum_k (-1)^{P_k}\, \psi[j^{n-1}_{1,\ldots,k-1,k+1,\ldots,n-1}(\alpha_2 J_2)\, j_n(J_0)\, j_k J\, M]. \tag{26.26}$$

In order to bring (26.26) to the form (26.14), the nth particle should be coupled last. We thus apply the transformation (15.29) to change the order of coupling and obtain

$$\psi[j^{n-1}_{1,\ldots,n-1}(\alpha_0 J_0)\, j_n J\, M] - \sum_{\alpha_2 J_2} [j^{n-2}(\alpha_2 J_2)\, j\, J_0 \| j^{n-1}\alpha_0 J_0]$$

$$\times \sum_{kJ_1} (-1)^{P_k} \langle J_2 j(J_0) j\, J | J_2 j(J_1) j\, J\rangle\, \psi[j^{n-2}_{1,\ldots,k-1,k+1,\ldots,n-1}(\alpha_2 J_2)\, j_k(J_1)\, j_n J\, M]. \tag{26.27}$$

On the antisymmetric wave functions of $n-2$ particles in (26.27) we now perform the permutations which bring the $(n-1)$-particle to the position of the missing k-particle (if $k \neq n-1$) and in this way obtain

$$\psi[j^{n-1}_{1,\ldots,n-1}(\alpha_0 J_0)\, j_n J\, M] - \sum_{\alpha_2 J_2} [j^{n-2}(\alpha_2 J_2)\, j\, J_0 \| j^{n-1}\alpha_0 J_0]$$

$$\times \sum_{J_1} \langle J_2 j(J_0) j\, J | J_2 j(J_1) j\, J\rangle$$

$$\times \Big[\psi[j^{n-2}_{1,\ldots,n-2}(\alpha_2 J_2)\, j_{n-1}(J_1)\, j_n J\, M]$$

$$- \sum_k [j^{n-2}_{1,\ldots,k-1,n-1,k+1,\ldots n-2}(\alpha_2 J_2)\, j_k(J_1)\, j_n J\, M]\Big]. \tag{26.28}$$

The expression in the square brackets is precisely the result of the operation $\mathscr{A}_1 \psi[j^{n-2}_{1\ldots n-2}(\alpha_2 J_2) j_{n-1}(J_1) j_n J\, M]$ where $\mathscr{A}_1$ is the operator defined by (26.19) for the first $n-1$ particles. We can therefore rewrite (26.28) in a more concise form

$$\mathscr{A}\psi[j^{n-1}(\alpha_0 J_0)j_n J\,M] = \psi[j^{n-1}(\alpha_0 J_0)j_n J\,M] - \sum_{\alpha_2 J_2 J_1'} [j^{n-2}(\alpha_2 J_2)j\,J_0|\}j^{n-1}\alpha_0 J_0]$$

$$\times \langle J_2 j(J_0)j\,J|J_2 j(J_1')j\,J\rangle\, \mathscr{A}_1\psi[j^{n-2}(\alpha_2 J_2)j_{n-1}(J_1')j_n J\,M]. \qquad (26.29)$$

Multiplying both sides by $\psi^*[j^{n-1}(\alpha_1 J_1)j_n JM]$ and integrating, we obtain, by using (26.23) and the explicit form of the transformation coefficients (15.30),

$$n[j^{n-1}(\alpha_0 J_0)j\,J|\}j^n[\alpha_0 J_0]J]\,[j^{n-1}(\alpha_1 J_1)j\,J|\}j^n[\alpha_0 J_0]J]$$

$$= \delta_{\alpha_1\alpha_0}\,\delta_{J_1 J_0} + (n-1)\sum_{\alpha_2 J_2}(-1)^{J_0+J_1}\sqrt{(2J_0+1)(2J_1+1)}\begin{Bmatrix} J_2 & j & J_1 \\ J & j & J_0 \end{Bmatrix}$$

$$\times [j^{n-2}(\alpha_2 J_2)j\,J_0|\}j^{n-1}\alpha_0 J_0]\,[j^{n-2}(\alpha_2 J_2)j\,J_1|\}j^{n-1}\alpha_1 J_1]. \qquad (26.30)$$

Equation (26.30) can be used to obtain the c.f.p. in the j^n configuration in terms of the c.f.p. in the j^{n-1} configuration. It can therefore serve as a recursion formula for the c.f.p. It should be remembered that the wave functions $\psi(j^n[\alpha_0 J_0]J\,M)$ with different principal parents are generally not independent. Therefore a better scheme of wave functions, with additional quantum numbers α replacing $[\alpha_0 J_0]$, is required. This problem will be subsequently treated.

We shall now discuss the use of antisymmetric wave functions (26.14) as expressed with coefficients of fractional parentage. It turns out to be convenient to work with antisymmetric wave functions like (26.14) in which a definite particle (the nth in our case) is singled out and coupled last. As the first example let us calculate matrix elements of single particle operators $\sum f_i$.

Due to the symmetry of $\sum f_i$ and of the product of two antisymmetric wave functions it follows that

$$\langle j^n\alpha\,J\,M\Big|\sum f_i\Big|j^n\alpha' J' M'\rangle = n\langle j^n\alpha\,J\,M|f_n|j^n\alpha' J' M'\rangle. \qquad (26.31)$$

In (26.31) we choose to calculate the contribution of f_n, since the nth particle is in a special position in the wave functions (26.14). Using c.f.p. we obtain from (26.31)

$$\langle j^n\alpha\,J\,M\Big|\sum f_i\Big|j^n\alpha' J' M'\rangle$$

$$= n\sum_{\alpha_1 J_1 \alpha_1' J_1'}[j^n\alpha\,J\{|j^{n-1}(\alpha_1 J_1)j\,J]\,[j^{n-1}(\alpha_1' J_1')j\,J'|\}j^n\alpha' J']$$

$$\times \langle j^{n-1}(\alpha_1 J_1)j_n J\,M|f_n|j^{n-1}(\alpha_1' J_1')j_n J' M'\rangle. \qquad (26.32)$$

Since wave functions of $n - 1$ particles with different values of $\alpha_1 J_1$ and $\alpha_1' J_1'$ are orthogonal, the matrix elements of f_n in (26.32) vanish unless $\alpha_1' = \alpha_1$ and $J_1' = J_1$. The matrix elements of f_n can be easily calculated. If f is a component of a tensor operator $f_\kappa^{(k)}$, we can use (15.27) and obtain the final result

$$\left(j^n \alpha\, J \middle\| \sum \mathbf{f}^{(k)}(i) \middle\| j^n \alpha' J'\right)$$

$$= n(j||\mathbf{f}^{(k)}||j) \sum_{\alpha_1 J_1} [j^n \alpha\, J\{| j^{n-1}(\alpha_1 J_1) j\, J]\, [j^{n-1}(\alpha_1 J_1) j\, J' |\} j^n \alpha' J']$$

$$\times (-1)^{J_1+j+J+k} \sqrt{(2J+1)(2J'+1)} \begin{Bmatrix} j & J & J_1 \\ J' & j & k \end{Bmatrix}. \qquad (26.33)$$

The appearance of the W-coefficient in (26.33) implies that this expression vanishes unless both (j, j, k) and (J, J', k) satisfy the triangular condition. If the single particle operator is a scalar $f_0^{(0)} = f$ this expression can be further simplified. Thus, for $k = 0$, there can be no change in the total angular momentum, i.e., $J' = J$. Putting the values of the Racah coefficients for this case in (26.33) we obtain the expression

$$\left(j^n \alpha\, J \middle\| \sum \mathbf{f}^{(0)}(i) \middle\| j^n \alpha' J\right)$$

$$= n \sqrt{\frac{2J+1}{2j+1}} (j||\mathbf{f}^{(0)}||j) \sum_{\alpha_1 J_1} [j^n \alpha\, J\{| j^{n-1}(\alpha_1 J_1) j\, J]\, [j^{n-1}(\alpha_1 J_1) j\, J |\} j^n \alpha' J].$$

Using the orthogonality relation (26.15) of the c.f.p. we obtain

$$\left(j^n \alpha\, J \middle\| \sum \mathbf{f}^{(0)}(i) \middle\| j^n \alpha' J\right) = n \sqrt{\frac{2J+1}{2j+1}} (j||\mathbf{f}^{(0)}||j)\, \delta_{\alpha\alpha'}. \qquad (26.34)$$

Since the matrix elements of $f_0^{(0)}$ are independent of m, we obtain from (26.34)

$$\left\langle j^n \alpha\, J M \middle| \sum f_0^{(0)}(i) \middle| j^n \alpha' J M \right\rangle = n \langle j\, m | f_0^{(0)} | j\, m \rangle\, \delta_{\alpha\alpha'} \qquad (26.35)$$

as, of course, should have been expected.

Equation (26.33) is the expression of matrix elements of single particle operators in the j^n configuration in terms of single particle matrix elements. The case of two-particle operators is slightly more complicated since we do not obtain directly the result in terms of two-particle matrix elements. Let us consider matrix elements of $\sum_{i<k} g_{ik}$ taken between two wave functions (26.14)

$$\langle j^n \alpha\, J M \Big| \sum_{i<k}^{n} g_{ik} \Big| j^n \alpha' J' M' \rangle$$

$$= \sum_{\alpha_1 J_1 \alpha_1' J_1'} [j^n \alpha\, J \{| j^{n-1}(\alpha_1 J_1) j\, J] [j^{n-1}(\alpha_1' J_1') j\, J' |\} j^n \alpha' J']$$

$$\times \langle j^{n-1}(\alpha_1 J_1) j_n J\, M \Big| \sum_{i<k}^{n} g_{ik} \Big| j^{n-1}(\alpha_1' J_1') j_n J' M' \rangle. \qquad (26.36)$$

We take g_{ik} to be a *scalar* operator and thus, its only nonvanishing matrix elements are diagonal in J and M. Because of the symmetry of $\sum g_{ik}$ and of the product of two antisymmetric wave functions, all $[n(n-1)/2]$ terms g_{ik} contribute equally to (26.36). However, unlike the previous case, it is not possible to choose some of the g_{ik} so that integration over the first $n-1$ particles could be carried out. Instead, we can choose the g_{ik} which *do not* operate on the coordinates of the nth particle and integrate (26.36) over the coordinates of *that* particle. We can thus replace the $[n(n-1)]/2$ terms of $\sum_{i<k}^{n} g_{ik}$ by the $[(n-1)(n-2)]/2$ terms of $\sum_{i<k}^{n-1} g_{ik}$, provided we multiply the latter sum by $n/(n-2)$. We thus obtain

$$\langle j^n \alpha\, J M \Big| \sum_{i<k}^{n} g_{ik} \Big| j^n \alpha' J\, M \rangle = \frac{n}{n-2} \langle j^n \alpha\, J M \Big| \sum_{i<k}^{n-1} g_{ik} \Big| j^n \alpha' J\, M \rangle$$

$$= \frac{n}{n-2} \sum_{\alpha_1 J_1 \alpha_1' J_1'} [j^n \alpha\, J \{| j^{n-1}(\alpha_1' J_1') j\, J] [j^{n-1}(\alpha_1 J_1) j\, J |\} j^n \alpha' J]$$

$$\times \langle j^{n-1}(\alpha_1 J_1) j_n J\, M \Big| \sum_{i<k}^{n-1} g_{ik} \Big| j^{n-1}(\alpha_1' J_1') j_n J\, M \rangle.$$

The integration over the coordinates of the nth particle can now be carried out. Since $\sum_{i<k}^{n-1} g_{ik}$ is a scalar operator, its matrix elements do not vanish only if $J_1 = J_1'$ and are independent of M_1. The final result is therefore

$$\langle j^n \alpha\, J M \Big| \sum_{i<k}^{n} g_{ik} \Big| j^n \alpha' J\, M \rangle$$

$$= \frac{n}{n-2} \sum_{\alpha_1 \alpha_1' J_1} [j^n \alpha\, J \{| j^{n-1}(\alpha_1 J_1) j\, J] [j^{n-1}(\alpha_1' J_1) j\, J |\} j^n \alpha' J]$$

$$\times \langle j^{n-1} \alpha_1 J_1 M_1 \Big| \sum_{i<k}^{n-1} g_{ik} \Big| j^{n-1} \alpha_1' J_1 M_1 \rangle. \qquad (26.37)$$

Equation (26.37) gives the expression of matrix elements of two-particle operators in the n-particle configuration as linear combinations of such matrix elements in the $n-1$ particle configuration. We can proceed in this reduction from $n-1$ to $n-2$ particles, etc., until we finally arrive at the two-particle configuration. In this way we obtain matrix elements in the n-particle configuration as linear combinations of expectation values in the two-particle configuration.

A simple application of (26.37) can be used in order to check numerical values of c.f.p. For this purpose we take a two-body operator whose matrix elements are known. The simplest case is the operator

$$\sum_{i<k}^{n} 2(\mathbf{j}_i \cdot \mathbf{j}_k) = \Big(\sum_{i}^{n} \mathbf{j}_i\Big)^2 - \sum_{i}^{n} \mathbf{j}_i^2 = \mathbf{J}^2 - \sum_{i}^{n} \mathbf{j}^2 = J(J+1) - nj(j+1). \tag{26.38}$$

Since this operator is diagonal in our scheme, we obtain from (26.37), using (26.15) the following expression

$$J(J+1) + \frac{n}{n-2} j(j+1) = \frac{n}{n-2} \sum_{\alpha_1 J_1} [j^{n-1}(\alpha_1 J_1) j\, J | \} j^n \alpha\, J]^2\, J_1(J_1+1). \tag{26.39}$$

We shall now obtain matrix elements of Σg_{ik} directly in terms of those in the two-particle configuration. To do it we have to express the antisymmetric wave functions in a form where two, rather than one, particles are separated from the rest. Starting from (26.14) we can use, c.f.p. in the j^{n-1} configuration to separate the $(n-1)$ particle from the others

$$\psi(j^n \alpha\, J\, M) = \sum_{\alpha_2 J_2 \alpha_1 J_1} [j^{n-2}(\alpha_2 J_2) j\, J_1 | \} j^{n-1} \alpha_1 J_1]\, [j^{n-1}(\alpha_1 J_1) j\, J | \} j^n \alpha\, J] \times \psi[j^{n-2}(\alpha_2 J_2) j_{n-1}(J_1) j_n J\, M].$$

We now recouple j_{n-1} and j_n to J' using the transformation

$$\psi[j^{n-2}(\alpha_2 J_2) j_{n-1}(J_1) j_n J\, M] = \sum_{J'} \langle J_2 j_{n-1}(J_1) j_n J | J_2 j_{n-1} j_n(J') J \rangle\, \psi[j^{n-2}(\alpha_2 J_2), j_{n-1} j_n(J') J\, M]. \tag{26.40}$$

In the new wave functions $\psi[j^{n-2}(\alpha_2 J_2) j^2(J') J\, M]$, the first $n-2$ particles are coupled to form the state $\alpha_2 J_2$, $\mathbf{j}_{n-1}$ and $\mathbf{j}_n$ are coupled to $\mathbf{J}'$ and $\mathbf{J}_2$ and $\mathbf{J}'$ are coupled to form the total $\mathbf{J}$. The transformation coefficients in (26.40) are very similar to (15.30). They can be easily obtained as a special case of (14.37) and are given in the Appendix to be

$$\langle J_2 j(J_1) j\, J | J_2, j\, j(J') J \rangle = (-1)^{2j+J_2+J} \sqrt{(2J_1+1)(2J'+1)} \begin{Bmatrix} J_2 & j & J_1 \\ j & J & J' \end{Bmatrix}. \tag{26.41}$$

Carrying out this transformation we obtain

$$\psi(j^n\alpha\, J\, M) = \sum_{\alpha_2 J_2 \alpha_1 J_1 J'} [j^{n-2}(\alpha_2 J_2)\, j\, J_1 | \} j^{n-1}\alpha_1 J_1]\, [j^{n-1}(\alpha_1 J_1)\, j\, J | \} j^n\alpha\, J]$$

$$\times \langle J_2 j(J_1)\, j\, J | J_2, j j(J')J\rangle\, \psi[j^{n-2}(\alpha_2 J_2)\, j^2_{n-1,n}(J')J\, M]. \qquad (26.42)$$

It is clear that all terms in (26.42) with different values of $\alpha_2 J_2$ or J' are orthogonal to each other. Since $\psi(j^n\alpha JM)$ is antisymmetric in all particle coordinates, including the $(n-1)$th and n particles, it follows that in (26.42) only wave functions with allowed (i.e. even) values of J' are multiplied by nonvanishing coefficients. In (26.42) we have the expression of an antisymmetric wave function as a linear combination of terms in which the last *two* particles are singled out and coupled together. The expansion coefficients thus obtained are normalized and are also called coefficients of fractional parentage. They are denoted by a similar notation

$$\psi(j^n\alpha\, J\, M) = \sum_{\alpha_2 J_2, J'} [j^{n-2}(\alpha_2 J_2)\, j^2(J')J | \} j^n\alpha\, J]\, \psi[j^{n-2}_{1,\ldots,n-2}(\alpha_2 J_2)\, j^2_{n-1,n}(J')J\, M]. \qquad (26.43)$$

Equation (26.42) gives the expression of these $n \to n-2$ c.f.p. in terms of products of ordinary c.f.p. and Racah coefficients. Using (26.41) we obtain explicitly

$$[j^{n-2}(\alpha_2 J_2)\, j^2(J')J | \} j^n\alpha\, J] = \sum_{\alpha_1 J_1} [j^{n-2}(\alpha_2 J_2)\, j\, J_1 | \} j^{n-1}\alpha_1 J_1]\, [j^{n-1}(\alpha_1 J_1)\, j\, J | \} j^n\alpha\, J]$$

$$\times \sqrt{(2J_1+1)(2J'+1)}\, (-1)^{J_2+J+2j} \begin{Bmatrix} J_2 & j & J_1 \\ j & J & J' \end{Bmatrix}. \qquad (26.44)$$

The matrix elements of $\sum_{i<k}^n g_{ik}$ can be now directly calculated by using the expansion (26.43). We simply evaluate the contribution of $g_{n-1,n}$ and multiply it by $[n(n-1)]/2$. In the matrix elements we integrate over all particle coordinates with $i, k \leq n-2$, and due to the fact that g_{ik} is a scalar operator, we obtain

$$\langle j^n\alpha\, J\, M \left| \sum_{i<k}^n g_{ik} \right| j^n\alpha' J\, M\rangle = \frac{n(n-1)}{2} \sum_{\alpha_2 J_2 J'} [j^n\alpha\, J \{ | j^{n-2}(\alpha_2 J_2)\, j^2(J')J]$$

$$\times [j^{n-2}(\alpha_2 J_2)\, j^2(J')J | \} j^n\alpha' J]\, \langle j^2 J' M' | g_{n-1,n} | j^2 J' M'\rangle. \qquad (26.45)$$

Equation (26.45) is the desired relation between matrix elements in the n-particle configuration and those in the two-particle configuration.

In the cases of interest, g_{ik} is the sum of scalar products of two irreducible tensors operating on the coordinates of the i- and k-particles. In this case the calculation of matrix elements of Σg_{ik} can be transformed into a calculation of matrix elements of single particle operators. Let us consider the case in which $g_{12} = (\mathbf{f}_1^{(k)} \cdot \mathbf{f}_2^{(k)})$ where $\mathbf{f}^{(k)}$ is an irreducible tensor operator of degree k. In the evaluation of matrix elements of $\Sigma_{i<k} g_{ik}$ within the j^n configuration, we actually take matrix elements of all the $f_i^{(k)}$ between single particle states with the same j. In fact, consider for instance the matrix element

$$\langle j^n\alpha\, J\, M| f_\kappa^k(1) f_{-\kappa}^{(k)}(2) | j^n\alpha' J\, M\rangle$$

$$\sum_{\alpha'' J' M'} \langle j^n\alpha\, J\, M| f_\kappa^{(k)}(1) | j^n\alpha'' J' M'\rangle \langle j^n\alpha'' J' M'| f_{-\kappa}^{(k)}(2) | j^n\alpha' J\, M\rangle.$$

The configuration of the intermediate states is the *same j^n configuration* as that of the initial and final states. The operator $f_\kappa^{(k)}(1)$ could have changed the orbit of particle 1 leading to a different configuration. However, the operator $f_{-\kappa}^{(k)}(2)$, operating on *another* particle could not then connect this different configuration with the original one. We can therefore replace $f_\kappa^{(k)}$ by $(j||\mathbf{f}^{(k)}||j) u_\kappa^{(k)}$ where the unit tensor operator $\mathbf{u}^{(k)}$ was defined in (21.54) by $(j||\mathbf{u}^{(k)}||j') = \delta_{jj'}$. Using this device we can express Σg_{ih} within the j^n configuration by

$$\sum_{i<h}^{n} g_{ih} = \sum_{i<h}^{n} (\mathbf{f}_i^{(k)} \cdot \mathbf{f}_h^{(k)}) = (j||\mathbf{f}^{(k)}||j)^2 \sum_{i<h}^{n} (\mathbf{u}_i^{(k)} \cdot \mathbf{u}_h^{(k)})$$

$$= (j||\mathbf{f}^{(k)}||j)^2\, \tfrac{1}{2} \left\{ \left(\sum_{i=1}^{n} \mathbf{u}_i^{(k)} \cdot \sum_{h=1}^{n} \mathbf{u}_j^{(k)} \right) - \sum_{i=1}^{n} (\mathbf{u}_i^{(k)} \cdot \mathbf{u}_i^{(k)}) \right\}$$

$$= (j||\mathbf{f}^{(k)}||j)^2\, \tfrac{1}{2} \left\{ (\mathbf{U}^{(k)} \cdot \mathbf{U}^{(k)}) - n \frac{1}{2j+1} \right\}. \tag{26.46}$$

In this expression we have used the definition $\mathbf{U}^{(k)} = \Sigma_{i=1}^{n} \mathbf{u}_i^{(k)}$ and (21.54). Equation (26.46) can be used to obtain matrix elements of two-particle operators in terms of those of the single particle operators $\mathbf{U}^{(k)}$. The use of the unit tensors in the derivation of (26.46) is intended to stress the fact that it is not the operator $(\mathbf{F}^{(k)} \cdot \mathbf{F}^{(k)})$ [where $\mathbf{F}^{(k)} = \Sigma_i^n \mathbf{f}_i^{(k)}$] which appears in (26.46). Similarly, it is generally *not true* that

$$(\mathbf{f}^{(k)} \cdot \mathbf{f}^{(k)}) = (j||\mathbf{f}^{(k)}||j)^2 \frac{1}{2j+1}.$$

The operators $\mathbf{f}^{(k)}$ and $\mathbf{F}^{(k)}$ may generally have nonvanishing matrix elements between the j^n configuration and other configurations as well as between states in the j^n configuration. In the multiplication of $U_\kappa^{(k)}$ and $U_{-\kappa}^{(k)}$, in (26.46), the intermediate states are automatically restricted to the j^n configuration.

27. Pairs of Particles Coupled to $J = 0$

The interaction energy in the j^3 configuration in the state with $J = j$ can be easily calculated using the results of the preceding section. If there are several such states we shall refer to the particular state (26.13) built from the principal parent with $J_0 = 0$. Inserting the c.f.p. (26.12) in (26.37) we obtain the result

$$\langle j^3 J = j \left| \sum_{i<k}^{3} V_{ik} \right| j^3 J = j \rangle = 3 \left\{ \frac{2j-1}{3(2j+1)} \langle j^2 0 | V_{12} | j^2 0 \rangle \right.$$

$$\left. + \frac{4}{3(2j+1)(2j-1)} \sum_{J_1 > 0} (2J_1 + 1) \langle j^2 J_1 | V_{12} | j^2 J_1 \rangle \right\}. \tag{27.1}$$

Since the matrix elements of $\sum V_{ik}$ are independent of M, we avoid explicit reference to it.

It is interesting to evaluate (27.1) for some simple interactions considered in Part II. We saw there that in the case of odd tensor interactions, there are simple relations between expectation values in the various states. These relations, which hold also for short range δ-type interactions, were given by (21.44)

$$\sum_{J_1 > 0,\ \text{even}} (2J_1 + 1) \langle j^2 J_1 | V_{12} | j^2 J_1 \rangle = \frac{2j-1}{2} \langle j^2 0 | V_{12} | j^2 0 \rangle.$$

Using this relation in (27.1) we obtain for odd tensor interactions the simple result

$$\langle j^3 J = j \left| \sum_{i<k}^{3} V_{ik} \right| j^3 J = j \rangle = \langle j^2 0 | V_{12} | j^2 0 \rangle = V_0. \tag{27.2}$$

This means that for such interactions the energy in this specific state of the j^3 configuration is equal to the interaction energy in the $J = 0$ state of the j^2 configuration. We shall say that an interaction V_{ik} has the *pairing property* if it satisfies (27.2) (a more general definition will be given below). For such interactions, the interaction energy in the state $\psi(j^3[J_0 = 0] J M)$ seems to be entirely due to the pair coupled to $J = 0$. The third, unpaired particle does not contribute anything to it.

Explicit calculations in each case, as we shall soon see, show that, if in the j^n configuration, with n even, there is only one state with $J = 0$, then for odd tensor interactions

$$\langle j^n\, J = 0 \left| \sum_{i<k}^{n} V_{ik} \right| j^n\, J = 0 \rangle = \frac{n}{2} V_0 \qquad n \text{ even.} \tag{27.3}$$

Similarly, in j^n configurations with n odd, if there is only one state with $J = j$, then for such interactions

$$\langle j^n\, J = j \left| \sum_{i<k}^{n} V_{ik} \right| j^n\, J = j \rangle = \frac{n-1}{2} V_0 \qquad n \text{ odd.} \tag{27.4}$$

It seems as if, for odd tensor interactions, the last odd nucleon, the unpaired one, does not contribute to the interaction energy. Thus, the expectation values of odd tensor interactions measure, in some sense, the number of pairs coupled to $J = 0$. For example, in the j^4 configuration, there are sometimes *two* states with $J = 0$. One of them may be obtained from two pairs each coupled to $J = 0$ and thus satisfy (27.3). The other state is obtained by coupling two pairs, each coupled to a $J_1 \neq 0$. This way we could attempt to distinguish between various states with the same value of J according to the number of pairs coupled to $J = 0$ in each of them.

A more precise definition of the proposed scheme of states can be obtained by requiring that an interaction which has the pairing property (odd tensor interaction) be diagonal in our scheme. Obviously, if there is only one state with a given J, the expectation value of this interaction, as well as any other interaction in this state, will be an eigenvalue of the interaction matrix (limited to the given configuration). However, if there are several states with the same value of J, this requirement defines a certain scheme. The eigenvalues of one specific interaction which is diagonal in this scheme may serve as additional quantum numbers for the various states which have the same value of J.

In order to prove (27.3) and (27.4) and define their generalizations, we shall use the identity (26.46) which gives the interaction energy in terms of the single particle operators $\mathbf{U}^{(k)}$. It is clear that if $U_\kappa^{(k)}$ itself is diagonal in the scheme which we are trying to define, the same will hold for $(\mathbf{U}^{(k)} \cdot \mathbf{U}^{(k)})$. Thus, we shall consider single particle operators $\mathbf{U}^{(k)} = \sum \mathbf{u}_i^{(k)}$ where k is an odd number. The specific form of $\mathbf{u}^{(k)}$ is of no importance and we may use a linear combination of any single particle tensor operators provided they are all of odd degrees. The important property of odd tensor operators is that they satisfy the following equation

$$\langle j^2 J\, M | u_\kappa^{(k)}(1) + u_\kappa^{(k)}(2) | j^2 0\, 0 \rangle = 0. \tag{27.5}$$

Thus, the matrix elements of $u_\kappa^{(k)}(1) + u_\kappa^{(k)}(2)$, for odd k, between the state $J = 0$ ($M = 0$) and *any* other state of the j^2 configuration are all zero. It is clear that according to the Wigner-Eckart theorem, the matrix elements in (27.5) vanish unless $J = k$ $M = \kappa$. However, since only even values of J are allowed by the Pauli principle, it is clear that this condition cannot hold if k is odd and this proves (27.5).

As a matter of fact, (27.5) is even more general and holds even if the two j-particles are not identical so that the Pauli principle does not limit J to even values. This is so because a state of the j^2 configuration with an odd value of J is *symmetric* under a permutation of the particles. The same is, of course, true of the tensor $\mathbf{u}^{(k)}(1) + \mathbf{u}^{(k)}(2)$. On the other hand, the state j^2 with $J = 0$ is antisymmetric. Therefore, the matrix element in (27.5) must vanish. The property (27.5) is obviously shared by any linear combination of odd tensor operators. It is worthwhile to mention that *even* unit tensors have always nonvanishing matrix elements of the form (27.5). In fact, using (15.26) and (15.27) we readily obtain

$$(j^2J||\mathbf{u}^{(k)}(1) + \mathbf{u}^{(k)}(2)||j^20) = \frac{2}{\sqrt{2j+1}} \qquad \text{for even} \qquad k = J.$$

We can interpret (27.5) by saying that there is no contribution to $\mathbf{U}^{(k)}$ from pairs coupled to $J = 0$. In order to express this property in the j^n configuration, we shall use $n \to n - 2$ c.f.p. Let us start from $\psi(j^{n-2}\alpha\ J\ M)$ and multiply it by the wave function of two additional particles coupled to $J = 0$. The total wave function of n particles obtained this way, $\psi[j^{n-2}_{1,\ldots,n-2}(\alpha\ J)j^2_{n-1,n}(0)J\ M]$, has the same value of J but is not antisymmetric. The antisymmetrization of this wave function is a projection operator. It projects the space of wave functions $\psi[j^{n-2}_{1,\ldots,n-2}(\alpha_1 J_1)j^2_{n-1,n}(J_2)J\ M]$ which are antisymmetric in the first $n - 2$ particles and in the last two particles on its subspace of fully antisymmetric wave functions. As in the case of ordinary $n \to n - 1$ c.f.p., we use the antisymmetrizing operator $\mathscr{A}$ which is a sum $\Sigma_P\,(-1)^P P$ over permutations P which exchange one or both of the last two particles with one or two of the others. There are altogether $[n(n-1)]/2$ such permutations, and therefore the projection operator $\mathscr{P}$, which satisfies $\mathscr{P}^2 = \mathscr{P}$ is equal in this case to $2\mathscr{A}/[n(n-1)]$.

The matrix elements of $U_\kappa^{(k)}$ can now be calculated between the normalized antisymmetric wave function $N_{\alpha J}\mathscr{A}\psi[j^{n-2}(\alpha\ J)j^2_{n-1,n}(0)J\ M]$ and another normalized antisymmetric wave function $\psi(j^n\alpha'\ J'M')$. Assuming that neither of these wave functions vanishes identically, we obtain

$$\int \psi^*(j^n\alpha'\ J'M')\left[\sum_{i=1}^{n} u_\kappa^{(k)}(i)\right] N_{\alpha J}\mathscr{A}\ \psi[j^{n-2}(\alpha\ J)\,j^2_{n-1,n}(0)J\ M]\ d(1)\ \ldots\ d(n). \quad (27.6)$$

We can now use the fact that $\mathscr{A}$ is a projection operator and evaluate (27.6) by shifting it to the other part of the integrand. Since $\sum u_\kappa^{(k)}(i)$ is symmetric under any permutation of the particle coordinates, and $\psi^*(j^n\alpha' J'M')$ is antisymmetric, the operation of $\mathscr{A}$ on their product is to multiply it by $[n(n-1)]/2$. We thus obtain the matrix element (27.6) in the form

$$\frac{n(n-1)}{2} N_{\alpha J} \int \psi^*(j^n\alpha' J'M') \left[\sum_{i=1}^{n} u_\kappa^{(k)}(i)\right] \psi[j^{n-2}(\alpha\, J) j^2_{n-1,n}(0) J\, M]\, d(1) \ldots d(n). \tag{27.7}$$

For tensors of odd degree k, (27.5) implies

$$\int \psi^*(j^n\alpha' J'M') [u_\kappa^{(k)}(n-1) + u_\kappa^{(k)}(n)] \psi[j^{n-2}(\alpha\, J) j^2_{n-1,n}(0) J\, M]\, d(n-1)\, d(n) = 0$$

for any values of $\alpha' J'$ and M'. We therefore expand $\psi^*(j^n\alpha' J'M')$ in (27.7) in terms of c.f.p. and obtain

$$\frac{n(n-1)}{2} N_{\alpha J} \sum_{\alpha_1 J_1 J_2} [j^{n-2}(\alpha_1 J_1) j^2(J_2) J' | \} j^n\alpha' J'] \int \psi^*[j^{n-2}(\alpha_1 J_1) j^2_{n-1,n}(J_2) J'M']$$

$$\times \left[\sum_{i=1}^{n-2} u_\kappa^{(k)}(i)\right] \psi[j^{n-2}(\alpha\, J) j^2_{n-1,n}(0) J\, M]\, d(1) \ldots d(n). \tag{27.8}$$

Here the summation over particle numbers does not include any more the particles $n-1$ and n. We can therefore carry out the integration over the coordinates of these two particles. The integrals in (27.8) do not vanish only for $J_2 = 0$, and $J_1 = J'$. Thus, the matrix element (27.8) vanishes unless the state $\psi(j^n\alpha' J'M')$ has a nonvanishing c.f.p. with $J_2 = 0$, namely, unless $[j^{n-2}(\alpha_1 J') j^2(0) J' | \} j^n\alpha' J'] \neq 0$. The nonvanishing integrals in (27.8) are thus matrix elements of $\mathbf{U}^{(k)}$ in the j^{n-2} configuration.

Specifying antisymmetric wave functions by the quantum numbers of the principal parents (written as before in square brackets) we can write (27.8) for odd tensors in the following form

$$\langle j^n\alpha' J'M' | \sum_{i}^{n} u_\kappa^{(k)}(i) | j^n[\alpha\, J\, 0] J\, M\rangle$$

$$= \sum_{\alpha_1} \langle j^{n-2}\alpha_1 J'M' | \sum_{i}^{n-2} u_\kappa^{(k)}(i) | j^{n-2}\alpha\, J\, M\rangle$$

$$\times \frac{n(n-1)}{2} N_{\alpha J} [j^{n-2}(\alpha_1 J') j^2(0) J' | \} j^n\alpha' J']. \tag{27.9}$$

If all the c.f.p. on the right-hand side of (27.9) vanish, $\psi(j^n\alpha' J'M')$ cannot be obtained from a principal parent $\psi(j^{n-2}J'M')$ by adding a pair coupled to $J = 0$ and antisymmetrizing. If there is a nonvanishing c.f.p. in (27.9), we can try to repeat the reduction of the matrix elements in the j^{n-2} configuration. We first find out by looking at the $n - 2 \rightarrow n - 4$ c.f.p. whether the wave functions $\psi(j^{n-2}\alpha\, JM)$ and $\psi(j^{n-2}\alpha_1 J'M')$ contain pairs with $J = 0$. If one of them does contain a pair $J = 0$, we can use again (27.9) and so on.

On the basis of this discussion, we can now specify the scheme in which odd tensor operators $\mathbf{U}^{(k)}$ are diagonal. For any given J, we look for the j^n configuration with the lowest value of n in which a state with this value of J appears. In this state of the smallest configuration there are no pairs coupled to $J = 0$. Otherwise there would have been a state with the same J in a configuration containing $n - 2$ particles. Let v be the number of particles in this "minimal configuration."

The state $\psi(j^v\alpha J)$ is said to have the seniority v if

$$[j^{v-2}(\alpha_1 J)j^2(0)J|\}j^v\alpha\, J] = 0 \quad \textit{for every } \alpha_1. \tag{27.10}$$

We can then write v as an additional quantum number. We now add to this state of the j^v configuration a pair coupled to $J = 0$ and antisymmetrize. We obtain an antisymmetric state with the same value of J in the j^{v+2} configuration. This state will also be said to have the seniority v. We can go on adding more pairs coupled to $J = 0$, and obtain after antisymmetrization corresponding states in the configuration j^v, j^{v+2}, j^{v+4}, etc. All these states have the same value of J and are defined to have the same seniority v. Thus,

The state $\psi(j^n J)$ will be said to have the seniority v if it can be obtained by adding successively pairs coupled to $J = 0$ to the principal parent $\psi(j^v v\, J)$ [defined by (27.10)] and antisymmetrizing. (27.11)

The notation v, introduced by Racah, comes from the Hebrew word for seniority, "vethek." We may note that, somewhat paradoxically, the "senior" of two states has a smaller seniority number v.

Having introduced the seniority scheme we should discuss its consistency. We shall now show that any two wave functions with different seniorities are orthogonal. We first observe that the wave function $\psi(j^v v\, J)$ is orthogonal to every wave function $\psi(j^v\alpha\, J)$ with lower seniority. In fact, such a wave function $\psi(j^v\alpha\, J)$ must be obtained, by definition, from a principal parent $\psi[j^{v-2}(\alpha_1 J)j^2(0)J]$. The integral

$$\int \psi^*(j^v v\, J)\mathscr{A}\, \psi[j^{v-2}(\alpha_1 J)\, j^2_{v-1,v}(0)J] \tag{27.12}$$

can be brought into the form

$$\frac{v(v-1)}{2}\int \psi^*(j^v v\, J)\,\psi[j^{v-2}(\alpha_1 J)\,j^2_{v-1,v}(0)J] = \frac{v(v-1)}{2}\,[j^{v-2}(\alpha_1 J)\,j^2(0)J|\}j^v v J] \tag{27.13}$$

by using the properties of the antisymmetrizer $\mathscr{A}$. The c.f.p. in (27.13) must however vanish by the definition (27.10) which proves the vanishing of (27.12) and the orthogonality.

Next we prove the theorem that given any two orthogonal functions in the j^n configuration, the wave functions obtained from them by adding to each a pair coupled to $J = 0$ and antisymmetrizing are also orthogonal. Let us therefore calculate the integral

$$\int \mathscr{A}\psi^*[j^n(\alpha' J)\,j^2_{n+1,n+2}(0)J]\,\mathscr{A}\psi[j^n(\alpha\, J)\,j^2_{n+1,n+2}(0)J]. \tag{27.14}$$

Using the properties of the antisymmetrizer $\mathscr{A}$, this integral can be written as

$$\frac{(n+1)(n+2)}{2}\int \psi^*\,[j^n(\alpha' J)\,j^2_{n+1,n+2}(0)J]\,\mathscr{A}\psi[j^n(\alpha\, J)\,j^2_{n+1,n+2}(0)J]. \tag{27.15}$$

We carry out the antisymmetrization in (27.15) explicitly and obtain

$$\begin{aligned}\mathscr{A}\psi[j^n(\alpha\, J)\,j^2_{n+1,n+2}(0)J] &= \psi[j^n(\alpha\, J)\,j^2_{n+1,n+2}(0)J]\\ &\quad -\sum_{k\neq n+1,n+2}\psi\,[j^n(\alpha\, J)\,j^2_{k,n+2}(0)J] - \sum_{k\neq n+1,n+2}\psi[j^n(\alpha\, J)\,j^2_{n+1,k}(0)J]\\ &\quad +\sum_{i<k\neq n+1,n+2}\psi[j^n(\alpha\, J)\,j^2_{ik}(0)J].\end{aligned} \tag{27.16}$$

Since particles $n+1$ and $n+2$ are coupled together in $\psi^*[j^n(\alpha' J)j_{n+1,n+2}(0)J]$, it is convenient to transform (27.16) and obtain functions of the type $\psi[j^{n-2}(\alpha_1 J_1)j^2(J')j^2(0)J]$ in which these two particles occur in either one of the pairs $j^2(J')$ or $j^2(0)$. To do this we apply a proper permutation P_k or P_{ik} on the functions appearing on the right-hand side of (27.16) with the proper phase, which brings particles $n+1$ and $n+2$ to the last places in $\psi(j^n\alpha\, J)$. We then expand in $n \to n-2$ c.f.p. and obtain

$$\begin{aligned}&\mathscr{A}\psi[j^n(\alpha\, J)\,j^2_{n+1,n+2}(0)J]\\ &= \psi[j^n(\alpha\, J)\,j^2_{n+1,n+2}(0)J] - \sum_k(-1)^{P_k}\sum_{\alpha_1 J_1 J'}[j^{n-2}(\alpha_1 J_1)\,j^2(J')J|\}j^n\alpha\, J]\\ &\quad\times\{\psi[j^{n-2}(\alpha_1 J_1)\,j^2_{p,n+1}(J')\,j^2_{k,n+2}(0)J] + \psi[j^{n-2}(\alpha_1 J_1)\,j^2_{p,n+2}(J')\,j^2_{n+1,k}(0)J]\}\\ &\quad+\sum_{i<k}(-1)^{P_{ik}}\sum_{\alpha_1 J_1 J'}[j^{n-2}(\alpha_1 J_1)\,j^2(J')J|\}j^n\alpha\, J]\,\psi[j^{n-2}(\alpha_1 J_1)\,j^2_{n+1,n+2}(J')\,j^2_{ik}(0)J].\end{aligned} \tag{27.17}$$

The index p in the second sum in (27.17) is equal to n if $k < n$ and if $k = n$, p is equal to $n - 1$. The first term and the last sum of terms in (27.17) have already the desired property.

To the terms of the second sum we apply a change of coupling from $\psi[j^2_{p,n+1}(J')j^2_{k,n+2}(0)J']$ to $\psi[j^2_{pk}(J'')j^2_{n+1,n+2}(J''')J']$. However, only the term with $J''' = 0$ will contribute to the integral (27.15) and therefore the only transformation coefficients involved are given by (14.38) and (15.3) to be simply

$$\langle j_p j_{n+1}(J') j_k j_{n+2}(0)J' | j_p j_k(J') j_{n+1} j_{n+2}(0)J' \rangle$$
$$= \langle j_p j_{n+2}(J') j_{n+1} j_k(0)J' | j_p j_k(J') j_{n+1} j_{n+2}(0)J' \rangle = \frac{1}{2j+1}.$$

After carrying out this transformation we can integrate in (27.15) over the coordinates of particles $n + 1$ and $n + 2$. We thus obtain for the integral (27.15) the form

$$\frac{(n+1)(n+2)}{2}\Big\{\int \psi^*(j^n\alpha' J)\psi(j^n\alpha J) - \frac{2}{2j+1}\sum_k (-1)^{P_k}\int \psi^*(j^n\alpha' J)$$
$$\times \sum_{\alpha_1 J_1 J'} [j^{n-2}(\alpha_1 J_1)j^2(J')J|\}j^n\alpha J]\,\psi[j^{n-2}(\alpha_1 J_1)j^2_{pk}(J')J]$$
$$+ \sum_{i<k}(-1)^{P_{ik}}\sum_{\alpha_1}[j^{n-2}(\alpha_1 J)j^2(0)J|\}j^n\alpha J]\int \psi^*(j^n\alpha' J)\,\psi[j^{n-2}(\alpha_1 J)j^2_{ik}(0)J]\Big\}. \tag{27.18}$$

We recall that

$$(-1)^{P_{ik}}\int \psi^*(j^n\alpha' J)\,\psi[j^{n-2}(\alpha_1 J)j^2_{ik}(0)J] = [j^{n-2}(\alpha_1 J)j^2(0)J|\}j^n\alpha' J]$$

and that

$$(-1)^{P_k}\sum_{\alpha_1 J_1 J'}[j^{n-2}(\alpha_1 J_1)j^2(J')J|\}j^n\alpha J]\,\psi[j^{n-2}(\alpha_1 J_1)j^2_{pk}(J')J] = \psi(j^n\alpha J).$$

Using these relations we can cary out the integration in (27.18) and thus obtain for the integral (27.14) the final form

$$\frac{(n+1)(n+2)}{2}\Big\{\delta_{\alpha\alpha'} - \frac{2n}{2j+1}\delta_{\alpha\alpha'} + \frac{n(n-1)}{2}$$
$$\times \sum_{\alpha_1}[j^{n-2}(\alpha_1 J)j^2(0)J|\}j^n\alpha J]\,[j^{n-2}(\alpha_1 J)j^2(0)J|\}j^n\alpha' J]\Big\}. \tag{27.19}$$

We shall now prove the orthogonality by induction using the expression (27.19). We assume that our theorem holds for principal parents in the j^{n-2} configuration and proceed to show that it holds then also in the j^n configuration. We first see that if the state α has seniority v and the α'-state has seniority v' then (27.19), as well as (27.14), vanish in the case $n = v$ (or $n = v'$) provided $\alpha \neq \alpha'$. In this case, according to the definition (27.10) all c.f.p. $[j^{v-2}(\alpha_1 J)j^2(0)J|\}j^v\alpha\ J]$ vanish. If, however, n is bigger than both v and v', then both the α-state and the α'-state must be obtained by adding pairs coupled to $J = 0$ to principal parents in the j^{n-2} configuration. Thus, we can write

$$\psi(j^n\ \alpha\ J) = N\mathscr{A}\psi(j^{n-2}(\alpha_0\ J)\ j^2_{n-1,n}(0)J).$$

Let us choose the complete set of orthonormal states $\psi(j^{n-2}\ \alpha_1\ J)$ in the j^{n-2} configuration, so that $\psi(j^{n-2}\ \alpha_0\ J)$ will be one of its functions. Since $\psi(j^{n-2}\ \alpha_0\ J)$ is orthogonal to all other $\psi(j^{n-2}\ \alpha_1\ J)$, it follows from the induction assumption that the wave function $\psi(j^n\ \alpha\ J)$ is orthogonal to all the functions $\mathscr{A}\psi(j^{n-2}\ (\alpha_1\ J)\ j^2_{n-1,n}(0)J)$ with $\alpha_1 \neq \alpha_0$. Therefore, due to (27.14) and (27.15), all c.f.p. $[j^{n-2}(\alpha_1\ J)j^2(0)J|\}\ j^n\ \alpha\ J]$ vanish unless $\alpha_1 \equiv \alpha_0$. On the other hand, $\psi(j^n\ \alpha'\ J)$ being orthogonal to $\psi(j^n\ \alpha\ J)$ is orthogonal also to its principal parent and therefore also $[j^{n-2}\ (\alpha_0\ J)\ j^2(0)J|\}\ j^n\ \alpha'\ J]$ vanishes. Thus, the summation in (27.19), and therefore also (27.14), vanishes if $\alpha' \neq \alpha$. Since the theorem holds for $n = v$ (or $n = v'$), it holds for every n.

We thus see that any two wave functions with different seniorities are orthogonal to each other. It is also true that the seniority scheme is complete. To prove the completeness we have to show that every wave function can be expanded in terms of wave functions with definite seniorities. In the j^n configuration we have to consider only wave functions orthogonal to all states with seniorities $v < n$. Such wave functions $\psi(j^n\alpha\ J)$ are also orthogonal to all principal parents $\psi[j^{n-2}(\alpha_1 J)j^2(0)J]$ and therefore all their c.f.p. $[j^{n-2}(\alpha_1 J)j^2(0)J|\}j^n\alpha\ J]$ vanish. According to the definition (27.10), such wave functions $\psi(j^n\alpha\ J)$ have seniority $v = n$.

The seniority scheme introduces an additional quantum number v for the classification of states. If there are several states with the same value of J in the j^n configuration, they may be distinguished by their seniorities. The seniority v can therefore replace sometimes the additional quantum numbers α. Still it may happen that in the j^v configuration there are several states with the same value of J and the seniority v. We shall have to distinguish between these states by some quantum number α, in addition to v, and denote them by $\psi(j^v v\ \alpha\ J)$. If the states $\psi(j^v v\ \alpha\ J)$ with different values of α are orthogonal, then, according to (27.19), also the functions obtained by adding pairs coupled to $J = 0$ and antisymmetrizing will be orthogonal. Thus, any orthogonal scheme of wave functions in the j^v configuration, with the same values of J and v, $\psi(j^v v\ \alpha\ J)$, defines uniquely an orthogonal scheme of wave functions $\psi(j^n v\ \alpha\ J)$ also in the j^n configuration. The same additional quantum numbers α defined in the j^v configuration can be used also in the j^n configuration. Therefore, when we shall discuss the relations between states of various configurations with the same values of v and J, we shall also assume that α is the same and often avoid explicit reference to this additional quantum number.

If the j^v configuration contains only one state with a given J and the j^n configuration contains two such states it is easy to build a state orthogonal to $\psi(j^n v\ J)$. Since such a state must have a different seniority, it cannot have a nonvanishing c.f.p. $[j^{n-2}(v\ J)j^2(0)J|\}j^n v'J]$. If we antisymmetrize two functions $\psi[j^{n-2}(\alpha_1 J_1)j^2(J_2)J]$ and $\psi[j^{n-2}(\alpha_1' J_1')j^2(J_2')J]$ we may obtain antisymmetric functions with nonvanishing c.f.p. $[j^{n-2}(v\ J)j^2(0)J|\}j^n J]$. A suitable linear combination of these may be chosen which does not have such a nonvanishing c.f.p. Such a linear combination is the desired wave function which is orthogonal to $\psi(j^n v\ J)$.

If there is only one state with a given J and a given seniority v, we can write v instead of α in order to specify the states completely. In this case, there is in the expansion of $\psi(j^n v\ J\ M)$ at most one nonvanishing c.f.p. $[j^{n-2}(v'J)j^2(0)J|\}j^n v\ J]$, namely the one for which $v' = v$. As discussed above, this is also the case if there are several states with the same values of v and J. For a given value of the additional quantum number α there is only one nonvanishing c.f.p. $[j^{n-2}(v'\alpha'J)j^2(0)J|\}j^n v\ \alpha\ J]$, namely the one for which $v' = v$ and $\alpha' \equiv \alpha$. For convenience of the notation we shall omit the α in the following. A successive application of the relation (27.9) (with v_1 and v replacing α_1 and α) separating $J = 0$ pairs from either $\psi(j^{n-2}v'J'M')$ or $\psi(j^{n-2}v\ J\ M)$ shows that matrix elements of $\mathbf{U}^{(k)}$ with an odd k vanish between any two states with different seniorities. The matrix elements (27.9) between states of the same seniority can be written in the form

$$\langle j^n v\ J'M' \mid \sum_i^n u_\kappa^{(k)}(i) \mid j^n v\ J\ M\rangle$$

$$= \langle j^{n-2}v\ J'M' \left| \sum_i^{n-2} u_\kappa^{(k)}(i) \right| j^{n-2}v\ J\ M\rangle \frac{n(n-1)}{2} N_{nvJ}[j^{n-2}(v\ J')j^2(0)J'|\}j^n v\ J']. \tag{27.20}$$

Equation (27.20) can be further simplified. Due to the properties of the projection operators $2\mathscr{A}/[n(n-1)]$, the same methods used to derive (26.24), give the relation

$$\frac{n(n-1)}{2} N_{nvJ}[j^{n-2}(v\ J)j^2(0)J|\}j^n v\ J] = 1. \tag{27.21}$$

Equation (27.21) cannot be used directly in (27.20) since in the latter case the normalization coefficient is connected with the value J, whereas the c.f.p. is for the state with J'. However, it can be easily shown that the normalization coefficient N_{nvJ} (as well as the corresponding c.f.p.) is independent of J (and α) and is a function of n and v only. In fact, from (27.21) follows

$$N_{nvJ}[j^{n-2}(v\ J)j^2(0)J|\}j^n v\ J] = N_{nvJ'}[j^{n-2}(v\ J')j^2(0)J'|\}j^n v\ J'].$$

On the other hand, the relation (27.20) could have been obtained also with the factor $\frac{1}{2}n(n-1)\ N_{nvJ'}[j^{n-2}(v\ J)j^2(0)J|\}j^n v\ J]$. Therefore, we obtain from the relations (27.20) (provided there is one such relation with some k and κ for which the matrix element does not vanish) the equation

$$N_{nvJ'}[j^{n-2}(v\ J)j^2(0)J|\}j^n v\ J] = N_{nvJ}[j^{n-2}(v\ J')j^2(0)J'|\}j^n v\ J'].$$

Dividing the last two equations we obtain that N_{nvJ} is equal to $N_{nvJ'}$ (apart from a possible phase which we fix as $+1$) and, in general, the normalization coefficient is independent of J. In particular it follows that also for $J \neq J'$

$$\frac{n(n-1)}{2} N_{nvJ}[j^{n-2}(v\ J')j^2(0)J'|\}j^n v\ J'] = 1.$$

Substituting this relation into (27.20) we obtain

$$\langle j^n v\ J'M' \Big| \sum_{i=1}^{n} u_\kappa^{(k)}(i) \Big| j^n v\ J\,M\rangle = \langle j^{n-2} v\ J'M' \Big| \sum_{i=1}^{n-2} u_\kappa^{(k)}(i) \Big| j^{n-2} v\ J\,M\rangle. \tag{27.22}$$

Thus, odd tensor operators are diagonal with respect to the seniority and their matrix elements in the seniority scheme do not depend on n. This property is a direct result of (27.5) which states that pairs coupled to $J = 0$ do not contribute to the matrix elements of odd operators.

In order to obtain the explicit value of N_{nvJ} (or of the related c.f.p.) we can use the expression (27.19) for the integral (27.15) in the case $\alpha \equiv \alpha'$. In this case, the integral (27.15) multiplied by N_{nvJ} is equal to

$$\frac{(n+1)(n+2)}{2}[j^n(v\ J)j^2(0)J|\}j^{n+2}v\ J].$$

Using (27.21) we can write the integral (27.15) in the form

$$\left\{\frac{(n+1)(n+2)}{2}[j^n(v\ J)j^2(0)J|\}j^{n+2}v\ J]\right\}^2.$$

On the other hand, taking $\alpha \equiv \alpha'$ in the seniority scheme there is only one nonvanishing term in the summation in (27.19) which assumes then the form

$$\frac{(n+1)(n+2)}{2}\left\{1 - \frac{2n}{2j+1} + \frac{n(n-1)}{2}[j^{n-2}(v\ J)j^2(0)J|\}j^n v\ J]^2\right\}.$$

Comparing these two expressions we obtain the following recursion relation for the c.f.p. considered

$$\frac{(n+1)(n+2)}{2}[j^n(v\ J)j^2(0)J|\}j^{n+2}v\ J]^2$$

$$= 1 - \frac{2n}{2j+1} + \frac{n(n-1)}{2}[j^{n-2}(v\ J)j^2(0)J|\}j^n v\ J]^2. \qquad (27.23)$$

This relation obviously holds only for $n \geqslant v$. The first conclusion that can be drawn from (27.23) is that the c.f.p. considered are independent of α and J (which was already mentioned). In fact, for $n = v$ we obtain

$$[j^{v-2}(v\ J)j^2(0)J|\}j^v v\ J] = 0$$

so that

$$[j^v(v\ J)j^2(0)J|\}j^{v+2}v\ J]^2 = \frac{2}{(v+1)(v+2)}\left(1 - \frac{2v}{2j+1}\right)$$

is independent of J (and α). The other c.f.p., for higher n, are obtained from (27.23) whose coefficients depend only on n and v and therefore are also independent of J (and α).

We also note that (27.23) is symmetric around the middle of the shell. In fact, considering

$$f(n) = \frac{n(n-1)}{2}[j^{n-2}(v\ J)j^2(0)J|\}j^n v\ J]^2$$

as a function of n we observe that it vanishes for $n = v$, becomes $[1 - 2v/(2j+1)]$ for $n = v+2$, $[1 - 2v/(2j+1)] + [1 - 2(v+2)/(2j+1)]$ for $n = v+4$, etc., until it obtains its maximum value at, or just before, the middle of the shell [depending on whether $\frac{1}{2}(2j+1) - v$ is even or odd]. Afterward it starts to decrease monotonically so that

$$f\left[\frac{2j+1}{2} + (k+2)\right] = f\left[\frac{2j+1}{2} - k\right],$$

until it vanishes again for $n - 2 = 2j + 1 - v$. Thus

$$[j^{2j+1-v}(v\ J)j^2(0)|\}j^{2j+3-v}v\ J] = 0$$

and hence

$$\psi(j^{2j+3-v}v\ J\ M) = 0.$$

It is clear that if $\psi(j^{n_0}v\ J\ M) = 0$ for $n_0 > v$ then $\psi(j^m v\ J\ M) = 0$ also for every $m \geqslant n_0$. Hence we see that if $\psi(j^v v\ J\ M)$ is not identically zero, there will be nonvanishing states $\psi(j^n v\ J\ M)$ with the same seniority v for every n differing from v by an even number and satisfying

$v \leq n \leq 2j + 1 - v$. Configurations with $n < v$ or $n > 2j + 1 - v$ do not contain states of seniority v. If we call a configuration with $2j + 1 - n$ particles a *configuration with n holes* we can also state this result by saying that states of seniority v occur only in configurations which do not have less than v particles and less than v holes. Equation (27.23) is seen to be completely symmetric with respect to particles and holes.

The state with lowest seniority is the $J = 0$ state of the j^2 configuration. As it contains a pair coupled to $J = 0$ its seniority is, by definition, $v = 0$. It therefore must appear in all j^n configurations with even n. Thus, if there is only one $J = 0$ state in a given j^n configuration it is obviously the state with $v = 0$. The state with $v = 1$ is the $J = j$ state of a single j particle. All j^n configurations with odd n contain therefore a state with $J = j$ and $v = 1$. In the cases where there is only one $J = j$ state in the j^n configuration it is the state with $v = 1$. In the j^2 configuration there appear, besides the state with $J = 0$ $v = 0$, states with even J, $0 < J < 2j$. These do not contain $J = 0$ pairs and thus have seniority 2.

From the property (27.22) of odd tensor operators follows the pairing property of odd tensor interactions. If

$$V_{12} = \sum_{k \text{ odd}} (\mathbf{f}_1^{(k)} \cdot \mathbf{f}_2^{(k)})$$

we can apply (26.46) to the calculation of expectation values of ΣV_{ik} in states with definite seniority. Thus, we obtain from (26.46), using (27.22)

$$\langle j^n v\, J \Big| \sum_{i<k}^{n} V_{ik} \Big| j^n v\, J\rangle$$

$$= \sum_{k \text{ odd}} (j||\mathbf{f}^{(k)}||j)^2 \left\{\langle j^n v\, J|(\mathbf{U}^{(k)} \cdot \mathbf{U}^{(k)})|j^n v\, J\rangle - n\frac{1}{2j+1}\right\}$$

$$= \sum_{k \text{ odd}} (j||\mathbf{f}^{(k)}||j)\,\tfrac{1}{2}\left\{\langle j^v v\, J|(\mathbf{U}^{(k)} \cdot \mathbf{U}^{(k)})|j^v v\, J\rangle - v\frac{1}{2j+1} - (n-v)\frac{1}{2j+1}\right\}$$

$$= \langle j^v v\, J \Big| \sum_{i<k}^{v} V_{ik} \Big| j^v v\, J\rangle - \frac{n-v}{2}\frac{1}{2j+1}\sum_{k \text{ odd}} (j||\mathbf{f}^{(k)}||j)^2. \tag{27.24}$$

For the j^2 configuration, it follows from (21.37) that

$$\langle j^2\ v = 0\ J = 0|V_{12}|j^2\ v = 0\ J = 0\rangle = V_0 = -\frac{1}{2j+1}\sum_{k \text{ odd}} (j||\mathbf{f}^{(k)}||j)^2.$$

Thus, (27.24) is the general expression for the pairing property of interactions V_{ik} which contain odd tensors only (and which include also the zero range δ-type interaction)

$$\langle j^n v\alpha\, JM \Big| \sum_{i<k}^{n} V_{ik} \Big| j^n v\alpha'\, JM\rangle = \langle j^v v\alpha\, JM \Big| \sum_{i<k}^{v} V_{ik} \Big| j^v v\alpha'\, JM\rangle + \frac{n-v}{2} V_0\, \delta_{\alpha\alpha'}. \tag{27.25}$$

All matrix elements between states with different seniorities vanish for such interactions as already mentioned. For the special cases $J = 0$ $v = 0$ and $J = j$ $v = 1$, (27.25) reduces to the relations (27.3) and (27.4).

Equation (27.25) shows the usefulness of the seniority, in particular for nuclear spectra. For attractive forces, as in the nuclear case, the order of levels is such that states with lower seniority are more tightly bound. The ground states of even-even nuclei have all $J = 0$ spin and those of odd-even nuclei mostly have the spin $J = j$ of the single nucleon. At least in the case where the configuration in question has no other states with these angular momenta, these states are characterized by the lowest possible seniorities. The existence of a rather strong pairing energy as demonstrated in the binding energy of nuclei is adequately represented by (27.3) and (27.4) and might indicate that the effective nuclear interaction has to a good approximation the pairing property. Equation (27.25) implies that, for odd tensor interactions, the order and spacing of levels with seniorities $v' \leq v$ in the j^n configuration are the same as in the j^v configuration.

Let us now use the recursion formula (27.23) in order to calculate explicitly the c.f.p. which appear in it. Equation (27.23) gives $\frac{1}{2}[n(n-1)]\,[j^{n-2}(vJ)j^2(0)| \} j^n vJ]^2$ as the sum of an arithmetical progression with $(n-v)/2$ terms. The first term is $1 - 2v/(2j+1)$ and the last is $1 - 2(n-2)/(2j+1)$ so that the sum is equal to

$$\frac{n(n-1)}{2}\,[j^{n-2}(v\, J)\, j^2(0) J| \} j^n v\, J]^2$$

$$= \frac{2}{n(n-1)}\,\frac{1}{N^2_{nvJ}} = \frac{n-v}{2}\,\frac{1}{2}\left[1 - \frac{2v}{2j+1} + 1 - \frac{2(n-2)}{2j+1}\right]$$

$$= \frac{1}{2}\,\frac{(n-v)\,(2j+3-n-v)}{2j+1}. \tag{27.26}$$

This expression vanishes indeed for $n = v$ and $n - 2 = 2j + 1 - v$ as mentioned before.

Equation (27.26) will be of great value for the actual calculation of c.f.p. in the seniority scheme as we shall soon see. Before proceeding with these calculations let us present an interesting interpretation of the quantities which appear in (27.26). We shall introduce a special two-particle interaction which is diagonal in the seniority scheme and its eigenvalues are given by (27.26). To do so we define a two-particle operator by the requirement that in the j^2 configuration it will vanish unless the pair is coupled to $J = 0$. It is obvious that such an operator is intimately connected with the seniority. For n particles we define $Q = \sum_{i<k} q_{ik}$ where

$$\langle j^2 J M | q_{12} | j^2 J' M' \rangle = (2j + 1)\delta_{JJ'}\delta_{MM'}\delta_{J0}. \tag{27.27}$$

The numerical factor $2j + 1$ was introduced only for convenience. We shall see that with this factor, Q has integral eigenvalues. We now calculate matrix elements of Q in the j^n configuration using $n \to n - 2$ c.f.p. Equation (26.45) taken in the seniority scheme, readily yields, in view of (27.27),

$$\langle j^n v\, J M \Big| \sum_{i<k}^{n} q_{ik} \Big| j^n v' J M \rangle$$

$$= \frac{n(n-1)}{2} [j^n v\, J \{| j^{n-2}(v\, J) j^2(0) J] \, [j^{n-2}(v' J) j^2(0) J |\} j^n v' J] \delta_{vv'} (2j + 1). \tag{27.28}$$

The operator Q is thus diagonal in the seniority scheme. Comparing (27.28) with (27.26) we see that Q is independent of J (and α) and its eigenvalues are given by

$$Q(n, v) = \tfrac{1}{2}(n - v)(2j + 3 - n - v). \tag{27.29}$$

In this way the seniority number v is given in terms of the eigenvalues of the two-particle operator Q. We now see the physical reason for the J independence of the c.f.p. in (27.26). All states with the same seniority v have the same amount of pairing to $J = 0$ as given by (27.28).

From its definition, or explicitly from (27.29) it follows that $Q(n, v)$ vanishes for $n = v$. We further see that Q does not possess the pairing property since besides terms linear in n (and v) its eigenvalues contain also terms quadratic in n(and v). We can see the source of these quadratic terms by expressing q_{ik} in terms of scalar products of tensor operators. The relation (15.16) can be written for the present case as

$$\sum_k (-1)^{2j+k}(2k + 1) \begin{Bmatrix} j & j & J \\ j & j & k \end{Bmatrix} = (2j + 1)\delta_{J0}.$$

Recalling (15.5) we can rewrite this identity in terms of scalar products of the unit tensor operators defined by (21.54). We thus obtain

$$\delta_{J0} = \frac{1}{2j + 1} \sum_{k=0}^{2j} (-1)^k (2k + 1) \langle j^2 J M | (\mathbf{u}_1^{(k)} \cdot \mathbf{u}_2^{(k)}) | j^2 J M \rangle. \tag{27.30}$$

This equation expresses q_{12} in terms of scalar products of unit tensor operators. However, the summation in (27.30) includes both odd and even values of k. Only the odd tensors are diagonal with respect to the seniority and it is therefore not clear why Q should also have this property. To understand this point we recall that the $2j+1$ scalar products of unit tensor operators are not independent in the j^2 configuration of identical particles. In fact, we obtained in Part II the identity (15.17) which can also be expressed, using (15.5), in terms of scalar products of unit tensor operators as follows:

$$-1 = \sum_{k=0}^{2j} (2k+1) \langle j^2 J M | (\mathbf{u}_1^{(k)} \cdot \mathbf{u}_2^{(k)}) | j^2 J M \rangle. \tag{27.31}$$

Equation (27.30) and (27.31) can now be combined to give

$$(2j+1)\delta_{J0} + 1 = -2 \sum_{k \text{ odd}} (2k+1) \langle j^2 J M | (\mathbf{u}_1^{(k)} \cdot \mathbf{u}_2^{(k)}) | j^2 J M \rangle. \tag{27.32}$$

It follows from this relation that q_{12} can, in fact, be expressed in terms of tensors with odd values of k and with $k = 0$. Thus

$$q_{12} = -1 - 2 \sum_{k \text{ odd}} (2k+1) (\mathbf{u}_1^{(k)} \cdot \mathbf{u}_2^{(k)}). \tag{27.33}$$

The operator q_{12} is therefore a sum of two terms. One of these terms is a sum of scalar products of odd tensors and is therefore diagonal with respect to the seniority and has the pairing property. The contribution of the other term (i.e., -1), which is independent of J (and M) to $Q = \Sigma_{i<k}^{n} q_{ik}$ is simply $-n(n-1)/2$. This term, being a plain number is diagonal in the seniority scheme, as well as in any other scheme, but it *does not* possess the pairing property. Thus, $Q(n, v) + \frac{1}{2} n(n-1)$ has the pairing property and (27.29) could therefore also be obtained directly from (27.25), recalling that $Q(v,v) = 0$, in the form

$$Q(n, v) + \frac{n(n-1)}{2} = \frac{v(v-1)}{2} + \frac{n-v}{2} [Q(2, 0) + 1]$$

$$= \frac{v(v-1)}{2} + \frac{n-v}{2} (2j+2). \tag{27.34}$$

The eigenvalues of the operator Q are closely related to the number of pairs coupled to $J = 0$ in the corresponding eigenstates. Actually, the number $(n-v)/2$ is the number of such pairs which have to be added to the principal parent in the j^v configuration. However, the eigenvalues of $Q = (2j+1) \ \Sigma_{i<k} \delta_{J_{ik}0}$ are not simply $(2j+1)(n-v)/2$ but rather $(2j+3-n-v)(n-v)/2 \le (2j+1)[(n-v)/2]$. The reason for this is that if there are more than two particles, the Pauli principle reduces the probability of the $(n-v)/2$ pairs to be each coupled to $J = 0$ simultaneously. Thus, although we started from adding $(n-v)/2$ pairs coupled to $J = 0$ to the principal parent in the j^v configuration, the antisymmetrization reduces the number of pairs coupled to $J = 0$ to $\frac{1}{2}(n-v)[(2j+3-n-v)/(2j+1)]$. Still, the number $(n-v)/2$ can be interpreted as the number of pairs coupled to $J = 0$ in the sense of (27.25) for interactions which have the pairing property.

Using (27.30) and (27.31) we could express q_{12} equally well by a sum of scalar products of *even* tensor operators. It is worthwhile to mention that we can define even tensor interactions, similar to q_{12}, which have a nonvanishing eigenvalue only for the state with J_0 of the j^2 configuration. We start from the identity (15.10) which can be written for our purpose as

$$(2J_0+1)\sum_k (2k+1) \begin{Bmatrix} j & j & k \\ j & j & J_0 \end{Bmatrix} \begin{Bmatrix} j & j & k \\ j & j & J \end{Bmatrix} = \delta_{JJ_0}.$$

The identity (15.11) can be written as

$$\sum_k (-1)^k (2k+1) \begin{Bmatrix} j & j & k \\ j & j & J_0 \end{Bmatrix} \begin{Bmatrix} j & j & k \\ j & j & J \end{Bmatrix} = (-1)^{J+J_0} \begin{Bmatrix} j & j & J_0 \\ j & j & J \end{Bmatrix}.$$

Combining these two equations we obtain

$$\delta_{JJ_0} = -(-1)^{J+J_0}(2J_0+1) \begin{Bmatrix} j & j & J_0 \\ j & j & J \end{Bmatrix} + \sum_k [1+(-1)^k](2J_0+1)(2k+1) \begin{Bmatrix} j & j & k \\ j & j & J_0 \end{Bmatrix} \begin{Bmatrix} j & j & k \\ j & j & J \end{Bmatrix}. \tag{27.35}$$

For the allowed states of the j^2 configuration with identical particles J_0 as well as J must be even. In this case (27.35) can be written down as an even tensor interaction. Using (15.5) we obtain for J and J_0 even

$$\begin{aligned} \delta_{JJ_0} &= (2J_0+1)\langle jjJM|(\mathbf{u}_1^{(J_0)} \cdot \mathbf{u}_2^{(J_0)})|jjJM\rangle - 2(2J_0+1) \\ &\times \sum_{k\ \text{even}} (2k+1) \begin{Bmatrix} j & j & k \\ j & j & J_0 \end{Bmatrix} \langle jjJM|(\mathbf{u}_1^{(k)} \cdot \mathbf{u}_2^{(k)})|jjJM\rangle \\ &= (2J_0+1) \sum_{k\ \text{even}} \left[\delta_{kJ_0} - 2(2k+1) \begin{Bmatrix} j & j & k \\ j & j & J_0 \end{Bmatrix}\right] \\ &\times \langle jjJM|(\mathbf{u}_1^{(k)} \cdot \mathbf{u}_2^{(k)})|jjJM\rangle. \end{aligned} \tag{27.36}$$

The existence of an even tensor interaction (27.36) which has a non-vanishing eigenvalue only for $J = J_0$ shows that the scalar products of even tensors are linearly independent in the j^2 configuration of identical particles. The matrix elements $\langle jj\,J\,M|(\mathbf{u}_1^{(k)}\cdot\mathbf{u}_2^{(k)})|jj\,J\,M\rangle$, for even values of J, can be considered as a vector whose $\frac{1}{2}(2j+1)$ components are defined by the values of J. Our statement means that the $\frac{1}{2}(2j+1)$ vectors, with k even, are linearly independent. We already saw that with any odd tensor interaction the $V_J = \langle jj\,J\,M|V_{12}|jj\,J\,M\rangle$ satisfy the relation (21.44). Such a relation implies a certain linear relation between the scalar products of tensors in the j^2 configuration of identical nucleons. No such relations can exist between the scalar products of even tensors. In fact, there are $\frac{1}{2}(2j+1)$ linear combinations (27.36), defined by the values of J_0, which span the whole space of $\frac{1}{2}(2j+1)$ dimensions.

It is of some interest to see the form of the scalar operator q_{ik} in the m-scheme, i.e., the form of the matrix elements $\langle j\,m_1\,j\,m_2|q_{12}|j\,m_3\,j\,m_4\rangle$. Since q_{12} vanishes in all states with $J > 0$, it follows from the sum method that it vanishes in all states with $M_J \neq 0$. The only matrix elements which do not vanish have therefore the form $\langle j\,m\,j-m|q_{12}|j\,m'j-m'\rangle$. Using Clebsch-Gordan coefficients we can expand this matrix element and obtain

$$\langle j\,m\,j-m|q_{12}|j\,m'\,j-m'\rangle$$
$$= \sum_J (j\,m\,j-m|jj\,J\,0)\,(j\,m'\,j-m'|jj\,J\,0)\,\langle j^2 J\,0|q_{12}|j^2 J\,0\rangle \qquad (27.37)$$

where all matrix elements between different values of J vanish since q_{12} is a scalar operator. Inserting the values (27.27) of the matrix elements of q_{12} into (27.37) we obtain

$$\langle j\,m\,j-m|q_{12}|j\,m'\,j-m'\rangle = (j\,m\,j-m|jj\,0\,0)\,(j\,m'\,j-m'|jj\,0\,0)\,(2j+1)$$
$$= \frac{(-1)^{j-m}}{\sqrt{2j+1}}\frac{(-1)^{j-m'}}{\sqrt{2j+1}}(2j+1) = -(-1)^{m+m'}. \qquad (27.38)$$

The interaction matrix of q_{12} for two particles in the m-scheme, given by (27.38) can be diagonalized very simply without the use of Clebsch-Gordan coefficients. This is a $2j+1$ by $2j+1$ matrix which is of rank 1, all its rows being proportional to each other. The proportionality factor is $+1$ or -1 and in fact by using a different phase convention (which would not change the physical contents) we can arrive at a matrix whose elements are all equal to 1. In either case there must be only one eigenvalue different from zero and its value is therefore equal to the trace of the matrix. As seen from (27.38) all the diagonal elements are $+1$ and the trace is equal to $2j+1$. Obviously, this eigenvalue is that of the state with $J = 0$ since the sum method implies that the eigenvalues of the states with $J > 0$ must be zero.

Interactions with similar properties appear in modern theories of superconductivity and were also considered in the case of nuclear structure. In the simple case of the j^n configuration such interactions are identical with q_{ik} and therefore diagonal in the seniority scheme and depend only on n and v.

28. C.f.p. and Matrix Elements in the Seniority Scheme

Restricting our attention to the scheme defined by the seniority quantum number, we can make more specific statements about the c.f.p. We shall first obtain the c.f.p. $[j^{n-2}(v_2 J_2) j^2(J') J \| j^n v\, J]$ for all values of n in terms of the c.f.p. in the case of one specific value of n. As a result of this reduction many expressions will become greatly simplified. These simplifications show the usefulness of the seniority scheme.

We first obtain a recursion formula for the c.f.p. considered (as a function of n). This coefficient is given by

$$[j^n(v_2 J_2)\, j^2(J') J \| j^{n+2} v\, J]$$

$$= \int \psi^*[j^n(v_2 J_2)\, j^2_{n+1,n+2}(J') J]\, \psi(j^{n+2} v\, J)$$

$$= \int N_{nv_2J_2} \mathscr{A}_n \psi^*[j^{n-2}(v_2 J_2)\, j^2_{n-1,n}(0)\, j^2_{n+1,n+2}(J') J]\, \psi(j^{n+2} v\, J). \qquad (28.1)$$

Here, $\mathscr{A}_n$ is the antisymmetrization operator which is the sum of $n(n-1)/2$ permutations of the coordinates of the first n particles, and $N_{nv_2J_2}$ is the corresponding normalization coefficient. Since $\psi(j^{n+2} v\, J)$ is fully antisymmetric in the $n+2$ particle coordinates, the operator $\mathscr{P} = 2\mathscr{A}_n/n(n-1)$, operating under the integral sign, does not change the integral at all. Thus (28.1) is equal to

$$\frac{n(n-1)}{2} N_{nv_2J_2} \int \psi^*[j^{n-2}(v_2 J_2)\, j^2_{n\;1,n}(0)\, j^2_{n+1,n+2}(J') J] \;\; (j^{n+2} v\, J).$$

Furthermore, since $\psi(j^{n+2} v\, J)$ is fully antisymmetric, we can apply to it or to any other part of the integrand any permutation of the particle coordinates without changing the value of the integral provided we multiply it by $(-1)^P$. We therefore exchange $n-1$, n with $n+1$, $n+2$ for the sake of convenience, and obtain for (28.1) the equivalent form

$$\frac{n(n-1)}{2} N_{nv_2J_2} \int \psi^*[j^{n-2}(v_2 J_2)\, j^2_{n-1,n}(J')\, j^2_{n+1,n+2}(0) J]\, \psi(j^{n+2} v\, J).$$

We now expand $\psi(j^{n+2} v\, J)$ in terms of c.f.p. and integrate over the coordinates of particles $n+1$ and $n+2$ obtaining for (28.1) the expression

$$\frac{n(n-1)}{2} N_{nv_2J_2} \sum_{v_2'J_2'J''} [j^n(v_2'J_2')j^2(J'')J|\}j^{n+2}v\,J]$$

$$\times \int \psi^*[j^{n-2}(v_2J_2)\,j^2_{n-1,n}(J')\,j^2_{n+1,n+2}(0)J]\,\psi[j^n(v_2'J_2')\,j^2_{n+1,n+2}(J'')J]$$

$$= \frac{n(n-1)}{2} N_{nv_2J_2}[j^n(v\,J)j^2(0)J|\}j^{n+2}v\,J] \int \psi^*[j^{n-2}(v_2J_2)\,j^2_{n-1,n}(J')J]\,\psi(j^nv\,J). \tag{28.2}$$

The integral on the right-hand side of (28.2) has the same form as (28.1) with n, rather than with $n+2$, particles. Therefore we obtain, using (27.21) and (27.26), the recursion formula

$$[j^n(v_2J_2)j^2(J')J|\}j^{n+2}v\,J]$$

$$= \frac{n(n-1)N_{nv_2J_2}}{(n+1)(n+2)N_{n+2,vJ}}[j^{n-2}(v_2J_2)j^2(J')J|\}j^nv\,J]$$

$$= \sqrt{\frac{n(n-1)(n+2-v)(2j+1-n-v)}{(n+1)(n+2)(n-v_2)(2j+3-n-v_2)}}[j^{n-2}(v_2J_2)j^2(J')J|\}j^nv\,J]. \tag{28.3}$$

The state of the j^{n+2} configuration with v and J, considered in the derivation of (28.3), has as a component the state obtained by adding a pair of particles coupled to J' to the principal parent with v_2 and J_2 of the j^n configuration and antisymmetrizing. The antisymmetrization of $\psi[j^n(v_2J_2)j^2_{n+1,n+2}(J')J]$ does not necessarily lead to a state with definite seniority. Only if the same state can also be obtained by adding a pair coupled to $J=0$ to the same principal parent and antisymmetrizing, will the state have a definite seniority $v=v_2$ (and $J=J_2$). In the general case, the resulting state will be a linear combination of states with a limited range of seniorities. We recall that the state with v_2 and J_2 in the j^n configuration is obtained from the principal parent (with v_2 and J_2) in the j^{v_2} configuration by adding $(n-v_2)/2$ pairs coupled to $J=0$ and antisymmetrizing. The order of the antisymmetrizations is irrelevant as can be seen best by considering them as projection operators. The number of pairs coupled to $J=0$ certainly cannot *decrease* when we add the pair coupled to J'. This number is $\frac{1}{2}(n-v_2)=\frac{1}{2}[(n+2)-(v_2+2)]$ and therefore the maximum seniority which can appear in the state of the j^{n+2} configuration is v_2+2. However, the state in the j^{n+2} configuration may contain states with seniority v which is *smaller* than v_2. We can couple first the pair coupled to J' to the principal parent $\psi(j^{v_2}v_2J_2)$ and only later the $(n-v_2)/2$ pairs coupled to $J=0$.

The principal parent $\psi(j^{v_2}v_2J_2)$ can be expressed, using c.f.p., as a linear combination of functions $\psi[j^{v_2-2}(v_3J_3)j^2(J'')J_2]$ where v_3 is equal to $v_2 - 2$ (it can be at most $v_2 - 2$ but as shown above v_2 can be at most $v_3 + 2$). Adding the pair coupled to J' we obtain the functions $\psi[j^{v_2-2}(v_3J_3)j^2(J'')J_2j^2(J')J]$. Changing the order of coupling we obtain the functions $\psi[j^{v_2-2}(v_3J_3), j^2(J'')j^2(J')J_0, J]$. We may now change the order of coupling of the particles in the two pairs (coupled to J'' and to J') as done in the derivation of (27.19). If there is a function with $J_3 = J$ and $J'' = J'$ we can obtain after the transformation a function $\psi[j^{v_2-2}(v_3J)j^2(0)j^2(0)J]$. Such a function will give after the antisymmetrization a wave function with seniority $v_3 = v_2 - 2$. Adding to it the other $(n - v_2)/2$ pairs coupled to $J = 0$ and antisymmetrizing we obtain a wave function in the j^{n+2} configuration with seniority $v = v_2 - 2$.

Thus we see that when we add a pair of particles to a state with a definite seniority v and antisymmetrize, the resulting state will contain states with seniorities which differ from v by $\Delta v = 0, \pm 2$. Therefore, the only values of v for which the c.f.p. in (28.3) do not vanish are $v = v_2$ and $v = v_2 \pm 2$.

Successive application of the recursion formula (28.3) gives the following results for the three cases with nonvanishing c.f.p.

$$[j^{n-2}(v_2J_2)j^2(J')J|\}j^nv\,J]$$

$$= \sqrt{\frac{v(v-1)(2j+3-n-v)(2j+5-n-v)}{n(n-1)(2j+3-2v)(2j+5-2v)}}\,[j^{v-2}(v-2, J_2)j^2(J')J|\}j^vv\,J]$$

$$\text{for} \quad v_2 = v - 2 \tag{28.4}$$

$$[j^{n-2}(v_2J_2)j^2(J')J|\}j^nv\,J]$$

$$= \sqrt{\frac{(v+1)(v+2)(n-v)(2j+3-n-v)}{2n(n-1)(2j+1-2v)}}\,[j^v(v\,J_2)j^2(J')J|\}j^{v+2}vJ]$$

$$\text{for} \quad v_2 = v \tag{28.5}$$

$$[j^{n-2}(v_2J_2)j^2(J')J|\}j^nv\,J]$$

$$= \sqrt{\frac{(v+3)(v+4)(n-v)(n-v-2)}{8n(n-1)}}\,[j^{v+2}(v+2, J_2)j^2(J')J|\}j^{v+4}v\,J]$$

$$\text{for} \quad v_2 = v + 2. \tag{28.6}$$

The relations (28.4), (28.5), and (28.6) are direct consequences of the more special result (27.26). We shall now see that similar relations follow from (27.26) also for ordinary, i.e., $n \to n - 1$, c.f.p. In order to derive these relations we shall follow a procedure similar to the one used in the derivation of (28.3). We start from the following integral which gives the $n+2 \rightarrow n+1$ c.f.p.

$$[j^{n+1}(v_1 J_1)\, j\, J|\} j^{n+2} v\, J]$$

$$= \int \psi^*[j^{n+1}(v_1 J_1)\, j_{n+2} J]\, \psi(j^{n+2} v\, J)$$

$$= N_{n+1,v_1 J_1} \int \mathscr{A}_{n+1}\, \psi^*[j^{n-1}(v_1 J_1)\, j^2_{n,n+1}(0)\, j_{n+2} J]\, \psi(j^{n+2} v\, J)$$

$$= \frac{n(n+1)}{2} N_{n+1,v_1 J_1} \int \psi^*[j^{n-1}(v_1 J_1)\, j^2_{n,n+1}(0)\, j_{n+2} J]\, \psi(j^{n+2} v\, J). \tag{28.7}$$

The last equality follows from the properties of the projection operator $\mathscr{P} = [2/n(n+1)]\mathscr{A}_{n+1}$. We now expand $\psi(j^{n+2} v\, J)$ in $n+2 \rightarrow n$ c.f.p. but remembering that this function is fully antisymmetric we first exchange the positions of particles n, $n+1$, and $n+2$ in (28.7) in order to simplify the notation. We thus obtain for (28.7) the following form

$$\frac{n(n+1)}{2} N_{n+1,v_1 J_1} \int \psi^*[j^{n-1}(v_1 J_1)\, j^2_{n+1,n+2}(0)\, j_n J]\, \psi(j^{n+2} v\, J)$$

$$= \frac{n(n+1)}{2} N_{n+1,v_1 J_1} \sum_{v_2 J_2 J'} [j^n(v_2 J_2)\, j^2(J') J|\} j^{n+2} v\, J]$$

$$\times \int \psi^*[j^{n-1}(v_1 J_1)\, j^2_{n+1,n+2}(0)\, j_n J]\, \psi[j^n(v_2 J_2)\, j^2_{n+1,n+2}(J') J]$$

$$= \frac{n(n+1)}{2} N_{n+1,v_1 J_1} [j^n(v\, J)\, j^2(0) J|\} j^{n+2} v\, J] \int \psi^*[j^{n-1}(v_1 J_1)\, j_n J]\, \psi(j^n v\, J). \tag{28.8}$$

The integral on the right-hand side is the $n \rightarrow n-1$ c.f.p. analogous to (28.7). Using the result (27.26) for the special $n+2 \rightarrow n$ c.f.p. and $N_{n+1,v_1 J_1}$ which appear in (28.8) we obtain the following recursion formula for the $n \rightarrow n-1$ c.f.p.

$$[j^{n+1}(v_1 J_1)\, j\, J|\} j^{n+2} v\, J]$$

$$= [j^{n-1}(v_1 J_1)\, j\, J|\} j^n v\, J] \sqrt{\frac{n(n+2-v)(2j+1-n-v)}{(n+2)(n+1-v_1)(2j+2-n-v_1)}}\,. \tag{28.9}$$

If we add another particle to the state $\psi(j^{n-1} v_1 J_1)$ and antisymmetrize, the resulting state need not have a definite seniority v. The state $\psi(j^{n-1} v_1 J_1)$ was obtained by successive coupling of $(n-1-v_1)/2$ pairs coupled to $J=0$ to the parent state $\psi(j^{v_1} v_1 J_1)$ and antisymmetrizing. The addition of another particle cannot decrease the number of pairs coupled to $J = 0$. Therefore, the number of such pairs in the wave function $\psi[j^{n-1}(v_1 J_1) j_n J]$ is *at least*

$$\frac{n-1-v_1}{2} = \frac{n-(v_1+1)}{2}.$$

Thus, the seniority of states contained in this function is *at most* $v_1 + 1$. However, the seniority can also decrease by the addition of another particle. As in the case of $n \to n-2$ c.f.p., we can change the order of antisymmetrizations and in $\psi[j^{n-1}(v_1 J_1) j_n J]$ couple first j_n to $\psi(j^{v_1} v_1 J_1)$ and only later the $(n-1-v_1)/2$ pairs coupled to $J = 0$. The principal parent $\psi(j^{v_1} v_1 J_1)$ can be expanded, using c.f.p., into a combination of the functions $\psi[j^{v_1-1}(v_2 J_2) j_{v_1} J_1]$, where v_2 must be equal to $v_1 - 1$. When j_n is coupled to such a function, yielding $\psi[j^{v_1-1}(v_2 J_2) j_{v_1}(J_1) j_n J]$, we can also change the order of coupling and obtain functions $\psi[j^{v_1-1}(v_2 J_2) j^2(J') J]$. If there are functions with $J_2 = J$ in the expansion we may obtain also functions with $J' = 0$. Such functions will give, after antisymmetrization, wave functions with seniority $v_2 = v_1 - 1$.

Thus we see that the addition of another particle to the state $\psi(j^{n-1} v_1 J_1)$ and antisymmetrization gives a function which is a linear combination of states with seniorities $v = v_1 + 1$ and $v = v_1 - 1$. Therefore the values of v_1 for which the c.f.p. in the recursion formula (28.9) do not vanish are $v_1 = v \pm 1$.

Successive applications of the recursion formula (28.9) give the following results for the two cases just mentioned

$$[j^{n-1}(v_1 J_1) j J | \} j^n v J]$$
$$= \sqrt{\frac{v(2j+3-n-v)}{n(2j+3-2v)}} [j^{v-1}(v-1, J_1) j J | \} j^v v J] \quad \text{for} \quad v_1 = v-1 \qquad (28.10)$$

$$[j^{n-1}(v_1 J_1) j J | \} j^n v J]$$
$$= \sqrt{\frac{(n-v)(v+2)}{2n}} [j^{v+1}(v+1, J_1) j J | \} j^{v+2} v J] \quad \text{for} \quad v_1 = v+1. \qquad (28.11)$$

A special case of (28.11) is worth mentioning. For the case $v = 0$ (and $J = 0$), v_1 can only be $v + 1 = 1$ (and $J_1 = j$). Since there is only one such state with $v_1 = 1$ (and $J_1 = j$), it is clear that only the c.f.p. with $v_1 = 1$ $J_1 = j$ does not vanish and therefore it must be equal to 1. In fact, inserting $v = 0$ in (28.11) we obtain that the c.f.p. for the various values of n are all equal to $[j j J = 0 | \} j^2 J = 0] = 1$. Thus the wave function for the $J = 0$ $v = 0$ state is given by the following simple expression

$$\psi(j^n v = 0 \; J = 0 \; M = 0) = \psi[j^{n-1}(v = 1 \; J = j) j_n \; J = 0 \; M = 0]. \qquad (28.12)$$

The wave function (28.12), although j_n is singled out in it, is fully antisymmetric in the coordinates of all particles.

In the relation (28.11) the seniorities which appear in the c.f.p. are $v+1$ and v. Yet on the right-hand side of (28.11) appears the c.f.p. of the j^{v+2} configuration which is not the minimal configuration for the seniorities v and $v+1$. Therefore, we shall try to express this c.f.p. in terms of the j^{v+1} configuration. We proceed to calculate directly this c.f.p. which is given by

$$[j^{v+1}(v+1, J_1)j\, J|\}j^{v+2}v\, J]$$
$$= \int \psi^*[j^{v+1}(v+1, J_1)j_{v+2}J]\, \psi(j^{v+2}v\, J)$$
$$= N \int \psi^*[j^{v+1}(v+1, J_1)j_{v+2}J]\, \mathscr{A}\psi[j^v(v\, J)j^2_{v+1,v+2}(0)J]. \tag{28.13}$$

Here $N = N_{v+2,vJ}$ is the appropriate normalization coefficient. When we carry out explicitly the antisymmetrization we obtain for the c.f.p. (28.13) the expression

$$N \int \psi^*[j^{v+1}(v+1, J_1)j_{v+2}J] \Big\{\psi[j^v(v\, J)j^2_{v+1,v+2}(0)J] - \sum_{k=1}^{v} \psi[j^v(v\, J)j^2_{k,v+2}(0)J]$$
$$- \sum_{k=1}^{v} \psi[j^v(v\, J)j_{v+1,k}(0)J] + \sum_{i<k}^{v} \psi[j^v(v\, J)j^2_{ik}(0)J]\Big\}. \tag{28.14}$$

Only the first two terms in the curly bracket in (28.14) contribute to the integral. In the other two terms, two of the first $v+1$ particles are coupled to $J=0$ whereas there are no such pairs in the wave function $\psi(j^{v+1}v+1, J_1)$. Since $\psi(j^{v+1}v+1, J_1)$ is antisymmetric, the contributions to (28.14) of each term $-\psi[j^v(v\, J)j_{k,v+2}(0)J]$ are equal to the contribution of the first term $\psi[j^v(v\, J)j_{v+1,v+2}(0)J]$. We now expand $\psi(j^{v+1}v+1, J_1)$ in terms of c.f.p. and apply a change of coupling. This yields

$$(v+1) \int \psi^*[j^{v+1}(v+1, J_1)j_{v+2}J]\, \psi[j^v(v\, J)j^2_{v+1,v+2}(0)J]$$
$$= (v+1) \sum_{J'} [j^v(v\, J')j\, J_1|\}j^{v+1}v+1, J_1]$$
$$\times \int \psi^*[j^v(v\, J')j_{v+1}(J_1)j_{v+2}J]\, \psi[j^v(v\, J)j^2_{v+1,v+2}(0)J]$$
$$= (v+1) \sum_{J'J''} [j(v\, J')j\, J_1|\}j^{v+1}v+1, J_1]$$
$$\times (-1)^{2j+J+J'} \sqrt{(2J_1+1)(2J''+1)} \begin{Bmatrix} J' & j & J_1 \\ j & J & J'' \end{Bmatrix}$$
$$\times \int \psi^*[j^v(v\, J')j^2_{v+1,v+2}(J'')J]\, \psi[j^v(v\, J)^2 j_{v+1,v+2}(0)J]. \tag{28.15}$$

When we carry out the integration, only the term with $J' = J$, $J'' = 0$ remains. Using the value of the Racah coefficient in this case, as well as the value of $N_{v+2,vJ}$ from (27.26), we obtain the result

$$[j^{v+1}(v+1, J_1)j\, J|\}j^{v+2}v\, J]$$

$$= (-1)^{J+j-J_1}\sqrt{\frac{(2J_1+1)\,2(v+1)}{(2J+1)(v+2)(2j+1-2v)}}\,[j^v(v\, J)j\, J_1|\}j^{v+1}v+1, J_1]. \tag{28.16}$$

The relations (28.4), (28.5), (28.6), (28.10), and (28.11) show the usefulness of the seniority scheme. In particular, (28.10), (28.11), together with (28.16) imply that when going from the j^n to the j^{n+1} configuration, only c.f.p. with $v = n + 1$ should be evaluated. All other c.f.p. are given in terms of the c.f.p. of preceding configurations.

We can make use of these recursion relations in order to calculate the c.f.p. of the state with $v = 1$ and $J = j$ in the j^n configuration. The parent states must have either $v_1 = 0$ or $v_1 = 2$. Using (28.10) and (28.11) we obtain in this case

$$[j^{n-1}(v_1 = 0\;\; J_1 = 0)j\;\; J = j|\}j^n\;\; v = 1\;\; J = j] = \sqrt{\frac{2j+2-n}{n(2j+1)}}$$

$$[j^{n-1}(v_1 = 2\;\; J_1 > 0)j\;\; J = j|\}j^n\;\; v = 1\;\; J = j]$$

$$= \sqrt{\frac{3(n-1)}{2n}}\,[j^2(v_1 = 2, J_1)j\;\; J = j|\}j^3\;\; v = 1\;\; J = j].$$

Using (28.16) in order to reduce further the $3 \to 2$ c.f.p. in the last equation we obtain the following explicit expressions

$$[j^{n-1}(J_1)j\;\; J = j|\}j^n\;\; v = 1\;\; J = j] = \begin{cases} \sqrt{\dfrac{2j+2-n}{n(2j+1)}} & \text{for}\quad J_1 = 0 \\[2ex] -\sqrt{\dfrac{2(n-1)}{n(2j+1)(2j-1)}}(2J_1+1) & \text{for}\quad J_1 > 0 \text{ even.} \end{cases} \tag{28.17}$$

In the case $n = 3$, (28.17) reduces to the result (26.12) obtained above.

All previous results were stated without explicit reference to additional quantum numbers which may be necessary if the seniority quantum number v does not specify uniquely the states with a given J in the j^n configuration. However, all these results, including (28.4), (28.5), and (28.6) as well as (28.10), (28.11), and (28.16) hold also in the general case when such additional quantum numbers are actually necessary. As already discussed, if there are two or more orthogonal states with the same J and v in the j^v configuration, we define α to be any additional quantum number necessary to distinguish between them. Starting from the states $\psi(j^v \alpha\, v\, J)$, adding successively pairs coupled to $J = 0$ and antisymmetrizing, we define a complete scheme of orthogonal wave functions $\psi(j^n \alpha\, v\, J)$ in the j^n configuration. It is therefore clear that all previous results hold if the additional quantum numbers are introduced. The relations (28.4), (28.5), and (28.6) then give the expression of the c.f.p. $[j^{n-2}(\alpha_2 v_2 J_2) j^2(J')J\}\, j^n \alpha\, v\, J]$ in terms of $[j^{v_2}(\alpha_2 v_2 J_2) j^2(J')J\}\, j^{v_2+2} \alpha\, v\, J]$. Similarly, (28.10) and (28.11) will give in this case the relations between $[j^{n-1}(\alpha_1 v_1 J_1) j\, J\}\, j^n \alpha\, v\, J]$ and $[j^{v_1}(\alpha_1 v_1 J_1) j\, J\}\, j^{v_1+1} \alpha\, v\, J]$. Therefore, we shall often continue to omit these additional quantum numbers for the sake of simplicity.

We can use (28.10) to obtain orthogonality relations of the c.f.p. which are more detailed than the general relation (26.15). In the present notation, (26.15) assumes the form

$$\sum_{\alpha_1 v_1 J_1} [j^n v\, \alpha\, J\{|j^{n-1}(v_1 \alpha_1 J_1) j\, J]\, [j^{n-1}(v_1 \alpha_1 J_1) j\, J|\}j^n v' \alpha' J] = \delta_{vv'} \delta_{\alpha\alpha'}. \tag{28.18}$$

We write now a summation similar to the one in (28.18), for $v = v'$, restricting it, however, to the value $v_1 = v - 1$. Using (28.10) we obtain

$$\sum_{\alpha_1 J_1} [j^n v\, \alpha\, J\{|\, j^{n-1}(v - 1, \alpha_1, J_1) j\, J]\, [j^{n-1}(v - 1, \alpha_1, J_1) j\, J|\}j^n v\, \alpha'\, J]$$

$$= \frac{v(2j + 3 - n - v)}{n(2j + 3 - 2v)} \sum_{\alpha_1 J_1} [j^v v\, \alpha\, J\{|\, j^{v-1}(v - 1, \alpha_1, J_1) j\, J]$$

$$\times [j^{v-1}(v - 1, \alpha_1, J_1) j\, J|\}j^v v\, \alpha'\, J]. \tag{28.19}$$

Since there are no states with seniority $v + 1$ in the j^{v-1} configuration, the summation on the right-hand side of (28.19) is exactly equal to the full orthogonality relation (28.18) in the j^v configuration. Thus, we obtain

$$\sum_{\alpha_1 J_1} [j^n v\, \alpha\, J\{|\, j^{n-1}(v - 1, \alpha_1, J_1) j\, J]\, [j^{n-1}(v - 1, \alpha_1, J_1) j\, J|\}j^n v\, \alpha'\, J]$$

$$= \frac{v(2j + 3 - n - v)}{n(2j + 3 - 2v)}\, \delta_{\alpha\alpha'}. \tag{28.20}$$

By subtracting this equation from the equation (28.18) (with $v = v'$) we obtain that the summation over $v_1 = v + 1$ only, has the value

$$\sum_{\alpha_1 J_1} [j^n v\, \alpha\, J\{|\, j^{n-1}(v + 1, \alpha_1, J_1) j\, J]\, [j^{n-1}(v + 1, \alpha_1, J_1) j\, J|\}j^n v\, \alpha'\, J]$$

$$= \frac{(n - v)\,(2j + 3 - v)}{n(2j + 3 - 2v)}\, \delta_{\alpha\alpha'}. \tag{28.21}$$

These relations can be used to check the values of c.f.p. Similar relations can be obtained for $n \to n-2$ c.f.p. However, we shall not do it here.

It is worthwhile to mention that the relations (28.20) and (28.21) can be obtained directly without the use of (28.10). If the summations on the left-hand sides of (28.20) and (28.21) will be denoted by x and y, respectively, we can write down two equations satisfied by them. The first is the orthogonality relation (28.18) (for $v = v'$)

$$x + y = \delta_{\alpha\alpha'}.$$

The second equation follows from the fact that Q is diagonal in the seniority scheme. Using (26.37) we obtain

$$\frac{n}{n-2} xQ(n-1, v-1) + \frac{n}{n-2} yQ(n-1, v+1) = Q(n, v)\delta_{\alpha\alpha'}.$$

It is clear that the determinant of the coefficients of x and y does not vanish. Therefore, if $\alpha \not\equiv \alpha'$, both x and y must vanish. In the case $\alpha \equiv \alpha'$, the solutions of the two equations are (28.20) and (28.21).

We can also use (28.10) to derive relations between matrix elements of single particle operators in different configurations. If in $\Sigma_{i=1}^{n} \mathbf{f}_i$ the $\mathbf{f}_i$ are tensor operators of odd degree, we already obtained the result that the matrix of $\Sigma\, \mathbf{f}_i^{(k)}$ within the j^n configuration is diagonal in the seniority scheme and independent of n. For such odd tensors we can therefore write

$$\left(j^n v\, \alpha\, J \left\| \sum_{i=1}^{n} \mathbf{f}_i^{(k)} \right\| j^n v\, \alpha'\, J'\right) = \left(j^v v\, \alpha\, J \left\| \sum_{i=1}^{v} \mathbf{f}_i^{(k)} \right\| j^v v\, \alpha'\, J'\right) \quad k \text{ odd}. \tag{28.22}$$

In the case of even tensors we can use (26.33) and insert for the c.f.p. their values given by (28.10) and (28.11). Equation (26.33) written in the seniority scheme is

$$\left(j^n v\, J \left\| \sum_{i=1}^{n} \mathbf{f}_i^{(k)} \right\| j^n v' J'\right)$$

$$= n \sum_{v_1 J_1} [j^n v\, J \{| j^{n-1}(v_1 J_1) j\, J] \, [j^{n-1}(v_1 J_1) j\, J' |\} j^n v' J']$$

$$\times (j \| \mathbf{f}^{(k)} \| j) (-1)^{J_1 + j + J + k} \sqrt{(2J+1)(2J'+1)} \begin{Bmatrix} j & J & J_1 \\ J' & j & k \end{Bmatrix}. \tag{28.23}$$

Since both v and v' differ by one from v_1, it is clear that either $v = v'$ or $v - v' = \pm 2$. If $v' = v - 2$ (or $v' = v + 2$ which is essentially the same case) there is only one possibility for v_1, namely $v_1 = v - 1 = v' + 1$. In this case the summation in (28.23) is only over J_1 (and α_1 in case such a quantum number is required). Using the relations (28.10) and (28.11) we can express the c.f.p. in (28.23) in terms of $v \to v-1$ c.f.p. as follows

$$n \sum_{J_1} [j^n v\, J\{| j^{n-1}(v-1, J_1) j\, J] [j^{n-1}(v-1, J_1) j\, J' |\} j^n\, v-2, J'] (j \| \mathbf{f}^{(k)} \| j)$$
$$\times (-1)^{J_1+j+J+k} \sqrt{(2J+1)(2J'+1)} \begin{Bmatrix} j & J & J_1 \\ J' & j & k \end{Bmatrix}$$
$$= n \sqrt{\frac{v(2j+3-n-v)(n-v+2)v}{n(2j+3-2v)2n}} \sum_{J_1} [j^v v\, J\{| j^{v-1}(v-1, J_1) j\, J]$$
$$\times [j^{v-1}(v-1, J_1) j\, J' |\} j^v\, v-2, J']$$
$$\times (j \| \mathbf{f}^{(k)} \| j) (-1)^{J_1+j+J+k} \sqrt{(2J+1)(2J'+1)} \begin{Bmatrix} j & J & J_1 \\ J' & j & k \end{Bmatrix}. \tag{28.24}$$

The expression under the summation sign on the right-hand side of (28.24) multiplied by v is exactly the value of the double-barred matrix element of $\Sigma\, \mathbf{f}_i^{(k)}$ in the j^v configuration (there are no states with $v+1$ in the j^{v-1} configuration!). We thus obtain

$$\left(j^n v \alpha\, J \middle\| \sum_{i=1}^{n} \mathbf{f}_i^{(k)} \middle\| j^n\, v-2, \alpha' J' \right)$$
$$= \sqrt{\frac{(n-v+2)(2j+3-n-v)}{2(2j+3-2v)}} \left(j^v v \alpha\, J \middle\| \sum_{i=1}^{v} \mathbf{f}_i^{(k)} \middle\| j^v\, v-2, \alpha' J' \right). \tag{28.25}$$

This result holds for even values of k as well as for the trivial cases of odd k or $k=0$, when both sides of (28.25) vanish.

The other possibility, to be treated now, is when v' is equal to v. In this case the summation in (28.23) includes terms with $v_1 = v-1$ as well as terms with $v_1 = v+1$. If we use again the relations (28.10) and (28.11) we cannot reduce (28.23) to the j^v configuration as before, since we still obtain $v+2 \to v+1$ c.f.p. as well as $v \to v-1$ c.f.p. Thus, such matrix elements in every j^n configuration can be expressed in terms of matrix elements in *two* configurations, the j^v and j^{v+2} configurations. We first evaluate the following matrix element, using (26.32) or (26.33):

$$\left(j^n v\, J \middle\| \sum_{i=1}^{n} \mathbf{f}_i^{(k)} \middle\| j^n v\, J' \right)$$
$$= n \sum_{v_1 J_1} [j^n v\, J\{| j^{n-1}(v_1 J_1) j\, J] [j^{n-1}(v_1 J_1) j\, J' |\} j^n v\, J'] (J_1 j_n J \| \mathbf{f}_n^{(k)} \| J_1 j_n J')$$
$$= n \sum_{J_1} [j^n v\, J\{| j^{n-1}(v-1, J_1) j\, J] [j^{n-1}(v-1, J_1) j\, J' |\} j^n v\, J']$$
$$\times (J_1 j_n J \| \mathbf{f}_n^{(k)} \| J_1 j_n J') + n \sum_{J_1} [j^n v\, J\{| j^{n-1}(v+1, J_1) j\, J]$$
$$\times [j^{n-1}(v+1, J_1) j\, J' |\} j^n v\, J'] (J_1 j_n J \| \mathbf{f}_n^{(k)} \| J_1 j_n J'). \tag{28.26}$$

The second sum on the right-hand side of (28.26) can be reduced, with the help of (28.11) to the corresponding sum in the j^{v+2} configuration. We can add to it an appropriate fraction of the first sum, reduced by (28.10) to a sum in the j^{v+2} configuration. Thus, we obtain the full expression of the matrix element in the j^{v+2} configuration. What remains of the first sum can be reduced, as before, to the matrix element in the j^v configuration. We thus obtain for (28.26) the expression

$$n\frac{(n-v)(v+2)}{2n}\sum_{J_1}[j^{v+2}v\,J\{|j^{v+1}(v+1,J_1)j\,J]$$

$$\times[j^{v+1}(v+1,J_1)j\,J'|\}j^{v+2}v\,J'](J_1j_nJ||\mathbf{f}_n^{(k)}||J_1j_nJ')$$

$$+n\frac{(n-v)(v+2)}{2n}\sum_{J_1}[j^{v+2}v\,J\{|j^{v+1}(v-1,J_1)j\,J]$$

$$\times[j^{v+1}(v-1,J_1)j\,J'|\}j^{v+2}v\,J'](J_1j_nJ||\mathbf{f}_n^{(k)}||J_1j_nJ')$$

$$+n\left(1-\frac{n-v}{2}\frac{2j+1-2v}{2j+3-n-v}\right)\sum_{J_1}[j^nv\,J\{|j^{n-1}(v-1,J_1)j\,J]$$

$$\times[j^{n-1}(v-1,J_1)j\,J'|\}j^nv\,J'](J_1j_nJ||\mathbf{f}_n^{(k)}||J_1j_nJ')$$

$$=\frac{n-v}{2}\left(j^{v+2}v\,J\left\|\sum_{i=1}^{v+2}\mathbf{f}_i^{(k)}\right\|j^{v+2}v\,J'\right)+\left(1-\frac{n-v}{2}\right)\left(j^vv\,J\left\|\sum_{i=1}^{v}\mathbf{f}_i^{(k)}\right\|j^vv\,J'\right). \tag{28.27}$$

This expression can be further reduced by expressing the matrix elements in the j^{v+2} configuration in terms of those in the j^v configuration. To do this we have to actually calculate the matrix elements in the j^{v+2} configuration making use of the explicit construction of $\psi(j^{v+2}v\,J\,M)$ from $\psi(j^vvJM)$. This procedure is similar to the derivation of (27.23). Before proceeding we would like to remark in passing that for scalars, i.e., for $k=0$, the result (28.27) is consistent with the general relation (26.34). This general relation for $k=0$ can be written in the present scheme as

$$\left(j^nv\,J\left\|\sum_{i=1}^{n}\mathbf{f}_i^{(0)}\right\|j^nv\,J\right)=n\sqrt{\frac{2J+1}{2j+1}}(j||\mathbf{f}^{(0)}||j)$$

$$=\frac{n}{v}\left(j^vv\,J\left\|\sum_{i=1}^{v}\mathbf{f}_i^{(0)}\right\|j^vv\,J\right)\qquad k=0. \tag{28.28}$$

In the following we shall consider only the case with $k>0$.

We now proceed to calculate the following matrix element

$$\langle j^{v+2}v\, J\, M \Big| \sum_{i=1}^{v+2} f_\kappa^{(k)}(i) \Big| j^{v+2}v\, J'M' \rangle$$

$$= N \int \psi^*(j^{v+2}v\, J\, M) \Big[\sum_{i=1}^{v+2} f_\kappa^{(k)}(i)\Big] \mathscr{A}\psi[j^v(v\, J')j^2_{v+1,v+2}(0)\, J'M'] \qquad (28.29)$$

where $N = N_{v+2,vJ}$ is the appropriate normalization coefficient. We use the properties of $\mathscr{A}$ as a projection and obtain for the integral (28.29) the form

$$\frac{(v+1)(v+2)}{2} N \int \psi^*(j^{v+2}v\, J\, M) \Big[\sum_{i=1}^{v+2} f_\kappa^{(k)}(i)\Big] \psi[j^v(v\, J')j^2_{v+1,v+2}(0)\, J'M']$$

$$= \frac{(v+1)(v+2)}{2} N \int \psi^*(j^{v+2}v\, J\, M) \Big[\sum_{i=1}^{v} f_\kappa^{(k)}(i)\Big] \psi[j^v(v\, J')j^2_{v+1,v+2}(0)\, J'M']$$

$$+ \frac{(v+1)(v+2)}{2} N \int \psi^*(j^{v+2}v\, J\, M) [f_\kappa^{(k)}(v+1) + f_\kappa^{(k)}(v+2)]$$

$$\times \psi[j^v(v\, J')j^2_{v+1,v+2}(0)\, J'M']. \qquad (28.30)$$

The first integral on the right-hand side of (28.30) is simply

$$\frac{(v+1)(v+2)}{2} N[j^v(v\, J)j^2(0)J|\}j^{v+2}v\, J] \int \psi^*(j^v v\, J\, M) \Big[\sum_{i=1}^{v} f_\kappa^{(k)}(i)\Big] \psi(j^v v\, J'M')$$

$$= \langle j^v v\, J\, M \Big| \sum_{i=1}^{v} f_\kappa^{(k)}(i) \Big| j^v v\, J'M' \rangle \qquad (28.31)$$

in view of (27.21). The second integral can be written as

$$\frac{(v+1)(v+2)}{2} N^2 \int \mathscr{A}\psi^*[j^v(v\, J)j^2_{v+1,v+2}(0)\, J\, M]\, [f_\kappa^{(k)}(v+1) + f_\kappa^{(k)}(v+2)]$$

$$\times \psi[j^v(v\, J')j^2_{v+1,v+2}(0)\, J'M']. \qquad (28.32)$$

In the case of odd k, (28.32) vanishes due to the property (27.5) of odd tensor operators. In the general case we carry out the antisymmetrization explicitly and obtain as in (27.16)

$$\mathscr{A}\psi^*[j^v(v\,J)\,j^2_{v+1,v+2}(0)\,J\,M]$$

$$= \psi^*[j^v(v\,J)\,j^2_{v+1,v+2}(0)\,J\,M] - \sum_{h\neq v+1,v+2} \psi^*[j^v(v\,J)\,j^2_{h,v+2}(0)\,J\,M]$$

$$- \sum_{h\neq v+1,v+2} \psi^*[j^v(v\,J)\,j^2_{v+1,h}(0)\,J\,M]$$

$$+ \sum_{i<h\neq v+1,v+2} \psi^*[j^v(v\,J)\,j^2_{ih}(0)\,J\,M]. \tag{28.33}$$

When this expression is inserted into (28.32), the integral over the first term in (28.33) vanishes since for $k > 0$ the triangular condition cannot be satisfied (j_{v+1} and j_{v+2} are coupled to zero on both sides). Integrating each term of the last summation of (28.33) over the ith and hth coordinates also gives zero since in $\psi(j^v v\,J'M')$ there are no pairs coupled to $J=0$

The nonvanishing terms in the integral (28.32) are thus

$$-\frac{(v+1)\,(v+2)}{2}N^2 \sum_{h\neq v+1,v+2} \int \{\psi^*[j^v(v\,J)\,j^2_{h,v+2}(0)\,J\,M]$$

$$+ \psi^*[j^v(v\,J)\,j^2_{v+1,h}(0)\,J\,M]\}\,[f^{(k)}_\kappa(v+1) + f^{(k)}_\kappa(v+2)]$$

$$\times\,\psi[j^v(v\,J')\,j^2_{v+1,v+2}(0)\,J'M']. \tag{28.34}$$

The contributions to the integral of the two functions in the curly brackets of (28.34) are equal. The integral over one of them is equal to the integral over the other, with the numbers of particles $v+1$ and $v+2$ interchanged. Therefore, we can keep only the second function and multiply the result by 2. It is also clear that only that part of $\psi^*[j^v(v\,J)j^2_{v+1,h}(0)J\,M]$ will contribute in which particle $v+1$ and $v+2$ are coupled in an antisymmetric state (the $J=k$ state, according to the triangular conditions). For even values of k, we can thus replace $\mathbf{f}^{(k)}(v+1) + \mathbf{f}^{(k)}(v+2)$ by $\mathbf{f}^{(k)}(v+2)$ multiplied by 2. Hence we can write (28.34) in the form

$$-\,2(v+1)\,(v+2)\,N^2 \sum_{h\neq v+1,v+2} \int \psi^*[j^v(v\,J)\,j^2_{v+1,h}(0)\,J\,M] f^{(k)}_\kappa(v+2)$$

$$\times\,\psi[j^v(v\,J')\,j^2_{v+1,v+2}(0)\,J'M']. \tag{28.35}$$

Since particle $v+2$ in $\psi^*[j^v(v\,J)j^2_{v+1,h}(0)J\,M]$ is one of the v particles in the antisymmetric wave function, we try to obtain the same situation in

$\psi[j^v(v\ J')j^2_{v+1,v+2}(0)J'M']$. To do this we expand this latter function in c.f.p.

$$\psi[j^v(v\ J')j^2_{v+1,v+2}(0)\ J'M'] = (-1)^{P_h} \sum_{v_1J_1J_2} [j^{v-2}(v_1J_1)j^2(J_2)\ J'|\}j^v v\ J']$$

$$\times\ \psi[j^{v-2}(v_1J_1)j^2_{p,h}(J_2)j^2_{v+1,v+2}(0)\ J'M'] \qquad (28.36)$$

where P_h is the cyclic permutation that brings particle h to the last position in $\psi(j^v v\ J')$ and p denotes particle $v - 1$ if $h = v$ and particle v if $h < v$. We now apply a change of coupling transformation so that j_{v+1} and j_h will be coupled to J''. The only contribution to (28.35) will come from the term with $J'' = 0$ and the corresponding transformation coefficient is simply $1/(2j + 1)$. The only contributing term is thus

$$\frac{1}{2j+1}(-1)^{P_h} \sum_{v_1J_1J_2} [j^{v-2}(v_1J_1)j^2(J_2)\ J'|\}j^v v\ J']$$

$$\times\ \psi[j^{v-2}(v_1J_1)j^2_{p,v+2}(J_2)j^2_{v+1,h}(0)\ J'M']$$

$$= \frac{1}{2j+1}\psi[j^v(v\ J')j^2_{v+1,h}(0)\ J'M']. \qquad (28.37)$$

We insert the function (28.37) into (28.35) and integrate over the coordinates of particles $v + 1$ and h. This way we obtain a sum of v identical integrals. Each of these integrals is equal to $\int \psi^*(j^v v\ J\ M) f^{(k)}_\kappa(i) \psi(j^v v\ J'M')$ where i refers to any of the particles which appear in the antisymmetric wave functions $\psi(j^v v\ J\ M)$. The integral (28.35) then becomes simply

$$-\frac{2(v+1)\,(v+2)}{2j+1} N^2 \left\langle j^v v\ J\ M \left| \sum_{i=1}^{v} f^{(k)}_\kappa(i) \right| j^v v\ J'M' \right\rangle$$

$$= -\frac{4}{2j+1-2v} \left\langle j^v v\ J\ M \left| \sum_{i=1}^{v} f^{(k)}_\kappa(i) \right| j^v v\ J'M' \right\rangle. \qquad (28.38)$$

This equality was obtained by using the value of N^2 for $n = v + 2$ taken from (27.26).

By adding the expression (28.38) to (28.31) we obtain the matrix elements of *even* tensor operators (with $k > 0$) in the j^{v+2} configuration as follows

$$\left\langle j^{v+2} v\ J\ M \left| \sum_{i=1}^{v+2} f^{(k)}_\kappa(i) \right| j^{v+2} v\ J'M' \right\rangle$$

$$= \left(1 - \frac{4}{2j+1-2v}\right) \left\langle j^v v\ J\ M \left| \sum_{i=1}^{v} f^{(k)}_\kappa(i) \right| j^v v\ J'M' \right\rangle. \qquad (28.39)$$

Using this value in the matrix elements in the j^n configuration (28.27) we obtain the final result

$$\left(j^n v\,\alpha\,J\right\|\sum_{i=1}^{n}\mathbf{f}_i^{(k)}\left\|j^n v\,\alpha'\,J'\right)=\frac{2j+1-2n}{2j+1-2v}\left(j^v v\,\alpha\,J\right\|\sum_{i=1}^{v}\mathbf{f}_i^{(k)}\left\|j^v v\,\alpha'\,J'\right)\quad k>0\text{ even} \tag{28.40}$$

We see from (28.40) that the value of these matrix elements for given v, J, and J' decreases linearly with n. It reverses its sign at the middle of the shell and for $n = 2j + 1 - v$ has the opposite sign and the same absolute value as for $n = v$. This should be contrasted with the odd tensor operators whose matrix elements (28.22) are independent of n in absolute magnitude as well as in sign.

Equation (28.40) shows in particular, that in the middle of the shell, for $n = (2j + 1)/2$, matrix elements of even tensor operators, with $k > 0$, vanish between states with the same seniority of the $j^{(2j+1)/2}$ configuration. This has been shown so far only for states with seniorities $v < \frac{1}{2}(2j + 1)$, since only for such values of v the relation (28.40) holds. However, this statement is true for $v = \frac{1}{2}(2j + 1)$ also, namely

$$\left(j^{(2j+1)/2}\,v=\tfrac{1}{2}(2j+1),J\right\|\sum_{i=1}^{(2j+1)/2}\mathbf{f}_i^{(k)}\left\|j^{(2j+1)/2}\,v=\tfrac{1}{2}(2j+1),J'\right)=0 \qquad k>0\text{ even.} \tag{28.41}$$

Equation (28.41) can be proved by a calculation similar to the derivation of (28.38). We make use of the fact that if we add to the function $\psi(j^{v_0}v_0J\,M)$ with $v_0 = \frac{1}{2}(2j + 1)$ a pair coupled to $J = 0$ and antisymmetrize, the result must vanish. We therefore calculate the vanishing matrix element

$$0=\int\mathscr{A}\psi^*[j^{v_0}(v_0J)\,j^2_{v_0+1,v_0+2}(0)J\,M]\,[f_\kappa^{(k)}(v_0+1)+f_\kappa^{(k)}(v_0+2)]$$
$$\times\,\psi[j^{v_0}(v_0J')\,j^2_{v_0+1,v_0+2}(0)J'M'].$$

The procedure used in the derivation of (28.38) yields in this case

$$-\frac{4}{2j+1}\langle j^{v_0}v_0J\,M\left|\sum_{i=1}^{v_0}f_\kappa^{(k)}(i)\right|j^{v_0}v_0J'M'\rangle=0 \qquad k>0\text{ even.}$$

This proves the statement (28.41).

It is worthwhile to consider (28.41) in the special case of $j = \frac{3}{2}$ where there is only one state with $v = 2$ and $J = 2$ in the $(\frac{3}{2})^2$ configuration. In this case there is also only one even value of $k > 0$, namely, $k = 2$. Due to the relation (28.41) in the $(\frac{3}{2})^2$ configuration the Racah coefficient $W(2\,2\,2|\frac{3}{2}\,\frac{3}{2}\,\frac{3}{2})$ must vanish. This vanishing is called "accidental" since it does not follow from violation of the triangular conditions. It is somewhat surprizing to realize that it vanishes because of the symmetry properties which give rise to (28.41).

Equation (28.40) can be applied to the case of the quadrupole moment which is the $\kappa = 0$ component of a tensor operator of degree $k = 2$. However, (28.40) holds if only identical nucleons are outside closed shells. Therefore, it is not possible to add any quantitative result beyond the statements made in Part II.

We shall now establish the connection with the considerations of Part II which were based on the m-scheme. We saw that in the m-scheme, matrix elements of even (odd) single particle tensors in the j^n configuration have the opposite (the same) signs and the same absolute values as those in the *complementary* j^{2j+1-n} configuration. In the m-scheme it was possible to assign to each state of the j^n configuration a state of the complementary configuration in a unique way. Thus, to the state characterized by $(m_1, m_2, ..., m_n)$, with all the m_i different, we could assign the state characterized by the $2j + 1 - n$ missing m-values, namely $(m_{n+1}, ..., m_{2j+1-n})$. Let us see what happens when we go over from the m scheme to the scheme characterized by the eigenvalues of $\mathbf{J}^2$. Each state $\psi(j^nJM)$ can be obtained by a linear transformation of the states of the same configuration in the m scheme. By applying the same linear transformation to the corresponding m-scheme states in the j^{2j+1-n} configuration we obtain states with the same value of J in this configuration. If the j^n configuration has more than one state with a definite J, then each of these states can be obtained by a different linear transformation applied to the states in the m-scheme. By applying these transformations to the corresponding states in the j^{2j+1-n} configuration we obtain the various states of the same J in that configuration. In this way a correspondence can be established between the states $\psi(j^n\alpha JM)$ and $\psi(j^{2j+1-n}\alpha' J, -M)$. The relations (28.40), (28.41) along with the relation (28.22) for odd tensors, show that in the seniority scheme, states with a definite v and J in the j^n configuration are those corresponding to the states with the same J and v in the complementary j^{2j+1-n} configuration. This conclusion could be obtained also by considering the symmetry of the relation (27.23) around the middle of the j-shell.

A word of caution should be added about phases. The relations (28.22) and (28.40) for the case $J = J'$ concern only diagonal matrix elements in the seniority scheme. Therefore, the statement just made is correct up to a phase only. We can further conclude by considering $J \neq J'$ that states with the same v have in the j^{2j+1-n} configuration the same relative phases as those in the j^n configuration. It is interesting to recall at this point the relation (28.25) dealing with matrix elements of even tensors between states with *different* seniorities. We see that the matrix elements between states with seniorities v and $v - 2$ have the same absolute values for n and $2j+1-n$ but we also see that they have the same sign for any value of n. This is different from the situation in the m-scheme. It shows that two states with seniorities v and $v-2$ have in the j^{2j+1-n} configuration a relative sign opposite to the one in the j^n configuration.

In other words, if we want to use for the j^{2j+1-n} configuration the matrices of operators obtained in the seniority scheme in the j^n configuration, we should remember that the phases will not always be the same. We shall now establish the relation between the phases in the seniority scheme and the phases in the m-scheme introduced in Part II. By coupling $\frac{1}{2}(2j+1)$ pairs with $J=0$ and antisymmetrizing we obtain the state with $J=0$ $v=0$ of the closed j^{2j+1} configuration. The difference in phase between this state and the one in the m scheme given by $(j, j-1, j-2, ..., -j+2, -j+1, -j)$ is found to be $(-1)^{(2j+1)/4}$ if $(2j+1)/4$ is an integer or $(-1)^{(2j-1)/4}$ if it is not. There is no phase difference for $j=\frac{1}{2}, j=\frac{7}{2}, j=\frac{9}{2}, j=\frac{15}{2}$, etc. and there is a difference in sign for $j=\frac{3}{2}, j=\frac{5}{2}, j=\frac{11}{2}$, etc. For the state with $J=j$ $v=1$ the phase difference is opposite to that in the $v=0$ case. It follows from (28.25) that the phase differences of states with even values of v are equal to the $v=0$ phase difference multiplied by $(-1)^{v/2}$. Similarly, states with odd values of v have a phase difference equal to that of the $v=1$ state multiplied by $(-1)^{(v-1)/2}$. In the following we shall continue to use the seniority scheme for all values of n, also beyond the middle of the shell.

We go over now to the calculation of matrix elements of two particle operators in the seniority scheme. The relations to be obtained can be calculated from the recursion relations (28.4), (28.5), and (28.6) of the $n \rightarrow n-2$ c.f.p. using (26.45). We shall use a somewhat different procedure in which we express the two-particle operators in terms of single particle operators according to (26.46) and then make use of the relations (28.22), (28.25), and (28.40) satisfied by the latter operators.

Let the two-particle scalar operator be given as a sum of scalar products of tensor operators

$$V_{12} = \sum_k (\mathbf{f}_1^{(k)} \cdot \mathbf{f}_2^{(k)})$$

where k goes over odd as well as even values. Equation (26.46) can then be written in the seniority scheme as

$$\langle j^n v J \left| \sum_{i<h}^{n} V_{ih} \right| j^n v' J \rangle$$

$$= \sum_k \tfrac{1}{2} (j||\mathbf{f}^{(k)}||j)^2 \langle j^n v J|(\mathbf{U}^{(k)} \cdot \mathbf{U}^{(k)})|j^n v' J\rangle - \delta_{vv'} \frac{n}{2(2j+1)} \sum_k (j||\mathbf{f}^{(k)}||j)^2 \tag{28.42}$$

where $U_\kappa^{(k)} = \sum_{i=1}^n u_\kappa^{(k)}(i)$. The matrix elements which appear in (28.42) are of the form

$$\langle j^n v\, J M \Big| \sum_\kappa (-1)^\kappa U_\kappa^{(k)} U_{-\kappa}^{(k)} \Big| j^n v' J M\rangle$$

$$= \sum_\kappa (-1)^\kappa \sum_{v_1 J_1 M_1} \langle j^n v\, J M | U_\kappa^{(k)} | j^n v_1 J_1 M_1\rangle \langle j^n v_1 J_1 M_1 | U_{-\kappa}^{(k)} | j^n v' J M\rangle$$

$$= \sum_{v_1 J_1} \frac{(-1)^{J-J_1}}{2J+1} (j^n v\, J || \mathbf{U}^{(k)} || j^n v_1 J_1)\, (j^n v_1 J_1 || \mathbf{U}^{(k)} || j^n v' J). \tag{28.43}$$

The selection rules on the matrix elements of single particle operators imply in this case $v_1 = v \pm 2$ and $v_1 = v' \pm 2$. Therefore, v' can be either equal to v or differ from it by 2 or 4. We shall first treat the cases $v' = v - 4$ and $v' = v - 2$. In these cases only tensors with even degree (with $k > 0$) contribute to (28.42) and we have therefore to use only (28.25) and (28.40). For $v \neq v'$ the second term of (28.42) vanishes and only the first term should be considered.

If $v' = v - 4$, v_1 must be equal to $v - 2$ and there is no summation over other values of v_1. We thus obtain from (28.42) using (28.43),

$$\langle j^n v\, J \Big| \sum_{i<h}^n V_{ih} \Big| j^n v - 4 J\rangle = \sum_k \tfrac{1}{2} (j || \mathbf{f}^k || j)^2$$

$$\times \sum_{J_1} \frac{(-1)^{J-J_1}}{2J+1} (j^n v\, J || \mathbf{U}^{(k)} || j^n\, v-2, J_1)\, (j^n\, v-2, J_1 || \mathbf{U}^{(k)} || j^n\, v-4, J).$$

We insert for the reduced matrix elements in this expression the corresponding terms of the j^v configuration taken from (28.25). We thus obtain the following relation between matrix elements in the j^n configuration and matrix elements in the j^v configuration

$$\langle j^n v\, \alpha J \Big| \sum_{i<h}^n V_{ih} \Big| j^n\, v-4, \alpha' J\rangle$$

$$= \sqrt{\frac{(n-v+2)\,(n-v+4)\,(2j+3-n-v)\,(2j+5-n-v)}{8(2j+3-2v)\,(2j+5-2v)}}$$

$$\times \langle j^v v\, \alpha J \Big| \sum_{i<h}^v V_{ih} \Big| j^v\, v-4, \alpha' J\rangle. \tag{28.44}$$

In the next case, $v' = v - 2$, the sum in (28.43) (for every k) splits into two sums according to the two possible values, v and $v - 2$, of v_1.

In both these sums we substitute for the reduced matrix elements their values in terms of the reduced matrix elements in the j^v configuration using (28.25) and (28.40). We find that each sum is equal to the corresponding sum in the j^v configuration multiplied by the same factor. Therefore, the two sums together are equal to the full matrix element in the j^v configuration multiplied by this factor. Thus, we obtain

$$\langle j^n v\, \alpha\, J \Big| \sum_{i<h}^{n} V_{ih} \Big| j^n\, v-2,\, \alpha'\, J\rangle$$

$$= \frac{2j+1-2n}{2j+1-2v}\sqrt{\frac{(n-v+2)(2j+3-n-v)}{2(2j+3-2v)}}\,\langle j^v v \alpha J \Big| \sum_{i<h}^{v} V_{ih} \Big| j^v v-2, \alpha' J\rangle \tag{28.45}$$

The matrix elements (28.45) vanish for $n = (2j+1)/2$. This vanishing, due to (28.40), is true for $v < (2j+1)/2$ in which case (28.40) holds. However, the matrix element (28.45) vanishes also for $n = v = (2j+1)/2$ due to (28.41). As will be shown below, this property gives rise to a great simplification in the spectra of $(\frac{7}{2})^n$ configurations.

We now consider the diagonal elements of two-particle operators by taking the case $v' = v$. We start again from (28.42), using (28.43) and substitute for the reduced matrix elements their values from (28.25) and (28.40) for even tensors, as well as from (28.22) for odd tensors. We obtain in this way

$$\langle j^n v\, J \Big| \sum_{i<h}^{n} V_{ih} \Big| j^n v\, J\rangle = \sum_{k\ \text{odd}} \tfrac{1}{2}(j||\mathbf{f}^{(k)}||j)^2 \langle j^v v\, | J(\mathbf{U}^{(k)} \cdot \mathbf{U}^{(k)}) | j^v v\, J\rangle$$

$$+ \sum_{k>0\ \text{even}} \tfrac{1}{2}(j||\mathbf{f}^{(k)}||j)^2 \frac{(n-v)(2j+1-n-v)}{2(2j-1-2v)} \sum_{J_1} \frac{(-1)^{J-J_1}}{2J+1}$$

$$\times (j^{v+2} v\, J||\mathbf{U}^{(k)}||j^{v+2}\, v+2,\, J_1)(j^{v+2}\, v+2,\, J_1||\mathbf{U}^{(k)}||j^{v+2} v\, J)$$

$$+ \sum_{k>0\ \text{even}} \tfrac{1}{2}(j||\mathbf{f}^{(k)}||j)^2 \left(\frac{2j+1-2n}{2j+1-2v}\right)^2$$

$$\times \sum_{J_1} \frac{(-1)^{J-J_1}}{2J+1} (j^v v\, J||\mathbf{U}^{(k)}||j^v v\, J_1)(j^v v\, J_1||\mathbf{U}^{(k)}||j^v v\, J)$$

$$+ \sum_{k>0\ \text{even}} \tfrac{1}{2}(j||\mathbf{f}^{(k)}||j)^2 \frac{(n-v+2)(2j+3-n-v)}{2(2j+3-2v)}$$

$$\times \sum_{J_1} \frac{(-1)^{J-J_1}}{2J+1} (j^v v\, J||\mathbf{U}^{(k)}||j^v\, v-2,\, J_1)(j^v\, v-2,\, J_1||\mathbf{U}^{(k)}||j^v v\, J)$$

$$+ \frac{n(n-1)}{2}\frac{(j||\mathbf{f}^{(0)}||j)^2}{2j+1} - \frac{n}{2(2j+1)}\sum_{k>0}(j||\mathbf{f}^{(k)}||j)^2. \tag{28.46}$$

Equation (28.46) can be transformed into a closed expression by using the same technique as in previous cases. However, we shall calculate this closed expression in a simpler way. We notice that the matrix element (28.46) is a quadratic (and linear) function of n. Therefore, we can write

$$\langle j^n v\, J \Big| \sum_{i<h}^{n} V_{ih} \Big| j^n v\, J\rangle$$

$$= \langle j^v v\, J \Big| \sum_{i<h}^{v} V_{ih} \Big| j^v v\, J\rangle + \frac{n-v}{2}\,\alpha(v, J) + \frac{n-v}{2}(2j+1-n-v)\,\beta(v, J). \tag{28.47}$$

The last two terms are proportional to $n - v$ and therefore vanish for $n = v$. The quadratic term in n was chosen to have that specific form which vanishes not only for $n = v$ but also for $n = 2j + 1 - v$. We can now determine $\alpha(v, J)$ by computing (28.47) for $n = 2j + 1 - v$. A glance at (28.46) shows that the diagonal elements in the j^n and j^{2j+1-n} configurations differ only by the value of the last two terms. This is also evident from the discussion in the m-scheme since (28.46) is a diagonal element. Thus,

$$\langle j^{2j+1-v} v\, J \Big| \sum_{i<h}^{2j+1-v} V_{ih} \Big| j^{2j+1-v} v\, J\rangle = \langle j^v v\, J \Big| \sum_{i<h}^{v} V_{ih} \Big| j^v v\, J\rangle$$

$$+ \frac{2j+1-2v}{2}\,\frac{2j}{2j+1}(j||\mathbf{f}^{(0)}||j)^2 - \frac{2j+1-2v}{2}\,\frac{1}{2j+1}\sum_{k>0}(j||\mathbf{f}^{(k)}||j)^2. \tag{28.48}$$

Comparing (28.48) and the relation (28.47) for $n = 2j + 1 - v$, we see that $\alpha(v, J)$ is independent of v and J and its value, E_0, is given by

$$E_0 = \frac{2j}{2j+1}(j||\mathbf{f}^{(0)}||j)^2 - \frac{1}{2j+1}\sum_{k>0}(j||\mathbf{f}^{(k)}||j)^2. \tag{28.49}$$

The physical meaning of E_0 is very simple. We can calculate the interaction energy of the closed j-shell by taking V_{ih} to be a two-body interaction and computing (28.47) for $n = 2j + 1$, $v = 0$. In this case the expectation value (28.47) is simply $[(2j + 1)/2]E_0$. We actually saw in Part II that this E_0 is given by (22.8) and (22.9) in terms of the matrix elements in the j^2 configuration $\langle j^2 J|V_{12}|j^2 J\rangle = V_J$, by the expression

$$E_0 = (j||\mathbf{f}^{(0)}||j)^2 - \frac{1}{2j+1}\sum_{k\geqslant 0}(j||\mathbf{f}^{(k)}||j)^2 = \frac{2}{2j+1}\sum_{J \text{ even}}(2J+1)\,V_J$$

which is identical to (28.49).

In order to calculate the value of $\beta(v, J)$, we write (28.47) for $n = v + 2$

$$\langle j^{v+2} v\, J \Big| \sum_{i<h}^{v+2} V_{ih} \Big| j^{v+2} v\, J \rangle$$

$$= \langle j^{v} v\, J \Big| \sum_{i<h}^{v} V_{ih} \Big| j^{v} v\, J \rangle + E_0 + (2j - 1 - 2v)\, \beta(v, J).$$

This equation gives $\beta(v, J)$ in terms of the diagonal elements in the j^v and j^{v+2} configurations. Using this value of $\beta(v, J)$ we obtain the general result

$$\langle j^{n} v\, \alpha\, J \Big| \sum_{i<h}^{n} V_{ih} \Big| j^{n} v\, \alpha'\, J \rangle$$

$$= \langle j^{v} v\, \alpha\, J \Big| \sum_{i<h}^{v} V_{ih} \Big| j^{v} v\, \alpha'\, J \rangle + \frac{n-v}{2} E_0\, \delta_{\alpha\alpha'} + \frac{(n-v)(2j+1-n-v)}{2(2j-1-2v)}$$

$$\times \left[\langle j^{v+2} v\, \alpha\, J \Big| \sum_{i<h}^{v+2} V_{ih} \Big| j^{v+2} v\, \alpha'\, J \rangle - \langle j^{v} v\, \alpha\, J \Big| \sum_{i<h}^{v} V_{ih} \Big| j^{v} v\, \alpha'\, J \rangle - E_0\, \delta_{\alpha\alpha'} \right]. \tag{28.50}$$

This equation gives any diagonal element in the j^n configuration as the sum of the corresponding diagonal element in the j^v configuration and terms linear and quadratic in n (in which the j^{v+2} configuration also appears). The simplicity of this expression is a direct consequence of using the seniority scheme.

It is interesting to compare this general expression (28.50) to the special result (27.25) which was obtained for interactions which include scalar products of odd tensors only. In that case E_0 is equal to V_0 and the square bracket in the last term of (28.50) vanishes due to (27.25). Thus, in the case of odd tensor interactions, which have the pairing property, (28.50) reduces to the simpler formula (27.25). The difference of a general interaction from an odd tensor interaction is demonstrated by the fact that (28.50) contains the matrix elements of another configuration (the j^{v+2} configuration) besides those of the j^v configuration.

The special cases of (28.50) for $v = 0$ and $v = 1$ will be considered in detail in the next section. Another special case, for $v = 2$ and $j = \frac{5}{2}$, is also simple. In addition to the value $n = v = 2$, there is, in that case, only one other possible n value for $v = 2$, namely, $n = v + 2 = 4$. However, the last term in (28.50) vanishes for $n = 4$, $v = 2$. Thus, any interaction has the pairing property in $(\frac{5}{2})^n$ configurations.

In the case of an interaction that is diagonal in the seniority scheme, (28.50) can be further simplified. To achieve this we have to evaluate matrix elements of such an interaction in the j^{v+2} configuration. Let us calculate the matrix element

$$\left\langle j^{v+2}\, v'\, J \middle| \sum_{i<h}^{v+2} V_{ih} \middle| j^{v+2}\, v\, J \right\rangle$$

$$= N \int \psi^*(j^{v+2}\, v'\, J) \Big(\sum_{i<h}^{v+2} V_{ih} \Big) \mathscr{A} \psi(j^v(v\, J) j^2_{v+1,v+2}(0)\, J)$$

$$= \frac{(v+1)(v+2)}{2} N \int \psi^*(j^{v+2}\, v'\, J) \Big(\sum_{i<h}^{v+2} V_{ih} \Big) \psi(j^v(v\, J) j^2_{v+1,v+2}(0)\, J). \quad (28.51)$$

For $v' = v$, (28.51) is the matrix element which appears in (28.50). For $v' \neq v$, (28.51) must vanish in the case of an interaction which is diagonal in the seniority scheme. By considering separately the terms for which, $i, h \leq v$, the $V_{v+1,v+2}$ term and the other terms, we obtain for (28.51) the expression

$$\frac{(v+1)(v+2)}{2} N[j^v(v\, J) j^2(0)\, J| \}j^{v+2}\, v'\, J] \left\langle j^v\, v\, J \middle| \sum_{i<h}^{v} V_{ih} \middle| j^v\, v\, J \right\rangle$$

$$+ \frac{(v+1)(v+2)}{2} N[j^v(v\, J) j^2(0)\, J| \}j^{v+2}\, v'\, J] \langle j^2\, J=0 \mid V_{v+1,v+2} \mid j^2\, J=0 \rangle$$

$$+ \frac{(v+1)(v+2)}{2} N \int \psi^*(j^{v+2}\, v'\, J) \Big[\sum_{i=1}^{v} (V_{i,v+1} + V_{i,v+2}) \Big]$$

$$\times \psi(j^v(v\, J) j^2_{v+1,v+2}(0)\, J)$$

$$= \left\langle j^v\, v\, J \middle| \sum_{i<h}^{v} V_{ih} \middle| j^v\, v\, J \right\rangle \delta_{vv'} + V_0 \delta_{vv'}$$

$$+ \frac{(v+1)(v+2)}{2} N \int \psi^*(j^{v+2}\, v'\, J) \Big[\sum_{i=1}^{v} (V_{i,v+1} + V_{i,v+2}) \Big]$$

$$\times \psi(j^v(v\, J) j^2_{v+1,v+2}(0)\, J). \quad (28.52)$$

In the derivation of the last equality, the relation (27.21) for $n = v + 2$ was used as well as the properties of the c.f.p. in the seniority scheme.

The first two terms in (28.52) are diagonal in the seniority scheme. A possible contribution for $v' \neq v$ may come only from the last term. We observe that the expression under the integral sign in that term is fully symmetric in the coordinates of the first v particles. Therefore, we can restrict the summation to the first $v - 1$ particles (for $v > 1$) and multiply the result by $v/(v - 1)$. We then obtain

$$\int \psi^*(j^{v+2}\, v'\, J) \left[\sum_{i=1}^{v} (V_{i,v+1} + V_{i,v+2}) \right] \psi(j^v(v\, J) j^2_{v+1,v+2}(0)\, J)$$

$$= \frac{v}{v-1} \sum_{v_1' J_1' v_1 J_1} [j^{v+2}\, v'\, J \{| j^{v+1}(v_1'\, J_1') j\, J] [j^{v-1}(v_1 J_1) j\, J |\} j^v\, v\, J]$$

$$\times \int \psi^*(j^{v+1}(v_1'\, J_1') j_v\, J) \left[\sum_{i=1}^{v-1} (V_{i,v+1} + V_{i,v+2}) \right] \psi(j^{v-1}(v_1\, J_1) j^2_{v+1,v+2}(0) j_v\, J). \tag{28.53}$$

We now integrate over the coordinates of particle v, and since V_{ik} is a scalar operator, J_1' must be equal to J_1 and we obtain for (28.53) the expression

$$\frac{v}{v-1} \sum_{v_1' v_1 J_1} [j^{v+2}\, v'\, J \{| j^{v+1}(v_1'\, J_1) j\, J] [j^{v-1}(v_1\, J_1) j\, J |\} j^v\, v\, J]$$

$$\times \int \psi^*(j^{v+1}\, v_1'\, J_1) \left[\sum_{i=1}^{v-1} (V_{i,v} + V_{i,v+1}) \right] \psi(j^{v-1}(v_1\, J_1) j^2_{v,v+1}(0)\, J). \tag{28.54}$$

In the integrals appearing in (28.54) the numbers of particles $v + 1$ and $v + 2$ were conveniently changed to v and $v + 1$. The seniorities v_1' and v_1 may be equal to each other even if v' is different from v. For instance, if $v' = v - 2$, there is a term with $v_1' = v_1 = v - 1$ in (28.54).

The integral (28.53) in the j^{v+2} configuration is reduced in (28.54) into a sum of similar integrals in the j^{v+1} configuration. Any one of these integrals can be further reduced into a sum of such integrals in the j^v configuration and so on. Thus, if for a certain interaction all integrals (28.53) vanish in a given j^{v_0+2} configuration for all values of J and v' (including $v' = v_0$), the analogous integrals vanish also in every j^{v+2} configuration with $v > v_0$. As a result, the matrix elements (28.51) will vanish in all such configurations for $v' \neq v$. Due to (28.44) and (28.45) all matrix elements of the interaction between different seniorities v' and v ($v' \leq v + 2, v \geq v_0$) will vanish in all j^n configurations (with $n > v_0$).

From this discussion it follows that a sufficient condition for an interaction to be diagonal in the seniority scheme in all j^n configurations is that all integrals (28.53) vanish for $v = v_0 = 1$ in the j^3 configuration. The possible values of v' in this case are $v' = 3$ and $v' = 1$. We shall show that *any* given interaction can be expressed as a sum of an interaction for which (28.53) in the j^3 configuration vanishes for $v' = v = 1$ and a scalar term whose only contribution is a $\frac{1}{2}n(n-1)$ term in the j^n configuration. Therefore, the only real sufficient condition is the vanishing of (28.53) in the j^3 configuration between states with $v' = 3$ and $v = 1$. This, however, is also a necessary condition for the interaction to be diagonal in the seniority scheme. Thus, we obtain

The necessary and sufficient conditions for an interaction to be diagonal in the seniority scheme are

$$\int \psi^*(j^3 v' = 3 J = j)(V_{12} + V_{13})\psi(j_1 j_{23}^2(0) J = j) = 0 \qquad (28.55)$$

for all possible states with $v' = 3$ $J = j$ in the j^3 configuration.

As a special case we see that *any* two-body interaction in j^n configurations with $j \leq \frac{7}{2}$ is diagonal in the seniority scheme. In the j^3 configurations with $j \leq \frac{7}{2}$, there is only one state with $J = j$, namely, the one with $v = 1$. The condition (28.55) is thus trivially satisfied irrespective of the nature of the interaction.

Another result of the reduction (28.54) of the integrals (28.53) is that whenever the integrals appearing in (28.55) do not vanish, the matrix elements of the interaction which connect different seniorities are linear conbinations of these integrals. The integrals (28.53) are obtained by repeating (28.54) as linear combinations of the integrals appearing in (28.55). Therefore, this is also true for the matrix element (28.52) with $v' \neq v$. Due to the relations (28.44) and (28.45), every nondiagonal matrix element in the seniority scheme is a linear combination of the integrals appearing in (28.55).

We have still to consider the integrals (28.53) with $v' = v = 1$ in the j^3 configuration. Let us expand the interaction V_{12} in terms of scalar products of irreducible tensors as was done in (28.42):

$$V_{12} = \sum_k (j \| \mathbf{f}^{(k)} \| j)^2 (\mathbf{u}_1^{(k)} \cdot \mathbf{u}_2^{(k)}) = \sum_k F^k(\mathbf{u}_1^{(k)} \cdot \mathbf{u}_2^{(k)}). \qquad (28.56)$$

Substituting this expansion in (28.53), we obtain

$$\int \psi^*(j^3 v = 1 J = j)(V_{12} + V_{13})\psi(j_1 j_{23}^2(0) J = j)$$

$$= \int \psi^*(j^3 v = 1 J = j)\left[\sum_k F^k \mathbf{u}_1^{(k)} \cdot (\mathbf{u}_2^{(k)} + \mathbf{u}_3^{(k)})\right]\psi(j_1 j_{23}^2(0) J = j). \quad (28.57)$$

Expanding $\psi(j^3 v = 1 J = j)$ in c.f.p., we can carry out the integration in (28.57) and obtain for it, using (15.5) as well as (15.26) and (15.27), the expression

$$\sum_{k \text{ even}} F^k[j, j^2(k)\, J = j|\}j^3\, J = 1\, J = j]$$

$$\times (-1)^{k+2j} \begin{Bmatrix} j & k & j \\ 0 & j & k \end{Bmatrix} (-1)^{2j+k}\, 2\sqrt{2k+1} \begin{Bmatrix} j & j & k \\ 0 & k & j \end{Bmatrix}$$

$$= \frac{2}{2j+1} \sum_{k \text{ even}} \frac{F^k}{\sqrt{2k+1}} [j, j^2(k)\, J = j|\}j^3\, v = 1\, J = j].$$

Introducing into this expression the actual values of the c.f.p. given by (26.12), we obtain for the integral (28.57) the expression

$$\frac{2}{2j+1}\sqrt{\frac{2j-1}{3(2j+1)}}\left[F^0 - \frac{2}{2j-1}\sum_{k>0 \text{ even}} F^k\right]. \tag{28.58}$$

Thus, if in addition to satisfying (28.55), the interaction (28.56) makes (28.58) vanish, it follows that it is diagonal in the seniority scheme. However, it is rather easy to arrange the vanishing of (28.58) by changing, if necessary, the value of F^0. Starting from any given interaction (28.56), we write it as a sum of two interactions:

$$V_{12} = \left(F^0 - \frac{2}{2j-1}\sum_{k>0 \text{ even}} F^k\right)(\mathbf{u}_1^{(0)} \cdot \mathbf{u}_2^{(0)}) + V'_{12} = F^{0\prime}(\mathbf{u}_1^{(0)} \cdot \mathbf{u}_2^{(0)}) + V'_{12} \tag{28.59}$$

where

$$V'_{12} = \frac{2}{2j-1}\left(\sum_{k>0 \text{ even}} F^k\right)(\mathbf{u}_1^{(0)} \cdot \mathbf{u}_2^{(0)}) + \sum_{k>0} F^k(\mathbf{u}_1^{(k)} \cdot \mathbf{u}_2^{(k)}). \tag{28.60}$$

The definition (28.60) makes the integral (28.57) with V'_{12} vanish [as seen from (28.58)]. The other interaction is proportional to the product of two scalars, with $k = 0$, and is therefore diagonal in any orthogonal scheme in the j^n configuration; it has in this case the eigenvalue

$$\frac{n(n-1)}{2}\frac{F^{0\prime}}{2j+1} = \frac{n(n-1)}{2}\frac{1}{2j+1}\left(F^0 - \frac{2}{2j-1}\sum_{k>0 \text{ even}} F^k\right). \tag{28.61}$$

Therefore, if ΣV_{ih} is diagonal in the seniority scheme, so also is $\Sigma V'_{ih}$ and vice versa.

Any given interaction V_{12} that satisfies the condition (28.55) can thus be decomposed into V'_{12} and a scalar interaction according to (28.59). The interaction V'_{12} satisfies, in addition to (28.55), the vanishing of

(28.57), and thus $\Sigma V'_{ih}$ is diagonal in the seniority scheme. Therefore, the original interaction ΣV_{ih} is also diagonal in the seniority scheme [although it does not necessarily lead to the vanishing of (28.57)]. The interaction V'_{12} also has the pairing property. In fact, the vanishing of (28.53) for V'_{12} in the j^3 configuration leads to the vanishing of these integrals in any j^{v+2} configuration. Therefore, we obtain from (28.52)

$$\left\langle j^{v+2} v J \middle| \sum_{i<h}^{v+2} V'_{ih} \middle| j^{v+2} v J \right\rangle = \left\langle j^{v} v J \middle| \sum_{i<h}^{v} V'_{ih} \middle| j^{v} v J \right\rangle + V'_0. \tag{28.62}$$

We go now back to the general expression (28.50) and make use of the decomposition (28.59) and the properties (28.61) and (28.62) of the scalar interaction and the interaction V'_{ih}, respectively. E_0, given by (28.49), is expressed in the notation of (28.56) by

$$\frac{2j+1}{2} E_0 = \sum_{J \text{ even}} (2J+1) V_J = \tfrac{1}{2}\Big(2j F^0 - \sum_{k>0} F^k\Big). \tag{28.63}$$

On the other hand, we obtain from (28.56) the following expression for V_0:

$$\frac{2j+1}{2} V_0 = \tfrac{1}{2}\sum_k (-1)^k F^k = \tfrac{1}{2} F^0 + \tfrac{1}{2} \sum_{k>0 \text{ even}} F^k - \tfrac{1}{2} \sum_{k \text{ odd}} F^k. \tag{28.64}$$

Subtracting (28.64) form (28.63), we obtain the following expression for F'_0:

$$\frac{1}{2j+1} F'_0 = \frac{1}{2j+1}\Big(F^0 - \frac{2}{2j-1} \sum_{k>0 \text{ even}} F^k\Big) = \frac{1}{2j-1}(E_0 - V_0)$$

$$= \frac{2}{(2j-1)(2j+1)}\Big[\sum_{J \text{ even}} (2J+1) V_J - \frac{(2j+1)}{2} V_0\Big]. \tag{28.65}$$

Using (28.62) we obtain

$$\left\langle j^{v+2} v J \middle| \sum_{i<h}^{v+2} V_i^h \middle| j^{v+2} v J \right\rangle - \frac{(v+2)(v+1)}{2} \frac{F^{0\prime}}{2j+1}$$

$$= \left\langle j^{v} v J \middle| \sum_{i<h}^{v} V_{ih} \middle| j^{v} v J \right\rangle - \frac{v(v-1)}{2} \frac{F^{0\prime}}{2j+1} + V_0 - \frac{F^{0\prime}}{2j+1}.$$

Substituting from this into (28.50), we obtain, using (28.65),

$$\left\langle j^{n} v J \middle| \sum_{i<h}^{n} V_{ih} \middle| j^{n} v J \right\rangle$$

$$= \left\langle j^{v} v J \middle| \sum_{i<h}^{v} V_{ih} \middle| j^{v} v J \right\rangle + \frac{n-v}{2} V_0 + \frac{(n-v)(n+v-2)}{2(2j-1)} (E_0 - V_0). \tag{28.66}$$

This is the form which (28.50) assumes in the case of an interaction which is diagonal in the seniority scheme. Equation (28.66) can also be written in the following way which demonstrates the pairing property of V'_{ih} and the $n(n-1)$ behavior of the scalar interaction:

$$\left\langle j^n v J \middle| \sum_{i<h}^{n} V_{ih} \middle| j^n v J \right\rangle - \frac{n(n-1)}{2} \frac{F^{0\prime}}{2j+1}$$

$$= \left\langle j^v v J \middle| \sum_{i<h}^{v} V_{ih} \middle| j^v v J \right\rangle - \frac{v(v-1)}{2} \frac{F^{0\prime}}{2j+1} + \frac{n-v}{2}\left(V_0 - \frac{F^{0\prime}}{2j+1}\right). \tag{28.67}$$

We can also summarize the contents of (28.67) in words as follows:

Every interaction which is diagonal in the seniority scheme is equal to the sum of a scalar interaction and an interaction that has the pairing property. (28.68)

We have seen in Section 27 that any odd tensor interaction is diagonal in the seniority scheme and has the pairing property. The result (28.67) indicates that the only interaction that is diagonal in the seniority scheme and does not have the pairing property is the scalar interaction. We shall now show that

Every interaction that has the pairing property can be expressed as an odd tensor interaction. (28.69)

In order to prove the statement (28.69) we have to consider, along with the vanishing of (28.58) also the condition (28.55). Instead, we can use other relations which are obtained from (28.55) and the vanishing of (28.58). We shall consider the pairing property in the j^4 configuration for the $v = 2$ levels with $J > 0$ even. We shall compute the interaction energies

$$\left\langle j^4 v=2\, J \middle| \sum_{i<h}^{4} V_{ih} \middle| j^4 v=2\, J \right\rangle = 6 \sum_{J_1 J_2} [j^2(J_1) j^2(J_2) J| \}j^4 v=2\, J]^2 V_{J_2} \tag{28.70}$$

and compare them with the V_J. The c.f.p. which appear in (28.70) can be calculated in a straightforward manner. We start by adding to $\psi(j^2 J)$, with $J > 0$ even, a pair coupled to $J = 0$ and antisymmetrizing:

$$\mathscr{A}\psi(j^2_{12}(J)\, j^2_{34}(0)\, J) = \psi(j^2_{12}(J)\, j^2_{34}(0)\, J) - \psi(j^2_{13}(J)\, j^2_{24}(0)\, J) - \psi(j^2_{14}(J)\, j^2_{32}(0)\, J)$$

$$- \psi(j^2_{32}(J)\, j^2_{14}(0)\, J) - \psi(j^2_{42}(J)\, j^2_{31}(0)\, J) + \psi(j^2_{34}(J)\, j^2_{12}(0)\, J).$$

Using change of coupling transformations, we obtain for this function

$$\psi(j_{12}^2(J)\,j_{34}^2(0)\,J)+\psi(j_{12}^2(0)\,j_{34}(J)\,J)$$

$$-\sum_{J_1J_2}(1+(-1)^{J_1})\,(1+(-1)^{J_2})\,\langle j_1j_3(J)\,j_2j_4(0)\,J\mid j_1j_2(J_1)\,j_3j_4(J_2)\,J\rangle$$

$$\times\,\psi(j_{12}^2(J_1)\,j_{34}^2(J_2)\,J)$$

$$=\sum_{J_1J_2\text{ even}}\left[\delta_{J_1J}\,\delta_{J_20}+\delta_{J_10}\,\delta_{J_2J}+4\sqrt{\frac{(2J_1+1)\,(2J_2+1)}{2j+1}}\begin{Bmatrix}j & J_1 & j\\ J_2 & j & J\end{Bmatrix}\right]$$

$$\times\,\psi(j_{12}^2(J_1)\,j_{34}^2(J_2)\,J).\tag{28.71}$$

The latter equality was obtained by using the actual values (14.37) of the transformation matrix elements as well as their special values given by (15.3). The normalization of the function (28.71) can be carried out by using (27.21) to obtain the normalization factor. We obtain in this way the following values of the c.f.p. for any value of $j>\frac{3}{2}$:

$$[j^2(J_1)\,j^2(J_2)\,J|\}\,j^4\,v=2\,J]$$

$$=\sqrt{\frac{2j+1}{6(2j-3)}}\left(\delta_{J_1J}\,\delta_{J_20}+\delta_{J_10}\,\delta_{J_2J}+4\sqrt{\frac{(2J_1+1)\,(2J_2+1)}{2j+1}}\begin{Bmatrix}j & J_1 & j\\ J_2 & j & J\end{Bmatrix}\right).\tag{28.72}$$

Substituting the c.f.p. (28.72) in (28.70), we obtain

$$\left\langle j^4\,v=2\,J\left|\sum_{i<h}^{4}V_{ih}\right|j^4\,v=2\,J\right\rangle$$

$$=\frac{2j-3}{2j+1}V_0+\frac{2j-7}{2j-3}V_J+\frac{16}{2j-3}$$

$$\times\sum_{J_1\text{ even}}\sum_{J_2>0\text{ even}}(2J_1+1)\,(2J_2+1)\begin{Bmatrix}j & J_1 & j\\ J_2 & j & J\end{Bmatrix}^2V_{J_2}$$

$$=\frac{2j-3}{2j+1}V_0+\frac{2j-7}{2j-3}V_J+\frac{8}{(2j+1)\,(2j-3)}\sum_{J_2>0\text{ even}}(2J_2+1)\,V_{J_2}$$

$$-\frac{8}{2j-3}\sum_{J_2>0\text{ even}}(2J_2+1)\,V_{J_2}\begin{Bmatrix}j & j & J_2\\ j & j & J\end{Bmatrix}.\tag{28.73}$$

The last equality in (28.73) was obtained by using (15.10) and (15.11).

The last term in (28.73) contains expressions that were considered before. In (27.35) we obtained an expression for the operator δ_{J_0J}, which has in it the term $(2J_0 + 1)\, W(jj\,J_0 \,|\, jj\,J)$. We can obtain for δ_{J_0J}, in the same way that (27.35) was obtained, an expression analogous to it, which will contain, in addition to that term, scalar products of odd tensors only. This expansion is

$$\delta_{J_2J} = (2J_2 + 1) \begin{Bmatrix} j & j & J_2 \\ j & j & J \end{Bmatrix} + 2(2J_2 + 1) \sum_{k \text{ odd}} (2k + 1) \begin{Bmatrix} j & j & k \\ j & j & J_2 \end{Bmatrix} \begin{Bmatrix} j & j & k \\ j & j & J \end{Bmatrix}. \tag{28.74}$$

Using this expansion we obtain the last summation in (28.73) in the form

$$\sum_{J_2>0 \text{ even}} (2J_2 + 1)\, V_{J_2} \begin{Bmatrix} j & j & J_2 \\ j & j & J \end{Bmatrix}$$

$$= \sum_{J_2>0 \text{ even}} V_{J_2}\, \delta_{J_2J} - 2 \sum_{\substack{k \text{ odd} \\ J_2>0 \text{ even}}} (2J_2 + 1)\, V_{J_2} (2k + 1) \begin{Bmatrix} j & j & k \\ j & j & J_2 \end{Bmatrix} \begin{Bmatrix} j & j & k \\ j & j & J \end{Bmatrix}. \tag{28.75}$$

We make use of the identity (28.75) in (28.73) and recall that in this case $J > 0$ so that $\Sigma V_{J_2}\, \delta_{J_2J} = V_J$. We obtain

$$\left\langle j^4\, v = 2\, J \left| \sum_{i<h}^{4} V_{ih} \right| j^4\, v = 2\, J \right\rangle$$

$$= \frac{2j - 3}{2j + 1} V_0 + \frac{2j - 15}{2j - 3} V_J + \frac{8}{(2j + 1)(2j - 3)} \sum_{J_2>0 \text{ even}} (2J_2 + 1)\, V_{J_2}$$

$$+ \frac{16}{2j - 3} \sum_{\substack{k \text{ odd} \\ J_2>0 \text{ even}}} (2J_2 + 1)\, V_{J_2} (2k + 1) \begin{Bmatrix} j & j & k \\ j & j & J_2 \end{Bmatrix} \begin{Bmatrix} j & j & k \\ j & j & J \end{Bmatrix}. \tag{28.76}$$

If the interaction V_{12}, considered above, has the pairing property, the scalar interaction contained in it, which gives rise to a $n(n - 1)$ term, must vanish. For any given interaction V_{12} diagonal in the seniority scheme, this condition is satisfied by the interaction V'_{12} defined in (28.59). Therefore, we can continue the discussion for the interaction V_{12} and show that the corresponding V'_{12} can be expressed as an odd tensor interaction. The requirement on the interaction matrix elements (28.76) is thus that they satisfy the relation (28.66), namely

$$\left\langle j^4\, v = 2\, J \left| \sum_{i<h}^{4} V_{ih} \right| j^4\, v = 2\, J \right\rangle$$

$$= V_J + V_0 + \frac{8}{(2j - 1)(2j + 1)} \left[\sum_{J>0 \text{ even}} (2J + 1)\, V_J - \frac{2j - 1}{2} V_0 \right]. \tag{28.77}$$

Using the result (28.76) in (28.77), we obtain the following equation for $V_J (J > 0)$:

$$V_J = \langle j^2 J | V_{12} | j^2 J \rangle = \frac{4}{3(2j+1)(2j-1)} \sum_{J_2 > 0 \text{ even}} (2J_2 + 1) V_{J_2}$$

$$+ \tfrac{4}{3} \sum_{\substack{k \text{ odd} \\ J_2 > 0 \text{ even}}} (2J_2 + 1) V_{J_2} (2k+1) \begin{Bmatrix} j & j & k \\ j & j & J_2 \end{Bmatrix} \begin{Bmatrix} j & j & k \\ j & j & J \end{Bmatrix} \quad \text{for } J > 0. \quad (28.78)$$

The expansion (28.78) is indeed in terms of only odd tensors in addition to the scalar term. However, this expansion is not yet complete since it holds only for $J > 0$. The inclusion of the $J = 0$ state does not present any difficulty since there exists the operator q_{12} which vanishes for the states with $J > 0$ in the j^2 configuration. We first calculate the contribution of the interaction which appears in (28.78) for $J = 0$. The difference between this contribution and V_0, multiplied by $q_{12}/(2j + 1)$, can then be added to the interaction without affecting (28.78) for $J > 0$. Putting $J = 0$ in (28.78) we obtain

$$\frac{4}{3(2j+1)(2j-1)} \sum_{J_2 > 0 \text{ even}} (2J_2 + 1) V_{J_2}$$

$$+ \frac{4}{3(2j+1)} \sum_{J_2 > 0 \text{ even}} (2J_2 + 1) V_{J_2} \sum_{k \text{ odd}} (2k+1) \begin{Bmatrix} j & j & k \\ j & j & J_2 \end{Bmatrix}$$

$$= \frac{2}{3(2j-1)} \sum_{J_2 > 0 \text{ even}} (2J_2 + 1) V_{J_2}. \quad (28.79)$$

The last equality was obtained by using (15.16) and (15.17). We can now build an interaction that will have in the state with given J of the j^2 configuration the eigenvalues

$$\left[V_0 - \frac{2}{3(2j-1)} \sum_{J_2 > 0 \text{ even}} (2J_2 + 1) V_{J_2} \right] \frac{q_{12}}{2j+1}$$

$$+ \frac{4}{3(2j+1)(2j-1)} \sum_{J_2 > 0 \text{ even}} (2J_2 + 1) V_{J_2}$$

$$+ \tfrac{4}{3} \sum_{\substack{k \text{ odd} \\ J_2 > 0 \text{ even}}} (2J_2 + 1) V_{J_2} (2k+1) \begin{Bmatrix} j & j & k \\ j & j & J_2 \end{Bmatrix} \begin{Bmatrix} j & j & k \\ j & j & J \end{Bmatrix}. \quad (28.80)$$

This interaction has the eigenvalues V_J for $J > 0$ according to (28.78) and the eigenvalue V_0 for $J = 0$. Obviously, this is equivalent to an expansion of the original interaction V_{12} in the j^2 configuration.

We saw in (27.33) that q_{12} can be expanded in terms of odd tensors in addition to a scalar term. Making use of this expansion we can rewrite (28.80) as follows:

$$V_{12} = \frac{2}{(2j-1)(2j+1)} \Big[\sum_{J_2>0 \text{ even}} (2J_2+1) V_{J_2} - \frac{2j-1}{2} V_0 \Big]$$

$$- \frac{2}{2j+1} \Big[V_0 - \frac{2}{3(2j-1)} \sum_{J_2>0 \text{ even}} (2J_2+1) V_{J_2} \Big] \sum_{k \text{ odd}} (2k+1)(\mathbf{u}_1^{(k)} \cdot \mathbf{u}_2^{(k)})$$

$$- \tfrac{4}{3} \sum_{k \text{ odd}} \Big[\sum_{J_2>0 \text{ even}} (2J_2+1) V_{J_2} \begin{Bmatrix} j & j & k \\ j & j & J_2 \end{Bmatrix} \Big] (2k+1)(\mathbf{u}_1^{(k)} \cdot \mathbf{u}_2^{(k)}). \tag{28.81}$$

The first term in (28.81) is the scalar term of the interaction. If V_{12} has the pairing property, this term vanishes. The rest of the interaction is therefore V'_{12} which has the pairing property. We see from (28.81) that such an interaction can indeed be expressed in terms of scalar products of odd tensors only. We can thus summarize (28.68) and (28.69) by the following statement.

Every interaction which is diagonal in the seniority scheme can be expressed as a sum of a scalar interaction and an odd tensor interaction. (28.82)

29. Average Interaction Energies in the Seniority Scheme

In the preceding section we obtained the expression (28.50) for the expectation values of two-particle operators. The most important application of this formula is the calculation of the interaction energy in the j^n configuration. Equation (28.50) is rather complicated if the j^v configuration contains many states. However, for the two simple cases of $v = 0$ and $v = 1$, where there is only one state in each case with $J = 0$ and $J = j$, respectively, (28.50) can be written down explicitly. In both cases there is no interaction energy in the j^v configuration, while the j^{v+2} configuration is rather simple.

For the $J = 0$ state with $v = 0$ we obtain directly from (28.50)

$$\langle j^n\ v=0\ J=0 \Big| \sum_{i<k}^{n} V_{ik} \Big| j^n\ v=0\ J=0\rangle$$

$$= \frac{n}{2} E_0 + \frac{n(2j+1-n)}{2(2j-1)} [\langle j^2\ J=0 | V_{12} | j^2\ J=0\rangle - E_0]$$

$$= \frac{n(2j+1-n)}{2(2j-1)} V_0 + \frac{n(n-2)}{2(2j-1)} E_0$$

$$= \frac{n(2j+3-n)}{2(2j+1)} V_0 + \frac{n(n-2)}{(2j+1)(2j-1)} \sum_{J>0\ \text{even}} (2J+1) V_J. \qquad (29.1)$$

The last equality was obtained by inserting the value of E_0 from (22.8). Equation (29.1) is the expression of the energy in the j^n configuration in terms of the energies in the j^2 configuration. It also demonstrates the fact that the expectation value in the state with $v = 0$ is a linear combination of V_0 and of the average of V_{ik} over all states with seniority $v = 2$. The factor $2J + 1$ comes from the fact that $\langle J\ M | V_{12} | J\ M\rangle$ is independent of M and, hence, each V_J appears $2J + 1$ times when the above average is taken. The important point is that the coefficients of the linear combination in (29.1) are independent of the particular interaction considered. They are functions of n, v, and j only. Since the $J = 0$ state with the lowest seniority is the ground state of all even-even nuclei, (29.1) will turn out to be a very useful relation. We may remark that, in the case of an interaction which has the pairing property, where $E_0 = V_0$, (29.1) is reduced to (27.3).

In the case of $v = 1$, the energy of the only state, with $J = j$, is given by

$$\langle j^n\ v=1\ J=j \Big| \sum_{i<k}^{n} V_{ik} \Big| j^n\ v=1\ J=j\rangle = \frac{n-1}{2} E_0 + \frac{(n-1)(2j-n)}{2(2j-3)}$$

$$\times \left[\langle j^3\ v=1\ J=j \Big| \sum_{i<k}^{3} V_{ik} \Big| j^3\ v=1\ J=j\rangle - E_0\right]. \qquad (29.2)$$

We have already calculated the expectation value of the interaction energy in the $J = j$ state with $v = 1$ of the j^3 configuration. Inserting this value, given by (27.1), into (29.2) and using the value (22.8) for E_0, we obtain the explicit formula

$$\langle j^n\ v=1\ J=j \Big| \sum_{i<k}^{n} V_{ik} \Big| j^n\ v=1\ J=j\rangle$$

$$= \frac{n-1}{2} E_0 + \frac{(n-1)(2j-n)}{2(2j-3)}$$

$$\times \left[\frac{2j-1}{2j+1} V_0 + \frac{4}{(2j+1)(2j-1)} \sum_{J>0 \text{ even}} (2J+1) V_J - E_0\right]$$

$$= \frac{(n-1)(2j+2-n)}{2(2j+1)} V_0 + \frac{(n-1)^2}{(2j+1)(2j-1)} \sum_{J>0 \text{ even}} (2J+1) V_J. \qquad (29.3)$$

The $J = j$ states with $v = 1$ are the ground states of most odd even nuclei and therefore (29.3) will be used extensively in the following. Also in this case, the energies of the $v = 1$ states in the j^n configurations are linear combinations of V_0 and the average interaction energy of all $v = 2$ states in the j^2 configuration. This average will be denoted by

$$\bar{V}_2 = \sum_{J>0 \text{ even}} (2J+1) V_J \Big/ \sum_{J>0 \text{ even}} (2J+1)$$

$$= \frac{1}{(j+1)(2j-1)} \sum_{J>0 \text{ even}} (2J+1) V_J. \qquad (29.4)$$

The fact that both (29.1) and (29.3) are linear functions of only two interaction energies, V_0 and $\bar{V}_2$, is specific to the lowest seniorities $v = 0$ and $v = 1$. Obviously, this cannot hold in cases where there is more than one state with the same v. A trivial example is the case $v = 2$ in the j^2 configuration where the V_J are certainly not proportional to each other (with factors which are independent of the interaction).

However, even in this example, the average interaction energy of all states with $v = 2$, taken with the weights $(2J + 1)$, has this property (it is equal by definition to $\bar{V}_2$). In general,

the average interaction energy of all states with the same v in the j^n configuration is a linear function of V_0 and $\bar{V}_2$ only. (29.5)

In fact, (29.1) and (29.3) offer special cases of this statement since there is only one state with lowest seniority ($v = 0$ or $v = 1$) in the j^n configuration so that the average interaction energy coincides with the actual energy of this state.

Before we prove the general theorem (29.5) about the averages, we shall make use of it to derive some simple expressions for the average interaction energy of the states with a definite v in the j^n configuration. Once we know that these averages are linear functions of V_0 and $\bar{V}_2$ we can determine them right away. We first observe that the factor of V_0 in every state with the given v in the j^n configuration is given by (27.26) as $Q(n, v)/(2j + 1)$. Next we recall that, if all interaction energies V_J in the j^2 configuration are equal to V, the interaction energy of each state in the j^n configuration is simply $[n(n - 1)/2]\, V$. Hence the sum of the coefficients of V_0 and $\bar{V}_2$ in the expression of the average interaction energy is simply $n(n - 1)/2$. We thus obtain the general result

$$\langle j^n v \Big| \sum_{i<k}^{n} V_{ik} \Big| j^n v \rangle$$

$$\equiv \frac{\sum_{\alpha J} (2J+1) \langle j^n v \alpha J \Big| \sum_{i<k}^{n} V_{ik} \Big| j^n v \alpha J \rangle}{\sum_{\alpha J} (2J+1)}$$

$$= \frac{n(n-1)}{2} \bar{V}_2 + \frac{(n-v)(2j+3-n-v)}{2(2j+1)} (V_0 - \bar{V}_2). \tag{29.6}$$

In the special cases $v = 0$ and $v = 1$, this result reduces to (29.1) and (29.3), respectively. In the case $n = v$, the second term in (29.6) vanishes and the only term which remains is $[n(n - 1)/2]\, \bar{V}_2$. Obviously, V_0 should not appear in this case since there are no pairs coupled to $J = 0$. Equation (29.6) holds for any two-particle interaction. In the special case of an odd tensor interaction, where $\bar{V}_2 = V_0/2(j + 1)$, it reduces to the average of (27.25), according to the pairing property of such interactions. The physical meaning of (29.6) is straightforward. There are altogether $[n(n - 1)/2]$ interactions in the j^n configuration. In a state with the seniority v, there are exactly $Q(n, v)/(2j + 1)$ interactions in the $J = 0$ state (this is exactly the probability to find a pair coupled to $J = 0$). The other $[n(n-1)/2] - [Q(n, v)/(2j+1)]$ interactions are in $v = 2$ states of the two interacting particles.

We shall now derive (29.6) in a somewhat different method which can be used also in other cases. In order to illustrate this method we shall first discuss the simple case of the $(\frac{5}{2})^n$ configuration.

There are three states (with $J = 0, 2, 4$) in the $(\frac{5}{2})^2$ configuration of identical particles. As we often mentioned, the interaction energy in the $(\frac{5}{2})^n$ configuration is a linear combination of the energies in the $(\frac{5}{2})^2$ configuration. Therefore, *any* two interactions which have the same expectation values V_J, in the $(\frac{5}{2})^2$ configuration, have also the same expectation values in all $(\frac{5}{2})^n$ configurations. Two such interactions need not, of course, coincide in other configurations. Thus, they will generally be two different interactions. However, the fact that their expectation values coincide in the one specific j^2 configuration assures us that they will also coincide in all j^n configurations with the same value of j (implying also the same values of l and the radial quantum numbers), but different values of n. We may, therefore, try to find an interaction, as simple as possible, which will reproduce the energies in the $(\frac{5}{2})^2$ configuration. If we find such an interaction, whose expectation values are known for every state $\psi(j^n v\, J\, M)$, we shall have a closed expression for the energies in all states of the $(\frac{5}{2})^n$ configurations for the case of the original interaction.

Actually, we have already considered three such interactions. The simplest is $V_{ik} = 1$, for any i and k; its eigenvalues in any state of the j^n configuration are simply $n(n-1)/2$. Another interaction is $V_{ik} = 2(\mathbf{j}_i \cdot \mathbf{j}_k)$. Its eigenvalues in any state with a definite J in the j^n configuration are given by

$$\langle j^n J \left| \sum_{i<k}^{n} V_{ik} \right| j^n J \rangle = J(J+1) - nj(j+1).$$

The third interaction is diagonal in the seniority scheme and is simply $V_{ik} = q_{ik}$. Its eigenvalues in the states with definite v in the j^n configuration are given by (27.29) as $Q(n, v) = [(n-v)/2]\,(2j+3-n-v)$.

We now construct the interaction

$$V'_{ik} = a + 2b(\mathbf{j}_i \cdot \mathbf{j}_k) + cq_{ik} \tag{29.7}$$

and adjust the coefficients a, b, and c in such a way that (29.7) will reproduce the interaction energies V_J of the original interaction V_{ik} in the three states of the $(\frac{5}{2})^2$ configuration. To do this we need only write the three linear equations, for $J = 0$, $J = 2$, and $J = 4$,

$$V_J = a + b[J(J+1) - \tfrac{35}{2}] + c\frac{2-v}{2}(6-v) \tag{29.8}$$

and by solving them obtain a, b, and c in terms of the various V_J. The solutions of the three equations (29.8) are

$$a = \tfrac{1}{28}(5V_2 + 23V_4),\; b = \tfrac{1}{14}(V_4 - V_2),\text{ and } c = \tfrac{1}{6}V_0 - \tfrac{1}{42}(10V_2 - 3V_4). \tag{29.9}$$

Using now the interaction (29.7) with these values of a, b, and c, we obtain for the original interaction the result

$$\langle(\tfrac{5}{2})^n v\, J \Big| \sum_{i<k}^{n} V_{ik} \Big| (\tfrac{5}{2})^n v\, J\rangle$$

$$= a\frac{n(n-1)}{2} + b[J(J+1) - n\tfrac{35}{4}] + c\frac{n-v}{2}(8-n-v). \tag{29.10}$$

Equation (29.10), with the values (29.9) of a, b, and c, is a closed expression of the interaction energies of all states in the $(\frac{5}{2})^n$ configurations in terms of the interaction energies V_J in the $(\frac{5}{2})^2$ configuration. We notice that in this case the seniority v is a good quantum number since in no $(\frac{5}{2})^n$ configuration are there two states with the same value of J. Therefore (29.10) gives actually the *eigenvalues* of *any* interaction within the $(\frac{5}{2})^n$ configuration.

It may also be noted that, after the substitution of (29.9) in (29.10), the latter assumes the form

$$\langle(\tfrac{5}{2})^n v\, J|\sum_{i<k}^{n} V_{ik}|(\tfrac{5}{2})^n v\, J\rangle = \frac{n(n-1)}{2}\sum_{J'} A_{J'}\,\langle(\tfrac{5}{2})^2 J'|V_{12}|(\tfrac{5}{2})^2 J'\rangle.$$

Comparing this result with (26.45) we see that the $A_{J'}$ are given in terms of the $n \to n-2$ c.f.p. by

$$A_{J'} = \sum_{\alpha'' J''} [(\tfrac{5}{2})^{n-2}(\alpha'' J'')\,(\tfrac{5}{2})^2(J')J|\}(\tfrac{5}{2})^n v\, J]^2.$$

This method certainly works also in the simpler case of $j = \frac{3}{2}$, where we can put either b or c equal to zero. For $j > \frac{5}{2}$ we need, in general, more than three independent interactions to reproduce all levels of the j^2 configuration. However, if we are satisfied with the calculation of *averages of interaction energies* over groups of states with the same seniority in the j^n configuration, the above method can be applied after a slight modification.

We first treat the averages of states with the same seniority in the j^2 configuration. Since there are only two possible values of v, namely 0 and 2, we can express these averages in any case by using

$$V'_{ik} = a + bq_{ik}. \tag{29.11}$$

The requirement that the interaction (29.11) will have the same averages as the original interaction is

$$V_0 = V_0' = a + (2j+1)b$$

$$\bar{V}_2 = \bar{V}_2' = a.$$

From this follows that, to satisfy our conditions, a and b should be given by

$$a = \bar{V}_2 \qquad b = \frac{1}{2j+1}(V_0 - \bar{V}_2). \tag{29.12}$$

The interaction (29.11) with a and b given by (29.12) is a special two-particle interaction which has the same expectation values in all states of the j^2 configuration with $J > 0$ even, i.e., in all states with $v = 2$.

We can now calculate the expectation values of the interaction (29.11) in the j^n configuration. These values depend only on n and v and are given by

$$\langle j^n v\, J | \sum_{i<k}^{n} (a + bq_{ik}) | j^n v\, J \rangle$$

$$= \frac{n(n-1)}{2} a + Q(n, v) b$$

$$= \frac{n(n-1)}{2} \bar{V}_2 + \frac{(n-v)(2j+3-n-v)}{2(2j+1)} (V_0 - \bar{V}_2). \tag{29.13}$$

Since the expectation values of this particular interaction are independent of J, (29.13) gives directly the averages of this interaction over all states with the same seniority v. As stated in (29.5) the averages of *any* given interaction V_{ik} over all states with the same v in the j^n configuration are linear combinations of V_0 and $\bar{V}_2$ with coefficients which are *independent* of the interaction. Therefore, these averages are the same for any two interactions V_{ik} and V'_{ik} for which $V_0 = V_0'$ and $\bar{V}_2 = \bar{V}_2'$. If V_{ik} is any given interaction we can choose a and b to be given by V_0 and $\bar{V}_2$ of this interaction, according to (29.12). Then the averages of the interaction V_{ik} in the j^n configuration over all states with the same v will be also given by (29.13) whose right-hand side coincides with that of (29.6).

We shall now prove the statement (29.5) about the average interaction energy. We first note that

The average interaction energy of all states of the j^n configuration is a multiple of the average interaction energy of all states in the j^2 configuration. (29.14)

The coefficient of the two-particle average interaction in the average of the n-particle interaction is independent of the interaction considered. To see this we note that the sum

$$\sum_{vJ}(2J+1)\langle j^n v\, J|\sum_{i<k}^{n} V_{ik}|j^n v\, J\rangle = \sum_{vJM}\langle j^n v\, J M|\sum_{i<k}^{n} V_{ik}|j^n v\, J M\rangle \quad (29.15)$$

is the *trace* of the submatrix of the interaction energy $\sum_{i<k}^{n} V_{ik}$ in the j^n configuration. Therefore (29.15) is equal to the trace of this interaction matrix written in any scheme. Let us evaluate this trace in the m-scheme. The states in the m scheme are characterized by all sets of different $m_1\, m_2, \ldots, m_n$. As we saw in Section 25, the expectation value of any two-particle operator in such a state is the sum of all diagonal matrix elements $\langle m_i m_k|V_{ik}|m_i m_k\rangle$, where m_i and m_k go over all values $m_1, m_2, \ldots, m_n$ and the wave function $\psi(m_i, m_k)$ is antisymmetrized. In the group of all states of the j^n configuration all possible m_i appear in a symmetrical way. Therefore, it is clear that the trace of the matrix in the j^n configuration is equal to a multiple of the trace of V_{12} in the j^2 configuration. This latter trace is simply equal to $\sum_{\text{even } J=0}^{2j-1}(2J+1)V_J$. The factor of proportionality between the average interaction energy in the j^n configuration and this average in the j^2 configuration is simply $n(n-1)/2$.

In (29.15) V_0 plays the same role as all other V_J. It is only when we calculate the trace of the interaction submatrix characterized by a *definite* v, that V_0 has a special distinction. Clearly, this is due to the special role played by the $J=0$ state in the seniority scheme.

We shall prove (29.5) by induction with respect to n. First, (29.5) holds for $v=0$ and $v=1$ according to (29.1) and (29.3) for every n. It is clear that if (29.5) holds for all seniorities $v<n$ in the j^n configuration, it also holds for the average of $\langle j^n\, v=n\; J\, M|\sum V_{ik}|j^n\, v=n\; J\, M\rangle$. In fact, the sum over all states with seniority n is equal to the full sum (29.15) minus the sum over all states with seniorities $v<n$. Thus, the property (29.5) holds in the case $n=2$ both for $v=0$ and for $v=2$ and in the case $n=3$ both for $v=1$ and for $v=3$. We shall now prove that if (29.5) holds for $n-1$ and all possible seniorities then it holds also for n (and all possible seniorities).

We use (26.37), written in the seniority scheme, to calculate the matrix elements of $\sum V_{ik}$ and obtain for $n>2$

$$\langle j^n v\,\alpha\, J|\sum_{i<k}^{n} V_{ik}|j^n v\,\alpha\, J\rangle = \frac{n}{n-2}\sum_{v_1 v_1'\alpha_1\alpha_1' J_1}[j^{n-1}(v_1\alpha_1 J_1)j\, J|\}j^n v\,\alpha\, J]$$

$$\times [j^{n-1}(v_1'\alpha_1' J_1)j\, J|\}j^n v\,\alpha\, J]\,\langle j^{n-1} v_1\alpha_1 J_1|\sum_{i<k}^{n-1} V_{ik}|j^{n-1} v_1'\alpha_1' J_1\rangle. \quad (29.16)$$

The possible values of v_1 and v_1' are $v \pm 1$. We can therefore write (29.16) more explicitly as follows

$$\frac{n}{n-2} \sum_{\alpha_1 \alpha_1' J_1} [j^{n-1}(v+1, \alpha_1 J_1) j \, J| \} j^n v \alpha \, J] \, [j^{n-1}(v+1, \alpha_1' J_1) j \, J| \} j^n v \alpha \, J]$$

$$\times \langle j^{n-1} \, v+1, \alpha_1 \, J_1 \Big| \sum_{i<k}^{n-1} V_{ik} \Big| j^{n-1} \, v+1, \alpha_1' \, J_1 \rangle$$

$$+ \frac{n}{n-2} \sum_{\alpha_1 \alpha_1' J_1} [j^{n-1}(v-1, \alpha_1 J_1) j \, J| \} j^n v \alpha \, J] \, [j^{n-1}(v-1, \alpha_1' J_1) j \, J| \} j^n v \alpha \, J]$$

$$\times \langle j^{n-1} \, v-1, \alpha_1 \, J_1 \Big| \sum_{i<k}^{n-1} V_{ik} \Big| j^{n-1} \, v-1, \alpha_1' \, J_1 \rangle$$

$$+ \frac{2n}{n-2} \sum_{\alpha_1 \alpha_1' J_1} [j^{n-1}(v+1, \alpha_1 J_1) j \, J| \} j^n v \alpha \, J] \, [j^{n-1}(v-1, \alpha_1' J_1) j J \, | \} j^n v \alpha \, J]$$

$$\times \langle j^{n-1} \, v+1, \alpha_1 \, J_1 \Big| \sum_{i<k}^{n-1} V_{ik} \Big| j^{n-1} \, v-1, \alpha_1' J_1 \rangle. \qquad (29.17)$$

In order to obtain the average interaction, we have to multiply this expression by $(2J+1)$ and sum over all values of J which belong to the states with seniority v. We have then to evaluate sums of the form

$$\sum_{\alpha J} (2J+1) \, [j^{n-1}(v_1 \alpha_1 J_1) j \, J| \} j^n v \alpha \, J] \, [j^{n-1}(v_1' \alpha_1' J_1) j \, J| \} j^n v \alpha \, J].$$

As it stands, such a sum does not appear as an orthogonality relation. However, we can use (28.16) to transform such sums and then we can use the orthogonality relations (28.20) and (28.21).

In the case $v_1 = v_1' = v+1$, we first obtain by using (28.11)

$$\sum_{\alpha J} (2J+1) \, [j^{n-1}(v+1, \alpha_1 J_1) j \, J| \} j^n v \alpha \, J] \, [j^{n-1}(v+1, \alpha_1' J_1) j \, J| \} j^n v \alpha \, J]$$

$$= \frac{(n-v)(v+2)}{2n} \sum_{\alpha J} (2J+1) \, [j^{v+1}(v+1, \alpha_1 J_1) j \, J| \} j^{v+2} v \alpha \, J]$$

$$\times [j^{v+1}(v+1, \alpha_1' J_1) j \, J| \} j^{v+2} v \alpha \, J]. \qquad (29.18)$$

We now make use of (28.16) and then use the orthogonality relation (28.20) in the j^{v+1} configuration. Thus, we obtain for the sum (29.18) the following form

$$\frac{(n-v)(v+1)(2J_1+1)}{n(2j+1-2v)} \sum_{\alpha J} [j^v(v \alpha \, J) j \, J_1 | \} j^{v+1} \, v+1, \alpha_1 J_1]$$

$$\times [j^v(v \alpha \, J) j \, J_1 | \} j^{v+1} \, v+1, \alpha_1' J_1] = \frac{(n-v)(v+1)}{n(2j+1-2v)} (2J_1+1) \, \delta_{\alpha_1 \alpha_1'}.$$

This way we obtain a new type of orthogonality relations, namely

$$\sum_{\alpha J}(2J+1)\,[j^n v\,\alpha\,J\{|\,j^{n-1}(v+1,\alpha_1 J_1)j\,J]\,[j^{n-1}(v+1,\alpha_1' J_1)j\,J|\}j^n v\,\alpha\,J]$$
$$=\frac{(n-v)\,(v+1)}{n(2j+1-2v)}\,(2J_1+1)\,\delta_{\alpha_1\alpha_1'}. \tag{29.19}$$

In the next case, with $v_1 = v_1' = v - 1$, we first reduce the c.f.p. by using (28.10). We then use (28.16) and obtain with the help of the orthogonality relation (28.21) in the j^{v+1} configuration, the following relation

$$\sum_{\alpha J}(2J+1)\,[j^n v\,\alpha\,J\{|\,j^{n-1}(v-1,\alpha_1 J_1)j\,J]\,[j^{n-1}(v-1,\alpha_1' J_1)j\,J|\}j^n v\,\alpha\,J]$$
$$=\frac{(2j+3-n-v)\,(2j+4-v)}{n(2j+5-2v)}\,(2J_1+1)\,\delta_{\alpha_1\alpha_1'}. \tag{29.20}$$

In the last case, with $v_1 = v + 1$ and $v_1' = v - 1$, we first use (28.10) and (28.11). Making use now of (28.16) we obtain the ordinary orthogonality relation between states with seniorities $v + 1$ and $v - 1$ in the j^{v+1} configuration. As a result we obtain

$$\sum_{\alpha J}(2J+1)\,[j^n v\,\alpha\,J\{|\,j^{n-1}(v+1,\alpha_1 J_1)j\,J]\,[j^{n-1}(v-1,\alpha_1' J_1)j\,J|\}j^n v\,\alpha\,J]=0. \tag{29.21}$$

Multiplying (29.17) by $(2J + 1)$ and summing over J and α, we obtain, using (29.19), (29.20), and (29.21), the following result

$$\sum_{\alpha J}(2J+1)\left\langle j^n v\,\alpha\,J\left|\sum_{i<k}^{n}V_{ik}\right|j^n v\,\alpha\,J\right\rangle$$
$$=\frac{(n-v)\,(v+1)}{(n-2)\,(2j+1-2v)}\sum_{\alpha_1 J_1}(2J_1+1)$$
$$\times\left\langle j^{n-1}\,v+1,\alpha_1 J_1\left|\sum_{i<k}^{n-1}V_{ik}\right|j^{n-1}\,v+1,\alpha_1 J_1\right\rangle$$
$$+\frac{(2j+3-n-v)\,(2j+4-v)}{(n-2)\,(2j+5-2v)}\sum_{\alpha_1 J_1}(2J_1+1)$$
$$\times\left\langle j^{n-1}\,v-1,\alpha_1 J_1\left|\sum_{i<k}^{n-1}V_{ik}\right|j^{n-1}\,v-1,\alpha_1 J_1\right\rangle. \tag{29.22}$$

We thus see that the average interaction in the states with seniority v in the j^n configuration is a linear combination of the corresponding averages in the j^{n-1} configuration with seniorities $v + 1$ and $v - 1$. Therefore, if (29.5) holds for $n - 1$ and every possible seniority it holds also for n and every possible seniority. Since (29.5) holds for $n = 3$ it thus holds for every possible value of n and v.

A simple application of (29.22) for the case $n = v + 2$ may be of some interest. If we put $V_{ik} = q_{ik}$, the expectation values of this interaction, $Q(n, v)$, are the same for all states with the same v. Furthermore, we know that $Q(v + 1, v + 1) = 0$ so that we obtain directly

$$Q(v+2, v) \sum (2J+1)$$
$$= \frac{(2j+1-2v)(2j+4-v)}{v(2j+5-2v)} Q(v+1, v-1) \sum (2J_1+1).$$

Inserting the values of $Q(n, v)$ from (27.29) we obtain the result

$$\left[\sum (2J+1)\right]_v = \frac{(2j+4-v)(2j+3-2v)}{v(2j+5-2v)} \left[\sum (2J+1)\right]_{v-1}. \qquad (29.23)$$

In (29.23) we indicated the seniority of the states over which the summation has to be taken. Obviously these sums $[\Sigma(2J+1)]_v$ depend only on the seniority v but not on n. Thus, (29.23) gives the total number of states [every J value is counted as $(2J + 1)$ states] with the seniority v in terms of this number for the seniority $v - 1$.

30. Applications to Nuclear and Coulomb Energies

In Part II we have considered the order of levels in the j^2 configuration. It is pointed out there that for quite a general class of attractive interactions, the $J = 0$ state is the ground state and above it lie the levels with $J = 2, 4, 6, ..., 2j - 1$, in that order. We also saw that the shorter the range of the attractive potential, the larger the spacing between the $J = 0$ level and the other levels. We can summarize these results by saying that in these cases the $v = 0$ ($J = 0$) level lies considerably lower than the $v = 2$ levels. This situation should be contrasted with the case of a repulsive (ordinary) potential, like the Coulomb potential between electrons in atoms. Since in the $J = 0$ state there is maximum overlap of the spatial wave functions of the two particles, this $J = 0$ state will be the highest one in the case of a repulsive potential and the higher J levels will be lower.

Let us now see what can be said on the order of levels in the j^n configuration. Equation (29.6) gives all the necessary information on the center of mass of all levels with the same seniority. We immediately see that if the $J = 0$ level is lower than the center of mass of the $v = 2$ levels in the j^2 configuration, i.e., if $V_0 < \bar{V}_2$, the order of centers of mass in the j^n configuration is that of increasing seniority. The levels with $v = 0$ ($J = 0$) or $v = 1$ ($J = j$) are lower in this case than centers of mass characterized by higher values of v. The reason for this behavior is very simple: $Q(n, v)/(2j + 1)$, which is the number of pairs coupled to $J = 0$, is a monotonically decreasing function of v (for $v \leq n$ and $v \leq 2j + 1 - n$). Since $V_0 < \bar{V}_2$, centers of mass of levels with higher values of $Q(n, v)$ involve more pairs coupled to $J = 0$ and therefore lie lower [the total number of pairs is $n(n - 1)/2$]. On the other hand, if $V_0 > \bar{V}_2$, as in the case of a repulsive potential, the order of centers of mass will be reversed. If all levels with $v = 2$ in the j^2 configuration have approximately the same energy, our statement about the centers of mass might apply also to the individual levels. In the general case, however, nothing definite can be said about the positions of individual levels.

In special cases we can make more detailed statements on the order of levels. If the interaction considered can be written in terms of odd tensor operators and has therefore the pairing property, we can consider (27.25). This relation implies that the order and spacing of levels with seniority v in the j^n configuration is the same as that in the j^v configuration. It is clear that this is true also for the order and spacings of all levels with seniorities smaller than or equal to v. Therefore, in this case, the order of levels with $v = 0$ and $v = 2$ in the j^n configuration (with even n) is the same as that in the j^2 configuration. It should be pointed out that every interaction which is diagonal in the seniority scheme has this property. In fact, if we add to an odd tensor interaction a scalar term which is equal to the same constant for every i and k, the interaction will no more possess the pairing property. However, the contribution of the new term will be proportional to $n(n-1)/2$ and thus will not change the fact that the interaction is diagonal in the seniority scheme. Such a term will also not affect the order of levels and their spacings.

Such an interaction, although very specialized, is sufficient for expressing the energies of all j^n configurations with $j \leq \frac{7}{2}$. We saw that an interaction of the type (29.7) can replace any given interaction in the $(\frac{5}{2})^n$ configuration. This is also trivially true for the $(\frac{3}{2})^n$ and the $(\frac{1}{2})^n$ configurations. We also saw in Section 28 that any interaction in the $(\frac{7}{2})^n$ configuration can be replaced by one with only odd tensors and zero degree tensors.

It should be mentioned that, even for interactions possessing the pairing property, nothing can be said on the position of individual levels. The only statement which can be made is the one which follows from (29.6) on centers of mass of states with the same seniority. Groups of states with different seniorities (other than with $v = 0$ and $v = 2$) may very well overlap each other and thus it is impossible to say anything definite about the order of individual levels with different seniorities. For instance, in the j^3 configuration with an attractive interaction, some levels with $v = 3$ may be quite close to the $v = 1$ $J = j$ level and even lie lower than it even if the $J = 0$ level is well below all $v = 2$ levels in the j^2 configurations.

Let us discuss now in some more detail the energies of the $v = 0$ ($J = 0$) and $v = 1$ ($J = j$) states which are the ground states of j^n configurations in the case of short-range attractive potentials. Equations (29.1) and (29.3) which give the energies of the states with $v = 0$ and $v = 1$ in the j^n configurations, in terms of V_0 and $\bar{V}_2$, can be written as follows:

$$\langle j^n\ v=0\ J=0 \left| \sum V_{ik} \right| j^n\ v=0\ J=0 \rangle$$

$$= \frac{n(n-1)}{2} \left(\frac{2(j+1)\,\bar{V}_2 - V_0}{2j+1} \right) + \frac{n}{2}\,\frac{2(j+1)}{2j+1}\,(V_0 - \bar{V}_2) \qquad n \text{ even}$$

$$\langle j^n\ v=1\ J=j \left| \sum V_{ik} \right| j^n\ v=1\ J=j \rangle$$

$$= \frac{n(n-1)}{2} \left(\frac{2(j+1)\,\bar{V}_2 - V_0}{2j+1} \right) + \frac{n-1}{2}\,\frac{2(j+1)}{2j+1}\,(V_0 - \bar{V}_2) \qquad n \text{ odd.}$$

The two expressions for the energy, for even and odd n and minimum seniority in each case, can be combined in one formula

$$E_{\text{g.s.}}(j^n) = \frac{n(n-1)}{2}\left(\frac{2(j+1)\bar{V}_2 - V_0}{2j+1}\right) + \left[\frac{n}{2}\right]\frac{2(j+1)}{2j+1}(V_0 - \bar{V}_2) \quad (30.1)$$

where $[n/2]$ stands for the largest integer not exceeding $n/2$. The structure of the two terms in (30.1) is clear. If $V_{ik} = 1$ then $\bar{V}_2 = V_0$ and the second term vanishes while the first term simply counts the number of interactions $n(n-1)/2$. On the other hand, if V_{ik} is a pairing interaction then $\bar{V}_2 = V_0/2(j+1)$. The first term, quadratic in n, then vanishes and the second term is simply $[(n-v)/2]\,V_0$ in accordance with (27.25) (it should be remembered that for $v = 0$ or $v = 1$ there is no interaction in the j^v configuration). The factor multiplying $[n/2]$ is called the *pairing term*, since it gives the energy associated with saturated pairs in the j^n configuration. It is seen that the pairing term is bigger for bigger separations between the $J = 0$ state and the center of mass of the other states in the j^2 configuration.

We shall apply (30.1) to the evaluation of nuclear binding energies in the following. However, before going into it we shall use (30.1) for an approximate treatment of Coulomb energies of nuclei, i.e., for the electrostatic interaction energy of the protons in the nucleus.

We are confining ourselves to the discussion of Coulomb energies in nuclei rather than atoms, since the methods developed above are more appropriate for the former case. Thus, all the equations which relate properties of the j^n configurations to those of the j^2 configuration can be applied to actual cases only if the central field is the *same* for both configurations. This will turn out to be a rather good approximation for nuclei, whereas it is generally a poor approximation for atoms. The Coulomb energy of an electron in an atom, which is identical with its ionization energy, includes not only its interaction with the charge of the nucleus and with the other electrons in the atom. It also includes the change in energy of all other electrons which occurs when the atom is ionized, due to the change in the central field.

Another point which facilitates the treatment of Coulomb energies in nuclei is the nature of the ground state. In atoms, because of the repulsive Coulomb interaction, levels with high angular momentum lie lower. Apart from cases where there are no other states, seniorities $v = 0$ and $v = 1$ are never those of ground states.* Therefore (30.1) or its equivalent for LS coupling cannot be applied to the ground states of atomic configurations. In nuclei, on the other hand, the ground states are determined by the attractive *nuclear* interaction and have, as a rule,

(*) We have discussed so far only the jj-coupling scheme but also in LS coupling, which is the predominant coupling scheme for atoms, the situation is the same.

the lowest possible seniority. As a result, we can use (30.1) for the Coulomb energies. As we shall see, the Coulomb energy in nuclei shows different regularities than those observed in atoms. This difference is due mainly to the nature of the nuclear ground state. In fact, although it is due to the rather long-range Coulomb potential, the electrostatic energy in nuclei shows features which are remarkably similar to those of the attractive nuclear energy.

If we calculate the Coulomb energy in nuclei by using (30.1) we should remember that this is only an approximation. Nuclei contain both protons and neutrons and, as we shall see later, neither of them are in states with definite seniorities. Every state of the whole system of protons and neutrons outside closed shells can be expanded in a sum of products of wave functions of the protons and of the neutrons. Each product can be characterized by the values of the protons seniority and the neutrons seniority as well as by their other quantum numbers. Still, it turns out that in such an expansion of the ground states of nuclei, the predominant part is the one in which both protons and neutrons are in states of lowest seniority. We shall therefore attempt an approximate calculation of Coulomb energies in nuclei based on the use of (30.1).

The Coulomb energy of Z' protons in the j-orbit outside closed shells is equal to their electrostatic interaction with the protons in the closed shells plus their mutual electrostatic interaction. We discussed before the interaction of a single particle with closed shells and stated the fact that it is independent of the direction of the angular momentum of the single particle. Let C denote the electrostatic interaction of one proton in the j-orbit with the closed shells. The interaction of the Z' protons with the closed shells will then be simply $Z'C$. For the mutual electrostatic interaction of the Z' protons we shall use (30.1). Introducing the notation

$$a = \frac{[2(j+1)\bar{V}_2 - V_0]}{2j+1} \quad \text{and} \quad b = \frac{2(j+1)(V_0 - \bar{V}_2)}{2j+1},$$

where V_0 and $\bar{V}_2$ refer now only to the electrostatic interaction in the j-shell, the Coulomb energy considered is

$$E_c(Z') = Z'C + \frac{Z'(Z'-1)}{2}a + \left[\frac{Z'}{2}\right]b. \tag{30.2}$$

The experimental data on the Coulomb energy are mainly given by the energy differences between pairs of mirror nuclei. Since one nucleus of a mirror pair is obtained from the other by changing protons into neutrons and vice versa, and since the nuclear interaction is charge-independent, the difference in their binding energies is due only to electromagnetic interactions. Similar data can be gathered from the energy differences of other corresponding states, not necessarily ground states, in nuclei with the same mass number but different values of Z and N. In dealing with these cases we neglect the slight change in the nuclear wave functions due to the different Coulomb fields in the pairs of nuclei considered.

The difference in the Coulomb energy between two such nuclear states, one with Z' protons and the other with $Z' - 1$ protons, outside closed shells, is thus given by (30.2) as follows

$$E_c(Z') - E_c(Z' - 1) = \Delta_c(Z') = C + (Z' - 1)\,a + \tfrac{1}{2}[1 + (-1)^{Z'}]\,b. \quad (30.3)$$

If the effective central field in which the nucleons move is approximately the same for all nuclei in which the same j-shell is being filled, C, a, and b will be independent of Z' and can be determined by making a best fit of (30.3) to the experimental data.

Certain features of $E_c(Z')$ can be predicted directly from (30.3). From their definition we can see that, for a repulsive Coulomb potential, C, a, and b are all positive. This is clear for C. For b it follows from the fact that for the repulsive Coulomb potential the $J = 0$ state is above all states with $J > 0$ in the j^2 configuration ($V_2 < V_0$). The coefficient a vanishes in the short-range limit (δ-interaction). In the long-range limit, where all levels coincide ($V_0 = \bar{V}_2$), a is positive (for a repulsive force). The Coulomb potential is between these two limits and the coefficient a which belongs to it is also positive. Therefore, if we plot $\Delta_c(Z')$ as a function of Z' (or of Z) we should obtain *two* rising parallel straight lines, the upper line for even values of Z and the lower line for odd values of Z. At the beginning of a new j-shell a break in the lines can be expected since the values of C, a, and b do depend on the j-orbit considered.

The experimental values of $\Delta_c(Z)$ taken from current literature, are plotted as a function of Z in Fig. 30.1. It is clear that the points in any j-shell lie on two distinct and parallel straight lines—one for odd and the other for even values of Z. This behavior indicates that the potential well of the nucleus is approximately the same for all nuclei in the same j-shell. Furthermore, the straight lines for different j-shells have different slopes, clearly indicating the beginning and the end of the various shells.

More detailed results can be obtained by assuming a special form of the potential well, e.g., a harmonic oscillator, in which case C, a, and b can be calculated. We saw in Section 23 that the electrostatic interaction between particles moving in an oscillator well is given as a multiple of $e^2\sqrt{\nu/\pi}$ where ν is proportional to the "spring constant" of the oscillator. We can take the various experimental values of $\Delta_c(Z)$ and compare them to the theoretical expression. This way we can determine the parameter $e^2(\nu/\pi)^{1/2}$ in each case. If we do so, we find that $e^2(\nu/\pi)^{1/2}$ turns out to be fairly constant within the same j shell. For example, the six $\Delta_c(Z)$ of the $d_{5/2}$ shell determine an average $e^2(\nu/\pi)^{1/2} = 0.342$ Mev with all deviations within $\pm$ 0.003 Mev of this average. Similarly, the four $\Delta_c(Z)$ in the $d_{3/2}$ shell determine values of $e^2(\nu/\pi)^{1/2}$ within $\pm$ 0.004 Mev of an average value of 0.312 Mev.

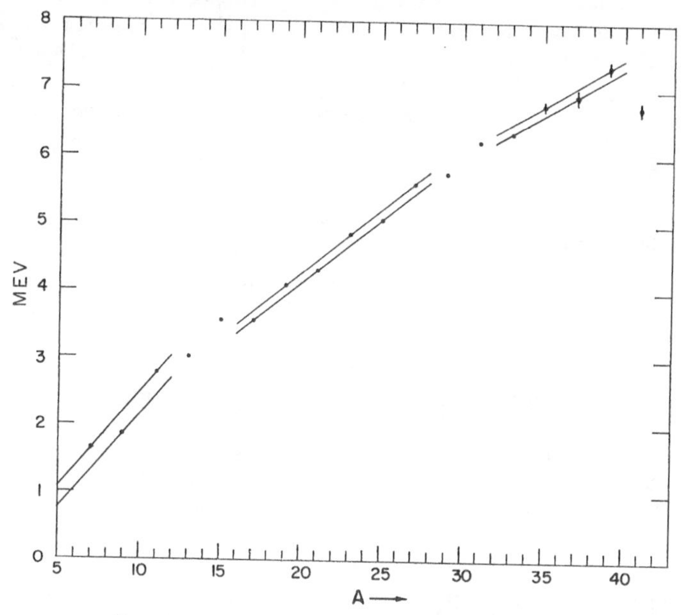

FIG. 30.1 Coulomb energy difference $\Delta_c(Z)$ in Mev.

A more careful analysis shows, however, that with a harmonic oscillator well it is impossible to reproduce exactly the straight lines of Fig. 30.1. In the $d_{5/2}$ shell the best fit to the experimental data is given by (30.3) with $C = 3.55$ Mev, $a = 0.375$ Mev, and $b = 0.15$ Mev. These constants calculated with harmonic oscillator wave functions, using the best value of $e^2(\nu/\pi)^{1/2} = 0.342$ Mev, turn out to be 3.54 Mev, 0.392 Mev, and 0.115 Mev respectively. In particular, the absolute value of the pairing term, thus obtained, is lower than that observed. In the $d_{3/2}$ shell the values of the constants which give the best fit to the experimental data are $C = 6.35$ Mev, $a = 0.275$ Mev, and $b = 0.125$ Mev. The values calculated with $e^2(\nu/\pi)^{1/2} = 0.312$ Mev turn out to be 6.27 Mev, 0.35 Mev, and 0.095 Mev, respectively. Again, the calculated pairing term is too small (in absolute value), and the calculated value of

a is too big. The good agreement, which is nevertheless obtained using the harmonic oscillator, is obviously due to the fact that by far the biggest contribution comes from C and relatively big changes in a and b have small influence on $E_c(Z')$. Thus, even though the harmonic oscillator potential may give a good agreement between theory and experiment for the total Coulomb energies, it is a rather poor potential for the determination of finer details, e.g., the Coulomb pairing term b.

Harmonic oscillator wave functions are frequently used in the nuclear shell model due to their great simplicity. It is important to realize that since the nuclear potential well seems to be the same within a j-shell, harmonic oscillator wave functions do not yield the right electrostatic energies. This fact should be kept in mind whenever harmonic oscillator wave functions are used in the calculation of the much more complicated nuclear interactions.

The Neutron and Proton $f_{7/2}$ Shell in Nuclei

We shall now treat in some detail energies of nuclei in which there are *either* neutrons or protons in the first $f_{7/2}$ shell outside closed shells. We shall consider here the nuclear energy proper and treat both ground states (binding energies) and excited levels of such nuclei. The treatment will be rather simple since in the cases to be considered, there are either only neutrons or only protons outside closed shells. This will, therefore, be a direct application of the results derived in the preceding sections.

As we saw in Part I, the first $f_{7/2}$ shell begins to be filled beyond $N = 20$ or $Z = 20$. Thus, nuclei with $f_{7/2}^n$ *neutron configurations*, with protons confined to closed shells only, are the Ca isotopes—Ca^{41} to Ca^{48}. The isotopes of Ni with $A \leq 56$ also belong to this group, but they have not yet been studied experimentally. Nuclei with $f_{7/2}^n$ *proton configurations* and closed neutron shells are those with 28 neutrons, i.e., Sc^{49}, Ti^{50}, V^{51}, Cr^{52}, Mn^{53}, Fe^{54}, Co^{55}, and Ni^{56}. There are sufficient data available so that we could attempt a comparison of the theoretical predictions with the experimental results.

As in the case of Coulomb energies, and also in the comparisons between theory and experiment in Part II, we shall try to derive the spectrum of energy levels and the binding energies in the $f_{7/2}$ shell in terms of a few parameters determined from the experimental data. In the case of the Coulomb energies we also tried to calculate these parameters from "first principles" using the harmonic oscillator well and the unmodified Coulomb potential between protons. This may be justified since the Coulomb force is a long-range force. Therefore, its expectation values are not expected to change much whether taken with the idealized shell model wave functions or with the real wave functions which probably contain strong correlations at small distances. The situation is quite different for the nuclear interaction. No reliable effective two-body interaction which can be used in the shell model has yet been obtained for actual nuclei. Therefore, we make no assumption on the effective interaction. We shall only assume that the effective nuclear forces are two-body forces and investigate whether this assumption is consistent with the experimental facts. In doing so we shall be using the basic assumption of the shell model, namely that the nuclear states can be approximately described by *jj*-coupling wave functions of nucleons in a central field. We shall also assume that the potential well, in which the nucleons move, is approximately the same for all nuclei in which the same shell is being filled. This is consistent with our previous findings on the relation between particle-particle and particle-hole configurations (Section 22) and on the behavior of Coulomb energies. However, we do not assume any specific form of the central potential well.

The fact that for two-particle interactions the matrix elements in the j^n configuration are linear combinations of the expectation values in the j^2 configuration, gives rise to relations between energy levels in various configurations. We investigate whether these theoretically predicted relations are satisfied by the experimental data. Whenever we find agreement with the experiment, we can express all experimental nuclear energies in a certain region in terms of a few parameters. These theoretical parameters are the expectation values of the effective interaction in two-particle configurations. In an ultimate theory these parameters should be calculable from the real free nucleon interaction. For the time being, we have to be satisfied with their values obtained from the analysis of the experimental data. Thus, we can only check the *consistency* of the experimental data with the shell model. If good agreement is obtained, the theoretical parameters, thus determined, can be used for the calculation of other, yet undetermined, nulcear energies.

We consider first energies of ground states, i.e., binding energies. All ground states observed in nuclei with either protons or neutrons in the $f_{7/2}$ shell have either $J = 0$ (in even-even nuclei) or $J = \frac{7}{2}$ (in odd-even nuclei). Since the $(\frac{7}{2})^2$ and $(\frac{7}{2})^4$ configurations have each only *one* $J = 0$ state, it is clear that every $J = 0$ state in a $(\frac{7}{2})^n$ configuration has necessarily $v = 0$. Similarly, since both $\frac{7}{2}$ and $(\frac{7}{2})^3$ configurations have only one $J = \frac{7}{2}$ state, any $J = \frac{7}{2}$ state in a $(\frac{7}{2})^n$ configuration has $v = 1$. Therefore, the seniorities of the ground states are $v = 0$ and $v = 1$ in even A and odd A nuclei respectively. The expectation values of the interaction energy in ground states of $f^n_{7/2}$ configurations are then given *exactly* in terms of (30.1). In order to obtain the binding energy, we should add to this mutual interaction energy the energy of the closed shells, the kinetic energy of the n $f_{7/2}$ nucleons and their interactions with the closed shells. This latter part is the sum of n equal single nucleon contributions. Let us denote by C the sum of the kinetic energy of each $f_{7/2}$ nucleon and its (nuclear) interaction with the closed shells. Then the total binding energy (B.E.) of a nucleus with n $f_{7/2}$ nucleons outside closed shells is equal to the B.E. of the closed shells plus the following terms

$$nC + \frac{n(n-1)}{2}\left(\frac{2(j+1)\bar{V}_2 - V_0}{2j+1}\right) + \left[\frac{n}{2}\right]\frac{2(j+1)}{2j+1}(V_0 - \bar{V}_2). \qquad (30.4)$$

According to our assumption of a constant potential well, the B.E. of the closed shells is the same for all nuclei considered and is equal to the B.E. of the nucleus with no $f_{7/2}$ nucleons outside the same closed shells. This closed shells nucleus, whose binding energy may thus serve as a reference energy in (30.4), is Ca^{40} if we consider $f^n_{7/2}$ neutron configurations. On the other hand, in nuclei with proton configurations the $f_{7/2}$ neutron shell is closed and the nucleus whose binding energy serves as reference is Ca^{48}. The energies C, V_0, and $\bar{V}_2$ in (30.4) are determined by the effective *nuclear* interaction between the nucleons (in the case of proton configurations this interaction contains also the electrostatic repulsion). We shall denote again the coefficients in (30.4) by a and b and remember that they have now a meaning that is different from the one in (30.2) and (30.3). The B.E. minus that of the closed shells nucleus, i.e., the part due to the $f_{7/2}$ nucleons only, is thus given by

$$\text{B.E.}\,(j^n) = nC + \frac{n(n-1)}{2}a + \left[\frac{n}{2}\right]b. \qquad (30.5)$$

It is worthwhile to mention that (30.5), in spite of its simplicity and intuitive meaning, is a rigorous result of the shell model. It is completely equivalent to the use of c.f.p. for each configuration with the appropriate expressions of the interaction energy given in the preceding sections. We shall not try to calculate C, a, and b, but rather see whether the experimental energies obey the relation (30.5). If we find good agreement, this will determine C, a, and b. The values of C for proton and neutron configurations should not be equal since in addition to the effect of the Coulomb interaction the closed shells are different in these two cases. The values of a and b must differ by the contribution of the mutual electrostatic interaction of the $f_{7/2}$ protons. In addition, any change in the *radial* wave functions of the neutrons and the protons, due to the electrostatic central field of protons in closed shells, may also cause differences in a and b in the two cases.

We have derived (30.5) from the theory by making the assumption on the constancy of the potential well. It is important to realize that, in the case of *slight* changes in the potential well, an expression that looks like (30.5) can give a good approximation. In fact, any change in the B.E. of the closed shells (which in our calculation was assumed to be constant), linear or quadratic in n, would be absorbed into the coefficients C and a respectively. Since this B.E. forms a considerable part of the total B.E., even slight changes may well affect C and a. Similarly, any linear change in C would be absorbed in a. Inasmuch as (30.5) reproduces the experimental data accurately, it means that the dependence of the B.E. of the closed shells on higher powers of n (n^3, n^4, etc.), or the dependence of C on higher powers of n (n^2, n^3, etc.) is very small, if it exists at all.

If we take now the expression (30.5) with the appropriate value of n for each nucleus and equate it to the experimental energy, we obtain linear equations for the three unknowns, C, a, and b. In order to check whether (30.5) gives a good description of the energies, we must have more than 3 such equations. Using the available experimental data given in current literature we have seven equations in the case of neutron configurations and six equations for proton configurations. If (30.5) with some specific values of C, a, and b would reproduce exactly all the data, the equations should be dependent. In practice, we determine C, a, and b so that the theoretical predictions would deviate from all experimental values as little as possible. This is done by requiring the sum of squares of these deviations to be as small as possible, i.e., by a *least squares fit*. The accuracy of the approximation can be measured by the smallness of the *root mean square* (*rms*) *deviation*, defined, as is well known, by

$$\text{rms deviation} \equiv \sqrt{\sum_{i=1}^{N} \Delta_i^2/(N-k)}. \tag{30.6}$$

Here, Δ_i is the deviation of the homogeneous part of the ith equation from the corresponding inhomogeneous part, N is the number of equations and, k is the number of unknowns. It is important to note that the rms deviation is proportional to $1/\sqrt{N-k}$. Therefore, if we increase the number of the parameters ("unknowns") we can make each Δ_i smaller but $1/\sqrt{N-k}$ will become bigger and the net result may be an increase of the rms deviation. Thus, the smallness of the rms deviation is not due to the fact that enough parameters were used, but is indicative of the actual fit between theory and experiment.

The results of the least squares fit for neutron and proton configurations are given in Table 30.1. The binding energies are given there for the sake of convenience, as positive quantities. The experimental values of the B.E. quoted are the total binding energies from which the B.E. of Ca^{40} was subtracted in the case of neutron configurations and the B.E. of Ca^{48} in the case of proton configurations. The calculated

TABLE 30.1

BINDING ENERGIES OF $f^n_{7/2}$ CONFIGURATIONS IN Mev

Nucleus	Binding energies[a]		Nucleus	Binding energies[b]	
	Experimental	Calculated		Experimental	Calculated
${}_{20}\mathrm{Ca}^{41}_{21}$	8.36	8.38	${}_{21}\mathrm{Sc}^{49}_{28}$	—	9.69
${}_{20}\mathrm{Ca}^{42}_{22}$	19.83	19.86	${}_{22}\mathrm{Ti}^{50}_{28}$	21.78	21.72
${}_{20}\mathrm{Ca}^{43}_{23}$	27.75	27.78	${}_{23}\mathrm{V}^{51}_{28}$	29.82	29.85
${}_{20}\mathrm{Ca}^{44}_{24}$	38.89	38.80	${}_{24}\mathrm{Cr}^{52}_{28}$	40.34	40.32
${}_{20}\mathrm{Ca}^{45}_{25}$	46.31	46.26	${}_{25}\mathrm{Mn}^{53}_{28}$	46.90	46.90
${}_{20}\mathrm{Ca}^{46}_{26}$	56.72	56.82	${}_{26}\mathrm{Fe}^{54}_{28}$	55.75	55.80
${}_{20}\mathrm{Ca}^{47}_{27}$	—	63.81	${}_{27}\mathrm{Co}^{55}_{28}$	60.85	60.81
${}_{20}\mathrm{Ca}^{48}_{28}$	73.95	73.93	${}_{28}\mathrm{Ni}^{56}_{28}$	—	68.17

([a]) From these the binding energy of Ca^{40} was subtracted.

([b]) From these the binding energy of Ca^{48} was subtracted.

values are the expressions (30.5) with the appropriate values of n and the values of C, a, and b determined from the least squares fit. It is seen that the agreement between the theoretical and experimental energies is very good. The rms deviation in the case of neutron configurations is 0.075 Mev which is only 0.12 % of the width of the energy range considered (64 Mev). In the case of proton configurations, the rms deviation is 0.056 Mev which is 0.14 % of the width of the corresponding energy range (39 Mev).

The best values of the parameters C, a, and b as determined from the least squares fit are given below, using the same sign convention used for binding energies, i.e., positive B.E. indicates real binding. These values are, in Mev,

for neutron configurations

$$C = 8.38 \pm 0.05 \qquad a = -0.23 \pm 0.01 \qquad b = 3.33 \pm 0.12$$

for proton configurations (30.7)

$$C = 9.69 \pm 0.04 \qquad a = -0.78 \pm 0.01 \qquad b = 3.11 \pm 0.09.$$

The statistical errors on these parameters are rather small which indicates their reliability. There is no simple way to compare the values of C since, as noted, even the closed shells are different in the two cases. The values of a and b in the case of proton configurations contain, as pointed out, the contribution of the electrostatic repulsion. The difference in the parameters in the two cases turns out to be in the right direction—less binding for the protons. The difference in the two values of b, about 0.2 Mev, is bigger than the expected value of the Coulomb pairing term (which is 0.15 Mev in the $d_{5/2}$ shell and only 0.125 in the $d_{3/2}$ shell). Similarly, the difference in the values of a, about 0.55 Mev, is bigger than expected (it is 0.375 in the $d_{5/2}$ shell and only 0.275 in the $d_{3/2}$ shell).

The analysis described above summarizes all that can be learned by considering energies of ground states. We therefore now turn our attention to excited levels in these $f^n_{7/2}$ configurations. As often stated, if the energy levels of $(f_{7/2})^2$ are given, all energy levels in $f^n_{7/2}$ configurations can be expressed as linear combinations of these. Our analysis of binding energies gives only the values of a and b (apart from C, which does not affect the spacings of excited states in a given $f^n_{7/2}$ configuration). According to (30.4) we therefore have the values of V_0 and $\bar{V}_2$. The only interesting parameter for dealing with excited states is $V_0 - \bar{V}_2$. This is given directly in terms of b by

$$V_0 - \bar{V}_2 = \frac{1}{27}[5(V_0 - V_2) + 9(V_0 - V_4) + 13(V_0 - V_6)] = \frac{8}{9}b. \qquad (30.8)$$

Because of our sign convention, (30.8) shows that the position of the center of mass of the $v = 2$ levels ($\bar{V}_2$) in the $(f_{7/2})^2$ configuration is *above* the $J = 0$ state. From the actual values of b given in (30.7) we deduce for (30.8) the values 2.96 ± 0.11 Mev and 2.77 ± 0.08 Mev for neutron and proton configurations respectively. In order to obtain more detailed information we have to consider more detailed level schemes of nuclei in which such $f^n_{7/2}$ configurations appear.

The states of the $f^2_{7/2}$ configuration, i.e., the states with $v = 0$, $J = 0$, and $v = 2$, $J = 2, 4, 6$ appear also in the $f^4_{7/2}$ and $f^6_{7/2}$ configurations. The order of levels and their spacings in the $f^6_{7/2}$ configuration should be the same as in the complementary $f^2_{7/2}$ configuration. However, it was mentioned in the preceding section that any two-body interaction is diagonal in the seniority scheme in $(\frac{7}{2})^n$ configurations. Therefore, also in the $f^4_{7/2}$ configuration the levels with $v = 0$ and $v = 2$ have the same order and spacings as in the $f^2_{7/2}$ configuration. Therefore, we should look in the even configurations for the same pattern of $v = 0$ and $v = 2$ energy levels. The case with $f^n_{7/2}$ configurations with odd values of n is even simpler. Apart from the trivial cases of the $f_{7/2}$ and $f^7_{7/2}$ configurations, where there is only the state $J = \frac{7}{2}$ $v = 1$, there are the two complementary configurations $f^3_{7/2}$ and $f^5_{7/2}$. We should therefore expect identical level schemes in these two cases. These predictions follow from the general principles of the shell model. The more detailed general relations of the type which should be satisfied are the relations giving the energy levels of the $f^3_{7/2}$ configuration and the $v = 4$ levels of the $f^4_{7/2}$ configuration in terms of the energy levels of the $f^2_{7/2}$ configuration.

Using the appropriate c.f.p. given in the Appendix, we can use either (26.37) or (26.45) (which are identical for $n = 3$) to express the interaction energies in the $f^3_{7/2}$ configuration in terms of those in the $f^2_{7/2}$ configuration. Denoting $\langle(\frac{7}{2})^3 J|\Sigma^3_{i<k} V_{ik}|(\frac{7}{2})^3 J\rangle$ by $V(J)$ (there is no danger of confusing these with V_J since the spins in the $(\frac{7}{2})^3$ configuration are half-integers), these expressions are found to be as follows

$$V(\tfrac{7}{2}) = \tfrac{3}{4} V_0 + \tfrac{1}{12}(5V_2 + 9V_4 + 13V_6)$$

$$V(\tfrac{3}{2}) = \tfrac{1}{14}(9V_2 + 33V_4)$$

$$V(\tfrac{5}{2}) = \tfrac{1}{66}(121V_2 + 12V_4 + 65V_6)$$

$$V(\tfrac{9}{2}) = \tfrac{1}{462}(143V_2 + 900V_4 + 343V_6)$$

$$V(\tfrac{11}{2}) = \tfrac{1}{66}(55V_2 + 39V_4 + 104V_6)$$

$$V(\tfrac{15}{2}) = \tfrac{1}{22}(15V_4 + 51V_6).$$

For the energies in the $f^2_{7/2}$ configuration, V_J, we can take the energies *above* the ground state, i.e., $V_J - V_0$, since the addition of a constant does not change the relative positions of energy levels. Since, in addition, the kinetic energy and the interaction with the closed shells are the same for all levels, $V_J - V_0$ is given simply by the spacing between the J-state and the ground state of the $f^2_{7/2}$ configuration.

Before considering the experimental data, let us look at the behavior of energy levels in various cases. In Fig. 30.2 we plotted the levels of the $f^2_{7/2}$ configuration (in heavy lines) as a function of a parameter x which changes the two-body interaction V_{ik} in some particular way. The dependence of V_{ik} on x is chosen to be such that for $x = 0$, $V_{ik} = \delta(r_i - r_k)$. For $x = 1$, V_{ik} is an interaction proportional to $(\mathbf{j}_i \cdot \mathbf{j}_k)$, that yields $V_J - V_0$ proportional to $J(J + 1)$. Between these two limits $V_2 - V_0$ remains constant, $V_4 - V_0$ changes linearly and $V_6 - V_0$ is chosen to vary in such a way that for a certain value $x = x_0$, V_{ik} reproduces the experimental values for both $(V_4 - V_0)/(V_2 - V_0)$ and $(V_6 - V_0)/(V_2 - V_0)$. Since the relative positions of levels in any j^2 configuration are independent of each other, and any order can be obtained by a proper choice of the interaction, it is clear that a V_{ik} as a function of x with the above properties can be constructed. It is convenient to introduce this "variable" interaction, since for each value of x it allows a comparison of the spectrum of the j^2 and j^3 configurations, etc.

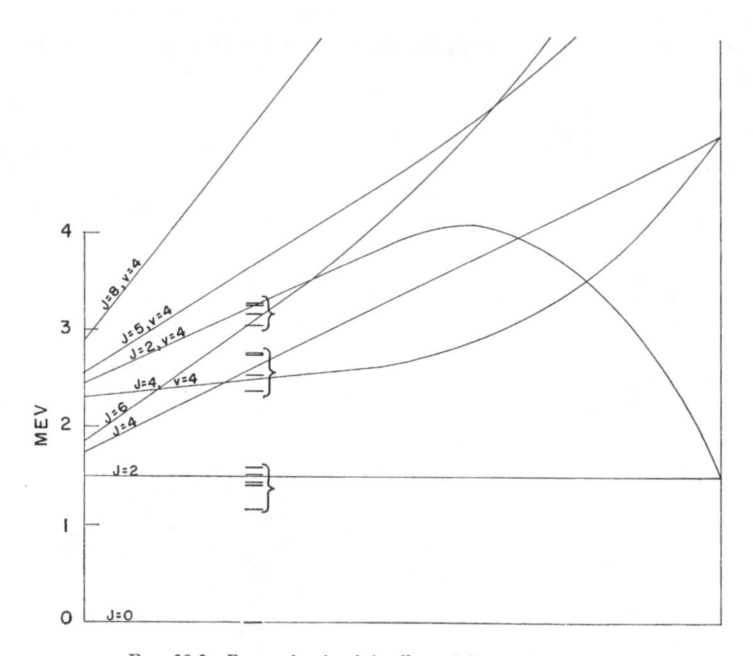

FIG. 30.2 Energy levels of the $f^2_{7/2}$ and $f^4_{7/2}$ configurations.

Using the same scale and limits, Fig. 30.3 gives the level spacings of the $f^3_{7/2}$ (and $f^5_{7/2}$) configuration. In the δ-limit all levels with $v = 3$ are grouped high above the $v = 1$ $J = \frac{7}{2}$ level. In the $J(J+1)$ limit the order of levels is according to their values of J, the $J = \frac{3}{2}$ and $J = \frac{5}{2}$ lying below the $J = \frac{7}{2}$ level. More interesting is the behavior of the levels between these two limits. When x changes from 0 to 1 the $J = \frac{5}{2}$ level is moving toward the $J = \frac{7}{2}$ ground state while all other levels remain well above it. Even the $J = \frac{3}{2}$ level, which becomes the ground state in the $J(J+1)$ limit, crosses the $J = \frac{7}{2}$ level only very close to the limit $x = 1$. In Fig. 30.2 the levels with $v = 4$ are also plotted. While they are above the $v = 2$ levels in the δ-limit ($x = 0$), the energy levels in the $J(J+1)$ limit ($x = 1$) depend only on J. Still, the $v = 4$ $J = 4$ level approaches relatively fast the corresponding $v = 2$ $J = 4$ level when x changes from 0 to 1 and even becomes lower than it in the middle of the range. On the other hand, the $v = 4$ $J = 2$ is always higher than the $v = 2$ $J = 2$ level up to the $J(J+1)$ limit.

This $v = 4$, $J = 2$ level crosses the $J = 4$ levels (with $v = 2$ and $v = 4$) only very close to the $J(J+1)$ limit ($x = 1$).

We shall now turn our attention to the experimentally established level schemes of nuclei in which $f^n_{7/2}$ configurations appear. The spins of the first excited states, both in the even and odd configurations are

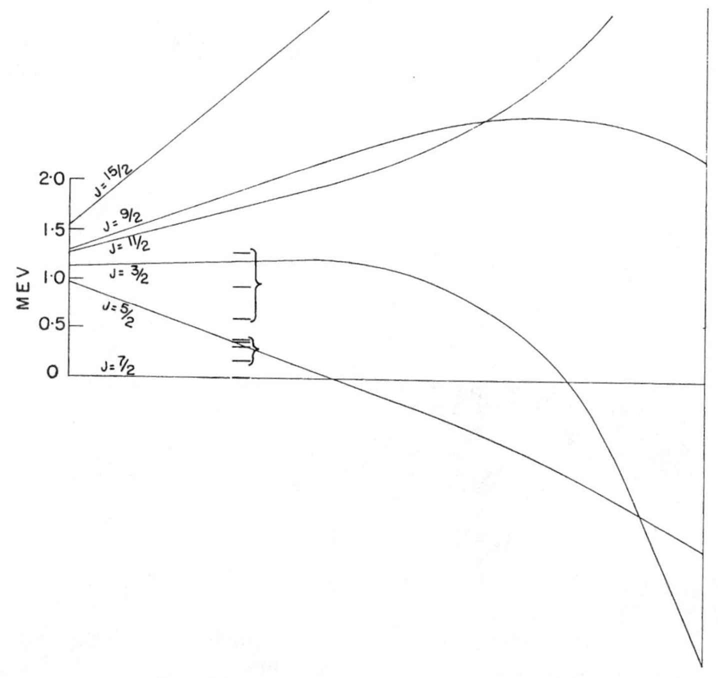

Fig. 30.3 Energy levels of the $f^3_{7/2}$ configuration.

known. They are $J = 2$ and $J = \frac{5}{2}$ respectively. Furthermore, these levels are sufficiently isolated from other levels so that their interpretation is straightforward in terms of $f^n_{7/2}$ configurations. The $J = 2$ levels are known in five of the six possible even configurations. Except for the Ca^{44} case, they all have roughly the same position above the ground state. Since the separation between the $v = 2$ $J = 2$ and the $J = 0$ levels in Ca^{44} should be the same as in the other cases, irrespective of the nuclear interaction, we conclude that the discrepancy is due to configuration interaction. The $J = \frac{5}{2}$ levels lie rather close to the $J = \frac{7}{2}$ ground states. In three of the four possible odd configurations, the $J = \frac{5}{2}$ states are roughly in the same position above the ground states. In Ca^{45} the $J = \frac{5}{2}$ lies lower, probably due to configuration interaction. Both in Ca^{44} and Ca^{45} the deviations are only a few tenths of Mev. This might indicate an order of magnitude of the maximum effects of configuration interaction on energy levels in this region.

In order to obtain the $J = \frac{5}{2} - J = \frac{7}{2}$ separation from the energy levels of the $f^2_{7/2}$ configuration, we need information on the positions of the $J = 4$ and $J = 6$ levels in this configuration. This is somewhat more difficult since in the region where these levels occur, there are already many levels, some of which certainly belong to other configurations. Yet, in some even configurations, levels with $J = 4$ and $J = 6$ were identified which probably belong to $f^n_{7/2}$ configurations. On the basis of this information we were able to draw the lines in Fig. 30.2. Such $J = 4$ and $J = 6$ levels can be tentatively identified also in other cases according to their position above the ground states. The $v = 2$ levels, thus adopted, give a $J = \frac{5}{2} - J = \frac{7}{2}$ separation, which agrees very well with the experimental values. The experimental values of the relative positions of the $v = 2$ and the $v = 0$ levels are indicated in Fig. 30.2. The experimental $J = \frac{5}{2} - J = \frac{7}{2}$ spacings are indicated in Fig. 30.3. It is seen that the best fits in the two figures are obtained at the same point of the abscissa. The center of mass of the adopted $v = 2$ levels, at 1.5, 2.7, and 3.2 Mev for $J = 2$, $J = 4$, and $J = 6$ respectively, lies at 2.72 Mev above the $J = 0$ level. This agrees rather well with the values predicted from the analysis of the binding energies, namely 2.77 ± 0.08 and 2.96 ± 0.11 Mev. The information on higher states in odd configurations is rather scarce. The only two identified states are both with $J = \frac{3}{2}$, in V^{51} at 0.93 Mev and in Ca^{43} at 0.59 Mev above the ground states. The position of this state taken from Fig. 30.3 is about 1.2 Mev above the ground state. The Ca^{43} value is certainly too low (probably due to interaction with the neighboring $p_{3/2}$ state); however, the V^{51} value agrees fairly well with the theoretical value. Nothing is yet known experimentally about the $v = 4$, $J = 4$, and other $v = 4$ levels in Ca^{44} and Cr^{52}.

The analysis presented above should be considered as a rather preliminary one. It indicates the power of the methods developed above and their general validity. More experimental data are required before we can obtain a clearer picture. It is evident, however, in which direction such experimental evidence has to be looked for and how the consistency of the shell model can be tested without going into too many details of the nuclear interaction.

31. Antisymmetric States in *LS*-Coupling

In this section we shall develop the method of constructing antisymmetric wave functions in LS-coupling. As in jj-coupling we shall introduce coefficients of fractional parentage, establish the seniority scheme, and calculate matrix element of single particle and two-particle operators. Almost all derivations, *mutatis mutandis*, are the same as in the somewhat simpler case of jj-coupling treated in the previous sections. Therefore, we shall not present here proofs of the theorems stated, but rather indicate all points in which there is something different in either the results or the derivations. As we saw in Section 25 the construction of antisymmetric states is straightforward in all cases but those of equivalent particles. Therefore, we shall consider here only the case of n particles in the same orbits, i.e., all having the same radial and angular momentum quantum numbers. Such configuration will be called, in the present case, l^n configurations.

We shall begin the construction of antisymmetric states in the l^n configuration by simply coupling ("geometrically") the state of the nth particle to the antisymmetric states of the l^{n-1} configuration. We should only remember that each particle carries now *two* independent angular momenta $\mathbf{s}$ and $\boldsymbol{l}$. The antisymmetric states of the l^{n-1} configuration are characterized by the values of S_1 and L_1. The vectors $\mathbf{S}_1$ $\mathbf{L}_1$ are coupled to form a $\mathbf{J}_1$ with a definite value of J_1. However, when we add another particle we have to couple $\mathbf{S}_1$ and $\mathbf{s}_n$ to form a total $\mathbf{S}$, $\mathbf{L}_1$, and $\boldsymbol{l}$ to a total $\mathbf{L}$, and only then couple $\mathbf{S}$ and $\mathbf{L}$ to the total $\mathbf{J}$. This implies that for the states of the l^{n-1} configuration we have to use the scheme characterized by $S_1L_1M_{S_1}M_{L_1}$ rather than that characterized by $S_1L_1J_1M_{J_1}$. The states of the l^n configuration obtained by direct coupling will be characterized by $S\,L\,M_SM_L$. Since the transformation from this scheme to the $S\,L\,J\,M$ scheme is very simple and straightforward we shall leave it out from our considerations for the time being and characterize the wave functions only by the values of S and L (and any additional necessary quantum number α). Only when it will become important, shall we refer to the values of J explicitly.

By coupling $\mathbf{s}_n$ to $\mathbf{S}_1$, $\boldsymbol{l}_n$ to $\mathbf{L}_1$ we obtain states of the n particles system with definite S and L, which are antisymmetric in the coordinates of the first $n-1$ particles. Therefore, any wave function with the same values of S and L, antisymmetric in *all* particle coordinates, must be a linear combination of these wave functions. The process of taking the appropriate linear combinations is a *projection* from the space, characterized by S and L, of wave functions antisymmetric in the first $n - 1$ particles to the subspace of fully antisymmetric wave functions.

Thus we obtain the following expression, analogous to (26.14), for the antisymmetric wave functions, in LS coupling

$$\psi(l^n \alpha\, S\, L) = \sum_{\alpha_1 S_1 L_1} [l^{n-1}(\alpha_1 S_1 L_1) l\, S\, L | \} l^n \alpha\, S\, L]\, \psi[l^{n-1}(\alpha_1 S_1 L_1) l_n S\, L]. \tag{31.1}$$

In this expression, α_1 and α stand for additional quantum numbers necessary to distinguish between different possible states with the same values of S_1, L_1 and S, L respectively. In the *coefficients of fractional parentage* $[l^{n-1}(\alpha_1 S_1 L_1) l\, S\, L | \} l^n \alpha\, S\, L]$ the obvious fact that $s_n = \frac{1}{2}$ is not mentioned for the sake of convenience. Had we specified the states in (31.1) in more detail, we should have written $\psi(l^n \alpha\, S\, L\, M_S M_L)$ instead of $\psi(l^n \alpha\, S\, L)$, etc. Obviously, the c.f.p. in (31.1) are independent of M_S and M_L. Therefore, if we transform the wave functions (31.1) from the $M_S M_L$ scheme to the JM_J scheme, using the Clebsch-Gordan coefficients $(S\, M_S L\, M_L | S\, L\, J\, M)$, we obtain wave functions $\psi(l^n \alpha\, S\, L\, J\, M)$ given in terms of $\psi[l^{n-1}(\alpha_1 S_1 L_1) l_n S\, L\, J\, M]$ by the expression (31.1) with the same c.f.p.

Like in jj-coupling, the c.f.p. are normalized to unity. The c.f.p. of two orthogonal wave functions with the *same* values of S and L are orthogonal [cf. (26.15)]. Thus we have

$$\sum_{\alpha_1 S_1 L_1} [l^n \alpha\, S\, L \{ | l^{n-1}(\alpha_1 S_1 L_1) l\, S\, L]\, [l^{n-1}(\alpha_1 S_1 L_1) l\, S\, L | \} l^n \alpha' S\, L] = \delta_{\alpha\alpha'}. \tag{31.2}$$

The c.f.p. can be calculated by coupling particle n to a principal parent in the l^{n-1} configuration and antisymmetrizing. In the case of three particles this is a simple matter and we can obtain a closed formula in the same way used to obtain (26.11). After antisymmetrizing (with respect to the spin as well as the space coordinates) we restore the same order of coupling by a change-of-coupling transformation which in this case is

$$\psi[l_1 l_3(S_2 L_2) l\, S\, L] = \sum_{S_1 L_1} \langle l_1 l_3(S_2 L_2) l_2 S\, L | l_1 l_2(S_1 L_1) l_3 S\, L \rangle\, \psi[l_1 l_2(S_1 L_1) l_3 S\, L]. \tag{31.3}$$

Obviously, $(-1)^{S_2 + L_2} = 1$, since we started from a wave function antisymmetric in the two equivalent particles 1 and 3 [cf. (18.14)]. The change of coupling transformation (31.3) is carried out simultaneously on the two independent sets of spin and orbital angular momenta. Thus, it is given as a *product* of two transformations (15.29), one for the $\mathbf{S}$ vectors and the other for the $\mathbf{L}$ vectors.

Hence, we obtain

$$\langle l_1 l_3(S_2L_2)l_2 S\,L | l_1 l_2(S_1L_1)l_3 S\,L\rangle = (-1)^{2l+1+S_1+L_1+S_2+L_2}$$

$$\times [(2S_1+1)(2S_2+1)(2L_1+1)(2L_2+1)]^{1/2} \begin{Bmatrix} l & l & L_1 \\ L & l & L_2 \end{Bmatrix} \begin{Bmatrix} \frac{1}{2} & \frac{1}{2} & S_1 \\ S & \frac{1}{2} & S_2 \end{Bmatrix}. \tag{31.4}$$

The analog of (26.7) for the nonnormalized antisymmetric wave function in LS-coupling is

$$\frac{1}{N}\psi(l^3[S_2L_2]S\,L)$$

$$= \sum_{S_1+L_1 \text{ even}} [\delta_{S_1S_2}\delta_{L_1L_2} - 2\langle l\,l(S_2L_2)l\;S\,L | l\,l(S_1L_1)l\;S\,L\rangle]\,\psi[l^2(S_1L_1)l_3 S\,L]$$

$$= \sum_{S_1+L_1 \text{ even}} \Big(\delta_{S_1S_2}\delta_{L_1L_2} + 2[(2S_1+1)(2S_2+1)(2L_1+1)(2L_2+1)]^{1/2}$$

$$\times \begin{Bmatrix} l & l & L_1 \\ L & l & L_2 \end{Bmatrix} \begin{Bmatrix} \frac{1}{2} & \frac{1}{2} & S_1 \\ S & \frac{1}{2} & S_2 \end{Bmatrix}\Big)\,\psi[l^2(S_1L_1)l_3 S\,L]. \tag{31.5}$$

In (31.5) we use the notation, introduced before, of indicating the principal parent in square brackets. It is rather a simple matter to normalize (31.5) since all functions which appear there are orthogonal to each other. It is, however, even simpler if we consider the antisymmetrization as a projection operation and use the analog of (26.24). The inverse of the normalization factor N is equal to $n(n = 3)$ times the *normalized* c.f.p. of the principal parent. Therefore N^{-2} is equal to 3 times the coefficient of the principal parent in (31.5).

Denoting the S and L values of the principal parent by S_0 and L_0, we obtain, using (31.5)

$$N^2 = \left[3\left(1 + 2(2S_0+1)(2L_0+1)\begin{Bmatrix} l & l & L_0 \\ L & l & L_0 \end{Bmatrix} \begin{Bmatrix} \frac{1}{2} & \frac{1}{2} & S_0 \\ S & \frac{1}{2} & S_0 \end{Bmatrix}\right)\right]^{-1}. \tag{31.6}$$

From this we obtain the following expressions for the c.f.p.

$$[l^2(S_1L_1)l\;S\,L|\}l^3[S_0L_0]S\,L]$$

$$= \Big[\delta_{S_1S_0}\delta_{L_1L_0} + 2\sqrt{(2S_0+1)(2S_1+1)(2L_0+1)(2L_1+1)}$$

$$\times \begin{Bmatrix} l & l & L_1 \\ L & l & L_0 \end{Bmatrix} \begin{Bmatrix} \frac{1}{2} & \frac{1}{2} & S_1 \\ S & \frac{1}{2} & S_0 \end{Bmatrix}\Big] \left[3 + 6(2S_0+1)(2L_0+1)\begin{Bmatrix} l & l & L_0 \\ L & l & L_0 \end{Bmatrix} \begin{Bmatrix} \frac{1}{2} & \frac{1}{2} & S_0 \\ S & \frac{1}{2} & S_0 \end{Bmatrix}\right]^{-1/2}. \tag{31.7}$$

Equation (31.7) assumes a particularly simple form for the c.f.p. deriveable from the principle parent $S_0 = 0$, $L_0 = 0$, i.e., for the c.f.p. of the state $l^3(^2l)$. Inserting the appropriate values of the Racah coefficients we obtain for this particular state

$$[l^2(^1S)l\,^2l\{\!|l^3\,^2l] = \sqrt{\frac{2l}{3(2l+1)}}$$

$$[l^2(^{2S_1+1}L_1)l\,^2l\{\!|l^3\,^2l] = -\sqrt{\frac{(2S_1+1)(2L_1+1)}{6l(2l+1)}} \quad L \neq 0, \quad (-1)^{S_1+L_1} = 1. \tag{31.8}$$

Here a state with S_1 and L_1 is denoted by $^{2S_1+1}L_1$. These equations are the analogs of (26.12).

The matrix elements in the l^n configuration can be expressed simply in terms of c.f.p. The analog of (26.32) is

$$\langle l^n\alpha\, S\, L\, M_S M_L \Big| \sum f_i \Big| l^n\alpha' S' L' M'_S M'_L\rangle$$

$$= n \sum_{\alpha_1 S_1 L_1} [l^n\alpha\, S\, L\{\!|l^{n-1}(\alpha_1 S_1 L_1)l\, S\, L]\, [l^{n-1}(\alpha_1 S_1 L_1)l\, S' L'|\!\}l^n\alpha' S' L']$$

$$\langle (S_1L_1)l_n S\, L\, M_S M_L | f_n | (S_1L_1)l_n S' L' M'_S M'_L\rangle. \tag{31.9}$$

Equation (31.9) can be simplified if f_n is a product of a certain component of an irreducible tensor of degree k_1 operating on the spin coordinates (of the nth particle) and a component of a tensor of degree k_2 operating on the space coordinates (of the same particle). Let us denote such a *double tensor* operator by $f^{(k_1k_2)}_{\kappa_1\kappa_2}$. In this case we can use the Wigner-Eckart theorem with respect to both the spin coordinates and the space coordinates to obtain from (31.9) the analog of (26.33)

$$\left(l^n S\, L \left\| \sum_i \mathbf{f}_i^{(k_1k_2)} \right\| l^n\alpha' S' L'\right)$$

$$= n(\tfrac{1}{2}\, l \| \mathbf{f}^{(k_1k_2)} \| \tfrac{1}{2}\, l)\sqrt{(2S+1)(2S'+1)(2L+1)(2L'+1)}$$

$$\times \sum_{\alpha_1 S_1 L_1} [l^n\alpha\, S\, L\{\!|l^{n-1}(\alpha_1 S_1 L_1)l\, S\, L]\, [l^{n-1}(\alpha_1 S_1 L_1)l\, S' L'|\!\}l^n\alpha' S' L']$$

$$\times (-1)^{S_1+1/2+S+k_1+L_1+l+L+k_2} \begin{Bmatrix} \frac{1}{2} & S & S_1 \\ S' & \frac{1}{2} & k_1 \end{Bmatrix} \begin{Bmatrix} l & L & L_2 \\ L' & l & k_2 \end{Bmatrix}. \tag{31.10}$$

No such simple expression can be found for a reduced matrix element of $\mathbf{f}_i$ in the state with a definite J. The reason is that a general double tensor need not be an *irreducible* tensor operator with respect to simultaneous rotations of both ordinary and spin coordinates. If, however,

f is a component $f_\kappa^{(k_1k_2)k}$ of an irreducible tensor of degree k with respect to the total J, i.e.

$$f_\kappa^{(k_1k_2)k} = \sum_{\kappa_1\kappa_2} (k_1\kappa_1k_2\kappa_2|k_1k_2k\ \kappa) f_{\kappa_1\kappa_2}^{(k_1k_2)}$$

then we have according to (14.26)

$$\left(l^n\alpha\ S\ L\ J \left\| \sum_i \mathbf{f}_i^{(k_1k_2)k} \right\| l^n\alpha' S'L'J'\right)$$

$$= \left(l^n\alpha\ S\ L \left\| \sum_i \mathbf{f}_i^{(k_1k_2)} \right\| l^n\alpha'S'L'\right) \sqrt{(2J+1)(2J'+1)(2k+1)} \begin{Bmatrix} S & L & J \\ S' & L' & J' \\ k_1 & k_2 & k \end{Bmatrix}. \tag{31.11}$$

When dealing with two-particle operators $\Sigma_{i<k}^n g_{ik}$ we shall assume that g_{ik} is a scalar operator with respect to the total $\mathbf{J}$. If g_{ik} is, in addition, a scalar with respect to both $\mathbf{S}$ and $\mathbf{L}$, then the analog of (26.37) assumes a very simple form. In this case, the matrix elements of $\Sigma\, g_{ik}$ are diagonal in S and L and independent of M_S and M_L (or of J and M_J). Two-body interactions with this property were called scalar forces in Part II. For *such operators* we obtain easily

$$\langle l^n\alpha\ S\ L \left| \sum_{i<k}^n g_{ik} \right| l^n\alpha' S\ L\rangle = \frac{n}{n-2} \sum_{\alpha_1\alpha_1' S_1 L_1} [l^n\alpha\ S\ L\{|l^{n-1}(\alpha_1 S_1 L_1)l\ S\ L]$$

$$\times [l^{n-1}(\alpha_1' S_1 L_1)l\ S\ L|\}l^n\alpha' S\ L]\ \langle l^{n-1}\alpha_1 S_1 L_1 \left| \sum_{i<k}^{n-1} g_{ik} \right| l^{n-1}\alpha_1' S_1 L_1\rangle. \tag{31.12}$$

Successive applications of (31.12) give the interaction in the l^n configuration in terms of the interactions in the l^2 configuration.

In the case of more complicated interactions, like vector forces or tensor forces, the expression of matrix elements is more involved. In the general case, as we saw in Part II, g_{ij} is the sum of products of two double tensors $\mathbf{f}_i^{(k_1k_2)}$ and $\mathbf{f}_j^{(k_1'k_2')}$ operating on the coordinates of particle i and j.

Let us now separate the spin and space dependence of these double tensors by writing

$$f_{\kappa_1\kappa_2}^{(k_1k_2)} = t_{\kappa_1}^{(k_1)} r_{\kappa_2}^{(k_2)}.$$

Here $\mathbf{t}^{(k_1)}$ is an irreducible tensor operator of degree k_1 operating on the spin coordinates and $\mathbf{r}^{(k_2)}$ a tensor of degree k_2 operating on the space coordinates. Since the total interaction g_{ij} is a scalar with respect to simultaneous rotations in the spin and ordinary spaces, it is clear that by using change of coupling transformations we can write g_{ij} in the form

$$g_{ij} = \sum_k g_{ij}^{(k)} \tag{31.13}$$

where

$$g_{ij}^{(k)} = \sum_{k_1k_1'k_2k_2'} \lambda_{k_1k_1'k_2k_2'k} \left([\mathbf{t}_i^{(k_1)} \times \mathbf{t}_j^{(k_1')}]^{(k)} \cdot [\mathbf{r}_i^{(k_2)} \times \mathbf{r}_j^{(k_2')}]^{(k)} \right). \tag{31.14}$$

Since g_{ij} is symmetric with respect to the exchange of i and j, i.e., $g_{ij} = g_{ji}$, it follows from (31.14) that we must have

$$\lambda_{k_1k_1'k_2k_2'k} = (-1)^{k_1+k_1'+k_2+k_2'} \lambda_{k_1'k_1k_2'k_2k}.$$

Otherwise there are no further general relations which are satisfied by the λ. They are determined by the interaction g_{ij} on the one hand and the normalization of the irreducible tensors $\mathbf{t}^{(k)}$ and $\mathbf{r}^{(k)}$ on the other hand. Thus, $\Sigma\, g_{ij}$ can be expressed as a sum of terms (31.14), each of which is a scalar product of a tensor of degree k with respect to $\mathbf{S}$ and a tensor of the same degree k with respect to $\mathbf{L}$. It should be kept in mind that $g_{ij}^{(k)}$ defined by (31.14) is thus a *scalar* with respect to $\mathbf{J}$.

The three possible values of k are 0, 1, and 2 which correspond to scalar, vector, and tensor interactions respectively. According to (15.5) [which can be derived also from (31.11) by putting $k = 0$] we obtain

$$\begin{aligned} &\langle l^n\alpha\, S\, L\, J \Big| \sum_{i<j}^{n} g_{ij}^{(k)} \Big| l^n\alpha' S' L' J \rangle \\ &= (-1)^{S'+L-J} \left(l^n\alpha\, S\, L \Big\| \sum_{i<j}^{n} g_{ij}^{(k)} \Big\| l^n\alpha' S' L' \right) \begin{Bmatrix} S & L & J \\ L' & S' & k \end{Bmatrix}. \end{aligned} \tag{31.15}$$

The reduced matrix element which appears in (31.15) can be calculated by using c.f.p. Restricting the summation to $i, j \leq n-1$ and multiplying by $n/(n-2)$, we first obtain

$$\begin{aligned} &\left(l^n\alpha\, S\, L \Big\| \sum_{i<j}^{n} g_{ij}^{(k)} \Big\| l^n\alpha' S' L' \right) \\ &= \frac{n}{n-2} \left(l^n\alpha\, S\, L \Big\| \sum_{i<j}^{n-1} g_{ij}^{(k)} \Big\| l^n\alpha\, S\, L \right) \\ &= \frac{n}{n-2} \sum_{\alpha_1\alpha_1'S_1S_1'L_1L_1'} [l^n\alpha\, S\, L\{| l^{n-1}(\alpha_1 S_1 L_1) l\, S\, L]\, [l^{n-1}(\alpha_1' S_1' L_1') l\, S' L' |\} l^n\alpha' S' L'] \\ &\qquad \times \left(l^{n-1}(\alpha_1 S_1 L_1) l\, S\, L \Big\| \sum_{i<j}^{n-1} g_{ij}^{(k)} \Big\| l^{n-1}(\alpha_1' S_1' L_1') l\, S' L' \right). \end{aligned} \tag{31.16}$$

Since the operator in the reduced matrix element on the right-hand side of (31.16) is independent of the nth particle we can evaluate it by using (15.26) for the spins and for the orbital angular momenta. We thus obtain the final result

$$\left(l^n\alpha\, S\, L \left\| \sum_{i<j}^{n} g_{ij}^{(k)} \right\| l^n\alpha' S' L'\right)$$

$$= \frac{n}{n-2}[(2S+1)(2S'+1)(2L+1)(2L'+1)]^{1/2}$$

$$\sum_{\alpha_1\alpha_1' S_1 S_1' L_1 L_1'} [l^n\alpha\, S\, L\{|l^{n-1}(\alpha_1 S_1 L_1) l\, S\, L]$$

$$\times [l^{n-1}(\alpha_1' S_1' L_1') l\, S' L'|\} l^n\alpha' S' L'] \left(l^{n-1}\alpha_1 S_1 L_1 \left\| \sum_{i<j}^{n-1} g_{ij}^{(k)} \right\| l^{n-1}\alpha_1' S_1' L_1'\right)$$

$$\times (-1)^{S_1-1/2+S'+L_1-l+L'} \begin{Bmatrix} S_1 & S & \frac{1}{2} \\ S' & S_1' & k \end{Bmatrix} \begin{Bmatrix} L_1 & L & l \\ L' & L_1' & k \end{Bmatrix}. \qquad (31.17)$$

In the case of a scalar interaction, $k = 0$, (31.17) reduces to the simpler case of (31.12).

Another way to obtain the reduced matrix element in (31.15) is to use the procedure which led to (26.46) in the case of jj-coupling. For this purpose we write the operators which appear in (31.14) in terms of double tensors, using the notation

$$\sum_{i<j}^{n} [\mathbf{t}_i^{(k_1)}\mathbf{r}_i^{(k_2)} \times \mathbf{t}_j^{(k_1')}\mathbf{r}_j^{(k_2')}]^{(kk)}. \qquad (31.18)$$

We replace $\mathbf{t}_i^{(k_1)}\mathbf{r}_i^{(k_2)}$ by $(\frac{1}{2}||\mathbf{t}^{(k_1)}||\frac{1}{2})(l||\mathbf{r}^{(k_2)}||l)\,\mathbf{u}^{(k_1k_2)}$ where $\mathbf{u}^{(k_1k_2)}$ is the *unit double tensor* defined by $(\frac{1}{2}\, l||\mathbf{u}^{(k_1k_2)}||\frac{1}{2}\, l') = \delta_{ll'}$. The sum in (31.18) can then be written as

$$(\tfrac{1}{2}||\mathbf{t}^{(k_1)}||\tfrac{1}{2})(\tfrac{1}{2}||\mathbf{t}^{(k_1')}||\tfrac{1}{2})(l||\mathbf{r}^{(k_2)}||l)(l||\mathbf{r}^{(k_2')}||l)$$

$$\times \sum_{i<j}^{n} [\mathbf{u}_i^{(k_1k_2)} \times \mathbf{u}_j^{(k_1'k_2')}]^{(kk)}. \qquad (31.19)$$

The summation over $i < j$ can be written as follows

$$\sum_{i<j}^{n} [\mathbf{u}_i^{(k_1k_2)} \times \mathbf{u}_j^{(k_1'k_2')}]^{(kk)}$$

$$= \tfrac{1}{2}\left(\left\{\left[\sum_i^n \mathbf{u}_i^{(k_1k_2)}\right] \times \left[\sum_j^n \mathbf{u}_j^{(k_1'k_2')}\right]\right\}^{(kk)} - \sum_i^n \left[\mathbf{u}_i^{(k_1k_2)} \times \mathbf{u}_i^{(k_1'k_2')}\right]^{(kk)}\right). \qquad (31.20)$$

The last sum in (31.20) is a single particle operator $\Sigma_i \mathbf{f}_i^{(kk)}$ and its reduced matrix elements are given by (31.10). The first sum in (31.20) can be written as

$$[\mathbf{U}^{(k_1k_2)} \times \mathbf{U}^{(k_1'k_2')}]^{(kk)} \tag{31.21}$$

where we write $\mathbf{U}^{(k_1k_2)}$ for the single particle operator $\mathbf{U}^{(k_1k_2)} = \Sigma_i^n \mathbf{u}_i^{(k_1k_2)}$. The reduced matrix elements of (31.21) are given according to (15.23) by

$$(l^n\alpha\, S\, L||[\mathbf{U}^{(k_1k_2)} \times \mathbf{U}^{(k_1k_2)}]^{(kk)}||l^n\alpha' S' L')$$

$$= (-1)^{S+S'+L+L'}(2k+1) \sum_{\alpha''S''L''} (l^n\alpha\, S\, L||\mathbf{U}^{(k_1k_2)}||l^n\alpha'' S'' L'')$$

$$\times\, (l^n\alpha'' S'' L''||\mathbf{U}^{(k_1'k_2')}||l^n\alpha' S' L') \begin{Bmatrix} S & k_1 & S'' \\ k_1' & S' & k \end{Bmatrix} \begin{Bmatrix} L & k_2 & L'' \\ k_2' & L' & k \end{Bmatrix}. \tag{31.22}$$

Using this procedure it is possible to express matrix elements of two-particle operators in terms of matrix elements of single particle operators in the l^n configuration. All these procedures are not so convenient and are much more complicated that in the jj-coupling scheme. In that case, there are only scalar interactions and the calculations are simpler.

As in jj-coupling, we can introduce $n \to n-2$ c.f.p. defined by the analog of (26.43). Such c.f.p. have properties similar to these in jj-coupling. Since we shall not make explicit use of them we shall not write down in detail all the relations involving them.

The Seniority Scheme in LS-Coupling

In analogy with jj-coupling, we obtain the seniority scheme by making use of pairs coupled to $J = 0$. However, in the present case there are *two* $J = 0$ states in the l^2 configuration, i.e., the $^1S_0$ and the $^3P_0$ states. We shall make use only of ^1S pairs since in this case $L = 0$ as well as $S = 0$. Two states, in the l^n configuration and l^{n-2} configuration, will have the same seniority if the wave function $\psi(l^n\alpha\, S\, L)$ can be obtained by adding a ^1S pair to $\psi(l^{n-2}\alpha\, S\, L)$ and antisymmetrizing. Such two states will clearly have the same values of S and L. We can continue this procedure of "subtracting" ^1S pairs until we arrive at a wave function $\psi(l^v\alpha\, S\, L)$ which cannot be obtained by adding a ^1S pair to $\psi(l^{v-2}S\, L)$ and antisymmetrizing. The number v is then called the *seniority* of the state $\psi(l^n\alpha\, S\, L)$.

The ^1S state of the l^2 configuration has seniority $v = 0$ and it appears in all l^n configurations with even n. The state of a single particle ^2l has seniority $v = 1$ and it appears in all l^n configurations with odd n.

Certain operators turn out to be diagonal in the seniority scheme. The single particle operators analogous to the odd tensors in jj coupling are in the present case *odd double tensors.* These are double tensors $\mathbf{u}^{(k_1k_2)}$, for which $k_1 + k_2$ is *odd.* As mentioned before, if we construct of such double tensors irreducible tensors of degree k with respect to $\mathbf{j}$, only those tensors with *odd* k have nonvanishing expectation values. This follows simply from (14.26) which gives the expectation value $(\frac{1}{2}\,lj||\mathbf{u}^{(k_1k_2)k}||\frac{1}{2}\,lj)$ in terms of the $9-j$ symbol

$$\begin{Bmatrix} \frac{1}{2} & l & j \\ \frac{1}{2} & l & j \\ k_1 & k_2 & k \end{Bmatrix}$$

which has two identical rows and vanishes according to (14.32) unless $k_1 + k_2 + k$ is even. The important property of such odd double tensors, i.e.,

$$\langle l^2 S\,L|u^{(k_1k_2)}_{\kappa_1\kappa_2}(1) + u^{(k_1k_2)}_{\kappa_1\kappa_2}(2)|l^2\,{}^1S\rangle = 0, \qquad k_1 + k_2 \text{ odd} \tag{31.23}$$

is obtained in the same way as (27.5). The triangular conditions require that $S = k_1$ and $L = k_2$ in order for the left-hand side of (31.23) not to vanish. However, in this case $(-1)^{S+L} = (-1)^{k_1+k_2} = -1$ so that $\psi(l^2S\,L)$ is *symmetric* in the two particles. Since $\mathbf{u}(1) + \mathbf{u}(2)$ is also symmetric but $\psi(l^2\,{}^1S)$ is *antisymmetric*, (31.23) vanishes identically.

From (31.23) follows that the odd double tensors $\mathbf{U}^{(k_1k_2)} = \sum_i^n \mathbf{u}_i^{(k_1k_2)}$ are diagonal in the seniority scheme and their matrix elements in the l^n configurations are independent of n. This result is obtained by using $n \to n-2$ c.f.p. in the same way as (27.22) was obtained in the case of jj-coupling. The analog of (27.22) is

$$\langle l^n v\,\alpha\,S\,L\,M_SM_L\left|\sum_{i=1}^{n} u^{(k_1k_2)}_{\kappa_1\kappa_2}(i)\right| l^n v'\alpha' S'L'M_S'M_L'\rangle$$

$$= \langle l^v v\,\alpha\,S\,L\,M_SM_L\left|\sum_{i=1}^{v} u^{(k_1k_2)}_{\kappa_1\kappa_2}(i)\right| l^v v\,\alpha'\,S'L'\,M_S'M_L'\rangle\,\delta_{vv'} \qquad k_1 + k_2 \text{ odd.} \tag{31.24}$$

Using (31.24) for the case of two-particle interactions built of odd double tensors, we obtain that the first term of (31.20) is independent of n. In the case of scalar forces, $k = 0$, the second term in (31.20) is a sum of n equal single particle terms. In this case, the interaction has the pairing property, in analogy with (27.25),

$$\langle l^n v\, \alpha\, S\, L\, J \Big| \sum_{i<j}^{n} V_{ij} \Big| l^n v\, \alpha' S' L' J \rangle$$

$$= \langle l^v v\, \alpha\, S\, L\, J \Big| \sum_{i<j}^{v} V_{ij} \Big| l^v v\, \alpha\, S' L' J \rangle + \frac{n-v}{2} V_0\, \delta_{SS'}\, \delta_{\alpha\alpha'}\, \delta_{LL'}. \qquad (31.25)$$

In (31.25), V_0 is the expectation value of V_{ij} in the ${}^1S$ state of the l^2 configuration. The general interaction V_{ij} in (31.25) may be noncentral and thus may have nonvanishing matrix elements between states with different values of S and L. In particular, equation (31.25) holds also for tensor forces in which case $V_0 = 0$. The second term in (31.20) vanishes for $k = 2$ since there is no second degree tensor built of a single particle spin vector $\mathbf{s}$. In other words, the triangular conditions cannot be satisfied with $\frac{1}{2}$, $\frac{1}{2}$, and 2.

Also in LS-coupling there is an operator $Q = \sum_{i<j}^{n} q_{ij}$ which is intimately connected with the seniority scheme. The two-particle operator q_{ij} is defined, in analogy with (27.27) by

$$\langle l^2 L\, M | q_{12} | l^2 L' M' \rangle = (2l+1)\, \delta_{LL'}\, \delta_{MM'}\, \delta_{L0}. \qquad (31.26)$$

We shall calculate the eigenvalues of Q in the seniority scheme, directly from the definition (31.26) by trying to express q_{12} in terms of odd (double) tensors. The identity (15.16) can be written by using (15.5) as

$$(2l+1)\, \delta_{L0} = \sum_{k=0}^{2l} (-1)^k (2k+1) \langle l^2 L\, M | (\mathbf{u}_1^{(k)} \cdot \mathbf{u}_2^{(k)}) | l^2 L\, M \rangle. \qquad (31.27)$$

In (31.27) the $\mathbf{u}^{(k)}$ are the unit tensors, defined by (21.54), operating on the space coordinates only. Another identity, satisfied by scalar products of unit tensors, is obtained from (15.17)

$$(-1)^{2l-L} = (-1)^L = \sum_{k=0}^{2l} (2k+1) \langle l^2 L\, M | (\mathbf{u}_1^{(k)} \cdot \mathbf{u}_2^{(k)}) | l^2 L\, M \rangle. \qquad (31.28)$$

Equations (31.27) and (31.28) can now be combined to give the analog of (27.32)

$$(2l+1)\, \delta_{L0} - (-1)^L = -2 \sum_{k \text{ odd}} (2k+1) \langle l^2 L\, M | (\mathbf{u}_1^{(k)} \cdot \mathbf{u}_2^{(k)}) | l^2 L\, M \rangle. \qquad (31.29)$$

The term $(-1)^L$ is equal, for antisymmetric states, to $(-1)^S$ and this latter number can be expressed, as we saw in Part II, by

$$(-1)^S = -\tfrac{1}{2}[1 + 4(\mathbf{s}_1 \cdot \mathbf{s}_2)]. \tag{31.30}$$

Introducing (31.30) into (31.29) we obtain q_{12} in the following form

$$q_{12} = -\tfrac{1}{2} - 2(\mathbf{s}_1 \cdot \mathbf{s}_2) - 2 \sum_{k \text{ odd}} (2k+1)(\mathbf{u}_1^{(k)} \cdot \mathbf{u}_2^{(k)}). \tag{31.31}$$

We see that q_{12} is a scalar interaction which can be expressed in terms of the odd tensors $\mathbf{u}_i^{(k)}$ and $\mathbf{s}_i$ and in addition a constant ($k = 0$) term. The odd part has the pairing property (31.25) while the contribution of the constant, $-\frac{1}{2}$, to $Q = \Sigma_{i<j}^n q_{ij}$ is simply $-\frac{1}{2}n(n-1)/2$. It is worthwhile to mention that in the analogous expression (27.33) the constant is -1 instead of the $-\frac{1}{2}$ which appears here. Therefore, the explicit dependence of $Q(n, v)$ on n and v will be different in the two cases. In LS-coupling we obtain, using (31.25),

$$Q(n, v) + \tfrac{1}{2}\frac{n(n-1)}{2} = Q(v, v) + \tfrac{1}{2}\frac{v(v-1)}{2} + \frac{n-v}{2}[Q(2, 0) + \tfrac{1}{2}]. \tag{31.32}$$

Putting $Q(v, v) = 0$ and $Q(2, 0) = 2l + 1$ in (31.32) yields the final result

$$\begin{aligned} Q(n, v) &= -\tfrac{1}{2}\frac{n(n-1)}{2} + \tfrac{1}{2}\frac{v(v-1)}{2} + \frac{n-v}{2}(2l + \tfrac{3}{2}) \\ &= \tfrac{1}{4}(n-v)(4l + 4 - n - v). \end{aligned} \tag{31.33}$$

We shall see later that some relations obtained in jj-coupling, can be translated into their corresponding expressions in LS-coupling by simply replacing $2j + 1$ in the former case by $2(2l + 1)$ in the latter. This, however, is not true of the expression (31.33) and its analog (27.29). The reason for the difference between these two expressions is the appearance of the different constants in (31.31) and (27.33).

Using $n \to n - 2$ c.f.p. to calculate $Q(n, v)$ we obtain the expression analogous to (27.28)

$$Q(n, v) = \frac{n(n-1)}{2}[l^{n-2}(v\,S\,L)l^2(^1S)S\,L|\}l^n v\,S\,L]^2(2l + 1). \tag{31.34}$$

This relation gives the value of the important c.f.p. in analogy with (27.26) as follows

$$\frac{n(n-1)}{2}[l^{n-2}(v\,S\,L)l^2(^1S)S\,L|\}l^n v\,S\,L]^2 = \tfrac{1}{4}\frac{(n-v)(4l + 4 - n - v)}{2l + 1}. \tag{31.35}$$

Equation (31.35) can be used to obtain recursion relations for the c.f.p. in the seniority scheme. We shall use the same procedure as that leading to (28.9). This will give the following recursion relation, similar to (28.9),

$$[l^{n-1}(v_1S_1L_1)l\,S\,L|\}l^n v\,S\,L]$$

$$= [l^{n-3}(v_1S_1L_1)l\,S\,L|\}l^{n-2}v\,S\,L]\sqrt{\frac{(n-2)(n-v)(4l+4-n-v)}{n(n-v_1-1)(4l+5-n-v_1)}}. \quad (31.36)$$

Successive applications of (31.36) give the following results for the two possible cases $v_1 = v - 1$ and $v_1 = v + 1$.

$$[l^{n-1}(v_1S_1L_1)l\,S\,L|\}l^n v\,S\,L]$$

$$= \sqrt{\frac{v(4l+4-n-v)}{n(4l+4-2v)}}[l^{v-1}(v-1,S_1L_1)l\,S\,L|\}l^v v\,S\,L] \quad \text{for} \quad v_1 = v-1 \quad (31.37)$$

$$[l^{n-1}(v_1S_1L_1)l\,S\,L|\}l^n v\,S\,L]$$

$$= \sqrt{\frac{(n-v)(v+2)}{2n}}[l^{v+1}(v+1,S_1L_1)l\,S\,L|\}l^{v+2}v\,S\,L] \quad \text{for} \quad v_1 = v+1. \quad (31.38)$$

These relations are the analogs of the relations (28.10) and (28.11) for the case of jj-coupling. We note that (31.37) and (31.38) can be obtained from (28.10) and (28.11) by replacing $2j + 1$ by $4l + 2$. As we remarked earlier, the expressions for $Q(n, v)$ in jj-coupling and LS-coupling [(27.29) and (31.33) respectively] differ by a factor $\frac{1}{2}$. Since the factors in (31.37) and (31.38) involve *ratios* of eigenvalues of $Q(n, v)$, the extra factors disappear and the simpler relation between the two coupling schemes shows up. We shall meet similar situations in the following expressions.

The relations (31.37) and (31.38) can be augmented by another useful relation which expresses the c.f.p. on the right-hand side of (31.38) in terms of a $v \to v + 1$ c.f.p. This relation is the analog of (28.16), and is obtained in a similar way, yielding

$$[l^{v+1}(v+1,S_1L_1)l\,S\,L|\}l^{v+2}v\,S\,L]$$

$$= (-1)^{S+L+l+1/2-S_1-L_1}\sqrt{\frac{(2S_1+1)(2L_1+1)2(v+1)}{(2S+1)(2L+1)(v+2)(4l+2-2v)}}$$

$$\times [l^v(v\,S\,L)l\,S_1L_1|\}l^{v+1}\,v+1,S_1L_1]. \quad (31.39)$$

Using the relations (31.37) and (31.38) we can obtain orthogonality relations between c.f.p. with definite seniorities. These relations, the analogs of (28.20) and (28.21), are as follows

$$\sum_{\alpha_1 S_1 L_1} [l^n v \alpha S L\{|l^{n-1}(v-1, \alpha_1, S_1 L_1) l S L] [l^{n-1}(v-1, \alpha_1, S_1 L_1) l S L|\} l^v v \alpha' S L] = \frac{v(4l+4-n-v)}{n(4l+4-2v)} \delta_{\alpha\alpha'} \tag{31.40}$$

$$\sum_{\alpha_1 S_1 L_1} [l^n v \alpha S L\{|l^{n-1}(v+1, \alpha_1, S_1 L_1) l S L] [l^{n-1}(v+1, \alpha_1, S_1 L_1) l S L|\} l^v v \alpha' S L] = \frac{(n-v)(4l+4-v)}{n(4l+4-2v)} \delta_{\alpha\alpha'}. \tag{31.41}$$

Equation (31.39) can be used to obtain another type of orthogonality relations which are analogous to the relations (29.19), (29.20), and (29.21). These are

$$\sum_{SL} (2S+1)(2L+1) [l^n v S L\{|l^{n-1}(v+1, S_1 L_1) l S L] \times [l^{n-1}(v-1, S_1 L_1) l S L|\} l^n v S L] = 0 \tag{31.42}$$

$$\sum_{SL} (2S+1)(2L+1) [l^{n-1}(v+1, S_1 L_1) l S L|\} l^n v S L]^2 = \frac{(n-v)(v+1)}{n(4l+2-2v)} (2S_1+1)(2L_1+1) \tag{31.43}$$

$$\sum_{SL} (2S+1)(2L+1) [l^{n-1}(v-1, S_1 L_1) l S L|\} l^n v S L]^2 = \frac{(4l+4-n-v)(4l+5-v)}{n(4l+6-2v)} (2S_1+1)(2L_1+1). \tag{31.44}$$

In case other quantum numbers α are required to specify uniquely the states with the same values of v, S, and L, the summations in (31.42), (31.43), and (31.44) include also summation over α and the relations assume the form of (29.19), (29.20), and (29.21).

We shall now consider matrix elements of single particle operators in the seniority scheme in the case of LS-coupling. We already mentioned the fact that odd double tensor operators are diagonal in the seniority scheme and their expectation values are independent of n. Thus, in analogy with (28.22),

$$\left(l^n v S L \left\| \sum_{i=1}^{n} \mathbf{f}_i^{(k_1 k_2)} \right\| l^n v S' L'\right) = \left(l^v v S L \left\| \sum_{i=1}^{v} \mathbf{f}_i^{(k_1 k_2)} \right\| l^v v S' L'\right) \qquad k_1 + k_2 \text{ odd}. \tag{31.45}$$

In the case of even double tensor operators there may be nonvanishing matrix elements between states with different seniorities v and v'. It follows from (31.37), (31.38) that v and v' can differ at most by two. As in the derivation of (28.25) we can use (31.37) to express matrix elements between states with seniorities v and $v' = v - 2$ in terms of the l^v configuration. We obtain

$$\left(l^n v\, S\, L \left\| \sum_{i=1}^{n} \mathbf{f}_i^{(k_1 k_2)} \right\| l^n\, v-2,\, S'L'\right)$$

$$= \sqrt{\frac{(n-v+2)\,(4l+4-n-v)}{2(4l+4-2v)}} \left(l^v v\, S\, L \left\| \sum_{i=1}^{v} \mathbf{f}_i^{(k_1 k_2)} \right\| l^v\, v-2,\, S'L'\right). \tag{31.46}$$

The case $v' = v$ is slightly more complicated but can be handled exactly as in jj-coupling. A calculation similar to the one carried out in deriving (28.40) from (28.26) yields

$$\left(l^n v\, S\, L \left\| \sum_{i=1}^{n} \mathbf{f}_i^{(k_1 k_2)} \right\| l^n v\, S'L'\right) = \frac{4l+2-2n}{4l+2-2v} \left(l^v v\, S\, L \left\| \sum_{i=1}^{v} \mathbf{f}_i^{(k_1 k_2)} \right\| l^v v\, S'L'\right)$$

$$k_1 + k_2 > 0 \text{ even.} \tag{31.47}$$

The matrix elements of a scalar, i.e., $k_1 = 0\; k_2 = 0$, are much simpler. In the l^n configuration only the diagonal elements do not vanish and they are equal to the matrix element in the single l-particle configuration multiplied by n. Thus,

$$\left(l^n v\, S\, L \left\| \sum_{i=1}^{n} \mathbf{f}_i^{(00)} \right\| l^n v\, S\, L\right) = \frac{n}{v} \left(l^v v\, S\, L \left\| \sum_{i=1}^{v} \mathbf{f}_i^{(00)} \right\| l^v v\, S\, L\right)$$

$$= n \sqrt{\frac{(2S+1)\,(2L+1)}{2(2l+1)}}\, (\tfrac{1}{2}\, l \| \mathbf{f}^{(00)} \| \tfrac{1}{2}\, l). \tag{31.48}$$

The behavior of even double tensor operators given by (31.47) is similar to that in jj-coupling. There is a linear change of the matrix element with n, and a reversal of the sign at the middle of the shell (where $n = 2l + 1$). All this agrees with the observations made when discussing the m-scheme (Part II). As can be seen from (31.46), the phases of states defined by the seniority scheme in LS-coupling differ from the phases defined by the convention we made in the m-scheme for the corresponding states of holes. Obviously, if one use only the seniority scheme, and does not go over to the m-scheme for the holes, everything is fully consistent. The phase relations between the two conventions are quite simple. The states with $v = 0$ and $v = 1$ have the *same phases* in both LS-coupling and the m scheme. States with even $v > 0$ differ by a phase $(-1)^{v/2}$, and states with odd v have the phase difference $(-1)^{(v-1)/2}$.

Using the relations (31.45), (31.46), (31.47), and (31.48) in the expression (31.22), it is possible to obtain relations expressing matrix elements of two-particle operators in the l^n configuration in terms of simpler configurations. These relations are analogous to those obtained in jj-coupling and we shall, therefore, not derive them in the present case.

32. The Group Theoretical Classification of States

Many of the considerations presented until now can be introduced more generally by using methods of group theory. Although this approach is not essential for the actual understanding of the structure of spectra, it is instructive to look at them from this point of view. In this section, we shall give a short description of the group theoretical meaning of the concepts introduced so far. We shall not develop the mathematical machinery of the theory of representations. Neither shall we devote time to the rigorous proofs of various statements. Rather, we shall try to convey the physical ideas underlying the use of these methods. This section can therefore be skipped without hampering the understanding of the following sections.

In Part II we considered the wave functions of a Hamiltonian which is invariant under rotations. We saw there, that all wave functions which transform among themselves under rotations must belong to the same eigenvalue. We could therefore group all wave functions into irreducible sets. The functions in each such set transform among themselves under rotations. These sets can thus form bases for irreducible representations of the rotations. This enabled us to obtain quantum numbers characterizing these irreducible representations (i.e., the eigenvalues of the total angular momentum). We could study many properties of the wave functions (i.e., those connected with the behavior under rotations) in a general way without further specifying the Hamiltonian. The only information we used was the invariance of the Hamiltonian under rotations. The strength of this approach lies in the fact that instead of studying in detail complicated physical systems, we can use to the full extent the general mathematical theory of the rotation group and its irreducible representations. This method of obtaining quantum numbers can be extended by the study of other groups of transformations which leave the Hamiltonian invariant. The eigenstates of the Hamiltonian considered can then be characterized by the irreducible representations of these other groups.

If a group of transformations leaves the Hamiltonian invariant (i.e., commutes with the Hamiltonian), then the set of eigenfunctions belonging to the same eigenvalue of H generally forms a basis of an irreducible representation of the group. In fact, let P be any transformation of the group considered. Let ψ_{ni} be the set of eigenfunctions belonging to the eigenvalue E_n. Since $PH = HP$ it follows that $\varphi = P\psi_{ni}$ is also an eigenfunction of H belonging to E_n. Thus, φ can be expressed as a linear combination of ψ_{ni} with different values of i but the same value of n

$$P\psi_{ni} = \sum_j P_{ij}\psi_{nj}.$$

The matrices P_{ij} form a representation of the group of transformations P, which is generally irreducible. Alternatively, we can consider sets of functions φ_{ni}, such that under the transformations P all functions φ_{ni} with the same index n transform irreducibly among themselves. It then follows that the Hamiltonian H, which commutes with the group, is diagonal in this scheme. Moreover, all the function φ_{ni}, with the same n, belong to the *same* eigenvalue E_n of the Hamiltonian (Schur's lemma). An example of this is the $2J + 1$ degeneracy of the eigenvalues of Hamiltonians invariant under rotations.

In some cases a group of transformations may commute with a part of the Hamiltonian only. Nevertheless, if the energy matrix is approximately diagonal in the scheme of this group, this scheme may be very useful. The quantum numbers which characterize the irreducible representations of the group may be referred to, in such cases, as *approximately good quantum numbers*. In any case, such a scheme of wave functions is well-defined mathematically and can be used in computations even if the energy matrix is not diagonal. An example of such a scheme is the seniority scheme introduced above which will be shown to be connected with the irreducible representations of a certain group of transformations.

Every Hamiltonian which governs a system of identical particles is invariant under permutations of the particle numbers. In particular, the Hamiltonian (24.1) of identical particles in a central field interacting with each other, is invariant under permutations of the coordinates of the particles. Therefore, if P is a permutation applied to the space and spin coordinates of the particles and ψ is an eigenfunction of H which belongs to an eigenvalue E, then $P\psi$ should also be an eigenfunction which belongs to the same eigenvalue E. As we saw, this apparent *equivalence degeneracy* of the $n!$ wave functions $P\psi$ is removed by the Pauli principle. It states that for fermions only one linear combination of all these wave functions is physically admissible, namely the antisymmetric combination (24.3). Obviously, every wave function, antisymmetric in all n particle coordinates, can be the basis function of an irreducible representation of the group of permutations of the n particles. This representation is very simple. To each permutation P corresponds the number $(-1)^P$ which is $+1$ for even and -1 for odd permutations.

The restriction to wave functions which are antisymmetric with respect to interchange of all the coordinates of any two particles, thus leads to one simple irreducible representation of the permutations group. A wider class of representations is introduced by considering Hamiltonians which do not contain the spin operators of the particles explicitly. Such Hamiltonians do not contain spin-orbit interactions (either single particle or mutual) or tensor forces. For identical particles, to which we limit our considerations now, Bartlett forces are equivalent to Majorana forces. Since the Majorana exchange operator P_x does not contain the spins, any scalar interaction can be included in a spin-independent Hamiltonian. Obviously, the eigenfunctions of such Hamiltonians are given by the LS-coupling scheme. The somewhat simpler case of jj-coupling will be considered later on.

The antisymmetric wave functions of two particles in LS-coupling, discussed in Part II, are products of spin functions and wave functions of the space coordinates. There are two possibilities: either the spin functions are symmetric ($S = 1$) and the spatial functions antisymmetric (odd L), or the spin functions antisymmetric ($S = 0$) and the spatial functions symmetric (even L). This is not the case for the wave functions of systems with more than two particles. Such wave functions can clearly be expanded as a linear combination of products of spin functions and spatial functions. However, unlike the case of two particles, the spin functions or the spatial functions appearing in these combinations are in general neither symmetric nor antisymmetric. Only the whole combination must be antisymmetric under an exchange of the spin and space coordinates of any two particles.

If the Hamiltonian does not contain the spin operators, it is clear that the space functions, which appear in the expansion of the complete wave function, must all be eigenfunctions of the Hamiltonian which belong to the same eigenvalue. These eigenfunctions, which depend only on the spatial coordinates of the particles, are not restricted directly by the antisymmetry requirement. The physical reason for this is that particles can now be distinguished from each other by their spin orientation. Each wave function should be antisymmetric with respect to exchange of the space coordinates of two particles only if they have the same spin orientation.

The Hamiltonians we now consider are invariant under permutations of the space coordinates of the particles. Their spatial eigenfunctions can therefore be classified according to the irreducible representations of the permutations group. Let us assume that there is no accidental degeneracy in the Hamiltonian. If ψ is an eigenfunction of H, all functions $P\psi$, where P is any permutation, will also be eigenfunctions which belong to the same eigenvalue. We can choose a basis of independent functions so that all functions $P\psi$ (including ψ) can be written as linear combinations of the basis functions. These functions obviously form the basis of an irreducible representation of the permutations group. Every irreducible representation, like the symmetric and antisymmetric ones, defines a certain symmetry character of the eigenfunctions.

There are usually several independent eigenfunctions in each irreducible representation. However, this fact does not give rise to any real physical degeneracy. As will be discussed later all these independent functions should be multiplied by appropriate spin functions and added together in order to form a complete antisymmetric wave function. We shall first discuss the spatial eigenfunctions of the various irreducible representations and then see under what conditions they can be augmented by appropriate spin functions to form an antisymmetric wave function.

The interaction energy depends strongly on the space symmetry of the wave functions. If the wave function is antisymmetric with respect to the interchange of the space coordinates of two particles, it vanishes when these particles are close together. Therefore, if the forces are short-ranged and attractive, then the wave function with maximum possible symmetry (in the space coordinates) will give the lowest energy. In the case of repulsive forces, the situation will be reversed. The higher the symmetry of the wave function the higher the energy. Thus we see that for Hamiltonians with no strong spin-dependent forces, the quantum numbers obtained from the group of permutations may be of great value for the classification of states.

In order to describe the irreducible representations of the group of permutations we shall start from a spatial wave function in the m-scheme. Let us consider

$$\psi_{m_1 m_2 \cdots m_n} = \psi_{lm_1}(x_1)_{lm_2}(x_2)\dots_{lm_n}(x_n) \tag{32.1}$$

which is a function of the space coordinates of n particles, all with orbital angular momentum l. We shall first require all m_i to be different. In $\psi_{m_1 m_2 \cdots m_n}$ it is understood that the first m is the magnetic quantum number of particle number 1, the second m is that of particle number 2, etc. If we perform a permutation on the particle coordinates in (32.1) we shall obtain a new function $\psi_{m_1' m_2' \cdots m_n'}$ where the set $m_1' m_2' \dots m_n'$ is a permutation of $m_1 m_2 \dots m_n$. It is therefore possible to consider permutations applied to the quantum numbers m_i rather than to the particle coordinates.

By applying all $n!$ permutations to (32.1) we obtain $n!$ different wave functions (since all m_i are different). These $n!$ functions form a basis of a representation of the permutations group. The operation of any permutation on one of these functions, or on any linear combination thereof, again gives a linear combination of them. However, this representation is not irreducible. We can form sets of linear combinations of these wave functions which transform among themselves under all permutations. In fact, one such linear combination is the antisymmetric wave function (24.3).

$$\psi_a = \sum (-1)^P P \psi_{m_1 m_2 \cdots m_n}. \tag{32.2}$$

As already remarked the antisymmetric function (32.2) is in itself the basis of an irreducible representation of order 1 of the permutations group. The matrices which correspond to the permutations are the numbers ± 1. The correspondence is simply

$$P \to (-1)^P \qquad \text{for all } P.$$

It should be mentioned that the antisymmetric function (32.2) vanishes if two (or more) of the m_i are equal. An even simpler one-dimensional representation is given by the *symmetric* combination. This is defined by

$$\psi_s = \sum P\psi_{m_1 m_2 \cdots m_n}. \tag{32.3}$$

Clearly, ψ_s is invariant under all permutations and therefore all representation matrices are equal to $+1$, namely

$$P \to +1 \qquad \text{for all } P.$$

The irreducible representations given by the symmetric and the antisymmetric combinations are the only ones which are one-dimensional. All other irreducible representations have other, more complicated types of symmetry of the wave functions. The various symmetry types have a simple graphical description in terms of *Young's diagrams* or tableaux. Each diagram contains n squares arranged in k rows. The rows are arranged according to their length from top to bottom. An example of such a diagram is the following:

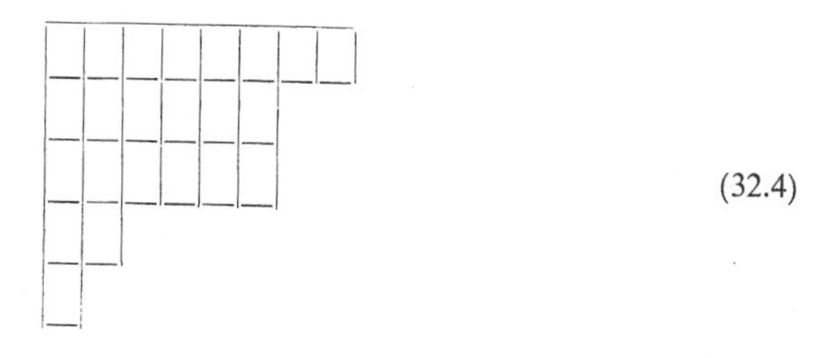

(32.4)

Each diagram is uniquely defined by the lengths of its rows. If n_i stands for the number of squares in the ith row, the diagram is defined by the set $n_1, \ldots, n_k$ satisfying

$$n_1 + n_2 + \ldots + n_k = n \qquad n_1 \geqslant n_2 \geqslant \ldots \geqslant n_k > 0. \tag{32.5}$$

We can get all the possible diagrams by considering the various *partitions* of n into a sum of nonnegative numbers. It will be shown soon that one irreducible representation of the group of permutations corresponds to each diagram. All irreducible representations can be obtained from such diagrams and different diagrams give rise to inequivalent representations.

The dimensionality of the irreducible representation that belongs to a definite diagram, which is the number of independent functions that transform among themselves under all permutations, can be calculated in a simple way. Write in each square one number from 1 to n. This can be done in $n!$ different ways. Among these, we distinguish the arrangements in which the numbers *in each row and each column* are written in increasing order (from left to right and from top to bottom respectively). These arrangements are called *standard arrangements.* For example, among the $3! = 6$ possible arrangements of 1, 2, and 3 in the diagram

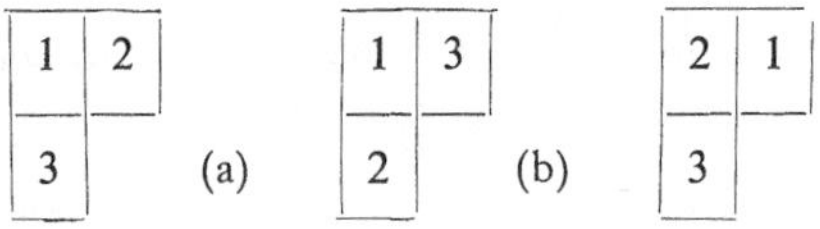

$$\begin{array}{|c|c|}\hline 1 & 2 \\ \hline 3 & \\ \cline{1-1}\end{array}\ \text{(a)} \qquad \begin{array}{|c|c|}\hline 1 & 3 \\ \hline 2 & \\ \cline{1-1}\end{array}\ \text{(b)} \qquad \begin{array}{|c|c|}\hline 2 & 1 \\ \hline 3 & \\ \cline{1-1}\end{array}\ \text{(c)} \tag{32.6}$$

only the arrangements (a) and (b) are standard but not that in (c). We state without proof that

The number of standard arrangements in a diagram is the dimensionality of the irreducible representation which corresponds to this diagram. (32.7)

To construct the irreducible representation defined by a given diagram T, we build a symmetry operation related to this diagram. First we take one of the standard arrangements of the n numbers in this diagram. We now consider all permutations p which permute only numbers written in the same row. Similarly, let q denote a permutation in which only numbers within the same column are permuted. We apply to the wave function (32.1) the following operation.

$$\psi^{(T)} = \sum_{qp} (-1)^q qp\psi = \left[\sum_q (-1)^q q\right] \left(\sum_p p\right) \psi. \tag{32.8}$$

This involves symmetrization of ψ with respect to the particles within rows, followed by antisymmetrization with respect to particles within columns. The order of operations in (32.8) is important since the antisymmetrization in the columns generally destroys the symmetry within rows. Special cases of this operation are the total symmetrization (32.3) which corresponds to the diagram $n = n + 0 + \dots + 0$ and the total antisymmetrization (32.2) given by $n = 1 + 1 + \dots + 1$.

The symmetric (or the antisymmetric) function is itself invariant under all permutations. In the general case, operating on $\psi^{(T)}$ with a permutation P we may obtain the same function or a different one. It follows from (32.8) that permutations of numbers within the same column can change at most the sign of $\psi^{(T)}$. Furthermore if $\psi^{(T)}$ is derived, for example, from the diagram

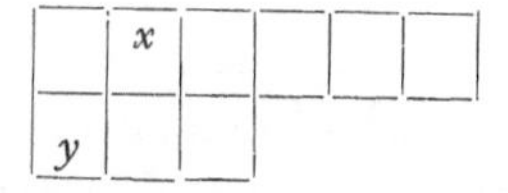

the permutation P which exchanges the numbers in the last two squares of the first row leaves $\psi^{(T)}$ invariant. On the other hand, if we exchange the numbers in the squares x and y we usually get a different function $\psi^{(T')}$ with the same symmetry. It turns out that all the functions $P\psi^{(T)}$ obtained by operating with all permutations P on $\psi^{(T)}$ derived from one given standard arrangement in the tableau T can be expressed as linear combinations of only g_T independent functions. According to (32.7) g_T is the number of possible standard arrangements in the given tableau T. The matrices constructed from these linear combinations form an irreducible representation of the permutation group of dimensionality g_T. Starting from different standard arrangements in the same tableau T we usually get different and independent functions leading to different representations. They are, however, all equivalent to each other. Thus, up to equivalence, a diagram determines an irreducible representation uniquely. In this way, the $n!$ functions $P\psi$ can be written in terms of linear combinations that belong to sets which transform among themselves under permutation. It is also true that along with any set with g_T independent functions there are additional $g_T - 1$ similar sets which give rise to the same irreducible representation. The total number of independent functions is $n!$ and we therefore obtain

$$\sum_T g_T^2 = n! \tag{32.9}$$

where the summation is over all possible different diagrams with n numbers. The relation (32.9) is the special case for the permutations group of a well-known relation in the theory of representations.

If two or more quantum numbers m_i in (32.1) are equal, the $n!$ functions $P\psi$ are no more independent We can still choose independent linear combinations and group them in sets which transform irreducibly among themselves under permutations. These sets do not give rise to all the irreducible representations obtained before. For instance, the antisymmetric combination clearly vanishes in this case, and we cannot obtain the antisymmetric representation from a basis in which two or more of the m_i are equal to each other. When all the m_i are different we can obtain g_T different sets of basis functions, each containing g_T

such functions, and all of them giving rise to the same representation (up to equivalence). Here, when two or more of the m_i are equal, there will generally be less than g_T sets (each containing however g_T basis functions). The representations generated by each of these sets will again be equivalent to each other.

The case in which not all the m_i are different is particularly important for the representations constructed from the spin functions. As shall be seen later, such functions are required for the construction of totally antisymmetric states. The preceding considerations can be applied to this case where now the m_i are restricted to the *two* values $+\frac{1}{2}$ and $-\frac{1}{2}$. If $n > 2$, the same values of m must occur several times in the wave function. When we construct the basis functions for the irreducible representations with the help of the Young tableaux, we antisymmetrize with respect to particles in the same column. Therefore, if there are two or more identical m_i in the same column, the function vanishes. Thus, all spin functions which give rise to irreducible representations correspond to diagrams having no more than *two* squares in their columns. In other words, the only irreducible representations of the permutations group constructed from the spin functions correspond to diagrams with at most *two* rows.

Let us now discuss in some more detail the spin functions and the representations to which they belong. We note first that starting from any spin function

$$\psi_{m_1 m_2 \cdots m_n} = \chi_1(m_1)\, \chi_2(m_2) \ldots \chi_n(m_n) \qquad m_i = \pm\tfrac{1}{2} \tag{32.10}$$

we can construct one nonvanishing *symmetric* combination. The corresponding diagram has one row only. It is clear that the symmetric function is in this case completely determined by giving the numbers n_+ and n_- specifying the number of m_i which have the values $+\frac{1}{2}$ and $-\frac{1}{2}$ respectively ($n_+ + n_- = n$). For a given set of values m_i, the value of $M_S = \Sigma\, m_i$ is also fixed and the corresponding wavefunction has therefore a well-defined value of S_z. Moreover, if it is a completely symmetric function it also has a well-defined value of $\mathbf{S}^2$. In fact, if all m_i are equal to $+\frac{1}{2}$, M_S obtains its maximum value $M_S = (n/2)$. The corresponding function is then also an eigenfunction of $\mathbf{S}^2$ with $S = (n/2)$. Operating now on this symmetric function with the symmetric operator $S_- = \Sigma_k\, [s_x^{(k)} - is_y^{(k)}]$ we obtain a symmetric function which still belongs to the same value of S but has $M_S = (n/2) - 1$. However, M_S and n determine n_+ and n_- uniquely. There is therefore only *one* symmetric spin function for a given value of n and M_S. From the foregoing we see that this function also has a definite value of S, namely $S = n/2$.

To obtain spin functions with different values of S we have to make use of symmetry types characterized by two rowed diagrams (three and more rowed diagrams give rise to vanishing spin functions, as was mentioned before). Let us consider a typical diagram

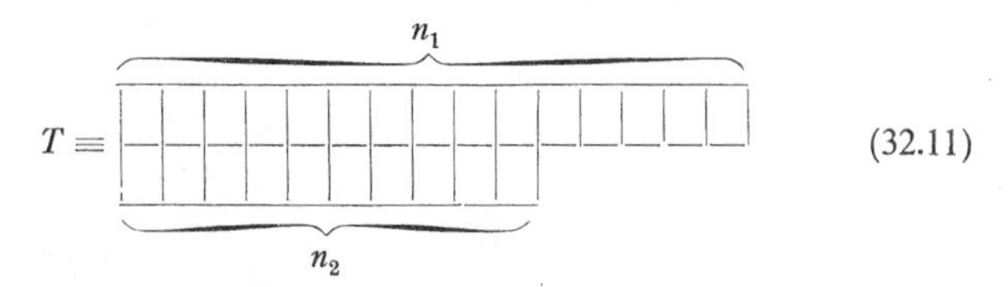

(32.11)

with n_1 squares in the first row and n_2 squares in the second row, where $n_1 + n_2 = n$. Starting from a spin function $\psi_{m_1 \cdots m_n}$, as defined by (32.10), this diagram generates, according to the prescription (32.8), a function $\psi^{(T)}$ with a definite symmetry. By applying all different permutations P to $\psi^{(T)}$ and taking appropriate linear combinations, we can construct a basis of g_T independent functions. All these functions have the same value of $M_S = \frac{1}{2}(n_+ - n_-)$. The functions $\psi^{(T)}$, and therefore also $P\psi^{(T)}$, vanish if $M_S > \frac{1}{2}(n_1 - n_2)$. Indeed, in this case there must be at least two particles in one column which are in the same m-state leading, upon antisymmetrization, to a vanishing result.

We have seen that the completely symmetric function is uniquely determined by n and M_S. We shall now see how many independent functions can be obtained by applying a definite tableau (32.11) to a function $\psi_{m_1 m_2 \cdots m_n}$ with a given $M_S = \Sigma\, m_i$. In other words, we are looking for the number of independent spin functions of a given symmetry and M_S. If all the m_i in $\psi_{m_1 m_2 \cdots m_n}$ are different from each other, each standard arrangement in the diagram T gives rise to g_T independent functions. Furthermore different standard arrangements give rise to independent sets of basis functions. Thus, for the case $m_i \neq m_k$ for every $i \neq k$, we obtain altogether g_T^2 independent functions of a given symmetry. However with $m_i = \pm \frac{1}{2}$, and $n > 2$ all the m_i cannot be different from each other and the number of independent functions of a given symmetry is reduced.

To find this number, we first prove a lemma. If $\psi_{M_S} = \psi_{m_1 \cdots m_n}$ is a given spin function and T is the operator corresponding to a given diagram, then

$$TP\psi_{M_S} = \pm\, T\psi_{M_S} \tag{32.12}$$

for all permutations P. To prove this lemma we note first that, if we apply to ψ_{M_S} a permutation p_0 of the numbers within the rows of the diagram T, then

$$Tp_0\psi_{M_S} = T\psi_{M_S}. \tag{32.13}$$

This follows from the fact that T can be decomposed into the product of two operators (Σp) and $[\Sigma (-1)^q q]$ as in (32.8). Obviously $(\Sigma p) = (\Sigma p p_0)$. For every function ψ_{M_S} we can find a permutation p_0 such that in $\psi'_{M_S} = p_0 \psi_{M_S}$ the m-value of every particle in the first n_2 squares of the *first* row will be different from the m value of the particles in the squares just below them. We can therefore limit our considerations to such special functions ψ'_{M_S}.

Let us see what is the result of applying Σp to ψ'_{M_S}. Some permutations p_1 exchange numbers only within the last $n_1 - n_2$ squares of the first row. All functions $p\psi'_{M_S}$ which have equal m values of particles in the same column will vanish after applying $\Sigma (-1)^q q$. All functions which will not vanish will be equal either to $p_1 \psi'_{M_S}$ or to $p_2 \psi'_{M_S}$. Here p_1 exchanges numbers only within the last $n_1 - n_2$ squares in the first row. Permutations p_2 exchange simultaneously numbers written in the same column. We can therefore apply to ψ'_{M_S} the operator $(\Sigma p_1)(\Sigma p_2)$ instead of the general $\Sigma_p p$. Before doing it we can further modify ψ'_{M_S} without changing the results of the operation (32.8). We apply to ψ'_{M_S} a permutation q_0 such that in $q_0 \psi'_{M_S}$ the first n_2 particles in the first row will have $m = +\frac{1}{2}$ and the particles in the second row will have $m = -\frac{1}{2}$. Operating with T on $\psi''_{M_S} = q_0 \psi'_{M_S}$ gives the same result, apart from a possible change of sign, as operating on ψ'_{M_S}. First q_0 commutes with all permutations p_1 since they operate on different sets of numbers. It is also clear that $p_2 q_0 \psi'_{M_S} = q'_0 p_2 \psi'_{M_S}$, $(-1)^{q_0} = (-1)^{q'_0}$, i.e., we can first permute columns by p_2 and then exchange the particles within the columns in their new positions. Therefore, $\Sigma(-1)^q q q'_0 = \Sigma(-1)^q q q_0$. As a result

$$\begin{aligned}
\left[\sum (-1)^q q p\right] q_0 \psi'_{M_S} &= \left[\sum (-1)^q q\right] \left(\sum p_2\right) \left(\sum p_1\right) q_0 \psi'_{M_S} \\
&= \sum (-1)^q q\, q_0 \left(\sum p_2\right) \left(\sum p_1\right) \psi'_{M_S} \\
&= (-1)^{q_0} \left[\sum (-1)^q q\right] \left(\sum p_2\right) \left(\sum p_1\right) \psi'_{M_S} \\
&= (-1)^{q_0} \left[\sum (-1)^q q p\right] \psi'_{M_S}.
\end{aligned} \tag{32.14}$$

It therefore follows that

$$T\psi''_{M_S} = T q_0 \psi'_{M_S} = \pm\, T\psi'_{M_S} = \pm\, T p_0 \psi_{M_S} = \pm\, T\psi_{M_S}.$$

Starting from the function $P\psi_{M_S}$, for any P, we arrive again at the *same* function ψ''_{M_S} (apart from a possible rearrangement of the last $n_1 - n_2$ numbers in the first row which is irrelevant). Since n and M_S determine uniquely n_+ and n_-, all functions ψ_{M_S} with a given M_S are obtained from each other by permutations. Hence, operating with a given T on any spin function ψ_{M_S} (with a given n and M_S) we always obtain the same result $T\psi''_{M_S}$. This proves our lemma.

We turn back to the original problem of determining the number of independent spin functions for given n and M_S with a given symmetry. We have to find the number of independent functions in the set

$$P\psi^{(T)} = PT\psi_{M_S} \tag{32.15}$$

where we take all g_T standard arrangements of the given diagram T and all possible permutations P. We have

$$PT\psi_{M_S} = PTP^{-1}P\psi_{M_S} = T'P\psi_{M_S}$$

where T' is an operator corresponding to the same diagram but not necessarily in a standard arrangement. Using the lemma (32.12) we therefore obtain

$$PT\psi_{M_S} = \pm\, T'\psi_{M_S}. \tag{32.16}$$

If in (32.16) T stands for one standard arrangement and P runs over all permutations, T' will run over all possible arrangements in the given diagram (both standard and nonstandard). We know that with one definite arrangement T and with P running over all permutations, the left-hand side of (32.16) yields exactly g_T independent functions. The functions on the right-hand side of (32.16) should therefore contain only g_T independent linear combinations. Since the right-hand side of (32.16) gives the same set of functions for *any* arrangement on the left-hand side, we conclude that *the number of independent spin functions of a given symmetry and a given M_S is exactly equal to the number of standard arrangements of the same symmetry*. For example, in the special case of symmetric functions there is only one standard arrangement, and there is therefore only one symmetric spin function, as was shown before.

We shall now show that $\psi^{(T)}$ is also an eigenstate of $\mathbf{S}^2$. We first choose a basis of g_T independent functions out of the functions $P\psi^{(T)}$ (P running over all $n!$ possible permutations). These are all the independent functions of the given symmetry which have the same value of M_S. $\mathbf{S}^2$ is a scalar operator and therefore must transform any function with a given M_S into a function of the same M_S. Since $\mathbf{S}^2$ is also a symmetric operator, it transforms a function with a given symmetry into a function with the same symmetry. Therefore $\mathbf{S}^2$ transforms the functions $P\psi^{(T)}_{M_S}$ among themselves. Furthermore, $\mathbf{S}^2$ commutes with all the permutations P. The g_T independent functions obtained from $P\psi^{(T)}_{M_S}$ form a basis for an irreducible representation of the permutation group. It follows therefore that $\mathbf{S}^2$ is a multiple of the unit matrix in the subspace of the functions considered. Hence $\psi^{(T)}$, and therefore also all functions $P\psi^{(T)}$, are eigenfunctions of $\mathbf{S}^2$.

It is now a simple matter to find the value of S for the spin functions $\psi^{(T)}_{M_S}$. If n_1 and n_2 have their previous meaning, $n_1 \geqslant n_2$, the maximum possible value of M_S for a spin function which does not vanish after the application of (32.8) is obtained when $n_+ = n_1$ and $n_- = n_2$. This maximum value $M_S = (n_1 - n_2)/2$ must be equal to S. The other sets of functions with lower values of M_S can be obtained from the one with $M_S = S$ by applying to it the symmetric operator S_-. We therefore see that every spin function obtained from a diagram $n = n_1 + n_2$ is uniquely characterized by the values of S and n. Thus

$$n_1 + n_2 = n \qquad \frac{n_1 - n_2}{2} = S. \tag{32.17}$$

We see that the spin S characterizes the symmetry type of the spin function for n particles as well as in the case of two particles discussed earlier. This result is due to the fact that m_s can have only two values.

We must now multiply spatial wave functions by spin functions and combine them so as to obtain a wave function, totally antisymmetric with respect to simultaneous permutations of the *space and spin* coordinates. Loosely speaking the antisymmetry implies that the spin functions must be antisymmetric where the spatial functions are symmetric and vice versa. Since the spatial wave functions, which belong to the same eigenvalue of H, have a definite symmetry type, we have to consider only those which belong to a definite diagram T. The spin functions by which they should be multiplied must belong to a definite diagram T' to obtain a totally antisymmetric function. For identical nucleons this is equivalent to the requirement that the complete function be an eigenfunction of $\mathbf{S}^2$ (we are considering LS-coupling!). The diagram T' is simply related to T. It is the *dual* of T, namely the rows of T' are the columns of T and vice versa. Furthermore, a space function obtained by (32.8) from a definite arrangement of the n numbers in T should be multiplied by a spin function obtained from the dual arrangement of the n numbers in the dual diagram T'. The antisymmetric combination must contain all $g_{T'} = g_T$ spin functions multiplied by the appropriate $g_T = g_{T'}$ basis functions of one irreducible representation. If there are N independent sets of g_T *spatial* basis functions there are also N independent totally antisymmetric combinations.

We see that the symmetry type of the space functions in the complete antisymmetric function uniquely determines the symmetry type of the spin functions and vice versa. The symmetry type of the spin functions is uniquely determined by S. Therefore, the symmetry type of the complete antisymmetric wave function in the space coordinates is uniquely determined by the spin S. As was seen before, there is only one set of $g_{T'} = g_T$ spin functions which must all be included in every antisymmetric combination. This means that there cannot be two (or more) independent antisymmetric functions with the same value of S constructed from the same spatial wave functions. On the other hand, there can generally be several independent sets of spatial wave functions of the same symmetry type which have either different values of L or even the same value of L. In the latter case, additional quantum numbers (like the seniority) are necessary.

In the case of maximum spin $S = n/2$, the spin function is symmetric, $g_T = 1$. In this case the total antisymmetric function is simply a product of the symmetric spin function and an antisymmetric space function. In general, for a given n, the higher the spin the higher the symmetry of the wave function in the spin coordinates. Therefore, the higher the spin the lower the symmetry of the wave functions in the space coordinates. States with higher spin should therefore lie higher for central attractive forces. On the other hand, for repulsive forces, like the Coulomb force, states with higher spin should lie lower. This is the basis of *Hund's rule* observed in atomic spectra. The positions of atomic energy levels are mainly determined by their spin values, the higher the spin the lower the energy.

As an illustration we may consider the case of a potential which can be approximated by a constant (an "infinite range square well" potential). Let the two-body interaction be $a + bP_{ik}$ where P_{ik} is the Majorana space exchange operator. For antisymmetric functions of identical particles, P_{ik} is given by

$$P_{12} = -\frac{1 + 4(\mathbf{s}_1 \cdot \mathbf{s}_2)}{2}.$$

We then obtain

$$\sum_{i<k}^{n} (a + bP_{ik}) = \frac{n(n-1)}{2} a - b \sum_{i<k}^{n} [\tfrac{1}{2} + 2(\mathbf{s}_i \cdot \mathbf{s}_k)]$$

$$= \frac{n(n-1)}{2} a - \frac{b}{2} \frac{n(n-1)}{2} - b \sum_{i<k}^{n} 2(\mathbf{s}_i \cdot \mathbf{s}_k)$$

$$= \frac{n(n-1)}{2} \left(a - \frac{b}{2}\right) - b \left(\sum_i \mathbf{s}_i\right)^2 + b \sum_i \mathbf{s}_i^2$$

$$= \frac{n(n-1)}{2} \left(a - \frac{b}{2}\right) - bS(S+1) + \tfrac{3}{4} nb. \qquad (32.18)$$

Thus, the energy of a state depends in this case only on its symmetry and therefore only on S (and n).

As an example of the preceding considerations on the symmetry of wave functions let us consider the p^3 configuration. Using the m-scheme we find that there are three independent states in this case. They are $^2D\,(S=\frac{1}{2}, L=2)$ and $^2P\,(S=\frac{1}{2}, L=1)$ and $^4S\,(S=\frac{3}{2}, L=0)$. Let us construct the wave functions of these states in the m scheme. We shall first construct wave functions with the appropriate S and L values and only then antisymmetrize them. For $M_S=\frac{1}{2}$ the maximum value of L is 2. This value belongs for example to the function $\psi^D_{M=2}=(1^+\,1^-\,0^+)$. On this function we operate with $L_-=\sum_{i=1}^3 l_-^{(i)}$ and obtain

$$L_-\psi^D_{M=2}=2\psi^D_{M=1}=\sqrt{2}\,(1^+\;0^-\;0^+)+\sqrt{2}\,(1^+\;1^-\;-1^+).$$

Therefore

$$\psi^D_{M=1}=\frac{1}{\sqrt{2}}[(1^+\;0^-\;0^+)+(1^+\;1^-\;-1^+)].$$

The function with $M=1$ $M_S=\frac{1}{2}$ which is orthogonal to it is the function of the P state. Choosing the phases arbitrarily, we obtain

$$\psi^P_{M=1}=\frac{1}{\sqrt{2}}[(1^+\;0^-\;0^+)-(1^+\;1^-\;-1^+)].$$

Operating on these two functions with L_- we obtain

$$\psi^D_{M=0}=\frac{1}{\sqrt{6}}[(1^+\;-1^-\;0^+)+2(1^+\;0^-\;-1^+)+(0^+\;1^-\;-1^+)]$$

$$\psi^P_{M=0}=\frac{1}{\sqrt{2}}[(1^+\;-1^-\;0^+)\qquad\qquad-(0^+\;1^-\;-1^+)].$$

The linear combination, orthogonal to these two, is the function of the ^4S state, with $M_S=\frac{1}{2}$, namely

$$\psi^S_{M=0}=\frac{1}{\sqrt{3}}[(1^+\;-1^-\;0^+)-(1^+\;0^-\;-1^+)+(0^+\;1^-\;-1^+)].$$

We now antisymmetrize all these functions with respect to both spin and space coordinates. Antisymmetrizing $\psi^S_{M=0}$ we obtain 18 terms which can be written as a product of spin functions and space functions. Let us write separately the spin and space functions so that $(1^+\,0^-\,-1^+)$, for instance, will be written as $(+\,-\,+)\,(1\;0\;-1)$ where the first bracket is a spin function with $m_1=+\frac{1}{2}$, $m_2=-\frac{1}{2}$, $m_3=+\frac{1}{2}$ and the second bracket is a space function. We can then write the antisymmetric $\psi^S_{M=0}$, apart from a normalization factor, as

$$[(+\ +\ -)+(+\ -\ +)+(-\ +\ +)]$$
$$\times\,[(1\,0\,-1)-(-1\,0\,1)-(0\,1\,-1)-(1\,-1\,0)+(-1\,1\,0)+(0\,-1\,1)].$$

This is simply a product of a symmetric spin function and an antisymmetric space function. Writing it in the form of diagrams it will be given by

$$\begin{array}{|c|c|c|}\hline 1 & 2 & 3 \\ \hline\end{array}\,(+\ +\ -) \qquad \begin{array}{|c|}\hline 1 \\ \hline 2 \\ \hline 3 \\ \hline\end{array}\,(1\,0\,-1).$$

The 12 terms of the antisymmetrized $\psi^P_{M=0}$ function can be expressed as a sum of two products of spin functions and space functions. Apart from a normalization, we obtain in this case

$$-\,[(+\ +\ -)-(-\ +\ +)]\,[(1\,0\,-1)+(-1\,0\,1)-(0\,1\,-1)-(0\,-1\,1)]$$
$$+\,[(+\ -\ +)-(-\ +\ +)]\,[(1\,-1\,0)+(-1\,1\,0)-(0\,1\,-1)-(0\,-1\,1)].$$

The spin function in the first term can be obtained by applying the diagram

$$\begin{array}{|c|c|}\hline 1 & 2 \\ \hline 3 & \multicolumn{1}{c}{} \\ \cline{1-1}\end{array}$$

to the function $(+\ +\ -)$. It is multiplied by a space function obtained from the dual diagram

$$\begin{array}{|c|c|}\hline 1 & 3 \\ \hline 2 & \multicolumn{1}{c}{} \\ \cline{1-1}\end{array}$$

and the function (1 0 − 1). Similar considerations hold for the second part. Using diagrams this function can be written as

$$-\begin{array}{|c|c|}\hline 1 & 2 \\ \hline 3 \\ \cline{1-1}\end{array}\;(+\;+\;-)\;\begin{array}{|c|c|}\hline 1 & 3 \\ \hline 2 \\ \cline{1-1}\end{array}\;(1\;0\;-1)$$

$$+\begin{array}{|c|c|}\hline 1 & 3 \\ \hline 2 \\ \cline{1-1}\end{array}\;(+\;-\;+)\;\begin{array}{|c|c|}\hline 1 & 2 \\ \hline 3 \\ \cline{1-1}\end{array}\;(1\;-1\;0).$$

The antisymmetrization of $\psi^{D}_{M=0}$ yields a similar result. We write here the somewhat simpler expression obtained by antisymmetrizing the combination

$$\frac{1}{\sqrt{2}}\,(\psi^{P}_{M=0} + \sqrt{3}\,\psi^{D}_{M=0}) = [(1^{+}\;-1^{-}\;0^{+}) + (1^{+}\;0^{-}\;-1^{+})].$$

The result is

$$[(+\;+\;-) - (-\;+\;+)]\;[(-1\;1\;0) + (0\;1\;-1) - (1\;-1\;0) - (1\;0\;-1)] -$$
$$[(+\;-\;+) - (-\;+\;+)]\;[(0\;-1\;1) + (-1\;0\;1) - (1\;0\;-1) - (1\;-1\;0)].$$

The spin functions appearing in this expression are the same as those in the previous case. The space functions are different but have the same symmetry properties.

The cases where two equal m-values occur in the space functions are even simpler. For example, antisymmetrization of $\psi^{D}_{M=2} = (1^{+}\;1^{-}\;0^{+})$ yields, apart from normalization,

$$-\,[(+\;+\;-) - (-\;+\;+)]\;[(1\;0\;1) - (0\;1\;1)]$$
$$+\,[(+\;-\;+) - (-\;+\;+)]\;[(1\;1\;0) - (0\;1\;1)].$$

In terms of diagrams it can be written as

$$\begin{array}{|c|c|}\hline 1 & 2 \\ \hline 3 \\ \cline{1-1}\end{array}\;(+\;+\;-)\;\begin{array}{|c|c|}\hline 1 & 3 \\ \hline 2 \\ \cline{1-1}\end{array}\;(1\;1\;0) - \begin{array}{|c|c|}\hline 1 & 3 \\ \hline 2 \\ \cline{1-1}\end{array}\;(+\;-\;+)\;\begin{array}{|c|c|}\hline 1 & 2 \\ \hline 3 \\ \cline{1-1}\end{array}\;(1\;0\;1).$$

We shall now discuss the quantum numbers which, along with the symmetry type (given by n and S) characterize the spatial wave functions.

If H is invariant also under rotations in ordinary space, L is a good quantum number. However, except for trivial cases, there are several states with the same L and the same symmetry type (i.e., n and S). For example, the antisymmetric states of the d^3 configuration are

$$S = \tfrac{3}{2} \qquad {}^4P, {}^4F$$

$$S = \tfrac{1}{2} \qquad {}^2P, {}^2D, {}^2D, {}^2F, {}^2G, {}^2H.$$

In this case there are two states with $S = \frac{1}{2}$ and $L = 2$. We therefore have to look for additional quantum numbers in order to specify uniquely the various states.

The L quantum number specifies the wave functions of n particles which transform according to the corresponding irreducible representations of the rotations group. Under rotations the $2L + 1$ wave functions ψ_{LM} transform among themselves. If there are two (or more) states with the same value of L (and M) we must find some means to label the different states. The symmetry type is indeed such a label. The rotations transform the coordinates of all particles in the same way and therefore commute with the permutations. Hence the conclusions that rotations do not change the symmetry type of the wave functions. However, rotations certainly do admix wave functions with different values of M_L. Thus they admix functions which belong to different representations of the permutations group. The important point however, is that they admix only functions which belong to *equivalent* representations (i.e., of the *same* symmetry type).

It is therefore useful to consider spatial wave functions with the same symmetry type irrespective of their M_L values. We can make another step and consider all spatial wave functions with the same symmetry type irrespective of either their M_L values or their L values. From the physical point of view the symmetry type then offers a rather coarse classification of states which can be made more precise by considering subsequently the three-dimensional rotations.

This approach has a well-defined mathematical formulation. All functions with a definite symmetry type (with all values of L and M_L) transform irreducibly under a certain group of transformations which is called the *linear group*. To introduce this group, we recall that under a three-dimensional rotation, the wave functions ψ_{lm} of a single particle undergo a linear transformation with the D-matrix according to

$$\psi'_{lm} = \sum_{m_1} \psi_{lm_1} D^{(l)}_{m_1 m}(R). \tag{32.19}$$

This transformation induces a linear transformation of the n-particle wave functions. As mentioned above, rotations do not change the symmetry of the n-particle wave function since the wave functions of *every particle* undergo the same transformation (32.19). A more general transformation, which also commutes with permutations, is obtained by subjecting the wave function of every particle to the same general linear transformation

$$\psi'_{lm} = \sum_{m_1} \psi_{lm_1} a_{m_1 m}. \tag{32.20}$$

The matrix elements $a_{m_1 m}$ in (32.20) are not restricted to be those of the D matrix but can be any complex numbers. The only requirement is that the transformation matrix $a_{m_1 m}$ is not singular (i.e., it must have a nonvanishing determinant).

The transformations (32.20) which belong to the linear group, induce in the spatial wave functions of n particles, linear transformation which transform functions with the same symmetry type irreducibly among themselves. The sets of all n particle functions with the same symmetry type can thus furnish the bases of the various inequivalent irreducible representations of the linear group.

From a mathematical point of view, the transformation (32.20) [as well as (32.19)] can be looked upon as taking place in an abstract space of $2l + 1$ dimensions. The ψ_{lm} are unit vectors in this space. In this language, each of the n-particle functions in the m-scheme is a component of a Cartesian tensor of rank n. Under the transformations (32.20) the components of the tensor undergo linear transformations. However, this tensor is reducible. We can make linear combinations of its components so as to obtain groups of functions which will transform among themselves under (32.20). These linear combinations are precisely the functions with definite symmetry type as explained above. These functions are obtained from all functions in the m-scheme by the operation of (32.8) with a definite diagram. Therefore, we see that the Young symmetrizer, operating on ψ in (32.8) is a projection operator into a subspace invariant under linear transformations.

The three-dimensional rotations are given by the transformations (32.19). The D matrices form a subgroup of the group of the general linear (nonsingular) matrices. The group of transformations induced in the ψ_{lm} by the three-dimensional rotations, R_3, is a subgroup of the linear group in the $2l + 1$-dimensional space. Consequently, an irreducible representation of the linear group in the n-particle functions may become reducible when we limit ourselves to the three-dimensional rotations. The set of functions with a definite symmetry type is thus split into subsets, each with $2L + 1$ functions, which transform irreducibly under rotations. However, an irreducible representation of R_3 may appear several times in this decomposition.

Before dealing with the problem of additional quantum numbers, it is worthwhile to make the following remarks. Although the transformation (32.20) and the rotations are treated here in the same manner, there is a considerable difference between the two. The mathematical transformation (32.20) has no simple physical meaning. It can be viewed as an operation on the Hamiltonian if H is written as a matrix in the m-scheme. However, the Hamiltonian is generally ***not invariant*** under these operations.* It is true that the Hamiltonian does not have nonzero matrix elements between states which belong to different irreducible representations of the linear group. However, this is due to the fact that such states have different symmetry types and H is invariant under permutations (of the space coordinates). The situation can be clarified by considering the invariance of H under rotations. In this case the energy matrix commutes with all rotation matrices. Therefore, H must have the same eigenvalue E in its submatrix corresponding to a given irreducible representation of the rotations group. This leads to the $(2L + 1)$-fold degeneracy of the energy in all states with the same value of L. There is generally no such degeneracy of the energy in all states with the same symmetry type. Only in very few cases is there such a degeneracy of H due to its invariance under the transformation (32.20). A simple example for this latter case is offered by a Hamiltonian which contains the kinetic energy and a long-range interaction like (32.18). The kinetic energy has the same value in all states of the same configuration and it follows from (32.18) that the potential energy depends only on S, and therefore only on the symmetry type of the wave functions.

Until now we have achieved a unified, group theoretical treatment of both the symmetry types of wave functions and their L quantum numbers. This approach can be used also to obtain additional quantum numbers. The irreducible representations of the unitary group $U(2l + 1)$ (or the linear group) are characterized by the symmetry type and thus offer an important quantum number. The restriction to the unitary transformations (32.19) which are induced by three-dimensional rotations, introduces the L quantum number. If we can find a subgroup $R(2l + 1)$ of the unitary group $U(2l + 1)$ which contains R_3 as a subgroup

$$U(2l +) \supset R(2l + 1) \supset R_3$$

then its irreducible representations may serve to label different states with the same value of L. If we go over from $U(2l + 1)$ to $R(2l + 1)$, an irreducible representation of $U(2l + 1)$, characterized by n and S,

(*) For the transformed Hamiltonian to have the same eigenvalues as the original we restrict the transformation matrix in (32.20) to be *unitary*. However, the irreducible representations of the linear group are the same as those of its unitary subgroup $U(2l + 1)$ which itself contains R_3 as a subgroup.

may be reduced to two (or more) irreducible representations of $R(2l+1)$. Let two such irreducible representations be characterized by v_1 and v_2. By going over from $R(2l+1)$ to R_3, the same value of L may appear twice. However, it may happen that one set of $2L+1$ functions is obtained in the reduction of v_1 and the other in that of v_2. Thus, at least in some cases, the irreducible representations of $R(2l+1)$ may distinguish between states with the same value of L. Even if such labels are not good quantum numbers, they still may serve to define a convenient scheme for the wave functions.

A group $R(2l+1)$ with the required properties does actually exist. It is the group of mathematical (real) rotations in the $2l+1$-dimensional space of the functions ψ_{lm}. A rotation is defined as a linear transformation which leaves the scalar product of two vectors invariant, and whose determinant is $+1$. It is, in fact, easy to find out which bilinear form must be invariant under $R(2l+1)$. Since this group has to contain R_3 as a subgroup,* also the transformations of R_3 must leave this bilinear form invariant. We have only one such bilinear form built of two vectors ψ_{lm}. It is the $L=0$ wave function of two particles (since it has only one component $M=0$, it is invariant under three-dimensional rotations). This $L=0$ function is given, apart from normalization, by

$$\psi^S = \sum_m \psi^*_{lm}(1)\psi_{lm}(2) = \sum_m (-1)^m \psi_{lm}(1)\psi_{l-m}(2). \tag{32.21}$$

In order to see that $R(2l+1)$, which leaves (32.21) invariant, is the group of real rotations in the $2l+1$-dimensional space it is more convenient to use *real* single particle wave functions. Let us define

$$\psi_0 = \psi_{l0} \qquad \psi_{m+} = \frac{1}{\sqrt{2}}(\psi_{lm} + \psi^*_{lm}) \qquad \psi_{m-} = \frac{-i}{\sqrt{2}}(\psi_{lm} - \psi^*_{lm})$$
$$\text{for} \quad 1 \leq m \leq l. \tag{32.22}$$

In this system the functions contain $\cos m\varphi$ and $\sin m\varphi$ instead of $e^{im\varphi}$. In terms of these wave functions, ψ^S can be written as

$$\psi^S = \psi_0(1)\psi_0(2) + \sum_{m=1}^{l} \psi_{m+}(1)\psi_{m+}(2) + \sum_{m=1}^{l} \psi_{m-}(1)\psi_{m-}(2). \tag{32.23}$$

Thus, a transformation which leaves ψ^S invariant (when applied simultaneously to the functions of the two particles) is equivalent to a real rotation in the $2l+1$-dimensional space of the real functions (32.22).

(*) It is not trivial that the rotation group in the $2l+1$-dimensional space can have R_3 as a subgroup. This will not be the case in the $2j+1$-dimensional space to be treated below in the case of jj-coupling.

We can now classify wave functions according to the irreducible representation of $R(2l+1)$. It should be kept in mind that the Hamiltonian is generally not invariant under the transformations of $R(2l+1)$. These transformations do, in fact, admix wave functions with different L values. In the case of $U(2l+1)$, H has nonvanishing matrix elements only between states which belong to the same irreducible representation. This is so, only because of the invariance of H under the permutations group and the relation between the irreducible representations of the permutations group and those of $U(2l+1)$. This is not the case for $R(2l+1)$. The Hamiltonian will generally *have* nonvanishing nondiagonal elements between states which belong to *different* irreducible representations of $R(2l+1)$. Clearly, such matrix elements will connect only states with the same symmetry type, and in the case of rotational invariance only states with the same value of L. In spite of this inconvenience, the scheme of wave functions, defined by the irreducible representation of $R(2l+1)$, is very useful and simple to work with. For certain interactions, the submatrix of the energy in one configuration may be diagonal in this scheme. If the energy submatrix is nearly diagonal, the irreducible representation of $R(2l+1)$ may give us approximately good quantum numbers.

We shall now consider the irreducible representations of $R(2l+1)$ in the space of wave functions of n particles. Sets of functions which transform among themselves under $U(2l+1)$ will certainly do so under $R(2l+1)$. Let us therefore confine ourselves to all the wave functions of the same symmetry type.

These can be written in the form of $2l+1$-dimensional tensors of rank n constructed from the ψ_{lm} (i.e., wave functions of n particles in an l-orbit). We can sometimes form linear combinations of these n-particle wave functions, such that two particles will be in the state ψ^S. Since ψ^S is invariant under $R(2l+1)$, it follows that the transformation of such tensors is the same as that of tensors constructed with $n-2$ particle wave functions.* We have thus established a relation between wave functions of $n-2$ particles and a special set of wave functions of n particles. Both sets transform in the same way under $R(2l+1)$ and thus give rise to the same representation. This relation between some wave functions of n particles and those of $n-2$ particles, obtained by removing a pair coupled to zero, is precisely that established in Section 31 by the seniority scheme. From now on we can make use of the formalism developed there.

Let us choose a set of functions of v particles, of the same symmetry, which contain no two particles in the special combination ψ^S. *These*

(*) It is essential that all single particle wave functions undergo the *same* transformation. The expression (32.21) is invariant also under $U(2l+1)$, if the wave functions of particle 1 transform with the matrix conjugate to that of particle 2.

functions will transform irreducibly among themselves under $R(2l+1)$. They are the wave functions with seniority v. If we start from wave functions of n particles with some symmetry type and remove pairs of ψ^S functions, arriving at wavefunctions with v particles, the value of the spin S, which depends only on the symmetry, does not change. In fact the $L=0$ pairs removed must all have $S=0$ because of the antisymmetry requirement. We can thus characterize the irreducible representations of $R(2l+1)$ by the symmetry type of the wave functions of v particles, i.e., in terms of S and v.

The symmetry types of wave functions of v particles with seniority v can also be obtained by the use of Young diagrams. These diagrams have in this case the property that the sum of the lengths of their first two columns cannot exceed $2l+1$. This is obvious since there are only two columns of lengths v_1 and v_2, and $v_1+v_2=v$, which is the seniority, cannot exceed $2l+1$ as we already found. The second column in any diagram is at most as long as the first one. From $v_1+v_2 \leq 2l+1$ follows $v_2 \leq 2l+1-v_1$. Therefore, if we define $v_1' = 2l+1-v_1$, there always exists a diagram with columns containing v_1' and v_2 squares (also $v_1'+v_2 \leq 2l+1$). Two such diagrams are called *associate diagrams*. The irreducible representations characterized by associate diagrams (which are always different) are inequivalent representations of $U(2l+1)$. However, the irreducible representations of $R(2l+1)$, characterized by associate diagrams, can be shown to be equivalent. We can therefore characterize the irreducible representations of $R(2l+1)$ by only one of the two associate diagrams. It is customary to choose as the representing diagram the one with the shorter first column. In this case $v_1 \leq l$ and the diagram can be described by l numbers in a decreasing order $w_1, w_2, \ldots, w_l$. The w_1 is the length of the first row, w_2 the length of the second row, etc. Some of the last w_i may be zero, but there are never more than l rows.

Since the rows of the diagrams under consideration contain at most 2 squares, the irreducible representations of $R(2l+1)$ are given by the number a of rows with 2 squares and the number b of rows with one square. Thus

$$w_1 = w_2 = \ldots = w_a = 2 \qquad w_{a+1} = w_{a+2} = \ldots = w_{a+b} = 1$$
$$w_{a+b+1} = w_{a+b+2} = \ldots + w_l = 0. \tag{32.24}$$

The numbers in (32.24) characterize both associate diagrams. If $v_1 \leq l$, the representing diagram will be that with column lengths v_1 and v_2. In this case

$$v = v_1 + v_2 = 2a + b$$
$$2S = v_1 - v_2 = b \tag{32.25}$$

so that $a = \frac{1}{2}v - S$ and $b = 2S$. However, if $v_1 > l$, the representing diagram will have column lengths $v_1' = 2l + 1 - v_1$ and v_2, i.e., $2l + 1 - v_1 = a + b$ and $v_2 = a$. In this case

$$\begin{aligned} v &= v_1 + v_2 = 2l + 1 - b \\ 2S &= v_1 - v_2 = 2l + 1 - 2a - b \end{aligned} \tag{32.26}$$

so that again $a = \frac{1}{2}v - S$ but now $b = 2l + 1 - v$. According to our convention b in (32.24) must be the smaller number of the two possible cases (32.25) and (32.26). We can thus express the quantum numbers a and b in terms of v and S by

$$a = \frac{v}{2} - S \qquad b = \min(2S, 2l + 1 - v). \tag{32.27}$$

Thus we see that the wave functions of a single particle with $v = 1$ $S = \frac{1}{2}$ (and $l > 0$) are characterized by $a = 0$ $b = 1$ so that $W \equiv (w_1 w_2 \ldots w_l) = (1\,0 \ldots 0)$. The ψ^S wave function of the l^2 configuration with $v = 0$ $S = 0$ has $a = 0$ and $b = 0$ and is thus characterized by $W = (0\,0 \ldots 0)$. If $l > 0$ there are $v = 2$ states with either even L and $S = 0$ or odd L and $S = 1$. For $S = 0$ we obtain from (32.27) $a = 1$ $v = 0$, i.e., $w = (2\,0 \ldots 0)$. For $S = 1$, the quantum numbers turn out to be $a = 0$ $b = 2$ (for $l > 1$) and therfore $W = (110 \ldots 0)$. For $l = 1$ the ^3P state is characterized according to (32.27) by $a = 0$ $b = 1$ and the only number w_1 is $w_1 = 1$.

If we return to the d^3 configuration, we see that under $U(2l + 1) = U(5)$ the space functions of ^4P and ^4F transform among themselves and the space functions of ^2P, ^2D, ^2D, ^2F, ^2G, and ^2H transform among themselves. When going over from $U(5)$ to $R(5)$, ^4P and ^4F still transform irreducibly among themselves but the other set is not irreducible. The functions of one of the ^2D states (with seniority $v = 1$) transform among themselves and the space functions of all other states ^2P, ^2D ($v = 3$), ^2F ^2G and ^2H transform among themselves. The states with $S = \frac{1}{2}$ $v = 3$ are given by $a = 1$, $b = 1$ and therefore by $(w_1 w_2) = (2\,1)$. On the other hand the diagram which corresponds to the states with $S = \frac{3}{2}$ $v = 3$ has $v_1 = 3v_2 = 0$ and therefore a and b are determined by the associate diagram. From (32.27) we find $a = 0$ $b = 2$, so that the state with $S = \frac{3}{2}$ $v = 3$ is characterized by $(w_1 w_2) = (1\,1)$. In the case of $l > 2$ the states of the l^3 configuration with $S = \frac{3}{2}$ $v = 3$ have $a = 0$, $b = 3$ and are therefore characterized by

$$(w_1 w_2 \ldots w_l) = (1\,1\,1\,0 \ldots 0).$$

The case of jj-coupling with identical particles is much simpler than that of LS-coupling. The Hamiltonian in this case is not invariant under permutations of the space coordinates alone. The only permutations which commute with H are those of spin and space coordinates of the particles. The Pauli principle removes the equivalence degeneracy by stating that only the antisymmetric combination of m-scheme wave functions is admitted.

We shall discuss in the following the case of n identical particles in the same j orbit. The m_i in the m-scheme are now the eigenvalues of the z-components of the $\mathbf{j}_i$. Every antisymmetric combination is the basis of a one-dimensional irreducible representation of the permutations group. It is characterized by the partition $n = 1 + 1 + ... + 1$, so that the corresponding diagram has only one column. The length of this column cannot exceed $2j + 1$.

All antisymmetric wave functions of the j^n configuration transform irreducibly among themselves when all the single particle wave functions undergo transformations of the linear (or unitary) group. These are given in the present case by

$$\psi'_{jm} = \sum_{m_1} \psi_{jm_1} a_{m_1 m}. \tag{32.28}$$

Here the matrix elements $a_{m_1 m}$ are those of a nonsingular matrix (or better still a unitary matrix). We can go over from $U(2j + 1)$ to R_3 by restricting the matrices $a_{m_1 m}$ to be the D-matrices $D^{(j)}_{m_1 m}(R)$. By doing this the antisymmetric representation of $U(2j + 1)$ is reduced to irreducible representations of R_3 characterized by the values of J. Until now it is all the same as for LS-coupling. The necessity of distinguishing between several states in the configuration j^n with the same J values occurs also in jj-coupling. Therefore, in order to find additional quantum numbers we look for a subgroup of $U(2j + 1)$ which will contain R_3 as a subgroup. Led by analogy with LS-coupling, we consider the group which leaves invariant the bilinear form $\psi_{J=0}$ of two particles. Since $\psi_{J=0}$ is invariant under three-dimensional rotations, R_3 is a subgroup of this new group. However, this new intermediate group is no longer the group of rotations in the $2j + 1$-dimensional space. The scalar product of two vectors which is invariant under rotations is a *symmetric* bilinear form (with respect to the exchange of the two particles). This was the case with $\psi^S = \psi(l^2 L = 0)$. However, $\psi_{J=0} = \psi(j^2 J = 0)$ is *antisymmetric* in the two particles. The group which leaves invariant a bilinear *antisymmetric* form of two vectors (in $2j + 1$-dimensional space) is called the *symplectic group*, $\mathrm{Sp}(2j + 1)$.

The irreducible representations of $\mathrm{Sp}(2j + 1)$ are analogous to those of $R(2l + 1)$. Again we build linear combinations of the wave functions of n particles with a definite symmetry (in our case antisymmetric). Some of them will have two particles in a $\psi_{J=0}$ state. Since $\psi_{J=0}$ is invariant under $\mathrm{Sp}(2j + 1)$, such linear combinations have the same transformation properties, with respect to the symplectic group, as wave functions of $n - 2$ particles. The relation thus established is precisely that introduced in Section 27 by the seniority scheme. All the results obtained there apply to the present discussion. The wave functions of v particles which contain no bilinear combinations $\psi_{J=0}$, transform irreducibly under Sp$(2j + 1)$. These are the functions with seniority v. They can also be obtained by using Young diagrams. The only diagrams in this case which give rise to nonvanishing wave functions with seniority v are those with not more than $\frac{1}{2}(2j + 1) = j + \frac{1}{2}$ rows. This is obvious in the present case since the diagrams contain only one column of length v and the seniority cannot be higher than half the number of particles in a shell. This is simpler than the corresponding restriction in the case of $R(2l + 1)$. As a result, all permissible diagrams are uniquely characterized by giving the lengths of $j + \frac{1}{2}$ rows, some of which may be zero. We therefore characterize these diagrams by

$$w_1 w_2 \dots w_{j+1/2} \qquad w_1 \geqslant w_2 \geqslant \dots \geqslant w_{j+1/2}. \tag{32.29}$$

Since in the present case, w_i can at most be equal to 1, we have

$$w_1 = w_2 = \dots = w_a = 1 \qquad w_{a+1} = w_{a+2} = \dots = w_{j+1/2} = 0. \tag{32.30}$$

The length of the column is simply equal to the seniority, $a = v$.

The single particle ψ_{jm} states, as well as all other states with $v = 1$, are thus characterized by $W = (1\ 0 \dots 0)$. The state $\psi_{J=0}$ of the j^2 configuration has $v = 0$ and is thus given by $W = (0\ 0 \dots 0)$. The other states, with even J and $v = 2$, of the j^2 configuration, have $a = 2$ and are characterized by $W = (1\ 1\ 0 \dots 0)$.

When we go over from Sp$(2j + 1)$ to R_3 we obtain a further specification of the states in terms of the quantum number J. We remark again that H is generally not invariant under either $U(2j + 1)$ or Sp$(2j + 1)$. However, in any case, the scheme of the irreducible representations of Sp$(2j + 1)$, which is the seniority scheme, is a simple and convenient scheme to work with. When H contains two-body interactions with only odd tensors we saw in Section 27 that all matrix elements of H between states of different irreducible representations of Sp$(2j + 1)$ (i.e., with different seniorities) vanish.

33. Many-Nucleon Wave Functions with Definite Isospin

In Part II we introduced the idea of charge independence. We mentioned there that the nuclear interaction between two neutrons, or two protons is the same as the one between a proton and a neutron. Furthermore, certain nuclear interactions involve charge exchange between a proton and a neutron. The electrostatic forces acting between two protons are different in this respect from the nuclear forces. They tend to destroy charge independence. However, these forces being weaker than the nuclear forces, become important only if there are many protons present. In other cases they, as well as the proton-neutron mass difference, can be disregarded in the first approximation. It is therefore worthwhile to consider the proton and neutron as two different states of the same particle—the *nucleon*. These states can be defined in terms of a charge coordinate and conveniently described by the formalism of the isospin introduced in Section 19. In the present section we shall extend this formalism to systems with many nucleons and consider the properties of their wave functions.

In the two-nucleon system, the introduction of the charge coordinates does not introduce any equivalence degeneracy. The total wave function can be written, in this case, as a product of two wave functions, one of the space and spin coordinates and the other of the charge coordinates. The symmetric charge functions necessarily have $T = 1$ and should be multiplied only by antisymmetric space and spin wave functions. The antisymmetric charge function, with $T = 0$ must be multiplied only by symmetric space and spin functions. This way of removing the spurious equivalence degeneracy is very convenient since it leads to a generalized exclusion principle. It allows for wave functions of two nucleons which are antisymmetric in the space, spin, and isospin (charge) coordinates.

We keep the isospin formalism for systems with many nucleons. We require the wave function of the system to be antisymmetric under exchange of the space, spin, and isospin coordinates of any two nucleons. This requirement completely removes the equivalence degeneracy since we saw in the preceding section that there is only *one* antisymmetric combination of any set of functions obtained from one another by permutations of the particle numbers.

Charge independence requires that the two-nucleon interactions be independent of the m_t values of the nucleons. Therefore, these interactions must be scalars in isospin space. As a result, also the Hamiltonian is invariant under rotations in isospin space. This implies that the total isospin T, defined by the eigenvalue of $\mathbf{T}^2 = T(T + 1)$, is a good quantum number. The vector $\mathbf{T}$ is the *total isospin* vector and is defined by

$$\mathbf{T} = \sum_{i=1}^{n} \mathbf{t}_i$$

where $\mathbf{t}_i$ is the isospin vector of the ith nucleon.

We can use the generalized exclusion principle to eliminate all charge coordinates from the Hamiltonian. In fact, the only non trivial scalars built of the isospin vectors of two nucleons are of the form $(\mathbf{t}_i \cdot \mathbf{t}_k)$. Such terms can always be expressed in terms of the charge exchange operators P_τ by using

$$P_{ik}^{\tau} = \frac{1 + 4(\mathbf{t}_i \cdot \mathbf{t}_k)}{2}. \tag{33.1}$$

If we denote the spin exchange operator by P_{ik}^{σ} and that of the spatial coordinates by P_{ik}^{x}, then the operator $P_{ik}^{\tau} P_{ik}^{\sigma} P_{ik}^{x}$ exchanges simultaneously all coordinates (charge, spin, and space). It therefore multiplies admissible wave functions by -1. Therefore, as long as we are working with such admissible wave functions (i.e., properly antisymmetrized), we can replace P_{ik}^{τ} by $-P_{ik}^{\sigma}P_{ik}^{x}$. The Hamiltonian will then become invariant under exchange of the space and spin coordinates of the nucleons only, without an exchange of the charge coordinates. The situation is similar to that of LS-coupling in the case of identical particles. The role played there by S is given here to T. We can conclude from the discussion of the last section that the complete antisymmetric wave function of n nucleons will be a linear combination of isospin functions multiplied by space and spin functions. The functions appearing in the products will all have a definite symmetry type (not necessarily symmetric or antisymmetric) characterized by n and T. We shall discuss the analogy between the two cases in more detail further on. We shall first make use of the isospin vectors in a straightforward method of obtaining totally antisymmetric wave functions for nucleons. This method is similar to the one used in Section 31 for the case of LS-coupling and identical particles.

We consider first the problem of n nucleons in the same j orbit. This case, of jj-coupling, is simpler than LS-coupling since with every nucleon is associated just *one* angular momentum vector $\mathbf{j}$ (instead of $\mathbf{s}$ and $\mathbf{l}$ in LS-coupling). These vectors should be combined to give a definite eigenvalue J of the total angular momentum operator $\mathbf{J}$. Each nucleon has also an isospin vector $\mathbf{t}$ in isospin space. These vectors should be combined to give a definite eigenvalue T so that $\mathbf{T}^2\psi = T(T+1)\psi$. A state with a definite T and with a definite value of $T_3 = M_T$ describes a system of Z protons and N neutrons so that

$(Z - N)/2 = M_T$. Obviously the values that T can assume in such a system are limited by

$$\tfrac{1}{2}|Z - N| \leq T \leq \tfrac{1}{2}(Z + N). \tag{33.2}$$

This condition on T could also be interpreted in the following way. The system of Z protons has *one* definite value of T namely $T_1 = Z/2$, similarly the neutron system is in a state with $T_2 = N/2$. The total isospin $\mathbf{T}$ is given by $\mathbf{T} = \mathbf{T}_1 + \mathbf{T}_2$ so that (33.2) is simply the triangular condition. If we consider for a given value of A a definite value of T, there are $(2T + 1)$ states with the various values of M_T for which $-T \leq M_T \leq T$. These $(2T + 1)$ states form a *charge multiplet* but they are found in *different nuclei*, determined by the values of $A = Z + N$ and $M_T = (Z - N)/2$. We shall now consider the problem of constructing wave functions with definite values of J and T, fully antisymmetric in the ordinary (space and spin) coordinates and the isospin (charge) coordinates of the nucleons.

The wave function of a single nucleon is characterized by the value of j, the eigenvalue m of j_z, the value of $t = \frac{1}{2}$, and the eigenvalue of t_3. According to our convention, the value $+\frac{1}{2}$ of t_3 characterizes a proton and the value $-\frac{1}{2}$ a neutron. The j^2 configuration was discussed in detail in Part II. The wave functions of the various states of the j^2 configuration are products of functions of space and spin coordinates and isospin functions. The former are characterized by J and the latter by T. A space-spin function is symmetric if J is odd and antisymmetric if J is even (similarly, an isospin function is symmetric for odd values of T and antisymmetric when T is even). Therefore, the wave functions admitted by the generalized exclusion principle with $T = 1$ are those with even values of J (as in the case of two identical nucleons where of course $T_z = \pm 1$). The wave functions with odd values of J have $T = 0$ (and therefore also $M_T = 0$).

The case of three j-nucleons is more complicated. If T has its maximum value ($T = \frac{3}{2}$), the wave functions are still simply products of a space-spin function and an isospin function. The space-spin functions are those admitted by the (ordinary) Pauli principle and are the antisymmetric functions discussed in Section 26. The isospin functions are the same as the symmetric functions of three spins $\frac{1}{2}$, i.e., the "configuration" $(\frac{1}{2})^3$, with total spin $\frac{3}{2}$. They can be built by simply coupling $\mathbf{t}_1$ and $\mathbf{t}_2$ to form $\mathbf{T}_{12}$ with $T_{12} = 1$, then coupling $\mathbf{t}_3$ and $\mathbf{T}_{12}$ to form $\mathbf{T}$ with $T = \frac{3}{2}$, and symmetrizing. We shall not discuss these functions in detail. Rather, we shall derive general formula for constructing the fully antisymmetric wave functions of three j-nucleons, valid also for the case of $T = \frac{1}{2}$.

We start from a wave function of two nucleons with allowed values of T_0 and J_0, where $\mathbf{T}_0 = \mathbf{t}_1 + \mathbf{t}_2$ and $\mathbf{J}_0 = \mathbf{j}_1 + \mathbf{j}_2$. Using vector addition coefficients, we form the sums of products of these wave functions with those of the third nucleon to obtain a state with definite values of T and J. The wave function thus obtained will be denoted by

$$\psi[j_{12}^2(T_0J_0)\,j_3T\,J\,M_TM]. \tag{33.3}$$

For convenience we shall omit in the following the eigenvalues of the z-projections, M_T and M, from the wave functions. The wave function (33.3), even though it may have values of T and J allowed by the exclusion principle, is generally not antisymmetric in all particle coordinates. To perform the antisymmetrization we use the operator $\Sigma(-1)^PP$ where the permutations P exchange both ordinary and isospin coordinates. Since $\psi(j^2T_0J_0)$ is already antisymmetric we obtain after antisymmetrization the function

$$2\{\psi[j_{12}^2(T_0J_0)\,j_3T\,J] - \psi[j_{13}^2(T_0J_0)\,j_2T\,J] + \psi[j_{23}^2(T_0J_0)\,j_1T\,J]\}. \tag{33.4}$$

The wave function (33.4) is both antisymmetric and has definite values of T and J. However, this wave function is not so easy to use, as explained in detail at the beginning of Section 26, and we would rather bring it to a form similar to (26.10). Therefore, we change the order of coupling in the last two terms of (33.4) so that $\mathbf{j}_1$ and $\mathbf{j}_2$ will be coupled to $\mathbf{J}_1$ and $\mathbf{t}_1$ and $\mathbf{t}_2$ will be coupled to $\mathbf{T}_1$. This change of coupling is carried out by using the transformation (15.29). We thus obtain from (33.4)

$$\begin{aligned}
&\mathscr{A}\psi[j_{12}^2(T_0J_0)\,j_3T\,J] \\
&= \psi[j_{12}^2(T_0J_0)\,j_3T\,J] - \psi[j_{13}^2(T_0J_0)\,j_2T\,J] + \psi[j_{23}(T_0J_0)\,j_1T\,J] \\
&= \psi[j_{12}^2(T_0J_0)\,j_3T\,J] - \sum_{T_1J_1} \langle t_1t_3(T_0)t_2T|t_1t_2(T_1)t_3T\rangle \\
&\times \langle j_1j_3(J_0)\,j_2J|\,j_1j_2(J_1)\,j_3J\rangle\,\psi[j_{12}^2(T_1J_1)\,j_3T\,J] \\
&+ \sum_{T_1J_1} \langle t_2t_3(T_0)t_1T|t_2t_1(T_1)t_3T\rangle\,\langle j_2j_3(J_0)\,j_1J|\,j_2j_1(J_1)\,j_3J\rangle \\
&\times \psi[j_{21}^2(T_1J_1)\,j_3T\,J].
\end{aligned} \tag{33.5}$$

The antisymmetrizer $\mathscr{A}$, introduced in (33.5), is analogous to the antisymmetrizer used in Section 26. It is a linear combination of permutations which exchange the space and spin as well as the isospin coordinates of the particles. We make use of the fact that $t_1 = t_2 = t_3 = \frac{1}{2}, j_1 = j_2 = j_3 = j$ to combine the last two terms in (33.5). In doing so we notice that all terms with T_1J_1 which yield symmetric functions $\psi(j^2T_1J_1)$

necessarily drop out and only terms with $T_1 J_1$ which yield antisymmetric two-particle functions remain. Such T_1 and J_1 satisfy $(-1)^{J_1+T_1} = -1$. Introducing the values of the transformation coefficients from (15.30) we obtain for (33.5) the expression

$$\psi[j_{12}^2(T_0J_0)j_3T\,J] - 2 \sum_{T_1+J_1 \text{ odd}} \sqrt{(2T_0+1)(2T_1+1)(2J_0+1)(2J_1+1)}$$

$$\times \begin{Bmatrix} \frac{1}{2} & \frac{1}{2} & T_1 \\ T & \frac{1}{2} & T_0 \end{Bmatrix} \begin{Bmatrix} j & j & J_1 \\ J & j & J_0 \end{Bmatrix} \psi[j_{12}^2(T_1J_1)j_3T\,J]$$

$$= \sum_{T_1+J_1 \text{ odd}} \Big(\delta_{T_1T_0}\delta_{J_1J_0} - 2\sqrt{(2T_0+1)(2T_1+1)(2J_0+1)(2J_1+1)}$$

$$\times \begin{Bmatrix} \frac{1}{2} & \frac{1}{2} & T_1 \\ T & \frac{1}{2} & T_0 \end{Bmatrix} \begin{Bmatrix} j & j & J_1 \\ J & j & J_0 \end{Bmatrix}\Big) \psi[j_{12}^2(T_1J_1)j_3T\,J]. \tag{33.6}$$

The wave function (33.6) is already of the desired form. The coefficients in this expansion are not yet coefficients of fractional parentage since (33.6) is not normalized to unity. If the particular values of T and J are not allowed by the exclusion principle, all the coefficients of the wave function (33.6) vanish. If however (33.6) does not vanish we can normalize it. To do so we should only recall the fact that antisymmetrization is a projection operation. From this we concluded [cf. (26.24)] that the inverse of the normalization factor N is equal to n (= 3 in the present case) times the *normalized* c.f.p. of the principal parent. Therefore N^{-2} is equal to 3 times the coefficient of the principal parent in (33.6), this parent being the state $\psi(j^2T_0J_0)$ from which we started. Thus, we obtain

$$N^2 = \left(3\left[1 - 2(2T_0+1)(2J_0+1)\begin{Bmatrix} \frac{1}{2} & \frac{1}{2} & T_0 \\ T & \frac{1}{2} & T_0 \end{Bmatrix} \begin{Bmatrix} j & j & J_0 \\ J & j & J_0 \end{Bmatrix}\right]\right)^{-1}. \tag{33.7}$$

The normalized wave function is expressed as

$$\psi(j^3T\,J) = \sum_{T_1+J_1 \text{ odd}} [j^2(T_1J_1)j\,T\,J|\}j^3T\,J]\,\psi[j_{12}^2(T_1J_1)j_3T\,J]. \tag{33.8}$$

The coefficients of fractional parentage (c.f.p.), as obtained from the antisymmetrization of (33.3), are given by

$$[j^2(T_1J_1)j\,T\,J|\}j^3[T_0J_0]T\,J]$$

$$= \Big[\delta_{T_1T_0}\delta_{J_1J_0} - 2\sqrt{(2T_1+1)(2T_0+1)(2J_1+1)(2J_0+1)}$$

$$\times \begin{Bmatrix} \frac{1}{2} & \frac{1}{2} & T_1 \\ T & \frac{1}{2} & T_0 \end{Bmatrix} \begin{Bmatrix} j & j & J_1 \\ J & j & J_0 \end{Bmatrix}\Big] \left[3 - 6(2T_0+1)(2J_0+1)\begin{Bmatrix} \frac{1}{2} & \frac{1}{2} & T_0 \\ T & \frac{1}{2} & T_0 \end{Bmatrix} \begin{Bmatrix} j & j & J_0 \\ J & j & J_0 \end{Bmatrix}\right]^{-1/2}. \tag{33.9}$$

In (33.9) the quantum numbers of the principal parent are written in the square brackets within the c.f.p. This result is similar to (31.7) in the case of LS-coupling (with identical particles). The correspondence is obtained by $l \rightarrow j$ and $s \rightarrow t$. The difference of sign between the two expressions (33.9) and (31.7) arises from the fact that l is integral whereas j is half-integral.

We first consider (33.9) in the special case of identical particles. In this case $T = \frac{3}{2}$ and consequently we must have $T_0 = 1$. The Racah coefficient $W(\frac{1}{2}\ \frac{1}{2}\ T_1 \mid \frac{3}{2}\ \frac{1}{2}\ 1)$ vanishes unless $T_1 = 1$ because of the triangular conditions. Inserting these values of T_1, T_0, and T as well as the value of $W(\frac{1}{2}\ \frac{1}{2}\ 1 \mid \frac{3}{2}\ \frac{1}{2}\ 1) = -\frac{1}{3}$ into (33.9) we obtain our previous result (26.11). This serves only as a check on the formalism since the results for a system of protons only, should, of course, be independent of the existence of neutrons. In general, if we calculate for n particles with maximum isospin $T = n/2$ the c.f.p. of the various states of $n-1$ particles, these c.f.p. will be different from zero only for the maximum isospin $T_1 = [(n-1)/2]$. This fact is due to the triangular conditions on the coupling $\mathbf{T}_1 + \mathbf{t} = \mathbf{T}$ and is obviously necessary since in this case (which can be obtained with identical particles) the isospin functions are not required.

Equation (33.9) gives new results only for the case $T = \frac{1}{2}$. Even if $T_0 = 1$, T_1 can be either 1 or 0 and there are nonvanishing c.f.p. with both of these possibilities. As an example, we calculate the c.f.p. for the case $T_0 = 1$ $J_0 = 0$ so that $J = j$. For $T = \frac{1}{2}$ we obtain from (33.9), using the values $W(\frac{1}{2}\ \frac{1}{2}\ 1 \mid \frac{1}{2}\ \frac{1}{2}\ 1) = \frac{1}{6}$ and

$$\begin{Bmatrix} j & j & J_1 \\ j & j & 0 \end{Bmatrix} = (-1)^{2j+J_1} \frac{1}{2j+1},$$

the following results for the c.f.p.

$$[j^2(T_1 J_1) j\ T = \tfrac{1}{2}\ J = j | \} j^3 [T_0 = 1\ J_0 = 0]\ T = \tfrac{1}{2}\ J = j]$$

$$= \begin{cases} \sqrt{\dfrac{2(j+1)}{3(2j+1)}} & \text{for} \quad T_1 = 1 \quad J_1 = 0 \\[2ex] \sqrt{\dfrac{2J_1+1}{6(j+1)(2j+1)}} & \text{for} \quad T_1 = 1 \quad J_1 > 0 \text{ even} \\[2ex] \dfrac{-\sqrt{2J_1+1}}{\sqrt{2(j+1)(2j+1)}} & \text{for} \quad T_1 = 0 \quad J \text{ odd.} \end{cases} \tag{33.10}$$

The coefficients of fractional parentage in the case of n nucleons are defined in a similar way. We consider all wave functions

$\psi[j^{n-1}(\alpha_1 T_1 J_1) j \; T \; J]$, with definite values of T and J, for all values of allowed α_1 T_1 and J_1. The α_1 are the additional quantum numbers necessary to define uniquely the states of $n-1$ nucleons. This is a complete set of functions, antisymmetric in the coordinates of the first $n-1$ nucleons. Therefore, a wave function with the given T and J, antisymmetric in the coordinates of *all* nucleons, can be expressed as a linear combination of these functions. Provided the antisymmetric function does not vanish identically, we normalize it and write

$$\psi(j^n \alpha \; T \; J) = \sum_{\alpha_1 T_1 J_1} [j^{n-1}(\alpha_1 T_1 J_1) j \; T \; J \| \} j^n \alpha \; T \; J] \, \psi[^{n-1}(\alpha_1 T_1 J_1) j_n T \; J]. \quad (33.11)$$

The quantum numbers α are those necessary to distinguish the different antisymmetric states of n nucleons with the same given T and J, if there are several of them. As before, we do not specify the z-projections M_T and M since the c.f.p. are independent of them. The requirement that the functions (33.11) are normalized and orthogonal to each other leads to the following relations between the c.f.p.

$$\sum_{\alpha_1 T_1 J_1} [j^n \alpha' T \; J \{ | j^{n-1}(\alpha_1 T_1 J_1) j \; T \; J] \, [j^{n-1}(\alpha_1 T_1 J_1) j \; T \; J | \} j^n \alpha \; T \; J] = \delta_{\alpha\alpha'}. \quad (33.12)$$

We can similarly define $n \rightarrow n-2$ c.f.p. by the following expansion of a normalized antisymmetric function

$$\psi(j^n \alpha \; T \; J) = \sum_{\alpha_1 T_1 J_1 T' J'} [j^{n-2}(\alpha_1 T_1 J_1) j^2(T' J') T \; J | \} j^n \alpha \; T \; J]$$
$$\times \psi[j^{n-2}(\alpha_1 T_1 J_1) j^2_{n-1,n}(T' J') T \; J]. \quad (33.13)$$

The actual construction of the wave functions (33.11) can be carried out in the same fashion as in the case of 3 nucleons. We start from an antisymmetric wave function of $n-1$ nucleons $\psi(j^{n-1}\alpha_0 T_0 J_0)$. To this state we add the nth nucleon so that its spin $\mathbf{j}$ is added to $\mathbf{J}_0$ to form $\mathbf{J}$ and its isospin $\mathbf{t}$ is combined with $\mathbf{T}_0$ to form $\mathbf{T}$. The resulting wave-function $\psi[j^{n-1}(\alpha_0 T_0 J_0) j_n T \; J]$ is now antisymmetrized by applying the operator $\mathscr{A}$ as defined in (33.14) below. In each term we single out the nth nucleon, the coordinates of which are in the function $\psi(j^{n-1}\alpha_0 T_0 J_0)$, by using $n-1 \rightarrow n-2$ c.f.p. obtaining

$$\mathscr{A}\psi[j^{n-1}(\alpha_0 T_0 J_0) j_n T \; J]$$
$$= \psi[j^{n-1}(\alpha_0 T_0 J_0) j_n T \; J] - \sum_k^{n-1} \psi[j^{n-1}(\alpha_0 T_0 J_0) j_k T \; J]$$
$$= \psi[j^{n-1}(\alpha_0 T_0 J_0) j_n T \; J] - \sum_k^{n-1} (-1)^{P_k} \sum_{\alpha_2 T_2 J_2} [j^{n-2}(\alpha_2 T_2 J_2) j \; T_0 J_0 | \} j^{n-1} \alpha_0 T_0 J_0]$$
$$\times \psi[j^{n-2}(\alpha_2 T_2 J_2) j_n (T_0 J_0) j_k T \; J]. \quad (33.14)$$

The P_k in (33.14) denote permutations, applied to $\psi(j^{n-1}\alpha_0 T_0 J_0)$, which bring particle n to the last place. Nucleons k and n are now interchanged by a change of coupling transformation which is a product of two transformations (15.29). One such transformation changes the order of coupling of the **j**-vectors and the other affects the coupling of the isospin vectors.

We thus obtain for (33.14) the expression

$$\psi[j^{n-1}(\alpha_0 T_0 J_0)\, j_n T\ J] - \sum_k^{n-1} (-1)^{P_k} \sum_{\alpha_2 T_2 J_2 T_1' J_1'} [j^{n-2}(\alpha_2 T_2 J_2)\, j\ T_0 J_0 | \} j^{n-1} \alpha_0 T_0 J_0]$$

$$\times \langle T_2\, \tfrac{1}{2}(T_0)\, \tfrac{1}{2}\, T | T_2\, \tfrac{1}{2}(T_1')\, \tfrac{1}{2}\, T \rangle \langle J_2 j(J_0) j\ J | J_2 j(J_1') j\ J \rangle$$

$$\times \psi[j^{n-2}(\alpha_2 T_2 J_2)\, j_k(T_1' J_1')\, j_n T\ J]. \tag{33.15}$$

We multiply (33.15) by $\psi^*[j^{n-1}(\alpha_1 T_1 J_1) j_n T\ J]$ and integrate. According to the relations (26.24), which follow from the antisymmetrization being a projection operation, this gives

$$N^{-1}[j^{n-1}(\alpha_1 T_1 J_1)\, j\ T\ J | \} j^n [\alpha_0 T_0 J_0] T\ J]$$

$$= n[j^{n-1}(\alpha_0 T_0 J_0)\, j\ T\ J | \} j^n[\alpha_0 T_0 J_0] T\ J]\, [j^{n-1}(\alpha_1 T_1 J_1)\, j\ T\ J | \} j^n[\alpha_0 T_0 J_0] T\ J]$$

$$= \delta_{\alpha_0 \alpha_1} \delta_{T_0 T_1} \delta_{J_0 J_1} - \sum_{\alpha_2 T_2 J_2} [j^{n-2}(\alpha_2 T_2 J_2)\, j\ T_0 J_0 | \} j^{n-1} \alpha_0 J_0 T_0]$$

$$\langle T_2\, \tfrac{1}{2}(T_0)\, \tfrac{1}{2}\, T | T_2\, \tfrac{1}{2}(T_1)\, \tfrac{1}{2}\, T \rangle \langle J_2 j(J_0) j\ J | J_2 j(J_1) j\ J \rangle$$

$$\times \int \psi^*(j^{n-1} \alpha_1 T_1 J_1) \sum_k (-1)^{P_k} \psi[j^{n-2}(\alpha_2 T_2 J_2)\, j_k T_1 J_1]$$

$$= \delta_{\alpha_0 \alpha_1} \delta_{T_0 T_1} \delta_{J_0 J_1} - \sum_{\alpha_2 T_2 J_2} \langle T_2\, \tfrac{1}{2}(T_0)\, \tfrac{1}{2}\, T | T_2\, \tfrac{1}{2}(T_1)\, \tfrac{1}{2}\, T \rangle \langle J_2 j(J_0) j\ J | J_2 j(J_1) j\ J \rangle$$

$$\times [j^{n-2}(\alpha_2 T_2 J_2)\, j T_0 J_0 | \} j^{n-1} \alpha_0 T_0 J_0]\, (n-1)\, [j^{n-2}(\alpha_2 T_2 J_2)\, j T_1 J_1 | \} j^{n-1} \alpha_1 T_1 J_1].$$

Introducing the actual values for the transformation coefficients (15.30) we obtain

$$n[j^{n-1}(\alpha_0 T_0 J_0)\, j\ T\ J | \} j^n[\alpha_0 T_0 J_0] T\ J]\, [j^{n-1}(\alpha_1 T_1 J_1)\, j\ T\ J | \} j^n[\alpha_0 T_0 J_0] T\ J]$$

$$= \delta_{\alpha_0 \alpha_1} \delta_{T_0 T_1} \delta_{J_0 J_1} - (-1)^{T_0 + T_1 + J_0 + J_1} (n-1)$$

$$\times [(2T_0 + 1)(2T_1 + 1)(2J_0 + 1)(2J_1 + 1)]^{1/2} \sum_{\alpha_2 T_2 J_2} \begin{Bmatrix} T_2 & \tfrac{1}{2} & T_0 \\ T & \tfrac{1}{2} & T_1 \end{Bmatrix} \begin{Bmatrix} J_2 & j & J_0 \\ J & j & J_1 \end{Bmatrix}$$

$$\times [j^{n-2}(\alpha_2 T_2 J_2)\, j\ T_0 J_0 | \} j^{n-1} \alpha_0 T_0 J_0]\, [j^{n-2}(\alpha_2 T_2 J_2)\, j\ T_1 J_1 | \} j^{n-1} \alpha_1 T_1 J_1]. \tag{33.16}$$

Equation (33.16) is a recursion relation for the c.f.p. It gives the c.f.p.

of the state $\psi(j^n[\alpha_0 T_0 J_0]T\, J)$ [obtained from the principal parent $\psi(j^{n-1}\alpha_0 T_0 J_0)$] in terms of the c.f.p. in the j^{n-1} configuration. The suqare of the c.f.p. of the principal parent is given by (33.16) on putting $\alpha_1 = \alpha_0$ $T_1 = T_0$ and $J_1 = J_0$. Once its sign is fixed by some convention, all the other c.f.p. including their signs are given by (33.16). It is worthwhile to mention that the $n - 1 \to n - 2$ c.f.p. which appear in (33.16) are referred to an arbitrary scheme of independent wave functions. On the other hand, the $n \to n - 1$ c.f.p. given by (33.16) define wave functions, obtained from principal parents by antisymmetrizing.

The c.f.p. defined above are very useful for the evaluation of matrix elements of operators involving both protons and neutrons. We give below the expressions, in terms of c.f.p., of matrix elements of single nucleon and two nucleon operators. These expressions are very similar to those previously obtained for jj-coupling and LS-coupling with identical particles.

We first consider single particle operators $\Sigma_i^n f_i$. Their matrix elements are given by

$$\langle j^n\alpha\, T\, J\, M_T M_J \Big| \sum_i^n f_i \Big| j^n\alpha' T' J' M_T' M_J' \rangle$$

$$= n \sum_{\alpha_1 T_1 J_1} [j^n\alpha\, T\, J\{\!|\, j^{n-1}(\alpha_1 T_1 J_1)\, j\, T\, J]\, [j^{n-1}(\alpha_1 T_1 J_1)\, j\, T\, J|\!\}\, j^n\alpha' T' J']$$
$$\times \langle j^{n-1}(\alpha_1 T_1 J_1)\, j_n T\, J\, M_T M_J |\, f_n\, |\, j^{n-1}(\alpha_1 T_1 J_1)\, j_n T' J' M_T' M_J' \rangle. \qquad (33.17)$$

This relation is the analog of (26.32) or (31.9). Equation (33.17) becomes considerably simpler when f is a product of a component of an irreducible tensor operator in ordinary space and a component of an irreducible tensor operator in isospin space. Let us, in fact, consider the single particle operator $f_\kappa^{(k)} h_\varrho^{(r)}$ where $f_\kappa^{(k)}$ is the κ-component of a tensor of degree k in ordinary space (including spin space). The other term $h_\varrho^{(r)}$, is the ρ-component of a tensor of degree r in isospin space. There are nonvanishing matrix elements of $h_\varrho^{(r)}$ between single nucleon isospin wave functions only if r is either 1 or 0 (only these values satisfy the triangular conditions with $\frac{1}{2}$ and $\frac{1}{2}$). Using (15.27) we obtain the following result for the reduced matrix element of the single particle operator

$$\Big(j^n\alpha\, T\, J \Big\| \sum_i^n \mathbf{f}^{(k)}(i)\mathbf{h}^{(r)}(i) \Big\| j^n\alpha' T' J'\Big)$$

$$= n(j\|\mathbf{f}^{(k)}\|j)\,(\tfrac{1}{2}\|\mathbf{h}^{(r)}\|\tfrac{1}{2})\,\sqrt{(2T+1)(2T'+1)(2J+1)(2J'+1)}$$
$$\times \sum_{\alpha_1 T_1 J_1} [j^n\alpha\, T\, J\{\!|\, j^{n-1}(\alpha_1 T_1 J_1)\, j\, T\, J]\, [j^{n-1}(\alpha_1 T_1 J_1)\, j\, T' J'|\!\}\, j^n\alpha' T' J']$$
$$\times (-1)^{T_1+1/2+T+r+J_1+j+J+k} \begin{Bmatrix} \frac{1}{2} & T & T_1 \\ T' & \frac{1}{2} & r \end{Bmatrix} \begin{Bmatrix} j & J & J_1 \\ J' & j & k \end{Bmatrix}. \qquad (33.18)$$

This result is the analog of (26.33) in the case of jj-coupling and of (31.10) for LS-coupling with identical particles.

We shall make use of (33.18) to derive an expression for the magnetic moment in the j^n configuration of protons and neutrons. The g-factor for a j proton will be denoted by g_P and that of a j neutron by g_N. The following relation holds in the j^n configuration

$$g\mathbf{J} = \sum_i^n g_i \mathbf{j}_i = \frac{g_P}{2}\sum_i^n [1 + \tau_3(i)]\mathbf{j}_i + \frac{g_N}{2}\sum_i^n [1 - \tau_3(i)]\mathbf{j}_i$$

$$= \frac{g_P + g_N}{2}\mathbf{J} + \frac{g_P - g_N}{2}\sum_i^n \tau_3(i)\,\mathbf{j}_i\ . \tag{33.19}$$

The second term on the right-hand side of (33.19) is the zeroth component of a tensor of degree 1 (vector) in the isospin space. The other terms (proportional to $\mathbf{J}$) are scalars in isospin space. Obviously, such an equation cannot hold for all states if the value of g is independent of these states. Equation (33.19) should therefore be understood to determine a value of g for each value of M_T, as well as of T, J, and other necessary quantum numbers. Using (33.18) we can further evaluate the g-factor, but before we do this we can draw directly some simple conclusions from (33.19).

The expectation values of the second term on the right-hand side of (33.19) are proportional, according to the Wigner-Eckart theorem, to $\langle TM_T|T_3|TM_T\rangle = M_T$. Thus, we conclude immediately that this term vanishes in all states with $M_T = 0$. Hence, in all self-conjugate nuclei, with $Z = N$, the g-factor is given simply by

$$g = \frac{g_P + g_N}{2} \qquad \text{if} \qquad Z = N. \tag{33.20}$$

This result is directly applicable to self-conjugate odd-odd nuclei and also to excited states in self-conjugate even-even nuclei. In the ground states of the latter, $J = 0$ and g has no meaning.

Another simple result can be obtained for the g-factors of mirror nuclei. A pair of mirror nuclei is characterized by $M_T = +\frac{1}{2}$ and $M_T = -\frac{1}{2}$. All other quantum numbers of corresponding states are the same. Thus, if we add the g-factors of two mirror nuclei, the contributions of the second term on the right-hand side of (33.19) cancel each other and the average g-factor is also given by (33.20). Measured magnetic moments of self-conjugate odd-odd nuclei satisfy rather well the relation (33.20). No comparison with experiment can yet be made with mirror nuclei since at least one of each pair is an unstable nucleus.

The magnetic moment operator (33.19) is closely related to the operator whose matrix elements determine the probability of magnetic dipole

($M1$) transitions. From (17.10) or (17.12) it follows that this operator, for a single particle, is given by $(e\hbar/2m_pc)\boldsymbol{l} + (e\hbar/2m_pc)\mu_p\boldsymbol{\sigma}$ for a proton and by $(e\hbar/2m_pc)\mu_n\boldsymbol{\sigma}$ for a neutron. These operators can be conveniently written, in terms of nuclear magnetons, as

$$\tfrac{1}{2}(1+\tau_3)\boldsymbol{l} + \tfrac{1}{2}(1+\tau_3)g_{sp}\mathbf{s} + \tfrac{1}{2}(1-\tau_3)g_{sn}\mathbf{s}. \qquad (33.21)$$

Due to the Landé formula, this operator in the j^n configuration can be written simply as

$$\tfrac{1}{2}(1+\tau_3)g_P\mathbf{j} + \tfrac{1}{2}(1-\tau_3)g_N\mathbf{j}. \qquad (33.22)$$

The sum of the operators (33.22) over all nucleons is identical with (33.19).

General considerations can be now made on transition probabilities within a given configuration in self-conjugate nuclei. In such nuclei, with $M_T = 0$, the matrix elements between two states with the same value of T can be easily shown to vanish. In fact, the first term on the right-hand side of (33.19) is proportional to $\mathbf{J}$ and therefore has no nonvanishing nondiagonal elements. The matrix elements of the second term are proportional, according to the Wigner-Eckart theorem, to $\langle T\,M_T|T_3|T\,M_T\rangle = M_T = 0$. Thus, we obtain the result

In the j^n configuration in self-conjugate nuclei no $M1$ transitions occur between states with the same value of $T(\Delta T = 0)$ to the approximation used in deriving (17.10) (the long wave approximation). (33.23)

It is worthwhile to mention that the attenuation of $M1$ transitions with $\Delta T = 0$ in self-conjugate nuclei is not a very good check on the validity of jj-coupling. Quite generally, summing (33.21) over all nucleons we obtain

$$\tfrac{1}{2}\sum_i^n [\boldsymbol{l}_i + (g_{sp}+g_{sn})\mathbf{s}_i] + \tfrac{1}{2}\sum_i^n \tau_3(i)\,[\boldsymbol{l}_i + (g_{sp}-g_{sn})\mathbf{s}_i]$$

$$= \tfrac{1}{2}\mathbf{J} + \tfrac{1}{2}(g_{sp}+g_{sn}-1)\sum_i^n \mathbf{s}_i + \tfrac{1}{2}\sum_i^n \tau_3(i)\,[\boldsymbol{l}_i + (g_{sp}-g_{sn})\mathbf{s}_i]. \qquad (33.24)$$

For $\Delta T = 0$ transitions in self-conjugate nuclei (with $M_T = 0$) the contribution of the last term on the right-hand side of (33.24) vanishes due to the Wigner-Eckart theorem. Also the first term there does not have any nonvanishing nondiagonal elements. Thus, the only contribution in this case comes from the operator $\mathbf{S}$ in the middle term. However, this operator is multiplied by $\frac{1}{2}(g_{sp} + g_{sn} - 1) = 0.38$. This factor should be compared, for instance, to the factor $\frac{1}{2}(g_{sp} - g_{sn}) = 4.70$ in the last term in (33.24) which is effective in other $M1$ transitions. The ratio 0.38/4.70 squared can give an indication of the attenuation of the rates of $M1$ transitions with $\Delta T = 0$ in self-conjugate nuclei. The attenuation turns out to be by a factor of about 10^{-2}. This result, due to Morpurgo, was derived without assuming any particular coupling scheme. The attenuation of $M1$ transitions with $\Delta T = 0$ in self-conjugate nuclei is observed experimentally in many cases.

Another result on $M1$ transitions in jj-coupling is of some interest. The operator for $M1$ transitions within a j^n configuration of identical particles is simply $\Sigma g\mathbf{j}_i = g\mathbf{J}$. Thus, there are no nondiagonal elements different from zero in this case. We therefore obtain that

within a j^n configuration of identical particles, no $M1$ transitions occur in the long wave approximation. (33.25)

The attenuation of $M1$ transitions in j^n configuration with identical nucleons is observed experimentally in several cases (e.g., in V^{51} and O^{19}).

There is another electromagnetic transition which is forbidden in $\Delta T = 0$ transitions in self-conjugate nuclei. The operator for electric dipole ($E1$) transitions is given according to (17.9) as

$$\sum_i^n e_i\mathbf{r}_i = \sum_i^n \tfrac{1}{2}[1 + \tau_3(i)]\mathbf{r}_i = \tfrac{1}{2}\sum_i^n \mathbf{r}_i + \tfrac{1}{2}\sum_i^n \tau_3(i)\mathbf{r}_i. \qquad (33.26)$$

The matrix elements of the second term on the right-hand side of (33.26) are proportional to $\langle T\, M_T|T_3|T\, M_T\rangle = M_T$ for $\Delta T = 0$ transitions. Therefore, these matrix elements vanish in self-conjugate nuclei (with $M_T = 0$) as was the case in the previous example. The first term on the right-hand side of (33.26) is proportional to the vector of the center of mass of the nucleus. Clearly, this operator has vanishing matrix elements between any two different internal states of the nucleus. It can be easily seen that even taking into account the proton-neutron mass difference does not change the situation. We thus obtain

No $E1$ transitions take place in self-conjugate nuclei between states with the same value of $T(\Delta T = 0)$ in the long wave approximation. (33.27)

We shall now come back to the evaluation of (33.19). Since (33.19) is a combination of tensors of different degrees (in isospin space), we must be a little careful with the use of the Wigner-Eckart theorem.

Therefore, we take first the expectation value of the z component of (33.19) in a state defined by αT, M_T, J, and M which gives

$$\left(g - \frac{g_P + g_N}{2}\right)(J||\mathbf{J}||J)(-1)^{J+M-1}\begin{pmatrix} J & J & 1 \\ -M & M & 0 \end{pmatrix}$$

$$= \left(\frac{g_P - g_N}{2}\right)(-1)^{J+M-1}\begin{pmatrix} J & J & 1 \\ -M & M & 0 \end{pmatrix}(-1)^{T+M_T-1}\begin{pmatrix} T & T & 1 \\ -M_T & M_T & 0 \end{pmatrix}$$

$$\times \left(j^n\alpha\, T\, J\left\|\sum_i^n \boldsymbol{\tau}_i \mathbf{j}_i\right\|j^n\alpha\, T\, J\right). \qquad (33.28)$$

Taking the case $M \neq 0$ we can divide by $\begin{pmatrix} J & J & 1 \\ -M & M & 0 \end{pmatrix}$ and then apply (33.18) to the right-hand side of (33.28). We thus obtain

$$g - \frac{g_P + g_N}{2} = \frac{g_P - g_N}{2}(-1)^{T+M_T-1}\begin{pmatrix} T & T & 1 \\ -M_T & M_T & 0 \end{pmatrix} n$$

$$\times 2(\tfrac{1}{2}||\mathbf{t}||\tfrac{1}{2})\frac{(j||\mathbf{j}||j)}{(J||\mathbf{J}||J)}(2T+1)(2J+1)$$

$$\times \sum_{\alpha_1 T_1 J_1}(-1)^{T_1+1/2+T+J_1+j+J}\begin{Bmatrix} \frac{1}{2} & T & T_1 \\ T & \frac{1}{2} & 1 \end{Bmatrix}\begin{Bmatrix} j & J & J_1 \\ J & j & 1 \end{Bmatrix}$$

$$\times [j^{n-1}(\alpha_1 T_1 J_1)\, j\, T\, J|\}j^n\alpha\, T\, J]^2. \qquad (33.29)$$

This can be further simplified by introducing the actual values of the following reduced matrix elements and Wigner coefficients:

$$(J||\mathbf{J}||J) = \sqrt{J(J+1)(2J+1)}, \qquad \begin{pmatrix} T & T & 1 \\ -M & M_T & 0 \end{pmatrix} = \frac{(-1)^{T+M_T-1}M_T}{\sqrt{T(T+1)(2T+1)}}.$$

The final result is

$$g - \frac{g_P + g_N}{2}$$

$$= nM_T(g_P - g_N)\sqrt{\tfrac{3}{2}}\sqrt{j(j+1)(2j+1)}\sqrt{\frac{2T+1}{T(T+1)}}\sqrt{\frac{2J+1}{J(J+1)}}$$

$$\times \sum_{\alpha_1 T_1 J_1}(-1)^{T_1+1/2+T+J_1+j+J}\begin{Bmatrix} \frac{1}{2} & T & T_1 \\ T & \frac{1}{2} & 1 \end{Bmatrix}\begin{Bmatrix} j & J & J_1 \\ J & j & 1 \end{Bmatrix}[j^{n-1}(\alpha_1 T_1 J_1)\, j\, T\, J|\}j^n\alpha\, T\, J]^2. \qquad (33.30)$$

The result (33.30) gives the g factor in terms of c.f.p. in the j^n configuration. We can introduce into this formula the c.f.p. for the case $n = 3$ given by (33.10). We thus obtain for the $T = \frac{1}{2}$ $J = j$ state (obtained from the principal parent with $T_0 = 1$ $J_0 = 0$) the following expression for the g factor

$$g = \frac{g_P + g_N}{2} + 6M_T(g_P - g_N)(2j+1)\Big[-\frac{2(j+1)}{3(2j+1)}\begin{Bmatrix}\frac{1}{2} & \frac{1}{2} & 1\\ \frac{1}{2} & \frac{1}{2} & 1\end{Bmatrix}\begin{Bmatrix}j & j & 0\\ j & j & 1\end{Bmatrix}$$

$$-\frac{\begin{Bmatrix}\frac{1}{2} & \frac{1}{2} & 1\\ \frac{1}{2} & \frac{1}{2} & 1\end{Bmatrix}}{6(j+1)(2j+1)}\sum_{J_1>0,\ \text{even}}(2J_1+1)\begin{Bmatrix}j & j & J_1\\ j & j & 1\end{Bmatrix}$$

$$-\frac{\begin{Bmatrix}\frac{1}{2} & \frac{1}{2} & 0\\ \frac{1}{2} & \frac{1}{2} & 1\end{Bmatrix}}{2(j+1)(2j+1)}\sum_{J_1\ \text{odd}}(2J_1+1)\begin{Bmatrix}j & j & J_1\\ j & j & 1\end{Bmatrix}\Big].$$

Using (15.16) and (15.17) we can sum the terms in the square brackets and obtain

$$g = \frac{g_P + g_N}{2} - M_T\frac{g_P - g_N}{3}\frac{j+4}{j+1} \quad \text{for} \quad n = 3,\ T = \tfrac{1}{2},\ J = j \tag{33.31}$$

In the case of one proton and two neutrons, $M_T = -\frac{1}{2}$ and (33.31) has the value

$$g = \frac{(4j+7)g_P + (2j-1)g_N}{6(j+1)} \qquad \begin{matrix}\text{for one proton and two neutrons}\\ \text{in the state with } T = \frac{1}{2}\ J = j.\end{matrix} \tag{33.32}$$

In the case $M_T = \frac{1}{2}$ the g-factor is obtained from (33.32) by interchanging g_P and g_N. For $j = \frac{1}{2}$ we obtain from (33.32) $g = g_P$. Obviously, in this case the two neutrons are coupled always to $J = 0$ and the magnetic moment is that of the remaining single proton. In general, the g factor (33.32) gives a magnetic moment which is different from the Schmidt value of the odd particle (i.e., g_P in this case). Since J is equal to j in the state considered, the results (33.31) and (33.32) hold also for the magnetic moments μ if we replace g_P and g_N by the single j-nucleon magnetic moments μ_P and μ_N. The deviation from the Schmidt value given by (33.32) is in a qualitative agreement with the experimental findings.

Coefficients of fractional parentage can be used also for evaluating matrix elements of two-particle operators. We shall discuss here two-nucleon operators g_{ik} which are scalars with respect to both J and T. Since the charge of the nucleus is determined by M_T, and there are no states of mixed charge, there is no physical meaning to a state in which $\mathbf{T}$ is coupled to $\mathbf{J}$. Therefore, in the case of nucleons in jj-coupling there are no analogs to the complications due to vector and tensor forces in LS-coupling with identical particles. Furthermore, as already mentioned, the exclusion principle makes it possible to eliminate entirely the isospin vectors from a two-nucleon charge-independent Hamiltonian. We shall thus consider only interactions which do not depend explicitly on the $\boldsymbol{\tau}_i$.

The matrix elements of g_{ik} are equal, for all pairs (i, k), because of the antisymmetry of the wave functions. We therefore have

$$\langle j^n \Big| \sum_{i<k}^{n} g_{ik} \Big| j^n \rangle = \frac{n(n-1)}{2} \langle j^n | g_{12} | j^n \rangle$$

$$= \frac{n(n-1)}{2} \left[\frac{(n-1)(n-2)}{2}\right]^{-1} \langle j^n \Big| \sum_{i<k}^{n-1} g_{ik} \Big| j^n \rangle$$

$$= \frac{n}{n-2} \langle j^n \Big| \sum_{i<k}^{n-1} g_{ik} \Big| j^n \rangle .$$

This way we obtain, exactly as in the case of identical particles,

$$\langle j^n \alpha\, T\, J \Big| \sum_{i<k}^{n} g_{ik} \Big| j^n \alpha' T\, J \rangle$$

$$= \frac{n}{n-2} \sum_{\alpha_1 \alpha_1' T_1 J_1} [j^n \alpha\, T\, J \{| j^{n-1}(\alpha_1 T_1 J_1) j\, T\, J] \, [j^{n-1}(\alpha_1' T_1 J_1) j\, T\, J |\} j^n \alpha' T\, J]$$

$$\langle j^{n-1} \alpha_1 T_1 J_1 \Big| \sum_{i<k}^{n-1} g_{ik} \Big| j^{n-1} \alpha_1' T_1 J_1 \rangle. \tag{33.33}$$

Since the interactions g_{ik} in (33.33) are taken to be diagonal in T and J, and independent of M_T and M, these latter quantum numbers were omitted from (33.33). Equation (33.33) is the analog of (26.37) in the case of jj-coupling and of (31.12) in the case of scalar forces in LS-coupling with identical particles.

If g_{ih} is the scalar product of two irreducible tensor operators of degree k, we can express matrix elements of g_{ih} by means of single nucleon operators. Any irreducible tensor operator $f_\kappa^{(k)}$ can be replaced within the j^n configuration by $(j||\mathbf{f}^{(k)}||j) u_\kappa^{(k)}$ where $\mathbf{u}^{(k)}$ is the unit tensor operator. Doing this we obtain

$$\sum_{i<h}^{n} g_{ih} = \sum_{i<h}^{n} (\mathbf{f}_i^{(k)} \cdot \mathbf{f}_h^{(k)}) = (j||\mathbf{f}^{(k)}||j)^2 \sum_{i<h}^{n} (\mathbf{u}_i^{(k)} \cdot \mathbf{u}_h^{(k)})$$

$$= (j||\mathbf{f}^{(k)}||j)^2 \, \tfrac{1}{2} \left\{ \left(\sum_i^n \mathbf{u}_i^{(k)} \cdot \sum_h^n \mathbf{u}_h^{(k)} \right) - \sum_i^n (\mathbf{u}_i^{(k)} \cdot \mathbf{u}_i^{(k)}) \right\}$$

$$= (j||\mathbf{f}^{(k)}||j)^2 \, \tfrac{1}{2} \left\{ (\mathbf{U}^{(k)} \cdot \mathbf{U}^{(k)}) - n \frac{1}{2j+1} \right\} . \tag{33.34}$$

Equation (33.34) gives the matrix elements of g_{ih} for the special case considered, in terms of matrix elements of the single nucleon operator $\mathbf{U}^{(k)} = \Sigma_i^n \mathbf{u}_i^{(k)}$. It is the analog of (26.46) in jj-coupling and of (31.20) in LS-coupling with identical particles.

In analogy with the case of identical particles, $n \to n-2$ c.f.p. can also be defined by

$$\psi(j^n\alpha\, T\, J) = \sum_{\alpha_1 T_1 J_1 T' J'} [j^{n-2}(\alpha_1 T_1 J_1)\, j^2(T'J')T\, J | \} j^n\alpha\, T\, J] \times \psi[j^{n-2}(\alpha_1 T_1 J_1)\, j^2_{n-1,n}(T'J')T\, J]. \qquad (33.35)$$

Equation (33.35) gives the expansion of a normalized antisymmetric wavefunction of n nucleons by means of $n \to n-2$ c.f.p. The properties of these coefficients are very similar to those in the case of identical particles. We shall not discuss them here in more detail. We only recall that such c.f.p. are useful for the calculation of matrix elements of two particle operators. For example, in the case of a charge independent interaction we obtain

$$\left\langle j^n\,\alpha'\,T\,J \left| \sum_{i<k}^{n} V_{ik} \right| j^n\,\alpha\,T\,J \right\rangle = \frac{n(n-1)}{2} \sum_{\alpha_1 T_1 J_1 T' J'} [j^n\alpha' T\, J\{| j^{n-2}(\alpha_1 T_1 J_1) j^2(T'J')T\, J] [j^{n-2}(\alpha_1 T_1 J_1) j^2(T'J')T\, J | \} j^n\alpha\, T\, J]\, V(j^2 T' J'). \qquad (33.36)$$

34. The Seniority Scheme in the Nuclear j^n Configuration

We have seen that in configurations of n identical particles we are often confronted with a situation in which J, the total angular momentum, is not sufficient to characterize a state uniquely. In nuclear configurations where each particle can be either a proton or a neutron, the situation is even more complicated. For identical particles, we introduced a further characterization of states by using the concept of seniority v. States in the j^n configuration were obtained from states in the j^v configuration by the addition of $(n - v)/2$ pairs coupled to $J = 0$ and antisymmetrization. This characterization of states turned out to be very convenient for calculations, and we shall therefore introduce a similar scheme for the nuclear j^n configuration.

We start from the principal parent $\psi(j^v \alpha_1 T_1 J)$. To go from this configuration to the configuration with $v + 2$ particles, we add a pair of particles coupled to $J = 0$. Such a pair, coupled to $J = 0$, must have $T = 1$. The resulting wave function of the j^{v+2} configuration has therefore the same total angular momentum J as that of the principal parent but *not* necessarily the same isospin. We denote this wave function by $\psi[j^v(\alpha_1 T_1 J) j^2_{v+1,v+2}(1\ 0) T\ J]$. We now antisymmetrize this wave function and obtain an antisymmetric wave function of $v + 2$ nucleons coupled to a total isospin T and total angular momentum J. We are interested, in particular, in states of the j^v configuration which cannot be obtained by this procedure from any parent state of the j^{v-2} configuration. In analogy with (27.10), and (27.11), we define

A state $\psi(j^v \alpha\ T\ J)$ will be said to have the seniority v if all its $n \rightarrow n-2$ c.f.p. $[j^{v-2}(\alpha_1 T_1 J) j^2(1\ 0) T J | \} j^v \alpha\ T\ J]$ vanish. All states $\psi(j^n \alpha\ T\ J)$ obtained from a state $\psi(j^v\ v\ T_1\ J)$ by adding $(n - v)/2$ pairs coupled to $J = 0$ (and $T = 1$) and antisymmetrizing are defined to have the seniority v. (34.1)

All states with the same seniority obtained according to (34.1) from $\psi(j^v \alpha\ T\ J)$ have the same value of J. This is analogous to the case of identical particles where such states have the same value of J in jj-coupling and the same values of L in LS-coupling. However, there is a fundamental difference between the two cases. In the case of LS-coupling with identical particles, the $l^2\ L = 0$ state must have $S = 0$. Hence, it follows that all states with the same seniority v obtained from

$\psi(l^v\, S\, L)$ have also the same value of S. In the present case, as mentioned before, the j^2 $J = 0$ state has $T = 1$. Whereas states in the nuclear j^n configuration with seniority v have the same value of J as the principal parent, the value of T need not be the same. Thus, we have to characterize the states $\psi(j^n\, \alpha\, T\, J)$, constructed according to the above procedure from $\psi(j^v\, \alpha_1\, T_1\, J)$, not only by the seniority v but also by the isospin T_1 of the principal parent. This isospin, denoted usually by t, is called the *reduced isospin*. Thus, a state $\psi(j^n\, \alpha\, v\, t\, T\, J)$ with seniority v and reduced isospin t is a state obtainable from $\psi(j^v\, \alpha_1\, T_1 = t,\, J)$ by the addition of $(n - v)/2$ pairs, each coupled to $T = 1$ $J = 0$, and antisymmetrization. Because of the triangular condition on the addition of isospins it follows that

$$\psi(j^n \alpha\, v\, t\, T\, J) = 0 \qquad \text{if} \qquad |T - t| > \frac{n - v}{2}.$$

The state of a single j nucleon is characterized by $v = 1$ and $t = \frac{1}{2}$. The $J = 0$ state of the j^2 configuration has, by definition, $v = 0$ and therefore $t = 0$. The other states with *even* values of J in the j^2 configuration have $v = 2$ and $t = T = 1$. On the other hand, the states with *odd* values of J have also $v = 2$ but $t = T = 0$.

Before studying the consistency of the scheme just defined, we wish to point out another aspect of the complication which occurs in the seniority scheme for nucleons. There is only one way of adding $(n - v)/2$ pairs coupled to $J = 0$ to a state $\psi(j^v\, J)$ in order to obtain a state $\psi(j^n J)$. There are, however, many ways of adding $(n - v)/2$ pairs coupled to $T = 1$ which lead from $\psi(j^v\, T_1 = t\, J)$ to $\psi(j^n\, T\, J)$. Thus, if we first construct the antisymmetric state $\psi(j^{n-v},\ v = 0,\ t = 0,\ T_0,\ J = 0)$ from the $(n - v)/2$ pairs coupled to $T = 1$ $J = 0$, we can obtain an n nucleon state $\psi(j^n\, v\, t\, T\, J)$ by the antisymmetrization of

$$\psi[j^v(\alpha\, v\, t\ \ T = t\ \ J)\, j^{n-v}(0, 0, T_0, 0) T\, J]. \tag{34.2}$$

Any allowed value of T_0, which is consistent with $|T - t| \leq T_0 \leq T + t$, can be chosen in (34.2). Although, in most practical cases different values of T_0 lead to the same state after antisymmetrization, this is not always the case. Thus, an additional quantum number must be introduced to differentiate between states in the j^n configuration with the same values of T, J, seniority v, and reduced isospin t. This new quantum number is different from the quantum number α which may be necessary to distinguish between states with the same values of v, $T = t$, and J in the j^v configuration. It appears only when going from the j^v to the j^n configuration. This additional quantum number is evidently related to the various possibilities for choosing T_0 in (34.2). It cannot be taken to be T_0 itself, since different choices of T_0 may lead to the *same* state after antisymmetrization. No simple definition can be given for this extra quantum number. We shall later make explicit reference to this additional quantum number.

We shall now prove the consistency of the seniority scheme in the nuclear j^n configuration. The proof, by induction, will be similar to that in the case of identical particles. We shall first show that any two wave functions with different seniorities and reduced isospins are orthogonal to each other. We notice that the wave function $\psi(j^v\ v\ T\ J)$ is orthogonal to any wave function $\psi(j^v\ \alpha\ T\ J)$ with lower seniority. The latter function is obtained, by the definition (34.1), from a principal parent $\psi[j^{v-2}(\alpha_1\ T_1\ J)j^2(1\ 0)T\ J)$. Therefore, we calculate the integral

$$\int \psi^*(j^v v\ T\ J)\,\mathscr{A}\psi[j^{v-2}(\alpha_1 T_1 J)\,j^2_{v-1,v}(1\ 0)T\ J]. \tag{34.3}$$

Using the properties of the antisymmetrizer $\mathscr{A}$, (34.3) can be brought into the form

$$\frac{v(v-1)}{2}\int \psi^*(j^v v\ T\ J)\,\psi[j^{v-2}(\alpha_1 T_1 J)\,j^2_{v-1,v}(1\ 0)T\ J]$$

$$= \frac{v(v-1)}{2}\,[j^{v-2}(\alpha_1 T_1 J)j^2(1\ 0)T\ J|\}j^v v\ T\ J]. \tag{34.4}$$

The $v \to v-2$ c.f.p. in (34.4) vanishes due to the definition (34.1). This proves the vanishing of the integral (34.3) and the orthogonality of the functions considered.

We further observe that any two wave functions in the j^v configuration both having the seniority v but different reduced isospins t_1 and t_2 are orthogonal. In fact, in this case also the total isospins $T_1 = t_1$ and $T_2 = t_2$ are different and the wave functions are orthogonal to each other.

We shall now consider two states in the j^{n+2} configuration with seniorities v and v' which are both smaller than $n+2$. These two wave functions can be obtained from principal parents in the j^n configuration by adding a pair coupled to $T=1$ $J=0$ and antisymmetrizing. We therefore calculate the value of the integral

$$\int \mathscr{A}\psi^*[j^n(v't'T_1'J)\,j^2_{n+1,n+2}(1\ 0)T\ J]\,\mathscr{A}\psi[j^n(v\,t\,T_1 J)\,j^2_{n+1,n+2}(1\ 0)T\ J]$$

$$= \frac{(n+1)(n+2)}{2}\int \psi^*[j^n(v't'T_1'J)\,j^2_{n+1,n+2}(1\ 0)T\ J]$$

$$\times\,\mathscr{A}\psi[j^n(v\,t\,T_1 J)\,j^2_{n+1,n+2}(1\ 0)T\ J]. \tag{34.5}$$

The right-hand side of (34.5) is analogous to the expression (27.15) in the case of identical particles. In that case we carried out explicitly the antisymmetrization and restored the order of the particles by using recoupling transformations. Due to the fact that the added pair has $J = 0$, the recouplings were extremely simple and we could therefore use the orthogonality of the principal parents in order to prove the vanishing of (27.15). Since the $J = 0$ pairs in the case of the nuclear j^n configuration carry nonvanishing isospin vectors, the recoupling transformations are much more complicated. Thus, the procedure used in the case of identical particles cannot be applied directly to (34.5). In the present case it is not even true that any two orthogonal principal parents which appear in (34.5) give rise to orthogonal wave functions.

In order to make full use of the fact that the total spin of the pair is $J = 0$, we have to consider specifically the space and spin parts of the wave functions. The antisymmetric wave functions (with a given value of T) can always be written as a linear combination of products of isospin functions and functions of the space and spin coordinates of the nucleons. As explained in the beginning of Section 33, the space and spin functions have all the same symmetry type and transform irreducibly among themselves under permutations of the nucleon coordinates. Similarly, the isospin functions have all the same symmetry type given by the Young diagram which is dual to the one which determines the symmetry of the space and spin functions. We shall now show that any two space and spin functions which belong to states with different seniorities or reduced isospins are orthogonal to each other.

We first demonstrate this property in the j^v configuration. Any two space and spin functions in the expansion of $\psi(j^v\, v\, t_1\, T_1 = t_1\, J)$ and $\psi(j^v\, v\, t_2\, T_2 = t_2\, J)$ are orthogonal if $t_1 \neq t_2$. In fact, since in this case the total isospins T_1 and T_2 are different, the space and spin functions (as well as the isospin functions) have different symmetry types and hence are orthogonal to each other. Furthermore, any space and spin function in the expansion of $\psi(j^v\, v\, t\, T\, J)$ is orthogonal to any such function in the expansion of $\psi(j^v\, v'\, t'\, T\, J)$ for $v' < v$. This can best be seen by expanding each space and spin function in terms of the complete set of functions $\varphi[j^{v-2}(J_1)j^2_{v+1,v+2}(J')J]$. The space and spin functions of $\psi(j^v\, v\, t\, T\, J)$ cannot contain in their expansions functions $\varphi(j^{v-2} J) \times \varphi(j^2_{v+1,v+2} J' = 0)$ with an admissible function $\varphi(j^{v-2} J)$ having the symmetry type given by $T_1 = T$ or $T_1 = T \pm 1$. Otherwise, the $v \to v - 2$ c.f.p. which appears in the definition (34.1) will not vanish. We recall that $\psi(j^v\, v'\, t'\, T\, J)$ can be obtained by antisymmetrizing

$$\psi[j^{v-2}(v'\, t'\, T_1'\, J)j^2_{v+1,v+2}(1\,0)T\, J].$$

Hence, the space and spin functions in its expansion can be obtained by starting from the function $\varphi(j^{v-2}\, J)\varphi(j^2_{v+1,v+2}\, J' = 0)$ and operating on it by projection operators, each of which is a linear combination of permutations. These are the operators which project this function onto the space of functions $\varphi(j^v\ J)$ with the symmetry type which is determined by a total isospin T. Since the space and spin functions of $\psi(j^v\ v\ t\ T\ J)$ have this same symmetry type, we can omit this projection operator in the integral

$$\int \varphi^*(j^v v\ t\ J) P\, \varphi(j^{v-2} J)\, \varphi(j^2_{v+1,v+2}\ J' = 0).$$

By integrating without the projection operator P over the (space and spin) coordinates of nucleons $v + 1$ and $v + 2$, we see that this integral vanishes.

We shall now show this orthogonality in the j^n configuration with $n > v, v'$. As explained above, the space and spin functions of $\psi(j^{n+2}\ v\ t\ T\ J)$ can be obtained by operating with projection operators P on $\varphi(j^n\ v\ t\ J)\varphi(j^2_{n+1,n+2}\ J' = 0)$. We can thus calculate

$$\int P\varphi^*(j^n v' t' J)\, \varphi^*(j^2_{n+1,n+2}\ J' = 0)\, P\varphi(j^n v\ t\ J)\, \varphi(j^2_{n+1,n+2}\ J' = 0)$$

$$= \int \varphi^*(j^n v' t' J)\, \varphi^*(j^2_{n+1,n+2}\ J' = 0)\, P\varphi(j^n v\ t\ J)\, \varphi(j^2_{n+1,n+2}\ J' = 0). \quad (34.6)$$

We now carry out explicitly the projection operation which is taking linear combinations of permutations. In each permuted function we restore the original order of nucleon coordinates by change of coupling transformations, as in the derivation of (27.19). As shown in detail in that case, the fact that the $n + 1$, $n + 2$ pair is originally coupled to $J' = 0$ greatly simplifies matters. We thus obtain that (34.6) vanishes provided any $\varphi(j^n\ v\ t\ J)$ and $\varphi(j^n\ v'\ t'\ J)$ are orthogonal and obtained by a projection from orthogonal functions $\varphi(j^{n-2}\ v\ t\ J)\varphi(j^2_{n-1,n}\ J' = 0)$ and $\varphi(j^{n-2}\ v'\ t'\ J)\varphi(j^2_{n-1,n}\ J' = 0)$. Thus, the orthogonality of any two space and spin functions with different seniorities or isospins is demonstrated by induction. Therefore, any two functions $\psi(j^n\ v\ t\ T\ J)$ and $\psi(j^n\ v'\ t'\ T'\ J)$ with different seniorities or reduced isospins are *a fortiori* orthogonal to each other.

The orthogonality of wave functions with different seniorities or reduced isospins is thus shown. This orthogonality can also be expressed in the following statement. In the expansion of a function $\psi(j^n\ v\ t\ T\ J)$ in terms of $n \to n - 2$ c.f.p., the only parents which contain pairs coupled to $T = 1$ $J = 0$ are those with the same seniority v and reduced isospin t. The completeness of the seniority scheme follows directly from the definition (34.1). Any function in the j^n configuration which is orthogonal to all functions with seniorities $v < n$ has no nonvanishing c.f.p. $[j^{n-2}(\alpha_1\ T_1\ J)j^2(1\ 0)T\ J|\}j^n\ \alpha\ T\ J]$, and thus, it has by definition the seniority n.

There is an important property of the seniority scheme that we shall now proceed to derive. It is analogous to (27.22) in the case of identical particles, namely

Matrix elements of odd tensor operators vanish between states with different seniorities or different reduced isospins. The nonvanishing matrix elements are independent of n and T. (34.7)

The theorem (34.7) follows from the analog of (27.5). We first observe that the equation

$$\langle j^2 T\ J\ M_T\ M | u_\kappa^{(k)}(1) + u^{(k)}(2) | j^2\ T' = 1\ \ J' = 0\ \ M_T'\ \ M' = 0 \rangle = 0 \qquad \text{for} \quad k \text{ odd} \tag{34.8}$$

holds for all values of T and J (as well as of M_T, M_T', and M). The $\mathbf{u}^{(k)}$ in (34.8) is an irreducible tensor operator of odd degree k with respect to rotations in ordinary (spin and Cartesian) space. It is *independent* of the isospin (charge) coordinates. According to the triangular conditions (the Wigner-Eckart theorem) T in (34.8) must be equal to 1 for the matrix element not to vanish. On the other hand, J must be equal to k. However, there is no antisymmetric state with $J = k$ for k odd and $T = 1$ in the j^2 configuration. Thus all matrix elements (34.8) vanish for odd values of k.

We shall now use (34.8) to calculate matrix elements of $\mathbf{U}^{(k)} = \sum_i^n \mathbf{u}^{(k)}$ between states in the j^n configuration. Let us consider the matrix element

$$\langle j^n v' t' T' J' \left| \sum_i^n u_\kappa^{(k)}(i) \right| j^n v\ t\ T\ J \rangle \tag{34.9}$$

where the projections M_T, M_T', M, and M' were dropped out to simplify the notation. Since the odd tensor in (34.9) is charge independent, the matrix element will vanish unless $T' = T$ and $M_T' = M_T$. Let us assume $n > v' \geqslant v$. In this case, $\psi(j^n\ v\ t\ T\ J)$ can be obtained by antisymmetrizing $\psi[j^{n-2}(v\ t\ T_1\ J) j^2_{n-1,n}(1\ 0) T\ J]$ with a proper choice of T_1. Let us denote by N_J the factor with which the antisymmetrized function should be multiplied in order to be normalized. Obviously, N_J may well depend on n, v, t, T_1, as well as on T. We can then write the matrix element (34.9) as

$$N_J \int \psi^*(j^n\ v'\ t'\ T\ J') \Big[\sum_i^n u_\kappa^{(k)}(i) \Big] \mathscr{A} \psi[j^{n-2}(v\ t\ T_1\ J) j^2_{n-1,n}(1\ 0) T\ J]. \tag{34.10}$$

Using the properties of the antisymmetrizer $\mathscr{A}$ we can write (34.10) in the form

$$\frac{n(n-1)}{2} N_J \int \psi^*(j^n v' t' T\, J') \left[\sum_i^{n-2} u_\kappa^{(k)}(i)\right] \psi[j^{n-2}(v\, t\, T_1 J) j^2_{n-1,n}(1\ 0) T\, J]$$

$$+ \frac{n(n-1)}{2} N_J \int \psi^*(j^n v' t' T\, J') \left[u_\kappa^{(k)}(n-1) + u_\kappa^{(k)}(n)\right]$$

$$\times\, \psi[j^{n-2}(v\, t\, T_1 J) j^2_{n-1,n}(1\ 0) T\, J]. \qquad (34.11)$$

Due to (34.8) the second term in (34.11) vanishes, for odd k, when the integration over nucleons $n-1$ and n is carried out.

In the first term we expand $\psi(j^n\ v'\ t'\ T\ J')$ in terms of c.f.p. and obtain for (34.11) the expression

$$\frac{n(n-1)}{2} N_J \sum_{v_1' t_1' T_1' J_1 T'' J''} [j^{n-2}(v_1'\, t_1'\, T_1'\, J_1') j^2(T'' J'') T\, J' | \} j^n v' t' T\, J']$$

$$\times \int \psi^*[j^{n-2}(v_1'\, t_1'\, T_1'\, J_1') j^2_{n-1,n}(T'' J'') T\, J'] \left[\sum_i^{n-2} u_\kappa^{(k)}(i)\right]$$

$$\times\, \psi[j^{n-2}(v\, t\, T_1 J) j^2_{n-1,n}(1\ 0) T\, J]. \qquad (34.12)$$

Since $\sum_{i=1}^{n-2} u_\kappa^{(k)}(i)$ is independent of the coordinates of nucleons $n-1$ and n, we can integrate over them and obtain a factor $\delta_{J''0}\delta_{T''1}$. It therefore follows, by examining the c.f.p. in (34.12), that we must have $v_1' = v'$ and $t_1' = t'$ for the matrix element not to vanish. Since, furthermore, $u_\kappa^{(k)}(i)$ is independent of the charge coordinates, we must have also $T_1' = T_1$. Equation (34.12) therefore reduces to

$$\frac{n(n-1)}{2} N_J [j^{n-2}(v' t' T_1 J') j^2(1\ 0) T\, J' | \} j^n v' t' T\, J']$$

$$\times\, \langle j^{n-2} v' t' T_1 J' \left| \sum_i^{n-2} u_\kappa^{(k)}(i) \right| j^{n-2} v\, t\, T_1\, J \rangle. \qquad (34.13)$$

We thus see that a nonvanishing result is obtained only if $\psi(j^n\ v'\ t'\ T\ J')$ has a nonvanishing c.f.p. of the form

$$[j^{n-2}(v' t' T_1 J') j^2(1\ 0) T\, J' | \} j^n v' t' T\, J'].$$

We can continue this process going down to the j^{n-4}, j^{n-6}, ... configurations until the $j^{v'}$ configuration is reached (since $v' \geqslant v$). If v is actually smaller than v' we can continue this process. The matrix element (34.9) then vanishes since there is no nonvanishing c.f.p. of the form

$$[j^{v'-2}(v'' t'' T_1'' J) j^2(1\ 0) T'' J | \} j^{v'} v' t' T'' J].$$

Therefore, (34.9) can be different from zero only if $v' = v$. This proves the first part of the theorem (34.7).

To prove that (34.9) is also diagonal in the reduced isospin t, we note that, for $v = v'$, (34.13) becomes, after going down to the j^v configuration, proportional to

$$\langle j^v v\, t' T_1' = t' J' \left| \sum_i^v u_\kappa^{(k)}(i) \right| j^v v\, t\, T_1 = t\, J \rangle.$$

Since the isospin in the j^v configuration is equal to the reduced isospin, this matrix element will vanish unless $t = t'$. We thus see that the matrix element (34.9) for odd k between two states in the j^n configuration with different reduced isospins vanishes even if these states have the same seniority.

Let us now consider the nonvanishing matrix elements of odd tensor operators. From (34.13) we have

$$\langle j^n v\, t\, T\, J' \left| \sum_i^n u_\kappa^{(k)}(i) \right| j^n v\, t\, T\, J \rangle$$

$$= \frac{n(n-1)}{2} N_J [j^{n-2}(v\, t\, T_1 J') j^2(1\ 0) T\, J' | \} j^n v\, t\, T\, J']$$

$$\times \langle j^{n-2} v\, t\, T_1 J' \left| \sum_i^{n-2} u_\kappa^{(k)}(i) \right| j^{n-2} v\, t\, T_1 J \rangle. \qquad (34.14)$$

We can obtain an equivalent expression by writing

$$\psi(j^n v\, t\, T\, J') = N_{J'} \mathscr{A} \psi [j^{n-2}(v\, t\, T_1 J') j^2_{n-1,n}(1\ 0) T\, J']$$

and expanding $\psi(j^n\, v\, t\, T\, J)$ in terms of c.f.p. This way we arrive at the following result for the same matrix element*

$$\langle j^n v\, t\, T\, J' \left| \sum_i^n u_\kappa^{(k)}(i) \right| j^n v\, t\, T\, J \rangle$$

$$= \frac{n(n-1)}{2} N_{J'} [j^{n-2}(v\, t\, T_1\, J) j^2(1\ 0) T\, J | \} j^n v\, t\, T\, J]$$

$$\times \langle j^{n-2} v\, t\, T_1\, J' \left| \sum_i^{n-2} u_\kappa^{(k)}(i) \right| j^{n-2} v\, t\, T_1 J \rangle. \qquad (34.15)$$

(*) If the set of states with given v, t, and T and different values of J occurs only once in the j^n configuration, it is clear that both $\psi(j^n\, v\, t\, T\, J)$ and $\psi(j^n\, v\, t\, T\, J')$ can be obtained by antisymmetrization from a principal parent with the same value of T_1. If several such sets of states occur, these considerations apply only to two states with J and J' obtained from principal parents in the j^{n-2} configuration with the same value of T_1. The more complicated cases will not be discussed here. It should be only pointed out that if no such value of T_1 exists, (34.12) vanishes.

Provided we can find at least one value of k and κ, such that the matrix element under consideration does not vanish, we obtain from (34.14) and (34.15) the result

$$N_J[j^{n-2}(v\,t\,T_1J')j^2(1\;0)T\,J'|\}j^nv\,t\,T\,J']$$
$$= N_{J'}[j^{n-2}(v\,t\,T_1J)j^2(1\;0)T\,J|\}j^nv\,t\,T\,J]. \tag{34.16}$$

On the other hand, we obtain from the analog of (27.21), modified to our present case, that

$$N_J[j^{n-2}(v\,t\,T_1J)j^2(1\;0)T\,J|\}j^nv\,t\,T\,J]$$
$$= N_{J'}\,[j^{n-2}(v\,t\,T_1J')j^2(1\;0)T\,J'|\}j^nv\,t\,T\,J']. \tag{34.17}$$

Using (34.16) and (34.17) we obtain $N_J = N_{J'}$ as well as the equality of the two corresponding c.f.p. Using this result, (34.14) is greatly simplified. The product of $\{[n(n-1)]/2\}\,N_J$ and the c.f.p. becomes simply 1 due to (27.21). Our result assumes now the final form

$$\langle j^nv't'T'J' \Big| \sum_i^n u_\kappa^{(k)}(i) \Big| j^nv\,t\,T\,J\rangle$$
$$= \langle j^{n-2}v't'T_1J' \Big| \sum_i^{n-2} u_\kappa^{(k)}(i) \Big| j^{n-2}v\,t\,T_1J\rangle\,\delta_{vv'}\delta_{tt'}\delta_{TT'} \quad \text{for} \quad k \text{ odd} \tag{34.18}$$

or the equivalent form

$$\langle j^nv\,t\,T\,J' \Big| \sum_i^n u_\kappa^{(k)}(i) \Big| j^nv\,t\,T\,J\rangle$$
$$= \langle j^vv\,t\;T_1 = t\;J' \Big| \sum_i^v u_\kappa^{(k)}(i) \Big| j^vv\,t\;T_1 = t\;J\rangle \quad \text{for} \quad k \text{ odd.} \tag{34.19}$$

The form (34.19) displays more clearly the fact that the value of this matrix element is independent not only of n but also of T, thus completing the proof of the theorem (34.7).

Before proceeding with the utilization of the property (34.19), let us point out another result obtained above. The normalization factors N_J used for constructing the wave function from principal parents do not depend on J. This factor, or the corresponding c.f.p., may be a function of v, t, T_1, and T, but is the same for all states with the same v, t, and T and various J values. We shall make use of this fact in the following.

The relation (34.19) gives rise to the *pairing property* of two nucleon interactions which are sums of scalar products of odd tensor operators. If V_{ik} is the sum of scalar products $(\mathbf{f}_i^{(r)} \cdot \mathbf{f}_k^{(r)})$ of tensor operators with odd degrees, we can write it in the form (33.34). We then reduce the interaction matrix in the j^n configuration to the j^v configuration using (34.19). In analogy with (27.25) in jj-coupling and (31.25) in LS-coupling for identical particles, we obtain in the present case that such a V_{ik} is diagonal in v and t and its nonvanishing matrix elements are given by

$$\langle j^n v\, t\, T\, J \left| \sum_{i<k}^{n} V_{ik} \right| j^n v\, t\, T\, J\rangle$$

$$= \langle j^v v\, t\ \ T_1 = t\ \ J \left| \sum_{i<k}^{v} V_{ik} \right| j^v v\, t\ \ T_1 = t\ \ J\rangle + \frac{n-v}{2} V_0. \qquad (34.20)$$

In this equation V_0 is defined by

$$V_0 = -\frac{1}{2j+1} \sum_{r\ \mathrm{odd}} (j \| \mathbf{f}^{(r)} \| j)$$

$$= \langle j^2\ v = 0\ t = 0\ T = 1\ J = 0\ | V_{12} | j^2\ v = 0\ t = 0\ T = 1\ J = 0\rangle. \qquad (34.21)$$

The last equality was already obtained in Part II. It can also be directly obtained form (34.20) in the case $n = 2\ J = 0$ by using (34.8).

As in the previous cases, we shall calculate matrix elements of a special interaction which is intimately connected with the seniority scheme. In analogy with (27.27) and (31.26) we define a two nucleon operator $Q = \Sigma\, q_{ik}$ by

$$\langle j^2 T\, J\, M_T\, M | q_{12} | j^2 T' J' M'_T\, M'\rangle = (2j+1)\, \delta_{TT'} \delta_{JJ'} \delta_{M_T M'_T} \delta_{MM'} \delta_{J0}. \qquad (34.22)$$

This operator measures the number of pairs coupled to $T = 1\ J = 0$ in a given state. The operator q_{12} can be expanded into scalar products of tensor operators as in the previous cases [(Eqs. (27.30) or (31.27)]. However, we shall try, as before, to expand q_{12} in terms of odd tensors so that we will be able to make use of the pairing property in evaluating its matrix elements. Equations (15.16) and (15.17) can be combined with (15.5) to give the analog of (27.33) or (31.29)

$$(2j+1)\, \delta_{J0} = -(-1)^J - 2 \sum_{r\ \mathrm{odd}} (2r+1) (\mathbf{u}_1^{(r)} \cdot \mathbf{u}_2^{(r)}). \qquad (34.23)$$

Now, unlike the case of identical particles, $(-1)^J$ is not always equal to $+1$, but is a simple function of T. Due to the exclusion principle, we have in the j^2 configuration $(-1)^{T+J} = -1$. Thus, $-(-1)^J$ can be replaced by $(-1)^T$ which in turn is simply equal to $-P^\tau$. Using (33.1) we can write $(-1)^T = -P^\tau = -\{[1 + 4(\mathbf{t}_1 \cdot \mathbf{t}_2)]/2\}$ and insert this value in (34.23). We thus obtain

$$q_{12} = -\tfrac{1}{2} - 2(\mathbf{t}_1 \cdot \mathbf{t}_2) - 2 \sum_{r \text{ odd}} (2r+1)(\mathbf{u}_1^{(r)} \cdot \mathbf{u}_2^{(r)}). \qquad (34.24)$$

We see that q_{12} can be expressed as the sum of two very simple interactions and an interaction with the pairing property. The operator Q is manifestly diagonal in the seniority scheme. Its eigenvalues can now be calculated with the use of (34.20). We obtain

$$\langle j^n v\, t\, T\, J | \sum_{i<k}^{n} q_{ik} | j^n v\, t\, T\, J \rangle$$

$$= -\tfrac{1}{2}\frac{n(n-1)}{2} - \sum_{i<k}^{n} 2(\mathbf{t}_i \cdot \mathbf{t}_k)$$

$$- \langle j^n v\, t\, T\, J | 2 \sum_{r \text{ odd}} \sum_{i<k}^{n} (2r+1)(\mathbf{u}_i^{(r)} \cdot \mathbf{u}_k^{(r)}) | j^n v\, t\, T\, J \rangle$$

$$= -\frac{n(n-1)}{4} - [T(T+1) - \tfrac{3}{4}n]$$

$$- \langle j^v v\, t\; T_1 = t\; J | 2 \sum_{r \text{ odd}} \sum_{i<k}^{v} (2r+1)(\mathbf{u}_i^{(r)} \cdot \mathbf{u}_k^{(r)}) | j^v v\, t\; T_1 = t\; J \rangle$$

$$+ \frac{n-v}{2} V_0$$

$$= -\frac{n(n-1)}{4} - [T(T+1) - \tfrac{3}{4}n]$$

$$+ \langle j^v v\, t\; T_1 = t\; J | \sum_{i<k}^{v} q_{ik} + \frac{v(v-1)}{4} + \sum_{i<k}^{v} 2(\mathbf{t}_i \cdot \mathbf{t}_k) | j^v v\, t\; T_1 = t\; J \rangle$$

$$+ \frac{n-v}{2} V_0. \qquad (34.25)$$

The result (34.25) can be now greatly simplified. First we notice that the expectation value of Q vanishes in all states with seniority v in the j^v configuration. The V_0 which appears in (34.25) is obtained from (34.24) in the state $T = 1$ $J = 0$ as $V_0 = 2j + 1 + 1 = 2j + 2$. Collecting the nonvanishing terms in (34.25), we see that Q is independent of J and is given by

$$Q(n, v, t, T) = -\frac{n(n-1)}{4} - T(T+1) + \tfrac{3}{4}n + \frac{v(v-1)}{4} + t(t+1)$$

$$- \tfrac{3}{4}v + \frac{n-v}{2}(2j+2)$$

$$= \frac{n-v}{4}(4j+8-n-v) - T(T+1) + t(t+1). \qquad (34.26)$$

For $n = v$, T must be equal to t, so that (34.26) vanishes as it should. For identical nucleons, the isospin has its maximum value $T = (n/2)$ and $t = (v/2)$. In this case (34.26) reduces to the expression (27.29) previously obtained.

Both in jj-coupling and LS-coupling with identical particles, Q is directly related to *one* $n \to n-2$ c.f.p. That particular c.f.p. could be used to obtain recursion relations for all other c.f.p. in the seniority scheme. The situation is much more complicated in the present case. Using $n \to n-2$ c.f.p. to evaluate Q we readily obtain from (33.36) the following result

$$Q(n, v, t, T) = \frac{n(n-1)}{2}(2j+1)\sum_{T_1} [j^{n-2}(v\,t\,T_1J)j^2(1\;0)T\,J|\}j^n v\,t\,T\,J]^2. \qquad (34.27)$$

In the previous cases only one c.f.p. appeared in the analog of (34.27). But here, due to the value $T = 1$ of the pair coupled to a $J = 0$ state, T_1 can be generally equal to $T+1$, $T-1$, or T. Therefore, apart from special cases, the c.f.p. cannot be determined from (34.27) and no simple recursion relations follow from it. We shall not discuss here the possible methods that could be used to solve this problem. We shall only mention that it is impossible to find other two nucleon operators apart from Q, that will furnish further relations between the c.f.p. which appear in (34.27). This is obvious since in the expansion of $\psi(j^n\,v\,t\,T\,J)$ in $n \to n-2$ c.f.p., the last two nucleons are coupled to the *same* $T = 1$ $J = 0$ state in all terms which contribute to (34.27). It is clear that one must look for three nucleon operators with simple eigenvalues. However, as simple as these operators may be, a new quantum number must be introduced in order to express their eigenvalues. This quantum number is related to the additional quantum number which was mentioned earlier. We shall not pursue further this discussion since this difficulty has not yet been fully resolved.

At the end of Section 28, in the case of identical particles, we saw that any interaction which is diagonal in the seniority scheme can be expressed as the sum of an odd tensor interaction and a scalar term. We shall now see which is the most general interaction that is diagonal in the seniority scheme in nuclear j^n configurations.

We deal with charge independent interactions and consider their expansions into a sum of scalar products of space and spin irreducible tensors. In the same way that (28.55) was obtained we find that

necessary conditions for an interaction to be diagonal in the seniority scheme are

$$\int \psi^*(j^3\, v = 3, t = T, T, J = j)$$

$$\times \left[\sum_k F^k(\mathbf{u}_1^{(k)} \cdot (\mathbf{u}_2^{(k)} + \mathbf{u}_3^{(k)}))\right] \psi(j_1 j_{23}^2(10)T, J = j) = 0$$

for all possible states with $v = 3$, $J = j$, $T = \frac{3}{2}$ *and* $T = \frac{1}{2}$, *in the* j^3 *configuration.* (34.28)

These conditions will turn out to be also sufficient conditions as we shall see a little later.

Expanding the $v = 3$ states appearing in (34.28) in terms of c.f.p., we obtain the conditions (34.28) in the following form

$$\sum_{k \text{ even}} \frac{F^k}{\sqrt{2k+1}} [j, j^2(1k)\, T, J = j \| j^3\, v = 3, t = T, T, J = j] = 0. \qquad (34.29)$$

Odd tensors do not appear in (34.29) because of (34.8). As we saw above, every odd tensor interaction is diagonal in the seniority scheme and has the pairing property (34.20). The summation in (34.29) does not include the value $k = 0$ since for $v = 3$ there is no nonvanishing c.f.p. with a pair coupled to $T = 1$ $J = 0$. A scalar term, with $k = 0$, is obviously diagonal in the seniority scheme.

For every possible $v = 3$ $J = j$ state in the j^3 configuration there is one homogeneous equation (34.29) in which the F^k (or $F^k/\sqrt{2k+1}$) appear as the unknowns. It is a simple matter to find the number of these states. We first consider the total number of $J = j$ states with both $T = \frac{3}{2}$ and $T = \frac{1}{2}$. For this purpose we take the j^3 configuration with $M_T = \frac{1}{2}$ which has two protons and one neutron. The two protons must be coupled to antisymmetric j^2 states with J_1 even where $0 \leqslant J_1 \leqslant 2j - 1$. Every such state, with a given value of J_1, can be coupled to the j neutron to form a state with $J = j$. Thus, there are altogether $j + \frac{1}{2}$ states with $J = j$ in the j^3 configuration. Two of these are the $v = 1$ $J = j$

states with $T = \frac{1}{2}$ and $T = \frac{3}{2}$ (for $j > \frac{1}{2}$). Therefore, there are $j + \frac{1}{2} - 2 = j - \frac{3}{2}$ states with $v = 3$ $J = j$ in the j^3 configuration and accordingly $j - \frac{3}{2}$ equations (34.29). In particular, we see that for $j \leqslant \frac{3}{2}$ every two body interaction is diagonal in the seniority scheme. This fact, for the $j = \frac{3}{2}$ case, will be used explicitly in Section 36.

We now notice that all equations (34.29) are linearly independent. In fact, the c.f.p. appearing in (34.29) satisfy the orthogonality relations (33.12). [These relations hold for the c.f.p. of states with the same value of T. However, they hold, in the present case, also between states with $T = \frac{3}{2}$ and $T = \frac{1}{2}$ since these $v = 3$ states have $t = \frac{3}{2}$ and $t = \frac{1}{2}$ and thus different seniority quantum numbers.] As a result, the set of equations (34.29) is a set of $j - \frac{3}{2}$ linearly independent equations with $j - \frac{1}{2}$ unknowns F^k for $k > 0$ even. Therefore, there is only one set of these F^k, uniquely determined up to a common multiplicative factor, which satisfy these equations.

To find this set of F^k it is not necessary to compute the c.f.p. which appear in (34.29). The operator Σq_{ih} is diagonal in the seniority scheme and hence the F^k in the expansion of q_{12} satisfy the equations (34.29). We recall that due to (27.30) we have the following expansion of q_{12}

$$q_{12} = \sum_{k=0}^{2j} (-1)^k (2k+1)(\mathbf{u}_1^{(k)} \cdot \mathbf{u}_2^{(k)})$$

$$= \sum_{k \text{ even}} (2k+1)(\mathbf{u}_1^{(k)} \cdot \mathbf{u}_2^{(k)}) - \sum_{k \text{ odd}} (2k+1)(\mathbf{u}_1^{(k)} \cdot \mathbf{u}_2^{(k)}). \quad (34.30)$$

We thus conclude that

the scalar products of even tensors, with $k > 0$, in the expansion of any interaction which is diagonal in the seniority scheme must appear only as multiples of the combination

$$\sum_{k>0 \text{ even}} (2k+1)(\mathbf{u}_1^{(k)} \cdot \mathbf{u}_2^{(k)}). \quad (34.31)$$

Unlike the case of identical particles, the coefficients of all F^k are linearly independent due to (22.51). Therefore, no further simplification in the expansion of the interaction is possible if we use only space and spin irreducible tensors.

An alternative expansion can be obtained if we make use of the isospin vectors. Instead of using the combination (34.31) of even tensors we can use the operator q_{12} itself. In (34.24), q_{12} is expressed as a sum of a scalar term, a term proportional to $(\mathbf{t}_1 \cdot \mathbf{t}_2)$ and an odd tensor interaction.

We can, therefore, formulate the conditions on an interaction which is diagonal in the seniority scheme as follows.

Every interaction which is diagonal in the seniority scheme in the j^n configurations can be expressed as a sum of a scalar term (proportional to $n(n-1)/2$), *a* $(\mathbf{t}_1 \cdot \mathbf{t}_2)$ *term* (proportional to $T(T+1) - \frac{3}{4}n$) *and an odd tensor interaction* (which has the pairing property). (34.32)

The conditions (34.31) and (34.32) were obtained from the *necessary* conditions (34.28). However, the interaction described in (34.32) is certainly diagonal in the seniority scheme. Thus, the conditions (34.28), (34.29), (34.31), and (34.32) are also *sufficient* conditions for an interaction to be diagonal in the seniority scheme.

The Group Theoretical Description of the Seniority Scheme

We shall now discuss the group theoretical meaning of the quantum numbers v and t. The situation is slightly more complicated than in the case of jj-coupling for identical particles. On the other hand, there is some limited similarity to the case of LS-coupling with identical particles.

The antisymmetric wave functions of a charge-independent Hamiltonian (which can be made not to contain the isospin vectors) are taken, according to the exclusion principle, to be antisymmetric in *all* particle coordinates. Every wave function can be written as a sum of products of isospin functions and functions of the space and spin coordinates. As the Hamiltonian is charge-independent, each of the latter functions must be an eigenfunction of H which belongs to the same eigenvalue E. Since the Hamiltonian is invariant under permutations of the spin and space coordinates of the nucleons, these eigenfunctions must all have the same symmetry type and transform irreducibly among themselves under such permutations. These functions can therefore be obtained from a certain Young diagram T. The requirement that the complete wave function be antisymmetric implies that the isospin functions which multiply these space and spin functions must have a definite symmetry type which belongs to the dual diagram T'.

The diagrams which give rise to irreducible representations of the permutations group in the space of the isospin functions can have, at most, two rows. This is due to the limitation of the projection of an isospin vector of a single nucleon to the two values $+\frac{1}{2}$ and $-\frac{1}{2}$. Any column with more than two squares must contain nucleon numbers with at least two identical m_i values so that the antisymmetrization of the wave function gives zero. These isospin functions have the same form as the spin functions discussed above. The functions obtained from a given diagram are eigenfunctions of $\mathbf{T}^2$. The eigenvalue T to which they belong is simply given by the diagram. If n_1 and n_2 are the lengths of the two rows we obtain as before

$$n_1 + n_2 = n \qquad \frac{n_1 - n_2}{2} = T. \tag{34.33}$$

This means that T is a good quantum number of the complete wave function. Furthermore, T determines the symmetry type of the space and spin part of the wave function. The lower the T, the more symmetric the space and spin wave function. Therefore, for two nucleon forces that are stronger in space and spin symmetric states, the states with different isospin T values will differ widely in energy. The lower the T value, the lower the energy of the state.

We now turn to the space and spin wave functions characterized by the irreducible representations of the permutations group (symmetry types). The diagrams to which these functions belong must be dual to those of the isospin functions in order to make the complete wave function antisymmetric. This means that the diagrams under consideration have rows with no more than two squares. The length of the two columns can be at most $2j + 1$ which is the number of different m values of a single nucleon. We consider again as one set *all* spin and space wave functions with the same symmetry type in the j^n configuration, irrespective of their J or M values. All these functions transform irreducibly among themselves if the single nucleon wave functions undergo a general linear (or unitary) transformation (32.20) with a nonsingular matrix. In the case of identical particles the isospin T has its maximum value $n/2$ and there is only one isospin function which is fully symmetric. Therefore, the only space and spin function possible is the fully antisymmetric one. In the general case there are functions with other symmetry type which are likewise bases of irreducible representations of the linear or unitary group in the $2j + 1$-dimensional space $U(2j + 1)$.

If we restrict now the unitary transformation to be the $D^{(j)}_{mm'}$ matrices which represent rotations in three-dimensional space, every irreducible representation of $U(2j + 1)$ will be decomposed into the irreducible representations of R_3. In other words, only the $2J + 1$ components of a state with a given J transform among themselves under rotations. There are no nonvanishing matrix elements of the Hamiltonian between two states, even with the same J and M, if they belong to two different symmetry types. On the other hand, there may be several states of the same symmetry type (i.e., with the same value of T) with the same value of J (and M). As in the case of identical particles, we shall therefore try to characterize these various states by going from $U(2j + 1)$ to R_3 in two steps.

We first restrict the unitary matrices to be those which belong to the symplectic group Sp$(2j + 1)$. The transformations of Sp$(2j + 1)$ leave invariant the $\psi_{J=0}$ wave function of two nucleons (therefore, the transformations of R_3 are a subgroup of this symplectic group). The states with the same symmetry type are grouped into sets which transform irreducibly under Sp$(2j + 1)$. These sets of states will be characterized by the corresponding irreducible representations of Sp$(2j + 1)$. The energy matrix *will* in general have nonvanishing nondiagonal elements connecting states which belong to different irreducible representations of Sp$(2j + 1)$ provided they have the same J value. This is clearly so, since generally the Hamiltonian is not invariant under the transformations of Sp$(2j + 1)$ [and *a fortiori* under those of $U(2j + 1)$], as explained in detail in the case of identical particles. However, these nondiagonal elements may, for some Hamiltonians, be small, in which case the quantum numbers furnished by Sp$(2j + 1)$ will be approximately good quantum numbers. In any case, the scheme of wave functions which transform irreducibly under Sp$(2j + 1)$ is a definite mathematical scheme which can be used for practical calculations.

In the second step we restrict Sp$(2j + 1)$ to its subgroup R_3. The bases of the irreducible representations of Sp$(2j + 1)$ will be decomposed into sets of $2J + 1$ functions with the same value of J, which are the bases of the irreducible representations of R_3. States with the same value of J which originate in sets of functions of different representations of Sp$(2j + 1)$ will be distinguished by the quantum numbers which characterize these representations. These quantum numbers are just those of the seniority scheme.

We have only to see how the irreducible representations of Sp$(2j + 1)$ are characterized in the present case. *The wave functions with a certain symmetry type in the j^v configuration which do not contain a $J = 0$ state of two nucleons transform irreducibly under* Sp$(2j + 1)$. The various symmetry types of these wave functions, which have the seniority v, can be obtained by using Young diagrams. The only diagrams in this case which give rise to nonvanishing wave functions are those which have no more than $\frac{1}{2}(2j + 1) = j + \frac{1}{2}$ rows. Since in our case the diagrams have, as we saw, at most two columns, this requirement implies that the number of nucleons v is, at most, $2j + 1$. This simply means that the seniority cannot exceed half of the maximum number of nucleons in the shell which is $2(2j + 1) = 4j + 2$. This is analogous to the case of identical particles where the seniority v cannot exceed $\frac{1}{2}(2j + 1)$.

All permissible diagrams are uniquely characterized by giving the lengths of the possible $j + \frac{1}{2}$ rows. To do this we introduce the numbers w_i, $i = 1, 2, ..., j + \frac{1}{2}$ which give the length of each row and satisfy

$$w_1 \geq w_2 \ldots \geq w_{j+1/2} \geq 0. \tag{34.34}$$

In the present case, the w_i can be equal to either 2, 1, or 0. Let us define a and b by

$$\begin{aligned} w_1 = w_2 = ... = w_a &= 2 \\ w_{a+1} = w_{a+2} = ... = w_{a+b} &= 1 \\ w_{a+b+1} = w_{a+b+2} = ... = w_{j+1/2} &= 0. \end{aligned} \tag{34.35}$$

The length of the first column is therefore $a + b$ and that of the second column a. If we make use of (34.33) in the j^v configuration for the symmetry type considered we obtain

$$2a + b = v \qquad \frac{b}{2} = t. \tag{34.36}$$

Equation (34.36) is the group theoretical definition of v and t.

The state with $v = 0$ $t = 0$ $J = 0$ is thus given by the empty diagram and in terms of the w_i by (000 ... 0), or simply (0). The state with $v = 1$ $t = \frac{1}{2}$ and $J = j$ belongs, according to (34.36), to $a = 0$ $b = 1$ or (100 ... 0). The states with $v = 2$ belong to *two* irreducible representations of the symplectic group. Those with $J > 0$ even have $t = 1$, and therefore $a = 0$ $b = 2$ or (110 ... 0). Those with J odd have $t = 0$. According to (34.36), they belong to $a = 1$ $b = 0$ which in terms of the w_i corresponds to (200 ... 0).

We saw the important role played by the odd tensor operators in the seniority scheme due to the relation (34.8). We shall now discuss the group theoretical meaning of this relation. The transformations of Sp$(2j + 1)$ can be written as*

$$\psi'_{jm_1} = \sum_m a_{m_1 m} \psi_{jm}. \tag{34.37}$$

In (34.37) the matrix elements $a_{mm'}$ are so defined that the transformation applied to nucleons 1 and 2 leaves their $\psi_{J=0}$ wave function invariant. Let us find the form of the matrices $||a_{mm'}||$ which leave invariant $\psi_{J=0}$. In order to do this it is easier to consider *infinitesimal transformations* as was done in the case of the rotation group. We define the infinitesimal operator $\alpha_{mm'}$ by

$$a_{mm'} = \delta_{mm'} + \epsilon\alpha_{mm'} \tag{34.38}$$

where ϵ is considered infinitesimal. The condition of the invariance of $\psi_{J=0}$ under (34.37) becomes, if we write the transformation as an operator,

$$a(1)a(2)\psi_{J=0} = \psi_{J=0}. \tag{34.39}$$

(*) We write here the transformation as a matrix multiplying a column wave function rather than a row wave function multiplying a matrix as in (32.20). This makes no difference whatsoever and is adopted only for the sake of convenience.

Using (34.38) we have

$$a(1)a(2) = 1 + \epsilon[\alpha(1) + \alpha(2)] + \epsilon^2\alpha(1)\alpha(2) \tag{34.40}$$

where the unity on the right-hand side stands for the unit operator in the space of wave functions of the two nucleons. The term proportional to ϵ^2 in (34.40) can be neglected. Therefore, the condition (34.39) of invariance of $\psi_{J=0}$ becomes

$$[\alpha(1) + \alpha(2)]\, \psi_{J=0} = 0. \tag{34.41}$$

This equation, written in the JM scheme of the j^2 configuration, is

$$\langle j^2 J\, M|\alpha(1) + \alpha(2)|j^2\ J = 0\ M = 0\rangle = 0 \quad \text{for} \quad \text{all } J, M. \tag{34.42}$$

We now realize that (34.42) is exactly the relation (34.8) which is satisfied by all odd tensor operators and by *no even* unit tensor operator. This means that the odd tensor operators $u_\kappa^{(k)}$ can be taken for the α in (34.41). Thus, the $u_\kappa^{(k)}$ with odd k are the infinitesimal operators of the symplectic group in $2j + 1$-dimensional space of the ψ_{jm}.

This result was obtained without making use of the half-integral character of j. Therefore, in LS coupling the corresponding $u_\kappa^{(k)}$ with odd k are the infinitesimal operators of the rotation group $R(2l + 1)$ in the $2l + 1$-dimensional space of the ψ_{lm}. The following considerations will therefore be also applicable *mutatis mutandis* to the l^n configuration.

In the j^n configuration the transformations induced by $\mathrm{Sp}(2j + 1)$ are direct products of operators [like (34.40) for $n = 2$]. The infinitesimal operators in this space are thus the *sums* of the individual infinitesimal operators. Thus, $U_\kappa^{(k)} = \Sigma_i^n u_\kappa^{(k)}(i)$ with odd k, are the infinitesimal operators of the representation of $\mathrm{Sp}(2j + 1)$ in the space of wave functions of the j^n configuration. It is therefore clear that these operators do not have nonvanishing matrix elements connecting states which belong to different irreducible representations of $\mathrm{Sp}(2j + 1)$ (i.e., with different values of v or t). It is also clear that the matrices of all $U_\kappa^{(k)}$ with k odd cannot be reducible in the subspace of states which belong to the same irreducible representation.

We consider next the operator Q which can be built according to (34.24) from odd tensors $\mathbf{U}^{(k)}$. The explicit dependence of Q on the $\mathbf{U}^{(k)}$ is according to (34.24) and (33.34) the following

$$Q = \sum_{i<k}^{n} q_{ik} = -\frac{n(n-1)}{4} - 2\sum_{i<k}^{n} (\mathbf{t}_i \cdot \mathbf{t}_k) + \frac{n}{2j+1}(j+1)(2j+1)$$
$$- \sum_{r \text{ odd}} (2r+1)(\mathbf{U}^{(r)} \cdot \mathbf{U}^{(r)}). \tag{34.43}$$

The operator $\Sigma_{r \text{ odd}}\, (2r + 1)\, (\mathbf{U}^{(r)} \cdot \mathbf{U}^{(r)})$ plays in the groups $\mathrm{Sp}(2j + 1)$

and $R(2l+1)$ the same role as $\mathbf{L}^2$ in R_3. In fact, for $l = 1$ this operator is equal to a multiple of $\mathbf{L}^2$. The operator

$$G = 2 \sum_{r \text{ odd}} (2r+1)(\mathbf{U}^{(r)} \cdot \mathbf{U}^{(r)})$$

is called *Casimir's operator* of the group considered (note the factor 2 in its definition). Like $\mathbf{L}^2$ commuting with L_x, L_y, and L_z, G commutes with all $U_\rho^{(r)}$ as follows from (34.43) due to the properties of Q. Like Q it is diagonal in the seniority scheme and its eigenvalues are the same for all states with the same v and t. Its explicit expression in terms of these quantum numbers is given by

$$G = 2 \langle j^n v\, t\, T\, J \left| \sum_{r \text{ odd}} (2r+1)(\mathbf{U}^{(r)} \cdot \mathbf{U}^{(r)}) \right| j^n v\, t\, T\, J \rangle$$

$$= -2Q(n, v, t, T) - \frac{n(n-1)}{2} - 2[T(T+1) - \tfrac{3}{4}n] + 2(j+1)n$$

$$= \frac{v}{2}(4j+8-v) - 2t(t+1). \tag{34.44}$$

Thus, $G = G(v, t)$ is independent of n and T and its eigenvalues, like the $L(L+1)$ eigenvalues of $\mathbf{L}^2$, are determined only by the quantum numbers of the irreducible representations of the group to which it belongs. In terms of the G function (34.44), we can give Q the following form which is sometimes found in the literature

$$Q(n, v, t, T) = \left[\frac{n}{4}(4j+8-n) - T(T+1)\right] - \left[\frac{v}{4}(4j+8-v) - t(t+1)\right]$$

$$= \tfrac{1}{2}[G(n, T) - G(v, t)]. \tag{34.45}$$

In terms of the w_i which can be used to characterize the irreducible representations of $\mathrm{Sp}(2j+1)$, the eigenvalues of G are given by the equivalent expression

$$G(v, t) = G(W) = w_1(w_2 + 2j + 1) + w_2(w_2 + 2j - 1) + \ldots + w_{j+1/2}(w_{j+1/2} + 2) \tag{34.46}$$

where $W \equiv w_1 w_2 \ldots w_{j+1/2}$. This expression for the eigenvalues of Casimir's operator can be obtained directly by group theoretical methods and is often given in the literature. It is of some interest to write down the analogous expression for Casimir's operator of the rotation group $R(2l+1)$. Its irreducible representations are given by l numbers w_i, as already mentioned in Section 32. In terms of these, the eigenvalues of the corresponding Casimir's operator are given by a similar expression

$$w_1(w_1 + 2l - 1) + w_2(w_2 + 2l - 3) + \ldots + w_l(w_l + 1). \tag{34.47}$$

For $l = 1$ there is only one w and (34.47) is then identical with $L(L+1)$.

35. Coefficients of Fractional Parentage in the Seniority Scheme

In the case of identical particles we were able to make use of the seniority scheme in reducing the c.f.p. in the j^n configuration to those in lower configurations. For example, using (28.9) we obtained (28.10) which gives some c.f.p. in the j^n configuration in terms of those in the j^v configuration. The other c.f.p. in the j^n configuration are given by (28.11) and (28.16) in terms of the j^{v+1} configuration. These reduction formulas are very useful for calculating the c.f.p. and in some simple cases they can be used to find a closed expression for the c.f.p. [e.g. (28.17)]. These reduction formulas also enabled us to derive the results obtained in Section 29 on the averages of the interaction in groups of states with the same seniority. The relations (28.10) and (28.11) show that the c.f.p. in the seniority scheme are factorized into products of two terms. The first term depends on n and v, but is independent of the angular momenta involved (i.e., J_1 and J). The second term depends on v, J_1, and J but is independent of n. This *factorization of the c.f.p.* is due to the fact that the coefficient for normalization of a wave function with a certain value of J, to which a pair with $J = 0$ was added and the result antisymmetrized, is independent of J. In the case of identical particles the value of this normalization coefficient was calculated and is given by (27.26). In the present case, with nucleons of both kinds, we were unable to calculate the corresponding normalization coefficient explicitly. Still, we obtained the important property of this coefficient, namely the fact that it is independent of J. This in itself is sufficient to insure many of the results obtained before in the case of identical particles. We shall now proceed to the discussion of the factorization of the c.f.p. in the present case, and the results that follow from it.

We first try to obtain recursion relations for the c.f.p. similar to (28.8) and (28.9) in the case of identical particles. We start from

$$[j^{n+1}(v_1t_1T_1J_1)j\,T\,J|\}j^{n+2}v\,t\,T\,J]$$

$$= \int \psi^*[j^{n+1}(v_1t_1T_1J_1)j_{n+2}T\,J]\,\psi(j^{n+2}v\,t\,T\,J)$$

$$= N_{J_1}\int \mathscr{A}\psi^*[j^{n-1}(v_1t_1T_2J_1)j_{n,n+1}(1\ \ 0)(T_1J_1)j_{n+2}T\,J]\,\psi(j^{n+2}v\,t\,T\,J). \quad (35.1)$$

The last equality is due to our method of constructing wave functions in the seniority scheme. The antisymmetrizer $\mathscr{A}$ exchanges nucleons n and $n+1$ with the first $n-1$ nucleons, and the normalization factor

N_{J_1} is a function of $n-1$, v_1, t_1, T_2 and T_1. Since $\psi(j^{n+2}\, v\, t\, T\, J)$ is antisymmetric, we can replace $\mathscr{A}$ by $[n(n+1)/2]$ under the integral sign in (35.1). We also permute nucleons n, $n+1$, and $n+2$ without affecting the result and obtain for (35.1) the form

$$\frac{n(n+1)}{2} N_{J_1} \int \psi^*[j^{n-1}(v_1 t_1 T_2 J_1)\, j^2_{n+1,n+2}(1\ 0)\,(T_1 J_1)\, j_n T\, J]\, \psi(j^{n+2} v\, t\, T\, J). \tag{35.2}$$

We now expand $\psi(j^{n+2}\, v\, t\, T\, J)$ in $n+2 \to n$ c.f.p. Upon the integration over $n+1$ and $n+2$, the only terms that will not vanish are those in which the parent wave functions have nucleons $n+1$ and $n+2$ coupled to $T=1$ $J=0$. Therefore, in these parents the first n nucleons will be in a state with the quantum numbers v, t, and J of the wave function $\psi(j^{n+2}\, v\, t\, T\, J)$. We thus obtain for (35.2)

$$\frac{n(n+1)}{2} N_{J_1} \sum_{T'} [j^n(v\, t\, T' J)\, j^2(1\ 0) T\, J |\} j^{n+2} v\, t\, T\, J]$$

$$\times \int \psi^*[j^{n-1}(v_1 t_1 T_2 J_1)\, j^2_{n+1,n+2}(1\ 0)\,(T_1 J_1)\, j_n T\, J]\, \psi[j^n(v\, t\, T' J)\, j^2_{n+1,n+2}(1\ 0) T\, J]. \tag{35.3}$$

In order to carry out the integration we have to change the order of coupling of the isospin vectors in ψ^* in (35.3). Carrying out this change of coupling transformation and integrating, we finally obtain

$$[j^{n+1}(v_1 t_1 T_1 J_1)\, j\, T\, J |\} j^{n+2} v\, t\, T\, J]$$

$$= \frac{n(n+1)}{2} N_{J_1} \sum_{T'} [j^n(v\, t\, T' J)\, j^2(1\ 0) T\, J |\} j^{n+2} v\, t\, T\, J]$$

$$\times \langle T_2 1(T_1) \tfrac{1}{2} T | T_2 \tfrac{1}{2}(T') 1\, T\rangle\, [j^{n-1}(v_1\, t_1\, T_2\, J_1)\, j\, T' J |\} j^n v\, t\, T' J]. \tag{35.4}$$

This result is certainly not a very handy formula. First of all, it contains a sum over T' which may in principle include three terms with $T' = T \pm 1$ and $T' = T$. We do not know the form of N_{J_1} and the $n+2 \to n$ c.f.p. which appear in (35.4). However, all terms, with the exception of the $n \to n-1$ c.f.p. on the right-hand side of (35.4), are independent of J_1 and J. The normalization factor N_{J_1} is independent of J_1 and is a certain function $N(n-1, v_1, t_1, T_2, T_1)$. This fact will enable us to establish the factorization of the c.f.p. although the factors as functions of the various quantum numbers will remain unknown.

We now use again the recursion relation (35.4) for the c.f.p. $[j^{n-1}(v_1\, t_1\, T_2\, J_1) j\, T'\, J |\} j^n\, v\, t\, T'\, J]$. This way we obtain more summations, and possibly the result will involve several $n-3 \to n-2$ c.f.p. However, as we repeat this procedure several times, the results become simpler. As in the case of identical particles, it is obvious that v can be equal to either $v_1 + 1$ or $v_1 - 1$. Let us consider first the case where $v_1 = v - 1$. In this case the process of repeating (35.4) will end when we obtain as the result of the reduction the $v \to v - 1$ c.f.p. In this case there will be only one such c.f.p. since the isospin in the j^v configuration of the state considered can be only equal to t, whereas the isospin of the j^{v-1} parent must be t_1. We thus obtain the c.f.p. on the left-hand side of (35.4) as a product of one $v \to v - 1$ c.f.p. and another factor. The other factor is a complicated sum over products of normalization coefficients and transformation matrix elements. The important property of this factor is that it depends on n, v_1, t_1, T_1, v, t, and T but is independent of J_1 and J. It also depends on the intermediate isospin vectors which were used in the successive couplings of pairs with $T = 1$ $J = 0$ to the parent state in the j^v configuration in order to obtain the state in the j^n configuration. Therefore, if there are two (or more) series of intermediate isospins that give independent states with the same values of v, t, and T in the j^n configuration, this factor will depend also on the quantum numbers which distinguish between these states. If there are a certain number of independent ways to arrive at the j^n configuration (for given v, t, and T), every state with the given values of v, t in the j^v configuration characterized by J and possibly by another necessary quantum number α, will appear the same number of times in the j^n configuration. Since the normalization coefficient is independent of J_1 and J, all states of the j^v configuration with v, t, T, and α will give rise to the same number of independent states (with the given T) in the j^n configuration. We can now distinguish between two types of additional quantum numbers necessary to specify uniquely states with the same values of v, t, T, and J. We shall denote by α the quantum numbers necessary to distinguish between states with the same values of v, t, and J in the j^v configuration. Such independent states with the same value of J belong to the *same* irreducible representation of $Sp(2j + 1)$ characterized by v and t. These quantum numbers were necessary also in some cases with identical particles. The other quantum numbers, to be denoted by β, distinguish between sets of states with the same values of v, t, and T which may appear for certain values of n and T in the j^n configuration (but never in the j^v configuration). Every such set of states belongs to a different irreducible representation of $Sp(2j + 1)$. All these representations are *equivalent* and are characterized by the same values of v and t. Using this notation we can summarize our result by

$$[j^{n-1}(\beta_1\alpha_1, v-1, t_1T_1J_1)j\,T\,J|\}j^n\beta\,\alpha\,v\,t\,T\,J]$$

$$= F(n\,\beta_1 t_1 T_1 \beta\,v\,t\,T)\,[j^{v-1}(\alpha_1, v-1, t_1t_1J_1)j\,t\,J|\}j^v\,\alpha\,v\,t\,t\,J]. \qquad (35.5)$$

This is the factorization of the c.f.p. in the case $v_1 = v - 1$. It will look far simpler once we drop, as will usually be done for the sake of conciseness, the additional quantum numbers. Equation (35.3) is the analog of (28.10) in the case of identical particles. In that particular case the value of the factor F is known.

Next we consider the case $v = v_1 + 1$ which is slightly more complicated. By using successively the relation (35.4), we can finally obtain the $n \to n - 1$ c.f.p. in terms of $v + 2 \to v + 1$ c.f.p. However, there may be two such c.f.p. $[j^{v+1}(v + 1, t_1 t_1 J_1)j\ T\ J \| j^{v+2}\ v\ t\ T\ J]$, since in principle T can be equal to either $t_1 + \frac{1}{2}$ or $t_1 - \frac{1}{2}$. The situation is more complicated than in the case of identical particles where there is only one such $v + 2 \to v + 1$ c.f.p., and the desired factorization is given directly by (28.11). In order to establish the factorization in the present case, we have to further evaluate this $v + 2 \to v + 1$ c.f.p. To do this we use a procedure similar to that used in arriving at (28.16) in the case of identical particles.

The $v + 2 \to v + 1$ c.f.p. considered can be evaluated by direct antisymmetrization. We begin by writing

$$[j^{v+1}(v + 1, t_1 t_1 J_1)\, j\ T\ J \| j^{v+2} v\ t\ T\ J]$$

$$= \int \psi^*[j^{v+1}(v + 1, t_1 t_1 J_1)\, j_{v+2} T\ J]\, \psi(j^{v+2} v\ t\ T\ J)$$

$$= N \int \psi^*[j^{v+1}(v + 1, t_1 t_1 J_1)\, j_{v+2} T\ J]\, \mathscr{A} \psi[j^v(v\ t\ t\ J))\, j^2_{v+1,v+2}(1\ \ 0) T\ J]. \qquad (35.6)$$

The antisymmetrizer $\mathscr{A}$ is a product of two antisymmetrizers $\mathscr{A}_1$ and $\mathscr{A}_2$ which exchange the first v nucleons with nucleons $v + 1$ and $v + 2$ respectively. Since $\psi[j^{v+1}(v + 1, t_1\, t_1\, J_1)j_{v+2}\ T\ J]$ is antisymmetric in the first $v + 1$ nucleons we can apply $\mathscr{A}_1$ to it rather than to the other function. This will multiply the integral by $v + 1$. We do it first, and then write down in detail the result of applying $\mathscr{A}_2$. We thus obtain for (35.6) the expression

$$(v + 1)N \int \psi^*[j^{v+1}(v + 1, t_1 t_1 J_1)\, j_{v+2} T\ J]$$

$$\times \{\psi[j^v(v\ t\ t\ J))\, j^2_{v+1,v+2}(1\ \ 0) T\ J] - \sum_k^v \psi[j^v(v\ t\ t\ J))\, j^2_{v+1,k}(1\ \ 0) T\ J]\}. \qquad (35.7)$$

Only the first term in the curly brackets contributes to (35.7), since in $\psi^*(j^{v+1}\ v + 1, t_1\, t_1\, J_1)$ no two nucleons (like nucleons k and $v + 1$) are coupled to $T = 1$ $J = 0$. We now expand $\psi^*(j^{v+1}\ v + 1, t_1\, t_1\, J_1)$ in $v + 1 \to v$ c.f.p. and integrate. The only parents in the j^v configuration that will contribute to the integral are those with $\psi^*(j^v\ v\ t\ t\ J)$. In order to carry out the integration we need to perform a transformation to couple nucleons $v+1$ and $v+2$ to a $T=1$ $J=0$ pair. Doing all this we obtain for (35.7) the expression

$$(v+1)N\,[j^v(v\,t\,t\,J)\,j\,t_1J_1|\}j^{v+1}\,v+1,\,t_1t_1J]$$
$$\times\,\langle t\,\tfrac{1}{2}(t_1)\,\tfrac{1}{2}\,T|t,\,\tfrac{1}{2}\,\tfrac{1}{2}(1)T\rangle\,\langle Jj(J_1)\,j\,J|J,jj(0)J\rangle. \qquad (35.8)$$

Inserting the actual value of the transformation coefficient of the J vectors and including the additional necessary quantum numbers, we obtain the final result

$$[j^{v+1}(\alpha_1,\,v+1,\,t_1t_1J_1)\,j\,T\,J|\}j^{v+2}\alpha\,v\,t\,T\,J]$$
$$=(v+1)\,N(v,\,v,\,t,\,t,\,T)\,\langle t\,\tfrac{1}{2}\,(t_1)\,\tfrac{1}{2}\,T|t,\,\tfrac{1}{2}\,\tfrac{1}{2}\,(1)T\rangle$$
$$\times\,\frac{1}{\sqrt{2j+1}}(-1)^{J+j-J_1}\sqrt{\frac{2J_1+1}{2J+1}}\,[j^v(\alpha\,v\,t\,t\,J)\,j\,t_1J_1|\}j^{v+1}\alpha_1,\,v+1,\,t_1t_1J_1]. \qquad (35.9)$$

This is the analog of (28.16) in the case of identical particles.

The result (35.9) shows that the $v+2\to v+1$ c.f.p. considered is equal in both cases, $T=t_1+\frac{1}{2}$ and $T=t_1-\frac{1}{2}$, to a constant which depends on T, but is independent of J_1 and J multiplied by the *same function* of J_1 and J. We can therefore continue further the reduction of the $n\to n-1$ c.f.p. in the case $v_1=v+1$ and finally obtain this c.f.p. in the factorized form

$$[j^{n-1}(\beta_1\alpha_1,\,v+1,\,t_1T_1J_1)\,j\,T\,J|\}j^n\beta\,\alpha\,v\,t\,T\,J]$$
$$=G(n\,\beta_1t_1T_1\beta\,v\,t\,T)\,(-1)^{J+j-J_1}\sqrt{\frac{2J_1+1}{2J+1}}$$
$$\times\,[j^v(\alpha\,v\,t\,t\,J)\,j\,t_1J_1|\}j^{v+1}\,\alpha_1,\,v+1,\,t_1t_1J_1]. \qquad (35.10)$$

This factorization (35.10), together with (35.5), are the desired general expressions. They both give the $n\to n-1$ c.f.p. as a product of a function which depends only on the structure of the irreducible representations of Sp$(2j+1)$ involved and a function of all other quantum numbers. In the following we shall make use of this factorization although we do not know the explicit forms of the functions F and G.

We can now use (35.10) to go over from the $v\to v+1$ c.f.p. to $v+1\to v+2$ c.f.p. All factors which depend on J_1 or J will disappear, and the result will be simply

$$[j^{n-1}(\beta_1\alpha_1,\,v+1,\,t_1T_1J_1)\,j\,T\,J|\}j^n\beta\,\alpha\,v\,t\,T\,J]$$
$$=F'(n\,\beta_1t_1T_1\beta\,v\,t\,T)\,[j^{v+1}(\alpha_1,\,v+1,\,t_1t_1J_1)\,j\,t\,J|\}j^{v+2}\alpha\,v\,t\,t\,J]. \qquad (35.11)$$

Using (35.5) and (35.11) we can obtain certain orthogonality relations obeyed by the c.f.p. which are analogous to (28.20) and (28.21) in the case of identical particles. We start with the expression

$$\sum_{\alpha_1 J_1} [j^n \beta' \alpha' v' t' T' J \{| j^{n-1}(\beta_1 \alpha_1, v-1, t_1 T_1 J_1) j\, T' J] \\ \times [j^{n-1}(\beta_1 \alpha_1, v-1, t_1 T_1 J_1) j\, T\, J |\} j^n \beta\, \alpha\, v\, t\, T\, J]. \qquad (35.12)$$

Using (35.5) it can be brought into the form

$$F(n\, \beta_1 t_1 T_1 \beta' v' t' T')\, F(n\, \beta_1 t_1 T_1 \beta\, v\, t\, T) \sum_{\alpha_1 J_1} [j^v \alpha' v' t' t' J \{| j^{v-1}(\alpha_1, v-1, t_1 t_1 J_1) j\, t' J] \\ \times [j^{v-1}(\alpha_1, v-1, t_1 t_1 J_1) j\, t\, J |\} j^v \alpha\, v\, t\, t\, J]. \qquad (35.13)$$

The summation in the last expression is a part of the orthogonality relation

$$\sum_{\alpha_1 J_1 t_1} [j^v \alpha' v' t' t' J \{| j^{v-1}(\alpha_1, v-1, t_1 t_1 J_1) j\, t' J] \\ \times [j^{v-1}(\alpha_1, v-1, t_1 t_1 J_1) j\, t\, J |\} j^v \alpha\, v\, t\, t\, J] = \delta_{\alpha\alpha'} \delta_{vv'} \delta_{tt'}. \qquad (35.14)$$

The summation over t_1 in (35.14) includes two values, namely $t_1 = t + \frac{1}{2}$ and $t_1 = t - \frac{1}{2}$. In order to separate the two t_1 values we calculate the matrix elements of the two-body operator $\Sigma\, 2(\mathbf{t}_i \cdot \mathbf{t}_k)$ in the j^v configuration. We obtain

$$\frac{v}{v-2} \sum_{\alpha_1 J_1 t_1} [j^v \alpha' v' t' t' J \{| j^{v-1}(\alpha_1, v-1, t_1 t_1 J_1) j\, t' J] \\ [j^{v-1}(\alpha_1, v-1, t_1 t_1 J_1) j\, t\, J |\} j^v \alpha\, v\, t\, t\, J]\, [t_1(t_1+1) - \tfrac{3}{4}(v-1)] \\ = [t(t+1) - \tfrac{3}{4} v]\, \delta_{\alpha\alpha'} \delta_{vv'} \delta_{tt'}. \qquad (35.15)$$

We can solve the two equations (35.14) and (35.15) and obtain the following result for the unknown sums

$$\sum_{\alpha_1 J_1} [j^v \alpha' v' t' t' J \{| j^{v-1}(\alpha_1, v-1, t_1 t_1 J_1) j\, t' J]\, [j^{v-1}(\alpha_1, v-1, t_1 t_1 J_1) j\, t\, J |\} j^v \alpha\, v\, t\, t\, J] \\ = \begin{cases} 0 & \text{if} \quad \alpha\, v\, t \not\equiv \alpha' v' t' \\ \text{independent of } \alpha \text{ and } J \text{ if } \alpha\, v\, t \equiv \alpha' v' t'. \end{cases} \qquad (35.16)$$

The exact dependence of (35.16) on t_1, t, and v is not important since we would like to use it only to prove that the expression (35.13), as well as (35.12), vanishes unless $\alpha v t \equiv \alpha' v' t'$, and is independent of α and J in this case. This result is the analog of (28.20), but in the present case we have not obtained the explicit form of the sum on the various quantum numbers.

The same result holds for the analogous expression

$$\sum_{\alpha_1 J_1} [j^n \beta' \alpha' v' t' T' J \{| j^{n-1}(\beta_1 \alpha_1, v+1, t_1 T_1 J_1) j\, T' J] \\ \times [j^{n-1}(\beta_1 \alpha_1, v+1, t_1 T_1 J_1) j\, T\, J |\} j^n \beta\, \alpha\, v\, t\, T\, J]. \qquad (35.17)$$

In order to see this, we use (35.11) and bring (35.17) to the form

$$F'[n\, \beta_1 t_1 T_1 \beta' v' t' T')\, F'(n\, \beta_1 t_1 T_1 \beta\, v\, t\, T] \\ \times \sum_{\alpha_1 J_1} [j^{v+2} \alpha' v' t' t' J \{| j^{v+1}(\alpha_1, v+1, t_1 t_1 J_1) j\, t' J] \\ \times [j^{v+1}(\alpha_1, v+1, t_1 t_1 J_1) j\, t\, J |\} j^{v+2} \alpha\, v\, t\, t\, J]. \qquad (35.18)$$

We now use the orthogonality relations in the j^{v+2} configuration and the properties of (35.12) for $n = v + 2$ [in the same way that (28.21) was obtained] to obtain that the expression

$$\sum_{\alpha_1 J_1 t_1} [j^{v+2} \alpha' v' t' t' J \{| j^{v+1}(\alpha_1, v+1, t_1 t_1 J_1) j\, t' J] \\ \times [j^{v+1}(\alpha_1, v+1, t_1 t_1 J_1) j\, t\, J |\} j^{v+2} \alpha\, v\, t\, t\, J] \qquad (35.19)$$

vanishes unless $\alpha v t \equiv \alpha' v' t'$, and in this case is independent of α and J. Since t_1 can be only equal to $t + \frac{1}{2}$ or $t - \frac{1}{2}$, we can use again the operator $\Sigma\, 2(\mathbf{t}_i \cdot \mathbf{t}_k)$ to prove that the summation (35.19) has this property also if there is no summation on t_1. As a result, the expression (35.17) has the same properties as (35.12). We summarize this by

$$\sum_{\alpha_1 J_1} [j^n \beta' \alpha' v' t' T' J \{| j^{n-1}(\beta_1 \alpha_1 v_1 t_1 T_1 J_1) j\, T' J] \\ \times [j^{n-1}(\beta_1 \alpha_1 v_1 t_1 T_1 J_1) j\, T\, J |\} j^n \beta\, \alpha\, v\, t\, T\, J] \\ = \begin{cases} 0 & \text{if} \quad \alpha\, v\, t \not\equiv \alpha' v' t' \\ \text{independent of } \alpha \text{ and } J & \text{if} \quad \alpha\, v\, t \equiv \alpha' v' t'. \end{cases} \qquad (35.20)$$

The relation (35.10) can now be used to obtain from (35.20) another type of orthogonality relations analogous to (29.19), (29.20), and (29.21) in the case of identical particles. These relations can be stated as follows

$$\sum_{\alpha J} (2J+1)\, [j^n \beta\, \alpha\, v\, t\, T\, J \{| j^{n-1}(\beta_1' \alpha_1' v_1' t_1' T_1' J_1) j\, T\, J] \\ \times [j^{n-1}(\beta_1 \alpha_1 v_1 t_1 T_1 J_1) j\, T\, J |\} j^n \beta\, \alpha\, v\, t\, T\, J] \\ = \begin{cases} 0 \quad \text{if} \quad \alpha_1 v_1 t_1 \not\equiv \alpha_1' v_1' t_1' \\ (2J_1 + 1) \text{ multiplied by a term independent of } \alpha_1 \text{ and } J_1 \\ \qquad \text{if} \quad \alpha_1 v_1 t_1 \equiv \alpha_1' v_1' t_1'. \end{cases} \qquad (35.21)$$

The states with the lowest seniority in the j^n configuration with n even are those with $v = 0$ $t = 0$ and $J = 0$. The T value of such a state $\psi(j^n\, 0\, 0\, T\, 0)$ can have integral values between 0 and $n/2$. However, not all such values correspond to antisymmetric states. The allowed values of T in these states are $(n/2)$, $(n/2) - 2$, ..., 0 for $n = 4m$ and are $(n/2)$, $(n/2) - 2$, ..., 1 for $n = 2m$ with m odd. We shall prove now, by induction, that all other values of T cannot belong to antisymmetric wave functions. The actual construction of wave functions with the allowed values of T will not be discussed here.

To find out which states with $v = 0$ $t = 0$ are not allowed, we shall build these states by coupling first all the $n/2$ pairs with $T = 1$ $J = 0$ to form a state with the given T and $J = 0$. The wave function ψ of the state thus obtained should then be antisymmetrized with respect to nucleons in different pairs. Let us carry out this antisymmetrization in two steps. We first consider all permutations which exchange the *two* nucleons in one pair with *two* nucleons in another pair. The operator $\Sigma(-1)^{P'}P'$, where P' is such a permutation, is equal to $\Sigma P'$ since all such permutations are even. If we apply first this operator to the function ψ, the result will become antisymmetric if we apply to it another operator which exchanges always one nucleon from one pair with a nucleon from another pair. The application of $\Sigma P'$ to ψ causes a *symmetrization* of ψ with respect to the $n/2$ pairs. Therefore, any such function ψ that will vanish as the result of applying $\Sigma P'$ cannot give rise to an antisymmetric wave function $\psi(j^n\, 0\, 0\, T\, 0)$. These pairs of fermions play in this case the role of bosons with isospin 1. We shall see that the requirement of *symmetry* of a function of pairs is sufficient to exclude all values of T which are not equal to $(n/2)$, $(n/2) - 2$, etc.

The simplest function is obtained for $(n/2) = 2$ pairs (each with $T = 1$). From the symmetry properties of the Clebsch-Gordan coefficients, it follows that a permutation of the two isospin 1 vectors multiplies the function by $(-1)^{2-T}$. Thus the states with $T = 0$ and $T = 2$ are symmetric and the state with $T = 1$ is antisymmetric. Therefore this state with $T = 1$ cannot give rise to an antisymmetric state of j^4 with $v = 0$ $t = 0$. We shall now prove our statement by induction. Assuming that the statement that the only symmetric states with $(n/2)$ pairs have isospins T given by $(n/2)$, $(n/2) - 2$, ... holds for n (and also for even numbers smaller than n) we shall show that it holds also for $n + 2$. We shall build these symmetric pair states by using appropriate coefficients of fractional parentage. The state with $(n/2) + 1$ pairs with a value of T different from $(n/2) + 1$, $(n/2) + 1 - 2$, ... can have only *one* parent state with $n/2$ pairs with the same value of T. This value of T is allowed according to our assumption. The state $\psi\{[j^2(1\,0)]^{n/2}\,(T)j^2(1\,0)T\}$ must therefore be symmetric with respect to the exchange of pairs. This state can be written as

$$a\psi(\{[j^2(1\;0)]^{n/2-1}(T-1)\,j^2(1\;0)T\}\,j^2(1\;0)T)$$
$$+\,b\psi(\{[j^2(1\;0)]^{n/2-1}(T+1)\,j^2(1\;0)T\}j^2(1\;0)T), \qquad (35.22)$$

where a and b play the role of c.f.p. According to our assumption, these are the only possible parents of $\psi\{[j^2(1\ 0)]^{n/2}T\}$. Exchanging now the last two pairs, and restoring their order by a change of coupling, we immediately see that the function obtained is not the original one. This means that the function considered could not be symmetric. That the function $\psi\{[j^2(1\ 0)]^m\ (T)\ j^2(1\ 0)T\}$ cannot be symmetric if $\psi\{[j^2(1\ 0)]^m T\}$ is symmetric can be demonstrated also in a simpler way. If we calculate $\sum_{\alpha<\beta}^{m+1} 2(\mathbf{T}_\alpha \cdot \mathbf{T}_\beta)$, where $\mathbf{T}_\alpha$ and $\mathbf{T}_\beta$ are the isospins of the α and β pairs, by evaluating $\sum_{\alpha<\beta}^{m} 2(\mathbf{T}_\alpha \cdot \mathbf{T}_\beta)$ and multiplying it by $(m+1)/(m-1)$ we obtain

$$T(T+1) - 2(m+1) = \frac{m+1}{m-1}\,[T(T+1) - 2m].$$

The fact that $T_\alpha(T_\alpha + 1) = 2$ as well as the assumed complete symmetry in the pairs was used in obtaining this result. This equation is therefore not satisfied unless $m = T(T+1) - 1$. This demonstration is therefore not sufficient to exclude some cases (with odd m). It is mentioned here only because of its simplicity.

On the other hand, it can be easily shown by actual construction with c.f.p. that it is possible to obtain states of $n/2$ pairs with T equal to $(n/2)$, $(n/2) - 2$, etc. which are symmetric under exchange of pairs. All such states when further antisymmetrized give rise to nonvanishing antisymmetric states of the j^n configuration with $v = 0$ and $t = 0$. However, we shall not deal here with the actual construction.

There is another important conclusion to be obtained from the construction of states which are symmetric in the pairs coupled to $T = 1$, $J = 0$. Each state with $v = 0$, $t = 0$, $J = 0$, and an allowed value of T in the j^n configuration has, in general, *two* possible parents, with $T_1 = T - 1$ and $T_1 = T + 1$, in the j^{n-2} configuration. However, there is only *one* such symmetric state with a given allowed value of T. Using the appropriate "c.f.p." a and b, as in (35.22), we evaluate again $\sum_{\alpha<\beta}^{m} 2(\mathbf{T}_\alpha \cdot \mathbf{T}_\beta)$, where $m = (n/2)$, and obtain

$$T(T+1) - 2m$$
$$= \frac{m}{m-2}\,\{a^2[(T-1)T - 2(m-1)] + b^2[(T+1)(T+2) - 2(m-1)]\}.$$

Making use of the normalization of the function, given by $a^2 + b^2 = 1$, we can solve and obtain unique values for a^2 and b^2 as follows

$$a^2 = \frac{T(m+T+1)}{m(2T+1)}; \qquad b^2 = \frac{(T+1)(m-T)}{m(2T+1)}.$$

Apart from the cases in which there is only one possible parent ($T = 0$ with $T_1 = 1$ and $a^2 = 0$ and $T = m$ with $T_1 = m - 1$ and $b^2 = 0$), both a and b do not vanish. Therefore, there is only one symmetric state having the given value of T with a^2 and b^2 as given above which can be constructed from the two given parents. (There cannot be two symmetric states differing by the relative phase of a and b. If these could have existed, it would be possible to make linear combinations with $a^2 = 0$ and $b^2 = 0$, respectively, which would also be symmetric. However, in a symmetric state both a and b must not vanish as shown above.) Starting with $m = 3$ we see that there is only one symmetric state with $T = 1$ and $T = 3$ since there is only one symmetric parent with $T_1 = 0$, or $T_1 = 2$. Similarly, in the case $m = 4$, it follows that there is only one symmetric state with $T = 0$, $T = 2$, and $T = 4$. Adding more pairs with $T = 1$, $J = 0$, and antisymmetrizing, we obtain unique states with given allowed values of T. Thus, we conclude that there is only one state of the j^n configuration with $v = 0$, $t = 0$, $J = 0$ and a given allowed value of T ($T = (n/2), (n/2) - 2, \ldots$).

The same reasoning as in the case of identical particles leads to the result that the only parents of the state $\psi(j^n 0\, 0\, T 0)$ are the two states $\psi[j^{n-1}\,(v = 1\; t = \frac{1}{2}\; T_1\; J = j)\, j_n T 0]$ with $T_1 = T \pm \frac{1}{2}$. Thus,

$$\begin{aligned}\psi(j^n 0\, 0\, T\, 0) = {} & [j^{n-1}(1\, \tfrac{1}{2}, T + \tfrac{1}{2}, j)\, j\, T\, 0| \}j^n 0\, 0\, T\, 0] \\ & \times \psi[j^{n-1}(1\, \tfrac{1}{2}, T + \tfrac{1}{2}, j)\, j_n T\, 0] \\ & + [j^{n-1}(1\, \tfrac{1}{2}, T - \tfrac{1}{2}, j)\, j\, T\, 0| \}j^n\, 0\, 0\, T\, 0] \\ & \times \psi[j^{n-1}(1\, \tfrac{1}{2}, T - \tfrac{1}{2}, j)\, j_n T\, 0]. \end{aligned} \qquad (35.23)$$

In the case of maximum isospin, only the term with $T - \frac{1}{2}$ is possible and (35.23) reduces to (28.12). Also in the case $n = 4m$ and $T = 0$ there is only one term in (35.23) with $T_1 = \frac{1}{2}$. In the general case, where both c.f.p. may be different from zero, we can calculate them using a simple trick. We shall evaluate the operator $\Sigma_{i<j}^{n} 2(\mathbf{t}_i \cdot \mathbf{t}_j)$ with the wave function (35.23), by evaluating $\Sigma_{i<j}^{n-1} 2(\mathbf{t}_i \cdot \mathbf{t}_j)$ and multiplying by $n/(n-2)$. We thus obtain

$$\begin{aligned} T(T+1) - \tfrac{3}{4}n = \frac{n}{n-2} & \{[j^{n-1}(1\, \tfrac{1}{2}, T + \tfrac{1}{2}, j)\, j\, T\, 0| \}j^n 0\, 0\, T\, 0]^2 \\ & \times [(T + \tfrac{1}{2})(T + \tfrac{3}{2}) - \tfrac{3}{4}(n-1)] \\ & + [j^{n-1}(1\, \tfrac{1}{2}, T - \tfrac{1}{2}, j)\, j\, T\, 0| \}j^n 0\, 0\, T\, 0]^2 \\ & \times [(T - \tfrac{1}{2})(T + \tfrac{1}{2}) - \tfrac{3}{4}(n-1)]\}. \end{aligned} \qquad (35.24)$$

Using in addition to (35.24) the normalization of (35.23), we readily obtain the values of the squared c.f.p. as follows:

$$\begin{aligned} [j^{n-1}(1\, \tfrac{1}{2}, T + \tfrac{1}{2}, j)\, j\, T\, 0| \}j^n 0\, 0\, T\, 0]^2 &= (n - 2T)(T+1)/n(2T+1) \\ [j^{n+1}(1\, \tfrac{1}{2}, T - \tfrac{1}{2}, j)\, j\, T\, 0| \}j^n 0\, 0\, T\, 0]^2 &= (n + 2T + 2)T/n(2T+1). \end{aligned} \qquad (35.25)$$

If there were several states with $v = 1$ $t = \frac{1}{2}$ and the same isospin $T + \frac{1}{2}$ or $T - \frac{1}{2}$) in the j^{n-1} configuration the expressions on the right-hand side of (35.25) would have been equal to the sum over β of the square of the $n \to n - 1$ c.f.p. We shall now see that the uniqueness of the state with $v = 0$, $t = 0$, $J = 0$ and a given allowed value of T implies that there is only *one* state with $v = 1$, $t = \frac{1}{2}$, $J = j$ and a given T in the j^n-configuration.

We consider next the states with $v = 1$ $t = \frac{1}{2}$ in the j^n configuration with n odd. Such states, obtained by coupling pairs with $T = 1$ $J = 0$ to a single j nucleon, have $J = j$ and some isospin T. The ground states of most odd-even nuclei are apparently these states with $v = 1$ $t = \frac{1}{2}$ $J = j$ or at least contain these states with large amplitudes. Therefore, we shall discuss these states in some detail.

There are some simple limitations on the c.f.p. of the states $\psi(j^n\, 1\, \frac{1}{2}\, T j)$. The parents $\psi[j^{n-1}(v\, t\, T_1\, J_1)\, j_n\, T j]$ can have either $v = 0$ or $v = 2$. The value of T_1 can be either $T + \frac{1}{2}$ or $T - \frac{1}{2}$. However, in the case $v = 0$ $t = 0$, T_1 is limited to the allowed values $(n-1)/2$, $(n-1)/2 - 2$ ... in the j^{n-1} configuration. Therefore, in the case $v = 0$ $t = 0$, for any value of T, T_1 can have only *one* value, either $T + \frac{1}{2}$ or $T - \frac{1}{2}$. We thus distinguish between two cases, Case I and Case II. In Case I we have $T = n/2$, $(n - 4)/2$, $(n - 8)/2$, ... and therefore only $T_1 = T - \frac{1}{2}$ can have the allowed values $(n - 1)/2$, $(n - 1)/2 - 2$, etc. On the other hand, in Case II we have $T = (n - 2)/2$, $(n - 6)/2$, $(n - 10)/2$, ... so that only $T_1 = T + \frac{1}{2}$ can have the allowed values.

By definition, the $v = 1$, $t = \frac{1}{2}$ state can be obtained by antisymmetrizing $\psi[j_1, j^{n-1}(0\, 0\, T_1\, 0)\, T\, J = j]$. We saw that there is only one possible allowed value of T_1 for any given T and that there is only one state with $v = 0$, $t = 0$, $J = 0$ with this value of T_1 in the j^{n-1} configuration. Therefore, there can be only *one* state with $v = 1$, $t = \frac{1}{2}$, $J = j$ and a given T in the j^n configuration.

The parents with $v = 2$ and $t = 1$ have even values of J_1. There is no additional limitation on the values of their T_1 and it can have both values $T + \frac{1}{2}$ and $T - \frac{1}{2}$. The states with $v = 2$ and $t = 0$ of the j^{n-1} configuration were obtained from the state $\psi(j^2\, v = 2\, t = 0\, T = 0\, J_1)$ with J_1 odd, by adding $(n - 3)/2$ pairs with $T = 1$ $J = 0$ coupled to form a state with isospin T_1. Therefore, the T_1 of the state

$$\psi(j^{n-1}\, v = 2\;\; t = 0\, T_1\, J_1)$$

is the same as the T_1 of the $(n - 3)/2$ pairs and can have only the values $(n - 3)/2$, $[(n - 3)/2] - 2$, etc. Accordingly, in Case I the only possible values of T_1 are those given by $T + \frac{1}{2}$, whereas in Case II they are given only by $T - \frac{1}{2}$. From the uniqueness of the state with $v = 0$, $t = 0$, $J = 0$ and an allowed value of T_1 in the j^{n-3} configuration, it follows that there is only one state with $v = 2$, $t = 0$, J_1 odd and the given possible value of T_1 in the j^{n-1} configuration.

Hence, in each case there are only four sets of values of v, t, and T_1 which belong to possible parents of the state $\psi(j^n\, 1\, \frac{1}{2}\, T\, j)$. In Case I the possible parents have $v = 0 \;\; t = 0 \;\; T_1 = T - \frac{1}{2} \;\; J_1 = 0$, $v = 2$ $t = 0 \;\; T_1 = T + \frac{1}{2} \;\; J_1$ odd, and $v = 2 \;\; t = 1 \;\; T_1 = T \pm \frac{1}{2} \;\; J_1 > 0$ even. In Case II the possible parents have the following quantum numbers: $v = 0 \; t = 0 \; T_1 = T + \frac{1}{2} \; J_1 = 0$, $v = 2 \; t = 0 \; T_1 = T - \frac{1}{2}$ J_1 odd, and $v = 2 \;\; t = 1 \;\; T_1 = T \pm \frac{1}{2} \;\; J_2 > 0$ even. In the following we shall calculate the c.f.p. which correspond to these parents.

The state obtained by antisymmetrizing $\psi[j^{n-1}(0\, 0\, T_1\, 0)\, j_n\, T\, j]$ has obviously seniority $v = 1$ and $t = \frac{1}{2}$ and is therefore, apart from a possible phase, the state $\psi(j^n\, 1\, \frac{1}{2}\, T\, j)$ considered. Since this latter state can be obtained from a principal parent, we can make use of the recursion relation (33.16). This relation in the present case is

$$n[j^{n-1}(0\,0\,T_1\,0)\,j\,T\,j|\}j^n\,1\,\tfrac{1}{2}\,T\,j]\,[j^{n-1}(v\,t\,T_1'J_1)\,j\,T\,j|\}j^n\,1\,\tfrac{1}{2}\,T\,j]$$

$$= \delta_{v0}\delta_{t0}\delta_{T_1T_1'}\delta_{J_10} + (-1)^{T_1+T_1'}(n-1)\sqrt{(2T_1+1)(2T_1'+1)}\,\frac{\sqrt{2J_1+1}}{2j+1}$$

$$\times \sum_{T_2} \begin{Bmatrix} T_2 & \frac{1}{2} & T_1 \\ T & \frac{1}{2} & T_1' \end{Bmatrix} [j^{n-2}(1\,\tfrac{1}{2}\,T_2\,j)\,j\,T_1 0|\}j^{n-1} 0\,0\,T_1 0]$$

$$\times\, [j^{n-2}(1\,\tfrac{1}{2}\,T_2\,j)\,j\,T_1'J_1|\}j^{n-1} v\,t\,T_1'J_1]. \qquad (35.26)$$

Using (35.26) for the case $v = 0 \;\; t = 0 \;\; T_1 = T_1'$ and $J_1 = 0$, we obtain

$$n[j^{n-1}(0\,0\,T_1 0)\,j\,T\,j|\}j^n 1\,\tfrac{1}{2}\,T\,j]^2$$

$$= 1 + (n-1)\frac{(2T_1+1)}{2j+1}\sum_{T_2}\begin{Bmatrix} T_2 & \frac{1}{2} & T_1 \\ T & \frac{1}{2} & T_1 \end{Bmatrix}[j^{n-2}(1\,\tfrac{1}{2}\,T_2\,j)\,j\,T_1 0|\}j^{n-1} 0\,0\,T_1 0]^2. \qquad (35.27)$$

Inserting the values of $W(T_2\,\frac{1}{2}\,T_1|T\,\frac{1}{2}\,T_1)$ and the values of the squares of the $n-1 \to n-2$ c.f.p. from (35.25), we obtain the values of the c.f.p. in the following cases

$$\text{Case I} \quad [j^{n-1}(0\,0,\,T-\tfrac{1}{2},\,0)\,j\,T\,j|\}j^n\,1\,\tfrac{1}{2}\,T\,j]^2 = \frac{4j+4-n-2T}{2n(2j+1)}$$

$$\text{Case II} \quad [j^{n-1}(0\,0,\,T+\tfrac{1}{2},\,0)\,j\,T\,j|\}j^n\,1\,\tfrac{1}{2}\,T\,j]^2 = \frac{4j+6-n+2T}{2n(2j+1)}. \qquad (35.28)$$

In order to calculate the other c.f.p. of the state $\psi(j^n\, 1\, \frac{1}{2}\, T\, j)$ we make use of certain relations that they satisfy. As in calculating the c.f.p. in (35.25) we shall make use of the normalization of the wave function and the expression of $\sum_{i<k}^n (\mathbf{t}_i \cdot \mathbf{t}_k)$ by means of the $n \to n-1$ c.f.p.

In addition, we shall make use of the property (34.19) which in this case is simply

$$\left(j^n\, 1\, \tfrac{1}{2}\, Tj \left\| \sum_i^n \mathbf{f}_i^{(k)} \right\| j^n\, 1\, \tfrac{1}{2} T j\right) = (j \| \mathbf{f}^{(k)} \| j) \sqrt{\frac{2T+1}{2}} \qquad \text{for} \qquad k \text{ odd.} \tag{35.29}$$

The factor $\sqrt{(2T+1)/2}$ appears in this equation because of the formal reason that the matrix element is reduced both with respect to J and to T.

Using (33.18) we can express this equation in terms of c.f.p. by

$$n(2j+1) \sum_{vtT_1J_1} [j^{n-1}(v\, t\, T_1 J_1)\, j\, Tj|\}j^n\, 1\, \tfrac{1}{2}\, Tj]^2 (-1)^{J_1} \begin{Bmatrix} j & j & J_1 \\ j & j & k \end{Bmatrix} = 1 \qquad \text{for} \qquad k \text{ odd.} \tag{35.30}$$

These equations, for odd values of k, along with the normalization

$$\sum_{vtT_1J_1} [j^{n-1}(v\, t\, T_1 J_1)\, j\, Tj|\}j^n\, 1\, \tfrac{1}{2}\, Tj]^2 = 1 \tag{35.31}$$

and the one derived from

$$\sum_{i<k}^n 2(\mathbf{t}_i \cdot \mathbf{t}_k) = T(T+1) - \tfrac{3}{4}n,$$

namely,

$$\frac{n}{n-2} \sum_{vtT_1J_1} [j^{n-1}(v\, t\, T_1 J_1)\, j\, Tj|\}j^n\, 1\, \tfrac{1}{2}\, Tj]^2 [T_1(T_1+1) - \tfrac{3}{4}(n-1)] = T(T+1) - \tfrac{3}{4}n \tag{35.32}$$

are sufficient to obtain the squares of the c.f.p. involved.

Before trying to solve Eqs. (35.30), (35.31), and (35.32) we can greatly simplify the procedure by factorizing the J_1 dependence of the c.f.p. involved. This factorization is important only for $v = 2$, since there is only one value of J_1, namely $J_1 = 0$, in the c.f.p. with $v = 0$ which was already calculated in (35.28). Since $v = 2$ is bigger by one unit than the seniority of the state considered ($v = 1$), the factorization of the c.f.p. is given by (35.10). We put $v = 1$ $t = \frac{1}{2}$ $J = j$ in (35.10) and obtain

$$[j^{n-1}(2\, t\, T_1 J_1)\, j\, Tj|\}j^n\, 1\, \tfrac{1}{2}\, Tj] = G(n\, t\, T_1, 1, \tfrac{1}{2}, T)(-1)^{2j-J_1} \sqrt{\frac{2J_1+1}{2j+1}} [j\, j\, t\, J_1|\}j^2 t\, J_1]. \tag{35.33}$$

The c.f.p. on the right-hand side of (35.33) is equal to 1. We change slightly the definition of G and absorb into it the phase $(-1)^{2j-J_1} = (-1)^t$ and the factor $1/\sqrt{2j+1}$. Dropping the n, 1, $\frac{1}{2}$, and T out of G, since they are the same for the one state we consider, we obtain

$$[j^{n-1}(2\,t\,T_1J_1)\,j\,T\,j|\}j^n\,1\,\tfrac{1}{2}\,T\,j] = G(t, T_1)\sqrt{2J_1+1}\,. \qquad (35.34)$$

Similarly, in accordance with (35.5) we define for the sake of conciseness

$$[j^{n-1}(0\,0\,T_1 0)\,j\,T\,j|\}j^n\,1\,\tfrac{1}{2}\,T\,j] = F(0, T_1). \qquad (35.35)$$

Introducing (35.34) into the equations (35.30) we see that in view of (15.16) and (15.17) they all reduce to the same equation

$$n(2j+1)\left\{\frac{F(0, T_1)^2}{2j+1} - \tfrac{1}{2}\,G(0, T_1')^2 + G(1, T_1)^2\,\frac{2j-1}{2(2j+1)}\right.$$
$$\left. + G(1, T_1')^2\,\frac{2j-1}{2(2j+1)}\right\} = 1. \qquad (35.36)$$

In (35.36), $T_1 = T - \frac{1}{2}$ and $T_1' = T + \frac{1}{2}$ in Case I whereas $T_1 = T + \frac{1}{2}$, $T_1' = T - \frac{1}{2}$ in Case II. Similarly, the normalization condition (35.31) becomes

$$F(0, T_1)^2 + (j+1)(2j+1)\,G(0, T_1')^2 + (j+1)(2j-1)\,G(1, T_1)^2$$
$$+ (j+1)(2j-1)\,G(1, T_1')^2 = 1. \qquad (35.37)$$

The last equation (35.32) becomes in the same way, using also (35.31),

$$\frac{n}{n-2}\{F(0, T_1)^2\,T_1(T_1+1) + (j+1)(2j+1)\,G(0, T_1')^2\,T_1'(T_1'+1)$$
$$+ (j+1)(2j-1)\,G(1, T_1)^2\,T_1(T_1+1)$$
$$+ (j+1)(2j-1)\,G(1, T_1')^2\,T_1'(T_1'+1)$$
$$- \tfrac{3}{4}(n-1)\} = T(T+1) - \tfrac{3}{4}n. \qquad (35.38)$$

Equations (35.36), (35.37), and (35.38) can be solved after introducing the value (35.28) for $F(0, T_1)$. We thus obtain for the squares of the c.f.p. the following results:

Case I $\qquad T = \dfrac{n}{2}, \quad \dfrac{n-4}{2}, \quad \ldots$

$$[j^{n-1}(0\,0,\,T-\tfrac{1}{2},\,0)\,j\,T\,j|\}j^n\,1\,\tfrac{1}{2}\,T\,j]^2$$
$$= (4j+4-n-2T)/2n(2j+1)$$

$$[j^{n-1}(2\,0, T+\tfrac{1}{2}, J_1)j\,Tj|\}j^n\,1\,\tfrac{1}{2}\,Tj]^2$$
$$= (2J_1+1)(n-2T)/4n(j+1)(2j+1) \qquad J_1 \text{ odd}$$

$$[j^{n-1}(2\,1, T-\tfrac{1}{2}, J_1)j\,Tj|\}j^n\,1\,\tfrac{1}{2}\,Tj]^2$$
$$= \frac{(2J_1+1)\{4(j+1)[T(n+2T)-1]+n-2T\}}{2n(j+1)(2j-1)(2j+1)(2T+1)} \qquad J_1 > 0 \text{ even}$$

$$[j^{n-1}(2\,1, T+\tfrac{1}{2}, J_1)j\,Tj|\}j^n\,1\,\tfrac{1}{2}\,Tj]^2$$
$$= (2J_1+1)(n-2T)(2T+3)/4n(j+1)(2j-1)(2T+1) \qquad J_1 > 0 \text{ even}) \tag{35.39}$$

Case II $\qquad T = \dfrac{n-2}{2}, \quad \dfrac{n-6}{2}, \quad \ldots$

$$[j^{n-1}(0\,0, T+\tfrac{1}{2}, 0)j\,Tj|\}j^n\,1\,\tfrac{1}{2}\,Tj]^2$$
$$= (4j+6-n+2T)/2n(2j+1)$$

$$[j^{n-1}(2\,0, T-\tfrac{1}{2}, J_1)j\,Tj|\}j^n\,1\,\tfrac{1}{2}\,Tj]^2$$
$$= (2J_1+1)(n+2T+2)/4n(j+1)(2j+1) \qquad J_1 \text{ odd}$$

$$[j^{n-1}(2\,1, T+\tfrac{1}{2}, J_1)j\,Tj|\}j^n\,1\,\tfrac{1}{2}\,Tj]^2$$
$$= \frac{(2J_1+1)\{4(j+1)[(T+1)(n-2T-2)+1]-(n+2T+2)\}}{2n(j+1)(2j-1)(2j+1)(2T+1)}$$
$$J_1 > 0 \text{ even}$$

$$[j^{n-1}(2\,1, T-\tfrac{1}{2}, T_1)j\,Tj|\}j^n\,1\,\tfrac{1}{2}\,Tj]^2$$
$$= (2J_1+1)(n+2T+2)(2T-1)/4n(j+1)(2j-1)(2T+1)$$
$$J_1 > 0 \text{ even.} \tag{35.40}$$

We can compare these results with previous ones in the case $n = 3$. In that case, the state with $v = 1$ $t = \frac{1}{2}$, and $T = \frac{3}{2}$, is Case I and the state with $v = 1$ $t = \frac{1}{2}$ and $T = \frac{1}{2}$ is Case II. We see that for $n = 3$ $T = \frac{3}{2}$ Eq. (35.39) reduces to (26.12), whereas (35.40) reduces for $n = 3$ $T = \frac{1}{2}$ to (33.10). In general, the case of identical particles in which $T = n/2$ is Case I. For this case, only the c.f.p. with $T - \frac{1}{2}$ do not vanish and (35.39) reduces to (28.17). In the present case we obtained only the absolute magnitude of the c.f.p., but not their phases. Nevertheless, in many applications only the squares of the c.f.p. enter, and we shall be able to make use of (35.39) and (35.40). If there are several states with the same $v = 2$, $t = 1$, and T_1 in the j^{n-1} configuration, it is clear from the way in which (35.39) and (35.40) were obtained that they give the sum of squares of such c.f.p. Even if this is the case, all expressions in the applications that we shall make of these results will remain the same. We shall, therefore, not refer explicitly to the additional quantum numbers β.

We shall now apply the c.f.p. calculated in (35.39) and (35.40) to the calculation of the magnetic moments in the states with $v = 1$ $t = \frac{1}{2}$ $J = j$. We use for this purpose the expression (33.30) obtained above, in which we put the values

$$W(\tfrac{1}{2}\,T, T - \tfrac{1}{2}|T\,\tfrac{1}{2}\,1) = \sqrt{(T + 1)/6T(2T + 1)}$$

and

$$W(\tfrac{1}{2}\,T, T + \tfrac{1}{2}|T\,\tfrac{1}{2}\,1) = \sqrt{T/6(T + 1)(2T + 1)}.$$

Substituting for the c.f.p. in (33.30) the values (35.39) or (35.40), we obtain in the two cases the following results:

$$g = \frac{g_P + g_N}{2} + M_T \frac{g_P - g_N}{2} \frac{4(j + 1)(T + 1) - n + 2T}{4(j + 1)\,T(T + 1)}$$

$$\text{Case I} \qquad T = \frac{n}{2}, \quad \frac{n - 4}{2}, \quad \ldots \tag{35.41}$$

$$g = \frac{g_P + g_N}{2} - M_T \frac{g_P - g_N}{2} \frac{4(j + 1)\,T + n + 2T + 2}{4(j + 1)\,T(T + 1)}$$

$$\text{Case II} \qquad T = \frac{n - 2}{2}, \quad \frac{n - 6}{2}, \quad \ldots. \tag{35.42}$$

In the special case $T = n/2$ of identical particles we obtain from (35.41) $g = g_P$ if $M_T = T$ and $g = g_N$ if $M_T = -T$. In the case $n = 3$, $T = \frac{1}{2}$ (35.42) reduces to our previous result (33.31).

We can make use of the results (35.41) and (35.42) for the calculation of magnetic moments of the ground states of odd even nuclei. In this case, $n = Z + N$ and $M_T = (Z - N)/2$ where Z and N are the numbers of protons and neutrons in the j shell. The isospin of the ground state has, as we shall discuss later, the minimum value $T = |Z - N|/2$. Let us take, for the sake of definiteness, Z to be odd and N to be even. The following expressions will be valid for the other possibility (i.e., Z even and N odd) if we interchange Z and N as well as g_P and g_N. We now distinguish between two cases.

(a) $Z > N$, so that $T = (Z - N)/2 = (n - 2N)/2$. Since N is even this corresponds to our Case I and the magnetic moment is obtained from (35.41), with M_T put equal to T, to be

$$g = g_P\left[1 - \frac{N}{4(j + 1)(T + 1)}\right] + g_N \frac{N}{4(j + 1)(T + 1)}. \tag{35.43}$$

(b) $Z < N$ so that $T = (N - Z)/2 = (n - 2Z)/2$. Since Z is odd this is Case II and the magnetic moment is given by (35.42), with $M_T = - T$, to be

$$g = g_P \left[1 - \frac{2j + 1 - N}{4(j + 1)(T + 1)}\right] + g_N \frac{2j + 1 - N}{4(j + 1)(T + 1)}. \qquad (35.44)$$

Since in the state considered, with $v = 1$ $t = \frac{1}{2}$, J is equal to j, the expressions (35.41) through (35.44) give also directly the magnetic moments μ if we replace g_P and g_N by the single j-nucleon magnetic moments μ_P and μ_N. We note in passing that (35.44) is obtained from (35.43) by taking, instead of N, the number $2j + 1 - N$ of neutron holes. Apparently we can couple the $2j + 1 - Z$ proton holes and $2j + 1 - N$ neutron holes, which have the same g factors as protons and neutrons respectively, to a state with the same values $v = 1$, $t = \frac{1}{2}$, and T. Equations (35.43) and (35.44) do not contain n and will thus be still valid. Obviously, from $Z < N$ follows $2j + 1 - Z > 2j + 1 - N$, so that (35.43) and (35.44) are interchanged when going over from nucleons to holes. If $2j + 1$ is a multiple of 4 (i.e., $j = \frac{3}{2}, \frac{7}{2}, \ldots$), the two expressions (35.43) and (35.44) become equal for $N = (2j + 1)/2$.

In both extreme cases, $N = 0$ and $N = 2j + 1$, we obtain $g = g_P$. The neutrons in these cases must be in the state with $J = 0$ and do not contribute to the magnetic moments. From (35.43) follows that for a fixed Z, the bigger the N (consistent with $N < Z$) the larger the deviation of g from g_P. From (35.44) we see that for a fixed Z, the smaller the N (consistent with $N > Z$) the larger the deviation of g from g_P. The maximum deviations are obtained for $N = Z - 1$ and $N = Z + 1$, i.e., in the case of mirror nuclei. For mirror nuclei, with $T = \frac{1}{2}$, in the case of *odd Z*, the magnetic moments are given by

$$\mu = \mu_P \left[1 - \frac{N}{6(j + 1)}\right] + \mu_N \frac{N}{6(j + 1)} \qquad Z = N + 1 \qquad (35.45)$$

$$\mu = \mu_P \left[1 - \frac{2j + 1 - N}{6(j + 1)}\right] + \mu_N \frac{2j + 1 - N}{6(j + 1)} \qquad Z = N - 1. \qquad (35.46)$$

The magnetic moments given by (35.45) and (35.46) are always between μ_P and μ_N and closer to μ_P (for *odd Z*). The magnetic moments given by (35.45) and (35.46) have the form $\mu_P + (\mu_N - \mu_P)X$ where X can be at most $(2j + 1)/12(j + 1)$. For $j = \frac{1}{2}$, $X = \frac{1}{9}$, and as j increases X goes down to its limiting values $\frac{1}{6}$. These features are in qualitative agreement with the experimentally measured magnetic moments of mirror nuclei.

36. Average Energies of States with Given T, v, and t

In the case of identical particles the average interaction energy in a group of states with the same seniority has the simple property (29.5). In this section we shall derive similar results for the case of protons and neutrons in the same j-orbit.

We first note that in the present case

The average interaction energy of all states of the j^n configuration (with all possible values of T and M_T) is a multiple of the average energy of all states of the j^2 configuration. (36.1)

The coefficient of the average energy of the j^2 configuration in (36.1) depends on n but not on the particular interaction considered. This is analogous to (29.14) in the case of identical particles. Also the proof of (36.1) is very similar to the one used in the case of identical particles. We again notice that the sum of all energies is the *trace* of the submatrix of the interaction energy defined by the j^n configuration. When evaluated in the m scheme, defined by both m_j and m_t of the individual nucleons, this trace is found to be equal to a multiple of the trace of V_{12} in the j^2 configuration. The coefficient of the average energy in the j^2 configuration is simply $n(n-1)/2$.

The property (36.1) holds for any two-body interaction. If V_{12} is charge-independent we can obtain more specific results for the average energies of states with the same value of T in the j^n configuration. We find that

The average charge-independent interaction energy of all states with a definite T (and M_T) in the j^n configuration is a linear combination of the average energies of the $T=1$ states and the $T=0$ states of the j^2 configuration. (36.2)

The coefficients of the linear combination in (36.2) depend on n and T but not on the charge-independent interaction considered. Clearly, the value of such an interaction is independent of M_T.

In order to prove (36.2) we first observe that, in the case $T=(n/2)$, $M_T=(n/2)$, in which there are only identical particles, the statement (36.2) reduces to (29.14). For this value of maximum T (and all values of M_T), the average interaction is simply a multiple of the average energy of the $T=1$ states in the j^2 configuration. We evaluate next the trace of the energy matrix defined by all states of the j^n configuration with $M_T=(n/2)-1$. We can characterize these states in the m scheme by fixing the m_t value of nucleons 1 to $n-1$ to be $m_t = +\frac{1}{2}$ whereas for the nth nucleon we choose $m_t = -\frac{1}{2}$. It follows that the trace considered will be a linear combination of the trace of the two-body interaction in the j^2 configuration with $m_{t_1} = m_{t_2} = +\frac{1}{2}$ (two protons) and the trace in the j^2 configuration with $m_{t_1} = +\frac{1}{2}$ $m_{t_2} = -\frac{1}{2}$ (a proton and a neutron). The first trace is the one for $T = 1$, and the second is a linear combination of the traces in the $T = 0$ and $T = 1$ states. The trace of the submatrix defined in the j^n configuration by $T = (n/2) - 1$ is obtained by subtracting the trace for $T = (n/2)$ $M_T = (n/2) - 1$, which is equal to the trace for $T = (n/2)$ $M_T = (n/2)$ discussed above. Thus, proceeding by induction the statement (36.2) is proved.

The coefficients of the linear combination described in (36.2) can be easily calculated. We denote by $\bar{V}_{T=0}$ and $\bar{V}_{T=1}$ the average interactions in the j^2 configuration in the states with $T = 0$ and $T = 1$ respectively. Since the coefficients are independent of the interaction chosen, we consider an interaction which has the same value in *all* states of the j^2 configuration. In this case $\bar{V}_{T=0} = \bar{V}_{T=1} = V$ and the average interaction in the j^n configuration is simply $[n(n-1)/2]\,V$. Thus, the sum of the coefficients of $\bar{V}_{T=0}$ and $\bar{V}_{T=1}$ in the linear combination of (36.2) is $n(n-1)/2$. Next we consider a two-body interaction which vanishes for states with $T = 0$ and is the same for all states with $T = 1$. Such an interaction is given by

$$[2(\mathbf{t}_1 \cdot \mathbf{t}_2) + \tfrac{3}{2}]\frac{V}{2} = \mathbf{T}^2 \frac{V}{2}$$

in which case $\bar{V}_{T=1} = V$. The coefficient of $\bar{V}_{T=1}$ must therefore be

$$\sum_{i<k}^{n} (\mathbf{t}_i \cdot \mathbf{t}_k) + \sum_{i<k}^{n} \tfrac{3}{4} = \tfrac{1}{2}[T(T+1) - \tfrac{3}{4}n] + \tfrac{3}{4}\frac{n(n-1)}{2}$$

As a result, we can write down the average interaction in all states with the same value of T in the j^n configuration as

$$\left[\frac{n(n+2)}{8} - \tfrac{1}{2}T(T+1)\right]\bar{V}_{T=0} + \left[\frac{3n(n-2)}{8} + \tfrac{1}{2}T(T+1)\right]\bar{V}_{T=1}$$

$$= \frac{\bar{V}_{T=0} + 3\bar{V}_{T=1}}{4}\,\frac{n(n-1)}{2} + \frac{\bar{V}_{T=1} - \bar{V}_{T=0}}{2}[T(T+1) - \tfrac{3}{4}n] \qquad (36.3)$$

Obviously, in the case of identical particles, for which $T = (n/2)$, the coefficient of $\bar{V}_{T=0}$ in (36.3) vanishes and the coefficient of $\bar{V}_{T=1}$ becomes $n(n-1)/2$.

We shall now go over to averages of groups of states with the same quantum numbers v and t of the seniority scheme. In the j^n configuration we consider all states with the same values of T, v, and t and the same additional quantum numbers β. As described above, the numbers β are necessary if several *sets* of states with the same values of T, v, and t occur which were all obtained from the same set of states, with t and v, in the j^v configuration. The interaction energies of these states are given by

$$\langle j^n\beta\,\alpha\,v\,t\,T\,J \,|\, \sum_{i<k}^{n} V_{ik} \,|\, j^n\beta\,\alpha\,v\,t\,T\,J\rangle$$

$$= \frac{n}{n-2} \sum_{\substack{\beta_1\alpha_1v_1t_1T_1J_1 \\ \beta_1'\alpha_1'v_1't_1'}} [j^{n-1}(\beta_1\alpha_1v_1t_1T_1J_1)\,j\,T\,J|\}j^n\beta\,\alpha\,v\,t\,T\,J]$$

$$\times [j^{n-1}(\beta_1'\alpha_1'v_1't_1'T_1J_1)\,j\,T\,J|\}j^n\beta\,\alpha\,v\,t\,T\,J]$$

$$\times \langle j^{n-1}\beta_1\alpha_1v_{11}tT_1J_1 \,|\, \sum_{i<k}^{n-1} V_{ik} \,|\, j^{n-1}\beta_1'\alpha_1'v_1't_1'T_1J_1\rangle \tag{36.4}$$

The expression (36.4) will be greatly simplified if we multiply it by $(2J+1)$ and sum over all the states with the various J and α and the same β, v, t, and T. According to (35.21) the only terms which will not vanish must have $\alpha_1 = \alpha_1'$ $v_1 = v_1'$ and $t_1 = t_1'$. Furthermore, the right-hand side of (36.4) will be a linear combination of the sums

$$\sum_{\alpha_1J_1} (2J_1+1)\langle j^{n-1}\beta_1\alpha_1v_1t_1T_1J_1 \,|\, \sum_{i<k}^{n-1} V_{ik} \,|\, j^{n-1}\beta_1'\alpha_1v_1t_1T_1J_1\rangle \tag{36.5}$$

The coefficients of the sums (36.5) will depend according to (35.21) only on n, β_1, β_1', v_1, t_1, and T_1 but not on α_1 and J_1. We can therefore prove by induction the following property

The average (charge-independent) interaction energy of all states with the same quantum numbers β,v, t, and T in the j^n configuration is a linear combination of the average interaction energies of states with the same values of v and t in the j^2 configuration. (36.6)

The coefficients of the linear combination in (36.6) are obviously independent of the particular interaction V_{ik}.

In fact, we see from (36.5) that the property (36.6) holds also for the sum of matrix elements of the interaction energy which are not diagonal in β. Therefore, if it holds for the j^{n-1} configuration it holds also in the j^n configuration. The property (36.6) holds trivially for $n = 2$, and this completes the proof by induction.

Once we know that (36.6) holds, it is possible to calculate explicitly the coefficients of the linear combinations described in (36.6). As in the case of identical particles, we construct a simple interaction V'_{ik} and calculate the coefficients in this special case. Since the coefficients are independent of the interaction, this procedure must give the desired result. Let us consider the two-body interaction

$$V'_{ik} = a + 2b(\mathbf{t}_i \cdot \mathbf{t}_k) + cq_{ik} \tag{36.7}$$

In the j^2 configuration there are three groups of states with the same values of v and t. These are the ground states with $v = 0$ $t = 0$, the states with even values of $J > 0$ with $v = 2$ $t = 1$, and the states with odd values of J with $v = 2$ $t = 0$. The average interaction energies of these groups for an arbitrary two-body interaction V_{ik} will be denoted by V_0, $\bar{V}_2$, and $\bar{V}_1$ respectively. The coefficients a, b, and c in (36.7) can be chosen so that V'_{ik} will have the same average values as the original interaction V_{ik}. In fact, given definite values of V_0, $\bar{V}_2$, and $\bar{V}_1$ we can always solve the three equations

$$\begin{aligned} V_0 &= a + \tfrac{1}{2}b + (2j+1)c \\ \bar{V}_2 &= a + \tfrac{1}{2}b \\ \bar{V}_1 &= a - \tfrac{3}{2}b \end{aligned}$$

and obtain a, b, and c in terms of V_0, $\bar{V}_2$, and $\bar{V}_1$. These parameters turn out to be

$$a = \frac{3\bar{V}_2 + \bar{V}_1}{4}; \qquad b = \frac{\bar{V}_2 - \bar{V}_1}{2}; \qquad c = \frac{V_0 - \bar{V}_2}{2j+1}. \tag{36.8}$$

With the coefficients in (36.7) thus chosen, the interaction V'_{ik} can replace the original interaction when calculating averages of groups of states with the same seniority. Therefore, in the j^n configuration these averages of V_{ik} (as well as of V'_{ik}) are given simply by

$$\frac{n(n-1)}{2}a + b[T(T+1) - \tfrac{3}{4}n] + cQ(n, v, t, T). \tag{36.9}$$

Introducing the expression (34.26) for the operator Q into (36.9) it becomes, after a slight rearrangement,

$$\frac{n(n-1)}{2}a + (b-c)\,[T(T+1) - \tfrac{3}{4}n] \\ + c\left[\frac{n-v}{4}(4j+5-n-v) + t(t+1) - \tfrac{3}{4}v\right]. \tag{36.10}$$

Substituting for a, b, and c their values (36.8) we finally obtain

$$\frac{\sum_{\alpha J}(2J+1)\langle j^n\beta\,\alpha\,v\,t\,T\,J|\sum_{i<k}^{n}V_{ik}|j^n\beta\,\alpha\,v\,t\,T\,J\rangle}{\sum_{\alpha J}(2J+1)}$$

$$=\frac{n(n-1)}{8}(3\bar{V}_2+\bar{V}_1)$$

$$+\frac{(2j+3)\bar{V}_2-2V_0-(2j+1)\bar{V}_1}{2(2j+1)}[T(T+1)-\tfrac{3}{4}n]$$

$$+\frac{V_0-\bar{V}_2}{2j+1}\left[\frac{n-v}{4}(4j+5-n-v)+t(t+1)-\tfrac{3}{4}v\right]. \quad (36.11)$$

Putting $T=(n/2)$, $t=(v/2)$, (36.11) reduces to (29.6) which is the analogous expression for identical particles.

The interaction (36.7) is diagonal in the seniority scheme. Furthermore, it has the same eigenvalues for all states with the same n, v, t, and T and is independent of β. The average interaction energy, given by (36.11), shares some of these properties. The averages (36.11) depend on n, v, t, and T but not on the additional quantum numbers β. Moreover, it follows that sum over α and J of the matrix elements of any interaction V_{ik} between two sets of states with the same values of α, v, t, T, and J [like the expression (36.5)] vanishes if the two sets are characterized by different values of β. In such a sum we can replace V_{ik} by the special interaction V'_{ik}, and the vanishing follows directly from the orthogonality of the wave functions with different quantum numbers β.

The expression (36.11) gives only the position of the center of mass of the group of levels with the same values of v, t, and T in the j^n configuration. However, for every value of n, there are certain values of v and t for which there is only one state with these quantum numbers. In these cases, (36.11) gives directly the interaction energy of the state considered. These special states are those with the lowest possible seniority. For even n, the state with $J=0$, $v=0$, $t=0$, and any allowed value of T is the only one in its group. For odd values of n, the state with $J=j$, $v=1$, $t=\frac{1}{2}$, and any possible T has the same property. In these cases we obtain from (36.11), after slight rearrangements, the following expressions

$$\langle j^n\ v=0\ t=0\ T\ J=0|\sum V_{ik}|j^n\ v=0\ T=0\ T\ J=0\rangle$$

$$=\frac{n(n-1)}{2}\frac{(6j+5)\bar{V}_2+(2j+1)\bar{V}_1-2V_0}{4(2j+1)}$$

$$+\frac{(2j+3)\bar{V}_2-2V_0-(2j+1)\bar{V}_1}{2(2j+1)}[T(T+1)-\tfrac{3}{4}n]+\frac{n}{2}\frac{2(j+1)}{2j+1}(V_0-\bar{V}_2)$$

for $\quad n$ even

$$\langle j^n\ v=1\ t=\tfrac{1}{2}\ T\ J=j \Big| \sum V_{ik} \Big| j^n\ v=1\ t=\tfrac{1}{2}\ T\ J=j\rangle$$
$$= \frac{n(n-1)}{2}\frac{(6j+5)\bar{V}_2+(2j+1)\bar{V}_1-2V_0}{4(2j+1)}$$
$$+\frac{(2j+3)\bar{V}_2-2V_0-(2j+1)\bar{V}_1}{2(2j+1)}[T(T+1)-\tfrac{3}{4}n]$$
$$+\frac{n-1}{2}\frac{2(j+1)}{2j+1}(V_0-\bar{V}_2) \qquad \text{for} \qquad n \text{ odd.}$$

These two expressions can be combined in one formula which gives the interaction energy in the state with lowest seniority in the j^n configuration

$$\frac{n(n-1)}{2}\frac{(6j+5)\bar{V}_2+(2j+1)\bar{V}_1-2V_0}{4(2j+1)}$$
$$+\frac{(2j+3)\bar{V}_2-2V_0-(2j+1)\bar{V}_1}{2(2j+1)}[T(T+1)-\tfrac{3}{4}n]$$
$$+\left[\frac{n}{2}\right]\frac{2(j+1)}{2j+1}(V_0-\bar{V}_2). \tag{36.12}$$

In (36.12) the function $[n/2]$ is equal to the largest integer not exceeding $n/2$. This formula reduces in the case of identical nucleons, i.e., for $T=(n/2)$, to (30.1). It exhibits a pairing term proportional to $[n/2]$ as well as a "symmetry term" proportional to $T(T+1)$. In the case of odd tensor interactions, according to (21.43) and (21.46), $\bar{V}_2 = V_0/2(j+1)$ and $\bar{V}_1 = -V_0/2(j+1)$. As a result, the first two terms in (36.12) vanish in agreement with the pairing property exhibited by such interactions.

The ground states of even-even nuclei have all spin $J=0$. Most of the ground states of odd-even nulcei have $J=j$ where j is the spin of the single nucleon orbit being filled. It is therefore tempting to identify these states as those of lowest seniority. If we consider the j^2 configuration of identical particles with attractive short-range forces, the $J=0$ state is considerably lower than the other states. This is also the case for the effective nuclear interaction observed in nuclei (cf. the discussion in Section 30 on the $f^n_{7/2}$ configurations). Therefore, it is plausible that the ground state of such configurations will have as many pairs coupled to $J=0$ as possible. Such a state is, as we saw in Section 27, that of lowest seniority. With both protons and neutrons in the same j-orbit, the situation is more complicated. The state with $J=0$ is the lowest only among the states with $T=1$ J even, in the j^2 configuration. On the other hand, states with $T=0$ J odd, may lie rather low and in some cases lower than the $J=0$ state. Therefore, it is not at all clear that the interaction energy in the j^n configuration of protons and neutrons will be diagonal in the seniority scheme.

Nevertheless, the expression (36.12) for the interaction energy can be used as an approximation in the case of ground states of nuclei. In the cases where there is only one state with a given T and $J = 0$ or $J = j$ in the j^n configurations, i.e., for $j \leq \frac{3}{2}$ the seniority is a good quantum number and (36.12) holds exactly for any two-body interaction. In addition to the nuclear interaction energy of the ground states, given by (36.12), the binding energy of a nucleus includes also the energy due to the closed shells, the kinetic energy of the j-nucleons, their interaction with the closed shells and the Coulomb energy of the j-protons. We shall make the same assumptions as made in Section 30 to obtain (30.4) and (30.5). We take the central field, in which the nucleons move, to be the same for all nuclei in which the same j-orbit is being filled. As a result, the energy due to closed shells is the same in all nuclei considered and is equal to the binding energy of the nucleus with closed shells in which there are no j-nucleons. We subtract this binding energy from all nuclear energies to be considered. Furthermore, the kinetic energy and the interaction with the closed shells of the j-nucleons is a sum of equal single nucleon terms. The interaction energy of the j-nucleons is given by (36.12) with constant values for the combinations of matrix elements V_0, $\bar{V}_1$, and $\bar{V}_2$.

The binding energy (B.E.) of a nucleus with the j^n configuration outside closed shells and an isospin $T = |N - Z|/2$ is given by

$$\text{B.E.}\,(n) - \text{B.E.}\,(n = 0)$$

$$= nC + \frac{n(n-1)}{2}\,a + [T(T+1) - \tfrac{3}{4}n]\,b + \left[\frac{n}{2}\right] c + \text{C.E.} \qquad (36.13)$$

The constants a, b, and c in (36.13) are the coefficients of $n(n - 1)/2$, of the symmetry term, and of the pairing term respectively in (36.12). They are different from, and should not be confused with the constants in (36.8). The last term in (36.13) is the Coulomb energy (C.E.). This latter energy is according to (30.2) the sum of the Coulomb interaction of the j-protons with the protons in the closed shells and the mutual j-proton Coulomb interaction. As in Section 30, we try to determine the constants in (36.13) with best possible accuracy. The constants thus determined should be interpreted as the matrix elements of the single nucleon and two nucleon effective interaction. The expression (36.13) with these values of the constants gives the calculated energies, to be compared to the experimental values.

The matrix elements are taken to be free parameters and are determined by a least squares fit. These theoretical parameters are the C, a, b, and c in (36.13) and some parameters of the Coulomb energy. This latter energy can be calculated with harmonic oscillator wave functions in which case there is only one parameter which enters the calculation. Alternatively, we can take one parameter for the Coulomb interaction of one j-proton with the protons in closed shells and calculate the rest, i.e., the mutual Coulomb interaction of the j-protons, in the oscillator model. Altogether there are 5 or 6 parameters for the binding energies of all nuclei with j^n configurations. In the case $j = \frac{1}{2}$, there are only two levels in the $(\frac{1}{2})^2$ configuration with $J = 0\,(T = 1)$ and $J = 1\,(T = 0)$. In this case the term proportional to c in (36.13) can be omitted. For $j = \frac{3}{2}$, the expression (36.13) is exact, as mentioned above. Only for $j \geq \frac{5}{2}$ is the expression (36.13) an approximation, valid to the extent that the seniority is a good quantum number for the effective nuclear two-body interaction.

We present here the results of such an analysis of binding energies for nuclei in the $d_{3/2}$ shell. The $d_{3/2}$ orbit is being filled between proton

TABLE 36.1

BINDING ENERGIES IN THE $d_{3/2}$ SHELL IN MEV

Nucleus	J of ground state	Binding energy minus B.E. (S^{32})	
		Experimental	Calculated
${}_{16}S^{33}_{17}$	$\frac{3}{2}$	8.64	8.63
${}_{16}S^{34}_{18}$	0	20.06	20.11
${}_{16}S^{35}_{19}$	$\frac{3}{2}$	27.05	27.09
${}_{16}S^{36}_{20}$	0	36.93	36.91
${}_{17}Cl^{33}_{16}$	$\frac{3}{2}$	2.29	2.35
${}_{17}Cl^{34}_{17}$	0	13.86	13.83
${}_{17}Cl^{35}_{18}$	$\frac{3}{2}$	26.43	26.35
${}_{17}Cl^{37}_{20}$	$\frac{3}{2}$	45.33	45.19
${}_{18}A^{35}_{17}$	$\frac{3}{2}$	19.67	19.50
${}_{18}A^{36}_{18}$	0	35.04	34.87
${}_{18}A^{37}_{19}$	$\frac{3}{2}$	43.73	43.89
${}_{18}A^{38}_{20}$	0	55.57	55.75
${}_{19}K^{37}_{18}$	$\frac{3}{2}$	36.83	37.09
${}_{19}K^{39}_{20}$	$\frac{3}{2}$	61.93	61.86
${}_{20}Ca^{39}_{19}$	$\frac{3}{2}$	54.55	54.49
${}_{20}Ca^{40}_{20}$	0	70.28	70.24

(or neutron) numbers 16 and 20. Thus, nuclei from S^{32} to Ca^{40} are considered. The binding energies of these nuclei are given in Table 36.1. As in Section 30 we take the binding energies with a minus sign so that they become positive quantities. From these binding energies, the binding energy of S^{32}, which has closed shells and no $d_{3/2}$ nucleons, was subtracted. The best values of the interaction constants, determined by a least squares fit are given below with their statistical errors

$$C = 8.634 \pm 0.082$$

$$a = 0.096 \pm 0.024 \qquad b = -1.848 \pm 0.054 \qquad c = 3.672 \pm 0.179 \qquad \text{Mev.}$$

The energies calculated using (36.13) with these best values of the theoretical parameters are given in the last column of Table 36.1 where they can be compared to the experimental energies.

We see that the agreement obtained is very good. The rms deviation, as defined in (30.6), turns out to be only 0.16 Mev. The quadratic term in (36.13) turns out to be positive (attractive) although the actual value of a is small. This is in no contradiction to the behavior of the binding energies found in Section 30. In the case of identical nucleons, $T = (n/2)$, the expression (36.13) reduces to the form of (30.5). The coefficient of $n(n-1)/2$ becomes then $a + (b/2)$ which is also in the present case negative (repulsive) and has the value -0.828 Mev.

The expressions (36.9) or (36.11) give only the positions of the centers of mass of groups of states with the same values of T, v, and t in the j^n configuration. These expressions were obtained by choosing a simple interaction to replace the actual interaction V_{ik}. The simplified interaction (36.7) is not sufficiently general to reproduce all energy levels in the j^2 configurations. It can only reproduce the averages V_0, $\bar{V}_1$, and $\bar{V}_2$. If a simple interaction (i.e., one expressed in terms of known simple operators) can be found to give all the energies V_J of the j^2 configuration, it will be possible to obtain the energy matrix in the j^n configuration in a straightforward manner. In the case $j = \frac{1}{2}$ it is clear that (36.7) itself, with $c = 0$, is sufficient to replace any two-body interaction. The next simple case is $j = \frac{3}{2}$ with four levels in the j^2 configuration ($J = 0, 2$ with $T = 1$ and $J = 1, 3$ with $T = 0$). Therefore, the interaction (36.7) is itself not sufficient. If, however, another simple operator, $(\mathbf{j}_i \cdot \mathbf{j}_k)$, is also added, the interaction obtained can reproduce all the energies V_J in the $(\frac{3}{2})^2$ configuration. The eigenvalues of the operators which appear in the simplified interaction can serve as good quantum numbers for the states. Therefore, the search for such operators in the cases with $j \geq \frac{5}{2}$ is connected with the difficult problem of a unique characterization of the various states. The situation for $j = \frac{3}{2}$ is simple since in this case T, v, and t along with J characterize uniquely all states of the $(\frac{3}{2})^n$ configuration.

Let us replace in $(\frac{3}{2})^n$ configurations the actual interaction V_{ik} by the simplified interaction

$$V'_{ik} = a + 2b(\mathbf{t}_i \cdot \mathbf{t}_k) + cq_{ik} + 2d(\mathbf{j}_i \cdot \mathbf{j}_k)$$

The constants a, b, c, and d can be chosen to satisfy the following relations

$$\begin{aligned} V_0 &= a + \tfrac{1}{2}b + 4c - \tfrac{15}{2}d \\ V_2 &= a + \tfrac{1}{2}b \qquad\quad - \tfrac{3}{2}d \\ V_1 &= a - \tfrac{3}{2}b \qquad\quad - \tfrac{11}{2}d \\ V_3 &= a - \tfrac{3}{2}b \qquad\quad + \tfrac{9}{2}d \end{aligned}$$

By solving these equations we obtain the following expressions for the various interaction parameters

$$a = \frac{15V_2 + 5V_3}{20} \qquad b = \frac{5V_2 - 3V_1 - 2V_3}{10} \qquad c = \frac{5V_0 - 5V_2 - 3V_1 + 3V_3}{20}$$

$$d = \frac{V_3 - V_1}{10}. \tag{36.14}$$

The interaction energy $\Sigma\, V'_{ik}$ in the $(\frac{3}{2})^n$ configuration is given by

$$\frac{n(n-1)}{2}a + b\,[T(T+1) - \tfrac{3}{4}n] + cQ(n, v, t, T) + d[J(J+1) - \tfrac{15}{4}n] \tag{36.15}$$

with a, b, c, and d given by (36.14). Inserting the value of Q from (34.26) with $j = \frac{3}{2}$ and using the values (36.14) we obtain after some rearrangements the expression (36.15) in the form

$$\frac{n(n-1)}{2}\left[\frac{-5V_0 + 35V_2 + 3V_1 + 7V_3}{40}\right].$$

$$+ \frac{n}{4}\left[\frac{65V_0 - 95V_2 + 9V_1 + 21V_3}{20}\right] - \frac{5V_0 - 15V_2 + 3V_1 + 7V_3}{20}T(T+1)$$

$$+ \frac{V_3 - V_1}{10}J(J+1) + \frac{5V_0 - 5V_2 - 3V_1 + 3V_3}{20}\left[\frac{v(v-14)}{4} + t(t+1)\right]. \tag{36.16}$$

The first two terms in (36.16) which depend only on n do not affect the energy spacings within a configuration. They are required, along with the single nucleon energies and the Coulomb energy, only for the determination of the binding energy of the nucleus considered. If we consider the relative positions of energy levels, we have to take into account only the last three terms in (36.16). Unfortunately, there is not enough experimental information on the spectra of $d^n_{3/2}$ configurations. Therefore, no actual example is available to check the validity of (36.16). Nevertheless, we shall discuss by using (36.16) some properties of the energy levels in the $d_{3/2}$ shell and wherever possible, make a comparison with the experimental data.

We first notice that the constants b and c in (36.13) furnish some information on the spacings of energy levels in the j^2 configuration. In fact, using (36.12), we obtain the following relations

$$V_0 - \bar{V}_2 = V_0 - \sum_{J>0 \text{ even}} (2J+1)V_J \Big/ \sum_{J>0 \text{ even}} (2J+1)$$

$$= V_0 - \sum_{J>0 \text{ even}} (2J+1)V_J \Big/ (j+1)(2j-1) = \frac{2j+1}{2(j+1)} c$$

$$V_0 - \bar{V}_1 = V_0 - \sum_{J \text{ odd}} (2J+1)V_J \Big/ \sum_{J \text{ odd}} (2J+1)$$

$$= V_0 - \sum_{J \text{ odd}} (2J+1)V_J \Big/ (j+1)(2j+1) = 2b + \frac{2j+3}{2(j+1)} c. \tag{36.17}$$

In the case of $j = \frac{3}{2}$, there is only one state with $J > 0$ even, namely the one with $J = 2$. Substituting the value of the interaction constant c given above in (36.17) we obtain $V_0 - V_2 = \frac{4}{5} c = 2.94$ Mev. This fixes the position of the $J = 2$ level of the $d_{3/2}^2$ configuration at 2.94 Mev above the $J = 0$ ground state. There are several $J = 2$ levels known in the $d_{3/2}$ region but it is not at all clear whether any of those can be assigned a pure $d_{3/2}^2$ configuration. For $V_0 - \bar{V}_1 = V_0 - (3V_1 + 7V_3)/10$ we find the value 0.73 Mev. We can now use these values of $V_0 - V_2$ and $V_0 - \bar{V}_1$ in (36.16). The last three terms of that expression (in Mev) become now equal to

$$- 1.85T(T+1) + \frac{V_3 - V_1}{10} J(J+1)$$

$$+ [0.72 + \tfrac{3}{20}(V_3 - V_1)] \left[\frac{v(v-14)}{4} + t(t+1)\right].$$

Some information on the $V_3 - V_1$ spacing can be obtained by comparing the energies of the $J = 0$ and $J = 3$ states of the $d_{3/2}^2$ configuration. From the observed spectra of Cl^{34} and K^{38} we know that these two energies are roughly equal. From this we obtain that $V_3 - V_1$ is about 1.5-2.5 Mev. This quantity cannot be compared with the experimental data since the $J = 1$ level of the $d_{3/2}^2$ configuration has not been identified. With a value of $V_3 - V_1$ taken to be 2 Mev, in which case we have $V_3 = V_0$, we obtain for the interaction energies in the $d_{3/2}^2$ configuration the following values

$$V_0 = 2.84 \qquad V_1 = 0.84 \qquad V_2 = -0.09 \qquad V_3 = 2.84 \qquad \text{Mev.}$$

Taking the value $V_3 - V_1 = 2$ Mev we obtain that the spacings of energy levels, determined by the last three terms of (36.16), are given by

$$-1.85\, T(T+1) + 0.2\, J(J+1) + \left[\frac{v(v-14)}{4} + t(t+1)\right].$$

We can apply this expression for a rough determination of order and spacing of levels in $d^n_{3/2}$ configurations. In particular, we consider the $d^4_{3/2}$ configuration in Cl^{36}. The levels with $T = 1$ have $J = 2$ $v = 2$ $t = 1$ and $J = 3,1$ $v = 2$ $t = 0$. The $J = 0$ level which can occur in Cl^{36} has $T = 2$ (it is the analog of the S^{36} ground state). With our values of the interaction constants it follows that the lowest state has $J = 2$. The $J = 3$ level lies about 0.8 Mev above it while the $J = 1$ level lies 2 Mev above the $J = 3$ level. The $J = 0$ level lies, because of its $T = 2$, much higher at about 4.5 Mev above the ground state.

The ground state of Cl^{36} is known experimentally to have spin $J = 2$ (and a known excited state at 0.79 Mev may well have spin $J = 3$). Since this ground state is the only one having $T = 1$ $v = 2$ $t = 1$ for $n = 4$ we can calculate its energy by using (36.11). This binding energy of Cl^{36}, minus that of S^{32}, turns out to be given by $4c + 6a - b + 1.2c$ + C.E. Using our values of the interaction constants, we obtain for this binding energy the value 35.08 Mev. This is in excellent agreement with the experimental value which is 35.01 Mev.

Nuclear l^n Configurations

We shall now consider the wave functions of n nucleons in LS-coupling. With charge-independent interactions, T, S, and L are then good quantum numbers. As in the case of identical nucleons, if S and L commute with the Hamiltonian the energies of all the states with given values of S and L are independent of the value of J. The $2S + 1$ states (if $S \leq L$) form a multiplet denoted by ^{2S+1}L. If the value of T should be explicitly indicated, the notation $^{2T+1,2S+1}L$ is often used. All wave functions with definite values of T, S, and L, admissible by the generalized Pauli principle, are antisymmetric under exchange of the isospin, spin, and space coordinates of any two nucleons. We shall not go in detail into the construction of these antisymmetric functions. Rather, we shall discuss their general properties. To do this it is convenient to consider Hamiltonians which do not contain the spin or isospin vectors of the nucleons.

Every Hamiltonian of physical interest is invariant under any permutations of the nucleon coordinates. As a result, as described in detail in Section 32, the wave functions can be classified according to their symmetry properties. In the general case, this does not furnish any additional information. The wave functions must be antisymmetric with respect to exchange of all coordinates of any two nucleons. If, however, the isospin and spin vectors of the nucleons do not appear in the Hamiltonian, much more can be said about the wave functions. In this case, the Hamiltonian is invariant under any permutation of only the *space coordinates* of the nucleons. This invariance can be utilized in the same fashion as in the case of identical particles. Every complete antisymmetric wave function can be expressed as a sum of products of functions of the isospin and spin coordinates and spatial functions. The spatial eigenfunctions of the Hamiltonian have definite values of L and definite symmetry properties and the energy eigenvalues depend strongly on these symmetries. The spatial functions which appear in the expansion of a given antisymmetric wave function have all the same symmetry type and each of them is multiplied by an isospin and spin function with definite T and S and a definite symmetry type. If the symmetry of the functions of the space coordinates is defined by a certain Young diagram, the symmetry of the isospin and spin functions is given by the *dual* Young diagram. We shall now discuss the diagrams which appear in the present case.

Consider the symmetry types of the isospin and spin functions. The Young diagrams indicate symmetrization with respect to particle numbers within rows followed by antisymmetrization within the columns. There are only four different single nucleon functions of the isospin and spin coordinates. Therefore, the columns of the diagrams cannot contain more than four nucleon numbers. Thus, the Young diagrams of the isospin and spin functions have at most *four rows*. The dual diagrams of the spatial functions cannot then have more than four columns (the maximum length of which is obviously $2l + 1$). The admissible diagrams are thus characterized by the partition

$$n = n_1 + n_2 + n_3 + n_4 \qquad 2l + 1 \geq n_1 \geq n_2 \geq n_3 \geq n_4 \geq 0. \tag{36.18}$$

As in the case of identical particles in LS-coupling, we can consider among the states with the symmetry (36.18) the ones with maximum M_T and among them the one with maximum M_S. We can further arrange that all n_1 nucleons whose numbers appear in the first row will have $m_t = +\frac{1}{2}$, $m_s = +\frac{1}{2}$. In the second row the n_2 nucleons will have $m_t = +\frac{1}{2}$ but $m_s = -\frac{1}{2}$. In the third row they will have $m_t = -\frac{1}{2}$, $m_s = +\frac{1}{2}$ and in the fourth $m_t = -\frac{1}{2}$, $m_s = -\frac{1}{2}$. We can now attach a physical interpretation to the numbers n_i, which will, however, be more complicated than in the case of identical particles.

We first remark that, due to (36.18), only three numbers, in addition to n, are required to characterize the symmetry type. These can be conveniently chosen as follows. In the special state considered, we can write down immediately the values of M_T and M_S. In fact,

$$M_T = \tfrac{1}{2}(n_1 + n_2 - n_3 - n_4)$$
$$M_S = \tfrac{1}{2}(n_1 - n_2 + n_3 - n_4).$$

The total number of particles is given by

$$n = n_1 + n_2 + n_3 + n_4$$

and therefore only one more number is required. Along with the operators $\sum_i^n t_3^{(i)}$ and $\sum_i^n s_z^{(i)}$ with the eigenvalues M_T and M_S, we consider the operator

$$Y = 2\sum_i^n t_3^{(i)} s_z^{(i)}. \tag{36.19}$$

This is a symmetric operator which commutes with both T_3 and S_z. The eigenvalue of this operator in the state considered is

$$Y = \tfrac{1}{2}(n_1 - n_2 - n_3 + n_4).$$

Thus, the eigenvalues M_T, M_S, and Y, along with n, determine uniquely the partition (36.18) and, therefore, the corresponding symmetry type. Clearly, the operators **T** and **S** play identical roles in the present formalism. They could be interchanged without affecting the discussion. Moreover, the operators T_3, S_z, and Y are also interchangeable in this way (instead of the definition $y = 2t_3s_z$ we could start from y and define $s_z = 2yt_3$ or $t_3 = 2ys_z$). Therefore we describe the partition (36.18) more generally, by the three quantum numbers P, P', and P'' defined by

$$\begin{aligned} P &= \tfrac{1}{2}(n_1 + n_2 - n_3 - n_4) \\ P' &= \tfrac{1}{2}(n_1 - n_2 + n_3 - n_4) \\ P'' &= \tfrac{1}{2}(n_1 - n_2 - n_3 + n_4). \end{aligned} \tag{36.20}$$

It follows from (36.18) that $P \geq P' \geq P''$. The physical meaning of these operators is that P is the largest eigenvalue which any one of the operators T_3, S_z, and Y can have for a state with the symmetry defined by (36.20). Similarly, P' is the largest eigenvalue that another one of these operators can have under the condition that P is the eigenvalue of the first operator. With two of these operators having the eigenvalue P and P', P'' is the largest eigenvalue that the third operator can have.

As a result, the largest eigenvalues that M_T, and therefore T, can have in states with the partitions (36.20) is P. This is also the largest eigenvalue of S for such states. With $T = P$, S can have at most the value P' and vice versa. These interpretations are consistent with the definition (36.20) since it follows that both P and P' cannot be negative (and cannot exceed $n/2$). It is also clear that P, P', and P'' are all integers if n is even and are all half-integers if n is odd. P'' can be also negative but this does not lead to an inconsistency since Y, unlike T and S, can have negative values (like M_T and M_S).

The interactions occurring in the Hamiltonians considered above do not contain the spin and isospin vectors. Therefore, the eigenvalues of the Hamiltonian are determined by the spatial parts of the wave functions. For finite range forces it is expected that the more symmetric the wave function, the stronger the expectation value of the interaction. This is expected to hold for Wigner forces, and *a fortiori* for Majorana forces. These latter forces are attractive in space symmetric states of two nucleons and repulsive in antisymmetric states. If the potential is taken at the limit of a square well with infinite range, all states will have the same energy for Wigner forces, irrespective of their symmetry. In the case of Majorana forces, however, even in this limit the interaction energy will be stronger for states with higher spatial symmetry.

The energies of states, for the Hamiltonians we consider, are independent of S and T. These quantum numbers enter only implicitly through the partition (36.20). Thus, let us take a given spatial wave function with a definite symmetry and a definite value of L. Consider all antisymmetric states built from this function (and those obtained from it by permutations) and isospin and spin functions with the symmetry defined by the dual Young diagram. All these states will be degenerate in the case of the Hamiltonians considered. In the case of identical particles, this degeneracy was merely the $2S + 1$ degeneracy in M_S, since in that case a partition defines a definite value of S. In the present case, however, apart from the degeneracy in M_T and M_S, there may be further degeneracy. A partition (36.18) or (36.20) may well determine more than one value of T and of S. This degeneracy of states with different values of T and S is removed only if interactions which contain explicitly the spins are also considered. These interactions may be scalar forces, like Bartlett and Heisenberg forces, and therefore still have S and L as good quantum numbers. In the atomic case, the degeneracy considered here leads to $2S + 1$ states with different values of J (for $S \leq L$). These states form a multiplet. In analogy, all states with the same value of L and the same symmetry type, with the various values of T and S, form a nuclear *supermultiplet*. The states of a given supermultiplet may appear in several nuclei (if $T > 0$). In order to observe all states of a given supermultiplet the nucleus with the lowest value of $|M_T| = \frac{1}{2}|Z - N|$ should be looked at.

We shall now describe the supermultiplets with maximum symmetry in various nuclei. The ground states of the nuclei considered are expected to belong to these supermultiplets in the case of the Hamiltonians considered here.

From (36.20) follows that P, P', and P'', depending only on the differences of the n_i, remain unchanged if we add to every n_i the same number. Therefore, if n is increased by 4 the same supermultiplets occur also in the l^{n+4} configuration (if the middle of the shell is not reached). The bigger configuration may still contain more supermultiplets. The isospin and spin structure of the wave functions which belong to the same supermultiplet are the same in the l^n and l^{n+4} configurations. However, the bigger configuration may well have more spatial functions with the same symmetry type, and therefore more states of the given supermultiplet.

The maximum symmetry of the spatial functions occurs for maximum antisymmetry of the isospin and spin functions. This is the case when the n_i are as equal to each other as possible for the given value of n. The highest symmetry can thus be obtained for $n = 4n_1$ and $n_1 = n_2 = n_3 = n_4$. In this case $P = P' = P'' = 0$ and there is only one isospin value $T = 0$ and one spin value $S = 0$. All states in this supermultiplet, denoted by (0, 0, 0), are thus ^{11}L. The possible values of L depend on the configuration. For instance, in the p^4 configuration the possible states are ^{11}S, ^{11}D, and ^{11}G, whereas in the p^{12} configuration, which is a closed shell, there is only one state, namely ^{11}S.

In the case of even n which is *not* a multiple of 4, not all the n_i can be equal. The highest symmetry is obtained in this case for $n_1 = n_2$ and $n_3 = n_4 = n_1 - 1$. Thus, the supermultiplet considered is characterized by $P = 1$, $P' = P'' = 0$ and denoted by (1, 0, 0). The possible values of the isospin and spin are accordingly $T = 1$, $S = 0$ or $T = 0$, $S = 1$ and the states are thus either ^{31}L or ^{13}L. For example, for two nucleons with $l = 1$, the states of this supermultiplet, which can be simply defined by symmetric spatial wave functions, are ^{31}S, ^{13}S, ^{31}D, and ^{13}D. In the case of the p^6 configuration, in addition to these, there are also other states namely ^{31}F, ^{13}F, ^{31}G, ^{13}G as well as other independent ^{31}D and ^{13}D states.

The highest symmetry for odd n nuclei is different for $n = 4m + 1$ and $n = 4m + 3$. In the case $n = 4m + 1$, the highest symmetry is obtained with $n_1 = m + 1$ and $n_2 = n_3 = n_4 = m$. The corresponding supermultiplet is characterized by $P = P' = P'' = \frac{1}{2}$ or $(\frac{1}{2}, \frac{1}{2}, \frac{1}{2})$. On the other hand, in the case $n = 4m + 3$, the highest symmetry is obtained with $n_1 = n_2 = n_3 = m + 1$ $n_4 = m$. Thus, the lowest supermultiplet is given by $P = P' = \frac{1}{2}$, $P'' = -\frac{1}{2}$ or $(\frac{1}{2}, \frac{1}{2}, -\frac{1}{2})$. In both cases, however, the possible values of the isospin and spin are $T = \frac{1}{2}$ $S = \frac{1}{2}$ and the states are denoted by ^{22}L. In the case of a single nucleon outside closed shells, or missing from closed shells, the only state is ^{22}L with $L = l$. In other cases, more states occur. For example, in the p^3 configuration, the states of the $(\frac{1}{2}, \frac{1}{2}, -\frac{1}{2})$ supermultiplet are ^{22}P and ^{22}F. In the p^5 configuration the states ^{22}P, ^{22}D, ^{22}F and ^{22}G belong to the $(\frac{1}{2}, \frac{1}{2}, \frac{1}{2})$ supermultiplet.

It was mentioned before that for attractive Wigner and Majorana forces the energy of a state is largely determined by the symmetry properties of its spatial wave functions. It also depends on other properties of the spatial functions, e.g. on their L values. However, for the case mentioned above, of long-range square well (or Gaussian) potential, all states with different values of L within the same supermultiplet have the same energy. The Wigner force gives in this limit the same energy to the states of all supermultiplets. However, the interaction of the Majorana force, which is in this case proportional to $\sum_{i<k}^{n} P_{ik}^{(x)}$, has a larger value for states with higher symmetry. The operator $M = \sum P_{ik}^{(x)}$ is diagonal in the scheme of the supermultiplets and we shall now calculate its eigenvalues. This calculation does not have immediate applications, since the nuclear forces are certainly not long-range Wigner and Majorana forces. However, the eigenvalues of the operator M furnish some measure of the spatial symmetry of a state. They can, in fact, be considered as the number of symmetrically coupled pairs minus the number of antisymmetrically coupled pairs.

We first replace the operator $P_{ik}^{(x)}$ by

$$P_{ik}^{(x)} = -P_{ik}^{(\tau)} P_{ik}^{(\sigma)} = -\left[\frac{1 + 4(\mathbf{t}_i \cdot \mathbf{t}_k)}{2}\right]\left[\frac{1 + 4(\mathbf{s}_i \cdot \mathbf{s}_k)}{2}\right]. \tag{36.21}$$

This will enable us to study the symmetry of the spatial functions by looking at the corresponding isospin and spin functions which are simpler to handle. Thus, we have to calculate the eigenvalues of the operator

$$M = \sum_{i<k}^{n} P_{ik}^{(x)} = -\sum_{i<k}^{n}\left[\frac{1 + 4(\mathbf{t}_i \cdot \mathbf{t}_k)}{2}\right]\left[\frac{1 + 4(\mathbf{s}_i \cdot \mathbf{s}_k)}{2}\right]$$

$$= -\sum_{i<k}^{n} \tfrac{1}{4} - \tfrac{1}{2}\sum_{i<k}^{n} 2(\mathbf{t}_i \cdot \mathbf{t}_k) - \tfrac{1}{2}\sum_{i<k}^{n} 2(\mathbf{s}_i \cdot \mathbf{s}_k) - 2\sum_{i<k}^{n} 2(\mathbf{t}_i \cdot \mathbf{t}_k)(\mathbf{s}_i \cdot \mathbf{s}_k). \tag{36.22}$$

We shall now calculate the eigenvalue of M, as given in (36.22), for the supermultiplet characterized by the partition $n = n_1 + n_2 + n_3 + n_4$. We consider among all isospin and spin functions those with the maximum value of M_T, and choose from these the one with the maximum value of M_S. This particular function belongs to definite values of $T = P$ and $S = P'$.

We thus obtain from (36.22) in this case

$$M = -\tfrac{1}{4}\frac{n(n-1)}{2} - \tfrac{1}{2}[T(T+1) - \tfrac{3}{4}n] - \tfrac{1}{2}[S(S+1) - \tfrac{3}{4}n]$$

$$-2\sum_{i,k}^{n} (\mathbf{t}_i \cdot \mathbf{t}_k)(\mathbf{s}_i \cdot \mathbf{s}_k) + 2\sum_{i}^{n} (\mathbf{t}_i \cdot \mathbf{t}_i)(\mathbf{s}_i \cdot \mathbf{s}_i)$$

$$= \frac{n(16-n)}{8} - \tfrac{1}{2}[P(P+1) + P'(P'+1)] - 2\sum_{i,k}^{n} (\mathbf{t}_i \cdot \mathbf{t}_k)(\mathbf{s}_i \cdot \mathbf{s}_k). \quad (36.23)$$

The last term on the right-hand side of (36.23) can be calculated explicitly. By expanding the scalar products we obtain

$$\sum_{i,k}^{n} (\mathbf{t}_i \cdot \mathbf{t}_k)(\mathbf{s}_i \cdot \mathbf{s}_k)$$

$$= \sum_{i,k}^{n} [t_3^{(i)} t_3^{(k)} + \tfrac{1}{2} t_+^{(i)} t_-^{(k)} + \tfrac{1}{2} t_-^{(i)} t_+^{(k)}]\,[s_z^{(i)} s_z^{(k)} + \tfrac{1}{2} s_+^{(i)} s_-^{(k)} + \tfrac{1}{2} s_-^{(i)} s_+^{(k)}]. \quad (36.24)$$

Therefore, (36.24) can be expressed as the sum of single nucleon operators multiplied by their Hermitean conjugates. Such single nucleon operators are, for example, $\Sigma_i\, t_3^{(i)} s_z^{(i)}$ $\Sigma_i\, t_3^{(i)} s_+^{(i)}$ and $\Sigma_i\, t_-^{(i)} s_+^{(i)}$. The effect of these operators on the isospin and spin function considered can be easily seen. First we observe that the operator $[\Sigma_i\, t_3^{(i)} s_z^{(i)}]^2$ is equal to $[Y/2]^2$. The eigenvalue of Y in this case is given by $\tfrac{1}{2}(n_1 - n_2 - n_3 + n_4)$. Some of the other single particle operators give zero when applied to the function considered. These are all operators that transform the given function into a function in which two nucleons have the same values of m_t and m_s. Such operators are

$$\sum t_3^{(i)} s_+^{(i)}, \qquad \sum t_+^{(i)} s_z^{(i)}, \qquad \sum t_+^{(i)} s_+^{(i)}, \qquad \text{and} \qquad \sum t_+^{(i)} s_-^{(i)}.$$

On the other hand, the remaining (symmetric) operators transform the given isospin and spin function χ into other functions, with the same symmetry type, with different values of M_T or M_S. Operating on these functions with the Hermitian conjugate operators, they are transformed back into the original function χ. For example,

$$\left[\sum t_-^{(i)} s_z^{(i)}\right] \chi = \tfrac{1}{2}\sqrt{n_1 - n_3}\,\chi_1 - \tfrac{1}{2}\sqrt{n_2 - n_4}\,\chi_2. \quad (36.25)$$

Here, χ_1 is the normalized function obtained from χ by putting one of the n_1 values of m_t in the first row equal to $-\tfrac{1}{2}$. Similarly, χ_2 has one of the n_2 values of m_t in the second row equal to $-\tfrac{1}{2}$. The factors

$\sqrt{n_1 - n_3}$ and $\sqrt{n_2 - n_4}$ arise from the normalization of these particular functions. From (36.25) it now follows that

$$\left[\sum t_+^{(i)} s_z^{(i)}\right] \left[\sum t_-^{(i)} s_z^{(i)}\right] \chi$$

$$= \tfrac{1}{2}\sqrt{n_1 - n_3}\left[\sum t_+^{(i)} s_z^{(i)}\right] \chi_1 - \tfrac{1}{2}\sqrt{n_2 - n_4}\left[\sum t_+^{(i)} s_z^{(i)}\right] \chi_2$$

$$= \tfrac{1}{4}(n_1 - n_3)\,\chi + \tfrac{1}{4}(n_2 - n_4)\,\chi \,. \qquad (36.26)$$

Collecting all the contributions we obtain

$$\left[\sum_{i,k}^{n} (\mathbf{t}_i \cdot \mathbf{t}_k)\,(\mathbf{s}_i \cdot \mathbf{s}_k)\right] \chi$$

$$= [\tfrac{1}{16}(n_1 - n_2 - n_3 + n_4)^2 + \tfrac{1}{8}(n_1 - n_3) + \tfrac{1}{8}(n_2 - n_4) + \tfrac{1}{8}(n_1 - n_2)$$
$$+ \tfrac{1}{8}(n_3 - n_4) + \tfrac{1}{4}(n_1 - n_4) + \tfrac{1}{4}(n_2 - n_3)]\chi = \tfrac{1}{4}[P''^2 + 3P + P']\chi \,. \qquad (36.27)$$

Introducing (36.27) into (36.23) we obtain the expression of the eigenvalues of the Majorana exchange operator M in terms of the quantum numbers characterizing the symmetry of the state. Thus

$$M = \frac{n(16 - n)}{8} - \tfrac{1}{2}[P(P + 4) + P'(P' + 2) + P''^2]. \qquad (36.28)$$

We see that the smaller the numbers P, P', and P'', the bigger the symmetry of the state as given by M. It may be worthwhile to notice that the expression (36.28) can be equivalently written in terms of the n_i as

$$M = -\tfrac{1}{2}[n_1(n_1 - 1) + n_2(n_2 - 3) + n_3(n_3 - 5) + n_4(n_4 - 7)]$$
$$= [n_2 + 2n_3 + 3n_4]$$
$$- [\tfrac{1}{2}n_1(n_1 - 1) + \tfrac{1}{2}n_2(n_2 - 1) + \tfrac{1}{2}n_3(n_3 - 1) + \tfrac{1}{2}n_4(n_4 - 1)]. \qquad (36.29)$$

The expressions $(n_2 + 2n_3 + 3n_4)$ and $\tfrac{1}{2}\sum_i n_i(n_i - 1)$ in (36.29) can be looked upon as the numbers of antisymmetric and symmetric couplings in the isospin and spin function. It appears that the whole contribution to the number of antisymmetrically coupled pairs comes from pairs within the same columns. Similarly, the symmetric couplings come from pairs within rows of the Young diagram. However, it should be kept in mind that the isospin and spin wave function is not symmetric any more within rows after the antisymmetrization within columns is carried out. The result (36.29) is obtained only after collecting all contributions to the Majorana operator. The result (36.28) has been extensively used in the past in calculations of energies of ground states of nuclei (Wigner's supermultiplet theory). Since these calculations assume the validity of LS-coupling they will not be considered here.

37. Nonequivalent Particles; Configuration Interaction

In the preceding sections we discussed in detail configurations with equivalent particles (either identical or not). We shall now consider configurations of particles in several unfilled shells. It was shown in Section 25 that the antisymmetrization of the wave function with respect to particles in different orbits does not affect the reduction of the interaction energy into matrix elements in two-particle configurations. The only change introduced is that matrix elements in the two inequivalent particle configurations should be taken with properly antisymmetrized (two-particle) wave functions. This fact greatly simplifies the calculations. The considerations of Section 25 apply equally well for configurations with nonidentical nucleons. In this case, all wave functions should be antisymmetric if the charge coordinate is included. We shall therefore now summarize the results of Section 25, using the isospin formalizm.

Let us consider, for the sake of simplicity, two groups of equivalent nucleons. The generalization to more groups is straightforward, as shown in Section 25. Wave functions of the $j_1^{n_1} j_2^{n_2}$ configuration with given values of T and J can be constructed as follows. We take all antisymmetric states of the $j_1^{n_1}$ configuration and $j_2^{n_2}$ configuration. These states are characterized by $\alpha_1 T_1 J_1$ and $\alpha_2 T_2 J_2$ respectively, where α_1 and α_2 are any necessary additional quantum numbers. We then couple $\mathbf{T}_1$ with $\mathbf{T}_2$ to form the total $\mathbf{T}$ and $\mathbf{J}_1$ with $\mathbf{J}_2$ to form the total $\mathbf{J}$. The wave function thus obtained is denoted by

$$\psi[j_1^{n_1}(\alpha_1 T_1 J_1)\, j_2^{n_2}(\alpha_2 T_2 J_2) T\, J]. \tag{37.1}$$

The wave functions (37.1) are uniquely determined by the quantum numbers $\alpha_1 T_1 J_1 \alpha_2 T_2 J_2$ as well as T and J. Any two such functions, with different values of these quantum numbers are orthogonal to each other. The method of construction given by (37.1) exhausts all states of the $j_1^{n_1} j_2^{n_2}$ configuration. If we now antisymmetrize these wave functions with respect to particles in the j_1 orbit and the j_2 orbit we obtain from each wave function (37.1) a definite antisymmetric wave function (with the same values of T and J). The quantum numbers $\alpha_1 T_1 J_1$ $\alpha_2 T_2 J_2 T J$ thus characterize uniquely the antisymmetrized functions. Any two antisymmetric wave functions with different values of these quantum numbers are orthogonal to each other.

Matrix elements of single particle operators, $\Sigma_{i=1}^{n_1+n_2} f_i$, calculated with these normalized antisymmetrized wave functions, are exactly equal to those calculated with the original wave functions (37.1). Also matrix elements of any two-body operator $\Sigma_{i<k}^{n_1+n_2} V_{ik}$, taken with the antisymmetrized wave functions, can be simply obtained from those calculated with the wave functions (37.1). We first notice that

$$\langle j_1^{n_1}(\alpha_1 T_1 J_1)\, j_2^{n_2}(\alpha_2 T_2 J_2) T\, J \Big| \sum_{i<k}^{n_1+n_2} V_{ik} \Big| j_1^{n_1}(\alpha_1' T_1' J_1')\, j_2^{n_2}(\alpha_2' T_2' J_2') T\, J\rangle$$

$$= \langle j_1^{n_1}\alpha_1 T_1 J_1 \Big| \sum_{i<k}^{n_1} V_{ik} \Big| j_1^{n_1}\alpha_1' T_1' J_1'\rangle + \langle j_2^{n_2}\alpha_2 T_2 J_2 \Big| \sum_{i<k}^{n_2} V_{ik} \Big| j_2^{n_2}\alpha_2' T_2' J_2'\rangle$$

$$\langle j_1^{n_1}(\alpha_1 T_1 J_1)\, j_2^{n_2}(\alpha_2 T_2 J_2) T\, J \Big| \sum_{i=1}^{n_1} \sum_{k=n_1+1}^{n_1+n_2} V_{ik} \Big| j_1^{n_1}(\alpha_1' T_1' J_1')\, j_2^{n_2}(\alpha_2' T_2' J_2') T\, J\rangle. \tag{37.2}$$

The first two terms on the right-hand side of (37.2) can be calculated by the methods described in the preceding sections. The third term can always be expressed as a linear combination of the matrix elements $\langle j_1 j_2 T' J' | V_{12} | j_1 j_2 T' J'\rangle$. The coefficients of this linear combination when the antisymmetric wave functions are used are the same as those obtained with the wave functions (37.1). Thus, the wave functions (37.1) can be used in this calculation provided the matrix elements in the $j_1 j_2$ configuration are taken with the properly antisymmetric wave functions $\psi_a(j_1 j_2 T\, J)$. In the following we shall be mainly interested in the expression of the interaction between the $j_1^{n_1}$ and $j_2^{n_2}$ groups in terms of a linear combination of matrix elements in the $j_1 j_2$ configuration.

We notice that the matrix of the interaction $\Sigma\, V_{ik}$ is not diagonal in the scheme of wave functions (37.1), whether antisymmetrized or not. The interactions in the $j_1^{n_1}$ configuration and the $j_2^{n_2}$ configuration are diagonal in $T_1 J_1$ and $T_2 J_2$ respectively. These interactions may also be diagonal in the additional quantum numbers α_1 and α_2. However, the interaction between the two groups of inequivalent nucleons given by the last term in (37.2) is generally not diagonal in $T_1 J_1$ and $T_2 J_2$. The eigenstates of the interaction matrix *within the* $j_1^{n_1} j_2^{n_2}$ *configuration* are linear combinations of the (antisymmetrized) wave functions (37.1). We shall thus have to calculate diagonal as well as nondiagonal matrix elements of the interaction $\Sigma\, V_{ik}$, using the wave functions (37.1). Before starting this calculation, we shall discuss some properties of the diagonal elements.

Consider a given state of the $j_1^{n_1}$ configuration and a given state of the $j_2^{n_2}$ configuration. Let us consider, for simplicity, either only indentical particles or a case in which all j_1-nucleons are protons and all j_2-nucleons are neutrons. In these cases the two states will be characterized by the quantum numbers $\alpha_1 J_1$ and $\alpha_2 J_2$. We look at all states obtained by coupling $\mathbf{J}_1$ and $\mathbf{J}_2$ to $\mathbf{J}$. In order to calculate the expectation values of the interaction between the two groups of nucleons, we expand the interaction V_{ik} in the usual way

$$V_{ik} = \sum_r F^r(\mathbf{u}_i^{(r)} \cdot \mathbf{u}_k^{(r)}). \tag{37.3}$$

We include in (37.3) the direct terms as well as the exchange terms which can be written, as shown in Section 21, as the direct term of an appropriate interaction. The last term in (37.2) can now be written as

$$\langle j_1^{n_1}(\alpha_1 J_1)\, j_2^{n_2}(\alpha_2 J_2) J \left| \sum_r F^r \left[\sum_{i=1}^{n_1} \mathbf{u}_i^{(r)}\right] \cdot \left[\sum_{k=n_1+1}^{n_1+n_2} \mathbf{u}_k^{(r)}\right] \right| j_1^{n_1}(\alpha_1' J_1')\, j_2^{n_2}(\alpha_2' J_2') J \rangle. \tag{37.4}$$

According to (15.5) this can be written as

$$\sum_r F^r (-1)^{J_1+J_2+J} (j_1^{n_1}\alpha_1 J_1 || \mathbf{U}^{(r)}(1) || j_1^{n_1}\alpha_1' J_1')\, (j_2^{n_2}\alpha_2 J_2 || \mathbf{U}^{(r)}(2) || j_2^{n_2}\alpha_2' J_2') \begin{Bmatrix} J_1 & J_2 & J \\ J_2' & J_1' & r \end{Bmatrix} \tag{37.5}$$

where in (37.5) we used the definitions

$$\mathbf{U}^{(r)}(1) = \sum_{i=1}^{n_1} \mathbf{u}_i^{(r)} \qquad \text{and} \qquad \mathbf{U}^{(r)}(2) = \sum_{i=n_1+1}^{n_1+n_2} \mathbf{u}_i^{(r)}.$$

We now take the sum of the expressions (37.5) with $\alpha_1 = \alpha_1'$ $\alpha_2 = \alpha_2'$ $J_1 = J_1'$ $J_2 = J_2'$ over all states obtained by coupling definite J_1 and J_2 to all possible values of J (with the weight $2J + 1$, as usual). Due to (15.16) we then obtain

$$\sum_r F^r (j_1^{n_1}\alpha_1 J_1 || \mathbf{U}^{(r)}(1) || j_1^{n_1}\alpha_1 J_1)\, (j_2^{n_2}\alpha_2 J_2 || \mathbf{U}^{(r)}(2) || j_2^{n_2}\alpha_2 J_2)$$

$$\sum_J (-1)^{J_1+J_2+J}(2J+1) \begin{Bmatrix} J_1 & J_2 & J \\ J_2 & J_1 & r \end{Bmatrix}$$

$$= \sqrt{(2J_1+1)(2J_2+1)}\, F^0 (j_1^{n_1}\alpha_1 J_1 || \mathbf{U}^{(0)}(1) || j_1^{n_1}\alpha_1 J_1)$$
$$\times (j_2^{n_2}\alpha_2 J_2 || \mathbf{U}^{(0)}(2) || j_2^{n_2}\alpha_2 J_2). \tag{37.6}$$

Recalling (26.34) we obtain (37.6) in the simple form

$$F^0(2J_1+1)(2J_2+1) n_1 n_2 (j_1 || \mathbf{u}_1^{(0)} || j_1)\, (j_2 || \mathbf{u}_2^{(0)} || j_2) / \sqrt{(2j_1+1)(2j_2+1)}. \tag{37.7}$$

The number of states with all values of J and M is obviously $(2J_1 + 1)(2J_2 + 1)$. Therefore, it follows from (37.7) that the average interaction of all these states is

$$F^0 n_1 n_2 (j_1 || \mathbf{u}_1^{(0)} || j_1)\, (j_2 || \mathbf{u}_2^{(0)} || j_2) / \sqrt{(2j_1+1)(2j_2+1)}. \tag{37.8}$$

This average interaction is thus independent of both J_1 and J_2. This result is known as the *center-of-mass theorem*. It shows that if we use expectation values only, the position of the center of mass of all states with the same values of J_1 and J_2 relative to the positions of other such centers of mass is determined only by the interactions *within* the $j_1^{n_1}$ configuration and the $j_2^{n_2}$ configuration. Recalling (21.32) and (21.30) we can express the average interaction (37.8) in the form

$$n_1 n_2 \sum_J (2J+1) V(j_1 j_2 J) \Big/ \sum_J (2J+1)$$

$$= n_1 n_2 \sum_J (2J+1) V(j_1 j_2 J) \Big/ (2j_1+1)(2j_2+1). \tag{37.9}$$

In (37.9), $V(j_1 j_2 J)$ is the interaction energy of the $j_1 j_2$ configuration in the state with total spin J.

The appearance of nondiagonal matrix elements, defined in (37.2), severely limits the use of the center-of-mass theorem. An example where this theorem can be useful is the case where $j_1 \geq \frac{3}{2}$ and $j_2 = \frac{1}{2}$. Consider now the $j_1^2 j_2$ configuration. If the j_1-nucleons are identical, the states of the j_1^2 configuration have even values of the angular momenta $J_1 = 0, 2, \ldots$. The other angular momentum is $J_2 = j_2 = \frac{1}{2}$. Thus, the states of the system are uniquely defined by $J_1 = 0$ $J_2 = \frac{1}{2}$ $J = \frac{1}{2}$, $J_1 = 2$ $J_2 = \frac{1}{2}$ $J = \frac{3}{2}, \frac{5}{2}$ etc. In this case, within the $j_1^2 j_2$ configuration the interaction matrix is diagonal since there is only one state for each value of J. It follows from the center-of-mass theorem that the spacing between the center of mass of the levels with $J = \frac{3}{2}$ and $\mathrm{J} = \frac{5}{2}$ and the state with $J = \frac{1}{2}$ is equal to the spacing between the $J = 2$ $J = 0$ levels of the j_1^2 configuration. Similar relations hold in this case also for states with higher values of J.

A case which is often encountered in nuclei is when there are m protons in the j-orbit whereas the neutron j-shell is closed and n extra neutrons occupy the j'-orbit. If we disregard the j-neutrons as a closed shell, there is no need to use antisymmetrized wave functions in the matrix elements of the interaction in the jj' configuration. The j-nucleons, apart from the closed shell, are, in this case, definitely protons and the j'-nucleons are definitely neutrons. Therefore, the two-particle wave functions of the jj' configurations need not be antisymmetrized. As we shall see later, this does not hold in other cases where the matrix elements in the jj' configuration with $T = 0$ and $T = 1$ have to be considered separately. This fact can be also considered from the point of view of the isospin. The $2j+1+m$ nucleons in the j-shell are in a state with the isospin $T_1 = (2j+1-m)/2$. The n nucleons in the j'-orbit are in a state with $T_2 = n/2$. The *highest* value of T that such a state could have is $T = T_1 + T_2 = (2j+1-m+n)/2$. On the other hand, since the nucleus considered is in a state with $|M_T| = (2j+1+n-m)/2$, it cannot be in a state with a value of T *smaller* than

$(2j+1+n-m)/2$. Thus, the states of the $j^{2j+1+m}j'^n$ configuration with $M_T = -(2j+1+n-m)/2$ have all the same value of $T = (2j+1+n-m)/2$. The center-of-mass theorem described above can be applied also to this case.

A special case of the center-of-mass theorem may be of some interest. Consider the case in which $J_1 = 0$. There is, then, only one resulting state of the whole system, namely with $J = J_2$. The interaction energy between the j_1 particles and j_2 particles is then given directly by (37.8) or (37.9). Thus, we obtain for this interaction energy the form

$$V[j_1^{n_1}(J_1=0)j_2^{n_2}(J_2)\; J=J_2] = n_1n_2\sum_J (2J+1)V(j_1j_2J)\Big/\sum_J (2J+1)$$

$$= n_1n_2\sum_J (2J+1)V(j_2j_1J)\Big/(2j_1+1)(2j_2+1). \tag{37.10}$$

The expression (37.10) is seen to be independent of J_2. The same result holds also for $J_2 = 0$ for arbitrary values of J_1. In particular, we obtain from (37.10) for $n_1 = 2$ $n_2 = 1$ or $n_1 = 1$ $n_2 = 2$ the result

$$V[j_1^2(J_1=0)j_2\; J=j_2] = V[j_1, j_2^2(J_2=0)\; J=j_1]$$

$$= 2\sum (2J+1)V(j_1j_2J)\Big/(2j_1+1)(2j_2+1). \tag{37.11}$$

For $n_1 = 2j_1+1$ $n_2 = 1$ we obtain from (37.10) for the interaction of a particle with a closed shell the result (22.15) obtained above.

We turn now to the calculation of the j_1-j_2 interaction energy in the $j_1^{n_1}j_2^{n_2}$ configuration. We shall first consider either identical particles or protons and neutrons in different shells so that the isospin vectors are not required. The more general discussion will follow later. We calculate in the scheme of wave functions (37.1) the matrix elements of the interaction energy which are given by (37.4) and (37.5). We can use c.f.p. in the $j_1^{n_1}$ and in the $j_2^{n_2}$ configuration to express the reduced matrix elements of $\mathbf{U}^{(r)}$ in terms of single nucleon matrix elements. Recalling (26.33) we can rewrite (37.5) in the form

$$n_1n_2\sum_r F^r(-1)^{j_1+j_2}(j_1||\mathbf{u}^{(r)}||j_1)(j_2||\mathbf{u}^{(r)}||j_2)$$

$$\times \sum_{\alpha_{11}J_{11}\alpha_{22}J_{22}} [j_1^{n_1}\alpha_1J_1\{|j_1^{n_1-1}(\alpha_{11}J_{11})j_1J_1][j_1^{n_1-1}(\alpha_{11}J_{11})j_1J_1'|\}j_1^{n_1}\alpha_1'J_1']$$

$$\times [j_2^{n_2}\alpha_2J_2\{|j_2^{n_2-1}(\alpha_{22}J_{22})j_2J_2][j_2^{n_2-1}(\alpha_{22}J_{22})j_2J_2'|\}j_2^{n_2}\alpha_2'J_2'](-1)^{J_1+J_1'+2J_2+J+J_{11}+J_{22}}$$

$$\times [(2J_1+1)(2J_1'+1)(2J_2+1)(2J_2'+1)]^{1/2}$$

$$\times \begin{Bmatrix} j_1 & J_1 & J_{11} \\ J_1' & j_1 & r \end{Bmatrix} \begin{Bmatrix} j_2 & J_2 & J_{22} \\ J_2' & j_2 & r \end{Bmatrix} \begin{Bmatrix} J_1 & J_2 & J \\ J_2' & J_1' & r \end{Bmatrix}. \tag{37.12}$$

Equation (37.12) gives the matrix elements of the interaction energy in terms of the coefficients F^r of the tensor expansion of the two-body interaction. Instead, we can express these matrix elements in terms of $V(j_1 j_2 J)$, the matrix elements of the interaction in the $j_1 j_2$ configuration. Using the relation (22.51) we obtain (37.12) in the form

$$n_1 n_2 \sum_{\alpha_{11} J_{11} \alpha_{22} J_{22}} [j_1^{n_1}\alpha_1 J_1 \{| j_1^{n_1-1}(\alpha_{11}J_{11}) j_1 J_1] [j_1^{n_1-1}(\alpha_{11}J_{11}) j_1 J_1' |\} j_1^{n_1}\alpha_1' J_1']$$

$$\times [j_2^{n_2}\alpha_2 J_2 \{| j_2^{n_2-1}(\alpha_{22}J_{22}) j_2 J_2] [j_2^{n_2-1}(\alpha_{22}J_{22}) j_2 J_2' |\} j_2^{n_2}\alpha_2' J_2']$$

$$\times [(2J_1+1)(2J_1'+1)(2J_2+1)(2J_2'+1)]^{1/2} \sum_{J'} (2J'+1) V(j_1 j_2 J')$$

$$\times \sum_r (-1)^{J_1+J_1'+2J_2+J+J_{11}+J_{22}+J'} (2r+1) \begin{Bmatrix} j_1 & J_1 & J_{11} \\ J_1' & j_1 & r \end{Bmatrix} \begin{Bmatrix} j_2 & J_2 & J_{22} \\ J_2' & j_2 & r \end{Bmatrix}$$

$$\times \begin{Bmatrix} J_1 & J_2 & J \\ J_2' & J_1' & r \end{Bmatrix} \begin{Bmatrix} j_1 & j_2 & J' \\ j_2 & j_1 & r \end{Bmatrix}. \tag{37.13}$$

Another expression, equivalent to (37.13) can be obtained directly, without using as an intermediate step, the tensor expansion of the interaction. Due to the antisymmetry of the wave functions (37.1) in the first n_1 particles and in the last n_2 particles we have

$$\langle j_1^{n_1}(\alpha_1 J_1) j_2^{n_2}(\alpha_2 J_2) J \,|\, \sum_{i=1}^{n_1} \sum_{k=n+1}^{n_1+n_2} V_{ik} \,|\, j_1^{n_1}(\alpha_1' J_1') j_2^{n_2}(\alpha_2' J_2') J \rangle$$

$$= n_1 n_2 \langle j_1^{n_1}(\alpha_1 J_1) j_2^{n_2}(\alpha_2 J_2) J | V_{n_1, n_1+n_2} | j_1^{n_1}(\alpha_1' J_1') j_2^{n_2}(\alpha_2' J_2') J \rangle. \tag{37.14}$$

We can now use c.f.p. in the two functions (37.1) which appear in the matrix element (37.14). We obtain matrix elements of the form

$$\langle j_1^{n_1-1}(\alpha_{11}J_{11}) j_{1,n_1} J_1, j_2^{n_2-1}(\alpha_{22}J_{22}) j_{2,n_1+n_2} J_2, J | V_{n_1, n_1+n_2} |$$

$$| j_2^{n_2-1}(\alpha_{11}J_{11}) j_{1,n_1} J_1', j_2^{n_2-1}(\alpha_{22}J_{22}) j_{2,n_1+n_2} J_2', J \rangle \tag{37.15}$$

where α_{11}, J_{11}, α_{22}, and J_{22} must have the same values on both sides for a nonvanishing matrix element. In order to evaluate (37.15) we apply to both sides a change of coupling transformation to couple $\mathbf{j}_{1,n_1}$ and $\mathbf{j}_{2,n_1+n_2}$ to $\mathbf{J}'$ and $\mathbf{J}_{11}$ and $\mathbf{J}_{22}$ to $\mathbf{J}_{12}$. This will enable us to carry out the integration in (37.15) which will yield a matrix element in the $j_1 j_2$ configuration. According to (14.37) we have

$$\langle J_{11} j_1(J_1) J_{22} j_2(J_2) J | J_{11} J_{22}(J_{12}) j_1 j_2(J') J \rangle$$
$$= [(2J_1+1)(2J_2+1)(2J_{12}+1)(2J'+1)]^{1/2} \begin{Bmatrix} J_{11} & j_1 & J_1 \\ J_{22} & j_2 & J_2 \\ J_{12} & J' & J \end{Bmatrix}. \quad (37.16)$$

We thus obtain for the matrix element (37.14) the following expression

$$n_1 n_2 \sum_{\alpha_{11} J_{11} \alpha_{22} J_{22}} [j_1^{n_1}\alpha_1 J_1 \{| j_1^{n_1-1}(\alpha_{11} J_{11}) j_1 J_1] [j_1^{n_1-1}(\alpha_{11} J_{11}) j_1 J_1' |\} j_1^{n_1}\alpha_1' J_1']$$

$$\times [j_2^{n_2}\alpha_2 J_2 \{| j_2^{n_2-1}(\alpha_{22} J_{22}) j_2 J_2] [j_2^{n_2-1}(\alpha_{22} J_{22}) j_2 J_2' |\} j_2^{n_2}\alpha_2' J_2']$$

$$\times [(2J_1+1)(2J_1'+1)(2J_2+1)(2J_2'+1)]^{1/2}$$

$$\times \sum_{J'} (2J'+1) V(j_1 j_2 J') \sum_{J_{12}} (2J_{12}+1) \begin{Bmatrix} J_{11} & j_1 & J_1 \\ J_{22} & j_2 & J_2 \\ J_{12} & J' & J \end{Bmatrix} \begin{Bmatrix} J_{11} & j_1 & J_1' \\ J_{22} & j_2 & J_2' \\ J_{12} & J' & J \end{Bmatrix}. \quad (37.17)$$

This expression (37.17), is the same as (37.13) but for one difference. The sum of products of two 9-j coefficients in (37.17) is replaced by a sum of products of four 6-j coefficients in (37.13). The equality of the two sums can be verified also by using the expansion (15.19) of the 9-j coefficients in terms of the Racah coefficients and the properties of the latter.

In the special case $J_1 = J_1' = 0$ both (37.13) and (37.17) reduce to the result (37.10) obtained above. Another case, in which the expression of the interaction energy becomes simpler, is for $n_2 = 1$. In this case $J_2 = J_2' = j_2$ and there is only one value of J_{22}, namely $J_{22} = 0$, and the corresponding c.f.p. should be put equal to 1. We thus obtain for this special case, either from (37.13) or from (37.17), the following result

$$\langle j_1^n(\alpha_1 J_1) j_2 J \left| \sum_{i=1}^{n} V_{i,n+1} \right| j_1^n(\alpha_1' J_1') j_2 J \rangle$$

$$= n \sum_{\alpha_{11} J_{11}} [j_1^n \alpha_1 J_1 \{| j_1^{n-1}(\alpha_{11} J_{11}) j_1 J_1] [j_1^{n-1}(\alpha_{11} J_{11}) j_1 J_1' |\} j_1^n \alpha_1' J_1']$$

$$\times [(2J_1+1)(2J_1'+1)]^{1/2} \sum_{J'} (2J'+1) V(j_1 j_2 J') \begin{Bmatrix} J_{11} & j_1 & J_1 \\ j_2 & J & J' \end{Bmatrix} \begin{Bmatrix} J_{11} & j_1 & J_1' \\ j_2 & J & J' \end{Bmatrix}. \quad (37.18)$$

In the special case of $n = 2j_1$, using the appropriate c.f.p. given essentially by (26.12), Eq. (37.18) reduces to (22.52) giving the j_1-hole-j_2-particle interaction.

If we return to the form (37.5) of the matrix element we can obtain various relations between matrix elements in different configurations. In order to do it we can use the seniority scheme in the $j_1^{n_1}$ configuration and in the $j_2^{n_2}$ configuration. The matrix elements of the single particle tensor operators $\mathbf{U}^{(r)}$ are given in terms of those in the j^v configuration by (28.22), (28.25), (28.28), and (28.40). We shall not go here into the detailed derivation of these relations.

Let us consider, as an application of (37.18), the ground configuration of K^{42}. The proton configuration of this nucleus is taken to be, as in the K^{39} case, that of one $d_{3/2}$ proton missing from closed shells. There is only one state of this configuration, with $J_2 = j_2 = \frac{3}{2}$, which can be treated as a single $d_{3/2}$ hole state. The neutrons outside closed shells are most probably in the $f^3_{7/2}$ configuration. This is the ground configuration of Ca^{43} which was discussed in Section 30. There may be several states of the $f^3_{7/2}$ configuration with different values of J_1 which can couple with $j_2 = \frac{3}{2}$ to form independent states with any given value of J. All these states belong to the same configuration and therefore have the same expectation values of the kinetic energy and interaction energy with the closed shells. Only the expectation values of the interaction within the $f^3_{7/2}$ configuration, as well as of the $d^{-1}_{3/2}$-$f^3_{7/2}$ interaction, may well be different. These expectation values, as well as the nondiagonal matrix elements of the interaction matrix, are given by (37.2) and (37.18). The expectation values of

$$V(f^3_{7/2}J_1) = \langle f^3_{7/2}J_1 \Big| \sum_{i<k}^{3} V_{ik} \Big| f^3_{7/2}J_1 \rangle$$

were discussed in detail in Section 30. The matrix elements of the interaction between the $d_{3/2}$ proton hole and the $f_{7/2}$ neutrons are expressed by (37.18) in terms of $V(d^{-1}_{3/2}f_{7/2}\,J')$. The interaction between a $d_{3/2}$ proton hole and an $f_{7/2}$ neutron was discussed at the end of Section 22. We shall make use in the following of the actual values of the effective interaction considered there.

The state with $J_1 = \frac{7}{2}$ of the $f^3_{7/2}$ neutron configuration gives rise to states with $J = 2, 3, 4, 5$. For $J_1 = \frac{5}{2}$ we have states with $J = 1, 2, 3, 4$, etc. In order to find the energies of the various states, we have to write down the interaction matrices for each value of J and diagonalize them. The diagonal elements of these matrices contain the interaction in the $f^3_{7/2}$ neutron configuration. It is therefore clear that the lowest states of the $f^3_{7/2}$ configuration will have more weight in the lowest eigenstates for each value of J. In fact, for the states with $J = 2, 3, 4$ it is sufficient to consider the two lowest states, with $J_1 = \frac{7}{2}$ and $J_1 = \frac{5}{2}$, of the $f^3_{7/2}$ neutron configuration. The energies of these two states are only 0.37 Mev apart whereas the other states (of Ca^{43}) lie much higher. A more detailed calculation, including also other states, makes a negligible change in the following results.

If we are interested only in the level spacings in K^{42} (and not in the binding energy) we can ignore the kinetic energy and interaction with closed shells and consider the matrix defined by (37.2). We can further normalize $V(f^3_{7/2}\ J_1 = \frac{7}{2})$ to zero. We build the interaction matrices using (37.18) and the appropriate c.f.p. (given in the Appendix). For example, the interaction matrix for $J = 2$ is given in Table 37.1.

TABLE 37.1

THE INTERACTION MATRIX FOR $J = 2$ IN K^{42} (IN MEV)

J_1 \ J_1'	$\frac{7}{2}$	$\frac{5}{2}$
$\frac{7}{2}$	$\frac{1}{46200}(50325V_2 + 25025V_3 + 42075V_4 + 21175V_5)$	$\frac{1}{46200}\sqrt{\frac{15}{11}}(18150V_2 - 8470V_3 - 28314V_4 + 18634V_5)$
$\frac{5}{2}$	$\frac{1}{46200}\sqrt{\frac{15}{11}}(18150V_2 - 8470V_3 - 28314V_4 + 18634V_5)$	$\frac{1}{46200}(30360V_2 + 30030V_3 + 48642V_4 + 29568V_5) + 0.37$

The V_J in Table 37.1 are defined by $V_J = V(d^{-1}_{3/2} f_{7/2}\ J)$. On putting all the V_J equal, $V_J = V$, the diagonal elements in Table 37.1 contain simply $3V$ [and the $V(f^3_{7/2}\ J_1)$] whereas the nondiagonal elements vanish. This is also the case for the matrices for other values of J. It is therefore possible to consider the matrices for the various values of J in which we add (or subtract) the same number V from all the V_J. This will shift all eigenvalues by the same amount $3V$ but will not change their relative positions. We shall use this freedom to normalize the strongest interaction V_4 to zero. We then obtain from Table 22.1 the following values for the V_J: $V_2 = 0.80$, $V_3 = 0.03$, $V_4 = 0$, and $V_5 = 0.89$ Mev. Using these values, the interaction matrix becomes a numerical one given in Table 37.2. A glance at Table 37.2 shows that the interaction energy $V[d^{-1}_{3/2} f^3_{7/2}\ (J_1) J = 2]$ is stronger in the state with $J_1 = \frac{5}{2}$ than in the state with $J_1 = \frac{7}{2}$ by 0.19 Mev. Only when the difference $V(f^3_{7/2}\ J_1 = \frac{7}{2}) - V(f^3_{7/2}\ J_1 = \frac{5}{2}) = -0.37$ Mev is taken into account does the expectation value for $J_1 = \frac{7}{2}$ become lower (by 0.18 Mev). The difference between the diagonal matrix elements

TABLE 37.2

THE NUMERICAL MATRIX FOR $J = 2$ IN K^{42} (IN MEV)

J_1 \ J_1'	$\frac{7}{2}$	$\frac{5}{2}$
$\frac{7}{2}$	1.30	0.78
$\frac{5}{2}$	0.78	$1.11 + 0.37 = 1.48$

turns out to be smaller than the nondiagonal element in Table 37.2 so that a large admixture is expected. The diagonalization of the matrix for $J = 2$ shows that the lowest eigenvalue is lower by 0.69 Mev than the lowest diagonal element. The lowest eigenstate turns out to be

$$0.75\,\psi[f^3_{7/2}(\tfrac{7}{2})d^{-1}_{3/2}\ J = 2] - 0.66\,\psi[f^3_{7/2}(\tfrac{5}{2})d^{-1}_{3/2}\ J = 2]$$

and the per cent probabilities of the states with $J_1 = \frac{7}{2}$ and $J_1 = \frac{5}{2}$ are 56 % and 44 % respectively.

Looking at the matrices for other values of J, we find that the lowest diagonal element occurs for $J = 3$ (and $J_1 = \frac{7}{2}$). However, the nondiagonal element in the matrix for $J = 3$ is only 0.35 Mev and as a result the lowest eigenvalue with $J = 3$ turns out to be higher than the lowest eigenvalue for $J = 2$. This latter eigenvalue is lower than all other eigenvalues, with other J values, as well. We thus see that in general the spin of the ground state of an odd odd nucleus (like K^{42}) cannot be determined by considering the expectation values only. The fact that the ground-state spin of K^{42} is $J = 2$ and not $J = 3$ is due to the occurrence of a low-lying state with $J_1 = \frac{5}{2}$ in the $f^3_{7/2}$ neutron configuration. It is worthwhile to mention that the lowest state with $J = 4$ lies higher than the lowest states with $J = 2$ and $J = 3$. The fact that the ground-state spin of K^{42} is different from that of K^{40} (which has $J = 4$) is thus simply understood within the framework of the shell model by considering a pure $d^{-1}_{3/2}f^3_{7/2}$ configuration.

The general case of nucleons outside closed shells is slightly more complicated. Let us consider the case of n_1-nucleons in the j_1-orbit and n_2-nucleons in the j_2-orbit. We have to calculate the matrix elements (37.2) of the interaction energy. The first two sums on the right-hand side of (37.2) involve interactions within groups of equivalent nucleons and were dealt with before. It is the last term that we have to deal with explicitly. To do this, we can use the same procedure which was used in the derivation of (37.17). The change of coupling transformation (37.16) should be replaced in the present case by a product of two such transformations. The first is identical with (37.16) and the second has the same form but involves the isospin vectors rather than those of the angular momentum. The analog of (37.17) will be in the present case, for a charge-independent interaction, as follows

$$\langle j_1^{n_1}(\alpha_1 T_1 J_1)\, j_2^{n_2}(\alpha_2 T_2 J_2) T\ J \Big| \sum_{i=1}^{n_1} \sum_{k=n_1+1}^{n_1+n_2} V_{ik} \Big| j_1^{n_1}(\alpha_1' T_1' J_1')\, j_2^{n_2}(\alpha_2' T_2' J_2') T\ J \rangle$$

$$= n_1 n_2 \sum_{\alpha_{11} T_{11} J_{11} \alpha_{22} T_{22} J_{22}} [j_1^{n_1} \alpha_1 T_1 J_1 \{| j_1^{n_1-1}(\alpha_{11} T_{11} J_{11})\, j_1 T_1 J_1]$$

$$[j_1^{n_1-1}(\alpha_{11} T_{11} J_{11})\, j_1 T_1' J_1' |\} j_1^{n_1} \alpha_1' T_1' J_1'] \, [j_2^{n_2} \alpha_2 T_2 J_2 \{| j_2^{n_2-1}(\alpha_{22} T_{22} J_{22})\, j_2 T_2 J_2]$$

$$\times [j_2^{n_2-1}(\alpha_{22} T_{22} J_{22})\, j_2 T_2' J_2' |\} j_2^{n_2} \alpha_2' T_2' J_2']$$

$$\times [(2T_1+1)(2T_1'+1)(2T_2+1)(2T_2'+1)$$

$$\times (2J_1+1)(2J_1'+1)(2J_2+1)(2J_2'+1)]^{1/2}$$

$$\times \sum_{T'J'} (2T'+1)(2J'+1) V(j_1 j_2 T' J')$$

$$\times \sum_{T_{12}} (2T_{12}+1) \begin{Bmatrix} T_{11} & \frac{1}{2} & T_1 \\ T_{22} & \frac{1}{2} & T_2 \\ T_{12} & T' & T \end{Bmatrix} \begin{Bmatrix} T_{11} & \frac{1}{2} & T_1' \\ T_{22} & \frac{1}{2} & T_2' \\ T_{12} & T' & T \end{Bmatrix}$$

$$\times \sum_{J_{12}} (2J_{12}+1) \begin{Bmatrix} J_{11} & j_1 & J_1 \\ J_{22} & j_2 & J_2 \\ J_{12} & J' & J \end{Bmatrix} \begin{Bmatrix} J_{11} & j_1 & J_1' \\ J_{22} & j_2 & J_2' \\ J_{12} & J' & J \end{Bmatrix}. \tag{37.19}$$

In the special case of one j_2 nucleon, i.e., for $n_2 = 1$, $T_{22} = 0$, and $J_{22} = 0$, (37.19) reduces to a simpler expression which is the analog of (37.18).

Equation (37.19) gives the matrix elements of the interaction energy in terms of the matrix elements of the interaction in the $j_1 j_2$ configuration, namely $V(j_1 j_2 T\ J)$. In the general case, it is impossible to find a simple relationship between the coefficients of $V(j_1 j_2\ T=0\ J)$ and $V(j_1 j_2\ T=1\ J)$ which appear in (37.19). In the simpler case mentioned above, where $n_1 > 2j_1 + 1$ with a closed j_1 neutron shell and a $j_2^{n_2}$ neutron configuration, the interaction energy is a sum of two terms. One is the interaction of the n_2 extra neutrons with the closed j_1 neutron shell. The second is the interaction between the n_2 extra neutrons and the $j_1^{n_1-(2j_1+1)}$ proton configuration. In the latter term, only the combinations

$$V(j_1 j_2 \; T=0 \; J) + V(j_1 j_2 \; T=1 \; J) = 2V(j_1 j_2 \; J)$$

occur.

It was shown in Section 25 that the matrix elements of single particle operators evaluated with wave functions (37.1), antisymmetrized between the groups of inequivalent particles, are the same as those evaluated with the original functions (37.1). These matrix elements are thus the sum of matrix elements within the groups of equivalent particles. We shall now consider matrix elements of single particle operators between two different configurations. According to the considerations of Section 25, the matrix elements do not vanish only if the two configurations differ at most by the quantum number of one particle. Let us therefore consider the matrix element

$$\langle [j_1^{n-1}(\alpha_1 T_1 J_1) j_2]_a T' J' M' \left| \sum_{i=1}^{n} f_i \right| j_1^n \alpha \, T \, J \, M \rangle$$

$$= \int \psi_a^*[j_1^{n-1}(\alpha_1 T_1 J_1) j_2 T' J' M'] \left[\sum_{i=1}^{n} f_i \right] \psi(j_1^n \alpha \, T \, J \, M). \qquad (37.20)$$

This expression involves the wave function $\psi(j_1^n \alpha \, T \, J \, M)$ which is antisymmetric in the coordinates (including the charge coordinate) of the n particles. The other function ψ_a is the function

$$\psi[j_1^{n-1}(\alpha_1 T_1 J_1) j_2 T' J' M']$$

antisymmetrized in the coordinates of the first $n-1$ nucleons and the last nucleon. This antisymmetrization can be carried out simply by the operation

$$\mathscr{A}\psi[j_1^{n-1}(\alpha_1 T_1 J_1) j_2 T' J' M']$$

$$= \psi[j_1^{n-1}(\alpha_1 T_1 J_1) j_{2,n} T' J' M'] - \sum_{i=1}^{n-1} \psi[j_1^{n-1}(\alpha_1 T_1 J_1) j_{2,i} T' J' M']. \qquad (37.21)$$

Any two functions on the right-hand side of (37.21) are orthogonal to each other. Therefore, in order to normalize the wave function (37.21) we have to multiply it by $n^{-1/2}$. Introducing this expression into (37.20) and using the properties of the antisymmetrizer $\mathscr{A}$ we obtain

$$\frac{1}{\sqrt{n}} \int \mathscr{A}\psi^*[j_1^{n-1}(\alpha_1 T_1 J_1) j_2 T' J' M'] \left[\sum_{i=1}^{n} f_i\right] \psi(j_1^n \alpha\, T\, J\, M)$$

$$= \sqrt{n} \int \psi^*[j_1^{n-1}(\alpha_1 T_1 J_1) j_{2,n} T' J' M'] \left[\sum_{i=1}^{n} f_i\right] \psi(j_1^n \alpha\, T\, J\, M). \qquad (37.22)$$

Since the single nucleon wave functions in the j_1 and j_2 orbits are orthogonal, the only term that will give a nonvanishing contribution in the summation on the right-hand side of (37.22) is the one with $i = n$. Integrating (37.22) over the coordinates of the first $n - 1$ nucleons we obtain for the matrix element (37.20) the form

$$\sqrt{n}\,[j_1^{n-1}(\alpha_1 T_1 J_1) j_1 T\, J \rbrace j_1^n \alpha\, T\, J]\, \langle T_1 J_1, \tfrac{1}{2} j_{2,n}, T' J' M' | f_n | T_1 J_1, \tfrac{1}{2} j_{1,n}, T\, J\, M\rangle. \qquad (37.23)$$

If f is an irreducible tensor operator a more explicit expression can be obtained. We consider the case in which f is a product of an irreducible tensor of degree k operating on the space and spin coordinates $\mathbf{f}^{(k)}$ and another irreducible tensor $\mathbf{h}$ operating on the isospin coordinates. We can then use (15.27) to obtain from (37.23) the following expression for the matrix element, reduced with respect to the angular momentum only

$$\sqrt{n}\,[j_1^{n-1}(\alpha_1 T_1 J_1) j_1 T\, J \rbrace j_1^n \alpha\, T\, J]\, \langle T_1 \tfrac{1}{2} T'\, M_T | \mathbf{h}_n | T_1 \tfrac{1}{2} T\, M_T\rangle$$

$$\times (-1)^{J_1 + j_1 + J' + k} [(2J+1)(2J'+1)]^{1/2} \begin{Bmatrix} j_2 & j_1 & k \\ J & J' & J_1 \end{Bmatrix} (j_2 || \mathbf{f}^{(k)} || j_1). \qquad (37.24)$$

In (37.24) the M_T quantum number is explicitly indicated.

The dependence on n, for definite values of j_1, j_2, T_1, T, T', J_1, J, and J', of (37.23) and (37.24), is given by the factor

$$\sqrt{n}\,[j_1^{n-1}(\alpha_1 T_1 J_1) j_1 T\, J \rbrace j_1^n \alpha\, T\, J].$$

In some applications, the matrix elements (37.20) of the operator $\Sigma\, \mathbf{f}_i$ determine the rate of an electromagnetic or other transition. Such matrix elements appear also in the calculation of various reaction rates, e.g., stripping and pickup reactions. In all these cases the squares of the matrix elements appear. The dependence on n of these is given by

$$n[j_1^{n-1}(\alpha_1 T_1 J_1) j_1 T\, J \rbrace j_1^n \alpha\, T\, J]^2. \qquad (37.25)$$

We shall now consider the values of the factor (37.25) in some simple cases with lowest seniority.

The c.f.p. for the states with lowest seniority in the case of maximum isospin (or protons and neutrons in different shells*) are given by (28.12) for n even and by (28.17) for n odd. Using these values we obtain for the factor (37.25) the following results (in which j is written instead of j_1)

$$n[j^{n-1}(v=1\ J=j)j\ J=0|\}j^n\ v=0\ J=0]^2=n \qquad \text{for } n \text{ even} \tag{37.26}$$

$$n[j^{n-1}(v=0\ J=0)j\ J=j|\}j^n\ v=1\ J=j]^2=\frac{2j+2-n}{2j+1} \qquad \text{for } n \text{ odd.} \tag{37.27}$$

These expressions occur when the transition is between the state with lowest seniority of the j^n configuration and the state in which the j' particle (with j' replacing j_2) is coupled to the state with lowest seniority of the j^{n-1} configuration. In the cases where the seniority is a good quantum number, the states with the lowest seniorities are usually the ground states and (37.26) and (37.27) are expected to be applicable.

The behaviour of the factors (37.25), in the present case, for even values of n is quite different from the one for odd values. It follows from (37.26) that for even n this factor is a linear increasing function of n whereas the factor (37.27) for odd n starts at the single particle value for $n = 1$ and then decreases (linearly) with n. This difference in behaviour, verified experimentally in several cases, can be given a simple physical interpretation. Let us first consider the transition from the j^n configuration to the $j^{n-1}\,j'$ configuration. The factor (37.25) gives then the probability that one of the j-particles can be taken out leaving the other $n-1$ particles in the state with the given value of J_1. In the case of n even and $v=0\ J=0$ each of the n particles can be removed leaving the others in the $v=1\ J=j$ state. On the other hand, if n is odd and $v=1\ J=j$, only the removal of the "odd particle," which is not coupled to $J=0$, can give the $v=0\ J=0$ state. The factor in this case (37.27) is even smaller than unity due to the probability (increasing with n) that the removal of the odd particle leaves the j^{n-1} configuration in a state with $v=2$ (and $J_1>0$ even). Alternatively, we can consider the transition from the $j^{n-1}j'$ configuration to the j^n configuration. The factor (37.25) should then be interpreted as the

(*) In this case the j_1-neutron shell is closed and there are $n-1$ protons in the j_1-orbit and one j_2-neutron. We assign an $m_t=-\frac{1}{2}$ state to the j_2-nucleon and $m_t=+\frac{1}{2}$ to each of the $n-1$ nucleons in the j_1-orbit. We then have to antisymmetrize the resulting wave function according to (37.21). The antisymmetrization is essential if we consider processes, like beta decay, in which a neutron changes into a proton.

probability of putting another particle in the j-orbit and obtaining the state with the given J. In the case of n odd and $v = 1$ $J = j$, the number of available free orbits is decreased due to the Pauli principle (Fermi-Dirac statistics) by the presence of the $n - 1$ particles, by $[2j + 1 - (n - 1)]/(2j + 1)$ which is identical with (37.27). On the other hand, if n is even and $v = 0$ $J = 0$, the addition of another particle adds another *pair* to the existing $(n - 1)/2$ pairs coupled to $J = 0$. The factor n given by (37.26) for this case is highly suggestive of the effect of Bose-Einstein statistics obeyed by the fermion pairs (coupled to $J = 0$). Although the pairs coupled to $J = 0$ are not strictly speaking bosons, it seems as if they behave as such in determining the statistical factor in the present case.

In the general case of nuclear configurations we obtain the expressions analogous to (37.26) and (37.27) by using the c.f.p. given by (35.25) and (35.28). We shall not discuss these any further but rather present a simple example taken from actual nuclei in order to illustrate the possible use of such expressions. We consider electric quadrupole ($E2$) transitions in nuclei beyond O^{16} in which a single nucleon jumps from the $2s_{1/2}$ orbit into the $1d_{5/2}$ orbit. In F^{17}, for instance, with one proton outside the closed shells of O^{16}, the single $2s_{1/2}$ level lies 0.50 Mev above the $1d_{5/2}$ state. The transition between these two states is an electric quadrupole ($E2$) and its rate is given by the matrix elements of an irreducible tensor of degree two. The irreducible tensor operator of degree 2 is, according to (17.9), a product of the spherical harmonics of order 2 and a radial integral. Using some radial functions for the nucleons, e.g., harmonic oscillator wave functions, the matrix elements of this $E2$ transition can be calculated. If we make a reasonable choice of the radial extent of the wave function we obtain that the observed $E2$ transition is too fast. Far more difficult is the situation in the mirror transition (of 0.87 Mev) in O^{17}. In that case, an $E2$ transition could occur only due to the magnetic moment of the extra neutron making the rate very slow (in fact, the matrix elements for electric multipole radiation in the case of neutron transitions were completely ignored in Section 17). The observed transition is much faster and is comparable to the single proton transition. It is thus clear that in both cases there are effects which cannot be accounted for by the simple single nucleon wave functions. Applying to the transition rates the same philosophy we applied to the nuclear energies we can still try to use simple shell model wave functions with *modified* or *effective* single nucleon operators. In the case of $E2$ transitions we can use for example effective charges for the single proton and single neutron different from 1 and 0. If these effective single nucleon operators are the same for all transitions of the same type in a group of several nuclei, we can find simple relations between the rates of the transitions. Such relations will depend on geometrical factors which appear in (37.24) as well as on the statistical factors (37.25).

Two $E2$ transitions which are of the same type as those in F^{17} and O^{17} occur in another mirror pair — Al^{25} and Mg^{25}. We assume that the $T = \frac{1}{2}$ $J = \frac{5}{2}$ ground states of these nuclei belong to the $d^9_{5/2}$ configuration and have $v = 1$ $t = \frac{1}{2}$. The first excited $T = \frac{1}{2}$ $J = \frac{1}{2}$ states of these nuclei (at 0.45 Mev and 0.58 Mev respectively) are taken to belong to the $d^8_{5/2}s_{1/2}$ configuration. Clearly, the isospin T_1 of the $d^8_{5/2}$ configuration can be either $T_1 = 0$ or $T_1 = 1$ in order to give, when coupled to $t = \frac{1}{2}$, a total isospin $T = \frac{1}{2}$. However, since the $T_1 = 1$ states of the $d^8_{5/2}$ configuration lie much higher than those with $T_1 = 0$, we take T_1 to have only the value $T_1 = 0$. Similarly, we take the other quantum numbers of the $d^8_{5/2}$ configuration, to which the $s_{1/2}$ nucleon is coupled, to be $v = 0$, $t = 0$, and $J_1 = 0$. Under these assumptions, all geometrical factors which appear in (37.24) [or (37.23)] both for the angular momentum as well as for the isospin quantum numbers, for Al^{25} are the same as those for F^{17}. Also the same geometrical factors, with $M_T = -\frac{1}{2}$ this time, appear for Mg^{25} and O^{17}. Thus, the squares of the matrix elements for these $E2$ transitions for mass 25 nuclei differ from the corresponding ones for mass 17 only by the statistical factor (37.25). The present example, with $T = \frac{1}{2}$ in the $d^9_{5/2}$ configuration corresponds to Case I of Section 35 and the c.f.p. required is given by (35.28). Using the value of this c.f.p. (with $T_1 = 0 = T - \frac{1}{2}$) we obtain for the statistical factor (37.25) the value

$$\frac{4j + 4 - n - 2T}{2(2j + 1)} = \frac{2j - 3}{2j + 1} \text{ for } n=9 \ T=\tfrac{1}{2}, \text{ equal to } \tfrac{1}{3} \text{ for } j = \tfrac{5}{2}\cdot \quad (37.28)$$

The transition probability per unit time, given by (17.5), is proportional to $(\hbar\omega)^{2L+1}$, where $\hbar\omega$ is the energy of the emitted photon and L is the order of the transition. In the present example, of electric quadrupole radiation, we have $L = 2$. Thus, the lifetimes of the nuclear states are inversely proportional to $(\hbar\omega)^5$, where $\hbar\omega$ is the energy difference between the ground and excited states. Starting from the single nucleon transition rates given by the lifetimes of F^{17} and O^{17}, we can calculate the lifetimes in Al^{25} and Mg^{25} by dividing the former by the fifth power of the ratio of energy differences as well as by the statistical factor (37.28). The F^{17} (mean) lifetime is $(4.45 \pm 0.22)10^{-10}$ sec and that of O^{17} is $(2.55 \pm 0.13)10^{-10}$ sec. From these we obtain for the lifetimes of Al^{25} and Mg^{25}, respectively

$$(4.45 \pm 0.22)\ 10^{-10} \left(\frac{0.50}{0.45}\right)^5 \times 3 = (2.26 \pm 0.11)\ 10^{-9} \text{ sec}$$

$$(2.55 \pm 0.13)\ 10^{-10} \left(\frac{0.87}{0.58}\right)^5 \times 3 = (5.81 \pm 0.30)\ 10^{-9} \text{ sec.}$$

These calculated values agree fairly well with the measured lifetimes of $(2.6 \pm 0.4)10^{-9}$ sec and $(5.1 \pm 0.3)10^{-9}$ sec respectively.

It is a straightforward matter to generalize (37.23) or (37.24) to the case with several nucleons in the j_2-orbit. We consider the matrix element

$$\langle [j_1^{n_1}(\alpha_1' T_1' J_1')\, j_2^{n_2}(\alpha_2' T_2' J_2')]_a\, T' J' M_T' M' \,\Big|\, \sum_{i=1}^{n_1+n_2} f_i \,\Big|\, [j_1^{n_1-1}(\alpha_1 T_1 J_1)\, j_2^{n_2+1}(\alpha_2 T_2 J_2)]_a\, T\, J\, M_T M \rangle \qquad (37.29)$$

between antisymmetrized states (37.1) of the $j_1^{n_1} j_2^{n_2}$ configuration and the $j_1^{n_1-1} j_2^{n_2+1}$ configuration. The antisymmetrization of the two wave-functions involved can be carried out by antisymmetrizers $\mathscr{A}'$ and $\mathscr{A}$ respectively. These operators are defined by

$$\mathscr{A}\psi[j_{1,2,\ldots,m}^{m}(\alpha_1 T_1 J_1)\, j_{m+1,\ldots,m+n}^{\prime n}(\alpha_2 T_2 J_2) T\, J]$$
$$= \sum_P (-1)^P \psi[j_{i_1,i_2,\ldots,i_m}^{m}(\alpha_1 T_1 J_1)\, j^{\prime n}(\alpha_2 T_2 J_2) T\, J]. \qquad (37.30)$$

In (37.30) the nucleon numbers $i_1, i_2, \ldots, i_m$ stand for *one* choice of m numbers out of the numbers 1 to $m + n$. The summation is over permutations P each of which substitutes a different set of numbers in the first m places. Altogether the summation includes $\binom{m+n}{m} = \binom{m+n}{n}$ such permutations. All functions on the right-hand side of (37.30) are orthogonal so that the normalization factor of the wave function (37.30) is simply $\binom{m+n}{m}^{-1/2}$. Using these antisymmetrizers we obtain for (37.29) the expression

$$\binom{n_1+n_2}{n_1}^{-1/2} \binom{n_1+n_2}{n_1-1}^{-1/2} \int \mathscr{A}'\psi^*[j_1^{n_1}(\alpha_1' T_1' J_1')\, j_2^{n_2}(\alpha_2' T_2' J_2') T' J']$$
$$\times \Big(\sum_{i=1}^{n_1+n_2} f_i\Big)\, \mathscr{A}\psi[j_1^{n_1-1}(\alpha_1 T_1 J_1)\, j_2^{n_2+1}(\alpha_2 T_2 J_2) T\, J]. \qquad (37.31)$$

We can now let $\mathscr{A}'$ operate on the rest of the integrand which is fully antisymmetric, obtaining for (37.31) the form

$$\binom{n_1+n_2}{n_1}^{1/2} \binom{n_1+n_2}{n_1-1}^{-1/2} \int \psi^*[j_1^{n_1}(\alpha_1' T_1' J_1')\, j_2^{n_2}(\alpha_2' T_2' J_2') T' J']$$
$$\times \Big[\sum_{i=1}^{n_1+n_2} f_i\Big]\, \mathscr{A}\psi[j_1^{n_1-1}(\alpha_1 T_1 J_1)\, j_2^{n_2+1}(\alpha_2 T_2 J_2) T\, J]. \qquad (37.32)$$

If we carry out the integration over the coordinates of the first n_1 nucleons we see that the only terms in the summation in (37.32) which will contribute to the integral are the f_i for which $i = 1, 2, ..., n_1$. Due to the full antisymmetry of $\psi(j_1^{n_1}\alpha_1' T_1' J_1')$ the contributions of all these f_i are equal. We can therefore evaluate the integral over one of them, f_{n_1}, for instance, and multiply the result by n_1. The term with f_{n_1} is a sum over the various functions appearing in $\mathscr{A}\psi$, as given in (37.30). However, only one of these functions will give a nonvanishing contribution after the integration over the coordinates of the first $n_1 - 1$ nucleons. This is the function in which the first $n_1 - 1$ nucleons are in the j_1-orbit. We thus obtain for the matrix element (37.29) the form

$$n_1 \binom{n_1 + n_2}{n_1}^{1/2} \binom{n_1 + n_2}{n_1 - 1}^{-1/2}$$

$$\times \int \psi^*[j_{1;1,2,\dots,n_1}^{n_1}(\alpha_1' T_1' J_1') j_{2,n_1+1,\dots,n_1+n_2}^{n_2}(\alpha_2' T_2' J_2') T' J']$$

$$\times f_{n_1} \psi[j_{1;1,2,\dots,n_1-1}^{n_1-1}(\alpha_1 T_1 J_1) j_{2,n_1,n_1+1,\dots,n_1+n_2}^{n_2+1}(\alpha_2 T_2 J_2) T\, J]. \qquad (37.33)$$

The numerical factor multiplying the integral in (37.33) is equal to $\sqrt{n_1(n_2 + 1)}$. Integration over the coordinates of the first $n_1 - 1$ and the last n_2 nucleons yields two c.f.p., in th $j_1^{n_1}$ and in the $j_2^{n_2+1}$ configurations. The integration over the coordinates of nucleon n_1 gives (37.33) in terms of a matrix element of f_{n_1} as follows:

$$\sqrt{n_1(n_2 + 1)}\, [j_1^{n_1}\alpha_1' T_1' J_1' \{| j_1^{n_1-1}(\alpha_1 T_1 J_1) j_1 T_1' J_1'] (-1)^{n_2}$$

$$\times [j_2^{n_2}(\alpha_2' T_2' J_2') j_2 T_2 J_2 |\} j_2^{n_2+1} \alpha_2 T_2 J_2]$$

$$\times \langle T_1 J_1 \tfrac{1}{2} j_{1,n_1}(T_1' J_1'), T_2' J_2', T' J' M_T' M' | f_{n_1} | T_1 J_1, T_2' J_2' \tfrac{1}{2} j_{2,n_1}(T_2 J_2), T J M_T M \rangle.$$
$$(37.34)$$

If f is a product of a tensor operator of degree k with respect to the space and spin coordinates and a tensor operator of degree k' with respect to the isospin coordinates, (37.34) can be simplified. The reduced matrix element of (37.29) with respect to the angular momentum and isospin can be obtained by a change of coupling transformation in the matrix element in (37.34). By inspection, one would expect a 9-j symbol to appear where the reduced matrix element of (37.34) is expressed in terms of $(j_1 || \mathbf{f}^{(k)} || j_2)$. Another 9-$j$ symbol with the isospin vectors should also appear. To find out the correct phase we have to carry out the transformation in detail. We first interchange J_2' and j_2 on the right-hand side of the matrix element in (37.34) obtaining, due to

(13.56), the factor $(-1)^{J_2'+j_2-J}$. We then couple J_1 with j_2 to form some J_0 using (14.38) combined with (15.7). Using now (15.26) and (15.27) we obtain a sum over J_0 of a product of three Racah coefficients. This sum is equal to a 9-j symbol due to (15.19). The final result for the reduced matrix element is

$$\sqrt{n_1(n_2+1)}\,[j_1^{n_1}\alpha_1'T_1'J_1'\{|j_1^{n_1-1}(\alpha_1T_1J_1)j_1T_1'J_1']$$

$$\times(-1)^{n_2}\,[j_2^{n_2}(\alpha_2'T_2'J_2')j_2T_2J_2|\}j_2^{n_2+1}\alpha_2T_2J_2]$$

$$\times(-1)^{T_2'+1/2-T_2+J_2'+j_2-J_2}\,[(2T_1'+1)(2T_2+1)(2T+1)(2T'+1)$$

$$\times(2J_1'+1)(2J_2+1)(2J+1)(2J'+1)]^{1/2}$$

$$\times\begin{Bmatrix}T_1' & T_2' & T'\\ T_1 & T_2 & T\\ \frac{1}{2} & \frac{1}{2} & k'\end{Bmatrix}\begin{Bmatrix}J_1' & J_2' & J'\\ J_1 & J_2 & J\\ j_1 & j_2 & k\end{Bmatrix}(\tfrac{1}{2}j_1||\mathbf{f}^{(k'k)}||\tfrac{1}{2}j_2). \qquad (37.35)$$

In the special case of $n_2=0$, we have $T_2'=0$ and $J_2'=0$ and (37.35) reduces to the simpler expression (37.24).

Let us consider a special case of (37.35) in which the isospin vectors are not necessary (either maximum isospin or protons and neutrons in different shells) and where the states of the j^n configurations involved have all lowest seniority. If n_1 is odd and n_2 is even, we put $J_1'=j_1$, $J_2'=0$ and $J_1=0$, $J_2=j_2$. In this case (37.35) reduces, due to (15.7) and (15.15), to the simple expression

$$\sqrt{n_1(n_2+1)}\,[j_1^{n_1}\;J_1'=j_1\{|j_1^{n_1-1}(J_1=0)j_1\;\;J_1'=j_1]$$

$$\times[j_2^{n_2}(J_2'=0)j_2\;\;J_2=j_2|\}j_2^{n_2+1}\;\;J_2=j_2]\,(j_1||\mathbf{f}^{(k)}||j_2). \qquad (37.36)$$

In the other case, with n_1 even and n_2 odd, (37.35) reduces to a similar form with the appropriate c.f.p., differing only by the phase $(-)^{j_1-j_2+k+1}$.

The matrix elements of the tensor operator $\mathbf{f}^{(k)}$ may determine the rate of a certain transition between the $j^{n_1}j^{n_2}$ configuration and the $j_1^{n_1-1}j_2^{n_2+1}$ configuration. According to (15.35), the transition rate for n_1 odd and n_2 even is equal to that of the $j_1\rightarrow j_2$ single nucleon transition multiplied by the statistical factor

$$n_1(n_2+1)\,[j_1^{n_1}\{|j_1^{n_1-1}j_1]^2\,[j_2^{n_2}j_2|\}j_2^{n_2+1}]^2. \qquad (37.37)$$

In the other case, where n_1 is even and n_2 is odd, the same statistical factor appears and, in addition, the geometrical factor $(2j_1 + 1)/(2j_2 + 1)$. This latter factor arises from the fact that j_i [of (15.35)] is j_2 in the present case whereas it is equal to j_1 in the single nucleon $j_1 \to j_2$ transition. Using the c.f.p. given by (28.17) we obtain for the statistical factor in the first case

$$\left(\frac{2j_1 + 2 - n_1}{2j_1 + 1}\right)\left(\frac{2j_2 + 1 - n_2}{2j_2 + 1}\right) \qquad \text{for} \qquad n_1 \text{ odd} \quad n_2 \text{ even.} \tag{37.38}$$

In the other case, we obtain from (28.12) the following expression for the complete factor with which the single nucleon $j_1 \to j_2$ transition should be multiplied in order to obtain the transition considered. This factor is

$$\frac{2j_1 + 1}{2j_2 + 1} n_1(n_2 + 1) \qquad \text{for} \qquad n_1 \text{ even} \quad n_2 \text{ odd.} \tag{37.39}$$

In the special case $n_1 = 1$ (37.38) reduces to (37.27) with $n = n_2 + 1$. In the other case, with $n_2 = 0$, the statistical factor in (37.39) becomes identical with (37.26). In general, (37.38) and the statistical factor in (37.39) are products of the probabilities (37.27) and (37.26), respectively, of picking out a particle from the j_1-shell and putting it into the j_2-shell.

In Section 24 it was pointed out that, in general, the interaction Hamiltonian has matrix elements between different configurations. Such elements, nondiagonal in the scheme we use, give rise to configuration interaction. Only if the matrix elements connecting a state of a certain configuration to states in other configurations are small compared to the energy differences involved, we have an almost pure configuration. In other cases, nuclear (or atomic) states belong to admixtures of several configurations. If the admixtures are small, it may be worth while to consider the main configuration as pure. This case is much simpler and thus it is possible to make a direct comparison of the theory to the experiment. If several configurations are admixed, many more matrix elements, diagonal as well as nondiagonal, of the unknown effective nuclear interaction appear in the calculation. In order to see whether there is agreement with the experiment we need the energies of many nuclear states, which information is usually not available. There are several nuclear spectra, some of them mentioned above, where the use of an undetermined effective interaction along with the assumption of rather pure configurations is very successful. On the other hand, there are many cases where there is clear evidence for the effects of configuration interaction. In some of these latter cases there are enough experimental data so that it is possible to obtain good agreement with them on the assumption of *two* interacting configurations. As already discussed in Section 24, two configurations that get admixed by the action of (effective) two-body interactions can differ at most by the orbits of two particles. We shall first consider the case of two configurations which differ by the j values of two particles and later the other case in which they differ only by one orbit.

Let us first consider matrix elements of the two-body interaction between the j^n configuration and the $j^{n-2}j'j''$ configuration. The matrix element to be calculated has the form

$$\langle [j^{n-2}(\alpha_1 T_1 J_1) j'j''(T_2 J_2) T\, J]_a \left| \sum_{i<k}^{n} V_{ik} \right| j^n \alpha\, T\, J \rangle. \tag{37.40}$$

We take the interaction V_{ik} to be a charge-independent scalar operator, thus omitting in (37.40) the unnecessary M_T and M quantum numbers. The antisymmetric function on the left-hand side of the matrix element (37.40) can be built in a way similar to (37.30). In the case that the j' and j'' refer to the *same* orbit, the wave function $\psi(j'^2 T_2 J_2)$ is already antisymmetric for the allowed values of $T_2 J_2$. If the j'- and the j''-orbits are different (either by the angular momentum or by the radial quantum numbers) we consider the antisymmetrized wave function $\psi_a(j'j'' T_2 J_2)$. We define the following antisymmetrizer

$$\mathscr{A}\psi[j^{n-2}_{1,\ldots,n-2}(\alpha_1 T_1 J_1) j'j''(T_2 J_2) T\, J]$$

$$= \sum_P (-1)^P \psi[j^{n-2}_{i_1,\ldots,i_{n-2}}(\alpha_1 T_1 J_1) j'j''(T_2 J_2) T\, J]. \tag{37.41}$$

The summation in (37.41) is extended over $\binom{n}{2}$ permutations P, each of which substitutes for $i_1, \ldots, i_{n-2}$ a different set of numbers from 1 to n. All the functions on the right-hand side of (37.41) are orthogonal. Therefore the normalization factor of the function (37.41) is $\binom{n}{2}^{-1/2}$. In order to calculate the matrix element (37.40) we start with

$$\binom{n}{2}^{-1/2} \int \mathscr{A}\psi^*[j^{n-2}_{1,\ldots,n-2}(\alpha_1 T_1 J_1) j'j''(T_2 J_2) T\, J] \left(\sum_{i<k}^{n} V_{ik}\right) \psi(j^n \alpha\, T\, J). \tag{37.42}$$

Using the properties of $\mathscr{A}$, we can let it operate on the rest of the integrand obtaining

$$\binom{n}{2}^{1/2} \int \psi^*[j^{n-2}_{1,\ldots,n-2}(\alpha_1 T_1 J_1) j'j''(T_2 J_2) T\, J] \left(\sum_{i<k}^{n} V_{ik}\right) \psi(j^n_{\cdot} \alpha\, T\, J). \tag{37.43}$$

Since the single j-nucleon wave function is orthogonal to the single nucleon wave function in both the j'- and j''-orbits, only the term with $V_{n-1,n}$ will contribute to the integral (37.43). Carrying out the integration, we obtain the following expression of the matrix element (37.40) in terms of a nondiagonal matrix element of a two particle configuration

$$\sqrt{n(n-1)/2}\,[j^{n-2}(\alpha_1 T_1 J_1)j^2(T_2 J_2)T\,J|\}j^n\alpha\,T\,J]\,\langle(j'j''T_2J_2)_a|V_{n-1,n}|j^2T_2J_2\rangle. \tag{37.44}$$

Using nondiagonal elements like (37.44) along with the diagonal elements of the Hamiltonian, including two-body interactions, the energy matrix can be written down and diagonalized. In some special cases a simplified procedure can be adopted which consists of taking along with the diagonal elements, only the second-order correction due to nondiagonal matrix elements. The condition for the validity of using second-order perturbation theory is, as usual, that the nondiagonal matrix elements connecting the state considered with other states be small compared to the differences of the corresponding diagonal elements of the Hamiltonian. The contribution, to second order, of the matrix element (37.44) to the energy of the j^n configuration in the state characterized by α, T, and J is given by

$$\frac{n(n-1)}{2}[j^{n-2}(\alpha_1 T_1 J_1)j^2(T_2 J_2)T\,J|\}j^n\alpha\,T\,J]^2 \\ \times \frac{|\langle(j'j''T_2J_2)_a|V_{12}|j^2T_2J_2\rangle|^2}{E(j^n\alpha\,T\,J)-E[j^{n-2}(\alpha_1T_1J_1)j'j''(T_2J_2)T\,J]}. \tag{37.45}$$

The energies in the denominator of (37.45) are the total energies of the configuration involved in the particular states considered. The complete second-order contribution to the energy of the j^n configuration is obtained by summing (37.45) over all possible two-nucleon excitations and is given by

$$\frac{n(n-1)}{2}\sum_{\alpha_1T_1J_1T_2J_2}[j^{n-2}(\alpha_1 T_1 J_1)j^2(T_2 J_2)T\,J|\}j^n\alpha\,T\,J]^2 \\ \times \sum_{j'j''}\frac{|\langle(j'j''T_2J_2)_a|V_{12}|j^2T_2J_2\rangle|^2}{E(j^n\alpha\,T\,J)-E[j^{n-2}(\alpha_1T_1J_1)j'j''(T_2J_2)T\,J]}. \tag{37.46}$$

The form in which (37.46) is written displays the similarity with (33.36). Thus, the second-order contributions to the energy assume the form of an expectation value of a fictitious two-body interaction

$$\langle j^2T_2J_2|V'_{ik}|j^2T_2J_2\rangle=\sum_{j'j''}\frac{|\langle(j'j''T_2J_2)_a|V_{12}|j^2T_2J_2\rangle|^2}{E(j^n\alpha\,T\,J)-E[j^{n-2}(\alpha_1T_1J_1)j'j''(T_2J_2)T\,J]} \tag{37.47}$$

provided (37.47) is independent of α_1, T_1, and J_1. If, in addition, (37.47) is independent of α, T, and J, the fictitious interaction is the same in all states of the j^n configuration. If it is also independent of n, the same fictitious interaction can be used for all j^n configurations.

There may be several cases where (37.47) is approximately independent of α_1, T_1, J_1, α, T, J, and n. In general, if the diagonal as well as the nondiagonal elements of the interaction energy ΣV_{ik} are small compared to the differences of the eigenvalues of the zero-order Hamiltonian, we can replace the energy denominators in (37.47) by the differences of these eigenvalues. Since the zero-order Hamiltonian is a sum of single particle terms, the differences of eigenvalues are indeed independent of n as well as of the other quantum numbers. This situation is approximated in case the main contribution to (37.47) comes from highly excited $j^{n-2}j'j''$ configurations. Then, the energy denominators are determined principally by the differences in the single nucleon energies of the j^2 configuration and the $j'j''$ configuration. Thus, if the interaction is a small perturbation, the contributions of the second-order corrections (37.46) are equivalent to the first-order terms of a fictitious two-body interaction. By using a more elaborate perturbation expansion it can be shown that the contributions to all orders of certain excitations are also equivalent to a fictitious two-body interaction. This result is the theoretical justification for assuming that nucleons in nuclei are subject to an *effective* two-body interaction which can be used with shell model wave functions.

In actual cases of nuclear configurations the extreme conditions of perturbation theory are probably not satisfied. There may always be some low-lying configurations, the effect of which cannot be exactly expressed as an additional two-body interaction. However, if the second-order effects are not very large, it may be possible to find, even in such cases, an average fictitious two-body interaction to replace approximately the contributions (37.46). This is probably the reason why the idealized picture of a pure j^n configuration works so well in many cases of nuclear spectra (c.f. Section 30). In other cases, the average fictitious two-body interaction may be independent of n but it may well depend on α, T, and J. If this is the case, the effective two-body interaction obtained, for instance, from the analysis of ground states may not reproduce the energies of excited states. The simple shell model picture, with effective interaction, is clearly not applicable in such cases.

The behaviour of the second-order corrections in perturbation theory which was discussed above, can be seen quite generally by using the m scheme. However, we can complete the discussion in the TJ scheme by considering the second-order corrections to states of the $j_1^{n_1}j_2^{n_2}$ configuration. A two-body interaction can change at most the quantum numbers of two nucleons. Hence, the most complicated case is in which one j_1-nucleon is excited into the j_1'-orbit and a j_2-nucleon is excited into a j_2'-orbit. Therefore, all matrix elements of configuration interaction can be expressed in terms of (37.40) and the following matrix element

$$\langle j_1^{n_1-1}(\alpha_{11}T_{11}J_{11})\, j_2^{n_2-1}(\alpha_{22}T_{22}J_{22})\,(T_{12}J_{12})\, j_1'j_2'(T'J')T\,J\,|$$

$$|\sum_{i<k}^{n_1+n_2} V_{ik}\,\Big|\, j_1^{n_1}(\alpha_1T_1J_1)\, j_2^{n_2}(\alpha_2T_2J_2)T\,J\rangle. \qquad (37.48)$$

In (37.48) both j_1' and j_2' should be different from j_1 and j_2, otherwise we would have the case of one nucleon excitation (to be considered later). All states appearing in (37.48) where inequivalent nucleons are coupled to give a definite T and J are taken to be antisymmetric. No special notation is used in this case in order to simplify the expression. If there are several nucleons in either the j_1'- or the j_2'-orbits the expressions become more complicated. One should use in this case more expansions in c.f.p. and change of coupling transformations. We shall not consider this case here.

The evaluation of (37.48) is straightforward. We obtain

$$\psi_a[j_1^{n_1}(\alpha_1T_1J_1)j_2^{n_2}(\alpha_2T_2J_2)T\,J]$$

by applying an antisymmetrizer $\mathscr{A}$ to

$$\psi[j_1^{n_1}(\alpha_1T_1J_1)j_2^{n_2}(\alpha_2T_2J_2)T\,J]$$

and multiplying by the normalization factor $\binom{n_1+n_2}{n_1}^{-1/2}$. The other antisymmetric wave function, on the left-hand side of (37.48) is obtained by applying $\mathscr{A}'$ to

$$\psi[j_1^{n_1-1}(\alpha_{11}T_{11}J_{11})\; j_2^{n_2-1}(\alpha_{22}T_{22}J_{22})\;(T_{12}J_{12})j_{1,n_1}'j_{2,n_1+n_2}'(T'J')T\,J].$$

We take the wave function of the $j_1'j_2'$ configuration to be antisymmetric, $\psi_a(j_1'j_2'T'J')$, so that the normalization factor multiplying $\mathscr{A}'$ is equal to $\left[\binom{n_1+n_2-2}{n_1-1}\binom{n_1+n_2}{2}\right]^{-1/2}$. We let $\mathscr{A}'$ operate on the rest of the integrand obtaining a factor $\binom{n_1+n_2-n}{n_1-1}\binom{n_1+n_2}{2}$. The only term V_{ik} which contributes to the integral is V_{n_1,n_1+n_2}. We now expand $\psi(j_1^{n_1}\alpha_1T_1J_1)$ and $\psi(j_2^{n_2}\alpha_2T_2J_2)$ in terms of c.f.p. and use these expansions in $\psi[j_1^{n_1}(\alpha_1T_1J_1)\, j_2^{n_2}(\alpha_2T_2J_2)TJ]$. When the antisymmetrizer $\mathscr{A}$ is applied to the latter function we obtain $\binom{n_1+n_2}{n_1}$ different functions. Only two of these contribute to the integral over V_{n_1,n_1+n_2}. It is clear that the first n_1-1 nucleons must be in the j_1-orbit and nucleons n_1+1 to n_1+n_2-1 must be in the j_2-orbit for the integral not to vanish. The

n_1th nucleon may be either in the j_1-orbit and nucleon $n_1 + n_2$ in the j_2-orbit or vice versa. In order to carry out the integration we have to carry out a change of coupling transformation (37.16) on the right-hand side of (37.48) for the angular momenta as well as for the isospin vectors. We then integrate and obtain (37.48) in terms of the matrix element

$$\langle (j_1' j_2' T' J')_a | V_{n_1, n_1+n_2} | j_{1,n_1} j_{2,n_1+n_2} T' J' \rangle - \langle (j_1' j_2' T' J')_a | V_{n_1, n_1+n_2} | j_{1,n_1+n_2} j_{2,n_1} T' J' \rangle$$
$$= \sqrt{2} \, \langle (j_1' j_2' T' J')_a | V_{n_1, n_1+n_2} | (j_1 j_2 T' J')_a \rangle. \qquad (37.49)$$

Collecting all factors we obtain for (37.48) the form

$$\sqrt{n_1 n_2} \, [j_1^{n_1-1}(\alpha_{11} T_{11} J_{11}) j_1 T_1 J_1 | \} j_1^{n_1} \alpha_1 T_1 J_1]$$
$$\times [j_2^{n_2-1}(\alpha_{22} T_{22} J_{22}) j_2 T_2 J_2 | \} j_2^{n_2} \alpha_2 T_2 J_2]$$
$$\times [(2T_1+1)(2T_2+1)(2T_{12}+1)(2T'+1)$$
$$\times (2J_1+1)(2J_2+1)(2J_{12}+1)(2J'+1)]^{1/2}$$
$$\times \begin{Bmatrix} T_{11} & \frac{1}{2} & T_1 \\ T_{22} & \frac{1}{2} & T_2 \\ T_{12} & T' & T \end{Bmatrix} \begin{Bmatrix} J_{11} & j_1 & J_1 \\ J_{22} & j_2 & J_2 \\ J_{12} & J' & J \end{Bmatrix} \langle (j_1' j_2' T' J')_a | V_{12} | (j_1 j_2 T' J')_a \rangle. \qquad (37.50)$$

The nondiagonal matrix elements of the nuclear Hamiltonian can be written in terms of (37.50) as well as the simpler expression (37.44). We shall not discuss here actual cases of configuration interaction. We shall only point out that the second-order corrections due to matrix elements (37.50) assume the form of expectation values of a fictitous two-body interaction. In fact, we obtain by squaring (37.50), dividing by the corresponding energy difference and summing over j_1', j_2', α_{11}, T_{11}, J_{11}, α_{22}, T_{22}, J_{22} as well as T_{12}, J_{12}, T', and J' an expression which is identical to (37.19) if $V(j_1 j_2 T' J')$ is replaced by

$$\sum_{j_1' j_2'} \frac{|\langle (j_1' j_2' T' J')_a | V_{12} | (j_1 j_2 T' J')_a \rangle|^2}{\left(\begin{array}{l} E[j_1^{n_1}(\alpha_1 T_1 J_1) j_2^{n_2}(\alpha_2 T_2 J_2) T J] \\ \quad - E[j_1^{n_1-1}(\alpha_{11} T_{11} J_{11}) j_2^{n_2-1}(\alpha_{22} T_{22} J_{22}) (T_{12} J_{12}) j_1' j_2' (T' J') T J] \end{array} \right)} . \qquad (37.51)$$

In the case where the interaction V_{ik} is a small perturbation, the energy denominators in (37.51) can be approximated by the differences of single nucleon energies in the $j_1 j_2$ configuration and $j_1' j_2'$ configuration. If this is so, (37.51) is independent of n_1 and n_2 as well as of all other quantum numbers except T' and J', and can be considered as the expectation value of a fictitious two-body interaction $V'(j_1 j_2 T' J')$. The second-order contribution is then given by (37.19) with (37.51) replacing $V(j_1 j_2 T' J')$.

In some cases it may be more convenient to evaluate instead of (37.48), another matrix element, related to it by a change of coupling transformation. Let us consider the matrix element

$$\langle j_1^{n_1-1}(\alpha_{11} T_{11} J_{11}) j_1'(T_1' J_1') j_2^{n_2-1}(\alpha_{22} T_{22} J_{22}) j_2'(T_2' J_2') T\, J \,| \\ \Big| \sum_{i<h}^{n_1+n_2} V_{ih} \Big| j_1^{n_1}(\alpha_1 T_1 J_1) j_2^{n_2}(\alpha_2 T_2 J_2) T\, J \rangle. \tag{37.52}$$

It is more convenient to expand in this case the interaction V_{ih} in terms of scalar products of irreducible tensors, $(\mathbf{f}_i^{(k'k)} \cdot \mathbf{f}_h^{(k'k)})$, of degree k' in isospin space and degree k in ordinary and spin space. This expansion includes both direct and exchange terms of the two nucleon matrix element in (37.50). We shall now take V_{ih} to be equal to one such scalar product. By the same methods used in the evaluation of (37.48) we obtain for the matrix element (37.52) the expression

$$\langle [j_1^{n_1-1}(\alpha_{11} T_{11} J_{11}) j_1']_a (T_1' J_1') [j_2^{n_2-1}(\alpha_{22} T_{22} J_{22}) j_2']_a (T_2' J_2') T\, J \,| \\ \Big| \Big[\sum_{i=1}^{n_1} \mathbf{f}_i^{(k'k)} \Big] \cdot \Big[\sum_{h=n_1+1}^{n_1+n_2} \mathbf{f}_h^{(k'k)} \Big] \Big| j_1^{n_1}(\alpha_1 T_1 J_1) j_2^{n_2}(\alpha_2 T_2 J_2) T\, J \rangle. \tag{37.53}$$

The notation used in (37.53) implies that the wave functions with which the matrix element should be evaluated are antisymmetric with respect to the first n_1 nucleons and with respect to the other n_2 nucleons. No further antisymmetrization should be carried out in (37.53). This result looks as if no antisymmetrization is necessary between the two groups since (37.53) involves matrix elements of single particle operators. It should be pointed out, however, that the expansion in terms of scalar products contains already the required antisymmetrization in the two nucleon matrix elements. The further evaluation of (37.53) can be carried out by using the result (37.24) for matrix elements of single particle operators.

We turn now to discuss the interaction between two configurations that differ by the orbit of *one* nucleon. Let us first consider the following matrix element

$$\langle j^{n-1}(\alpha_1 T_1 J_1) j' T\, J \Big| \sum_{i<k}^{n} V_{ik} \Big| j^n \alpha\, T\, J \rangle \\ = \int \psi_a^*[j^{n-1}(\alpha_1 T_1 J_1) j' T\, J] \Big(\sum_{i<k}^{n} V_{ik} \Big) \psi(j^n \alpha\, T\, J). \tag{37.54}$$

The wave function on the left-hand side of the matrix element (37.54) is taken to be fully antisymmetric and thus given by (37.21) multiplied by $n^{-1/2}$. Using the properties of the antisymmetrizer $\mathscr{A}$ we obtain for (37.54)

$$n^{-1/2} \int \mathscr{A}\psi^*[j^{n-1}(\alpha_1 T_1 J_1) j'_n T\ J] \left(\sum_{i<k}^{n} V_{ik}\right) \psi(j^n \alpha\ T\ J)$$

$$= n^{1/2} \int \psi^*[j^{n-1}(\alpha_1 T_1 J_1) j'_n T\ J] \left(\sum_{i<k}^{n} V_{ik}\right) \psi(j^n \alpha\ T\ J). \qquad (37.55)$$

Due to the orthogonality of the single nucleon wave functions in the j- and j'-orbit, the only terms which contribute to the integral on the right-hand side of (37.55) are those in which $k = n$. Since both ψ^* and ψ in (37.55) are antisymmetric in the coordinates of the first $n - 1$ nucleons, the contributions of all terms V_{in} are equal. We can therefore evaluate one of them, $V_{n-1,n}$, and multiply the result by $n - 1$. In order to do this we expand $\psi(j^{n-1}\alpha_1 T_1 J_1)$ in terms of $n - 1 \to n - 2$ c.f.p. and $\psi(j^n \alpha\ T\ J)$ in $n \to n - 2$ c.f.p. We thus obtain for the matrix element (37.54) the form

$$(n - 1)\sqrt{n} \sum_{\alpha_2 T_2 J_2 T' J'} [j^{n-1}\alpha_1 T_1 J_1 \{| j^{n-2}(\alpha_2 T_2 J_2) j\ T_1 J_1]$$
$$\times [j^{n-2}(\alpha_2 T_2 J_2) j^2(T'J')T\ J |\} j^n \alpha\ T\ J]$$
$$\times \int \psi^*[j^{n-2}(\alpha_2 T_2 J_2) j_{n-1}(T_1 J_1) j'_n T\ J] V_{n-1,n} \psi[j^{n-2}(\alpha_2 T_2 J_2) j^2_{n-1,n}(T'J')T\ J]. \qquad (37.56)$$

In writing down (37.56) we made use of the orthogonality of the various wave functions $\psi(j^{n-2}\alpha_2 T_2 J_2)$. In order to carry out the integration in (37.56) we perform a change-of-coupling transformation (14.38) [combined with (15.7)]. We thus obtain from (37.56) the final form for the matrix element (37.54)

$$(n - 1)\sqrt{n} \sum_{\alpha_2 T_2 J_2 T' J'} [j^{n-1}\alpha_1 T_1 J_1 \{| j^{n-2}(\alpha_2 T_2 J_2) j\ T_1 J_1]$$
$$\times [j^{n-2}(\alpha_2 T_2 J_2) j^2(T'J')T\ J |\} j^n \alpha\ T\ J]$$
$$\times (-1)^{T_2+1+T+J_2+j+j'+J} [(2T_1 + 1)(2T' + 1)(2J_1 + 1)(2J' + 1)]^{1/2}$$
$$\times \begin{Bmatrix} T_2 & \frac{1}{2} & T_1 \\ \frac{1}{2} & T & T' \end{Bmatrix} \begin{Bmatrix} J_2 & j & J_1 \\ j' & J & J' \end{Bmatrix} \langle (jj'T'J')_a | V_{12} | j^2 T'J' \rangle / \sqrt{2}. \qquad (37.57)$$

The factor $\sqrt{2}$ in (37.57) arises from the fact that the wave function $\psi(j_{n-1} j'_n\ T'J')$ obtained from (37.56) by the transformation is not yet antisymmetric.

Instead of expressing (37.54) in terms of matrix elements in the two-nucleon configuration, we can use the tensor expansion of the interaction V_{ih}. We expand V_{ih} in terms of scalar products of irreducible tensors operating on the coordinates of nucleons i and h respectively. We consider one such product $(\mathbf{f}_i^{(k)} \cdot \mathbf{f}_h^{(k)})$ where the irreducible tensors can be assumed, in the case of a charge-independent interaction, to operate only on the space and spin coordinates. The expansion of V_{12} into a sum of products of irreducible tensors is the one which gives correctly the two-nucleon matrix elements in (37.57). We start from (37.55) and expand $\psi(j^n\alpha\, T\, J)$ in c.f.p. thereby obtaining

$$\sqrt{n} \sum_{\alpha_1' J_1'} [j^{n-1}(\alpha_1' T_1 J_1')\, j\, T\, J | \} j^n \alpha\, T\, J]$$

$$\times \langle j^{n-1}(\alpha_1 T_1 J_1)\, j_n' T\, J \Big| \Big[\sum_{i=1}^{n-1} \mathbf{f}_i^{(k)}\Big] \cdot \mathbf{f}_n^{(k)} \Big| j^{n-1}(\alpha_1' T_1 J_1')\, j_n T\, J \rangle. \qquad (37.58)$$

There is no summation over T_1' in (37.58) since for a tensor operator $\mathbf{f}^{(k)}$, scalar in T, we must have $T_1' = T_1$. The great similarity between (37.58) and (37.23) should be noted. In both matrix elements the change from the j-orbit to the j'-orbit (or the change $j_1 \rightarrow j_2$) of one nucleon is due to a single particle operator $\mathbf{f}^{(k)}$. The orbits of the other $n-1$ nucleons do not change. They remain in the same state [in (37.23)] or at most go over to another state (with α_1' and J_1') obtained by only a change of coupling [in (37.58)].

Let us consider a special case of (37.58) for $k = 0$. In this case, the value of j' must be equal to the value of j (the radial quantum numbers must then be different; they will be denoted here by N' and N to avoid confusion with n) and also J_1' must be equal to J_1. The other possible necessary quantum numbers α_1 and α_1' must also be equal since apart from a common radial integral, $\sum \mathbf{f}_i^{(0)}$ is just a number. Equation (37.58) assumes in this case the simpler form

$$\sqrt{n}\, [j^{n-1}(\alpha_1 T_1 J_1)\, j\, T\, J | \} j^n \alpha\, T\, J]$$

$$\times \langle j^{n-1} \alpha_1 T_1 J_1 \Big| \sum_{i=1}^{n-1} \mathbf{f}_i^{(0)} \Big| j^{n-1} \alpha_1 T_1 J_1 \rangle \langle N'j | \mathbf{f}^{(0)} | Nj \rangle. \qquad (37.59)$$

Equation (37.59), being very similar to (37.24) for $k = 0$, has essentially the form of the matrix element of a single particle operator. We can consider (37.59) as the nondiagonal matrix element of the central potential ($k = 0$) felt by one nucleon due to the action of the other $n - 1$ nucleons. This potential cannot change the value of the single nucleon angular momentum but can have nonvanishing matrix elements between states with different radial quantum numbers.

In the case where perturbation theory is valid, the amplitude of the admixture of $\psi_a[j^{n-1}(\alpha_1 T_1 J_1) j' T\, J]$ in the state which is mainly composed of $\psi(j^n \alpha\, T\, J)$ is given by (37.59) divided by $E(j^n \alpha\, T\, J) - E[j^{n-1}(\alpha_1 T_1 J_1) j' T\, J]$. If this can be approximated by the difference is single nucleon energies $\Delta = E(Nj) - E(N'j)$, we obtain for this amplitude the expression

$$\sqrt{n}\, [j^{n-1}(\alpha_1 T_1 J_1) j\, T\, J| \} j^n \alpha\, T\, J]\, F^0_{n-1} \frac{\langle N'j | \mathbf{f}^{(0)} | Nj \rangle}{\Delta}. \tag{37.60}$$

In (37.60), F^0_{n-1} stands for the matrix element

$$\langle j^{n-1} \alpha_1 T_1 J_1 \left| \sum_{i=1}^{n-1} \mathbf{f}_i^{(0)} \right| j^{n-1} \alpha_1 T_1 J_1 \rangle$$

which is independent of α_1, T_1, and J_1.

The change in the wave function $\psi(j^n \alpha\, T\, J)$ due to the admixtures given by (37.60) is equivalent to a modification of the radial parts of the single nucleon wave functions. This modification arises from the change in the central field acting on the Nj-nucleon due to its (central) interaction with the other $n - 1$ nucleons. In order to see it more clearly, we shall calculate the admixture of the function

$$\psi' = \sum_{\alpha_1 T_1 J_1} [j^{n-1}(\alpha_1 T_1 J_1) j\, T\, J| \} j^n \alpha\, T\, J]\, \psi_a[j^{n-1}(\alpha_1 T_1 J_1) j'_n T\, J]$$

in which $j' = j$ but $N' \neq N$. Multiplying (37.60) by $\psi_a[j^{n-1}(\alpha_1 T_1 J_1) j'_n T J]$ and summing over α_1, T_1, and J_1 we obtain ψ' multiplied by

$$\sqrt{n} \frac{F^{(0)}_{n-1} \langle N'j | \mathbf{f}^{(0)} | Nj \rangle}{\Delta}. \tag{37.61}$$

Let us consider the function ψ' in more detail. We first introduce the notation $\psi(N\, j\, m) = R_N \varphi(j\, m\, {}_t m)$ where R_N is a function of the radial coordinate r of the single nucleon and φ is a function of its angular, spin, and isospin coordinates. Using this notation, we can write

$$\psi(j^n \alpha\, T\, J) = R_N(1) R_N(2) \ldots R_N(n)\, \varphi(j^n \alpha\, T\, J).$$

Similarly, we obtain

$$\sum [j^{n-1}(\alpha_1 T_1 J_1) j\, T\, J| \} j^n \alpha\, T\, J]\, \psi[j^{n-1}(\alpha_1 T_1 J_1) j'_n T\, J]$$
$$= R_N(1) \ldots R_N(n-1)\, R_{N'}(n)\, \varphi(j^n \alpha\, T\, J). \tag{37.62}$$

When (37.60) is antisymmetrized by operating on it with $n^{-1/2} \mathscr{A}$, where $\mathscr{A}$ is defined by (37.21), it should be kept in mind that $\psi(j^n \alpha\, T\, J)$

is antisymmetric in the coordinates of all n nucleons (they all have the same angular, spin, and isospin functions since $j' = j$). We thus obtain

$$\psi' = n^{-1/2} \left[\sum_P R_N(i_1) \ldots R_N(i_{n-1}) R_{N'}(i_n) \right] \varphi(j^n \alpha\, T\, J). \qquad (37.63)$$

In (37.63) the products of the radial functions appear in the *symmetric* combination. The summation is over n permutations P of the numbers 1 to n each of which substitutes for i_n a different number. Thus, the admixture, to the first order, to $\psi(j^n \alpha\, T\, J)$ obtained by multiplying (37.63) by (37.61), is

$$\frac{F^{(0)}_{n-1} \langle N'j|\mathbf{f}^{(0)}|Nj\rangle}{\Delta} \left[\sum_P R_N(i_1) \ldots R_N(i_{n-1}) R_{N'}(i_n) \right] \varphi(j^n \alpha\, T\, J). \qquad (37.64)$$

The single nucleon function modified by the central interaction with the other $n - 1$ nucleons is given, to the first order, by

$$R'\varphi(j\, m_t m) = \left[R_N + \frac{F^{(0)}_{n-1} \langle N'j|\mathbf{f}^{(0)}|Nj\rangle}{\Delta} R_{N'} \right] \varphi(j\, m_t m). \qquad (37.65)$$

The $\psi(j^n \alpha\, T\, J)$, modified by (37.65), is given by

$$R'(1) \ldots R'(n)\, \varphi(j^n \alpha\, T\, J) = R_N(1) \ldots R_N(n)\, \varphi(j^n \alpha\, T\, J) + \frac{F^{(0)}_{n-1} \langle N'j|\mathbf{f}^{(0)}|Nj\rangle}{\Delta}$$

$$\times \left[\sum_P R_N(i_1) \ldots R_N(i_{n-1}) R_{N'}(i_n) \right] \varphi(j^n \alpha\, T\, J) + \ldots\, . \qquad (37.66)$$

We see that (37.66) is indeed equal, to the first order in $F^{(0)}_{n-1} \langle N'j|\mathbf{f}^{(0)}|Nj\rangle / \Delta$, to (37.64).

Thus, the interaction with configurations in which one nucleon is raised into an orbit with the same value of j but a different principal (radial) quantum number is equivalent ot a change in the radial parts of the single nucleon wave functions. Therefore, such interaction can be eliminated by an appropriate choice of the central potential which determines the single nucleon wave functions. If the central field is self-consistent, R_N is the radial part determined by the *total* central field acting on the nucleon considered. Therefore, we can assume that all first-order corrections to R_N vanish in this case.

The additional central field due to the $n - 1$ nucleons depends on n. In fact, we have

$$F^{(0)}_{n-1} = \langle j^{n-1} \alpha_1 T_1 J_1 \left| \sum_{i=1}^{n-1} \mathbf{f}^{(0)}_i \right| j^{n-1} \alpha_1 T_1 J_1 \rangle = (n-1) \langle j|\mathbf{f}^{(0)}|j\rangle.$$

However, if the main contribution to the central field comes from the closed shells, we may assume that the self-consistent central field is fairly independent of n. It should also be kept in mind that a linear change in the single nucleon energies leads to a quadratic term in the total energy which cannot be distinguished from a two-body interaction as already mentioned in Section 30.

In the general case, the admixtures given by the matrix elements (37.58) with $k \neq 0$ involve excitations to higher orbits with $j' \neq j$. A notable exception being the case with $J = 0$. In this case, J_1 as well as J_1' must be equal to j and therefore we must have $j' = J_1 = j$. Excitations with $j' \neq j$ cannot be attributed to a change in the central field. In fact, if we tried to build the many nucleon wave function from single nucleon wave functions, modified by the interaction, the latter would involve admixtures of wave functions with different values of j. The wave functions in a central field, which has spherical symmetry, have definite values of j. In order to admix functions with different values of j, we must consider *deformed* potential wells. The deformed potential well for a single nucleon serves as an idealization of the interaction with the other $n - 1$ nucleons which for $k \neq 0$ does not have spherical symmetry. We shall not enter here into a discussion of wave functions built this way, but only make a few comments. The wave functions obtained by multiplying several such single nucleon wave functions must first be projected into states with definite values of J. As long as the admixtures given by (37.58) are small, in the sense of perturbation theory, the deformed potential well is only a mathematical device. If, however, such admixtures are large, the deformed potential well may have a definite physical meaning. We may then succeed in building a wave function out of which we can project several wave functions with definite values of J which will be eigenstates of the Hamiltonian. If the energy differences of these states, which are said to belong to the same *rotational band*, are small compared to the interaction energy, we can use the adiabatic approximation. We can then look at the rotational states as due to a rotation of the system which is in a definite *intrinsic state*. The intrinsic state wave function is the one built from the wave functions of single nucleons in a deformed potential well.

Thus, the matrix elements (37.54), which involve single nucleon excitations, are those which lead to the breakdown of the spherical shell model. We may consider, in particular, excitations from the orbit given by $j = l + \frac{1}{2}$ to the higher orbit with $j = l - \frac{1}{2}$. The matrix elements of the two nucleon configuration appearing in (37.57) are linear combinations of the diagonal matrix elements $\langle l^2 T\, S\, L | V_{12} | l^2 T\, S\, L \rangle$. Therefore, the interaction with such configurations may be rather large. Configurations in which *two* nucleons are excited from the $j = l + \frac{1}{2}$ orbit into the one with $j = l - \frac{1}{2}$ are expected to lie much higher because of the strong spin orbit interaction. Hence, their admixtures are expected to be less important. Thus, the deviations of the nuclear coupling scheme from pure jj-coupling take the form of a deformation of the potential well which determines the single nucleon wave functions.

It may be worthwhile to mention in passing that single nucleon excitations do not occur in calculations involving "nuclear matter." If an infinite medium with translational invariance is considered, the single nucleon wave functions are characterized by the value of the *linear* momentum $\mathbf{k}$. In a state with a vanishing total linear momentum $\mathbf{K} = 0$ we cannot have one nucleon excited from a state $\mathbf{k}$ to another state $\mathbf{k}' \neq \mathbf{k}$. This is analogous to the case with $J = 0$ mentioned above.

Another type of matrix elements of the interaction between two configurations which differ by the orbit of one nucleon is the following

$$\langle j_1^{n_1}(\alpha_1 T_1 J_1)\, j_2^{n_2}(\alpha_2 T_2 J_2) T\, J \,\Big|\, \sum V_{ih} \,\Big|\, j_1^{n_1-1}(\alpha_1' T_1' J_1')\, j_2^{n_2+1}(\alpha_2' T_2' J_2') T\, J\rangle. \tag{37.67}$$

The matrix element (37.54) is a special case of (37.67) for $n_2 = 0$. The evaluation of this more general matrix element is straightforward. We shall not describe here the details but present the result for $V_{ih} = (\mathbf{f}_i^{(k)} \cdot \mathbf{f}_h^{(k)})$, as was done in the derivation of (37.58). We obtain for the matrix element (37.67) the form

$$\sqrt{n_1(n_2+1)} \sum_{\alpha_1'' J_1'' \alpha_2'' J_2''} [j_1^{n_1}\alpha_1 T_1 J_1 \{| j_1^{n_1-1}(\alpha_1'' T_1' J_1'')\, j_1 T_1 J_1]\, (-1)^{n_2}$$

$$\times [j_2^{n_2-1}(\alpha_2'' T_2 J_2'')\, j_2 T_2' J_2' |\}\, j_2^{n_2}\alpha_2' T_2' J_2']$$

$$\times \langle j_1^{n_1-1}(\alpha_1'' T_1' J_1'')\, j_{1,n_1}(T_1 J_1)\, j_2^{n_2}(\alpha_2 T_2 J_2) T\, J \,\Big|\, \Big[\sum_{i \neq n_1}^{n_1+n_2} \mathbf{f}_i^{(k)} \cdot \mathbf{f}_{n_1}^{(k)}\Big] \,\Big|$$

$$| j_1^{n_1-1}(\alpha_1' T_1' J_1')\, j_2^{n_2}(\alpha_2'' T_2 J_2'')\, j_{2,n_1}(T_2' J_2') T\, J\rangle. \tag{37.68}$$

The matrix element in (37.68) bears to the matrix element (37.34) the same relation as (37.58) to (37.23). The further evaluation of (37.68) can be carried out in the same fashion as in the evaluation of (37.34).

The matrix element (37.67) is not the most general one involving excitation of a single nucleon. The excitation of one j-nucleon into the j''-orbit may be due to the interaction of this nucleon with another nucleon in the j'-orbit. We shall consider here a special case of such a matrix element defined by

$$\langle j^{n-1}(\alpha_2 T_2 J_2)\, j''(T_1' J_1')\, j' T\, J | V_{ih} | j^n(\alpha_1 T_1 J_1)\, j' T\, J\rangle. \tag{37.69}$$

The wave functions which appear in (37.69) are taken to be fully antisymmetric. In the case with $T_1' = T_1$ and $J_1' = J_1$ there are contributions to (37.69) from terms V_{ih} between nucleons in the j and j''-orbits. These contributions can be expressed in terms of the matrix element (37.54). We shall consider here only the contributions of the matrix elements of V_{ih} between states of the jj' configuration and the $j''j'$ configuration. If T_1' and J_1' differ from T_1 and J_1, the only contribution to (37.69) comes from such matrix elements. Expanding V_{ih} in terms of scalar products of tensor operators we have to include both direct terms and exchange terms of these matrix elements since $j \neq j'$ and $j' \neq j''$. Therefore, the expression of the exchange terms as the direct terms of a modified interaction involves isospin vectors. We thus develop the complete interaction in a sum of scalar products of double tensor operators $(\mathbf{f}_i^{(k'k)} \cdot \mathbf{f}_h^{(k'k)})$. The operator $\mathbf{f}_i^{(k'k)}$ is an irreducible tensor in isospin space of degree k' (which is either 0 or 1) and an ordinary irreducible tensor of degree k. Considering the case with a given pair of values $k'k$, we obtain by the methods used above for the contribution to (37.69) the following expression

$$\sqrt{n}\,[j^{n-1}(\alpha_2 T_2 J_2)\,j\;T_1 J_1 |\} j^n \alpha_1 T_1 J_1]$$

$$\times \langle T_2 J_2 j_n''(T_1' J_1')\, j_{n+1}' T\; J | (\mathbf{f}_n^{(k,k)} \cdot \mathbf{f}_{n+1}^{(k,k)}) | T_2 J_2 j_n (T_1 J_1)\, j_{n+1}' T\; J \rangle. \tag{37.70}$$

The expression (37.70) can be evaluated by using (15.4) as well as (15.27).

If the admixtures of other configurations to a given configuration are small, they have a negligible effect on the energy. Wherever perturbation theory is applicable, the first-order correction to the wave function contributes only second-order corrections to the energy. The reason is that the energy has a stationary value at the exact eigenfunction of a state. Small deviations from this wave function do not affect the expectation value of the Hamiltonian. On the other hand, the expectation values of other operators may be very sensitive to the wave function used even if it is very close to the eigenfunction considered. In particular, this may be the case with the expectation values of single particle operators like the magnetic moments, quadrupole moments, etc., also the rates of electromagnetic and other transitions turn out to be very sensitive to the wave functions used.

Let us consider a single particle operator acting on a given eigenstate of the Hamiltonian. It produces a linear combination of states which belong to configurations differing from the given configuration at most by the orbit of one nucleon. If the single particle operator has the given state as an eigenstate it will have a stationary expectation value. If this is not the case, there will be first-order corrections to the expectation value if the eigenfunction has first-order corrections. These will be due to matrix elements between the given state and admixtures of configurations differing at most by the orbit of one nucleon. Therefore the interaction of configurations in which one nucleon is excited are most important to the evaluation of nuclear moments and transition probabilities.

The magnetic moment can serve as an example. Let us consider the magnetic moment of a group of identical particles. The operator from which the magnetic moment is calculated is, in this case, $g_s\mathbf{S} + g_l\mathbf{L}$. When this operator acts on a state of LS-coupling, with definite values of S and L, it produces a linear combination of states with the same values of S and L (these states may have different values of J). As a result, the expectation value of $g_s\mathbf{S} + g_l\mathbf{L}$ has a stationary value for a state with a definite value of J in LS-coupling. On the other hand, in the limit of jj-coupling the magnetic moment is far from being stationary. The operator $g_s\mathbf{S} + g_l\mathbf{L}$ has, in general, nonvanishing matrix elements connecting a given configuration to other configurations in which one nucleon is excited from the Nj-orbit to the Nj'-orbit if $j = l \pm \frac{1}{2}$ and $j' = l \mp \frac{1}{2}$. First-order admixtures of such configurations cause first-order changes in the value of the magnetic moment. In the case of nucleons of both kinds the magnetic moment is stationary neither in LS-coupling nor in jj-coupling.

Another example is offered by the quadrupole moments and $E2$ transition rates in nuclei. These quantities exceed in many cases the single particle values by several times. It is well known that these enhancements are one of the basic arguments for nuclear deformation. Moreover, there are several nuclei in which it is a single *neutron* that makes the $E2$ transition and yet the rate is of the same order of magnitude as for a proton transition. The case of O^{17}, mentioned above, offers an example of such a strong single neutron $s_{1/2} \rightarrow d_{5/2}$ transition. Also the ground state of O^{17}, which is assigned the configuration of a single $d_{5/2}$ neutron outside the closed shells of O^{16}, has a quadrupole moment comparable in size to that of a $d_{5/2}$ proton. We shall not attempt here a quantitative derivation of the O^{17} case but only sketch which configuration admixtures could contribute to the quadrupole moment and $E2$ transition rate.

The configurations which can be admixed to the ground-state function $\psi[O^{16}(T_1 = 0 \ \ J_1 = 0)d_{5/2} \ T = \frac{1}{2} \ J = \frac{5}{2}]$, of O^{17} and contribute in the first order to the quadrupole moment are those in which one nucleon form the closed shells is raised into a higher orbit. Let us therefore consider the wave function

$$\psi_{gs} = \psi[O^{16}(T_1 = 0 \ \ J_1 = 0)d_{5/2} \ T = \tfrac{1}{2} \ J = \tfrac{5}{2}] + \sum_{\alpha_1 T_1' J_1'} a_{\alpha_1 T_1' J_1'} \psi[O^{16}(\alpha_1 T_1' J_1')d_{5/2} \ T = \tfrac{1}{2} \ J = \tfrac{5}{2}]. \quad (37.71)$$

The only contributions to the quadrupole moment to first order (in the coefficients $a_{\alpha_1 T_1' J_1'}$) come from matrix elements of the quadrupole operator which is a component of a second-degree tensor. According to the Wigner-Eckart theorem, the only terms in (37.71) which will contribute in the first order must have $J_1' = 2$. The isospin T_1' can be either 0 or 1 since the single nucleon quadrupole operator has a factor $(1 + \tau_3)$. The expansion coefficients $a_{\alpha_1 T_1' J_1'}$ with $J_1' = 2$ in (37.71) are given by perturbation theory in terms of the matrix elements (37.69). Therefore, the only value of k which contributes to the corresponding matrix elements (37.70) is $k = 2$. The degree k' satisfies the triangular conditions with $T_1 = 0$ and T_1', and therefore, can be either 1 or 0.

Since the O^{16} nucleus in the $T = 0$ $J = 0$ ground state contains closed shells only, the c.f.p. in (37.70) are all equal to 1. The factor $n^{1/2}$ refers to the spin j of the nucleon which is raised to the higher j''-orbit. We have in this case $n = 2(2j + 1)$. The $p_{1/2}$- and $p_{3/2}$-nucleons must be excited to p- or f-orbits in order to obtain a state with the same parity. We neglect here possible excitations of the $1s_{1/2}$ nucleons into the $1d_{5/2}$-orbit and therefore take the j''-orbit to be different from the $1d_{5/2}$-orbit. We thus obtain, to first order, the expansion coefficients in (37.71) as follows

$$a_{\alpha_1 T_1'=0\ J_1'=2} = \sqrt{2(2j+1)}\, F_{\alpha_1}^{(02)} \times \langle T_2 = \tfrac{1}{2}\ J_2 = j\ j_n''(T_1' = 0\ J_1' = 2)d_{5/2,n+1}\ T = \tfrac{1}{2}\ J = \tfrac{5}{2}| \\ |(\mathbf{f}_n^{(02)} \cdot \mathbf{f}_{n+1}^{(02)})|T_2 = \tfrac{1}{2}\ J_2 = j\ j_n(T_1 = 0\ J_1 = 0)d_{5/2,n+1}\ T = \tfrac{1}{2}\ J = \tfrac{5}{2}\rangle/\Delta_0 \quad (37.72)$$

$$a_{\alpha_1 T_1'=1\ J_1'=2} = \sqrt{2(2j+1)}\, F_{\alpha_1}^{(12)} \times \langle T_2 = \tfrac{1}{2}\ J_2 = j\ j_n''(T_1' = 1\ J_1' = 2)d_{5/2}\ T = \tfrac{1}{2}\ J = \tfrac{5}{2}| \\ |(\mathbf{f}_n^{(12)} \cdot \mathbf{f}_{n+1}^{(12)})|T_2 = \tfrac{1}{2}\ J_2 = j\ j_n(T_1 = 0\ J_1 = 0)d_{5/2}\ T = \tfrac{1}{2}\ J = \tfrac{5}{2}\rangle/\Delta_1. \quad (37.73)$$

In (37.72) and (37.73) the Δ_0 and Δ_1 stand for the energy differences of the ground-state configuration and the admixed configuration. These energy differences depend on the orbits j and j'' which are denoted by α_1 as well as on the value of T_1' (this latter dependence is indicated by the subscript of Δ). The $F_{\alpha_1}^{(02)}$ and $F_{\alpha_1}^{(12)}$ are the coefficients of the expansion in products of irreducible tensors. They depend on the j-, j'-, and j''-orbits. In the case of a potential interaction these coefficients are simple linear combination of the direct and exchange radial integrals. Using (15.4) and (15.27) we obtain from (37.72) and (37.73) the simpler expression

$$a_{\alpha_1 T_1'\ J_1'=2} = F_{\alpha_1}^{(T_1' 2)} \frac{(-1)^{j''-j}}{\sqrt{5(2T_1'+1)}} (\tfrac{1}{2} j'' || \mathbf{f}^{(T_1' 2)} || \tfrac{1}{2} j)\, (\tfrac{1}{2}\, d_{5/2} || \mathbf{f}^{(T_1' 2)} || \tfrac{1}{2}\, d_{5/2}) / \Delta_{T_1'}. \quad (37.74)$$

Let us now calculate the expectation value, using the wave function (37.71), of the operator

$$\tfrac{1}{2}\sum_i (1 + \tau_{3i}) r_i^2 \mathbf{Y}^{(2)}(i) = \tfrac{1}{2}\sum_i r_i^2 \mathbf{Y}^{(2)}(i) + \tfrac{1}{2}\sum_i \tau_{3i} r_i^2 \mathbf{Y}^{(2)}(i). \qquad (37.75)$$

The first sum on the right-hand side of (37.75) is a scalar in isospin space and will have nonvanishing matrix elements only for $T_1' = T_1 = 0$. On the other hand, the second sum in (37.75) is a component of a vector in isospin space and will have no matrix elements for states with $T_1' = 0$. We obtain first the zero-order term

$$\langle \tfrac{1}{2}\, d_{5/2} M_T M | \tfrac{1}{2}(1 + \tau_3) r^2 Y_{2m} | \tfrac{1}{2}\, d_{5/2} M_T M \rangle$$

which vanishes for O^{17}(with $M_T = -\tfrac{1}{2}$) and has the single $d_{5/2}$ proton value for F^{17} (with $M_T = \tfrac{1}{2}$). We shall write down separately the first-order contributions for the isospin scalar and isospin vector in (37.75). We thus obtain by defining $r^2\mathbf{Y}^{(2)} = \mathbf{Y}^{(02)}$ and $\tau r^2 \mathbf{Y}^{(2)} = \mathbf{Y}^{(12)}$, the following expression

$$\left(\psi_{gs} || \tfrac{1}{2}\sum \mathbf{Y}^{(T_1'2)} || \psi_{gs}\right)$$

$$= 2\sum_{\alpha_1} a_{\alpha_1 T_1' J_1'=2} \Big(T_1 = 0\ J_1 = 0,\ t' = \tfrac{1}{2}\ j' = \tfrac{5}{2},\ T = \tfrac{1}{2}\ J = \tfrac{5}{2} \Big\|$$

$$\Big\| \tfrac{1}{2}\sum \mathbf{Y}^{(T_1'2)} \Big\| T_1'\ J_1' = 2,\ t' = \tfrac{1}{2}\ j' = \tfrac{5}{2},\ T = \tfrac{1}{2}\ J = \tfrac{5}{2}\Big)$$

$$= \sum_{\alpha_1} F_{\alpha_1}^{(T_1'2)} \frac{1}{5(2T_1' + 1)} \cdot \sqrt{\frac{6}{2j+1}}\, (\tfrac{1}{2} j'' || \mathbf{Y}^{(T_1'2)} || \tfrac{1}{2} j)\, (\tfrac{1}{2} j'' || \mathbf{f}^{(T_1'2)} || \tfrac{1}{2} j)$$

$$\times\, (\tfrac{1}{2}\, d_{5/2} || \mathbf{f}^{(T_1'2)} || \tfrac{1}{2}\, d_{5/2}) / \Delta_{T_1'}(\alpha_1). \qquad (37.76)$$

Due to the parity being a good quantum number, it follows that the j''- and j-orbits must have the same parity. We can therefore choose for the irreducible tensors $\mathbf{f}^{(k'k)}$ the spherical harmonics multiplied by an isospin operator, $\mathbf{Y}^{(k'2)}$. The expression (37.76) is then simplified into

$$\left(\psi_{gs} || \tfrac{1}{2}\sum \mathbf{Y}^{(T_1'2)} || \psi_{gs}\right)$$

$$= (\tfrac{1}{2}\, d_{5/2} || \tfrac{1}{2}\, \mathbf{Y}^{(T_1'2)} || \tfrac{1}{2}\, d_{5/2}) \sum_{\alpha_1} F_{\alpha_1}^{(T_1'2)} \frac{1}{5(2T_1' + 1)} \sqrt{\frac{24}{2j+1}}$$

$$\times\, (\tfrac{1}{2} j'' || \mathbf{Y}^{(T_1'2)} || \tfrac{1}{2} j)^2 \Delta / _{T_1'}(\alpha_1). \qquad (37.77)$$

From (37.77) follow that the first-order terms of the quadrupole moment operator are given by the single particle expression

$$\langle \tfrac{1}{2} d_{5/2} M_T M | \tfrac{1}{2}(a_0 + a_1\tau_3) r^2 Y_{2m} | \tfrac{1}{2} d_{5/2} M_T M \rangle. \tag{37.78}$$

In (37.78) the coefficients a_k are given by

$$a_k = \sum_{\alpha_1} F_{\alpha_1}^{(k2)} \frac{1}{5(2k+1)} \sqrt{\frac{24}{2j+1}} (\tfrac{1}{2} j'' || \mathbf{Y}^{(k2)} || \tfrac{1}{2} j)^2 / \Delta_k(\alpha_1). \tag{37.79}$$

The sum in (37.79) contains the squares of matrix elements reduced with respect to both isospin and angular momentum. The $\Delta_k(\alpha_1)$ are all negative since they are the energy differences between the ground configuration and excited configurations. Therefore, if the coefficients $F_{\alpha_1}^{(k2)}$ for all values of α_1 have the same sign, all terms in (37.79) contribute coherently to the quadrupole moment. For a potential interaction, in which case the $F_{\alpha_1}^{(k2)}$ are linear combinations of radial integrals, it is possible to find many situations where this is actually so. In the case of a potential interaction which is attractive, if the $F_{\alpha_1}^{(02)}$ have all the same sign they are expected to be negative. In this case the coefficient a_0 turns out to be positive and the O^{17} quadrupole moment is expected to have the same sign as the moment of a single $d_{5/2}$ proton. This agrees with the experimental findings.

A similar enhancement of the $E2$ transition in O^{17} (or F^{17}), mentioned above, can also occur as a result of configuration interaction. The wave function in the excited state may be given to the first order by

$$\psi_{1s} = \psi[O^{16}(T_1 = 0 \; J_1 = 0) s_{1/2} \; T = \tfrac{1}{2} \; J = \tfrac{1}{2}] + \sum_{\alpha_1 T_1' J_1'} b_{\alpha_1 T_1' J_1'} \psi[O^{16}(\alpha_1 T_1' J_1') d_{5/2} \; T = \tfrac{1}{2} \; J = \tfrac{1}{2}]. \tag{37.80}$$

The first-order contributions to the transition rate come from matrix elements of the operator (37.75) between the zero-order wave function in (37.71) and the wave functions multiplied by $b_{\alpha_1 T_1' J_1'}$ in (37.80). Therefore, we find again that we must have $J_1' = 2$ and T_1' equal to either 0 or 1. The configuration admixtures in (37.80), which contribute to the transition, are those in which one nucleon is excited from the closed shells of O^{16} to a higher orbit. The amplitudes $b_{\alpha_1 T_1' J_1'}$ are given to the first order in perturbation theory by matrix elements, like (37.52), divided by the appropriate energy differences. Looking at (37.53) we see that the first-order contributions to the matrix element can be expressed by

$$\langle \tfrac{1}{2} d_{5/2} M_T M | \tfrac{1}{2}(b_0 + b_1\tau_3) r^2 Y_{2m} | \tfrac{1}{2} s_{1/2} M_T M' \rangle$$

where b_0 and b_1 are given by expressions similar to (37.79). We shall not go here into the detailed derivation of these expressions. In addition to the contributions of the first order in $b_{\alpha_1 T_1' J_1'}$ to the transition rate, there are first-order contributions due to other possible admixtures to ψ_{gs}. These admixtures involve the functions $\psi[O^{16}(\alpha_1 T_1' J_1') s_{1/2}\ T = \frac{1}{2}\ J = \frac{5}{2}]$ which are connected by nonvanishing matrix elements of (37.75) to the zero-order function in (37.80). The enhancement due to these admixtures is quite similar to the one discussed above.

APPENDIX

Central Field Single Particle Hamiltonian and Wave Functions

Hamiltonian:

$$\left[\frac{p^2}{2M} + U(r)\right] \psi_{nlm}(r\theta\varphi) = \left[-\frac{\hbar^2}{2M}\Delta + U(r)\right] \psi_{nlm}(r\theta\varphi) = E_{nl}\psi_{nlm}(r\theta\varphi)$$

Wave function:

$$\psi_{nlm}(r\theta\varphi) = \frac{1}{r} R_{nl}(r)\, Y_{lm}(\theta\varphi)$$

Radial equation:

$$\frac{d^2R_{nl}(r)}{dr^2} + \left[\frac{2M}{\hbar^2}(E_{nl} - U(r)) - \frac{l(l+1)}{r^2}\right] R_{nl}(r) = 0$$

Square of orbital angular momentum (in units of $\hbar^2$):

$$l^2 = -\left[\frac{1}{\sin\theta}\frac{\partial}{\partial\theta}\left(\sin\theta\frac{\partial}{\partial\theta}\right) + \frac{1}{\sin^2\theta}\frac{\partial^2}{\partial\varphi^2}\right]$$

Spherical Harmonics

$$\boldsymbol{l}^2 Y_{lm}(\theta\varphi) = l(l+1)\, Y_{lm}(\theta\varphi)$$

$$l_z Y_{lm}(\theta\varphi) = m Y_{lm}(\theta\varphi)$$

$$(l_x \pm i l_y)\, Y_{lm}(\theta\varphi) = \sqrt{l(l+1) - m(m \pm 1)}\; Y_{lm\pm 1}(\theta\varphi)$$

$$= \sqrt{(l \mp m)\,(l \pm m + 1)}\; Y_{lm\pm 1}(\theta\varphi)$$

$$\Delta(r^l Y_{lm}(\theta\varphi)) = 0$$

$$\int_0^{2\pi}\int_0^{\pi} Y_{lm}^*(\theta\varphi)\, Y_{l'm'}(\theta\varphi) \sin\theta\, d\theta\, d\varphi = \delta(l, l')\, \delta(m, m')$$

$$Y_{lm}(\theta\varphi) = \frac{1}{\sqrt{2\pi}}\, \Theta_{lm}(\theta)\, e^{im\varphi}$$

$$\int_0^{\pi} \Theta_{lm}(\theta)\, \Theta_{l'm}(\theta) \sin\theta\, d\theta = \delta(l, l')$$

$$Y_{ll}(\theta\varphi) = (-1)^l \sqrt{\frac{(2l+1)!}{2}}\, \frac{1}{2^l l!} \sin^l\theta\, \frac{1}{\sqrt{2\pi}}\, e^{il\varphi}$$

$$Y_{lm}(\pi - \theta, \pi + \varphi) = (-1)^l\, Y_{lm}(\theta\varphi)$$

$$Y_{lm}(0\varphi) = \sqrt{\frac{2l+1}{4\pi}}\, \delta(m, 0), \qquad \Theta_{lm}(0) = \sqrt{\frac{2l+1}{2}}\, \delta(m, 0)$$

Some spherical harmonics:

$$Y_{0,0}(\theta\varphi) = \frac{1}{\sqrt{4\pi}}$$

$$Y_{1,0}(\theta\varphi) = \sqrt{\frac{3}{4\pi}} \cos\theta\,,$$

$$Y_{1,\pm 1}(\theta\varphi) = \mp \sqrt{\frac{3}{8\pi}} \sin\theta\, e^{\pm i\varphi}$$

$$Y_{2,0}(\theta\varphi) = \sqrt{\frac{5}{16\pi}} (3\cos^2\theta - 1),$$

$$Y_{2,\pm 1}(\theta\varphi) = \mp \sqrt{\frac{15}{8\pi}} \cos\theta \sin\theta\, e^{\pm i\varphi}$$

$$Y_{2,\pm 2}(\theta\varphi) = \sqrt{\frac{15}{32\pi}} \sin^2\theta\, e^{\pm 2i\varphi}$$

$$Y_{3,0}(\theta\varphi) = \sqrt{\frac{7}{16\pi}} (5\cos^3\theta - 3\cos\theta)$$

$$Y_{3,\pm 1}(\theta\varphi) = \mp\sqrt{\frac{21}{64\pi}}(4\cos^2\theta\sin\theta - \sin^3\theta)\,e^{\pm i\varphi}$$

$$Y_{3,\pm 2}(\theta\varphi) = \sqrt{\frac{105}{32\pi}}\cos\theta\sin^2\theta\,e^{\pm 2i\varphi}$$

$$Y_{3,\pm 3}(\theta\varphi) = \sqrt{\frac{35}{64\pi}}\sin^3\theta\,e^{\pm 3i\varphi}.$$

Legendre polynomial:

$$P_l(\cos\theta) = \sqrt{\frac{2}{2l+1}}\,\Theta_{l0}(\theta) = \sqrt{\frac{4\pi}{2l+1}}\,Y_{l0}(\theta\varphi) \qquad \text{(independent of } \varphi)$$

$$P_l(1) = 1$$

$$\int_0^\pi P_l(\cos\theta)\,P_{l'}(\cos\theta)\sin\theta\,d\theta = \frac{2}{2l+1}\,\delta(l,l')$$

$$\sum_l \frac{2l+1}{2}\,P_l(\cos\theta)\,P_l(\cos\theta') = \delta(\cos\theta - \cos\theta')$$

Addition theorem for spherical harmonics:

$$P_l(\cos\omega_{12}) = \frac{4\pi}{2l+1}\sum_m Y^*_{lm}(\Omega_1)\,Y_{lm}(\Omega_2)$$

where ω_{12} is the angle between the two directions determined by Ω_1 and Ω_2.

Coulomb Potential

$$U(r) = -\frac{Ze^2}{r}, \qquad a = \frac{\hbar^2}{Me^2}, \qquad \rho = \frac{2Z}{n+l}\,\frac{r}{a}$$

$$R_{nl} = \sqrt{\frac{Z(n-1)!}{(n+l)^2\,[(n+2l)!]^3\,a}}\;\rho^{l+1}\,e^{-\rho/2}\,L^{2l+1}_{n+2l}(\rho)$$

$$E_{nl} = -\frac{MZ^2e^4}{2\hbar^2(n+l)^2}$$

Associated Laguerre Polynomials $L_k^m(x)$

$$L_k^m(x) = (-1)^k\,\frac{k!}{(k-m)!}$$

$$\times\left\{x^{k-m} - \frac{k(k-m)}{1!}\,x^{k-m-1} + \frac{k(k-1)\,(k-m)\,(k-m-1)}{2!}\,x^{k-m-2} - + \ldots\right\}$$

$$\int_0^\infty x^m\, e^{-x}[L_k^m(x)]^2\, dx = \frac{(k!)^3}{(k-m)!}$$

$$\int_0^\infty x^{m+1}\, e^{-x}[L_k^m(x)]^2\, dx = (2k-m+1)\,\frac{(k!)^3}{(k-m)!}\,.$$

Harmonic Oscillator

$$U(r) = \tfrac{1}{2}\, M\omega^2 r^2, \qquad \nu = \frac{M\omega}{2\hbar}$$

$$R_{nl}(r) = \sqrt{\frac{2(2\nu)^{l+3/2}(n-1)!}{[\Gamma(n+l+\frac{1}{2})]^3}}\; r^{l+1} \exp(-\nu r^2)\, L_{n+l-1/2}^{l+1/2}(2\nu r^2)$$

$$E_{nl} = [2n + l - \tfrac{1}{2}]\,\hbar\omega$$

Single Nucleon States in the Nuclear Shell Model

Shell no.	1	2	3	4	5
s.p. states	$1s_{1/2}$	$1p_{3/2}, 1p_{1/2}$	$1d_{5/2}, 2s_{1/2}, 1d_{3/2}$	$1f_{7/2}$	$2p_{3/2}, 1f_{5/2}, 2p_{1/2}, 1g_{9/2}$
Magic no.	2	8	20	28	50

Shell no.	6	7
s.p. states	$2d_{5/2}, 1g_{7/2}, 3s_{1/2}, 2d_{3/2}, 1h_{11/2}$	$1h_{9/2}, 2f_{7/2}, 3p_{3/2}, 2f_{5/2}, 3p_{1/2}, 1i_{13/2}$
Magic no.	82	126

Transformation Properties

r: Coordinates of P in frame Σ

r$'$: Coordinates of same P in frame Σ'

$\| a \|$: Matrix for the rotation R from Σ to Σ'

$$x_i' = \sum a_{ik} x_k \qquad \sum_i a_{ik} a_{il} = \sum_i a_{ki} a_{li} = \delta(k, l)$$

$$Y_{lm'}(\theta'\varphi') = \sum_m Y_{lm}(\theta\varphi)\, D_{mm'}^{(l)}(R)$$

$$D^{(l)\dagger} D^{(l)} = 1$$

$$\int D_{m_1 m_2}^{(l)*}(R)\, D_{m_1' m_2'}^{(l')}(R)\, dR = \frac{8\pi^2}{2l+1}\, \delta(l, l')\, \delta(m_1, m_1')\, \delta(m_2, m_2')$$

Rotation around z by an angle φ:

$$D^{(l)}_{mm'}(R_z) = e^{-im\varphi}\delta(m, m')$$

Rotation around x by an angle π:

$$D^{(j)}_{mm'}(x_\pi) = (-1)^j\delta(m + m')$$

$$D^{(l)*}_{m,0}(\alpha\beta\gamma) = \sqrt{\frac{4\pi}{2l+1}}\, Y_{lm}(\beta\alpha)$$

$$D^{(l)*}_{0,m}(\alpha\beta\gamma) = (-1)^m \sqrt{\frac{4\pi}{2l+1}}\, Y_{lm}(\beta\gamma)$$

$$D^{(k)*}_{\kappa,\kappa'}(R) = (-1)^{\kappa-\kappa'} D^{(k)}_{-\kappa,-\kappa'}(R)$$

If U satisfies

$$U\left(\sum_j a_{ij}\sigma_j\right) U^{-1} = \sigma_i \qquad (\sigma_i\text{: Pauli spin matrices})$$

then

$$D^{(1/2)*}_{\kappa\kappa'}(R) = U_{\kappa'\kappa}(R)$$

where R is the rotation a_{ij}.

Wave Functions of Particles with Spin $\frac{1}{2}$

$$\psi(l, j = l + \tfrac{1}{2}, m)$$
$$= \sqrt{\frac{l + \frac{1}{2} + m}{2l+1}}\,\varphi(l, m - \tfrac{1}{2})\,\chi(\tfrac{1}{2}) + \sqrt{\frac{l + \frac{1}{2} - m}{2l+1}}\,\varphi(l, m + \tfrac{1}{2})\,\chi(-\tfrac{1}{2})$$

$$\psi(l, j = l - \tfrac{1}{2}, m)$$
$$= \sqrt{\frac{l + \frac{1}{2} - m}{2l+1}}\,\varphi(l, m - \tfrac{1}{2})\,\chi(\tfrac{1}{2}) - \sqrt{\frac{m + \frac{1}{2} + m}{2l+1}}\,\varphi(l, m + \tfrac{1}{2})\,\chi(-\tfrac{1}{2})$$

where φ is a space wave function and χ is a spin wave function.

Landé Formula

For any vector operator t:

$$j(j+1)\,\langle njm \mid \mathbf{t} \mid n'jm'\rangle = \langle jm \mid \mathbf{j} \mid jm'\rangle\,\langle njm' \mid (\mathbf{j}\cdot\mathbf{t}) \mid n'jm'\rangle$$

g-Factor of a single particle in a central field:

$$g = \frac{(2j - 1)\, g_l + g_s}{2j} \qquad \text{for } j = l + \tfrac{1}{2}$$

$$g = \frac{(2j + 3)\, g_l - g_s}{2(j + 1)} \qquad \text{for } j = l - \tfrac{1}{2}$$

Magnetic moment of a single particle:

$$\mu = l g_l + \tfrac{1}{2} g_s \qquad \text{for } j = l + \tfrac{1}{2}$$

$$\mu = \frac{j}{j + 1} \left[(l + 1)\, g_l - \tfrac{1}{2} g_s\right] \qquad \text{for } j = l - \tfrac{1}{2}$$

Quadrupole moment of a single proton:

$$Q = -\, e\langle r^2 \rangle \frac{2j - 1}{2j + 2}$$

Clebsch-Gordan Coefficients

Definition:

$$\psi(j_1 j_2 JM) = \sum_{m_1 m_2} (j_1 m_1 j_2 m_2 \mid j_1 j_2 JM)\, \psi(j_1 m_2)\, \psi(j_2 m_2)$$

$$(j_1 m_1 j_2 m_2 \mid j_1 j_2 JM) = 0 \quad \text{unless} \quad j_1 + j_2 \geq J \geq |j_1 - j_2| \ \text{ and } \ M = m_1 + m_2$$

Phase convention (Condon and Shortley):

$$(j_1 j_1 j_2 j_2 \mid j_1 j_2\, j_1 + j_2\, j_1 + j_2) = +1$$

$$\sum m_1 (j_1 m_1 j_2 m_2 \mid j_1 j_2 JM)\, (j_1 m_1 j_2 m_2 \mid j_1 j_2\, J - 1\, M) > 0$$

Reality:

$$(j_1 m_1 j_2 m_2 \mid j_1 j_2\, J\, M)^* = (j_1 m_1 j_2 m_2 \mid j_1 j_2\, J\, M)$$

Orthogonality:

$$\sum_{m_1 m_2} (j_1 m_1 j_2 m_2 \mid j_1 j_2\, J\, M)\, (j_1 m_1 j_2 m_2 \mid j_1 j_2\, J'\, M') = \delta(J, J')\, \delta(M, M')$$

$$\sum_{JM} (j_1 m_1 j_2 m_2 \mid j_1 j_2\, J\, M)\, (j_1 m_1' j_2 m_2' \mid j_1 j_2\, J\ \ M\,) = \delta(m_1, m_1')\, \delta(m_2, m_2')$$

Phase relations:

$$(j_1m_1j_2m_2 \mid j_1j_2JM)\,(j_1m_1'j_2m_2' \mid j_1j_2JM') = \frac{2J+1}{8\pi^2}\int D^{(j_1)}_{m_1m_1'}(R)\,D^{(j_2)}_{m_2m_2'}(R)D^{(J)*}_{MM'}(R)\,dR$$

$$(j_1m_1j_2m_2 \mid j_1j_2\,j_1+j_2\,M) \geq 0$$

$$(j_1m_1\,j_2m_2+1 \mid j_1j_2\,J\,J)\,(j_1m_1+1\,j_2m_2 \mid j_1j_2\,J\,J) \leq 0$$

$$(j_1m_1+1\,j_2m_2 \mid j_1j_2\,J\,J)\,(j_1m_1\,j_2m_2 \mid j_1j_2\,J-1\,J-1) \leq 0$$

$$(-1)^{j_1-m_1}(j_1m_1j_2m_2 \mid j_1j_2\,J\,J) \geq 0$$

Symmetry property:

$$(j_1m_1j_2m_2 \mid j_1j_2\,J\,M) = (-1)^{j_1+j_2-J}(j_2m_2j_1m_1 \mid j_2j_1\,J\,M)$$

Three-j Symbols

Definition:

$$\begin{pmatrix} j_1 & j_2 & j_3 \\ m_1 & m_2 & m_3 \end{pmatrix} = \frac{(-1)^{j_1-j_2-m_3}}{\sqrt{2j_3+1}}\,(j_1m_1j_2m_2 \mid j_1j_2j_3-m_3)$$

Symmetry properties:

$$\begin{pmatrix} j_1 & j_2 & j_3 \\ m_1 & m_2 & m_3 \end{pmatrix} = \begin{pmatrix} j_2 & j_3 & j_1 \\ m_2 & m_3 & m_1 \end{pmatrix} = \begin{pmatrix} j_3 & j_1 & j_2 \\ m_3 & m_1 & m_2 \end{pmatrix} = (-1)^{j_1+j_2+j_3}\begin{pmatrix} j_1 & j_3 & j_2 \\ m_1 & m_3 & m_2 \end{pmatrix}$$

$$\begin{pmatrix} j_1 & j_2 & j_3 \\ -m_1 & -m_2 & -m_3 \end{pmatrix} = (-1)^{j_1+j_2+j_3}\begin{pmatrix} j_1 & j_2 & j_3 \\ m_1 & m_2 & m_3 \end{pmatrix}$$

Orthogonality relations:

$$\sum_{m_1m_2}\begin{pmatrix} j_1 & j_2 & j_3 \\ m_1 & m_2 & m_3 \end{pmatrix}\begin{pmatrix} j_1 & j_2 & j_3' \\ m_1 & m_2 & m_3' \end{pmatrix} = \epsilon(j_1j_2j_3)\,\frac{1}{2j_3+1}\,\delta(j_3, j_3')\,\delta(m_3, m_3')$$

$$\sum_{m_1m_2m_3}\begin{pmatrix} j_1 & j_2 & j_3 \\ m_1 & m_2 & m_3 \end{pmatrix}\begin{pmatrix} j_1 & j_2 & j_3 \\ m_1 & m_2 & m_3 \end{pmatrix} = \epsilon(j_1j_2j_3)$$

$$\sum_{j_3m_3}(2j_3+1)\begin{pmatrix} j_1 & j_2 & j_3 \\ m_1 & m_2 & m_3 \end{pmatrix}\begin{pmatrix} j_1 & j_2 & j_3 \\ m_1' & m_2' & m_3 \end{pmatrix} = \epsilon(j_1j_2j_3)\,\delta(m_1, m_1')\,\delta(m_2, m_2')$$

where

$$\epsilon(j_1j_2j_3) = \begin{cases} 1 & \text{if } j_1+j_2 \geq j_3 \geq |\,j_1-j_2\,| \\ 0 & \text{otherwise} \end{cases}$$

Relation with *D*-matrices:

$$\begin{pmatrix} j_1 & j_2 & j_3 \\ m_1 & m_2 & m_3 \end{pmatrix} = \sum_{m_1'm_2'm_3'} \begin{pmatrix} j_1 & j_2 & j_3 \\ m_1' & m_2' & m_3' \end{pmatrix} D^{(j_1)}_{m_1'm_1}(R)\, D^{(j_2)}_{m_2'm_2}(R)\, D^{(j_3)}_{m_3'm_3}(R) \qquad \text{for any } R$$

9*j*-Symbols

Definition:

$$\begin{Bmatrix} j_{11} & j_{12} & j_{13} \\ j_{21} & j_{22} & j_{23} \\ j_{31} & j_{32} & j_{33} \end{Bmatrix} = \sum_{m_{ij}} \begin{pmatrix} j_{11} & j_{12} & j_{13} \\ m_{11} & m_{12} & m_{13} \end{pmatrix} \begin{pmatrix} j_{21} & j_{22} & j_{23} \\ m_{21} & m_{22} & m_{23} \end{pmatrix} \begin{pmatrix} j_{31} & j_{32} & j_{33} \\ m_{31} & m_{32} & m_{33} \end{pmatrix}$$
$$\begin{pmatrix} j_{11} & j_{21} & j_{31} \\ m_{11} & m_{21} & m_{31} \end{pmatrix} \begin{pmatrix} j_{12} & j_{22} & j_{32} \\ m_{12} & m_{22} & m_{32} \end{pmatrix} \begin{pmatrix} j_{13} & j_{23} & j_{33} \\ m_{13} & m_{23} & m_{33} \end{pmatrix}$$

Symmetry properties:

$$\begin{Bmatrix} s_1 & s_2 & s_3 \\ l_1 & l_2 & l_3 \\ j_1 & j_2 & j_3 \end{Bmatrix} = \begin{Bmatrix} s_1 & l_1 & j_1 \\ s_2 & l_2 & j_2 \\ s_3 & l_3 & j_3 \end{Bmatrix} = \begin{Bmatrix} j_1 & j_2 & j_3 \\ s_1 & s_2 & s_3 \\ l_1 & l_2 & l_3 \end{Bmatrix} = \begin{Bmatrix} l_1 & l_2 & l_3 \\ j_1 & j_2 & j_3 \\ s_1 & s_2 & s_3 \end{Bmatrix}$$

$$\begin{Bmatrix} s_1 & s_2 & s_3 \\ l_1 & l_2 & l_3 \\ j_1 & j_2 & j_3 \end{Bmatrix} = (-1)^{\Sigma} \begin{Bmatrix} l_1 & l_2 & l_3 \\ s_1 & s_2 & s_3 \\ j_1 & j_2 & j_3 \end{Bmatrix}, \quad \Sigma = s_1 + s_2 + s_3 + l_1 + l_2 + l_3 + j_1 + j_2 + j_3$$

$$\begin{Bmatrix} l_1 & l_2 & l_3 \\ l_1 & l_2 & l_3 \\ j_1 & j_2 & j_3 \end{Bmatrix} = 0 \quad \text{if} \quad 2(l_1 + l_2 + l_3) + j_1 + j_2 + j_3 \quad \text{is odd}$$

Orthogonality relations:

$$\sum_{J_{13}, J_{24}} (2J_{13} + 1)(2J_{24} + 1) \begin{Bmatrix} j_1 & j_2 & J_{12} \\ j_3 & j_4 & J_{34} \\ J_{13} & J_{24} & J \end{Bmatrix} \begin{Bmatrix} j_1 & j_2 & J_{12}' \\ j_3 & j_4 & J_{34}' \\ J_{13} & J_{24} & J \end{Bmatrix} = \frac{\delta(J_{12}, J_{12}')\,\delta(J_{34}, J_{34}')}{(2J_{12} + 1)(2J_{34} + 1)}$$

$$\sum_{J_{13}, J_{24}} (-1)^{(j_2+j_4+J_{24})+(j_2+j_3+J_{23})}(2J_{13} + 1)(2J_{24} + 1) \begin{Bmatrix} j_1 & j_3 & J_{13} \\ j_2 & j_4 & J_{24} \\ J_{12} & J_{34} & J \end{Bmatrix} \begin{Bmatrix} j_1 & j_4 & J_{14} \\ j_3 & j_2 & J_{23} \\ J_{13} & J_{24} & J \end{Bmatrix}$$
$$= (-1)^{j_3+j_4+J_{34}} \begin{Bmatrix} j_1 & j_4 & J_{14} \\ j_2 & j_3 & J_{23} \\ J_{12} & J_{34} & J \end{Bmatrix}$$

Relation with 3*j*-symbols:

$$\begin{pmatrix} J_{12} & J_{34} & J \\ M_{12} & M_{34} & M \end{pmatrix} \begin{Bmatrix} j_1 & j_2 & J_{12} \\ j_3 & j_4 & J_{34} \\ J_{13} & J_{24} & J \end{Bmatrix}$$

$$= \sum_{\substack{m_1m_2m_3m_4 \\ M_{13}M_{24}}} \begin{pmatrix} j_1 & j_2 & J_{12} \\ m_1 & m_2 & M_{12} \end{pmatrix} \begin{pmatrix} j_3 & j_4 & J_{34} \\ m_3 & m_4 & M_{34} \end{pmatrix} \begin{pmatrix} j_1 & j_3 & J_{13} \\ m_1 & m_3 & M_{13} \end{pmatrix} \begin{pmatrix} j_2 & j_4 & J_{24} \\ m_2 & m_4 & M_{24} \end{pmatrix} \begin{pmatrix} J_{13} & J_{24} & J \\ M_{13} & M_{24} & M \end{pmatrix}$$

Racah Coefficients (6*j*-Symbols)

Definition:

$$\begin{Bmatrix} j_1 & j_2 & J \\ j_2' & j_1' & k \end{Bmatrix} \equiv (-1)^{j_2+J+j_1'+k}\sqrt{(2J+1)(2k+1)} \begin{Bmatrix} j_1 & j_2 & J \\ j_1' & j_2' & J \\ k & k & 0 \end{Bmatrix}$$

$$\equiv W(j_1j_2J \mid j_2'j_1'k) \equiv (-1)^{j_1+j_2+j_1'+j_2'} W(j_1j_2j_1'j_2';\, Jk)$$

Symmetry:

$$\begin{Bmatrix} j_1 & j_2 & j_3 \\ l_1 & l_2 & l_3 \end{Bmatrix} = \begin{Bmatrix} j_2 & j_3 & j_1 \\ l_2 & l_3 & l_1 \end{Bmatrix} = \begin{Bmatrix} j_3 & j_1 & j_2 \\ l_3 & l_1 & l_2 \end{Bmatrix} = \begin{Bmatrix} j_2 & j_1 & j_3 \\ l_2 & l_1 & l_3 \end{Bmatrix} = \begin{Bmatrix} l_1 & l_2 & j_3 \\ j_1 & j_2 & l_3 \end{Bmatrix}$$

Orthogonality:

$$\sum (2j_3+1) \begin{Bmatrix} j_1 & j_2 & j_3 \\ l_1 & l_2 & l_3 \end{Bmatrix} \begin{Bmatrix} j_1 & j_2 & j_3 \\ l_1 & l_2 & l_3' \end{Bmatrix} = \frac{\delta(l_3, l_3')}{2l_3+1}$$

Special value:

$$\begin{Bmatrix} j_1 & j_1' & 0 \\ j_2 & j_2' & j_3 \end{Bmatrix} = \frac{(-1)^{j_1+j_2+j_3}}{\sqrt{(2j_1+1)(2j_2+1)}}\, \delta(j_1, j_1')\, \delta(j_2, j_2')$$

Sum rules:

$$\sum_j (-1)^{j_1+j_2+j}(2j+1) \begin{Bmatrix} j_1 & j_1 & j' \\ j_2 & j_2 & j \end{Bmatrix} = \sqrt{(2j_1+1)(2j_2+1)}\,\delta(j', 0)$$

$$\sum_j (-1)^{j+j'+j''}(2j+1) \begin{Bmatrix} j_1 & j_2 & j' \\ j_3 & j_4 & j \end{Bmatrix} \begin{Bmatrix} j_1 & j_3 & j'' \\ j_2 & j_4 & j \end{Bmatrix} = \begin{Bmatrix} j_1 & j_2 & j' \\ j_4 & j_3 & j'' \end{Bmatrix}$$

$$\sum_j (2j+1) \begin{Bmatrix} j_1 & j_2 & j \\ j_1 & j_2 & j' \end{Bmatrix} = (-1)^{2(j_1+j_2)}$$

$$\sum (-1)^{k_1+k_2+k}(2k+1) \begin{Bmatrix} j_1 & j_1' & k \\ j_2' & j_2 & j \end{Bmatrix} \begin{Bmatrix} k_1 & k_2 & k \\ j_1' & j_1 & j_2'' \end{Bmatrix} \begin{Bmatrix} k_1 & k_2 & k \\ j_2' & j_2 & j_1'' \end{Bmatrix}$$

$$= (-1)^{j_1+j_2+j_1'+j_2'+j_1''+j_2''+j} \begin{Bmatrix} j_1 & j_2 & j \\ j_1'' & j_2'' & k_1 \end{Bmatrix} \begin{Bmatrix} j_1' & j_2' & j \\ j_1'' & j_2'' & k_2 \end{Bmatrix}$$

Relations with 3*j*- and 9*j*-symbols:

$$\begin{pmatrix} j_1 & j_2 & j_3 \\ m_1 & m_2 & m_3 \end{pmatrix} \begin{Bmatrix} j_1 & j_2 & j_3 \\ l_1 & l_2 & l_3 \end{Bmatrix}$$

$$= \sum_{m_1'm_2'm_3'} (-1)^{l_1+l_2+l_3+m_1'+m_2'+m_3'} \begin{pmatrix} j_1 & l_2 & l_3 \\ m_1 & m_2' & -m_3' \end{pmatrix} \begin{pmatrix} l_1 & j_2 & l_3 \\ -m_1' & m_2 & m_3' \end{pmatrix} \begin{pmatrix} l_1 & l_2 & j_3 \\ m_1' & -m_2' & m_3 \end{pmatrix}$$

$$\begin{Bmatrix} j_1 & j_2 & J_{12} \\ j_3 & j_4 & J_{34} \\ J_{13} & J_{24} & J \end{Bmatrix} = \sum_{J'} (-1)^{2J'}(2J'+1) \begin{Bmatrix} j_1 & j_3 & J_{13} \\ J_{24} & J & J' \end{Bmatrix} \begin{Bmatrix} j_2 & j_4 & J_{24} \\ j_3 & J' & J_{34} \end{Bmatrix} \begin{Bmatrix} J_{12} & J_{34} & J \\ J' & j_1 & j_2 \end{Bmatrix}$$

$$= \sum_{J'} (-1)^{2J'}(2J'+1) \begin{Bmatrix} J_{12} & J_{34} & J \\ J_{13} & J_{24} & J' \end{Bmatrix} \begin{Bmatrix} j_1 & j_3 & J_{13} \\ J_{34} & J' & j_4 \end{Bmatrix} \begin{Bmatrix} j_2 & j_4 & J_{24} \\ J' & J_{12} & j_1 \end{Bmatrix}$$

$$= \sum_{J'} (-1)^{2J'}(2J'+1) \begin{Bmatrix} j_2 & j_4 & J_{24} \\ J & J_{13} & J' \end{Bmatrix} \begin{Bmatrix} J_{12} & J_{34} & J \\ j_4 & J' & j_3 \end{Bmatrix} \begin{Bmatrix} j_1 & j_3 & J_{13} \\ J' & j_2 & J_{12} \end{Bmatrix}$$

$$\sum_{J} (2J+1) \begin{Bmatrix} J_{12} & J_{34} & J \\ J_{13} & J_{24} & J' \end{Bmatrix} \begin{Bmatrix} j_1 & j_2 & J_{12} \\ j_3 & j_4 & J_{34} \\ J_{13} & J_{24} & J \end{Bmatrix} = (-1)^{2J'} \begin{Bmatrix} j_1 & j_3 & J_{13} \\ J_{34} & J' & j_4 \end{Bmatrix} \begin{Bmatrix} j_2 & j_4 & J_{24} \\ J' & J_{12} & j_1 \end{Bmatrix}$$

Change of Coupling Transformations

$$\psi(j_1j_3(J_{13})\, j_2j_4(J_{24})\, J) = \sum_{J_{12},J_{34}} \langle J_{13}J_{24}; J \mid J_{12}J_{34}; J\rangle\, \psi(j_1j_2(J_{12})\, j_3j_4(J_{34})\, J)$$

$$\langle J_{13}J_{24}; J \mid J_{12}J_{34}; J\rangle = \sqrt{(2J_{13}+1)(2J_{24}+1)(2J_{12}+1)(2J_{34}+1)} \times \begin{Bmatrix} j_1 & j_2 & J_{12} \\ j_3 & j_4 & J_{34} \\ J_{13} & J_{24} & J \end{Bmatrix}$$

$$\sum_{J_{12}J_{34}} \langle J_{13}J_{24}; J \mid J_{12}J_{34}; J\rangle \langle J_{12}J_{34}; J \mid J_{13}'J_{24}'; J\rangle = \delta(J_{13}, J_{13}')\,\delta(J_{24}, J_{24}')$$

$$\psi(j_1j_3(J_{13})\, j_2J) = \sum_{J_{12}} \langle j_1j_3(J_{13})\, j_2; J \mid j_1j_2(J_{12})\, j_3; J\rangle\, \psi(j_1j_2(J_{12})\, j_3J)$$

$$\langle j_1j_3(J_{13})\, j_2; J \mid j_1j_2(J_{12})\, j_3; J\rangle = (-1)^{j_2+j_3+J_{12}+J_{13}} \sqrt{(2J_{12}+1)(2J_{13}+1)} \begin{Bmatrix} j_1 & j_2 & J_{12} \\ J & j_3 & J_{13} \end{Bmatrix}$$

$$\langle j_1j_2(J_{12})\, j_3; J \mid j_1, j_2j_3(J_{23}); J\rangle = (-1)^{j_1+j_2+j_3+J} \sqrt{(2J_{12}+1)(2J_{23}+1)} \begin{Bmatrix} j_1 & j_2 & J_{12} \\ j_3 & J & J_{23} \end{Bmatrix}$$

Some Special Clebsch-Gordan Coefficients and 3*j*-Symbols

$(\frac{1}{2}m_s\, l\, m_l \mid \frac{1}{2}\, ljm)$

j \ m_s	$+\frac{1}{2}$	$-\frac{1}{2}$
$l+\frac{1}{2}$	$\sqrt{\left(\frac{l+\frac{1}{2}+m}{2l+1}\right)}$	$\sqrt{\left(\frac{l+\frac{1}{2}-m}{2l+1}\right)}$
$l-\frac{1}{2}$	$\sqrt{\left(\frac{l+\frac{1}{2}-m}{2l+1}\right)}$	$-\sqrt{\left(\frac{l+\frac{1}{2}+m}{2l+1}\right)}$

$(1\; m_s\, l\, m_l \mid 1\, ljm)$

j \ m_s	$+1$	0	-1
$l+1$	$\sqrt{\left(\frac{(l+m)(l+m+1)}{(2l+1)(2l+2)}\right)}$	$\sqrt{\left(\frac{(l-m+1((l+m+1)}{(2l+1)(l+1)}\right)}$	$\sqrt{\left(\frac{(l-m)(l-m+1)}{(2l+1)(2l+2)}\right)}$
l	$\sqrt{\left(\frac{(l+m)(l-m+1)}{2l(l+1)}\right)}$	$\frac{-m}{\sqrt{(l(l+1))}}$	$-\sqrt{\left(\frac{(l-m)(l+m+1)}{2l(l+1)}\right)}$
$l-1$	$\sqrt{\left(\frac{(l-m)(l-m+1)}{2l(2l+1)}\right)}$	$-\sqrt{\left(\frac{(l-m)(l+m)}{l(2l+1)}\right)}$	$\sqrt{\left(\frac{(l+m+1)(l+m)}{2l(2l+1)}\right)}$

$$(jm\, j' - m' \mid jj'00) = \frac{(-1)^{j-m}}{\sqrt{2j+1}}\,\delta(m, m')\,\delta(j, j')$$

$$\begin{pmatrix} j & 1 & j \\ -m & 0 & m \end{pmatrix} = (-1)^{j-m}\,\frac{m}{\sqrt{j(2j+1)(j+1)}}$$

$$\begin{pmatrix} j & 2 & j \\ -m & 0 & m \end{pmatrix} = (-1)^{j-m}\,\frac{3m^2 - j(j+1)}{\sqrt{(2j-1)\,j(2j+1)(j+1)(2j+3)}}$$

$$\begin{pmatrix} j_1 & j_2 & j_3 \\ 0 & 0 & 0 \end{pmatrix} = (-1)^g \sqrt{\frac{(2g-2j_1)!\,(2g-2j_2)!\,(2g-2j_3)!}{(2g+1)!}}$$

$$\times\,\frac{g!}{(g-j_1)!\,(g-j_2)!\,(g-j_3)!}$$

if $2g = j_1 + j_2 + j_3$ is even

$$\begin{pmatrix} j_1 & j_2 & j_3 \\ 0 & 0 & 0 \end{pmatrix} = 0 \qquad \text{if } j_1 + j_2 + j_3 \text{ is odd}$$

$$\begin{pmatrix} j_1 & j_2 & J \\ \frac{1}{2} & -\frac{1}{2} & 0 \end{pmatrix} = -\sqrt{(2l_1+1)(2l_2+1)}\begin{pmatrix} l_1 & l_2 & J \\ 0 & 0 & 0 \end{pmatrix}\begin{Bmatrix} j_1 & j_2 & J \\ l_2 & l_1 & \frac{1}{2} \end{Bmatrix}$$

where $l_1 = j_1 \pm \frac{1}{2}$ and $l_2 = j_2 \pm \frac{1}{2}$ are chosen so that $l_1 + l_2 + J$ is even.

$$\begin{pmatrix} l & k & l \\ -m & 0 & m \end{pmatrix} \approx \frac{(-1)^{l-m}}{2l+1} P_k\left(\frac{m}{l}\right) \qquad \text{for } l \to \infty \qquad (k \text{ finite})$$

Some Special Racah and 9j Coefficients

$$\begin{Bmatrix} j_1 & j_2 & j_3 \\ j_2 & j_1 & 0 \end{Bmatrix} = (-1)^{j_1+j_2+j_3} \frac{1}{\sqrt{(2j_1+1)(2j_2+1)}}$$

$$\begin{Bmatrix} j_1 & j_2 & j_3 \\ j_2+\frac{1}{2} & j_1+\frac{1}{2} & \frac{1}{2} \end{Bmatrix} = (-1)^{j_1+j_2+j_3+1} \left[\frac{(j_1+j_2+j_3+2)(j_1+j_2-j_3+1)}{(2j_1+1)(2j_1+2)(2j_2+1)(2j_2+2)}\right]^{1/2}$$

$$\begin{Bmatrix} j_1 & j_2 & j_3 \\ j_2+\frac{1}{2} & j_1-\frac{1}{2} & \frac{1}{2} \end{Bmatrix} = (-1)^{j_1+j_2+j_3} \left[\frac{(j_3+j_1-j_2)(j_2+j_3-j_1+1)}{2j_1(2j_1+1)(2j_2+1)(2j_2+2)}\right]^{1/2}$$

$$\begin{Bmatrix} j_1 & j_2 & j_3 \\ j_2-1 & j_1-1 & 1 \end{Bmatrix} = (-1)^{j_1+j_2+j_3}$$

$$\times \left[\frac{(j_1+j_2+j_3)(j_1+j_2+j_3+1)(j_1+j_2-j_3)(j_1+j_2-j_3-1)}{(2j_1-1)\,2j_1(2j_1+1)(2j_2-1)\,2j_2(2j_2+1)}\right]^{1/2}$$

$$\begin{Bmatrix} j_1 & j_2 & j_3 \\ j_2-1 & j_1 & 1 \end{Bmatrix} = (-1)^{j_1+j_2+j_3}$$

$$\times \left[\frac{2(j_1+j_2+j_3+1)(j_1+j_2-j_3)(j_2+j_3-j_1)(j_1-j_2+j_3+1)}{2j_1(2j_1+1)(2j_1+2)(2j_2-1)\,2j_2(2j_2+1)}\right]^{1/2}$$

$$\begin{Bmatrix} j_1 & j_2 & j_3 \\ j_2+1 & j_1-1 & 1 \end{Bmatrix} = (-1)^{j_1+j_2+j_3}$$

$$\times \left[\frac{(j_1-j_2+j_3-1)(j_1-j_2+j_3)(j_2+j_3-j_1+1)(j_2+j_3-j_1+2)}{(2j_2+1)(2j_2+2)(2j_2+3)(2j_1-1)\,2j_1(2j_1+1)}\right]^{1/2}$$

$$\begin{Bmatrix} j_1 & j_2 & j_3 \\ j_2 & j_1 & 1 \end{Bmatrix} = (-1)^{j_1+j_2+j_3+1} \frac{\frac{1}{2}[j_1(j_1+1)+j_2(j_2+1)-j_3(j_3+1)]}{\sqrt{j_1(j_1+1)(2j_1+1)j_2(j_2+1)(2j_2+1)}}$$

$$\begin{Bmatrix} S & S & 1 \\ l_1 & l_2 & L \\ j_1 & j_2 & L \end{Bmatrix}$$

$$= (-1)^{l_1+j_2+S+L} \frac{[l_1(l_1+1)-l_2(l_2+1)]-[j_1(j_1+1)-j_2(j_2+1)]}{2(2S+1)(2L+1)\sqrt{S(S+1)L(L+1)}} \begin{Bmatrix} l_1 & l_2 & L \\ j_2 & j_1 & S \end{Bmatrix}$$

Irreducible Tensors

Under a rotation δR by an angle $\delta\theta$ around the direction $\mathbf{n}$ an irreducible tensor is changed by

$$\delta T_\kappa^{(k)}(\mathbf{r}) = -i\,\delta\theta(\mathbf{n}\cdot\mathbf{J})\,T_\kappa^{(k)}(\mathbf{r})$$

where $\mathbf{J}$ is the sum of all angular momenta involved.

$$J_z T^{(k)}_\kappa(r) = \kappa T^{(k)}_\kappa(r)$$

$$(J_x \pm iJ_y)\, T^{(k)}_\kappa(r) = \sqrt{k(k+1) - \kappa(\kappa \pm 1)}\; T^{(k)}_{\kappa\pm1}(r)$$

Tensor products:

$$[\mathbf{T}^{(k_1)} \times \mathbf{T}^{(k_2)}]^{(k_3)}_{\kappa_3} = \sum_{\kappa_1\kappa_2} (k_1\kappa_1k_2\kappa_2 \mid k_1k_2k_3\kappa_3)\; T^{(k_1)}_{\kappa_1}\, T^{(k_2)}_{\kappa_2}$$

Scalar product:

$$(\mathbf{T}^{(k)} \cdot \mathbf{U}^{(k)}) = (-1)^k \sqrt{2k+1}\, [\mathbf{T}^{(k)} \times \mathbf{U}^{(k)}]^{(0)}_0$$

Wigner-Eckart theorem:

$$\langle JM \mid T^{(k)}_\kappa \mid J'M' \rangle = (-1)^{J-M} \begin{pmatrix} J & k & J' \\ -M & \kappa & M' \end{pmatrix} (J \,||\, \mathbf{T}^{(k)} \,||\, J')$$

If $T^{(k)}_0$ is a Hermitian operator, then

$$(J' \,||\, \mathbf{T}^{(k)} \,||\, J) = (-1)^{J-J'}\, \overline{(J \,||\, \mathbf{T}^{(k)} \,||\, J')}$$

Some special reduced matrix elements:

$$(J \,||\, 1 \,||\, J') = \sqrt{2J+1}\, \delta(J_1J')$$

$$(J \,||\, \mathbf{J} \,||\, J) \quad \sqrt{J(J+1)(2J+1)}$$

$$(l \,||\, \mathbf{Y}_k \,||\, l') = (-1)^l \sqrt{\frac{(2l+1)(2k+1)(2l'+1)}{4\pi}} \begin{pmatrix} l & k & l' \\ 0 & 0 & 0 \end{pmatrix}$$

$$(\tfrac{1}{2} lj \,||\, \mathbf{Y}_k \,||\, \tfrac{1}{2} l'j') = (-1)^{j-1/2} \sqrt{\frac{(2j+1)(2k+1)(2j'+1)}{4\pi}}$$

$$\times \begin{pmatrix} j & k & j' \\ -\frac{1}{2} & 0 & \frac{1}{2} \end{pmatrix} \tfrac{1}{2}[1 + (-1)^{l+k+l'}]$$

Reduced matrix elements of tensor products:

If $\mathbf{T}^{(k_1)}$ and $\mathbf{T}^{(k_2)}$ operate on the *same* set of coordinates and

$$\mathbf{T}^{(k)} = [\mathbf{T}^{(k_1)} \times \mathbf{T}^{(k_2)}]^{(k)}$$

then

$$(\alpha j \,\|\, \mathbf{T}^{(k)} \,\|\, \alpha' j') = (-1)^{j+k+j'} \sqrt{2k+1} \sum_{\alpha'' j''} (\alpha j \,\|\, \mathbf{T}^{(k_1)} \,\|\, \alpha'' j'')$$

$$\times (\alpha'' j'' \,\|\, \mathbf{T}^{(k_2)} \,\|\, \alpha' j') \begin{Bmatrix} k_1 & k_2 & k \\ j' & j & j'' \end{Bmatrix}$$

If $\mathbf{T}^{(k)} = [\mathbf{T}_1^{(k_1)} \times \mathbf{T}_2^{(k_2)}]^{(k)}$, 1 and 2 being two different systems, then

$$(\alpha_1 j_1 \alpha_2 j_2 J \,\|\, \mathbf{T}^{(k)} \,\|\, \alpha_1' j_1' \alpha_2' j_2' J') = \sqrt{(2J+1)(2k+1)(2J'+1)}$$

$$\times \begin{Bmatrix} j_1 & j_2 & J \\ j_1' & j_2' & J' \\ k_1 & k_2 & k \end{Bmatrix} (\alpha_1 j_1 \,\|\, \mathbf{T}_1^{(k_1)} \,\|\, \alpha_1' j_1')(\alpha_2 j_2 \,\|\, \mathbf{T}_2^{(k_2)} \,\|\, \alpha_2' j_2')$$

$$(\alpha_1 j_1 \alpha_2 j_1 J \,\|\, (\mathbf{T}_1^{(k)} \cdot \mathbf{T}_2^{(k)}) \,\|\, \alpha_1' j_1' \alpha_2' j_2' J') = (-1)^{j_2+J+j_1'} \sqrt{2J+1}$$

$$\times \begin{Bmatrix} j_1 & j_2 & J \\ j_2' & j_1' & k \end{Bmatrix} (\alpha_1 j_1 \,\|\, \mathbf{T}_1^{(k)} \,\|\, \alpha_1' j_1')(\alpha_2 j_2 \,\|\, \mathbf{T}_2^{(k)} \,\|\, \alpha_2' j_2')\, \delta(J, J')$$

$$\langle \alpha_1 j_1 \alpha_2 j_2 J M \mid (\mathbf{T}_1^{(k)} \cdot \mathbf{T}_2^{(k)}) \mid \alpha_1' j_1' \alpha_2' j_2' J' M' \rangle = (-1)^{j_2+J+j_1'} \begin{Bmatrix} j_1 & j_2 & J \\ j_2' & j_1' & k \end{Bmatrix}$$

$$\times (\alpha_1 j_1 \,\|\, \mathbf{T}_1^{(k)} \,\|\, \alpha_1' j_1')(\alpha_2 j_2 \,\|\, \mathbf{T}_2^{(k)} \,\|\, \alpha_2' j_2')\, \delta(J, J')\, \delta(M, M')$$

$$(\alpha_1 j_1 \alpha_2 j_2 J \,\|\, \mathbf{T}_1^{(k)} \,\|\, \alpha_1' j_1' \alpha_2' j_2' J') = (-1)^{j_1+j_2+J'+k} \sqrt{(2J+1)(2J'+1)}$$

$$\times (\alpha_1 j_1 \,\|\, \mathbf{T}_1^{(k)} \,\|\, \alpha_1' j_1') \begin{Bmatrix} j_1 & J & j_2 \\ J' & j_1' & k \end{Bmatrix} \delta(\alpha_2, \alpha_2')\, \delta(j_2, j_2')$$

$$(\alpha_1 j_1 \alpha_2 j_2 J \,\|\, \mathbf{T}_2^{(k)} \,\|\, \alpha_1' j_1' \alpha_2' j_2' J') = (-1)^{j_1+j_2'+J+k} \sqrt{(2J+1)(2J'+1)}$$

$$\times (\alpha_2 j_2 \,\|\, \mathbf{T}_2^{(k)} \,\|\, \alpha_2' j_2') \begin{Bmatrix} j_2 & J & j_1 \\ J' & j_2' & k \end{Bmatrix} \delta(\alpha_1, \alpha_1')\, \delta(j_1, j_1')$$

Calculation of transition probabilities:

$$\sum_{m_i m_f \kappa} | \langle j_i m_i \mid T_\kappa^{(k)} \mid j_f m_f \rangle |^2 = | (j_i \,\|\, \mathbf{T}^{(k)} \,\|\, j_f) |^2$$

$$\sum_{m_f \kappa} | \langle j_i m_i \mid T_\kappa^{(k)} \mid j_f m_f \rangle |^2 = \frac{1}{2j_i + 1} | (j_i \,\|\, \mathbf{T}^{(k)} \,\|\, j_f) |^2$$

Vector Spherical Harmonics

Wave functions for spin $s = 1$—three 3-dimensional vectors χ_κ, $\kappa = -1, 0, 1$, which satisfy:

$$\mathbf{s}^2\chi_\kappa = 2\chi_\kappa, \qquad s_z\chi_\kappa = \kappa\chi_\kappa$$

$$(s_x \pm is_y)\,\chi_\kappa = \sqrt{2 - \kappa(\kappa \pm 1)}\,\chi_{\kappa\pm 1}$$

Definition and properties:

$$\mathbf{Y}_m^{j(l1)}(\vartheta\varphi) = [\mathbf{Y}_l(\vartheta\varphi) \times \chi]_m^{(j)}$$

$$= \sum_{\lambda\kappa} (l\lambda 1\kappa \mid l1\,jm)\; Y_{l\lambda}(\vartheta\varphi)\,\chi_\kappa$$

$$\mathbf{s}^2\mathbf{Y}_m^{j(l1)} = 2\mathbf{Y}_m^{j(l1)}, \qquad \boldsymbol{l}^2\mathbf{Y}_m^{j(l1)} = l(l+1)\,\mathbf{Y}_m^{j(l1)}$$

$$\mathbf{j}^2\mathbf{Y}_m^{j(l1)} = j(j+1)\,\mathbf{Y}_m^{j(l1)}, \qquad j_z\mathbf{Y}_m^{j(l1)} = m\mathbf{Y}_m^{j(l1)}$$

If $j = l$,

$$\mathbf{Y}_m^{j(j1)}(\vartheta\varphi) = \frac{1}{\sqrt{j(j+1)}}\,\boldsymbol{l}\mathbf{Y}_{jm}(\vartheta\varphi)$$

Gradient formulae:

$$\text{grad}\,[r^l Y_{lm}(\vartheta\varphi)] = \sqrt{l(2l+1)}\,r^{l-1}\mathbf{Y}_m^{l(l-1,1)}(\vartheta\varphi)$$

For an arbitrary vector $\mathbf{v}$

$$(\text{grad}\,r^l Y_{lm}(\vartheta\varphi)\cdot\mathbf{v}) = \sqrt{l(2l+1)}\,r^{l-1}[\mathbf{Y}_{l-1}(\vartheta\varphi) \times \mathbf{v}]_m^{(l)}$$

Electromagnetic Transitions and Moments

Magnetic moment (without exchange current terms):

$$\mu = \langle J\,M = J \,\Big|\, \sum_i g_l^{(i)} l_{i,z} + g_s^{(i)} s_{i,z} \,\Big|\, J\,M = J\rangle$$

Quadrupole moment:

$$Q = \Big\langle J\,M = J \,\Big|\, \sqrt{\frac{16\pi}{5}} \sum e_i r_i^2 Y_{20}(\theta_i\varphi_i) \,\Big|\, J\,M = J\Big\rangle$$

$$\Big\langle j^{2j+1-n}\alpha JM \,\Big|\, \sum_{i=1}^{2j+1-n} T_0^{(k)}(i) \,\Big|\, j^{2j+1-n}\alpha JM\Big\rangle = (-1)^{k+1}\Big\langle j^n\alpha JM \,\Big|\, \sum_{i=1}^{n} T_0^{(k)}(i) \,\Big|\, j^n\alpha JM\Big\rangle$$

Transition rates in the long-wave approximation:

$$T(L) = 8\pi c \frac{e^2}{\hbar c} \frac{L+1}{L[(2L+1)!!]^2} k^{2L+1} B(L), \qquad k = \frac{\omega}{c} \ll \frac{1}{R}$$

$$B(\text{el}, L) = \frac{1}{2J_i+1} \left| \left(J_f \left\| \frac{1}{e} \sum_i e_i r_i^L \mathbf{Y}_L(\theta_i \varphi_i) \right\| J_i \right) \right|^2$$

$$B(\text{mag}, L) = \frac{1}{2J_i+1}$$

$$\times \left| \left(J_f \left\| \sum_i [\text{grad}\, r_i^L Y_{LM}(\theta_i \varphi_i)] \cdot \left[\frac{1}{ec(L+1)} \frac{e_i \hbar}{m_i} \mathbf{l}_i + \frac{1}{e} \mu_i \boldsymbol{\sigma}_i \right] \right\| J_i \right) \right|^2$$

For protons:

$$B_{\text{s.p.}}(\text{el}, L) = \frac{(2j_f+1)(2l_f+1)(2l_i+1)(2L+1)}{4\pi}$$

$$\times \begin{pmatrix} l_f & L & l_i \\ 0 & 0 & 0 \end{pmatrix}^2 \begin{Bmatrix} l_f & j_f & \frac{1}{2} \\ j_i & l_i & L \end{Bmatrix} \langle r^L \rangle^2$$

$$B_{\text{s.p.}}(\text{mag}, L) = \left(\frac{\hbar}{m_p c} \right)^2 \langle r^{L-1} \rangle^2 (l_f \| \mathbf{Y}_{L-1} \| l_i)^2 (2j_f+1) \frac{(2L+1)^2}{L}$$

$$\times \left[\frac{(-1)^{j_f+l_f+1/2} L}{L+1} \sqrt{(2j_i+1) j_i(j_i+1)} \begin{Bmatrix} l_f & j_f & \frac{1}{2} \\ j_i & l_i & L-1 \end{Bmatrix} \begin{Bmatrix} j_f & L & j_i \\ 1 & j_i & L-1 \end{Bmatrix} \right.$$

$$\left. + \sqrt{\tfrac{3}{2}} \left(L\mu_p - \frac{L}{L+1} \right) \begin{Bmatrix} l_f & \frac{1}{2} & j_f \\ l_i & \frac{1}{2} & j_i \\ L-1 & 1 & L \end{Bmatrix} \right]^2$$

If $L = |j_f - j_i|$,

$$B_{\text{s.p.}}(\text{mag}, L) = \left(\frac{\hbar}{m_p c} \right)^2 \langle r^{L-1} \rangle^2 (l_f \| \mathbf{Y}_{L-1} \| l_i)^2$$

$$\times (2j_f+1) \frac{(2L+1)^2}{L} \tfrac{3}{2} \left(L\mu_p - \frac{L}{L+1} \right)^2 \begin{Bmatrix} l_f & \frac{1}{2} & j_f \\ l_i & \frac{1}{2} & j_i \\ L-1 & 1 & L \end{Bmatrix}^2$$

or neutrons replace $m_p \to m_n$,

$$L\mu_p - \frac{L}{L+1} \to L\mu_n$$

In the formulae above,

$\hbar\omega$: energy of transition
$k = \omega/c \ll 1/R$, $\quad R =$ nuclear radius
L: multipolarity of transition ($|J_i - J_f| \leq L \leq J_i + J_f$)

μ_i: the nucleon's intrinsic magnetic moment in nuclear magnetons
el: electric radiation [change in parity between initial and final states $= (-1)^L$]
mag: magnetic radiation [change in parity between initial and final states $= (-1)^{L+1}$]
s.p.: single particle transition between states

$$(l_i j_i) \rightarrow (l_f j_f)$$

$$\langle r^L \rangle = \int_0^\infty R_{n_f l_f}(r) r^L R_{n_i l_i}(r)\, dr$$

m_p: proton mass
m_n: neutron mass

Matrix Elements of Interactions

In *LS*-coupling:

$$\Delta E_{SLJ} = \langle l_1 l_2 SLJM \mid V_{12} \mid l_1 l_2 SLJM \rangle$$

In *jj*-coupling:

$$\Delta E_J = \langle j_1 j_2 JM \mid V_{12} \mid j_1 j_2 JM \rangle$$

In antisymmetric states (identical particles):

LS-coupling:

$$\Delta E_{SLJ} = \langle l_1 l_2 SLJM \mid V_{12} \mid l_1 l_2 SLJM \rangle + (-1)^{l_1+l_2+L+S} \langle l_1 l_2 SLJM \mid V_{12} \mid l_2 l_1 SLJM \rangle$$

jj-coupling:

$$\Delta E_1 = \langle l_1 j_1 l_2 j_2 JM \mid V_{12} \mid l_1 j_1 l_2 j_2 JM \rangle - (-1)^{j_1+j_2-J} \langle l_1 j_1 l_2 j_2 JM \mid V_{12} \mid l_2 j_2 l_1 j_1 JM \rangle$$

Relation between matrix elements in *jj*-coupling and *LS*-coupling:

$$\Delta E_{j_1 j_2 J} = (2j_1 + 1)(2j_2 + 1) \sum_{S,L} (2S+1)(2L+1) \begin{Bmatrix} \frac{1}{2} & l_1 & j_1 \\ \frac{1}{2} & l_2 & j_2 \\ S & L & J \end{Bmatrix}^2 \Delta E_{SLJ}$$

For δ-forces (identical particles),

$$V_{12} = \delta(\mathbf{r}_1 - \mathbf{r}_2) = \frac{1}{r_1 r_2} \delta(r_1 - r_2)\, \delta(\cos\theta_1 - \cos\theta_2)\, \delta(\varphi_1 - \varphi_2)$$

$$\begin{aligned} \Delta E_{SLJ}(\delta) &= [1 + (-1)^S] \langle l_1 l_2 SLJM \mid \delta(\mathbf{r}_1 - \mathbf{r}_2) \mid l_1 l_2 SLJM \rangle \\ &= [1 + (-1)^S](2l_1 + 1)(2l_2 + 1) \begin{pmatrix} l_1 & L & l_2 \\ 0 & 0 & 0 \end{pmatrix}^2 \\ &\quad \times \frac{1}{4\pi} \int [R_{n_1 l_1}(r)\, R_{n_2 l_2}(r)]^2 \frac{1}{r^2}\, dr \end{aligned}$$

$$\Delta E_{j_1 j_2 J}(\delta) = \begin{cases} (2j_1+1)(2j_2+1)\begin{pmatrix} j_1 & j_2 & J \\ \frac{1}{2} & -\frac{1}{2} & 0 \end{pmatrix}^2 F_0 & \text{if } (-1)^{l_1+l_2-J} = +1 \\ 0 & \text{if } (-1)^{l_1+l_2-J} = -1 \end{cases}$$

where

$$F_0 = \frac{1}{4\pi}\int \frac{1}{r^2}[R_{n_1 l_1}(r)\, R_{n_2 l_2}(r)]^2\, dr$$

Expansion in Legendre polynomials:

$$V(|\mathbf{r}_1 - \mathbf{r}_2|) = \sum_{k=0}^{\infty} v_k(r_1, r_2)\, P_k(\cos\omega_{12})$$

$$v_k(r_1, r_2) = \frac{2k+1}{2}\int V(|\mathbf{r}_1 - \mathbf{r}_2|)\, P_k(\cos\omega_{12})\, d(\cos\omega_{12})$$

$$\langle l_1 l_2 LM \mid V(|\mathbf{r}_1 - \mathbf{r}_2|) \mid l_1 l_2 LM\rangle = \sum_{k\text{ even}} f_k F^k$$

where

$$F^k = \int [R_{n_1 l_1}(r)\, R_{n_2 l_2}(r)]^2\, v_k(r_1, r_2)\, dr_1\, dr_2$$

$$f_k = (-1)^{l_1+l_2+L}(l_1 \| \mathbf{C}^{(k)} \| l_1)(l_2 \| \mathbf{C}^{(k)} \| l_2)\begin{Bmatrix} l_1 & l_2 & L \\ l_2 & l_1 & k \end{Bmatrix}$$

$$C_\kappa^{(k)}(\Omega) = \sqrt{\frac{4\pi}{2k+1}}\, Y_{k\kappa}(\Omega)$$

Conversely:

$$(l_1 \| \mathbf{C}^{(k)} \| l_1)(l_2 \| \mathbf{C}^{(k)} \| l_2) F^k = \sum_L (-1)^{l_1+l_2+L}(2L+1)(2k+1)\begin{Bmatrix} l_1 & l_2 & L \\ l_2 & l_1 & k \end{Bmatrix} \times \langle l_1 l_2 LM \mid V_{12} \mid l_1 l_2 LM\rangle$$

If $V_{12} = V(|\mathbf{r}_1 - \mathbf{r}_2|)$, then, for identical particles,

$$\Delta E_{SLJ} = \sum f_k F^k + (-1)^{l_1+l_2+L+S}\sum g_k G^k$$

where

$$G^k = \int R_{n_1 l_1}(r_1)\, R_{n_2 l_2}(r_2)\, R_{n_1 l_1}(r_2)\, R_{n_2 l_2}(r_1)\, v_k(r_1, r_2)\, dr_1\, dr_2$$

$$g_k = (-1)^L (l_1 \| \mathbf{C}^{(k)} \| l_2)(l_2 \| \mathbf{C}^{(k)} \| l_1)\begin{Bmatrix} l_1 & l_2 & L \\ l_1 & l_2 & k \end{Bmatrix}$$

$$= (-1)^{l_1+l_2+L+k}(l_1 \| \mathbf{C}^{(k)} \| l_2)(l_2 \| \mathbf{C}^{(k)} \| l_1)$$

$$\langle l_1 l_2 LM \mid \sum_r \begin{Bmatrix} l_1 & l_2 & k \\ l_2 & l_1 & r \end{Bmatrix}(-1)^r(2r+1)(\mathbf{u}^{(r)}(1)\cdot\mathbf{u}^{(r)}(2)) \mid l_1 l_2 LM\rangle$$

Generally

$$V_{12} = \sum_{ss'kk'r} v_{ss',kk',r}(r_1, r_2)(\mathbf{T}^{(sk)r}(1)\cdot\mathbf{T}^{(s'k')r}(2))$$

where

$$\mathbf{T}^{(sk)r} = [\mathbf{\Sigma}^{(s)} \times \mathbf{u}^{(k)}(\mathbf{r})]^{(r)}$$

and $\mathbf{\Sigma}$ operates on spins only, $\mathbf{u}^{(k)}$ on space coordinates only.

$$\langle j_1 j_2 JM \mid V_{12} \mid j_1 j_2 JM \rangle = \sum f_r F^r \qquad [r \text{ stands for } (ss'kk', r)]$$

$$f_r = (-1)^{j_1+j_2+J} (l_1 j_1 \,\|\, \mathbf{T}^{(sk)r} \,\|\, l_1 j_1)(l_2 j_2 \,\|\, \mathbf{T}^{(s'k')r} \,\|\, l_2 j_2) \begin{Bmatrix} j_1 & j_2 & J \\ j_2 & j_1 & r \end{Bmatrix}$$

$$= (-1)^{j_1+j_2+J}(2r+1)(2j_1+1)(2j_2+1) \begin{Bmatrix} j_1 & j_2 & J \\ j_2 & j_1 & r \end{Bmatrix} \begin{Bmatrix} \frac{1}{2} & l_1 & j_1 \\ \frac{1}{2} & l_1 & j_1 \\ s & k & r \end{Bmatrix} \begin{Bmatrix} \frac{1}{2} & l_2 & j_2 \\ \frac{1}{2} & l_2 & j_2 \\ s' & k' & r \end{Bmatrix}$$

$$\times (\tfrac{1}{2} \,\|\, \mathbf{\Sigma}^{(s)} \,\|\, \tfrac{1}{2})(\tfrac{1}{2} \,\|\, \mathbf{\Sigma}^{(s')} \,\|\, \tfrac{1}{2})(l_1 \,\|\, \mathbf{u}^{(k)} \,\|\, l_1)(l_2 \,\|\, \mathbf{u}^{(k')} \,\|\, l_2)$$

$$\sum_{ss'kk'} F^{(ss',kk',r)} (l_1 j_1 \,\|\, \mathbf{T}^{(sk)r} \,\|\, l_1 j_1)(l_2 j_2 \,\|\, \mathbf{T}^{(s'k')r} \,\|\, l_2 j_2)$$

$$= (2r+1) \sum (-1)^{j_1+j_2+J} (2J+1) \begin{Bmatrix} j_1 & j_2 & J \\ j_2 & j_1 & r \end{Bmatrix} \langle j_1 j_2 JM \mid V_{12} \mid j_1 j_2 JM \rangle$$

Average jj' interaction energy:

$$\overline{\Delta E}(jj') = \frac{\Sigma (2J+1)\, \Delta E(jj'J)}{\Sigma\, 2J+1} = F^{(00)0} \qquad | j - j' | \leq J \leq j + j'$$

Interaction of j' particle with closed j shell:

$$\Delta E(j^{2j+1} j' \; J = j') = (2j+1)\, \overline{\Delta E(jj')}$$

Interaction of j hole with j' particle:

$$\Delta E(j^{2j} j' J) = (2j+1)\, \overline{\Delta E(jj')} - \sum_{J'} (2J'+1) \begin{Bmatrix} j & j' & J' \\ j & j' & J \end{Bmatrix} \Delta E(jj'J')$$

If $R_{nl}(r)$ is a radial function of the harmonic oscillator with frequency ω and mass $m/2$, and the particles, with mass m, move in a harmonic oscillator potential with frequency ω, then

$$\Delta E(l_1 l_2 L) = \sum_l \alpha_l^L I_l$$

where $0 \leq l \leq 2n_1 + l_1 + 2n_2 + l_2$,

$$I_l = \int | R_{1l}(r) |^2 \, V(r) \, dr$$

and the α_l^L are universal constants, independent of $V(r)$.

Coefficients of Fractional Parentage for Identical Particles

Definition:

$$\psi(j^n\alpha J) = \sum_{\alpha' J'} [j^{n-1}(\alpha' J')\, jJ| \} j^n \alpha J]\, \psi(j^{n-1}_{1\ldots n-1}(\alpha' J')\, j_n J)$$

Orthogonality:

$$\sum_{\alpha' J'} [j^{n-1}(\alpha' J')\, jJ| \} j^n \alpha_1 J]\, [j^{n-1}(\alpha' J')\, jJ| \} j^n \alpha_2 J] = \delta_{\alpha_1 \alpha_2}$$

In the j^3-configuration:

$$[j^2(J_1)\, jJ| \} j^3[J_0]\, J] = \left[\delta(J_1, J_0) + 2\sqrt{(2J_0+1)(2J_1+1)} \begin{Bmatrix} j & j & J_1 \\ J & j & J_0 \end{Bmatrix}\right] \times \left[3 + 6(2J_0+1) \begin{Bmatrix} j & j & J_0 \\ J & j & J_0 \end{Bmatrix}\right]^{-1/2}$$

For $J_0 = 0$:

$$[j^2(0)\, j\, J = j| \} j^3\, J = j] = \sqrt{\frac{2j-1}{3(2j+1)}}$$

$$[j^2(J_1)\, j\, J = j| \} j^3\, J = j] = -\frac{2\sqrt{2J_1+1}}{\sqrt{3(2j-1)(2j+1)}}, \qquad J_1 > 0,\ J_1 \text{ even}$$

Recursion formula:

$$n[j^{n-1}(\alpha_0 J_0)\, jJ| \} j^n[\alpha_0 j_0]\, J]\, [j^{n-1}(\alpha_1 J_1)\, jJ| \} j^n[\alpha_0 J_0]\, J]$$
$$= \delta(\alpha_1, \alpha_0)\, \delta(J_1, J_0) + (n-1) \sum_{\alpha_2 J_2} (-1)^{J_0+J_1} \sqrt{(2J_0+1)(2J_1+1)}$$
$$\times \begin{Bmatrix} J_2 & j & J_1 \\ J & j & J_0 \end{Bmatrix} [j^{n-2}(\alpha_2 j_2)\, jJ_0| \} j^{n-1} \alpha_0 J_0]\, [j^{n-2}(\alpha_2 J_2)\, jJ_1| \} j^{n-1} \alpha_1 J_1]$$

Applications of *c.f.p.*:

If $\mathbf{F}^{(k)} = \Sigma \mathbf{f}^{(k)}(i)$, then

$$(j^n \alpha J \,||\, \mathbf{F}^{(k)} \,||\, j^n \alpha' J')$$
$$= n \sum_{\alpha_1 J_1} [j^n \alpha J\{| j^{n-1}(\alpha_1 J_1)\, jJ]\, [j^{n-1}(\alpha_1 J_1)\, jJ'| \} j^n \alpha' J']\, (-1)^{J_1+j+J+k}$$
$$\times \sqrt{(2J+1)(2J'+1)} \begin{Bmatrix} j & J & J_1 \\ J' & j & k \end{Bmatrix} (j \,||\, \mathbf{f}^{(k)} \,||\, j)$$

$$(j^n \alpha J \,||\, \sum_i \mathbf{f}^{(0)}(i) \,||\, j^n \alpha' J') = n\sqrt{\frac{2J+1}{2j+1}}\, (j \,||\, \mathbf{f}^{(0)} \,||\, j)\, \delta(\alpha, \alpha')\, \delta(J, J')$$

If $G = \Sigma_{i<k} g_{ik}$, and G is a scalar, then

$$\langle j^n \alpha JM \mid G \mid j^n \alpha' JM \rangle$$

$$= \frac{n}{n-2} \sum_{\alpha_1 \alpha_1' J_1} [j^n \alpha J \{| j^{n-1}(\alpha_1 J_1) jJ] [j^{n-1}(\alpha_1' J_1) jJ |\} j^n \alpha' J]$$

$$\times \left\langle j^{n-1} \alpha_1 J_1 M_1 \mid \sum_{i<k}^{n-1} g_{ik} \mid j^{n-1} \alpha_1' J_1 M_1 \right\rangle$$

A possible check on values of *c.f.p.*:

$$\frac{n}{n-2} \sum_{\alpha_1 J_1} J_1(J_1+1) [j^{n-1}(\alpha_1 J_1) jJ |\} j^n \alpha J]^2 = J(J+1) + \frac{n}{n-2} j(j+1)$$

$n \to n-2$ *c.f.p.*:

$$[j^{n-2}(\alpha_2 J_2) j^2(J') J |\} j^n \alpha J]$$

$$= \sum_{\alpha_1 J_1} [j^{n-2}(\alpha_2 J_2) jJ_1 |\} j^{n-1} \alpha_1 J_1] [j^{n-1}(\alpha_1 J_1) jJ |\} j^n \alpha J] (-1)^{J_2+J+2j}$$

$$\times \sqrt{(2J_1+1)(2J'+1)} \begin{Bmatrix} J_2 & j & J_1 \\ j & J & J' \end{Bmatrix}$$

$$\left\langle j^n \alpha JM \left| \sum_{i<k}^{n} g_{ik} \right| j^n \alpha' JM \right\rangle$$

$$= \frac{n(n-1)}{2} \sum_{\alpha_2 J_2 J'} [j^n \alpha J \{| j^{n-2}(\alpha_2 J_2) j^2(J') J] [j^{n-2}(\alpha_2 J_2) j^2(J') J |\} j^n \alpha' J]$$

$$\times \langle j^2 J'M' \mid g_{12} \mid j^2 J'M' \rangle$$

Seniority Operator and Seniority Scheme

$$\langle j^2 JM \mid q_{12} \mid j^2 J'M' \rangle = (2j+1)\, \delta(J, J')\, \delta(M, M')\, \delta(J, 0)$$

$$Q = \sum_{i<k} q_{ik}$$

$$\langle j^n v JM \mid Q \mid j^n v JM \rangle = \tfrac{1}{2}(n-v)(2j+3-n-v)$$

$$[j^{n-2}(vJ) j^2(0) J |\} j^n v J]^2 = \frac{n-v}{n(n-1)} \frac{2j+3-n-v}{2j+1},$$

$$2j+1-v \geq n \geq v+2$$

Recursion relations for $n \to n-2$ *c.f.p.*:

$$[j^n(v_2 J_2) j^2(J') J |\} j^{n+2} v J] = \sqrt{\frac{n(n-1)(n+2-v)(2j+1-n-v)}{(n+1)(n+2)(n-v_2)(2j+3-n-v_2)}}$$

$$\times [j^{n-2}(v_2 J_2) j^2(J') |\} j^n v J]$$

$$[j^{n-2}(v_2J_2)j^2(J')\,J|\}\,j^nvJ] = \begin{cases} \sqrt{\dfrac{v(v-1)(2j+3-n-v)(2j+5-n-v)}{n(n-1)(2j+3-2v)(2j+5-2v)}} \\ \times [j^{v-2}(v-2,\,J_2)\,j^2(J')|\}\,j^vvJ] \quad \text{for } v_2 = v-2 \\ \sqrt{\dfrac{(v+1)(v+2)(n-v)(2j+3-n-v)}{2n(n-1)(2j+1-2v)}} \\ \times [j^v(v\,J_2)\,j^2(J')|\}\,j^{v+2}vJ] \quad \text{for } v_2 = v \\ \sqrt{\dfrac{(v+3)(v+4)(n-v)(n-v-2)}{8n(n-1)}} \\ \times [j^{v+2}(v+2,\,J_2)j^2(J')|\}\,j^{v+4}vJ] \quad \text{for } v_2 = v+2 \end{cases}$$

Recursion relations for ordinary ($n \to n-1$) *c.f.p.*:

$$[j^{n+1}(v_1J_1)\,jJ|\}\,j^{n+2}vJ] = \sqrt{\frac{n(n+2-v)(2j+1-n-v)}{(n+2)(n+1-v_1)(2j+2-n-v_1)}} \times [j^{n-1}(v_1J_1)\,jJ|\}\,j^nvJ]$$

$$[j^{n-1}(v_1J_1)\,jJ|\}\,j^nvJ] = \begin{cases} \sqrt{\dfrac{v(2j+3-n-v)}{n(2j+3-2v)}} \\ \times [j^{v-1}(v-1,\,J_1)\,jJ|\}\,j^vvJ] \quad \text{for } v_1 = v-1 \\ \sqrt{\dfrac{(n-v)(v+2)}{2n}} \\ \times [j^{v+1}(v+1,\,J_1)\,jJ|\}\,j^{v+2}vJ] \quad \text{for } v_1 = v+1 \end{cases}$$

$$[j^{v+1}(v+1,J_1)jJ|\}j^{v+2}vJ] = (-1)^{J+j-J_1}\sqrt{\frac{(2J_1+1)\,2(v+1)}{(2J+1)(v+2)(2j+1-2v)}} \times [j^v(vJ)\,jJ_1|\}\,j^{v+1}v+1\;J_1]$$

Special values of *c.f.p.*:

$$[j^{n-1}(v_1J_1 = j)\;j\;J = 0|\}\;j^n\;v = 0\;J = 0] = \delta(v_1, 1)$$

$$[j^{n-1}(v_1J_1)\;j\;J = j|\}\;j^n\;v = 1\;J = j] = \begin{cases} \sqrt{\dfrac{2j+2-n}{n(2j+1)}} \quad \text{for } J_1 = 0,\, v_1 = 0 \\ -\sqrt{\dfrac{2(n-1)(2J_1+1)}{n(2j+1)(2j-1)}} \quad \text{for } J_1 > 0,\, v_1 = 2 \end{cases}$$

Special orthogonality relations:

$$\sum_{\alpha_1 J_1} [j^n v\alpha J \{| j^{n-1}(v-1, \alpha_1, J_1)\, jJ]\, [j^{n-1}(v-1, \alpha_1, J_1)\, jJ |\} j^n v\alpha' J] = \frac{v(2j+3-n-v)}{n(2j+3-2v)}\, \delta(\alpha, \alpha')$$

$$\sum_{\alpha_1 J_1} [j^n v\alpha J \{| j^{n-1}(v+1, \alpha_1, J_1)\, jJ]\, [j^{n-1}(v+1, \alpha_1, J_1)\, jJ |\} j^n v\alpha' J] = \frac{(n-v)(2j+3-v)}{n(2j+3-2v)}\, \delta(\alpha, \alpha')$$

$$\sum_{\alpha J} (2J+1)\, [j^n v\alpha J \{| j^{n-1}(v+1, \alpha_1 J_1)\, jJ]\, [j^{n-1}(v+1, \alpha_1' J_1)\, jJ |\} j^n v\alpha J] = \frac{(n-v)(v+1)}{n(2j+1-2v)}\, (2J_1+1)\, \delta(\alpha_1, \alpha_1')$$

$$\sum_{\alpha J} (2J+1)\, [j^n v\alpha J \{| j^{n-1}(v-1, \alpha_1, J_1)\, jJ]\, [j^{n-1}(v-1, \alpha_1' J_1)\, jJ |\} j^n v\alpha J] = \frac{(2j+3-n-v)(2j+4-v)}{n(2j+5-2v)}\, (2J_1+1)\, \delta(\alpha_1, \alpha_1')$$

$$\sum_{\alpha J} (2J+1)\, [j^n v\alpha J \{| j^{n-1}(v+1, \alpha_1 J_1)\, jJ]\, [j^{n-1}(v-1, \alpha_1' J_1)\, jJ |\} j^n v\alpha J] = 0$$

Matrix Elements in the Seniority Scheme

Of single particle irreducible tensors:

$$(j^n v J \,||\, \textstyle\sum \mathbf{f}^{(k)}(i) \,||\, j^n v' J') = 0 \qquad \text{for odd values of } k \text{ and } v \neq v'$$

$$\left(j^n v J \,||\, \sum_{i=1}^{n} \mathbf{f}^{(k)}(i) \,||\, j^n v J'\right) = \left(j^v v J \,||\, \sum_{i=1}^{v} \mathbf{f}^{(k)}(i) \,||\, j^v v J'\right) \qquad \text{for } k \text{ odd}$$

$$= \frac{2j+1-2n}{2j+1-2v} \left(j^v v J \,||\, \sum_{i=1}^{v} \mathbf{f}^{(k)}(i) \,||\, j^v v J'\right) \qquad \text{for } k > 0 \text{ even}$$

$$= \frac{n}{v} \left(j^v v J \,||\, \sum_{i=1}^{v} \mathbf{f}^{(0)}(i) \,||\, j^v v J\right) \delta(J, J') \qquad \text{for } k = 0$$

$$\left(j^v v J \,||\, \sum_{i=1}^{n} \mathbf{f}^{(k)}(i) \,||\, j^n v-2\ J'\right) = \sqrt{\frac{(n-v+2)(2j+3-n-v)}{2(2j+3-2v)}} \times \left(j^v v J \,||\, \sum_{i=1}^{v} \mathbf{f}_i^{(k)} \,||\, j^v v-2\ J'\right)$$

Of two particle scalar operators V_{ik}:

$$\left\langle j^n v J \,|\, \sum_{i<k}^{n} V_{ik} \,|\, j^n v-4\, J \right\rangle$$

$$= \sqrt{\frac{(n-v+2)(n-v+4)(2j+3-n-v)(2j+5-n-v)}{8(2j+3-2v)(2j+5-2v)}}$$

$$\times \left\langle j^v v J \,|\, \sum_{i<k}^{n} V_{ik} \,|\, j^v v-4\, J \right\rangle$$

$$\left\langle j^n v J \,|\, \sum_{i<k}^{n} V_{ik} \,|\, j^n v-2\, J \right\rangle$$

$$= \frac{2j+1-2n}{2j+1-2v} \sqrt{\frac{(n-v+2)(2j+3-n-v)}{2(2j+3-2v)}} \left\langle j^v v J \,|\, \sum_{i<k}^{v} V_{ik} \,|\, j^v v-2\, J \right\rangle$$

$$\left\langle j^n v J \,|\, \sum_{i<k}^{n} V_{ik} \,|\, j^n v J \right\rangle = \left\langle j^v v J \,|\, \sum_{i<k}^{v} \,|\, j^v v J \right\rangle + \frac{n-v}{2} E_0$$

$$+ \frac{(n-v)(2j+1-n-v)}{2(2j-1-2v)}$$

$$\times \left[\left\langle j^{v+2} v J \,|\, \sum_{i<k}^{v+2} V_{ik} \,|\, j^{v+2} v J \right\rangle - \left\langle j^v v J \,|\, \sum_{i<k}^{v} V_{ik} \,|\, j^v v J \right\rangle - E_0 \right]$$

where

$$E_0 = \frac{2}{2j+1} \sum_{J \text{ even}} (2J+1) \langle j^2 J \,|\, V_{12} \,|\, j^2 J \rangle$$

Pairing property of odd tensor interactions:

$$\left\langle j^n v \alpha J \middle| \sum_{i<k}^{n} V_{ik} \middle| j^n v \alpha' J \right\rangle$$

$$= \left\langle j^v v \alpha J \middle| \sum_{i<k}^{v} V_{ik} \middle| j^v v \alpha' J \right\rangle + \frac{n-v}{2} \langle j^2 J=0 \,|\, V_{12} \,|\, j^2 J=0 \rangle \, \delta(\alpha, \alpha')$$

Average Interaction Energies

$$\sum_{\alpha J} (2J+1) \left\langle j^n v \alpha J \middle| \sum_{i<k}^{n} V_{ik} \middle| j^n v \alpha J \right\rangle \Big/ \sum_{\alpha J} (2J+1)$$

$$= \frac{n(n-1)}{2} \bar{V}_2 + \frac{(n-v)(2j+3-n-v)}{2(2j+1)} (V_0 - \bar{V}_2)$$

where

$$V_0 = \langle j^2 J = 0 \mid V_{12} \mid j^2 J = 0 \rangle$$

$$\bar{V}_2 = \frac{1}{(j+1)(2j-1)} \sum_{J>0 \text{ even}} (2J+1) \langle j^2 J \mid V_{12} \mid j^2 J \rangle$$

Interaction energy in states with lowest seniority $v = 0, J = 0$ for n even and $v = 1, J = j$ for n odd:

$$\frac{n(n-1)}{2} \left(\frac{2(j+1)\bar{V}_2 - V_0}{2j+1} \right) + \left[\frac{n}{2}\right] \frac{2(j+1)}{2j+1} (V_0 - \bar{V}_2)$$

where

$$\left[\frac{n}{2}\right] = \frac{n}{2} \quad \text{for even } n \text{ and} \quad \left[\frac{n}{2}\right] = \frac{n-1}{2} \quad \text{for odd } n.$$

Recursion formula for number of states with given seniority:

$$\left[\sum_{\alpha J} (2J+1)\right]_v = \frac{(2j+4-v)(2j+3-2v)}{v(2j+5-2v)} \left[\sum_{\alpha J} (2J+1)\right]_{v-1}$$

Wave Functions with Isotopic Spin

$$[j^2(T_1 J_1)\, jTJ| \} j^3[T_0 J_0]\, TJ]$$

$$= \Big[\delta(T_1, T_0)\, \delta(J_1, J_0) - 2\sqrt{(2T_1+1)(2T_0+1)(2J_1+1)(2J_0+1)}$$

$$\times \begin{Bmatrix} \frac{1}{2} & \frac{1}{2} & T_1 \\ T & \frac{1}{2} & T_0 \end{Bmatrix} \begin{Bmatrix} j & j & J_1 \\ J & j & J_0 \end{Bmatrix} \Big]$$

$$\times \left[3 - 6(2T_0+1)(2J_0+1) \begin{Bmatrix} \frac{1}{2} & \frac{1}{2} & T_0 \\ T & \frac{1}{2} & T_0 \end{Bmatrix} \begin{Bmatrix} j & j & J_0 \\ J & j & J_0 \end{Bmatrix} \right]^{-1/2}$$

For the state with $T = \frac{1}{2}$, $J = j$ obtained from $T_0 = 1$, $J_0 = 0$:

$$[j^2(T_1 J_1)\, jT = \tfrac{1}{2}\, J = j| \} j^3[10]\, T = \tfrac{1}{2}\, J = j]$$

$$= \begin{cases} \sqrt{\dfrac{2(j+1)}{3(2j+1)}} & \text{for } T_1 = 1, \quad J_1 = 0 \\[2ex] \sqrt{\dfrac{2J_1+1}{6(j+1)(2j+1)}} & \text{for } T_1 = 1, \quad J_1 > 0 \text{ even} \\[2ex] -\sqrt{\dfrac{2J_1+1}{2(j+1)(2j+1)}} & \text{for } T_1 = 0, \quad J_1 \text{ odd} \end{cases}$$

Recursion relation of *c.f.p.*:

$$n[j^{n-1}(\alpha_0 T_0 J_0)\, j\, T\, J|\}\, j^n[\alpha_0\, T_0\, J_0]\, T\, J]\, [j^{n-1}(\alpha_1 T_1 J_1)\, jTJ|\}\, j^n[\alpha_0 T_0 J_0]\, TJ]$$
$$= \delta(\alpha_0, \alpha_1)\, \delta(T_0, T_1)\, \delta(J_0, J_1) - (-1)^{T_0+T_1+J_0+J_1}(n-1)$$
$$\times [(2T_0+1)(2T_1+1)(2J_0+1)(2J_1+1)]^{1/2}$$
$$\times \sum_{\alpha_2 T_2 J_2} \begin{Bmatrix} T_2 & \frac{1}{2} & T_0 \\ T & \frac{1}{2} & T_1 \end{Bmatrix} \begin{Bmatrix} J_2 & j & J_0 \\ J & j & J_1 \end{Bmatrix} [j^{n-2}(\alpha_2 T_2 J_2)\, jT_0 J_0|\}\, j^{n-1}\alpha_0 T_0 J_0]$$
$$\times [j^{n-2}(\alpha_2 T_2 J_2)\, jT_1 J_1|\}\, j^{n-1}\alpha_1 T_1 J_1]$$

The g-factor of the state $|\, T\, M_T\, JM\rangle$ in the j^n-configuration is given by

$$g = \tfrac{1}{2}(g_p + g_n) + nM_T(g_p - g_n)$$
$$\times \sqrt{\tfrac{3}{2}}\sqrt{j(j+1)(2j+1)}\sqrt{\frac{2T+1}{T(T+1)}}\sqrt{\frac{2J+1}{J(J+1)}}$$
$$\times \sum_{\alpha_1 T_1 J_1} (-1)^{T_1+1/2+T+J_1+j+J} \begin{Bmatrix} \frac{1}{2} & T & T_1 \\ T & \frac{1}{2} & 1 \end{Bmatrix} \begin{Bmatrix} j & J & J_1 \\ J & j & 1 \end{Bmatrix} [j^{n-1}(\alpha_1 T_1 J_1)\, jTJ|\}\, j^n \alpha TJ]^2$$

The Seniority Scheme

Odd tensors are diagonal in the seniority v and reduced isospin t:

$$\left\langle j^n v' t' T' J' \left| \sum_{i=1}^{n} u_\kappa^{(k)}(i) \right| j^n v t T J \right\rangle$$

$$= \left\langle j^v vt\ T_1 = t\ J' \left| \sum_{i=1}^{v} u_\kappa^{(k)}(i) \right| j^n vt\ T_1 = t\ J \right\rangle \delta(v, v')\, \delta(t, t') \qquad \text{for } k \text{ odd}$$

The pairing property of odd tensor interactions:

$$\left\langle j^n\, vt\ T\ J \left| \sum_{i<k}^{n} V_{ik} \right| j^n\, vt\ T\ J \right\rangle$$

$$= \left\langle j^v\, vt\ T = t\ J \left| \sum_{i<k}^{v} V_{ik} \right| j^v\, vt\ T = t\ J \right\rangle$$

$$+ \frac{n-v}{2} \langle j^2\ T = 1\ J = 0 \mid V_{12} \mid j^2\ T = 1\ J = 0 \rangle.$$

The seniority operator defined by

$$\langle j^2 T M_T J M \mid q_{12} \mid j^2 T' M'_T J' M' \rangle$$
$$= (2j+1)\, \delta(T, T')\, \delta(M_T, M'_T)\, \delta(J, J')\, \delta(M, M')\, \delta(J, 0)$$

has, in the j^n configuration, the eigenvalues

$$Q(n, v, t, T) = \left\langle j^n vtTJ \middle| \sum_{i<k}^{n} q_{ik} \middle| j^n vtTJ \right\rangle$$

$$= \frac{n-v}{4}(4j + 8 - n - v) - T(T+1) + t(t+1)$$

C.f.p. of state with lowest seniority, n even, $v = 0$, $t = 0$, $J = 0$, and $T = n/2$, $n/2 - 2$, ...:

$$[j^{n-1}(v_1 = 1, t_1 = \tfrac{1}{2}, T_1 = T + \tfrac{1}{2}, J_1 = j)\, j\, T\, J = 0 | \} j^n\, v = 0\, t = 0\, T\, J = 0]^2$$
$$= (n - 2T)(T + 1)/n(2T + 1)$$

$$[j^{n-1}(v_1 = 1, t_1 = \tfrac{1}{2}, T_1 = T - \tfrac{1}{2}, J_1 = j)\, j\, T\, J = 0 | \} j^n\, v = 0\, t = 0\, T\, J = 0]^2$$
$$= (n + 2T + 2)\, T/n(2T + 1)$$

C.f.p. of state with lowest seniority, n odd, $v = 1$, $t = \frac{1}{2}$, T, and $J = j$:

Case I $\qquad T = \dfrac{n}{2}, \quad \dfrac{n-4}{2}, \quad \ldots$

$$[j^{n-1}(0\,0, T - \tfrac{1}{2}, 0)\, j\, T\, j | \} j^n\, 1\, \tfrac{1}{2}\, T\, j]^2$$
$$= (4j + 4 - n - 2T)/2n(2j + 1)$$

$$[j^{n-1}(2\,0, T + \tfrac{1}{2}, J_1)\, j\, T\, j | \} j^n\, 1\, \tfrac{1}{2}\, T\, j]^2$$
$$= (2J_1 + 1)(n - 2T)/4n(j + 1)(2j + 1), \qquad J_1 \text{ odd}$$

$$[j^{n-1}(2\,1, T - \tfrac{1}{2}, J_1)\, j\, T\, j | \} j^n\, 1\, \tfrac{1}{2}\, T\, j]^2$$
$$= \frac{(2J_1 + 1)\{4(j + 1)[T(n + 2T) - 1] + n - 2T\}}{2n(j + 1)(2j - 1)(2j + 1)(2T + 1)}, \qquad J_1 > 0 \text{ even}$$

$$[j^{n-1}(2\,1, T + \tfrac{1}{2}, J_1)\, j\, T\, j | \} j^n\, 1\, \tfrac{1}{2}\, T\, j]^2$$
$$= (2J_1 + 1)(n - 2T)(2T + 3)/4n(j + 1)(2j - 1)(2T + 1), \qquad J_1 > 0 \text{ even}$$

Case II $\qquad T = \dfrac{n-2}{2}, \quad \dfrac{n-6}{2}, \quad \ldots$

$$[j^{n-1}(0\,0, T + \tfrac{1}{2}, 0)\, j\, T\, j | \} j^n\, 1\, \tfrac{1}{2}\, T\, j]^2$$
$$= (4j + 6 - n + 2T)/2n(2j + 1)$$

$$[j^{n-1}(2\,0, T - \tfrac{1}{2}, J_1)\, j\, T\, j | \} j^n\, 1\, \tfrac{1}{2}\, T\, j]^2$$
$$= (2J_1 + 1)(n + 2T + 2)/4n(j + 1)(2j + 1), \qquad J_1 \text{ odd}$$

$$[j^{n-1}(2\,1,\,T+\tfrac{1}{2},\,J_1)\,j\,T\,j|\}j^n\,1\,\tfrac{1}{2}\,T\,j]^2$$
$$=\frac{(2J_1+1)\,\{4(j+1)\,[(T+1)\,(n-2T-2)+1]-(n+2T+2)\}}{2n(j+1)\,(2j-1)\,(2j+1)\,(2T+1)},\qquad J_1>0\text{ even}$$

$$[j^{n-1}(2\,1,\,T-\tfrac{1}{2},\,T_1)\,j\,T\,j|\}j^n\,1\,\tfrac{1}{2}\,T\,j]^2$$
$$=(2J_1+1)\,(n+2T+2)\,(2T-1)/4n(j+1)\,(2j-1)\,(2T+1),\qquad J_1>0\text{ even}$$

For $v=1, t=\frac{1}{2}$, and $J=j$ the g-factor of the state $|\,j^n 1\tfrac{1}{2}\,TM_T J=jM\rangle$ is

$$g=\tfrac{1}{2}(g_p+g_n)+\begin{cases} M_T\dfrac{g_p-g_n}{2}\,\dfrac{4(j+1)\,(T+1)-n+2T}{4(j+1)\,T(T+1)}, & T=\dfrac{n}{2},\dfrac{n-4}{2},\ldots \\[2ex] M_T\dfrac{g_n-g_p}{2}\,\dfrac{4(j+1)\,T+n+2T+2}{4(j+1)\,T(T+1)}, & T=\dfrac{n-2}{2},\dfrac{n-6}{2},\ldots \end{cases}$$

Average interaction energies:

$$\frac{\Sigma_{\alpha J}(2J+1)\,\langle j^n\beta\alpha vtTJ\mid\Sigma_{i<k}^n V_{ik}\mid j^n\beta\alpha vtTJ\rangle}{\Sigma_{\alpha J}(2J+1)}$$
$$=\frac{n(n-1)}{8}(3\bar{V}_2+\bar{V}_1)+\frac{(2j+3)\,\bar{V}_2-2V_0-(2j+1)\,\bar{V}_1}{2(2j+1)}[T(T+1)-\tfrac{3}{4}n]$$
$$+\frac{V_0-\bar{V}_2}{2j+1}\left[\frac{n-v}{4}(4j+5-n-v)+t(t+1)-\tfrac{3}{4}v\right]$$

where
$$\bar{V}_2=\sum_{J>0\text{ even}}(2J+1)\,V_J\Big/\sum_{J>0\text{ even}}(2J+1)$$
$$\bar{V}_1=\sum_{J\text{ odd}}(2J+1)\,V_J\Big/\sum_{J\text{ odd}}(2J+1)$$
$$V_J=\langle j^2TJ\mid V_{12}\mid j^2TJ\rangle\qquad (T=1\text{ for }J\text{ even},\ T=0\text{ for }J\text{ odd})$$

For the state of lowest seniority ($v=0,\ t=0,\ J=0$ for even n or $v=1,\ t=\frac{1}{2},\ J=j$ for odd n)

$$\left\langle j^nTJ\left|\sum_{i<k}^n V_{ik}\right|j^nTJ\right\rangle=\frac{n(n-1)}{2}\,\frac{(6j+5)\,\bar{V}_2+(2j+1)\,\bar{V}_1-2V_0}{4(2j+1)}$$
$$+\frac{(2j+3)\,\bar{V}_2-2V_0-(2j+1)\,\bar{V}_1}{2(2j+1)}[T(T+1)-\tfrac{3}{4}n]+\left[\frac{n}{2}\right]\frac{2(j+1)}{2j+1}(V_0-\bar{V}_2)$$

where

$$\left[\frac{n}{2}\right] = \frac{n}{2} \text{ for even } n \text{ and } \left[\frac{n}{2}\right] = \frac{n-1}{2} \text{ for odd } n.$$

Particles in Nonequivalent Orbits

Center-of-mass theorem:

$$\sum_{J=|J_1-J_2|}^{J_1+J_2} (2J+1)\left\langle j_1^{n_1}(\alpha_1 J_1)\, j_2^{n_2}(\alpha_2 J_2) J \left| \sum_{i=1}^{n_1} \sum_{k=n_1+1}^{n_1+n_2} V_{ik} \right| j_1^{n_1}(\alpha_1 J_1)\, j_2^{n_2}(\alpha_2 J_2) J\right\rangle$$

$$= n_1 n_2 \sum_{J=|j_1-j_2|}^{j_1+j_2} (2J+1)\, V(j_1 j_2 J)\Big/(2j_1+1)(2j_2+1)$$

For $J_1 = 0$:

$$\left\langle j_1^{n_1}(\alpha_1 J_1 = 0)\, j_2^{n_2}(\alpha_2 J_2)\, J = J_2 \left| \sum_{i=1}^{n_1} \sum_{k=n_1+1}^{n_1+n_2} V_{ik} \right| j_1^{n_1}(\alpha_1 J_1 = 0)\, j_2^{n_2}(\alpha_2 J_2)\, J = J_2 \right\rangle$$

$$= n_1 n_2 \sum (2J+1)\, V(j_1 j_2 J)\Big/(2j_1+1)(2j_2+1).$$

In the general case:

$$\left\langle j_1^{n_1}(\alpha_1 J_1)\, j_2^{n_2}(\alpha_2 J_2)\, J \left| \sum_{i=1}^{n_1} \sum_{k=n_1+1}^{n_1+n_2} V_{ik} \right| j_1^{n_1}(\alpha_1')\, j_2^{n_2}(\alpha_2')\, J \right\rangle$$

$$= n_1 n_2 \sum_{\alpha_{11} J_{11} \alpha_{22} J_{22}} [\, j_1^{n_1} \alpha_1 J_1 \{\!|\, j_1^{n_1-1}(\alpha_{11} J_{11}) j_1 J_1]\, [\, j_1^{n_1-1}(\alpha_{11} J_{11}) j_1 J_1' |\!\}\, j_1^{n_1} \alpha_1' J_1']$$

$$\times [\, j_2^{n_2} \alpha_2 J_2 \{\!|\, j_2^{n_2-1}(\alpha_{22} J_{22}) j_2 J_2]\, [\, j_2^{n_2-1}(\alpha_{22} J_{22}) j_2 J_2' |\!\}\, j_2^{n_2} \alpha_2' J_2']$$

$$\times [(2J_1+1)(2J_1'+1)(2J_2+1)(2J_2'+1)]^{1/2}$$

$$\times \sum_{J'} (2J'+1) V(j_1 j_2 J') \sum_{J_{12}} (2J_{12}+1) \begin{Bmatrix} J_{11} & j_1 & J_1 \\ J_{22} & j_2 & J_2 \\ J_{12} & J' & J \end{Bmatrix} \begin{Bmatrix} J_{11} & j_1 & J_1' \\ J_{22} & j_2 & J_2' \\ J_{12} & J' & J \end{Bmatrix}$$

In the case of a single j_2 particle:

$$\langle j_1^n(\alpha_1 J_1) j_2 J \left| \sum_{i=1}^{n} V_{i,n+1} \right| j_1^n(\alpha_1' J_1') j_2 J \rangle$$

$$= n \sum_{\alpha_{11} J_{11}} [\, j_1^n \alpha_1 J_1 \{\!|\, j_1^{n-1}(\alpha_{11} J_{11}) j_1 J_1]\, [\, j_1^{n-1}(\alpha_{11} J_{11}) j_1 J_1' |\!\}\, j_1^n \alpha_1' J_1']$$

$$\times [(2J_1+1)(2J_1'+1)]^{1/2} \sum_{J'} (2J'+1) V(j_1 j_2 J') \begin{Bmatrix} J_{11} & j_1 & J_1 \\ j_2 & J & J' \end{Bmatrix} \begin{Bmatrix} J_{11} & j_1 & J_1' \\ j_2 & J & J' \end{Bmatrix}$$

The general case with isotopic spin:

$$\langle j_1^{n_1}(\alpha_1 T_1 J_1)\, j_2^{n_2}(\alpha_2 T_2 J_2) T\, J \Big| \sum_{i=1}^{n_1} \sum_{k=n_1+1}^{n_1+n_2} V_{ik} \Big| j_1^{n_1}(\alpha_1' T_1' J_1') j_2^{n_2}(\alpha_2' T_2' J_2') T\, J \rangle$$

$$= n_1 n_2 \sum_{\alpha_{11} T_{11} J_{11} \alpha_{22} T_{22} J_{22}} [j_1^{n_1} \alpha_1 T_1 J_1 \{| j_1^{n_1-1}(\alpha_{11} T_{11} J_{11})\, j_1 T_1 J_1]$$

$$[j_1^{n_1-1}(\alpha_{11} T_{11} J_{11})\, j_1 T_1' J_1' |\} j_1^{n_1} \alpha_1' T_1' J_1'] \, [j_2^{n_2} \alpha_2 T_2 J_2 \{| j_2^{n_2-1}(\alpha_{22} T_{22} J_{22})\, j_2 T_2 J_2]$$

$$\times [j_2^{n_2-1}(\alpha_{22} T_{22} J_{22})\, j_2 T_2' J_2' |\} j_2^{n_2} \alpha_2' T_2' J_2']$$

$$\times [(2T_1+1)(2T_1'+1)(2T_2+1)(2T_2'+1)$$

$$\times (2J_1+1)(2J_1'+1)(2J_2+1)(2J_2'+1)]^{1/2}$$

$$\times \sum_{T'J'} (2T'+1)(2J'+1) V(j_1 j_2 T' J')$$

$$\times \sum_{T_{12}} (2T_{12}+1) \begin{Bmatrix} T_{11} & \frac{1}{2} & T_1 \\ T_{22} & \frac{1}{2} & T_2 \\ T_{12} & T' & T \end{Bmatrix} \begin{Bmatrix} T_{11} & \frac{1}{2} & T_1' \\ T_{22} & \frac{1}{2} & T_2' \\ T_{12} & T' & T \end{Bmatrix}$$

$$\times \sum_{J_{12}} (2J_{12}+1) \begin{Bmatrix} J_{11} & j_1 & J_1 \\ J_{22} & j_2 & J_2 \\ J_{12} & J' & J \end{Bmatrix} \begin{Bmatrix} J_{11} & j_1 & J_1' \\ J_{22} & j_2 & J_2' \\ J_{12} & J' & J \end{Bmatrix}$$

Matrix elements of single particle operators:

A special case

$$\left([j^{n-1}(\alpha_1 T_1 J_1)\, j_2]_a\, T' M_T' J' \,\Big\|\, \sum_{i=1}^{n} \mathbf{h}_i \mathbf{f}_i^{(k)} \,\Big\|\, j_1^n \alpha T M_T J \right)$$

$$= \sqrt{n}\, [j_1^{n-1}(\alpha_1 T_1 J_1)\, j_1 T\, J |\} j_1^n \alpha\, T\, J] \, \langle T_1 \tfrac{1}{2} T' M_T | \mathbf{h}_n | T_1 \tfrac{1}{2} T M_T \rangle$$

$$\times (-1)^{J_1+j_1+J'+k} [(2J+1)(2J'+1)]^{1/2} \begin{Bmatrix} j_2 & j_1 & k \\ J & J' & J_1 \end{Bmatrix} (j_2 \| \mathbf{f}^{(k)} \| j_1)$$

Some special statistical factors:

$$n[j^{n-1}(v=1 \;\; J=j)\, j \;\; J=0 |\} j^n \;\; v=0 \;\; J=0]^2 = n \qquad \text{for } n \text{ even}$$

$$n[j^{n-1}(v=0 \;\; J=0)\, j \;\; J=j |\} j^n \;\; v=1 \;\; J=j]^2 = \frac{2j+2-n}{2j+1} \qquad \text{for } n \text{ odd}$$

Two groups of equivalent nucleons:

$$\left([j_1^{n_1}(\alpha_1' T_1' J_1')\, j_2^{n_2}(\alpha_2' T_2' J_2')]_a\, T'J' \quad \left\| \sum_{i=1}^{n_1+n_2} \mathbf{f}_i^{(k'k)} \right\| [j_1^{n_1-1}(\alpha_1 T_1 J_1)\, j_2^{n_2+1}(\alpha_2 T_2 J_2)]_a\, TJ\right)$$

$$= \sqrt{n_1(n_2+1)}\, [\, j_1^{n_1}\alpha_1' T_1' J_1' \{\!| \, j_1^{n_1-1}(\alpha_1 T_1 J_1)\, j_1 T_1' J_1']$$

$$\times (-1)^{n_2}\, [\, j_2^{n_2}(\alpha_2' T_2' J_2')\, j_2 T_2 J_2 |\!\}\, j_2^{n_2+1}\alpha_2 T_2 J_2]$$

$$\times (-1)^{T_2'+1/2-T_2+J_2'+j_2-J_2}\, [(2T_1'+1)(2T_2+1)(2T+1)(2T'+1)$$

$$\times (2J_1'+1)(2J_2+1)(2J+1)(2J'+1)]^{1/2}$$

$$\times \begin{Bmatrix} T_1' & T_2' & T' \\ T_1 & T_2 & T \\ \frac{1}{2} & \frac{1}{2} & k' \end{Bmatrix} \begin{Bmatrix} J_1' & J_2' & J' \\ J_1 & J_2 & J \\ j_1 & j_2 & k \end{Bmatrix} (\tfrac{1}{2} j_1 || \mathbf{f}^{(k'k)} || \tfrac{1}{2} j_2)$$

Configuration Interaction

Two particle excitations:

$$\langle [\, j^{n-2}(\alpha_1 T_1 J_1)\, j'j''(T_2 J_2) T\, J]_a \left| \sum_{i<k}^{n} V_{ik} \right| j^n \alpha\, T\, J\rangle$$

$$= \sqrt{n(n-1)/2}\, [\, j^{n-2}(\alpha_1 T_1 J_1)\, j^2(T_2 J_2) T\, J |\!\}\, j^n\alpha\, T\, J]\, \langle (j'j'' T_2 J_2)_a | V_{n-1,n} |\, j^2 T_2 J_2\rangle$$

$$\langle\, j_1^{n_1-1}(\alpha_{11} T_{11} J_{11})\, j_2^{n_2-1}(\alpha_{22} T_{22} J_{22})\, (T_{12} J_{12})\, j_1' j_2' (T'J') T\, J \,|$$

$$\left| \sum_{i<k}^{n_1+n_2} V_{ik} \right| j_1^{n_1}(\alpha_1 T_1 J_1)\, j_2^{n_2}(\alpha_2 T_2 J_2) T\, J\rangle$$

$$= \sqrt{n_1 n_2}\, [\, j_1^{n_1-1}(\alpha_{11} T_{11} J_{11})\, j_1 T_1 J_1 |\!\}\, j_1^{n_1}\alpha_1 T_1 J_1]$$

$$\times [\, j_2^{n_2-1}(\alpha_{22} T_{22} J_{22})\, j_2 T_2 J_2 |\!\}\, j_2^{n_2}\alpha_2 T_2 J_2]$$

$$\times [(2T_1+1)(2T_2+1)(2T_{12}+1)(2T'+1)$$

$$\times (2J_1+1)(2J_2+1)(2J_{12}+1)(2J'+1)]^{1/2}$$

$$\times \begin{Bmatrix} T_{11} & \frac{1}{2} & T_1 \\ T_{22} & \frac{1}{2} & T_2 \\ T_{12} & T' & T \end{Bmatrix} \begin{Bmatrix} J_{11} & j_1 & J_1 \\ J_{22} & j_2 & J_2 \\ J_{12} & J' & J \end{Bmatrix} \langle (j_1' j_2' T'J')_a | V_{12} | (j_1 j_2 T'J')_a \rangle$$

Single particle excitations:

$$\langle j^{n-1}(\alpha_1 T_1 J_1) j' T\, J \left| \sum_{i<k}^{n} V_{ik} \right| j^n \alpha\, T\, J \rangle$$

$$= \int \psi_a^*[j^{n-1}(\alpha_1 T_1 J_1) j' T\, J] \left(\sum_{i<k}^{n} V_{ik} \right) \psi(j^n \alpha\, T\, J)$$

$$= (n-1)\sqrt{n} \sum_{\alpha_2 T_2 J_2 T' J'} [j^{n-1} \alpha_1 T_1 J_1 \{| j^{n-2}(\alpha_2 T_2 J_2) j\, T_1 J_1]$$

$$\times [j^{n-2}(\alpha_2 T_2 J_2) j^2 (T' J') T\, J |\} j^n \alpha\, T\, J]$$

$$\cdot (-1)^{T_2+1+T+J_2+j+j'+J} [(2T_1+1)(2T'+1)(2J_1+1)(2J'+1)]^{1/2}$$

$$\cdot \begin{Bmatrix} T_2 & \frac{1}{2} & T_1 \\ \frac{1}{2} & T & T' \end{Bmatrix} \begin{Bmatrix} J_2 & j & J_1 \\ j' & J & J' \end{Bmatrix} \langle (j j' T' J')_a | V_{12} | j^2 T' J' \rangle / \sqrt{2}$$

Coefficients of Fractional Percentage

$$[j^{n-1}(v_1, J_1), j; J |\} j^n\, v\, J] \qquad \text{maximum } T$$

1) $j = 3/2, \quad n = 3$

$$\left[\left(\tfrac{3}{2}\right)^2 (0, 0), \tfrac{3}{2}; \tfrac{3}{2} |\} \left(\tfrac{3}{2}\right)^3, 1, \tfrac{3}{2}\right] = 1/\sqrt{6}$$

$$\left[\left(\tfrac{3}{2}\right)^2 (2, 2), \tfrac{3}{2}; \tfrac{3}{2} |\} \left(\tfrac{3}{2}\right)^3, 1, \tfrac{3}{2}\right] = -\sqrt{(5/6)}$$

2a) $j = 5/2$, $n = 3$[a]

v	J \ v_1	0	2	
	J_1	0	2	4
1	$\frac{5}{2}$	$-\frac{\sqrt{2}}{3}$	$\frac{\sqrt{5}}{3\sqrt{2}}$	$\frac{1}{\sqrt{2}}$
3	$\frac{3}{2}$		$-\sqrt{\frac{5}{7}}$	$\sqrt{\frac{2}{7}}$
	$\frac{9}{2}$		$\sqrt{\frac{3}{14}}$	$-\sqrt{\frac{11}{14}}$

[a] A.R. Edmonds and B.H. Flowers, *Proc. Roy. Soc.* **A214**, 515 (1952).

2b) $j = 5/2, \; n = 4$[a]

v	J \ v_1	1	3	
	J \ J_1	$\frac{5}{2}$	$\frac{3}{2}$	$\frac{9}{2}$
0	0	1		
2	2	$-\frac{1}{2}$	$\sqrt{\frac{3}{7}}$	$\frac{3}{2\sqrt{7}}$
	4	$-\frac{1}{2}$	$-\sqrt{\frac{2}{21}}$	$-\frac{\sqrt{55}}{2\sqrt{21}}$

[a] A. R. Edmonds and B. H. Flowers, *Proc. Roy. Soc.* **A214**, 515 (1952).

3a) $j = 7/2, \quad n = 3$[a]

v	J \ v_1	0	2		
	J \ J_1	0	2	4	6
1	$\frac{7}{2}$	$-\frac{1}{2}$	$\frac{\sqrt{5}}{6}$	$\frac{1}{2}$	$\frac{\sqrt{13}}{6}$
3	$\frac{3}{2}$		$\sqrt{\frac{3}{14}}$	$-\sqrt{\frac{11}{14}}$	
	$\frac{5}{2}$		$\frac{\sqrt{11}}{3\sqrt{2}}$	$\sqrt{\frac{2}{33}}$	$-\frac{\sqrt{65}}{3\sqrt{22}}$
	$\frac{9}{2}$		$\frac{\sqrt{13}}{3\sqrt{14}}$	$-\frac{5\sqrt{2}}{\sqrt{77}}$	$\frac{7}{3\sqrt{22}}$
	$\frac{11}{2}$		$-\frac{\sqrt{5}}{3\sqrt{2}}$	$\sqrt{\frac{13}{66}}$	$\frac{2\sqrt{13}}{3\sqrt{11}}$
	$\frac{15}{2}$			$\sqrt{\frac{5}{22}}$	$\sqrt{\frac{17}{22}}$

[a] A. R. Edmonds and B. H. Flowers, *Proc. Roy. Soc.* **A214**, 515 (1952).

3b) $j = 7/2,\ n = 4$[a]

v \ v_1	J \ J_1	1	3				
		7/2	3/2	5/2	9/2	11/2	15/2
0	0	1					
2	2	$\frac{1}{\sqrt{3}}$	$\frac{3}{2\sqrt{35}}$	$-\frac{11}{2\sqrt{10}}$	$-\frac{\sqrt{13}}{2\sqrt{42}}$	$-\frac{1}{2}$	
	4	$\frac{1}{\sqrt{3}}$	$-\frac{\sqrt{11}}{2\sqrt{21}}$	$-\frac{1}{\sqrt{66}}$	$\frac{5\sqrt{5}}{\sqrt{462}}$	$\frac{\sqrt{13}}{2\sqrt{33}}$	$\sqrt{\frac{5}{33}}$
	6	$\frac{1}{\sqrt{3}}$		$\frac{\sqrt{5}}{2\sqrt{22}}$	$-\frac{7\sqrt{5}}{2\sqrt{858}}$	$\sqrt{\frac{2}{11}}$	$-\sqrt{\frac{51}{143}}$
4	2		$-\sqrt{\frac{11}{35}}$	$-\frac{1}{2\sqrt{10}}$	$-\frac{3\sqrt{39}}{2\sqrt{154}}$	$\frac{1}{\sqrt{11}}$	
	4		$\frac{\sqrt{13}}{2\sqrt{105}}$	$\sqrt{\frac{13}{30}}$	$\sqrt{\frac{3}{182}}$	$\frac{\sqrt{5}}{2\sqrt{3}}$	$-\frac{2}{\sqrt{39}}$
	5		$\frac{\sqrt{3}}{2\sqrt{5}}$	$-\frac{7}{2\sqrt{110}}$	$-\frac{3\sqrt{3}}{2\sqrt{22}}$	$\frac{\sqrt{65}}{2\sqrt{77}}$	$\sqrt{\frac{17}{77}}$
	8				$\frac{2\sqrt{5}}{\sqrt{143}}$	$\frac{3\sqrt{2}}{\sqrt{77}}$	$\sqrt{\frac{57}{91}}$

[a] A. R. Edmonds and B. H. Flowers, *Proc. Roy. Soc*, **A214**, 515 (1952).

4a) $j = 9/2 \quad n = 3$[a]

v \ v_1	J \ J_1	0	2			
		0	2	4	6	8
1	$\frac{9}{2}$	$-\frac{2}{\sqrt{15}}$	$\frac{1}{2\sqrt{3}}$	$\frac{\sqrt{3}}{2\sqrt{5}}$	$\frac{\sqrt{13}}{2\sqrt{15}}$	$\frac{\sqrt{17}}{2\sqrt{15}}$
3	$\frac{3}{2}$			$-\frac{2\sqrt{2}}{\sqrt{11}}$	$\sqrt{\frac{3}{11}}$	
	$\frac{5}{2}$		$-\frac{\sqrt{5}}{3\sqrt{2}}$	$\sqrt{\frac{13}{66}}$	$\frac{2\sqrt{13}}{3\sqrt{11}}$	
	$\frac{7}{2}$		$\frac{2\sqrt{13}}{3\sqrt{11}}$	$\frac{2\sqrt{5}}{\sqrt{143}}$	$\frac{1}{3\sqrt{55}}$	$-\sqrt{\frac{238}{715}}$
	$\frac{9}{2}$		$-\frac{\sqrt{13}}{6\sqrt{11}}$	$\frac{7\sqrt{5}}{2\sqrt{143}}$	$-\frac{31}{6\sqrt{55}}$	$\frac{3\sqrt{17}}{2\sqrt{715}}$
	$\frac{11}{2}$		$\frac{\sqrt{17}}{3\sqrt{11}}$	$-\sqrt{\frac{170}{429}}$	$-\frac{2\sqrt{14}}{3\sqrt{55}}$	$\frac{6\sqrt{19}}{\sqrt{2145}}$
	$\frac{13}{2}$		$\sqrt{\frac{10}{33}}$	$-\frac{3}{\sqrt{143}}$	$-\sqrt{\frac{17}{66}}$	$-\sqrt{\frac{323}{858}}$
	$\frac{15}{2}$			$\sqrt{\frac{19}{143}}$	$-\sqrt{\frac{7}{11}}$	$\sqrt{\frac{3}{13}}$
	$\frac{17}{2}$			$\frac{5\sqrt{5}}{\sqrt{429}}$	$-\sqrt{\frac{19}{110}}$	$-\sqrt{\frac{209}{390}}$
	$\frac{21}{2}$				$-\sqrt{\frac{7}{30}}$	$\sqrt{\frac{23}{30}}$

[a] A. de-Shalit, *Nuclear Phys.* **1**, 225 (1958).

4b) $j = 9/$

$(9/2)^4$ v	$(9/2)^3$ v' → / J' ↓ J	1				
		9/2	3/2	5/2	7/2	9/2
0	0	1				
2	2	$\frac{\sqrt{3}}{2\sqrt{2}}$		$\frac{1}{2\sqrt{3}}$	$\frac{2\sqrt{2\cdot 13}}{3\sqrt{5\cdot 11}}$	$\frac{\sqrt{13}}{2\cdot 3\sqrt{2\cdot 11}}$
	4	$\frac{\sqrt{3}}{2\sqrt{2}}$	$(-)\frac{2\sqrt{2}}{3\sqrt{11}}$	$(-)\frac{\sqrt{13}}{2\cdot 3\sqrt{11}}$	$\frac{2\sqrt{2\cdot 5}}{3\sqrt{11\cdot 13}}$	$(-)\frac{5\cdot 7}{2\cdot 3\sqrt{2\cdot 11\cdot 13}}$
	6	$\frac{\sqrt{3}}{2\sqrt{2}}$	$\frac{\sqrt{3}}{\sqrt{11\cdot 13}}$	$(-)\frac{\sqrt{2}}{\sqrt{3\cdot 11}}$	$\frac{\sqrt{2}}{3\sqrt{5\cdot 11\cdot 13}}$	$\frac{31}{2\cdot 3\sqrt{2\cdot 11\cdot 13}}$
	8	$\frac{\sqrt{3}}{2\sqrt{2}}$			$(-)\frac{2\sqrt{7}}{\sqrt{5\cdot 11\cdot 13}}$	$(-)\frac{3}{2\sqrt{2\cdot 11\cdot 13}}$
4	0					1
	2			$\frac{\sqrt{13}}{\sqrt{2\cdot 3\cdot 7}}$	$(-)\frac{\sqrt{11}}{3\sqrt{5\cdot 7}}$	$(-)\frac{\sqrt{7}}{3\sqrt{11}}$
	3		$\frac{\sqrt{3}}{2\sqrt{5}}$	$(-)\frac{\sqrt{2\cdot 13}}{\sqrt{5\cdot 7\cdot 11}}$	$\frac{\sqrt{3}}{2\sqrt{7\cdot 11}}$	$\frac{2\sqrt{3\cdot 5}}{\sqrt{11\cdot 13}}$
	4a		$(-)\frac{\sqrt{7}}{2\sqrt{3\cdot 23}}$	$\frac{8\sqrt{2\cdot 13}}{11\sqrt{3\cdot 7\cdot 23}}$	$(-)\frac{\sqrt{3\cdot 5\cdot 13}}{2\sqrt{7\cdot 23}}$	$\frac{8\sqrt{3\cdot 7}}{11\sqrt{13\cdot 23}}$
	4b		$\frac{\sqrt{13\cdot 17}}{3\sqrt{5\cdot 23}}$	$(-)\frac{43\sqrt{17}}{3\cdot 11\sqrt{2\cdot 5\cdot 23}}$	$(-)\frac{\sqrt{17}}{3\sqrt{23}}$	$(-)\frac{5\sqrt{5\cdot 17}}{3\cdot 11\sqrt{23}}$
	5		$\frac{\sqrt{3}}{\sqrt{5\cdot 11}}$	$\frac{\sqrt{7}}{\sqrt{5\cdot 11}}$	$\frac{\sqrt{3\cdot 7}}{\sqrt{2\cdot 11\cdot 13}}$	$\frac{\sqrt{2\cdot 3\cdot 5}}{\sqrt{11\cdot 13}}$
	6a		$\frac{\sqrt{11}}{\sqrt{5\cdot 13}}$	$\frac{13\sqrt{2}}{11\sqrt{5\cdot 11}}$	$\frac{5}{\sqrt{2\cdot 3\cdot 11\cdot 13}}$	$\frac{23\sqrt{2\cdot 5}}{11\sqrt{3\cdot 11\cdot 13}}$
	6b		0	$\frac{5\sqrt{17\cdot 19}}{11\sqrt{2\cdot 3\cdot 7\cdot 11}}$	$(-)\frac{\sqrt{2\cdot 5\cdot 17\cdot 19}}{3\sqrt{7\cdot 11\cdot 13}}$	$\frac{\sqrt{7\cdot 17\cdot 19}}{3\cdot 11\sqrt{2\cdot 11\cdot 13}}$
	7			$\frac{\sqrt{17}}{\sqrt{2\cdot 7\cdot 11}}$	$\frac{4\sqrt{3\cdot 17}}{\sqrt{5\cdot 7\cdot 11\cdot 13}}$	$(-)\frac{\sqrt{3\cdot 7}}{\sqrt{2\cdot 11\cdot 13}}$
	8				0	$\frac{\sqrt{19}}{\sqrt{2\cdot 11\cdot 13}}$
	9					$\frac{\sqrt{5}}{\sqrt{2\cdot 13}}$
	10					
	12					

[a] M. Sato, Institute of Industrial Science, University of Tokyo, Tokyo, Japan, privately circulated tables

$n = 4^a$

3				
11/2	13/2	15/2	17/2	21/2
$\frac{\sqrt{17}}{\sqrt{3\cdot5\cdot11}}$	$\frac{(-)\sqrt{7}}{\sqrt{3\cdot11}}$			
$\frac{(-)\sqrt{2\cdot5\cdot17}}{3\sqrt{11\cdot13}}$	$\frac{\sqrt{7}}{\sqrt{2\cdot11\cdot13}}$	$\frac{2\sqrt{19}}{3\sqrt{11\cdot13}}$	$\frac{(-)5\sqrt{5}}{\sqrt{2\cdot3\cdot11\cdot13}}$	
$\frac{(-)2\sqrt{2\cdot7}}{\sqrt{3\cdot5\cdot11\cdot13}}$	$\frac{\sqrt{7\cdot17}}{2\sqrt{3\cdot11\cdot13}}$	$\frac{(-)2\sqrt{7}}{\sqrt{11\cdot13}}$	$\frac{3\sqrt{19}}{2\sqrt{5\cdot11\cdot13}}$	$\frac{\sqrt{7\cdot11}}{2\sqrt{3\cdot5\cdot13}}$
$\frac{2\cdot3\sqrt{19}}{\sqrt{5\cdot11\cdot13\cdot17}}$	$\frac{\sqrt{7\cdot19}}{2\sqrt{3\cdot11\cdot13}}$	$\frac{2\sqrt{3}}{\sqrt{13\cdot17}}$	$\frac{\sqrt{3\cdot11\cdot19}}{2\sqrt{5\cdot13\cdot17}}$	$\frac{(-)\sqrt{11\cdot23}}{2\sqrt{3\cdot5\cdot17}}$
$\frac{(-)2\sqrt{2\cdot7\cdot17}}{\sqrt{3\cdot5\cdot11\cdot13}}$	$\frac{(-)\sqrt{11}}{\sqrt{2\cdot3\cdot13}}$			
$\frac{(-)\sqrt{17}}{\sqrt{11\cdot13}}$	$\frac{(-)\sqrt{2\cdot3}}{\sqrt{5\cdot11\cdot13}}$	$\frac{\sqrt{17\cdot19}}{\sqrt{2\cdot5\cdot11\cdot13}}$		
$\frac{(-)7\sqrt{5\cdot7\cdot17}}{11\sqrt{3\cdot13\cdot23}}$	$\frac{4\sqrt{3}}{\sqrt{13\cdot23}}$	$\frac{(-)17\sqrt{7\cdot19}}{11\sqrt{2\cdot3\cdot13\cdot23}}$	0	
$\frac{8}{3\cdot11\sqrt{23}}$	$\frac{\sqrt{7\cdot17}}{2\sqrt{5\cdot23}}$	$\frac{2\cdot41\sqrt{2\cdot19}}{3\cdot11\sqrt{5\cdot17\cdot23}}$	$\frac{\sqrt{23}}{2\sqrt{3\cdot17}}$	
$\frac{\sqrt{17}}{\sqrt{2\cdot11\cdot13}}$	$\frac{\sqrt{3\cdot7\cdot17}}{\sqrt{2\cdot5\cdot11\cdot13}}$	$\frac{\sqrt{19}}{\sqrt{5\cdot11\cdot13}}$	$\frac{(-)\sqrt{3\cdot19}}{\sqrt{2\cdot11\cdot13}}$	
$\frac{(-)55\sqrt{7}}{11\sqrt{2\cdot11\cdot13}}$	$\frac{(-)\sqrt{7\cdot17}}{2\sqrt{5\cdot11\cdot13}}$	$\frac{37\sqrt{3\cdot7}}{11\sqrt{5\cdot11\cdot13}}$	$\frac{\sqrt{3\cdot19}}{2\sqrt{11\cdot13}}$	$\frac{\sqrt{7}}{\sqrt{11\cdot13}}$
$\frac{8\sqrt{2\cdot5\cdot19}}{11\sqrt{3\cdot11\cdot13\cdot17}}$	$\frac{7\sqrt{19}}{4\sqrt{3\cdot11\cdot13}}$	$\frac{97}{11\sqrt{11\cdot13\cdot17\cdot19}}$	$\frac{9\cdot9\sqrt{7}}{4\sqrt{5\cdot11\cdot13\cdot17}}$	$\frac{(-)41\sqrt{13}}{2\sqrt{3\cdot5\cdot11\cdot17\cdot19}}$
$\frac{53}{\sqrt{5\cdot11\cdot13\cdot17}}$	$\frac{7\sqrt{19}}{4\sqrt{3\cdot11\cdot13}}$	$\frac{5\sqrt{3}}{\sqrt{11\cdot13\cdot17}}$	$\frac{\sqrt{7\cdot19}}{4\sqrt{5\cdot13\cdot17}}$	$\frac{\sqrt{11\cdot23}}{2\sqrt{3\cdot5\cdot17}}$
$\frac{9\sqrt{5}}{\sqrt{11\cdot13\cdot17}}$	$\frac{(-)\sqrt{7\cdot11}}{4\sqrt{3\cdot13}}$	$\frac{(-)3\cdot7\sqrt{3}}{\sqrt{13\ \ 17\cdot19}}$	$\frac{\sqrt{3\cdot11\cdot13}}{4\sqrt{5\cdot17}}$	$\frac{\sqrt{11\cdot23}}{2\sqrt{3\cdot5\cdot17\cdot19}}$
$\frac{3\sqrt{2}}{\sqrt{13\cdot17}}$	$\frac{\sqrt{7\cdot11}}{4\sqrt{3\cdot5\cdot13}}$	$\frac{(-,41\sqrt{3}}{\sqrt{5\cdot13\cdot17\cdot19}}$	$\frac{(-)\sqrt{3\cdot23}}{4\sqrt{17}}$	$\frac{(-)\sqrt{5\cdot7\cdot23}}{2\sqrt{3\cdot17\cdot19}}$
$\frac{\sqrt{23}}{\sqrt{13\cdot17}}$	$\frac{(-)\sqrt{11\cdot23}}{2\sqrt{3\cdot5\cdot13}}$	$\frac{\sqrt{2\cdot3\cdot11}}{\sqrt{5\cdot17\cdot19}}$	$\frac{\sqrt{11}}{2\sqrt{17}}$	$\frac{(-)\sqrt{5\cdot11\cdot13}}{\sqrt{2\cdot3\cdot17\cdot19}}$
		$\frac{5\sqrt{2}}{\sqrt{17\cdot19}}$	$\frac{(-)\sqrt{3\cdot13}}{\sqrt{2\cdot5\cdot17}}$	$\frac{(-)3\sqrt{13}}{\sqrt{2\cdot5\cdot19}}$

4c) $j = 9/2, \quad n = 5$

$(9/2)^5$ v	$(9/2)^4$ v' / J' — J	0	2							
		0	2	4	6	8	0	2	3	4a
1	9/2	$\frac{\sqrt{3}}{5}$	$\frac{(-)\,1}{\sqrt{2\cdot5}}$	$\frac{(-)\,3}{5\sqrt{2}}$	$\frac{(-)\sqrt{13}}{5\sqrt{2}}$	$\frac{(-)\sqrt{17}}{5\sqrt{2}}$				
	3/2			$\frac{4}{\sqrt{5\cdot11}}$	$\frac{(-)\sqrt{2\cdot3}}{\sqrt{5\cdot11}}$				$\frac{(-)\sqrt{3\cdot7}}{2\cdot5\sqrt{2}}$	$\frac{(-)\sqrt{3\cdot7}}{2\sqrt{2\cdot5\cdot23}}$
	5/2		$\frac{1}{3}$	$\frac{(-)\sqrt{13}}{\sqrt{3\cdot5\cdot11}}$	$\frac{(-)\,2\sqrt{2\cdot13}}{3\sqrt{5\cdot11}}$			$\frac{(-)\sqrt{13}}{3\sqrt{2\cdot7}}$	$\frac{(-)\sqrt{2\cdot13}}{5\sqrt{3\cdot11}}$	$\frac{(-)\,8\sqrt{2\cdot13}}{11\sqrt{5\cdot7\cdot23}}$
	7/2		$\frac{2\sqrt{2\cdot13}}{3\sqrt{5\cdot11}}$	$\frac{2\sqrt{2}}{\sqrt{11\cdot13}}$	$\frac{\sqrt{2}}{3\cdot5\sqrt{11}}$	$\frac{(-)\,2\sqrt{7\cdot17}}{5\sqrt{11\cdot13}}$		$\frac{\sqrt{11}}{2\cdot3\sqrt{5\cdot7}}$	$\frac{\sqrt{3}}{4\sqrt{5\cdot11}}$	$\frac{3\sqrt{3\cdot13}}{4\sqrt{7\cdot23}}$
	9/2		$\frac{\sqrt{13}}{3\sqrt{2\cdot5\cdot11}}$	$\frac{(-)\,7}{\sqrt{2\cdot11\cdot13}}$	$\frac{31}{3\cdot5\sqrt{2\cdot11}}$	$\frac{(-)\,3\sqrt{17}}{5\sqrt{2\cdot11\cdot13}}$	$\frac{(-)\,1}{5}$	$\frac{\sqrt{7}}{3\sqrt{5\cdot11}}$	$\frac{2\sqrt{3\cdot7}}{\sqrt{5\cdot11\cdot13}}$	$\frac{(-)\,8\cdot3\sqrt{3\cdot7}}{5\cdot11\sqrt{13\cdot23}}$
3	11/2		$\frac{\sqrt{2\cdot17}}{3\sqrt{5\cdot11}}$	$\frac{(-)\,2\sqrt{17}}{\sqrt{3\cdot11\cdot13}}$	$\frac{(-)\,4\sqrt{7}}{3\cdot5\sqrt{11}}$	$\frac{2\sqrt{2\cdot3\cdot19}}{5\sqrt{11\cdot13}}$		$\frac{2\sqrt{7\cdot17}}{3\sqrt{5\cdot11\cdot13}}$	$\frac{(-)\sqrt{7\cdot17}}{\sqrt{2\cdot3\cdot5\cdot11\cdot13}}$	$\frac{7\sqrt{7\cdot17}}{11\sqrt{2\cdot13\cdot23}}$
	13/2		$\frac{2}{\sqrt{3\cdot11}}$	$\frac{(-)\,3\sqrt{2}}{\sqrt{5\cdot11\cdot13}}$	$\frac{(-)\sqrt{17}}{\sqrt{3\cdot5\cdot11}}$	$\frac{(-)\sqrt{17\cdot19}}{\sqrt{3\cdot5\cdot11\cdot13}}$		$\frac{(-)\sqrt{11}}{\sqrt{2\cdot3\cdot7\cdot13}}$	$\frac{\sqrt{2\cdot3}}{5\sqrt{11\cdot13}}$	$\frac{4\cdot3\sqrt{3}}{\sqrt{5\cdot7\cdot13\cdot23}}$
	15/2			$\frac{\sqrt{2\cdot19}}{\sqrt{5\cdot11\cdot13}}$	$\frac{(-)\sqrt{2\cdot7}}{\sqrt{5\cdot11}}$	$\frac{\sqrt{2\cdot3}}{\sqrt{5\cdot13}}$			$\frac{\sqrt{7\cdot17\cdot19}}{4\cdot5\sqrt{11\cdot13}}$	$\frac{17\sqrt{3\cdot7\cdot19}}{4\cdot11\sqrt{5\cdot13\cdot23}}$

	[illegible]		$\sqrt{3\cdot11\cdot13}$	$5\sqrt{11}$	$5\sqrt{3\cdot13}$				
	21/2			$\frac{\sqrt{7}}{5\sqrt{3}}$	$\frac{(-)\sqrt{23}}{5\sqrt{3}}$				
	1/2								$\frac{\sqrt{2\cdot3\cdot7\cdot13}}{\sqrt{5\cdot11\cdot23}}$
	5/2						$\frac{2}{\sqrt{5\cdot7}}$	$\frac{1}{\sqrt{3\cdot5\cdot11}}$	$\frac{71}{11\sqrt{7\cdot23}}$
	7/2						$\frac{\sqrt{5\cdot17}}{2\sqrt{7\cdot11}}$	$\frac{(-)\sqrt{3\cdot5\cdot17}}{4\sqrt{11}}$	$\frac{(-)3\cdot43\sqrt{3\cdot17}}{4\cdot11\sqrt{7\cdot13\cdot23}}$
	9/2					$\frac{\sqrt{2}}{5}$	$\frac{\sqrt{7}}{\sqrt{2\cdot5\cdot11}}$	$\frac{(-)\sqrt{2\cdot3\cdot7}}{\sqrt{5\cdot11\cdot13}}$	$\frac{(-)4\cdot9\sqrt{2\cdot3\cdot7}}{5\cdot11\sqrt{13\cdot23}}$
5	11/2						$\frac{\sqrt{19}}{\sqrt{11\cdot13}}$	$\frac{\sqrt{19}}{2\sqrt{2\cdot3\cdot11\cdot13}}$	$\frac{(-)71\sqrt{19}}{2\cdot11\sqrt{2\cdot5\cdot13\cdot23}}$
	13/2						$\frac{3\sqrt{3\cdot17}}{\sqrt{2\cdot5\cdot7\cdot11\cdot13}}$	$\frac{\sqrt{2\cdot3\cdot17}}{\sqrt{5\cdot11\cdot13}}$	$\frac{(-)2\cdot8\sqrt{3\cdot17}}{11\sqrt{7\cdot13\cdot23}}$
	15/2							$\frac{\sqrt{5\cdot17}}{4\sqrt{11\cdot13}}$	$\frac{5\sqrt{3}}{4\cdot11\sqrt{13\cdot23}}$
	17/2								$\frac{4\sqrt{7\cdot19}}{\sqrt{5\cdot11\cdot13\cdot23}}$
	19/2								
	25/2								

4c) $j = 9/2, \quad n = 5$ *(continued)*

	4							
4b	5	6a	6b	7	8	9	10	12
$\frac{\sqrt{13\cdot 17}}{5\sqrt{2\cdot 23}}$	$\frac{(-)\sqrt{3}}{5\sqrt{2}}$	$\frac{\sqrt{11}}{5\sqrt{2}}$	0					
$\frac{43\sqrt{17}}{5\cdot 11\sqrt{2\cdot 3\cdot 23}}$	$\frac{\sqrt{7}}{5\sqrt{3}}$	$\frac{(-)13\sqrt{2\cdot 13}}{5\cdot 11\sqrt{3\cdot 11}}$	$\frac{(-)\sqrt{5\cdot 13\cdot 17\cdot 19}}{3\cdot 11\sqrt{2\cdot 7\cdot 11}}$	$\frac{\sqrt{17}}{\sqrt{2\cdot 7\cdot 11}}$				
$\frac{\sqrt{17}}{2\sqrt{5\cdot 23}}$	$\frac{\sqrt{3\cdot 7}}{2\sqrt{2\cdot 5\cdot 13}}$	$\frac{(-)\sqrt{5}}{2\sqrt{2\cdot 3\cdot 11}}$	$\frac{\sqrt{17\cdot 19}}{3\sqrt{2\cdot 7\cdot 11}}$	$\frac{2\cdot 3\sqrt{17}}{\sqrt{5\cdot 7\cdot 11\cdot 13}}$	0			
$\frac{\sqrt{5\cdot 17}}{11\sqrt{23}}$	$\frac{\sqrt{2\cdot 3}}{\sqrt{5\cdot 13}}$	$\frac{(-)23\sqrt{2}}{11\sqrt{3\cdot 5\cdot 11}}$	$\frac{(-)\sqrt{7\cdot 17\cdot 19}}{3\cdot 5\cdot 11\sqrt{2\cdot 11}}$	$\frac{(-)3\sqrt{7}}{\sqrt{2\cdot 5\cdot 11\cdot 13}}$	$\frac{(-)\sqrt{17\cdot 19}}{5\sqrt{2\cdot 11\cdot 13}}$	$\frac{\sqrt{19}}{\sqrt{2\cdot 5\cdot 13}}$		
$\frac{(-)4\sqrt{2}}{11\sqrt{3\cdot 5\cdot 23}}$	$\frac{\sqrt{17}}{2\sqrt{3\cdot 5\cdot 13}}$	$\frac{5\sqrt{5\cdot 7}}{2\cdot 11\sqrt{3\cdot 11}}$	$\frac{(-)8\sqrt{19}}{3\cdot 11\sqrt{11\cdot 17}}$	$\frac{53}{\sqrt{2\cdot 5\cdot 11\cdot 13\cdot 17}}$	$\frac{(-)3\sqrt{3}}{\sqrt{2\cdot 11\cdot 13}}$	$\frac{\sqrt{3\cdot 19}}{\sqrt{5\cdot 13\cdot 17}}$	$\frac{(-)\sqrt{7\cdot 23}}{\sqrt{2\cdot 5\cdot 13\cdot 17}}$	
$\frac{3\sqrt{17}}{2\cdot 5\sqrt{23}}$	$\frac{(-)\sqrt{3\cdot 17}}{5\sqrt{2\cdot 13}}$	$\frac{(-)\sqrt{17}}{2\cdot 5\sqrt{11}}$	$\frac{\sqrt{7\cdot 19}}{4\sqrt{3\cdot 5\cdot 11}}$	$\frac{(-)\sqrt{7\cdot 19}}{4\sqrt{11\cdot 13}}$	$\frac{(-)\sqrt{11\cdot 17}}{4\sqrt{3\cdot 5\cdot 13}}$	$\frac{(-)\sqrt{11\cdot 19}}{4\cdot 5\sqrt{3\cdot 13}}$	$\frac{(-)\sqrt{11\cdot 23}}{2\cdot 5\sqrt{13}}$	
$(-)41\sqrt{19}$	$\sqrt{19}$	$(-)37\sqrt{3\cdot 7}$	$(-)97$	$3\cdot 5$	$3\cdot 7\sqrt{3}$	$(-)41\sqrt{3}$	$(-)3\sqrt{7\cdot 11}$	$(-)5\sqrt{5}$

$2\ -3\cdot5\cdot17$	$\sqrt{2\cdot3\cdot5\cdot13}$	$2\sqrt{3\cdot5\cdot11}$	$4\cdot5\sqrt{11\cdot17}$	$4\sqrt{3\cdot5\cdot13\cdot17}$	$4\cdot5\sqrt{3}$	$4\sqrt{3\cdot5\cdot17}$	$2\sqrt{3\cdot5\cdot17}$	$\sqrt{2\cdot3\cdot17}$
		$\frac{(-)\sqrt{7}}{11\sqrt{5}}$	$\frac{13\cdot41}{25\cdot11\sqrt{3\cdot17\cdot19}}$	$\frac{\sqrt{23}}{2\sqrt{5\cdot17}}$	$\frac{(-)\sqrt{23}}{2\cdot5\sqrt{3\cdot19}}$	$\frac{(-)\sqrt{7\cdot23}}{2\sqrt{3\cdot11\cdot17}}$	$\frac{\sqrt{7\cdot13}}{\sqrt{2\cdot17\cdot19}}$	$\frac{3\sqrt{13}}{\sqrt{2\cdot11\cdot19}}$
$\frac{(-)3\sqrt{2\cdot17}}{5\sqrt{11\cdot23}}$	$\frac{(-)\sqrt{13}}{5}$							
$\frac{2\sqrt{3\cdot13\cdot17}}{11\sqrt{5\cdot23}}$	$\frac{\sqrt{2\cdot7}}{\sqrt{3\cdot5\cdot13}}$	$\frac{(-)7}{11\sqrt{3\cdot5\cdot11}}$	$\frac{(-)2\sqrt{17\cdot19}}{11\sqrt{7\cdot11}}$	$\frac{(-)8\sqrt{17}}{\sqrt{5\cdot7\cdot11\cdot13}}$				
$\frac{3\cdot19}{2\cdot11\sqrt{5\cdot23}}$	$\frac{\sqrt{3\cdot7\cdot17}}{2\sqrt{2\cdot5\cdot13}}$	$\frac{7\sqrt{3\cdot5\cdot17}}{2\cdot11\sqrt{2\cdot11}}$	$\frac{13\sqrt{19}}{11\sqrt{2\cdot7\cdot11}}$	$\frac{(-)3\sqrt{5}}{\sqrt{7\cdot11\cdot13}}$	$\frac{(-)\sqrt{7\cdot19}}{\sqrt{11\cdot13}}$			
$\frac{3\sqrt{5\cdot17}}{11\sqrt{2\cdot23}}$	$\frac{(-)\sqrt{3}}{\sqrt{5\cdot13}}$	$\frac{(-)23\sqrt{3}}{11\sqrt{5\cdot11}}$	$\frac{(-)\sqrt{7\cdot17\cdot19}}{2\cdot5\cdot11\sqrt{11}}$	$\frac{3\sqrt{7}}{2\sqrt{5\cdot11\cdot13}}$	$\frac{(-)3\sqrt{17\cdot19}}{2\cdot5\sqrt{11\cdot13}}$	$\frac{(-)\sqrt{19}}{2\sqrt{5\cdot13}}$		
$\frac{(-)\sqrt{3\cdot7\cdot17\cdot19}}{5\cdot11\sqrt{2\cdot23}}$	$\frac{\sqrt{7\cdot19}}{5\sqrt{3\cdot13}}$	$\frac{(-)\sqrt{17\cdot19}}{11\sqrt{3\cdot11}}$	$\frac{109}{11\sqrt{5\cdot7\cdot11}}$	$\frac{\sqrt{2\cdot19}}{\sqrt{7\cdot11\cdot13}}$	$\frac{\sqrt{2\cdot3\cdot7\cdot17}}{\sqrt{5\cdot11\cdot13\cdot19}}$	$\frac{(-)\sqrt{3\cdot7}}{5\sqrt{13}}$	$\frac{(-)4\sqrt{2\cdot23}}{5\sqrt{13\cdot19}}$	
$\frac{3\cdot31}{2\cdot11\sqrt{5\cdot23}}$	$\frac{(-)\sqrt{3}}{\sqrt{2\cdot5\cdot13}}$	$\frac{(-)37}{2\cdot11\sqrt{5\cdot11}}$	$\frac{(-)31\sqrt{3\cdot19}}{4\cdot11\sqrt{7\cdot11\cdot17}}$	$\frac{(-)43\sqrt{19}}{4\sqrt{5\cdot7\cdot11\cdot13\cdot17}}$	$\frac{9\sqrt{3}}{4\sqrt{11\cdot13}}$	$\frac{3\sqrt{3\cdot11\cdot19}}{4\sqrt{5\cdot13\cdot17}}$	$\frac{(-)\sqrt{11\cdot23}}{2\sqrt{5\cdot13\cdot17}}$	
$\frac{(-)3\cdot19\sqrt{7}}{11\sqrt{5\cdot17\cdot23}}$	$\frac{(-)3\sqrt{7}}{2\sqrt{2\cdot5\cdot13}}$	$\frac{(-)\sqrt{3\cdot5\cdot19}}{2\cdot11\sqrt{2\cdot11}}$	$\frac{3\cdot109}{2\cdot11\sqrt{2\cdot7\cdot11\cdot17}}$	$\frac{19\sqrt{5\cdot19}}{2\sqrt{2\cdot7\cdot11\cdot13\cdot17}}$	$\frac{(-)\sqrt{3\cdot7}}{2\sqrt{2\cdot13}}$	$\frac{\sqrt{3\cdot7\cdot19}}{2\sqrt{2\cdot5\cdot13\cdot17}}$	$\frac{\sqrt{11}}{2\sqrt{5\cdot17}}$	$\frac{(-)\sqrt{7}}{2\sqrt{17}}$
$\frac{9\sqrt{3\cdot19}}{2\cdot5\sqrt{11\cdot17\cdot23}}$	$\frac{\sqrt{11}}{5\sqrt{2\cdot3\cdot13}}$	$\frac{1}{2\cdot11\sqrt{3}}$	$\frac{(-)5\cdot13\sqrt{5\cdot7}}{4\cdot11\sqrt{17\cdot19}}$	$\frac{5\sqrt{7\cdot11}}{4\sqrt{3\cdot13\cdot17}}$	$\frac{3\sqrt{3\cdot5}}{4\sqrt{13\cdot19}}$	$\frac{(-)\sqrt{3\cdot23}}{4\cdot5\sqrt{11\cdot17}}$	$\frac{67\sqrt{7}}{2\cdot5\sqrt{3\cdot17\cdot19}}$	$\frac{5\sqrt{5\cdot13}}{\sqrt{2\cdot3\cdot11\cdot17\cdot19}}$
	$\frac{(-)\sqrt{23}}{2\sqrt{5\cdot13}}$	$\frac{\sqrt{3\cdot7\cdot23}}{2\cdot11\sqrt{5}}$	$\frac{2\cdot9\sqrt{23}}{5\cdot11\sqrt{17\cdot19}}$	$\frac{11\cdot23}{\sqrt{2\cdot5\cdot13\cdot17}}$	$\frac{\sqrt{3\cdot11}}{5\sqrt{2\cdot19}}$	$\frac{\sqrt{3\cdot7}}{\sqrt{5\cdot17}}$	$\frac{(-)5\sqrt{7\cdot11}}{\sqrt{2\cdot17\cdot19\cdot23}}$	$\frac{\sqrt{2\cdot3\cdot13}}{\sqrt{19.23}}$
					$\frac{\sqrt{5}}{\sqrt{2\cdot19}}$	$\frac{(-)\sqrt{3}}{\sqrt{2\cdot11}}$	$\frac{(-)3\sqrt{3\cdot7}}{\sqrt{2\cdot19\cdot23}}$	$\frac{3\sqrt{29}}{\sqrt{2\cdot11\cdot23}}$

Energies in Terms of Slater Integrals for Wigner Force in *LS*-Coupling*†

ls

$$^1L = F_0 + G_l \qquad G_l = \frac{1}{2l+1} G^l$$

$$^3L = F_0 - G_l$$

pp

$$^1S, {}^3S = F_0 + 10F_2 \pm (G_0 + 10\,G_2)$$

$$^1P, {}^3P = F_0 - 5F_2 \mp (G_0 - 5G_2) \qquad F_2 = \frac{F^2}{25}, \quad G_2 = \frac{G^2}{25}$$

$$^1D, {}^3D = F_0 + F_2 \pm (G_0 + G_2)$$

pd

$$^1P, {}^3P = F_0 + 7F_2 \pm (G_1 + 63G_3)$$

$$^1D, {}^3D = F_0 - 7F_2 \mp (3G_1 - 21G_3) \quad F_2 = \frac{F_2}{35}, \quad G_1 = \frac{G^1}{15}, \quad G_3 = \frac{G^3}{245}$$

$$^1F, {}^3F = F_0 + 2F_2 \pm (6G_1 + 3G_3)$$

dd

$$^1S, {}^3S = F_0 + 14F_2 + 126F_4 \pm (G_0 + 14G_2 + 126G_4)$$

$$^1P, {}^3P = F_0 + 7F_2 - 84F_4 \mp (G_0 + 7G_2 - 84G_4)$$

$$^1D, {}^3D = F_0 - 3F_2 + 36F_4 \pm (G_0 - 3G_2 + 36G_4)$$

$$^1F, {}^3F = F_0 - 8F_2 - 9F_4 \mp (G_0 - 8G_2 - 9G_4)$$

$$^1G, {}^3G = F_0 + 4F_2 + F_4 \pm (G_0 + 4G_2 + G_4)$$

$$F_2 = \frac{F^2}{49}, \quad F_4 = \frac{F^4}{441}$$

$$G_2 = \frac{G^2}{49}, \quad G_4 = \frac{G^4}{441}$$

pf

$$^1D, {}^3D = F_0 + 12F_2 \pm (3G_2 + 36G_4)$$

$$^1F, {}^3F = F_0 - 15F_2 \mp (15G_2 - 9G_4) \qquad F_2 = \frac{F_2}{75}, \quad G_2 = \frac{G^2}{175}, \quad G_4 = \frac{G^4}{189}$$

$$^1G, {}^3G = F_0 + 5F_2 \pm (45G_2 + G_4)$$

df

$$^1P, {}^3P = F_0 + 24F_2 + 66F_4 \pm (G_1 + 24G_3 + 330G_5)$$

$$^1D, {}^3D = F_0 + 6F_2 - 99F_4 \mp (3G_1 + 42G_3 - 165G_5)$$

$$F_2 = \frac{F^2}{105}, F_4 = \frac{F^4}{693}, G_1 = \frac{G^1}{35}$$

* The formulas for Majorana force are obtained by multiplying each term by $(-1)^{T+S+1}$.

† The formulas for the configurations l^2 with $T = 1$ are obtained by substituting $G_k = F_k$ in the ll-configurations and dividing by 2. For l^2 $T = 0$ one has to substitute $G_k = -F_k$ and to divide by 2.

$$^1F,\ ^3F = F_0 - 11F_2 + 66F_4 \pm (6G_1 + 19G_3 + 55G_5)$$

$$G_3 = \frac{G^3}{315}, \quad G_5 = \frac{G^5}{1524.6}$$

$$^1G,\ ^3G = F_0 - 15F_2 - 22F_4 \mp (10G_1 - 35G_3 - 11G_5)$$

$$^1H,\ ^3H = F_0 + 10F_2 + 3F_4 \pm (15G_1 + 10G_3 + G_5)$$

ff

$$^1S,\ ^3S = F_0 + 60F_2 + 198F_4 + 1716F_6 \pm (G_0 + 60G_2 + 198G_4 + 1716G_6)$$

$$F_2 = \frac{F^2}{225}$$

$$^1P,\ ^3P = F_0 + 45F_2 + 33F_4 - 1287F_6 \mp (G_0 + 45G_2 + 33G_4 - 1287G_6)$$

$$F_4 = \frac{F^4}{1089}$$

$$^1D,\ ^3D = F_0 + 19F_2 - 99F_4 + 715F_6 \pm (G_0 + 19G_2 - 99G_4 + 715G_6)$$

$$F_6 = \frac{F^6}{7361.64}$$

$$^1F,\ ^3F = F_0 - 10F_2 - 33F_4 - 286F_6 \mp (G_0 - 10G_2 - 33G_4 - 286G_6)$$

$$G_2 = \frac{G^2}{225}, \quad G_4 = \frac{G^4}{1089}$$

$$^1G,\ ^3G = F_0 - 30F_2 + 97F_4 + 78F_6 \pm (G_0 - 30G_2 + 97G_4 + 78G_6)$$

$$G_6 = \frac{G^6}{7361.64}$$

$$^1H,\ ^3H = F_0 - 25F_2 - 51F_4 - 13F_6 \mp (G_0 - 25G_2 - 51G_4 - 13G_6)$$

$$^1I,\ ^3I = F_0 + 25F_2 + 9F_4 + F_6 \pm (G_0 + 25G_2 + 9G_4 + G_6)$$

Energies in Terms of Slater Integrals for Wigner and Majorana Forces in *jj*-Coupling

ls

$$G_l = \frac{1}{2l+1} G^l$$

j	j'	T[a]	J	Wigner	Majorana
$l - 1/2$	$1/2$	1	$l - 1$	$F_0 - G_l$	$-F_0 + G_l$
			l	$F_0 - \frac{1}{2l+1} G_l$	$-\frac{1}{2l+1} F_0 + G_l$
$l + 1/2$	$1/2$		l	$F_0 + \frac{1}{2l+1} G_l$	$\frac{1}{2l+1} F_0 + G_l$
			$l + 1$	$F_0 - G_l$	$-F_0 + G_l$

[a] The formulas for two nonequivalent particles configurations with $T = 0$ are obtained by reversing the sign of the expressions for the Majorana force with $T = 1$.

s^2

j	j'	T^a	J	Wigner	Majorana
1/2	1/2	1	0	F_0	F_0
		0	1	F_0	F_0

pp

$$F_2 = \frac{F^2}{25}, \quad G_2 = \frac{G^2}{25}$$

j	j'	T^a	J	Wigner	Majorana
1/2	1/2	1	0	$F_0 + G_0$	$-\frac{1}{3}F_0 + \frac{20}{3}F_2 - \frac{1}{3}G_0 + \frac{20}{3}G_2$
			1	$F_0 - G_0$	$-\frac{5}{9}F_0 - \frac{20}{9}F_2 + \frac{5}{9}G_0 + \frac{20}{9}G_2$
1/2	3/2		1	$F_0 - 5G_2$	$-\frac{7}{9}F_0 - \frac{10}{9}F_2 - \frac{2}{9}G_0 + \frac{55}{9}G_2$
			2	$F_0 - G_2$	$-\frac{1}{3}F_0 + \frac{2}{3}F_2 + \frac{2}{3}G_0 + \frac{5}{3}G_2$
3/2	3/2		0	$F_0 + 5F_2 + G_0 + 5G_2$	$\frac{1}{3}F_0 + \frac{25}{3}F_2 + \frac{1}{3}G_0 + \frac{25}{3}G_2$
			1	$F_0 + F_2 - G_0 - G_2$	$\frac{1}{9}F_0 - \frac{59}{9}F_2 - \frac{1}{9}G_0 + \frac{59}{9}G_2$
			2	$F_0 - 3F_2 + G_0 - 3G_2$	$-\frac{1}{3}F_0 + \frac{11}{3}F_2 - \frac{1}{3}G_0 + \frac{11}{3}G_2$
			3	$F_0 + F_2 - G_0 - G_2$	$- F_0 - F_2 + G_0 + G_2$

[a] See footnote to table for ls.

p^2

$$F_2 = \frac{F^2}{25}$$

j	j'	T^a	J	Wigner	Majorana
1/2	1/2	1	0	F_0	$-\frac{1}{3}F_0 + \frac{20}{3}F_2$
		0	1	F_0	$\frac{5}{9}F_0 + \frac{20}{9}F_2$
1/2	3/2	1	1	$F_0 - 5F_2$	$- F_0 + 5F_2$
			2	$F_0 - F_2$	$\frac{1}{3}F_0 + \frac{7}{3}F_2$
		0	1	$F_0 + 5F_2$	$\frac{5}{9}F_0 + \frac{65}{9}F_2$
			2	$F_0 + F_2$	$F_0 + F_2$
3/2	3/2	1	0	$F_0 + 5F_2$	$\frac{1}{3}F_0 + \frac{25}{3}F_2$
		0	1	$F_0 + F_2$	$-\frac{1}{9}F_0 + \frac{59}{9}F_2$
		1	2	$F_0 - 3F_2$	$-\frac{1}{3}F_0 + \frac{11}{3}F_2$
		0	3	$F_0 + F_2$	$F_0 + F_2$

[a] See footnote to table for ls.

pd

$$F_2 = \frac{F^2}{35}, \quad G_1 = \frac{G^1}{15}, \quad G_3 = \frac{G^3}{245}$$

j	j'	T[a]	J	Wigner	Majorana
1/2	3/2	1	1	$F_0 + \frac{5}{3}G_1$	$-\frac{1}{3}F_0 + \frac{14}{3}F_2 - G_1 + 42G_3$
			2	$F_0 - 5G_1$	$-\frac{3}{5}F_0 - \frac{14}{5}F_2 + \frac{19}{5}G_1 + \frac{42}{5}G_3$
3/2	3/2		0	$F_0 + 7F_2 - G_1 - 63G_3$	$-F_0 - 7F_2 + G_1 + 63G_3$
			1	$F_0 + \frac{7}{5}F_2 + \frac{11}{15}G_1 - \frac{189}{5}G_3$	$-\frac{13}{15}F_0 - \frac{7}{15}F_2 - \frac{3}{5}G_2 + \frac{231}{5}G_3$
			2	$F_0 - \frac{21}{5}F_2 - \frac{1}{5}G_1 - \frac{63}{5}G_3$	$-\frac{3}{5}F_0 + \frac{7}{5}F_2 - G_1 + 21G_3$
			3	$F_0 + \frac{7}{5}F_2 - \frac{3}{5}G_1 - \frac{9}{5}G_3$	$-\frac{1}{5}F_0 + \frac{1}{5}F_2 + \frac{27}{5}G_1 + \frac{21}{5}G_3$
1/2	5/2		2	$F_0 - 35G_3$	$-\frac{11}{15}F_0 - \frac{28}{15}F_2 - \frac{4}{5}G_1 + \frac{203}{5}G_3$
			3	$F_0 - 5G_3$	$-\frac{1}{3}F_0 + \frac{4}{3}F_2 + 4G_1 + 7G_3$
3/2	5/2		1	$F_0 + \frac{28}{5}F_2 + \frac{3}{5}G_1 + \frac{84}{5}G_3$	$\frac{1}{5}F_0 + \frac{14}{5}F_2 + \frac{3}{5}G_1 + \frac{294}{5}G_3$
			2	$F_0 - \frac{4}{5}F_2 - \frac{9}{5}G_1 - \frac{92}{5}G_3$	$-\frac{1}{15}F_0 - \frac{86}{15}F_2 - G_1 + 38G_3$
			3	$F_0 - \frac{22}{5}F_2 + \frac{18}{5}G_1 - \frac{71}{5}G_3$	$-\frac{7}{15}F_0 + \frac{82}{15}F_2 - \frac{2}{5}G_1 + \frac{79}{5}G_3$
			4	$F_0 + 2F_2 - 6G_1 - 3G_3$	$-F_0 - 2F_2 + 6G_1 + 3G_3$

[a] See footnote to table for *ls*.

dd

$$F_2 = \frac{F^2}{49}, \quad F_4 = \frac{F^4}{441}, \quad G_2 = \frac{G^2}{49}, \quad G_4 = \frac{G^4}{441}$$

j	j'	T^a	J	Wigner
3/2	3/2	1	0	$F_0 + \frac{49}{5}F_2 + G_0 + \frac{49}{5}G_2$
			1	$F_0 + \frac{49}{25}F_2 - G_0 - \frac{49}{25}G_2$
			2	$F_0 - \frac{147}{25}F_2 + G_0 - \frac{147}{25}G_2$
			3	$F_0 + \frac{49}{25}F_2 - G_0 - \frac{49}{25}G_2$
3/2	5/2		1	$F_0 + \frac{196}{25}F_2 - \frac{21}{25}G_2 - 84G_4$
			2	$F_0 - \frac{28}{25}F_2 + \frac{47}{25}G_2 - 36G_4$
			3	$F_0 - \frac{154}{25}F_2 - \frac{46}{25}G_2 - 9G_4$
			4	$F_0 + \frac{14}{5}F_2 - \frac{6}{5}G_2 - G_4$
5/2	5/2		0	$F_0 + \frac{56}{5}F_2 + 42F_4 + G_0 + \frac{56}{5}G_2 + 42G_4$
			1	$F_0 + \frac{184}{25}F_2 - 6F_4 - G_0 - \frac{184}{25}G_2 + 6G_4$
			2	$F_0 + \frac{28}{25}F_2 - 21F_4 + G_0 + \frac{28}{25}G_2 - 21G_4$
			3	$F_0 - \frac{116}{25}F_2 + 19F_4 - G_0 + \frac{116}{25}G_2 - 19G_4$
			4	$F_0 - \frac{28}{5}F_2 - 7F_4 + G_0 - \frac{28}{5}G_2 - 7G_4$
			5	$F_0 + 4F_2 + F_4 - G_0 - 4G_2 - G_4$

j	j'	T^a	J	Majorana
3/2	3/2	1	0	$-\frac{1}{5}F_0+\frac{7}{5}F_2+\frac{504}{5}F_4-\frac{1}{5}G_0+\frac{7}{5}G_2+\frac{504}{5}G_4$
			1	$-\frac{7}{25}F_0+\frac{77}{25}F_2-\frac{1512}{25}F_4+\frac{7}{25}G_0-\frac{77}{25}G_2+\frac{1512}{25}G_4$
			2	$-\frac{11}{25}F_0+\frac{21}{5}F_2+\frac{504}{25}F_4-\frac{11}{25}G_0+\frac{21}{5}G_2+\frac{504}{5}G_4$
			3	$-\frac{17}{25}F_0-\frac{113}{25}F_2-\frac{72}{25}F_4+\frac{17}{25}G_0+\frac{113}{25}G_2+\frac{72}{25}G_4$
3/2	5/2		1	$-\frac{23}{25}F_0-\frac{182}{25}F_2-\frac{168}{25}F_4-\frac{2}{25}G_0+\frac{7}{25}G_2+\frac{2268}{25}G_4$
			2	$-\frac{19}{25}F_0+\frac{2}{5}F_2+\frac{216}{25}F_4+\frac{6}{25}G_0-\frac{13}{5}G_2+\frac{1116}{25}G_4$
			3	$-\frac{13}{25}F_0+\frac{58}{25}F_2-\frac{108}{25}F_4-\frac{12}{25}G_0+\frac{142}{25}G_2+\frac{333}{25}G_4$
			4	$-\frac{1}{5}F_0+\frac{2}{5}F_2+\frac{4}{5}F_4+\frac{4}{5}G_0+\frac{22}{5}G_2+\frac{9}{5}G_4$
5/2	5/2		0	$\frac{1}{5}F_0+\frac{28}{5}F_2+\frac{546}{5}F_4+\frac{1}{5}G_0+\frac{28}{5}G_2+\frac{546}{5}G_4$
			1	$\frac{3}{25}F_0+\frac{12}{25}F_2-\frac{2202}{25}F_4-\frac{3}{25}G_0-\frac{12}{25}G_2+\frac{2202}{25}G_4$
			2	$-\frac{1}{25}F_0-4F_2+\frac{1389}{25}F_4-\frac{1}{25}G_0-4G_2+\frac{1389}{25}G_4$
			3	$-\frac{7}{25}F_0-\frac{28}{25}F_2-\frac{637}{25}F_4+\frac{7}{25}G_0+\frac{28}{25}G_2+\frac{637}{25}G_4$
			4	$-\frac{3}{5}F_0+\frac{36}{5}F_2+\frac{37}{5}F_4-\frac{3}{5}G_0+\frac{36}{5}G_2+\frac{37}{5}G_4$
			5	$-F_0-4F_2-F_4+G_0+4G_2+G_4$

[a] See footnote to table for *ls*.

d^2

$$F_2 = \frac{F^2}{49} \quad F_4 = \frac{F^4}{441}$$

j	j'	T	J	Wigner	Majorana
3/2	3/2	1	0	$F_0 + \frac{49}{25}F_2$	$-\frac{1}{5}F_0 + \frac{7}{5}F_2 + \frac{504}{5}F_4$
		0	1	$F_0 + \frac{49}{5}F_2$	$\frac{7}{25}F_0 - \frac{77}{25}F_2 + \frac{1512}{25}F_4$
		1	2	$F_0 - \frac{147}{25}F_2$	$-\frac{11}{25}F_0 + \frac{21}{5}F_2 + \frac{504}{25}F_4$
		0	3	$F_0 + \frac{49}{25}F_2$	$\frac{17}{25}F_0 + \frac{113}{25}F_2 + \frac{72}{25}F_4$
3/2	5/2	1	1	$F_0 + 7F_2 - 84F_4$	$-F_0 - 7F_2 + 84F_4$
			2	$F_0 + \frac{19}{25}F_2 - 36F_4$	$-\frac{13}{25}F_0 - \frac{11}{5}F_2 + \frac{1332}{25}F_4$
			3	$F_0 - 8F_2 - 9F_4$	$-F_0 + 8F_2 + 9F_4$
			4	$F_0 + \frac{8}{5}F_2 - F_4$	$\frac{3}{5}F_0 + \frac{24}{5}F_2 + \frac{13}{5}F_4$
		0	1	$F_0 + \frac{217}{25}F_2 + 84F_4$	$\frac{21}{25}F_0 + \frac{189}{25}F_2 + \frac{2436}{25}F_4$
			2	$F_0 - 3F_2 + 36F_4$	$F_0 - 3F_2 + 36F_4$
			3	$F_0 - \frac{108}{25}F_2 + 9F_4$	$\frac{1}{25}F_0 + \frac{84}{25}F_2 + \frac{441}{25}F_4$
			4	$F_0 + 4F_2 + F_4$	$F_0 + 4F_2 + F_4$
5/2	5/2	1	0	$F_0 + \frac{56}{5}F_2 + 42F_4$	$\frac{1}{5}F_0 + \frac{28}{5}F_2 + \frac{546}{5}F_4$
		0	1	$F_0 + \frac{184}{25}F_2 - 6F_4$	$-\frac{3}{25}F_0 - \frac{12}{25}F_2 + \frac{2202}{25}F_4$
		1	2	$F_0 + \frac{28}{25}F_2 - 21F_4$	$-\frac{1}{25}F_0 - 4F_2 + \frac{1389}{25}F_4$
		0	3	$F_0 - \frac{116}{25}F_2 + 19F_4$	$\frac{7}{25}F_0 + \frac{28}{25}F_2 + \frac{637}{25}F_4$
		1	4	$F_0 - \frac{28}{5}F_2 - 7F_4$	$-\frac{3}{5}F_0 + \frac{36}{5}F_2 + \frac{37}{5}F_4$
		0	5	$F_0 + 4F_2 + F_4$	$F_0 + 4F_2 + F_4$

pf

$$F_2 = \frac{F^2}{75}, \quad G_2 = \frac{G^2}{175}, \quad G_4 = \frac{G^4}{189}$$

j	j'	T[a]	J	Wigner	Majorana
1/2	5/2	1	2	$F_0 + 7G_2$	$-\frac{1}{3}F_0 + 8F_2 - 5G_2 + 24G_4$
			3	$F_0 - 35G_2$	$-\frac{13}{21}F_0 - \frac{40}{7}F_2 + \frac{205}{7}G_2 + \frac{24}{7}G_4$
3/2	5/2		1	$F_0 + 12F_2 - 3G_2 - 36G_4$	$-F_0 - 12F_2 + 3G_2 + 36G_4$
			2	$F_0 - \frac{12}{7}F_2 + \frac{47}{7}G_2 - \frac{108}{7}G_4$	$-\frac{17}{21}F_0 + 4F_2 - \frac{43}{7}G_2 + \frac{156}{7}G_4$
			3	$F_0 - \frac{66}{7}F_2 - \frac{46}{7}G_2 - \frac{27}{7}G_4$	$-\frac{11}{21}F_0 + \frac{16}{7}F_2 - \frac{4}{7}G_2 + \frac{57}{7}G_4$
			4	$F_0 + \frac{30}{7}F_2 - \frac{30}{7}G_2 - \frac{3}{7}G_4$	$-\frac{1}{7}F_0 + \frac{300}{7}G_2 + \frac{9}{7}G_4$
1/2	7/2		3	$F_0 - 21G_4$	$-\frac{5}{7}F_0 - \frac{30}{7}F_2 - \frac{30}{7}G_2 + \frac{165}{7}G_4$
			4	$F_0 - \frac{7}{3}G_4$	$-\frac{1}{3}F_0 + \frac{10}{3}F_2 + 30G_2 + 3G_4$
3/2	7/2		2	$F_0 + \frac{75}{7}F_2 + \frac{9}{7}G_2 + \frac{45}{7}G_4$	$\frac{1}{7}F_0 + 3F_2 + \frac{15}{7}G_2 + \frac{243}{7}G_4$
			3	$F_0 - \frac{25}{7}F_2 - \frac{45}{7}G_2 - \frac{85}{7}G_4$	$-\frac{1}{7}F_0 - \frac{65}{7}F_2 - \frac{45}{7}G_2 + \frac{139}{7}G_4$
			4	$F_0 - \frac{65}{7}F_2 + \frac{135}{7}G_2 - \frac{131}{21}G_4$	$-\frac{11}{21}F_0 + \frac{35}{3}F_2 + \frac{15}{7}G_2 + \frac{47}{7}G_4$
			5	$F_0 + 5F_2 - 45G_2 - G_4$	$-F_0 - 5F_2 + 45G_2 + G_4$

[a] See footnote to table for *ls*.

df

$$F_2 = \frac{F^2}{105}, \quad F_4 = \frac{F^4}{693}, \quad G_1 = \frac{G^1}{35}, \quad G_3 = \frac{G^3}{315}, \quad G_5 = \frac{G^5}{1524.6}$$

j	j'	T^a	J	Wigner
3/2	5/2	1	1	$F_0 + \frac{84}{5}F_2 + \frac{7}{5}G_1 + \frac{108}{5}G_3$
			2	$F_0 - \frac{12}{5}F_2 - \frac{21}{5}G_1 - \frac{828}{35}G_3$
			3	$F_0 - \frac{66}{5}F_2 + \frac{42}{5}G_1 - \frac{639}{35}G_3$
			4	$F_0 + 6F_2 - 14G_1 - \frac{27}{7}G_3$
5/2	5/2		0	$F_0 + 24F_2 + 66F_4 - G_1 - 24G_3 - 330G_5$
			1	$F_0 + \frac{552}{35}F_2 - \frac{66}{7}F_4 + \frac{31}{35}G_1 + \frac{264}{35}G_3 - \frac{1650}{7}G_5$
			2	$F_0 + \frac{12}{5}F_2 - 33F_4 - \frac{23}{35}G_1 + \frac{348}{35}G_3 - \frac{825}{7}G_5$
			3	$F_0 - \frac{348}{35}F_2 + \frac{209}{7}F_4 + \frac{11}{35}G_1 - \frac{316}{35}G_3 - \frac{275}{7}G_5$
			4	$F_0 - 12F_2 - 11F_4 + \frac{1}{7}G_1 - \frac{76}{7}G_3 - \frac{55}{7}G_5$
			5	$F_0 + \frac{60}{7}F_2 + \frac{11}{7}F_4 - \frac{5}{7}G_1 - \frac{20}{7}G_3 - \frac{5}{7}G_5$
3/2	7/2	1	2	$F_0 + 15F_2 - \frac{15}{7}G_3 - 231G_5$
			3	$F_0 - 5F_2 + \frac{55}{7}G_3 - 77G_5$
			4	$F_0 - 13F_2 - \frac{85}{7}G_3 - \frac{77}{5}G_5$
			5	$F_0 + 7F_2 - 5G_3 - \frac{7}{5}G_5$
5/2	7/2		1	$F_0 + \frac{150}{7}F_2 + \frac{297}{7}F_4 + \frac{5}{7}G_1 + \frac{90}{7}G_3 + \frac{495}{7}G_T$
			2	$F_0 + 10F_2 - 33F_4 - \frac{15}{7}G_1 - \frac{190}{7}G_3 - \frac{253}{7}G_5$
			3	$F_0 - \frac{20}{7}F_2 - \frac{132}{7}F_4 + \frac{30}{7}G_1 + \frac{185}{7}G_3 - \frac{418}{7}G_5$
			4	$F_0 - 12F_2 + 36F_4 - \frac{50}{7}G_1 - \frac{15}{7}G_3 - \frac{1146}{35}G_5$
			5	$F_0 - \frac{74}{7}F_2 - \frac{123}{7}F_4 + \frac{75}{7}G_1 - \frac{190}{7}G_3 - \frac{311}{35}G_5$
			6	$F_0 + 10F_2 + 3F_4 - 15G_1 - 10G_3 - G_5$

j	j'	T^a	J	Majorana
3/2	5/2	1	1	$-\frac{1}{5}F_0 + \frac{12}{5}F_2 + \frac{264}{5}F_4 - \frac{3}{5}G_1 - \frac{12}{5}G_3 + 264G_5$
			2	$-\frac{11}{35}F_0 + \frac{228}{35}F_2 - \frac{2376}{35}F_4 + \frac{15}{7}G_1 - \frac{36}{7}G_3 + \frac{792}{7}G_5$
			3	$-\frac{17}{35}F_0 + \frac{264}{35}F_2 + \frac{1188}{35}F_4 - \frac{186}{35}G_1 + \frac{981}{35}G_3 + \frac{198}{7}G_5$
			4	$-\frac{5}{7}F_0 - \frac{72}{7}F_2 - \frac{44}{7}F_4 + \frac{78}{7}G_1 + \frac{97}{7}G_3 + \frac{22}{7}G_5$
5/2	5/2		0	$-F_0 - 24F_2 - 66F_4 + G_1 + 24G_3 + 330G_5$
			1	$-\frac{33}{35}F_0 - \frac{72}{5}F_2 + \frac{66}{5}F_4 - \frac{29}{35}G_1 - \frac{216}{35}G_3 + \frac{1782}{7}G_5$
			2	$-\frac{29}{35}F_0 - \frac{48}{35}F_2 + \frac{561}{35}F_4 + \frac{1}{7}G_1 - \frac{120}{7}G_3 + \frac{1023}{7}G_5$
			3	$-\frac{23}{35}F_0 + \frac{216}{35}F_2 - \frac{253}{35}F_4 + \frac{61}{35}G_1 + \frac{544}{35}G_3 + \frac{407}{7}G_5$
			4	$-\frac{3}{7}F_0 + \frac{24}{7}F_2 - \frac{11}{7}F_4 - \frac{41}{7}G_1 + \frac{216}{7}G_3 + \frac{99}{7}G_5$
			5	$-\frac{1}{7}F_0 + F_4 + \frac{95}{7}G_1 + \frac{80}{7}G_3 + \frac{11}{7}G_5$
3/2	7/2	1	2	$-\frac{31}{35}F_0 - \frac{501}{35}F_2 - \frac{396}{35}F_4 - \frac{12}{35}G_1 - \frac{93}{35}G_3 + \frac{1749}{7}G_5$
			3	$-\frac{5}{7}F_0 + \frac{13}{7}F_2 + \frac{132}{7}F_4 + \frac{12}{7}G_1 - \frac{17}{7}G_3 + \frac{649}{7}G_5$
			4	$-\frac{17}{35}F_0 + \frac{37}{7}F_2 - \frac{396}{35}F_4 - \frac{36}{7}G_1 + \frac{211}{7}G_3 + \frac{737}{35}G_5$
			5	$-\frac{1}{5}F_0 + F_2 + \frac{12}{5}F_4 + 12G_1 + 13G_3 + \frac{11}{5}G_5$
5/2	7/2		1	$\frac{1}{7}F_0 + 6F_2 + 33F_4 + \frac{3}{7}G_1 + \frac{102}{7}G_3 + \frac{2145}{7}G_5$
			2	$\frac{1}{35}F_0 - \frac{134}{35}F_2 - \frac{2409}{35}F_4 - \frac{33}{35}G_1 - \frac{562}{35}G_3 + \frac{1441}{7}G_5$
			3	$-\frac{1}{7}F_0 - \frac{46}{7}F_2 + \frac{528}{7}F_4 + \frac{6}{7}G_1 - \frac{71}{7}G_3 + \frac{748}{7}G_5$
			4	$-\frac{13}{35}F_0 + \frac{18}{7}F_2 - \frac{1744}{35}F_4 + \frac{6}{7}G_1 + \frac{169}{7}G_3 + \frac{1388}{35}G_5$
			5	$-\frac{23}{35}F_0 + 14F_2 + \frac{93}{5}F_4 - \frac{39}{7}G_1 + \frac{214}{7}G_3 + \frac{323}{35}G_5$
			6	$-F_0 - 10F_2 - 3F_4 + 15G_1 + 10G_4 + G_5$

[a] See footnote to table *ls*.

ff

$$F_2 = \frac{F^2}{225}, \quad F_4 = \frac{F^4}{1089}, \quad F_6 = \frac{F^6}{7361.64}, \quad G_2 = \frac{G^2}{225}, \quad G_4 = \frac{G^4}{1089}, \quad G_6 = \frac{G^6}{7361.64}$$

j	j'	T^a	J	Wigner
5/2	5/2	1	0	$F_0 + \frac{360}{7}F_2 + \frac{726}{7}F_4 + G_0 + \frac{360}{7}G_2 + \frac{726}{7}G_4$
			1	$F_0 + \frac{1656}{49}F_2 - \frac{726}{49}F_4 - G_0 - \frac{1656}{49}G_2 + \frac{726}{49}G_4$
			2	$F_0 + \frac{36}{7}F_2 - \frac{363}{7}F_4 + G_0 + \frac{36}{7}G_2 - \frac{363}{7}G_4$
			3	$F_0 - \frac{1044}{49}F_2 + \frac{2299}{49}F_4 - G_0 + \frac{1044}{49}G_2 - \frac{2299}{49}G_4$
			4	$F_0 - \frac{180}{7}F_2 - \frac{121}{7}F_4 + G_0 - \frac{180}{7}G_2 - \frac{121}{7}G_4$
			5	$F_0 + \frac{900}{49}F_2 + \frac{121}{49}F_4 - G_0 - \frac{900}{49}G_2 - \frac{121}{49}G_4$
5/2	7/2		1	$F_0 + \frac{2250}{49}F_2 + \frac{3267}{49}F_4 - \frac{45}{49}G_2 - \frac{1650}{49}G_4 - 1287G_6$
			2	$F_0 + \frac{150}{7}F_2 - \frac{363}{7}F_4 + \frac{17}{7}G_2 + \frac{330}{7}G_4 - 715G_6$
			3	$F_0 - \frac{300}{49}F_2 - \frac{1452}{49}F_4 \quad \frac{190}{49}G_2 - \frac{165}{49}G_4 - 286G_6$
			4	$F_0 - \frac{180}{7}F_2 + \frac{396}{7}F_4 + \frac{30}{7}G_2 - \frac{283}{7}G_4 - 78G_6$
			5	$F_0 - \frac{1110}{49}F_2 - \frac{1353}{49}F_4 - \frac{115}{49}G_2 - \frac{1146}{49}G_4 - 13G_6$
			6	$F_0 + \frac{150}{7}F_2 + \frac{33}{7}F_4 - \frac{25}{7}G_2 - \frac{30}{7}G_4 - G_6$
7/2	7/2	1	0	$F_0 + \frac{375}{7}F_2 + \frac{891}{7}F_4 + 429F_6 + G_0 + \frac{375}{7}G_2 + \frac{891}{7}G_4 + 429G_6$
			1	$F_0 + \frac{2125}{49}F_2 + \frac{2277}{49}F_4 - 143F_6 - G_0 - \frac{2125}{49}G_2 - \frac{2277}{49}G_4 + 143G_6$
			2	$F_0 + 25F_2 - \frac{297}{7}F_4 - 143F_6 + G_0 + 25G_2 - \frac{297}{7}G_4 - 143G_6$
			3	$F_0 + \frac{125}{49}F_2 - \frac{2727}{49}F_4 + 221F_6 - G_0 - \frac{125}{49}G_2 + \frac{2727}{49}G_4 - 221G_6$
			4	$F_0 - \frac{125}{7}F_2 + \frac{99}{7}F_4 - 143F_6 + G_0 - \frac{125}{7}G_2 + \frac{99}{7}G_4 - 143G_6$
			5	$F_0 - \frac{1375}{49}F_2 + \frac{2781}{49}F_4 + 53F_6 - G_0 + \frac{1375}{49}G_2 - \frac{2781}{49}G_4 - 53G_6$
			6	$F_0 - \frac{125}{7}F_2 - \frac{297}{7}F_4 - 11F_6 + G_0 - \frac{125}{7}G_2 - \frac{297}{7}G_4 - 11G_6$
			7	$F_0 + 25F_2 + 9F_4 + F_6 - G_0 - 25G_2 - 9G_4 - G_6$

j'	T^a	J	Majorana
/2	5/2 1	0	$-\frac{1}{7}F_0 + 66F_4 + \frac{10296}{7}F_6 - \frac{1}{7}G_0 + 66G_4 + \frac{10296}{7}G_6$
		1	$-\frac{9}{49}F_0 + \frac{144}{49}F_2 + \frac{2046}{49}F_4 - \frac{51480}{49}F_6 + \frac{9}{49}G_0 - \frac{144}{49}G_2 - \frac{2046}{49}G_4 + \frac{51480}{49}G_6$
		2	$-\frac{13}{49}F_0 + \frac{432}{49}F_2 - \frac{1023}{49}F_4 + \frac{25740}{49}F_6 - \frac{13}{49}G_0 + \frac{432}{49}G_2 - \frac{1023}{49}G_4 + \frac{25740}{49}G_6$
		3	$-\frac{19}{49}F_0 + \frac{744}{49}F_2 - \frac{3289}{49}F_4 - \frac{8580}{49}F_6 + \frac{19}{49}G_0 - \frac{744}{49}G_2 + \frac{3289}{49}G_4 + \frac{8580}{49}G_6$
		4	$-\frac{27}{49}F_0 + \frac{600}{49}F_2 + \frac{2981}{49}F_4 + \frac{1716}{49}F_6 - \frac{27}{49}G_0 + \frac{600}{49}G_2 + \frac{2981}{49}G_4 + \frac{1716}{49}G_6$
		5	$-\frac{37}{49}F_0 - \frac{1200}{49}F_2 - \frac{733}{49}F_4 - \frac{156}{49}F_6 + \frac{37}{49}G_0 + \frac{1200}{49}G_2 + \frac{733}{49}G_4 + \frac{156}{49}G_6$
/2	7/2	1	$-\frac{47}{49}F_0 - \frac{2160}{49}F_2 - \frac{3201}{49}F_4 - \frac{2574}{49}F_6 - \frac{2}{49}G_0 - \frac{45}{49}G_2 + \frac{1584}{49}G_4 + \frac{65637}{49}G_6$
		2	$\frac{43}{49}F_0 - \frac{936}{49}F_2 + \frac{1947}{49}F_4 + \frac{4290}{49}F_6 + \frac{6}{49}G_0 - \frac{5}{49}G_2 - \frac{2904}{49}G_4 + \frac{39325}{49}G_6$
		3	$-\frac{37}{49}F_0 + \frac{180}{49}F_2 + \frac{1056}{49}F_4 - \frac{3432}{49}F_6 - \frac{12}{49}G_0 + \frac{310}{49}G_2 + \frac{561}{49}G_4 + \frac{17446}{49}G_6$
		4	$-\frac{29}{49}F_0 + \frac{660}{49}F_2 - \frac{832}{49}F_4 + \frac{1560}{49}F_6 + \frac{20}{49}G_0 - \frac{810}{49}G_2 + \frac{3921}{49}G_4 + \frac{5382}{49}G_6$
		5	$-\frac{19}{49}F_0 + \frac{360}{49}F_2 - \frac{177}{49}F_4 - \frac{390}{49}F_6 - \frac{30}{49}G_0 + \frac{865}{49}G_2 + \frac{2676}{49}G_4 + \frac{1027}{49}G_6$
		6	$-\frac{1}{7}F_0 + 3F_4 + \frac{6}{7}F_6 + \frac{6}{7}G_0 + 25G_2 + 12G_4 + \frac{13}{7}G_6$
7/2	7/2 1	0	$\frac{1}{7}F_0 + 15F_2 + 99F_4 + \frac{10725}{7}F_6 + \frac{1}{7}G_0 + 15G_2 + 99G_4 + \frac{10725}{7}G_6$
		1	$\frac{5}{49}F_0 + \frac{305}{49}F_2 - \frac{495}{49}F_4 - \frac{62491}{49}F_6 - \frac{5}{49}G_0 - \frac{305}{49}G_2 + \frac{495}{49}G_4 + \frac{62491}{49}G_6$
		2	$\frac{1}{49}F_0 - \frac{275}{49}F_2 - \frac{2871}{49}F_4 + \frac{42757}{49}F_6 + \frac{1}{49}G_0 - \frac{275}{49}G_2 - \frac{2871}{49}G_4 + \frac{42757}{49}G_6$
		3	$-\frac{5}{49}F_0 - \frac{565}{49}F_2 + \frac{1275}{49}F_4 - \frac{23413}{49}F_6 + \frac{5}{49}G_0 + \frac{565}{49}G_2 - \frac{1275}{49}G_4 + \frac{23413}{49}G_6$
		4	$-\frac{13}{49}F_0 - \frac{205}{49}F_2 + \frac{2799}{49}F_4 + \frac{9815}{49}F_6 - \frac{13}{49}G_0 - \frac{205}{49}G_2 + \frac{2799}{49}G_4 + \frac{9815}{49}G_6$
		5	$-\frac{23}{49}F_0 + \frac{725}{49}F_2 - \frac{4107}{49}F_4 - \frac{2935}{49}F_6 + \frac{23}{49}G_0 - \frac{725}{49}G_2 + \frac{4107}{49}G_4 + \frac{2935}{49}G_6$
		6	$-\frac{5}{7}F_0 + 25F_2 + 45F_4 + \frac{79}{7}F_6 - \frac{5}{7}G_0 + 25G_2 + 45G_4 + \frac{79}{7}G_6$
		7	$-F_0 - 25F_2 - 9F_4 - F_6 + G_0 + 25G_2 + 9G_4 + G_6$

[a] See footnote to table for *ls*.

f^2

$$F_2 = \frac{F^2}{225}, \quad F_4 = \frac{F^4}{1089}, \quad F_6 = \frac{F^6}{7361.64}$$

j	j'	T	J	Wigner	Majorana
5/2	5/2	1	0	$F_0 + \frac{360}{7}F_2 + \frac{726}{7}F_4$	$-\frac{1}{7}F_0 + 66F_4 + \frac{10296}{7}$
		0	1	$F_0 + \frac{1656}{49}F_2 - \frac{726}{7}F_4$	$\frac{9}{49}F_0 - \frac{144}{49}F_2 - \frac{2046}{49}F_4 + \frac{51480}{49}$
		1	2	$F_0 + \frac{36}{7}F_2 - \frac{363}{7}F_4$	$-\frac{13}{49}F_0 + \frac{432}{49}F_2 - \frac{1023}{49}F_4 + \frac{25740}{49}$
		0	3	$F_0 - \frac{1044}{49}F_2 + \frac{2299}{49}F_4$	$\frac{19}{49}F_0 - \frac{744}{49}F_2 + \frac{3289}{49}F_4 + \frac{8580}{49}$
		1	4	$F_0 - \frac{180}{7}F_2 - \frac{121}{7}F_4$	$-\frac{27}{49}F_0 + \frac{600}{49}F_2 + \frac{2981}{49}F_4 + \frac{1716}{49}$
		0	5	$F_0 + \frac{900}{49}F_2 + \frac{121}{49}F_4$	$\frac{37}{49}F_0 + \frac{1200}{49}F_2 + \frac{733}{49}F_4 + \frac{156}{49}$
5/2	7/2	1	1	$F_0 + 45F_2 + 33F_4 - 1287F_6$	$-F_0 - 45F_2 - 33F_4 + 1287$
			2	$F_0 + \frac{167}{7}F_2 - \frac{33}{7}F_4 - 715F_6$	$-\frac{37}{49}F_0 - \frac{941}{49}F_2 - \frac{957}{49}F_4 + \frac{43615}{49}$
			3	$F_0 - 10F_2 - 33F_4 - 286F_6$	$-F_0 + 10F_2 + 33F_4 + 286$
			4	$F_0 - \frac{150}{7}F_2 + \frac{113}{7}F_4 - 78F_6$	$-\frac{9}{49}F_0 - \frac{150}{49}F_2 + \frac{3089}{49}F_4 + \frac{6942}{49}$
			5	$F_0 - 25F_2 - 51F_4 - 13F_6$	$-F_0 + 25F_2 + 51F_4 + 13F$
			6	$F_0 + \frac{125}{7}F_2 + \frac{3}{7}F_4 - F_6$	$\frac{5}{7}F_0 + 25F_2 + 15F_4 + \frac{19}{7}F$
		0	1	$F_0 + \frac{2295}{49}F_2 + \frac{4917}{49}F_4 + 1287F_6$	$\frac{45}{49}F_0 + \frac{2115}{49}F_2 + \frac{4785}{49}F_4 + \frac{68211}{49}F$
			2	$F_0 + 19F_2 - 99F_4 + 715F_6$	$F_0 + 19F_2 - 99F_4 + 715F$
			3	$F_0 - \frac{110}{49}F_2 - \frac{1287}{49}F_4 + 286F_6$	$\frac{25}{49}F_0 + \frac{130}{49}F_2 - \frac{495}{49}F_4 + \frac{20878}{49}F$
			4	$F_0 - 30F_2 + 97F_4 + 78F_6$	$F_0 - 30F_2 + 97F_4 + 78F$
			5	$F_0 - \frac{995}{49}F_2 - \frac{207}{49}F_4 + 13F_6$	$-\frac{11}{49}F_0 + \frac{505}{49}F_2 + \frac{2853}{49}F_4 + \frac{1417}{49}F$
			6	$F_0 + 25F_2 + 9F_4 + F_6$	$F_0 + 25F_2 + 9F_4 + F$

7/2	7/2	1	0	$F_0 + \frac{375}{7}F_2 + \frac{891}{7}F_4 + 429F_6$	$\frac{1}{7}F_0 + 15F_2 + 99F_4 + \frac{10725}{7}F_6$
		0	1	$F_0 + \frac{2125}{49}F_2 + \frac{2277}{49}F_4 - 143F_6$	$-\frac{5}{49}F_0 - \frac{305}{49}F_2 + \frac{495}{49}F_4 + \frac{62491}{49}F_6$
		1	2	$F_0 + 25F_2 - \frac{297}{7}F_4 - 143F_6$	$\frac{1}{49}F_0 - \frac{275}{49}F_2 - \frac{2871}{49}F_4 + \frac{42757}{49}F_6$
		0	3	$F_0 + \frac{125}{49}F_2 - \frac{2727}{49}F_4 + 221F_6$	$\frac{5}{49}F_0 + \frac{565}{49}F_2 - \frac{1275}{49}F_4 + \frac{23413}{49}F_6$
		1	4	$F_0 - \frac{125}{7}F_2 + \frac{99}{7}F_4 - 143F_6$	$-\frac{13}{49}F_0 - \frac{205}{49}F_2 + \frac{2799}{49}F_4 + \frac{9815}{49}F_6$
		0	5	$F_0 - \frac{1375}{49}F_2 + \frac{2781}{49}F_4 + 53F_6$	$\frac{23}{49}F_0 - \frac{725}{49}F_2 + \frac{4107}{49}F_4 + \frac{2935}{49}F_6$
		1	6	$F_0 - \frac{125}{7}F_2 - \frac{297}{7}F_4 - 11F_6$	$-\frac{5}{7}F_0 + 25F_2 + 45F_4 + \frac{79}{7}F_6$
		0	7	$F_0 + 25F_2 + 9F_4 + F_6$	$F_0 + 25F_2 + 9F_4 + F_6$

Slater Integrals F^k and G^k in Terms of Harmonic Oscillator Integrals I_l†

$(1s)^2$

$$F^0 = I_0$$

$(1p)^2$

$$F^0 = \frac{1}{12}[5(I_0 + I_2) + 2I_1]$$

$$F^2 = \frac{25}{12}[(I_0 + I_2) - 2I_1]$$

$(1d)^2$

$$F_0 = \frac{1}{240}[63(I_0 + I_4) + 28(I_1 + I_3) + 58I_2]$$

$$F^2 = \frac{7}{48}[9(I_0 + I_4) - 8(I_1 + I_3) - 2I_2]$$

$$F_4 = \frac{189}{80}[(I_0 + I_4) - 4(I_1 + I_3) + 6I_2]$$

$(1f)^2$

$$F^0 = \frac{1}{2240}[429(I_0 + I_6) + 198(I_1 + I_5) + 387(I_2 + I_4) + 212I_3]$$

$$F^2 = \frac{3}{448}[143(I_0 + I_6) - 66(I_1 + I_5) - 15(I_2 + I_4) - 124I_3]$$

$$F^4 = \frac{297}{2240}[13(I_0 + I_6) - 34(I_1 + I_5) + 19(I_2 + I_4) + 4I_3]$$

$$F^6 = \frac{5577}{2240}[(I_0 + I_6) - 6(I_1 + I_5) + 15(I_2 + I_4) - 20I_3]$$

† R. Thieberger, *Nuclear Phys.* **2**, 533 (1956/57).

$(1g)^2$

$$F^0 = \frac{1}{80640}[12155(I_0+I_8)+5720(I_1+I_7) + 10868(I_2+I_6) + 6248(I_3+I_5) + 10658I_4]$$

$$F^2 = \frac{11}{16128}[1105(I_0 + I_8) - 260(I_1 + I_7) + 52(I_2 + I_6) - 764(I_3 + I_5) - 266I_4]$$

$$F^4 = \frac{143}{26880}[255(I_0 + I_8) - 480(I_1 + I_7) + 68(I_2 + I_6) - 32(I_3 + I_5) + 378I_4]$$

$$F^6 = \frac{1859}{16128}[17(I_0 + I_8) - 76(I_1 + I_7) + 116(I_2 + I_6) - 52(I_3 + I_5) - 10I_4]$$

$$F^8 = \frac{41327}{16128}[(I_0 + I_8) - 8(I_1 + I_7) + 28(I_2 + I_6) - 56(I_3 + I_5) + 70I_4]$$

$(1h)^2$

$$F^0 = \frac{1}{709632}[88179(I_0 + I_{10}) + 41990(I_1 + I_9) + 78445(I_2 + I_8) + 46280(I_3 + I_7) + 76310(I_4 + I_6) + 47204I_5]$$

$$F^2 = \frac{13}{709632}[33915(I_0 + I_{10}) - 3230(I_1 + I_9) + 5695(I_2 + I_8) - 17960(I_3 + I_7) - 6842(I_4 + I_6) - 23156I_5]$$

$$F^4 = \frac{13}{78848}[6783(I_0 + I_{10}) - 9690(I_1 + I_9) - 765(I_2 + I_8) - 3480(I_3 + I_7) \quad 6270(I_4 + I_6) + 1764I_5]$$

$$F^6 = \frac{2873}{709632}[399(I_0 + I_{10}) - 1406(I_1 + I_9) + 1363(I_2 + I_8) - 8(I_3 + I_7) + 286(I_4 + I_6) - 1268I_5]$$

$$F^8 = \frac{71383}{709632}[21(I_0 + I_{10}) - 134(I_1 + I_9) + 337(I_2 + I_8) - 392(I_3 + I_7) + 154(I_4 + I_6) + 28I_5]$$

$$F^{10} = \frac{205751}{78848}[(I_0 + I_{10}) - 10(I_1 + I_9) + 45(I_2 + I_8) - 120(I_3 + I_7) + 210(I_4 + I_6) - 252I_5]$$

$(1i)^2$

$$F^0 = \frac{1}{12300288}[1300075(I_0 + I_{12}) + 624036(I_1 + I_{11}) + 1153110(I_2 + I_{10}) + 691220(I_3 + I_9) + 1116645(I_4 + I_8) + 711240(I_5 + I_7 + 1107636I_6]$$

$$F^2 = \frac{25}{12300288}[260015(I_0 + I_{12}) + 67830(I_2 + I_{10}) - 103360(I_3 + I_9) - 32895(I_4 + I_8) - 158784(I_5 + I_7) - 65612I_6]$$

$$F^4 = \frac{17}{4100096}[229425(I_0 + I_{12}) - 256956(I_1 + I_{11}) - 51870(I_2 + I_{10}) - 163020(I_3 + I_9) + 108735(I_4 + I_8) + 26760(I_5 + I_7) + 213852I_6]$$

$$F^6 = \frac{323}{946176}[4025(I_0 + I_{12}) - 11592(I_1 + I_{11}) + 7098(I_2 + I_{10}) + 1496(I_3 + I_9) + 6039(I_4 + I_8) - 6288(I_5 + I_7) - 1556I_6]$$

$$F^8 = \frac{5491}{1757184}[575(I_0 + I_{12}) - 3036(I_1 + I_{11}) + 5694(I_2 + I_{10}) - 3692(I_3 + I_9) - 303(I_4 + I_8) - 1464(I_5 + I_7) + 4452I_6]$$

$$F^{10} = \frac{52003}{585728}[25(I_0 + I_{12}) - 208(I_1 + I_{11}) + 730(I_2 + I_{10}) - 1360(I_3 + I_9 + 1335(I_4 + I_8) - 480(I_5 + I_7) - 84I_6]$$

$$F^{12} = \frac{4643125}{1757184}[I_0 + I_{12}) - 12(I_1 + I_{11}) + 66(I_2 + I_{10}) - 220(I_3 + I_9) + 495(I_4 + I_8) - 792(I_5 + I_7) + 924I_6]$$

$(2s)^2$

$$F^0 = \frac{41}{64}I_0 - \frac{79}{48}I_1 + \frac{385}{96}I_2 - \frac{175}{48}I_3 + \frac{105}{64}I_4$$

$(2p)^2$

$$F^0 = \frac{1}{1280}[363I_0 - 494I_1 + 2629I_2 - 5796I_3 + 9429I_4 - 7854I_5 + 3003I_6]$$

$$F^2 = \frac{1}{256}[363I_0 - 1986I_1 + 7125I_2 - 14364I_3 + 16485I_4 - 10626I_5 + 3003I_6]$$

$(2d)^2$

$$F^0 = \frac{1}{107520}[19737I_0 - 23024I_1 + 126556I_2 - 203984I_3 + 418806I_4 - 790416I_5 + 1158300I_6 - 926640I_7 + 328185I_8]$$

$$F^2 = \frac{1}{21504}[19737I_0 - 62492I_1 + 170452I_2 - 356324I_3 + 765702I_4 - 1452132I_5 + 1745172I_6 - 1158300I_7 + 328185I_8]$$

$$F^4 = \frac{1}{35840}[59211I_0 - 463752I_1 + 2228148I_2 - 6617592I_3 + 12376098I_4 - 14975928I_5 + 11505780I_6 - 5096520I_7 + 984555I_8]$$

$(2f)^2$

$$F^0 = \frac{1}{1290240}[175461I_0 - 189310I_1 + 1076897I_2 - 1623688I_3 + 3127562I_4 - 4533364I_5 + 7512362I_6 - 13104520I_7 + 18317585I_8 + 14318590I_9 + 4849845I_{10}]$$

$$F^2 = \frac{1}{258048}[175461I_0 - 413578I_1 + 121456I_2 - 2579800I_3 + 4363850I_4 - 6753340I_5 + 11603306I_6 - 20689240I_7 + 25318865I_8 - 17089930I_9 + 4849845I_{10}]$$

$$F^4 = \frac{1}{143360}[175461I_0 - 936870I_1 + 3100977I_2 - 6467208I_3 + 10190202I_4 - 19300644I_5 + 38686362I_6 - 55203720I_7 + 48461985I_8 - 23556390I_9 + 4849845I_{10}]$$

$$F^6 = \frac{13}{1290240}[175461I_0 - 1759186I_1 + 10492625I_2 - 40457560I_3 + 103656410I_4 - 182475436I_5 + 223872506I_6 - 188719960I_7 + 104083265I_8 - 33717970I_9 + 4849845I_{10}]$$

$1s1p$

$$F^0 = \frac{1}{2}(I_0 + I_1)$$

$$G^1 = \frac{3}{2}(I_0 - I_1)$$

$1p1d$

$$F^0 = \frac{1}{24}[7(I_0 + I_3) + 5(I_1 + I_2)]$$

$$F^2 = \frac{35}{24}[(I_0 + I_3) - (I_1 + I_2)]$$

$$G^1 = \frac{1}{8}[7(I_0 - I_3) - (I_1 - I_2)]$$

$$G^3 = \frac{49}{24}[(I_0 - I_3) - 3(I_1 - I_2)]$$

$1d1f$

$$F^0 = \frac{1}{160}[33(I_0 + I_5) + 21(I_1 + I_4) + 26(I_2 + I_3)]$$

$$F^2 = \frac{3}{32}[11(I_0 + I_5) - 5(I_1 + I_4) - 6(I_2 + I_3)]$$

$$F^4 = \frac{297}{160}[(I_0 + I_5) - 3(I_1 + I_4) + 2(I_2 + I_3)]$$

$$G^1 = \frac{1}{160}[99(I_0 - I_5) + 9(I_1 - I_4) + 38(I_2 - I_3)]$$

$$G^3 = \frac{21}{160}[11(I_0 - I_5) - 19(I_1 - I_4) + 2(I_2 - I_3)]$$

$$G^5 = \frac{363}{160}[(I_0 - I_5) - 5(I_1 - I_4) + 10(I_2 - I_3)]$$

$1f1g$

$$F^0 = \frac{1}{13440}[2145(I_0 + I_7) + 1287(I_1 + I_6) + 1749(I_2 + I_5) + 1539(I_3 + I_4)]$$

$$F^2 = \frac{11}{896}[65(I_0 + I_7) - 13(I_1 + I_6) - 15(I_2 + I_5) - 37(I_3 + I_4)]$$

$$F^4 = \frac{429}{4480}[15(I_0 + I_7) - 31(I_1 + I_6) + 3(I_2 + I_5) + 13(I_3 + I_4)]$$

$$F^6 = \frac{1859}{896}[(I_0 + I_7) - 5(I_1 + I_6) + 9(I_2 + I_5) - 5(I_3 + I_4)]$$

$1g1h$

$$F^0 = \frac{1}{32256}[4199(I_0 + I_9) + 2431(I_1 + I_8) + 3484(I_2 + I_7) + 2860(I_3 + I_6) + 3154(I_4 + I_5)]$$

$$F^2 = \frac{13}{32256}[1615(I_0 + I_9) - 85(I_1 + I_8) - 40(I_2 + I_7) - 736(I_3 + I_6) - 754(I_4 + I_5)]$$

$$F^4 = \frac{13}{3584}[323(I_0 + I_9) - 493(I_1 + I_8) - 92(I_2 + I_7) + 4(I_3 + I_6) + 258(I_4 + I_5)]$$

$$F^6 = \frac{2873}{32256}[19(I_0 + I_9) - 73(I_1 + I_8) + 80(I_2 + I_7) + 8(I_3 + I_6) - 34(I_4 + I_5)]$$

$$F^8 = \frac{71383}{32256}[(I_0 + I_9) - 7(I_1 + I_8) + 20(I_2 + I_7) - 28(I_3 + I_6) + 14(I_4 + I_5)]$$

$1s1d$

$$F^0 = \frac{1}{4}[(I_0 + I_2) + 2I_1]$$

$$G^2 = \frac{5}{4}[(I_0 + I_2) - 2I_1]$$

$1p1f$

$$F^0 = \frac{1}{16}[3(I_0 + I_4) + 4(I_1 + I_3) + 2I_2]$$

$$F^2 = \frac{15}{16}[(I_0 + I_4) - 2I_2]$$

$$G^2 = \frac{5}{48}[9(I_0 + I_4) - 8(I_1 + I_3) - 2I_2]$$

$$G^4 = \frac{27}{16}[(I_0 + I_4) - 4(I_1 + I_3) + 6I_2]$$

$1d1g$

$$F^0 = \frac{1}{960}[143(I_0 + I_6) + 154(I_1 + I_5) + 97(I_2 + I_4) + 172I_3]$$

$$F^2 = \frac{11}{192}[13(I_0 + I_6) + 2(I_1 + I_5) - 13(I_2 + I_4) - 4I_3]$$

$$F^4 = \frac{429}{320}[(I_0 + I_6) - 2(I_1 + I_5) - (I_2 + I_4) + 4I_3]$$

$1f1h$

$$F^0 = \frac{1}{1792}[221(I_0 + I_8) + 208(I_1 + I_7) + 156(I_2 + I_6) + 240(I_3 + I_5) + 142I_4)]$$

$$F^2 = \frac{13}{1792}[85(I_0 + I_8) + 20(I_1 + I_7) - 44(I_2 + I_6) - 20(I_3 + I_5) - 82I_4]$$

$$F^4 = \frac{117}{1792}[17(I_0 + I_8) - 24(I_1 + I_7) - 20(I_2 + I_6) + 24(I_3 + I_5) + 6I_4]$$

$$F^6 = \frac{2873}{1792}[I_0 + I_8) - 4(I_1 + I_7) + 4(I_2 + I_6) + 4(I_3 + I_5) - 10I_4]$$

$1s1f$

$$F^0 = \frac{1}{8}[(I_0 + I_3) + 3(I_1 + I_2)]$$

$$G^3 = \frac{7}{8}[(I_0 - I_3) - 3(I_1 - I_2)]$$

$2s1s$

$$F^0 = \frac{1}{8}[5(I_0 + I_2) - 2I_1]$$

$$G^0 = \frac{5}{8}[(I_0 + I_2) - 2I_1]$$

$2s1p$

$$F^0 = \frac{1}{48}[11I_0 + 37I_1 - 35I_2 + 35I_3]$$

$$G^1 = \frac{1}{16}[11I_0 - 41I_1 + 65I_2 - 35I_3]$$

$2s1d$

$$F^0 = \frac{1}{96}[15I_0 + 40I_1 + 34I_2 - 56I_3 + 63I_4]$$

$$G^2 = \frac{5}{96}[15I_0 - 44I_1 + 106I_2 - 140I_3 + 63I_4]$$

$2s1f$

$$F^0 = \frac{1}{64}[9I_0 + 9I_1 + 38I_2 - 10I_3 - 15I_4 + 33I_5]$$

$$G^3 = \frac{1}{64}[63I_0 - 147I_1 + 294I_2 - 630I_3 + 651I_4 - 231I_5]$$

INDEX

A

B

C

A CATALOG OF SELECTED
DOVER BOOKS
IN SCIENCE AND MATHEMATICS

Mathematics

FUNCTIONAL ANALYSIS (Second Corrected Edition), George Bachman and Lawrence Narici. Excellent treatment of subject geared toward students with background in linear algebra, advanced calculus, physics, and engineering. Text covers introduction to inner-product spaces, normed, metric spaces, and topological spaces; complete orthonormal sets, the Hahn-Banach Theorem and its consequences, and many other related subjects. 1966 ed. 544pp. 6⅛ x 9¼. 40251-7

ASYMPTOTIC EXPANSIONS OF INTEGRALS, Norman Bleistein & Richard A. Handelsman. Best introduction to important field with applications in a variety of scientific disciplines. New preface. Problems. Diagrams. Tables. Bibliography. Index. 448pp. 5⅜ x 8½. 65082-0

VECTOR AND TENSOR ANALYSIS WITH APPLICATIONS, A. I. Borisenko and I. E. Tarapov. Concise introduction. Worked-out problems, solutions, exercises. 257pp. 5⅜ x 8¼. 63833-2

THE ABSOLUTE DIFFERENTIAL CALCULUS (CALCULUS OF TENSORS), Tullio Levi-Civita. Great 20th-century mathematician's classic work on material necessary for mathematical grasp of theory of relativity. 452pp. 5⅜ x 8¼. 63401-9

AN INTRODUCTION TO ORDINARY DIFFERENTIAL EQUATIONS, Earl A. Coddington. A thorough and systematic first course in elementary differential equations for undergraduates in mathematics and science, with many exercises and problems (with answers). Index. 304pp. 5⅜ x 8½. 65942-9

FOURIER SERIES AND ORTHOGONAL FUNCTIONS, Harry F. Davis. An incisive text combining theory and practical example to introduce Fourier series, orthogonal functions and applications of the Fourier method to boundary-value problems. 570 exercises. Answers and notes. 416pp. 5⅜ x 8½. 65973-9

COMPUTABILITY AND UNSOLVABILITY, Martin Davis. Classic graduate-level introduction to theory of computability, usually referred to as theory of recurrent functions. New preface and appendix. 288pp. 5⅜ x 8½. 61471-9

ASYMPTOTIC METHODS IN ANALYSIS, N. G. de Bruijn. An inexpensive, comprehensive guide to asymptotic methods–the pioneering work that teaches by explaining worked examples in detail. Index. 224pp. 5⅜ x 8½ 64221-6

APPLIED COMPLEX VARIABLES, John W. Dettman. Step-by-step coverage of fundamentals of analytic function theory–plus lucid exposition of five important applications: Potential Theory; Ordinary Differential Equations; Fourier Transforms; Laplace Transforms; Asymptotic Expansions. 66 figures. Exercises at chapter ends. 512pp. 5⅜ x 8½. 64670-X

INTRODUCTION TO LINEAR ALGEBRA AND DIFFERENTIAL EQUATIONS, John W. Dettman. Excellent text covers complex numbers, determinants, orthonormal bases, Laplace transforms, much more. Exercises with solutions. Undergraduate level. 416pp. 5⅜ x 8½. 65191-6

Physics

OPTICAL RESONANCE AND TWO-LEVEL ATOMS, L. Allen and J. H. Eberly. Clear, comprehensive introduction to basic principles behind all quantum optical resonance phenomena. 53 illustrations. Preface. Index. 256pp. 5⅜ x 8½. 65533-4

QUANTUM THEORY, David Bohm. This advanced undergraduate-level text presents the quantum theory in terms of qualitative and imaginative concepts, followed by specific applications worked out in mathematical detail. Preface. Index. 655pp. 5⅜ x 8½. 65969-0

ATOMIC PHYSICS: 8th edition, Max Born. Nobel laureate's lucid treatment of kinetic theory of gases, elementary particles, nuclear atom, wave-corpuscles, atomic structure and spectral lines, much more. Over 40 appendices, bibliography. 495pp. 5⅜ x 8½. 65984-4

A SOPHISTICATE'S PRIMER OF RELATIVITY, P. W. Bridgman. Geared toward readers already acquainted with special relativity, this book transcends the view of theory as a working tool to answer natural questions: What is a frame of reference? What is a "law of nature"? What is the role of the "observer"? Extensive treatment, written in terms accessible to those without a scientific background. 1983 ed. xlviii+172pp. 5⅜ x 8½. 42549-5

AN INTRODUCTION TO HAMILTONIAN OPTICS, H. A. Buchdahl. Detailed account of the Hamiltonian treatment of aberration theory in geometrical optics. Many classes of optical systems defined in terms of the symmetries they possess. Problems with detailed solutions. 1970 edition. xv+360pp. 5⅜ x 8½. 67597-1

PRIMER OF QUANTUM MECHANICS, Marvin Chester. Introductory text examines the classical quantum bead on a track: its state and representations; operator eigenvalues; harmonic oscillator and bound bead in a symmetric force field; and bead in a spherical shell. Other topics include spin, matrices, and the structure of quantum mechanics; the simplest atom; indistinguishable particles; and stationary-state perturbation theory. 1992 ed. xiv+314pp. 6⅛ x 9¼. 42878-8

LECTURES ON QUANTUM MECHANICS, Paul A. M. Dirac. Four concise, brilliant lectures on mathematical methods in quantum mechanics from Nobel Prize–winning quantum pioneer build on idea of visualizing quantum theory through the use of classical mechanics. 96pp. 5⅜ x 8½. 41713-1

THIRTY YEARS THAT SHOOK PHYSICS: The Story of Quantum Theory, George Gamow. Lucid, accessible introduction to influential theory of energy and matter. Careful explanations of Dirac's anti-particles, Bohr's model of the atom, much more. 12 plates. Numerous drawings. 240pp. 5⅜ x 8½. 24895-X

ELECTRONIC STRUCTURE AND THE PROPERTIES OF SOLIDS: The Physics of the Chemical Bond, Walter A. Harrison. Innovative text offers basic understanding of the electronic structure of covalent and ionic solids, simple metals, transition metals and their compounds. Problems. 1980 edition. 582pp. 6⅛ x 9¼. 66021-4